Rare Earth Doped Silica Fiber and Its Applications

Lili Hu

Rare Earth Doped Silica Fiber and Its Applications

Springer

Lili Hu
Shanghai Institute of Optics and Fine Mechanics
Chinese Academy of Sciences
Shanghai, China

ISBN 978-981-96-5305-8 ISBN 978-981-96-5306-5 (eBook)
https://doi.org/10.1007/978-981-96-5306-5

The original submitted manuscript has been translated into English. The translation was done using artificial intelligence. A subsequent revision was performed by the author(s) to further refine the work and to ensure that the translation is appropriate concerning content and scientific correctness. It may, however, read stylistically different from a conventional translation.

Jointly published with Shanghai Scientific and Technical Publishers.

The print edition is not for sale in China (Mainland). Customers from China (Mainland) please order the print book from: Shanghai Scientific and Technical Publishers.

This Springer imprint is published by the registered company Springer Nature Singapore Pte Ltd.
The registered company address is: 152 Beach Road, #21-01/04 Gateway East, Singapore 189721, Singapore

Content Summary

This book, based on the years of research accumulation and related innovative achievements of the research team led by Prof. Dr. Hu Lili from the Shanghai Institute of Optics and Fine Mechanics, Chinese Academy of Sciences, summarizes the development history, composition, structure, properties, fabrication processing, and applications of rare earth-doped silica glass and fiber. This book comprehensively and systematically introduces the latest research results of rare earth-doped silica glasses and fibers to readers by integrating research reports from domestic and foreign peers. This book is divided into nine chapters. Chapter 1 systematically introduces the structure and properties of silica glass; Chap. 2 elaborates on the properties and structure of ytterbium-doped silica glass and analyzes the correlation between the local structure of ytterbium ions and the spectral properties of ytterbium ions; Chap. 3 elaborates on the preparation methods and performance characterization methods of rare earth-doped silica fiber; Chaps. 4 and 5 introduce the characteristics and applications of ytterbium-doped cladding structure and large mode area microstructure fibers, respectively; Chaps. 6 and 7 introduce the latest research progress and applications of 1.5 μm band erbium-doped silica fiber and 2 μm band thulium, holmium-doped silica fiber, respectively; Chap. 8 introduces the research progress of neodymium-doped silica fiber, focusing on the latest progress of 0.9 μm band laser applications; Chap. 9, facing the needs of fiber lasers for space application, elaborates on the preparation and application technology progress of radiation-resistant rare earth-doped silica fiber.

This book is the first professional book in China to systematically introduce rare earth-doped silica glass and fiber: on the one hand, it highlights application-oriented fundamental research, which is original, cutting-edge, and leading; on the other hand, the research team has outstanding scientific research background. Prof. Dr. Hu Lili and her team have long been engaged in the fundamental research and fabrication technology development of laser glass and laser fiber and have achieved many innovative results in the field of rare earth-doped silica glass and fiber research, providing core component support for domestic high-power fiber lasers. Therefore, this book has high academic value and great social benefits.

The readership of this book is broad, including engineers and students engaged in optics, photonics, laser fibers, and laser research and development in research institutes and universities, as well as practitioners in the fiber laser, laser fiber, and laser materials industry.

Preface

Since E. Snitzer first proposed the concept of fiber lasers in 1961, fiber lasers have gone through the initial research stage, the development of erbium-doped fiber amplifiers in the 1980s, and the beginning of optical communication applications, to the rapid development and widespread application of ytterbium-doped fiber lasers since the late 1990s. Fiber lasers, with their compact structure, high wall-plug efficiency, high beam quality, high reliability, easy operation and maintenance, and less environmental impact, have been widely used in advanced manufacturing, communication, measurement sensing, medical, defense, and scientific research fields.

Rare earth-doped silica fiber is the key component of fiber lasers and amplifiers. With the development of fiber lasers and amplifiers, the laser wavelength of rare earth-doped silica fiber has expanded from the initial 1 μm to the current visible light to 2 μm, the single-fiber output power has developed from the initial milliwatt level to the current single-mode kilowatt level, and the fiber structure has transitioned from the initial single cladding to double cladding, triple cladding, and microstructured fiber. Its fabrication technology has also developed rapidly.

As is well known, since the Corning Company in the United States first obtained low-loss silica fiber with a loss of 20 dB/km using MCVD technology in 1970, the University of Southampton in the UK successfully developed low-loss neodymium-doped and erbium-doped silica fiber using MCVD technology in 1985, which opened the application of erbium-doped fiber amplifiers in the field of optical communication. In 1988, E. Snitzer first proposed a double-clad neodymium-doped silica fiber structure, making it possible for a significant increase in the brightness and power of fiber lasers. In 1999, the international community first achieved a breakthrough in the output power of hundreds of watts from an ytterbium-doped silica fiber laser. In the following decade, with the development of laser diodes, ytterbium-doped silica fiber lasers achieved single-fiber single-mode kilowatt laser output. In 2000, the University of Bath in the UK first reported the research results of ytterbium-doped photonic crystal fibers. In 2003, the University of Jena in Germany first used large-mode-area ytterbium-doped photonic crystal fibers to achieve high average power laser output, providing a solution of active fibers for high peak power pulsed fiber lasers. The continuous research and development of rare earth-doped silica fibers not only promoted the application of fiber lasers and amplifiers in the fields of optical communication and measurement sensing but also greatly advanced the application of high-power fiber lasers represented by ytterbium-doped fiber lasers in the fields of industrial processing, medical treatment, and defense. At present, rare earth-doped silica fiber has become an indispensable gain material for communication fiber amplifiers, high-power lasers, and industrial fiber lasers.

So far, Chinese monographs containing professional knowledge content of rare earth-doped silica fiber mainly include: *New Type of Optically Functional Glass*, edited by Jiang Zhonghong (Chemical Industry Press, 2008), *High Power Fiber Lasers and Their Applications*, written by Lou Qihong (University of Science and Technology of China Press, 2009), *Industrial Fiber Lasers*, written by Yan Dapeng (Huazhong University of Science and Technology Press, 2022), and *Special Optical Fibers and Optical Fiber Communication*, written by Wang Tingyun (Shanghai Science and Technology Press, 2016). These four monographs focus on optically functional glass, high-power fiber lasers, industrial fiber lasers, and special optical fibers for communication, respectively. In addition, *Composite Glass Fibers* written by Yang Zhongmin and others (South China University of Technology Press, 2021) specifically introduces new types of fibers composed of multi-component glasses and composites of nanocrystals, crystals, semiconductors, metals, and organic materials with glass. Considering the important role of rare earth-doped silica fibers in fiber lasers and amplifiers, there is an urgent need for a book specifically introducing rare earth-doped silica fibers and their applications, to help everyone systematically understand the latest research and development and application progress of rare earth-doped silica fibers.

This book is based on the research accumulation and related innovative achievements of the team led by Prof. Dr. Hu Lili from the Shanghai Institute of Optics and Fine Mechanics, Chinese Academy of Sciences, in the field of rare earth-doped silica fiber over the past 10 years. It appropriately integrates research reports from domestic and foreign counterparts and provides a detailed introduction to the core glass composition, properties and structures of rare earth-doped silica fiber, the design and preparation of rare earth-doped silica fiber, parameter measurement, fiber properties, and applications, striving to introduce the latest research results and applications of rare earth-doped silica fiber comprehensively and systematically to readers. This book consists of nine chapters. Chapter 1 systematically introduces the basic properties of silica glass and the overview of rare earth-doped silica fiber; Chap. 2 focuses on the properties and structure of ytterbium-doped silica glass and analyzes the correlation between local structure of ytterbium ions and spectral properties of ytterbium ion; Chap. 3 focuses on the preparation methods and performance characterization methods of rare earth-doped silica fiber; Chaps. 4 and 5 introduce the structure, performance, and application of ytterbium-doped cladding structure and large mode area microstructure fiber, respectively; Chap. 6 introduces the latest research progress and application of erbium-doped silica fiber for 1.5 μm band application; Chap. 7 introduces the latest research progress and application of thulium and holmium-doped silica fiber for 2 μm band application; Chap. 8 focuses on the application of tunable lasers at 0.9 μm and 1 μm bands and introduces the latest research progress of neodymium and ytterbium-neodymium co-doped silica fiber; Chap. 9 focuses on the application needs of fiber lasers in space and elaborates on the radiation-resistant rare earth-doped silica fiber and its application.

The authors of this book mainly come from the Shanghai Institute of Optics and Fine Mechanics, Chinese Academy of Sciences. The whole book is edited and finalized by Hu Lili; Shao Chongyun is responsible for collecting manuscripts and liaising with authors and reviewers. The authors involved in the writing of this book are as follows: Hu Lili wrote Chap. 1, Guo Mengting and Shao Chongyun wrote Chap. 2, Wang Shikai and Wang Meng wrote Chap. 3, Yu Chunlei wrote Chap. 4, Feng Suya and Wang Meng wrote Chap. 5, Zhang Lei wrote Chap. 6, Wang Fan wrote Chap. 7, Wang Yafei and Wang Shikai wrote Chap. 8, and Shao Chongyun wrote Chap. 9.

Profs. Chen Danping and Ye Xisheng from the Shanghai Institute of Optics and Fine Mechanics, Chinese Academy of Sciences, reviewed all chapters of this book and provided valuable revision guidance. The Shanghai Science and Technology Publishing House carefully organized the editing and publishing of this book. We would like to express our heartfelt thanks to them.

This book is the first professional book in China that systematically introduces rare earth-doped silica fiber and its applications from a material perspective. It provides important references for scientific researchers, technicians, and students in research institutes and universities engaged in optics, photonics, laser fiber, and laser development, as well as technical personnel in the fiber laser and amplifier, laser fiber, and laser material industries. Due to the limitations of time and professional knowledge, there may be shortcomings in this book. We hope that readers will provide valuable suggestions for our continuous improvement.

Shanghai, China Lili Hu

Contents

Chapter 1
Silica Glass and Rare Earth-Doped Silica Fiber

Silica glass has excellent mechanical, thermal, optical, and stability properties, and is widely used in semiconductor manufacturing, chemical smelting, electric light sources, photovoltaic industry, aerospace, precision optical instruments, large laser facilities, optical communication, and fiber lasers. Rare earth doped silica fiber is the key component of fiber lasers, which usually consists of a rare earth doped silica glass core, a pure silica inner cladding, and an epoxy resin outer layer cladding. Silica glass is an important component of rare earth doped silica fiber. To develop high-performance rare earth doped silica fiber, it is necessary to first systematically understand the structure and properties of silica glass.

Theoretically, silica glass refers to an amorphous substance composed of a single compound, silicon dioxide. However, the chemical composition and properties of silica glass obtained by different preparation methods are different. In addition to silicon dioxide as the main component, due to different preparation process methods, silica glass may also contain impurities such as alkali metal ions, hydroxyl, chlorine ions, etc. at the ppm (one in a million, 10^{-6}) level.

The structure of silica glass is a three-dimensional continuous irregular network formed by the connection of silicon-oxygen tetrahedra in a corner-sharing. The properties of silica glass mainly depend on its structure. Any external factors that cause changes in the structure of silica glass, such as temperature, pressure, doping or impurity elements, will cause changes in the properties of silica glass. This chapter first focuses on the formation and types, structure, thermal properties, mechanical properties, optical properties, radiation resistance and laser damage characteristics of silica glass. Based on this, it briefly introduces the luminescent properties of rare earth ions and the development history of rare earth doped silica fiber.

L. Hu, *Rare Earth Doped Silica Fiber and Its Applications*,
https://doi.org/10.1007/978-981-96-5306-5_1

1.1 Formation and Types of Silica Glass

1.1.1 Formation of Silica Glass

From a thermodynamic point of view, the formation of silica glass is defined as a solid glass formed by cooling a melt above the melting point of quartz. As shown in Fig. 1.1, the formation of glass goes through two processes: supercooling and freezing of high-temperature melts. The glass transition can be described by the characteristics of thermodynamic properties such as enthalpy, free energy, and volume changing in a segmented manner with temperature. The first derivatives of these quantities are discontinuous. The temperature corresponding to the inflection point of the above thermodynamic properties or the breakpoint of the first derivative is called the glass transition temperature. The glass transition process is closely related to the cooling rate: the faster the cooling rate, the higher the glass transition temperature (T_{gf}); on the contrary, the slower the cooling rate, the lower the glass transition temperature (T_{gs}). The glass transition temperature is defined as the transition temperature corresponding to a cooling rate of 10 K/s.

Angell et al. [2] classified the ability of a liquid to form glass from high temperature cooling into strong and weak (brittle) categories. The liquids of silica and germanium oxide are defined as strong glass formers. The fragility of a liquid can be represented by the slope m of the $\lg(\eta)$ and T_g/T curve at the glass transition temperature: the larger the m value, the more brittle it is, the smaller the m value, the stronger the glass former. The m value of silica glass melt is about 20, and the m value of fragile liquids is usually greater than this value. As shown in Fig. 1.2, over a large temperature range above the glass transition temperature, their logarithmic values of viscosity

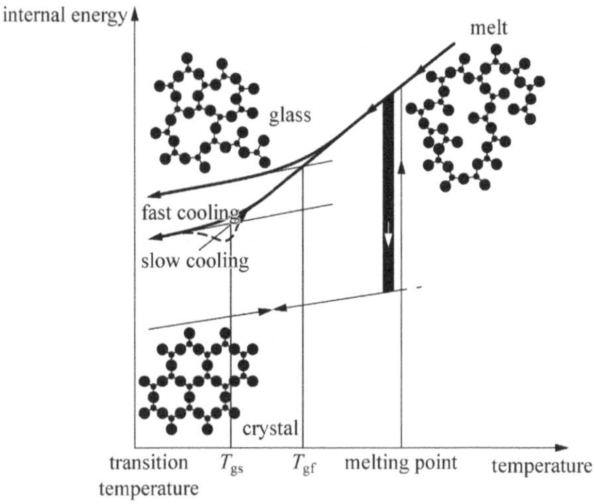

Fig. 1.1 Relationship between the internal energy E of a substance and temperature T. (Adapted from reference [1])

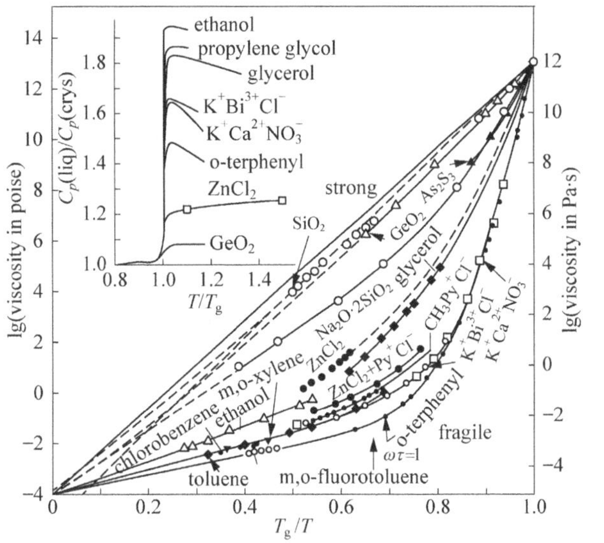

Fig. 1.2 The relationship between the logarithmic value of viscosity and T_g/T, the inset in the upper left corner is the relationship between the heat capacity of different substances and T/T_g [2]

and T_g/T values are linear or approximately linear. However, the logarithmic values of the viscosity of fragile liquids and T_g/T values deviate from the linear region. The insert of Fig. 1.2 shows the relationship between the heat capacity of different compounds before and after the glass transition temperature and T/T_g, it can be seen that the heat capacity change of strong glass formers near T_g temperature is relatively small, while the heat capacity change of fragile liquids before and after T_g temperature is much greater than that of strong glass formers.

Due to the structural characteristics of silica glass tetrahedra [SiO_2] connected at the vertices to form a three-dimensional network, its transition temperature is relatively high. Richet and Bottinga measured the highest glass transition temperature of silica glass to be 1480 K (heating rate of 20 K/min) [3], which is also the highest glass transition temperature. However, it is difficult to accurately test the glass transition temperature of silica glass using conventional differential scanning calorimetry (DSC) curves. This is mainly because the structural changes of silica glass during the process of forming glass from high temperature liquid phase cooling are relatively small, only the tetrahedra [SiO_2] changes in bond length and bond angle occur. Therefore, compared to ordinary glass, the configurational entropy change in the glass transition process of silica glass is smaller, and correspondingly, the configurational enthalpy is also smaller.

Yue et al. [4] evaluated the glass transition temperature of silica glass by measuring heat capacity. As shown in Fig. 1.3, by measuring the change in heat capacity with temperature during the heating process, a turning point can be seen near the glass transition temperature where the heat capacity rapidly increases with increasing temperature. The temperature at this turning point is the transition temperature

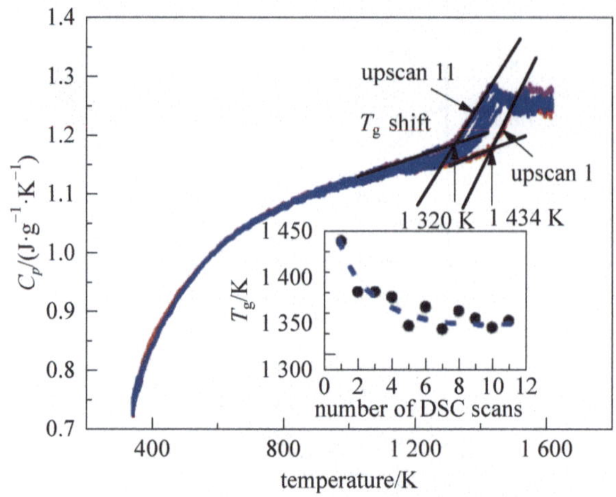

Fig. 1.3 Relationship between the heat capacity temperature curve of silica glass and the number of DSC heating process scanning tests (the inset in the lower right corner shows the change rule of the apparent glass transition temperature T_g with the number of DSC heating process scanning tests) [4]

of silica glass. However, even so, Yue believes that it is still difficult to accurately measure the glass transition temperature of silica glass. One reason is that the impurity content in silica glass greatly affects its glass transition temperature. The second reason is that the measurement of the glass transition temperature of silica glass by calorimetry is often affected by the number of heating-cooling cycles. The more heating-cooling cycles a silica glass containing only 1 ppm hydroxyl (OH^-) undergoes, the lower the glass transition temperature (T_g). The temperature T_g dropped from 1434 K in the first cycle test to 1320 K in the 11th cycle test, a decrease of 114 K was observed. This is mainly because during the heating-cooling cycle, the silica glass undergoes structural relaxation towards equilibrium. In the repeated heating-cooling measurement process, T_g decrease means that the bonding between the [$SiO_{4/2}$] tetrahedral structural groups in silica glass weakens. From the perspective of topological constraint theory, T_g decrease means that the topological constraints in silica glass decrease with the progress of the heating-cooling cycle. The decrease of T_g in silica glass with the number of heating-cooling differential thermal measurement cycles is its unique abnormal behavior, which is very different from conventional silicate glass. The T_g of ordinary silicate glass does not change with the number of differential thermal measurements. Yue believes that in order to obtain an accurate silica glass T_g temperature, it is necessary to exclude the influence of hydroxyl and differential thermal cycle measurement times, accurately test a series of silica glasses with different hydroxyl contents, and extrapolate to obtain the silica glass T_g with zero hydroxyl content. In addition, it is necessary to measure the T_g temperature of different differential thermal cycle times, and extrapolate to obtain the T_g[4] at zero cycle temperature measurement times. Of course, the viscosity temperature curve of silica glass can also be measured, and its T_g can be determined from the temperature corresponding to a viscosity of 10^{12} Pa·S. However, this requires a very high viscosity measurement temperature.

The main reason why quartz melt can form a glass state under a wide range of cooling speeds is that the bonding strength of silicon oxygen is relatively large, and the [SiO_2] tetrahedron presents a three-dimensional structure in the form of rings connected by apex angles in space.

1.1.2 Types of Silica Glass

Silica glass can be divided into the following four types according to different preparation methods:

1.1.2.1 Type I Silica Glass

Silica glass prepared from natural quartz crystal powder in a vacuum or inert atmosphere is called Type I silica glass. Type I silica glass contains a small amount of Al (100 ppm), Na (4 ppm), OH^- (within 5 ppm). The vacuum electric melting process of silica glass refers to loading quartz crystal powder into a graphite crucible, using a vacuum electric melting furnace to melt silica glass at 1800 ~ 2000 °C and 0.1 ~ 10 Pa vacuum degree. Using the vacuum electric melting process, low hydroxyl content, 1.8 m diameter and 650 mm thick silica glass can be produced. The main defect of the silica glass produced by this process is bubbles and stones. In order to reduce the defects in the vacuum electric melting silica glass, the two-step melting process of silica glass has been developed. Two-step melting refers to the secondary heating of electrically melted silica glass ingots or tubes to form the required silica glass rods, tubes or plates. The heating method of two-step melting is generally medium frequency induction heating.

1.1.2.2 Type II Silica Glass

Silica glass prepared from natural quartz crystal powder by hydrogen-oxygen flame heating method is called Type II silica glass. Type II silica glass contains 150 ~ 400 ppm of OH^-. The preparation of Type II silica glass uses hydrogen and oxygen as fuel sources, and quartz crystal powder as raw materials. It is melted on special equipment using a specially designed burner. This process was initially used to prepare transparent silica glass tubes and crucibles, which can be used to prepare transparent silica glass weights. The process is as follows: natural quartz crystal powder of 100 ~ 200 mesh is used as raw material, fed into a hydrogen-oxygen burner, sprayed on a silica glass target holder, the target surface continuously rotates and maintains a constant distance from the burner, so that the powder is melted to form silica glass. The silica glass prepared by this process has few bubbles; if the raw material quality is good, optical grade silica glass can be prepared. In addition, a plasma flame can also be used to replace the hydrogen-oxygen flame to heat the crystal powder for the preparation of low hydroxyl content Type II silica glass.

1.1.2.3 Type III Silica Glass

Silica glass obtained by hydrolysis of $SiCl_4$ under hydrogen-oxygen flame heating conditions is called Type III silica glass. Type III silica glass contains about 1000 ppm of OH^-, 100 ppm of Cl^-, but it does not contain metal ion impurities. A typical representative is Corning HPFS 7980 brand silica glass. Type III silica glass is prepared by the hydrogen-oxygen flame chemical vapor deposition method. The basic principle is as follows: the gasified $SiCl_4$ is sent into the combustion chamber with the carrier gas, under the action of the hydrogen-oxygen flame, $SiCl_4$ is oxidized to form amorphous silica, which is deposited on the rotating high-temperature target material. It then forms silica glass through a high-temperature heating process.

1.1.2.4 Type IV Silica Glass

Silica glass formed by $SiCl_4$ under plasma flame heating conditions is called Type IV silica glass. Type IV silica glass contains 0.4 ppm of OH^-, 200 ppm of Cl^-. A typical representative is Corning HPFS 8655 brand silica glass. Plasma chemical vapor deposition uses high-frequency induction equipment to excite the working gas oxygen, and poleless discharge generates high-frequency oxygen plasma, forming a high-temperature atmosphere above 1600 °C. The $SiCl_4$ gas phase raw material is brought into the plasma flame by the carrier gas, and directly reacts with the oxygen in the plasma to form silica and chlorine gas. The generated SiO_2 is deposited on the base rod, and is melted into silica glass by heating with a high-temperature plasma flame. The significant feature of the silica glass prepared by plasma chemical vapor deposition is its high purity, few impurities, and extremely low hydroxyl content. This process method is particularly beneficial for the preparation of ultra-high purity silica glass.

The ultraviolet transmittance of silica glass is mainly affected by metal impurities and oxygen vacancy defects, while the infrared transmittance is mainly affected by the hydroxyl content. Generally, the higher the hydroxyl content of high-purity silica glass, the less the oxygen vacancy defect content, so its ultraviolet transmittance will increase. Table 1.1 compares the raw materials, preparation methods, impurity content, light transmission range, appearance characteristics, and other parameters of four types of silica glass.

Table 1.1 Comparison of the characteristics of four types of silica glass [1, 5]

Category	Domestic Brand	Raw Material	Preparation Process	Metal Impurities /ppm	Cl^- Content /ppm	OH^- Content /ppm	Light Transmission range /nm	Appearance Characteristics
Type I	JGS3	Natural crystal	Vacuum electric melting	40 ~ 100	Very few	<5	300 ~ 3500	Few bubbles/stones
Type II	JGS2	Crystal powder	Hydrogen-oxygen flame heating	20 ~ 40	Very few	150 ~ 400	300 ~ 2000	Few bubbles
Type III	JGS1	$SiCl_4$	Hydrogen-oxygen flame heating	<1	>100	~1000	190 ~ 2000	No bubbles
Type IV	High purity JGS3	$SiCl_4$	Plasma heating	<1	~200	0.4	190 ~ 3500	No bubbles

1.2 Structure of Silica Glass

Compared with the regular structure of crystals, the structure of glass is very complex. The widely accepted theories of glass structure are the "crystallite theory" and the "random network theory". X-ray diffraction technology is an important method for studying crystal structure.

Figures 1.4a, b are the X-ray diffraction diagrams of silica glass powder and quartz crystal powder, respectively [6]. Among them, the quartz crystal is obtained by heat treating the silica glass at 1500 °C for several hours. The quartz crystal powder shows a intense, broad diffraction ring and three weak and narrow diffraction rings. The silica glass powder only shows a strong and broad diffraction ring, and the position of this diffraction ring is equivalent to the position of the strongest diffraction ring of the quartz crystal powder. For crystals, the smaller the grain size, the larger the diffraction ring width. Based on the above reasons, Randall et al. [7, 8] proposed in 1930 that silica glass is composed of countless tiny (~1.5 nm) "crystallites" and amorphous substances; the atomic arrangement within these "crystallites" is locally ordered, but their orientations are random, it is evenly dispersed in the amorphous substance, and there is no clear boundary between the "crystallites" and the amorphous substance. This is the origin of the "crystallite theory".

The "crystallite theory" is further confirmed by X-ray diffraction analysis. Fig. 1.5a, c are the X-ray diffraction curves of cristobalite powder and silica glass powder respectively, and Fig. 1.5b is the simulated X-ray diffraction curve when the average size of the cristobalite is assumed to be 1.5 nm [7]. It can be seen that the strongest diffraction peak of silica glass is comparable to the strongest diffraction peak of cristobalite. This strongest diffraction peak comes from the (111) crystallographic planes of the cristobalite. The experimental results show that as the grain size of the cristobalite gradually decreases, its strongest diffraction peak gradually widens (not shown in the Figure). Specifically, when the grain size of the quartz crystal is reduced to 1.5 nm, its diffraction curve is almost identical to the

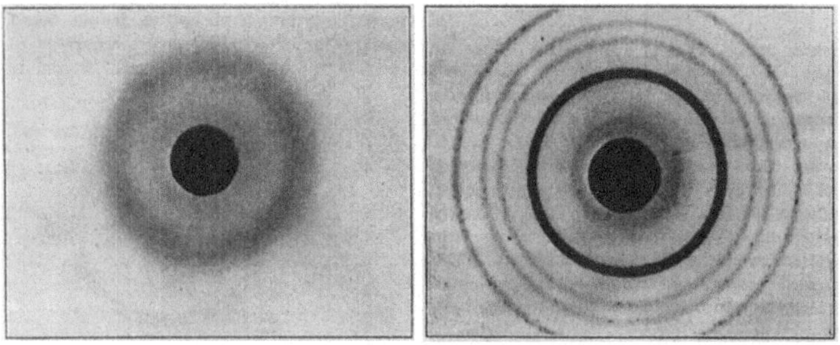

(a) (b)

Fig. 1.4 X-ray diffraction rings of silica glass powder (**a**) and quartz crystal powder (**b**)

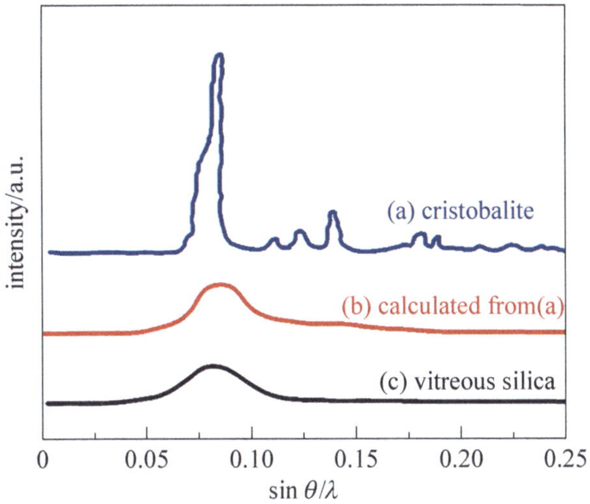

Fig. 1.5 (**a**) and (**c**) are the X-ray diffraction curves of cristobalite powder and silica glass powder respectively; (**b**) is the simulated X-ray diffraction curve when the average size of the cristobalite is assumed to be 1.5 nm

diffraction curve of silica glass. All of the above are strong evidence for the existence of "crystallites" in silica glass.

However, the "crystallite theory" also faces many unexplainable problems: ① lattice constant mismatch. According to the X-ray diffraction ring, the interplanar spacing of the cristobalite (111) crystal plane can be calculated to be 4.05 Å (1 Å = 10^{-10} m), and the interplanar spacing of the silica glass (111) crystal plane is 4.32 Å, with a lattice constant mismatch of up to 6%. ② The volume of the cristobalite will change significantly between 200 ~ 300 °C, but this phenomenon does not exist in silica glass. ③ When silica glass is heat-treated at 1500 °C, it can be expected that the "crystallites" in the silica glass will gradually grow, causing the width of its diffraction ring or diffraction peak to gradually narrow. However, the experimental result is that the width of its diffraction ring or diffraction peak suddenly narrows.

In 1932, Zachariasen [9] proposed the "random network" theory. This theory believes that the basic units of silica glass and quartz crystal are both silicon-oxygen tetrahedra [SiO_2], each silicon atom is surrounded by four oxygen atoms, each oxygen atom is connected to two silicon atoms, and adjacent silicon-oxygen tetrahedra [SiO_2] are connected by oxygen vertices to form a three-dimensional spatial network structure. Later, Warren and others [6, 10, 11] further confirmed the basic views of the "random network" theory through X-ray diffraction.

Figure 1.6 is the pair distribution function of silica glass and cristobalite [10, 12]. The first three peaks correspond to the first nearest neighbor Si—O_I (0.162 nm),

O—O_I (0.265 nm), Si—Si_I (0.312 nm) average bond length, the fourth peak corresponds to the overlap of the average bond length positions of the second nearest

Fig. 1.6 Pair distribution function of silica glass and quartz crystal [12]

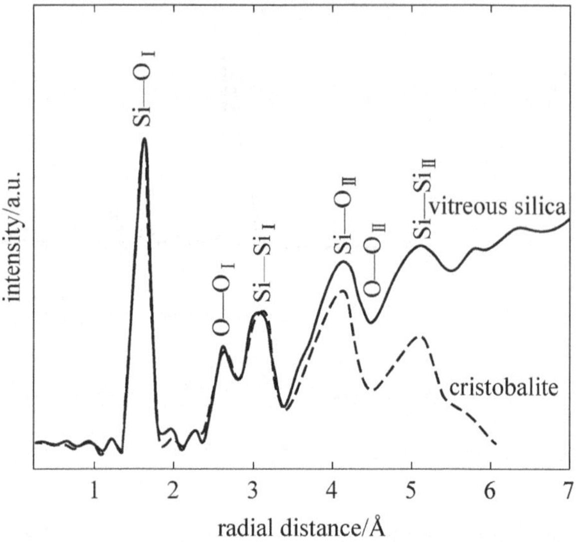

neighbor Si—O_{II} and O—O_{II}, the fifth peak corresponds to the average bond length of the second nearest neighbor Si—Si_{II}. It can be seen that: within the range of 3.5 Å, the pair distribution functions of silica glass and quartz crystal are basically consistent, which indicates the short-range order of the structure of silica glass. Beyond the range of 5 Å, the pair distribution function signal of quartz crystal is still clearly visible, but the pair distribution function signal of silica glass gradually weakens and even becomes indistinguishable, which is related to the long-range disorder of the structure of silica glass.

Figure 1.7 shows the theoretical model and experimental test results of the two-dimensional planar structure of silica glass. The theoretical model was first proposed by Zachariasen in 1932 [9], and later gradually confirmed by experimental means such as X-ray diffraction [13], neutron diffraction [14], nuclear magnetic resonance (NMR) [15], and annular dark field-scan transmission electron microscopy (ADF-STEM) [16]. Among them, the two-dimensional image of silica glass at the atomic scale taken by ADF-STEM is highly similar to the theoretical model proposed by Zachariasen, as shown in Fig. 1.7.

As shown in Figure 1.8a, the structural unit of silica glass is [SiO_2] tetrahedron, with Si atom at the center of the tetrahedron and O atoms at the four vertices. The [SiO_2] tetrahedra are connected by shared vertices to form a three-dimensional network structure, and the <Si—O—Si > bond angle α between adjacent [SiO_2] tetrahedra varies between 120 ~ 180°. In each [SiO_2] tetrahedron, the <O—Si—O > bond angle $\Phi = 109°28'$, the bond length between Si—O $d_1 \approx 1.6$ Å, and the distance between O—O $d_2 \approx 2.6$ Å.

As shown in Figure 1.8b, nuclear magnetic resonance (solid line), X-ray diffraction (dots), and spatial hindrance model (dashed line) calculations all indicate that the bond angle distribution between <Si—O—Si > in silica glass and silicate glass

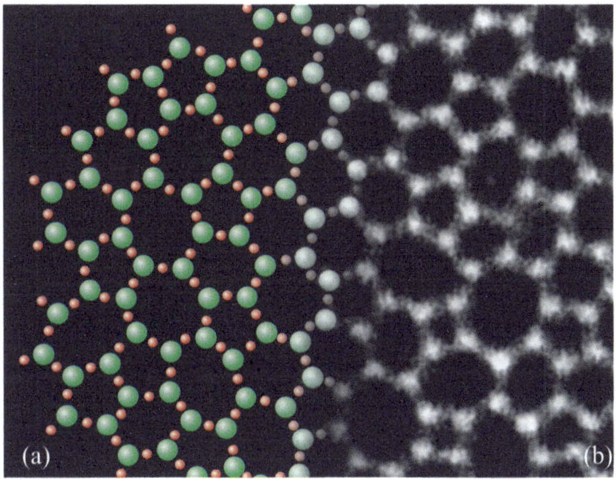

Fig. 1.7 Theoretical model of the two-dimensional planar structure of silica glass (**a**) and the experimental spectrum taken by ADF-STEM (**b**) [16]

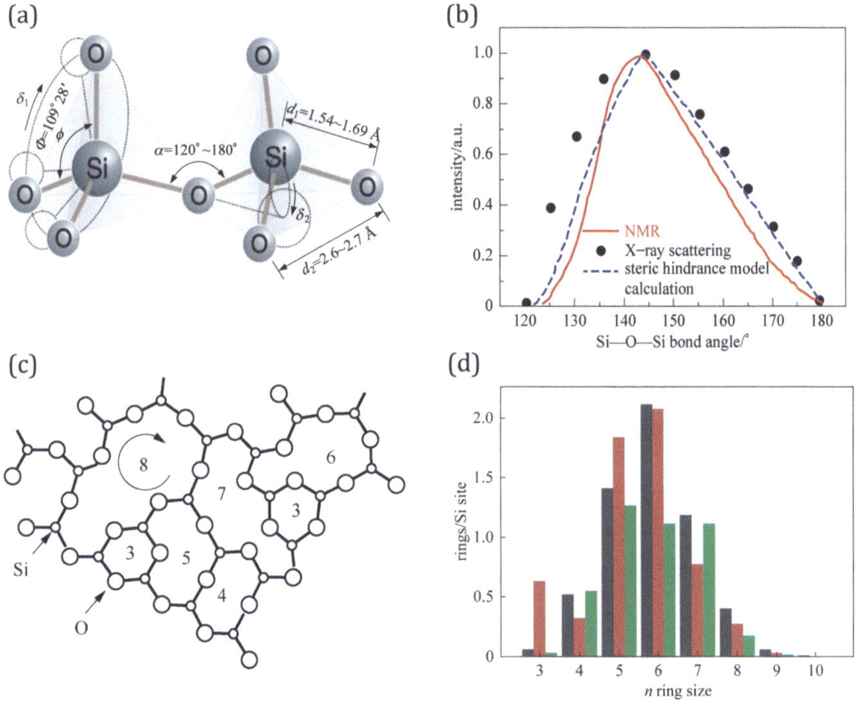

Fig. 1.8 Silica glass [SiO_2] tetrahedral structural unit [18] (**a**), Si—O—Si bond angle distribution [17] (**b**), planar ring structure schematic [19] (**c**) and ring structure distribution density [20] (**d**)

is relatively wide, varying between 120 ~ 180°, with an average value of about 145° [17]. Compared to silica glass, the bond angle distribution between <Si—O—Si > in quartz crystal is relatively narrow, with an average value of about 144°.

Figure 1.8c is a schematic diagram of the planar ring structure of silica glass. Due to the variation of <Si—O—Si > bond angle, it leads to the [SiO_2] tetrahedron is arranged in a disordered 2 ~ 10-membered ring structure in silica glass.

Figure 1.8d summarizes the distribution probabilities of 3 ~ 10-membered ring structures in silica glass obtained from three model simulations. Among them, the first model is classical molecular dynamics (black in the Figure); the second model is ab initio molecular dynamics (red in the Figure); the third model combines classical and ab initio molecular dynamics (green in the Figure). The statistical results of the three models all show that the probability of 5 ~ 7-membered ring structures is the highest, and the probability of 9 and 10-membered ring structures is the lowest. The occurrence probability of 3-membered and 4-membered ring structures depends on the simulation conditions and the thermal history of the sample.

Figures 1.9a, b are schematic diagrams of the planar 3-membered and 4-membered ring structures in silica glass, respectively. The Figure shows the changes in the <O—Si—O > and < Si—O—Si > bond angles (distributed as Φ and θ): the <O—Si—O > bond angle $\Phi = 109.5°$ is fixed, and the <Si—O—Si > bond angle θ value increases with the increase of the ring structure unit.

Figure 1.10a is a schematic diagram of the relationship between the formation energy (ΔE) of a single Si—O—Si bond and the <Si—O—Si > bond angle [21]. This result was calculated by Newton et al. [22] using ab initio self-consistent field molecular orbital methods. The formation energy of each planar ring structure $E_n = n \Delta E(\theta_n)$, where n is the number of [SiO_2] tetrahedra in the ring structure, θ_n is the <Si—O—Si > bond angle in the planar n -membered ring. Figure 1.10b is a schematic diagram of the relationship between the length of a single Si—O—Si bond and its bond angle. The larger the Si—O—Si bond angle, the shorter the corresponding bond length, so the more [SiO_2] in the planar k -membered ring, the shorter the Si—O—Si bond length in the ring. Table 1.2 gives the Si—O—Si bond angles of different planar membered rings, from which it can be seen that as the n value increases, the Si—O—Si bond angle significantly increases.

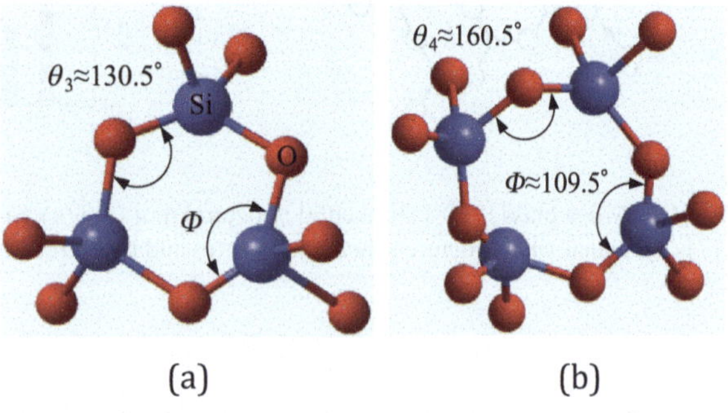

Fig. 1.9 Schematic diagram of planar 3-membered ring (**a**) and planar 4-membered ring (**b**) structures in silica glass

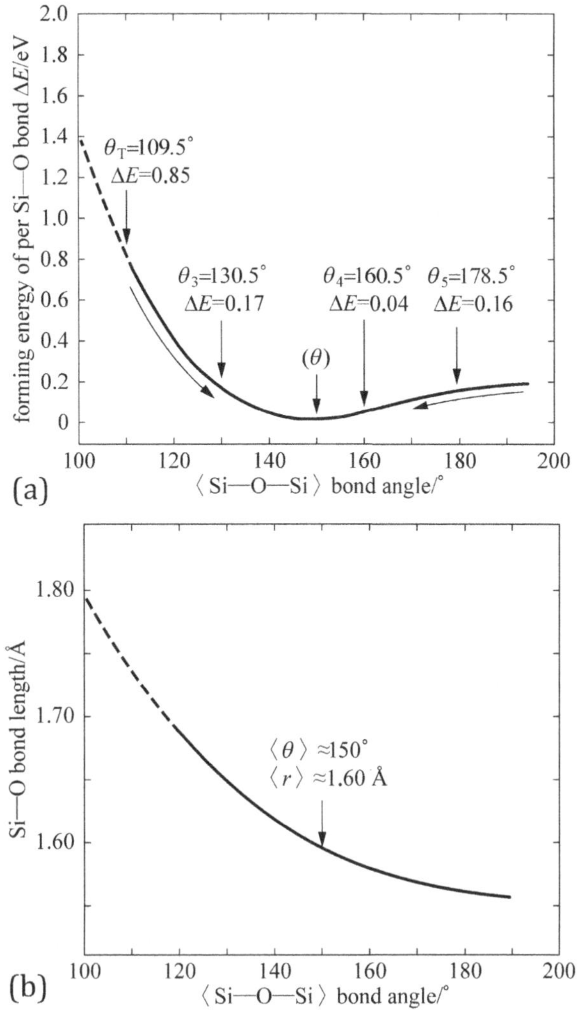

Fig. 1.10 Relationship between the formation energy (**a**) and length (**b**) of a single Si—O—Si bond and the <Si—O—Si > bond angle [21]

Table 1.2 Si—O—Si bond angles and total Si—O—Si bond formation energies in n -membered rings of silica glass [21]

n -membered ring	<Si—O—Si > bond angle θ_n/°	Formation energy E_n/eV
2	70.5	>5
3	130.5	0.51
4	160.5	0.16
5	178.5	0.80
6	190.5	1.08

Table 1.2 also lists the formation energies of n -membered rings in silica glass. As can be seen from the Table, the formation energy of the planar 4-membered ring is the smallest, followed by the planar 3-membered ring. The smaller the formation energy, the more stable the structure. Therefore, the content of 4-membered rings in silica glass is slightly higher than that of 3-membered rings. Studies have shown that the sum of the contents of planar 4-membered rings and planar 3-membered rings in silica glass often exceeds 1%. The formation energies of planar 4 ~ 9-membered ring structures can be greatly reduced by twisting the ring plane into a non-planar structure. However, the formation energies of planar 2-membered and 3-membered ring structures cannot be reduced by twisting the ring plane. Since the formation energy of the planar 2-membered ring is the largest, it is very unstable in the glass and has a small probability of existence. Planar 3-membered and 4-membered rings can reduce their formation energies by breaking bonds to form larger rings, and then twisting into non-planar structures.

Figure 1.11a, b are the Raman and infrared spectra of pure silica glass, respectively. The Raman spectrum was measured at an excitation wavelength of 633 nm, and the infrared spectrum was measured using the KBr pellet method. The corresponding peak assignments are shown in Table 1.3. The 487 cm $^{-1}$ and 603 cm $^{-1}$ in the Raman spectrum correspond to the vibrations of the Si—O—Si planar 4-membered ring (D_1) and planar 3-membered ring (D_2), respectively. D_1 and D_2 can be used as probe peaks for qualitatively evaluating the content of the planar 4-membered ring and planar 3-membered ring in silica glass.

Figure 1.11c, d are the magic angle spinning nuclear magnetic resonance (MAS NMR) spectra of ^{29}Si in pure silica glass and α-quartz crystal, respectively.

As can be seen from the Figure, both pure silica glass and quartz crystal consist of a symmetrical nuclear magnetic peak, which corresponds to the $Q^{(4)}$ structure of ^{29}Si, the superscript n of $Q^{(n)}$ indicates that there are n Si—O—Si bonds around each silicon, but the nuclear magnetic peak of pure silica glass is wider than that of quartz crystal, and its full width at half maxima, FWHM and the average Si—O—Si bond angle are listed in Table 1.4. Studies have shown that the full width at half maxima, FWHM of ^{29}Si MAS NMR is closely related to the distribution of Si—O—Si bond angles. The wider the distribution of Si—O—Si bond angles, the wider the corresponding NMR peak. Therefore, the full width at half maxima, FWHM of the NMR peak of pure silica glass is dozens of times that of quartz and cristobalite crystal.

Glass is a metastable substance, its structure is between the melt and the crystal. When the melt is quenched into glass, some structural features of the melt are preserved, resulting in the properties of the glass being equivalent to the properties of the melt at a certain temperature point during the cooling process. This characteristic temperature point was defined by Tool [24] as the fictive temperature of the glass (Fictive temperature, T_f). Above the glass softening temperature range, the fictive temperature of the glass is equal to the actual temperature. Below the transition temperature range, due to the very slow structural changes of the glass, the fictive temperature of the glass at room temperature is always higher than the actual temperature. In the range from the glass transition temperature to the softening temperature, the

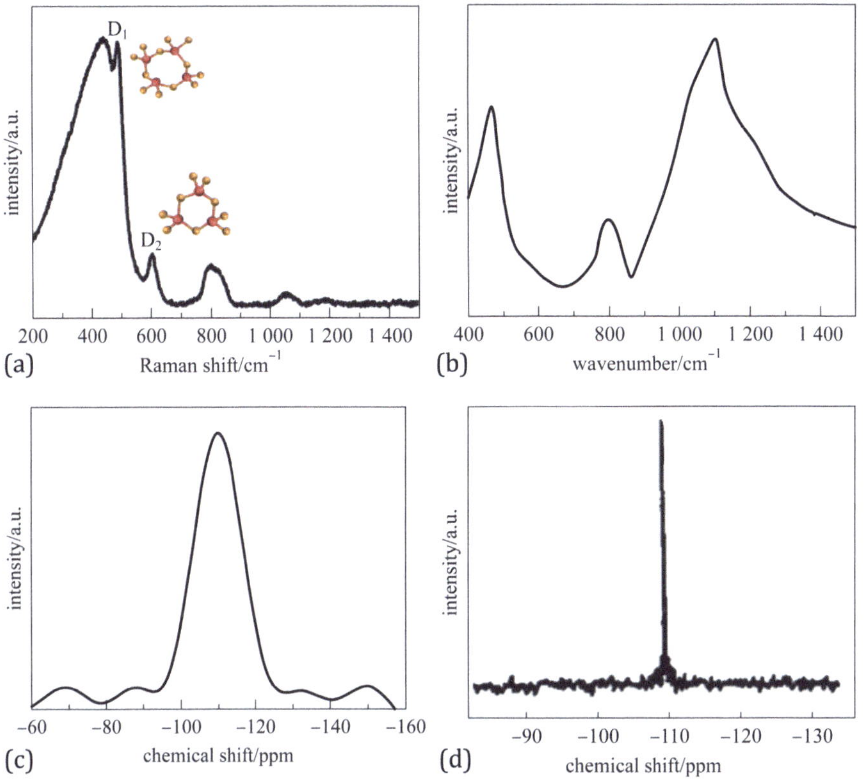

Fig. 1.11 Raman (**a**), infrared (**b**) of pure silica glass, and pure silica glass (**c**), quartz crystal (**d**) ^{29}Si MAS NMR [23]

Table 1.3 Assignment of Raman and infrared peaks of pure silica glass

Raman shift/cm $^{-1}$	Infrared absorption peak/cm $^{-1}$	Assignment
438	465	Si—O—Si bending vibration
487		Plane 4-ring vibration
603		Plane 3-ring vibration
800	800	Si—O—Si symmetric stretching vibration
1055, 1180	1000 ~ 1300	Si—O—Si antisymmetric stretching vibration

fictive temperature varies depending on the heating or cooling rate of the glass. Therefore, the fictive temperature of the glass is closely related to its thermal history. Usually, the faster the cooling rate of the melt, the higher the T_f of the glass, and the greater the structural disorder. Many properties of glass are related to its T_f, such as refractive index, density, ultraviolet transmittance, radiation resistance, etc.

Table 1.4 Chemical shift, full width at half maxima, FWHM and average Si—O—Si bond angle of ^{29}Si MAS NMR spectra of pure silica glass, quartz and cristobalite crystal

Sample	^{29}Si Chemical Shift/ppm	Full Width at Half Maxima, FWHM/ppm	Average Si—O—Si Bond Angle/°
Pure silica glass	−110.9	13.2	145.0
Quartz crystal	−107.2	0.20	143.6
Cristobalite	−108.5	0.20	146.4

Due to the large viscosity of silica glass melt at high temperature, silica glass can be formed at cooling rates ranging from hundreds of degrees per second to 1 °C per second. Therefore, the fictive temperature of silica glass can vary widely between 1000 ~ 1500 °C. Different fictive temperatures of silica glass correspond to different structures, including Si—O—Si bond angle distribution, Rayleigh scattering, X-ray small angle scattering, etc., which will change due to different fictive temperatures, resulting in changes in the density, refractive index and other properties of silica glass with the fictive temperature. The higher the fictive temperature of silica glass, the greater its density and refractive index.

The impact of thermal history on the structure of pure silica glass can be summarized as follows:

1. Bond angle changes. Figures 1.12a, b show the distribution of Si—O—Si bond angles in molten silica glass and the relationship between the average bond angle and the fictive temperature obtained by molecular dynamics simulation. The fictive temperature affects the distribution of Si—O—Si bond angles in silica glass, the higher T_f is, the wider the distribution of Si—O—Si bond angles, and the smaller the average bond angle of Si—O—Si (Si#1—O#3—Si#2, see the inset in Figure 1.12b), the denser the short-range structure of the glass.

   It has been reported that changes in thermal history can also cause changes in the distribution of O—Si—O, Si—Si—Si, O—O—O, Si—Si—O and O—O—Si bond angles. In the silicon-oxygen tetrahedron, the distribution of Si#1—O#3—Si#2 bond angles is wider than that of O1#—Si#1—O#3 and O1#—O#3—O#4 (see the inset in Figure 1.12b), indicating that the disorder in the short-range structure of silica glass mainly comes from the variability of Si—O—Si bond angles. Studies have found that a decrease in cooling rate will increase the probability of Si#1—O#3—Si#2, O1#—Si#1—O#3 and O1#—O#3—O#4 (see the inset in Figure 1.12b) bond angles at 143°, 109° and 60°, and the silicon-oxygen tetrahedron approaches a regular tetrahedron, as shown in Figures 1.12c, d. In addition, the distribution of O1#—O#3—O#2 bond angles formed by two silicon-oxygen tetrahedra sharing a vertex is wider. This is because this bond angle is related to both the relative position of the two connected silicon-oxygen tetrahedra and their relative direction. A decrease in cooling rate makes the distribution of O1#—O#3—O#2 bond angles wider and the average bond angle larger, leading to a looser medium-range structure, which may be one of the reasons for the decrease in density of silica glass. The

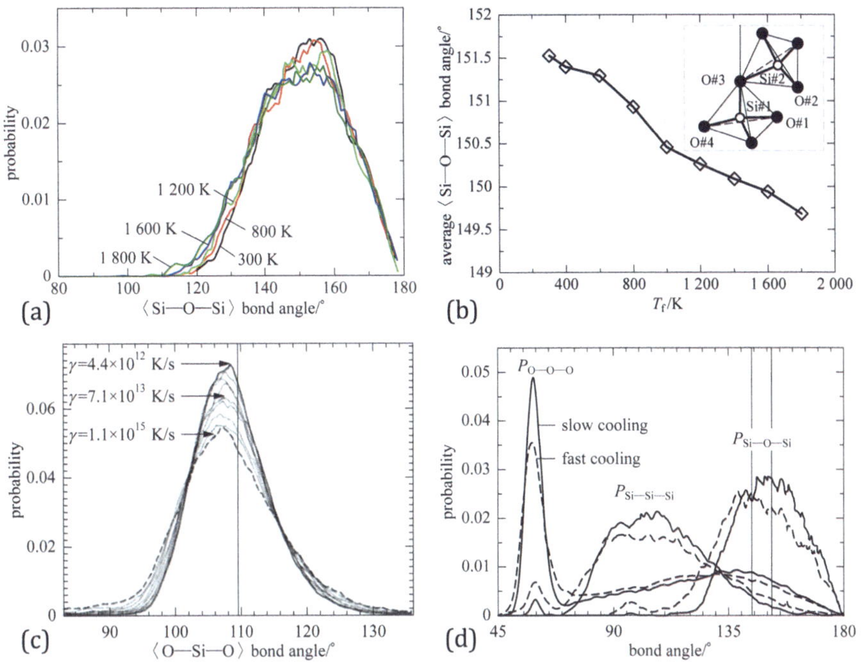

Fig. 1.12 Relationship between silica glass Si—O—Si bond angle distribution (**a**) and average Si—O—Si bond angle (**b**) with Fictive temperature, T_f [26]; Relationship between silica glass O—Si—O (**c**) and O—O—O, Si—Si—Si, Si—O—Si (**d**) bond angle distribution and cooling rate [27]

bond angle distribution of Si#1—Si#2—Si#3 bonds formed by three [SiO_2] can reflect the medium and long-range structure of the glass. The bond angle distribution of Si#1—Si#2—Si#3 consists of a small peak with an average bond angle of 60° and a wide peak with a bond angle distribution range of 90° ~ 120°. The former comes from planar 4-membered rings, and the latter comes from larger rings. A decrease in cooling rate will reduce the amount of the former and increase the amount of the latter, and the distribution becomes narrower.

2. Changes in ring structure distribution. The silicon-oxygen k ring structure can reflect the medium and long-range structure of silica glass. Figure 1.13a, b shows the relationship between the ring structure of pure silica glass and the cooling rate. Lowering the cooling rate will reduce the amount of small rings ($k < 5$) and large rings ($k > 8$), significantly increasing the amount of 6-membered rings. For example, the ring structures in the corresponding crystals of SiO_2 such as β-cristobalite, β-tridymite, and β-quartz formed at lower cooling rates are mostly 6-membered or 8-membered rings. Therefore, the slower the cooling rate, the lower the disorder of the glass structure, and the closer it is to the corresponding tridymite crystal structure.

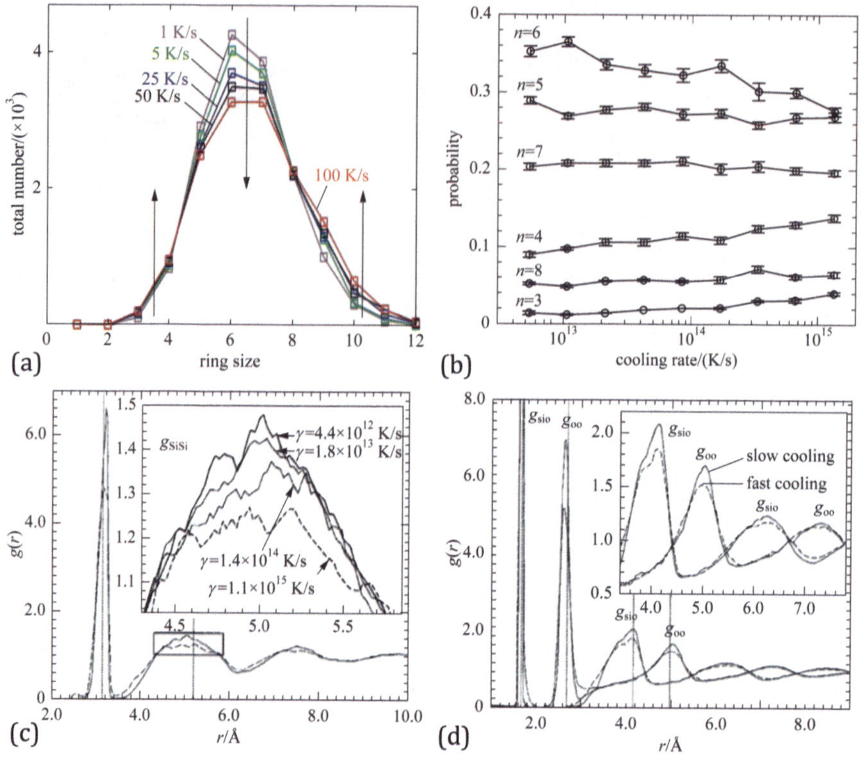

Fig. 1.13 The relationship between the radial distribution function curves of pure silica glass ring structure (**a**, **b**) [27], SiSi [28] (**c**), SiO and OO [28] (**d**) and the cooling rates

3. Changes in bond length. The atomic radial distribution function $g_{AB}(r)$ characterizes the probability of B atoms appearing at a distance of r from A atoms, reflecting the order of glass structure. As shown in Figures 1.13c, d, reducing the cooling rate will increase the probability of the distances between Si#1 and O#1 and between O#1 and O#3 in the silicon-oxygen tetrahedron being 1.6 Å and 2.62 Å, respectively, and will narrow the $g(r)$ curve, increasing the order of the near and medium-range structure ($r \le 8$ Å) of glass. The larger the peak distance between Si#1 and its first nearest neighbor Si#2 atoms and between O#1 and its second nearest neighbor O#2 atoms in the medium-range structure, the looser the medium-range structure of the glass.

The properties of glass are closely related to its thermal history, and the dependence of properties of different glasses on thermal history is different. For example, the density and refractive index of pure germanium dioxide and germanosilicate glass are negatively correlated with their fictive temperature, T_f [29], as shown in Fig. 1.14. In contrast, the density (Figure 1.14b, Fig. 1.40) and refractive index (Fig. 1.39) of pure silica glass are positively correlated with its fictive temperature, T_f [30]. From the above analysis, it can be seen that the lower the fictive

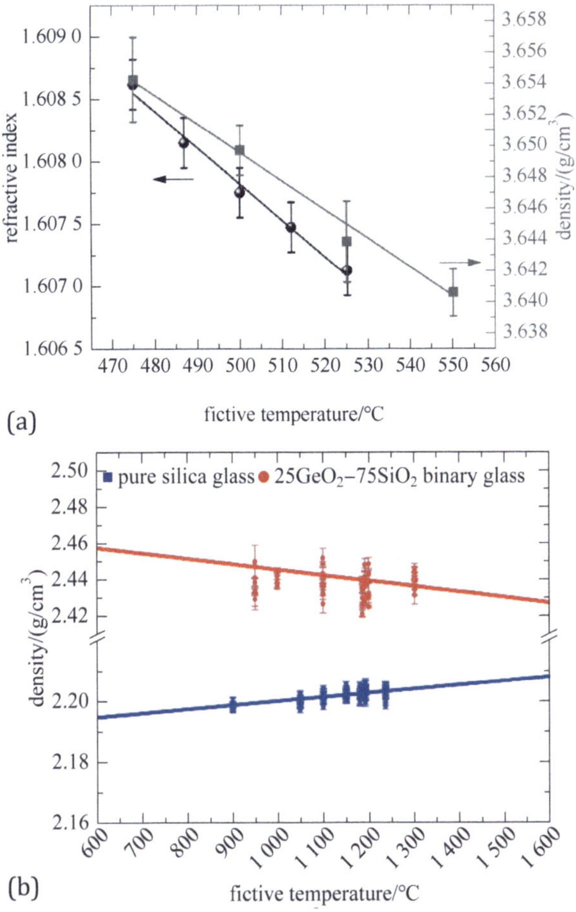

Fig. 1.14 The relationship between the refractive index of pure GeO_2 glass and the fictive temperature, T_f [29] (**a**); The relationship between the density of pure SiO_2 glass and $25GeO_2-75SiO_2$ (mol%) binary glass and the fictive temperature, T_f [30] (**b**)

temperature, the closer the short-range structure [SiO_2] of the glass is to the regular tetrahedron, and the medium and long-range structure is closer to the corresponding crystal. Compared with the short-range structure, the connection between silicon-oxygen tetrahedra is more closely related to the fictive temperature. The lower the fictive temperature, the larger the Si—O—Si and O—O—O bond angles of the two-bridge connected silicon-oxygen tetrahedra, the larger the Si—Si distance, which makes the structure of the glass looser, thereby leading to a decrease in the overall density of the glass. Table 1.5 summarizes the relationship between the physical and mechanical properties of silica glass and the fictive temperature (T_f).

Table 1.5 The impact of fictive temperature on the properties of silica glass [25]

Property	Increase in fictive temperature	Property	Increase in fictive temperature
Density	Increase	Compressibility	Decrease
Refractive index	Increase	Shear modulus	Decrease
Viscosity	Decrease	Young's modulus	Increase
Thermal expansion coefficient	Increase	Rayleigh scattering	Increase
Hydrofluoric acid erosion rate	Increase	Radiation resistance	Deteriorate
Moisture diffusion coefficient	Decrease	Density fluctuation	Increase
Hardness	Increase		

1.3 Thermal Properties of Silica Glass

As mentioned earlier, the formation and properties of silica glass are closely related to its thermal history and preparation process. The thermal properties of silica glass include glass transition temperature, thermal expansion coefficient, viscosity, thermal conductivity, and crystallization.

1.3.1 Glass Transition Temperature

Figure 1.15 is a curve of the volume change of Type I/II silica glass with temperature. It can be seen that the glass transition temperature of silica glass varies with the cooling rate. The faster the cooling rate, the higher the transition temperature ($t_{gf} > t_{gs}$), and the supercooled melt presents a temperature with the minimum specific volume during the cooling process. The impurity content in silica glass will also affect its glass transition temperature. Yue et al. [4] used differential thermal analysis to test the heat capacity method to study the glass transition temperature (T_g) of silica glass with different hydroxyl contents, as shown in Fig. 1.16. They found that the T_g temperature of silica glass with 1021 ppm OH^- is 98 K lower than that of silica glass with 1 ppm OH^- (the former is 1336 K, the latter is 1434 K).

In addition to the above-mentioned hydroxyl which can cause the transition temperature of silica glass to decrease, the trace amounts of other oxides in silica glass will significantly affect its inherent structure, thereby causing the glass transition temperature to decrease in different degrees. Figure 1.17 shows the trace amount of Na_2O, OH, Al_2O_3 and Ga_2O_3 affect the transition temperature of silica glass [31]. Using the characteristics that monovalent alkali metal oxide Na_2O can significantly reduce the glass transition temperature of silica glass, ultra-low loss communication silica fibers doped with trace amounts of Na_2O have been successfully developed.

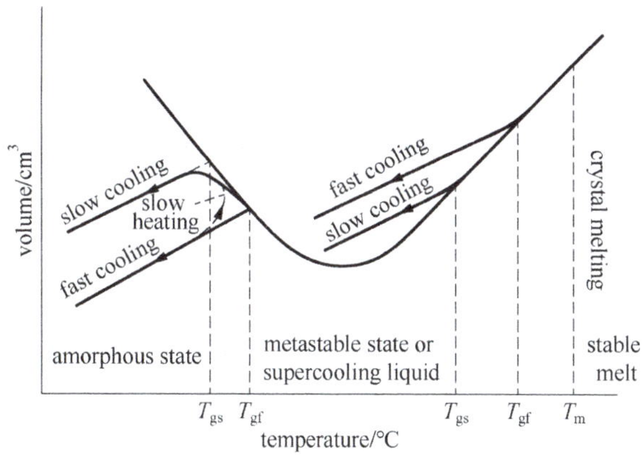

Fig. 1.15 Volume-Temperature curve of Type I/II silica glass

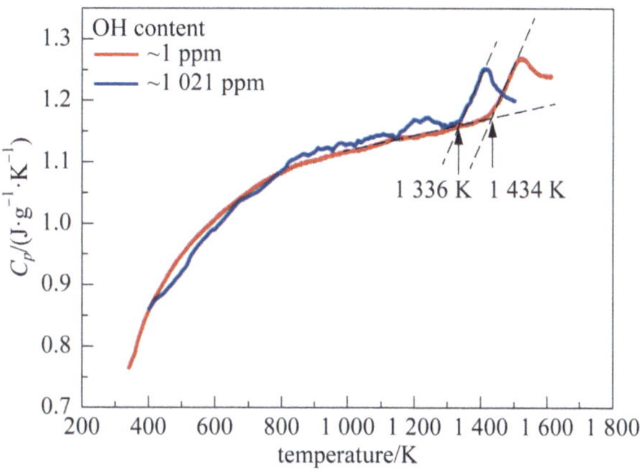

Fig. 1.16 Relationship between the isobaric heat capacity (C_p) and temperature of silica glass with different hydroxyl contents [4]

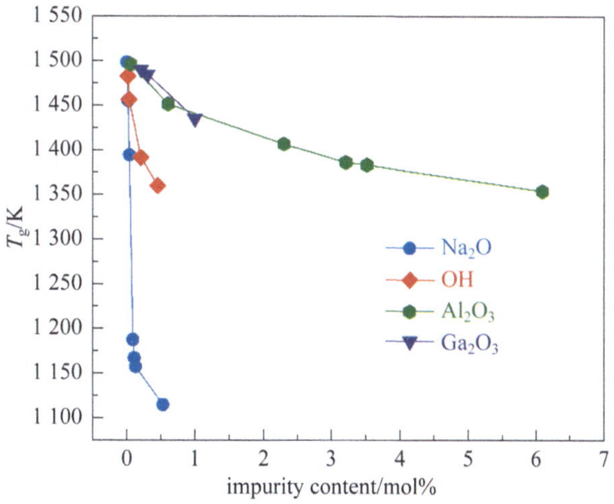

Fig. 1.17 Relationship between silica glass T_g and impurity contents

1.3.2 Thermal Expansion Coefficient

Figure 1.18 shows the thermal expansion coefficients of Type I/II/III silica glass at different fictive temperatures, T_f. It can be seen that thermal expansion coefficient of Type I/II silica glass with a fictive temperature, T_f of 1550 °C is larger. The Type III silica glass with a fictive temperature, T_f of 1460 °C shows the maximum thermal expansion coefficient.

As shown in Fig. 1.19, the thermal expansion curve test results of Type I/II silica glass are also affected by its thermal history. Quenched Type I/II silica glass samples show volume shrinkage in the first thermal expansion curve measurement.

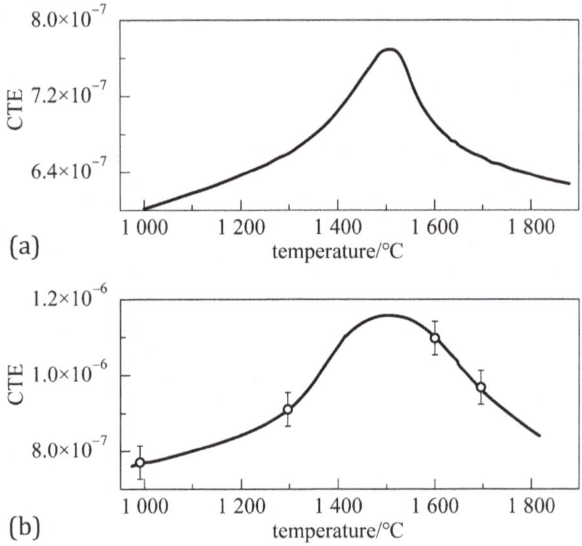

Fig. 1.18 Relationship between the thermal expansion coefficient and fictive temperature, T_f of Type I/II (**a**) and Type III (**b**) silica glass [1]

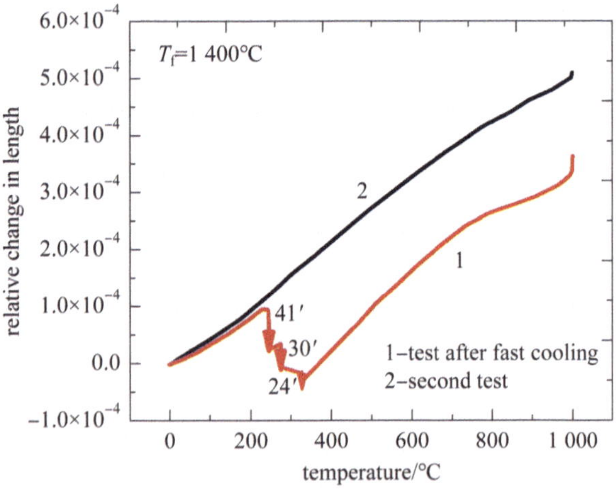

Fig. 1.19 Thermal expansion curve of Type I/II silica glass [1]

This may be related to the release of stress formed during the cooling process. No volume shrinkage was observed in the curve of the second test. This indicates that the stress was released during the first measurement, and the glass structure was also relaxed and repaired.

1.3.3 Viscosity

The melt of silica glass has a high viscosity, and its viscosity-temperature curve spans a large temperature range. The viscosity of silica glass is closely related to its thermal enthalpy, strain, volume, refractive index, and other properties of relaxation. Therefore, its relationship with temperature is widely studied. However, due to the following three reasons, it is difficult to accurately measure the viscosity of silica glass. First, the measurement temperature is high, the melt reacts with the crucible and atmosphere, affecting the measurement. Second, 1200 ~ 1700 °C is its crystallization temperature range. Third, below about 1400 °C, due to the viscosity of the silica glass melt, the time for the viscosity to reach equilibrium becomes longer. In addition, the impurity content in the silica glass obtained by different preparation processes will also affect its viscosity value.

In 1982, Urbain et al. [32] used high-purity silica glass (alkali metal less than 20 ppm, OH^- less than 10 ppm, Al less than 20 ppm) samples to test the viscosity data in the temperature range of 1200 ~ 2480 °C, the results are shown in Table 1.6. It was concluded that the logarithmic value of the viscosity of silica glass in this temperature range is linearly related to the reciprocal of temperature, that is, the change in viscosity with temperature satisfies the Arrhenian relationship, and the activation energy is 515.4 kJ/mol.

In 2002, Doremus [33] summarized the silica glass viscosity data published by different authors, believing that the test data of Urbain in the high temperature area of 1400 ~ 2500 °C is accurate, and the viscosity data tested by the fiber stretching method in the range of 1000 ~ 1400 °C is more accurate, as shown in Table 1.7. And in the temperature range of 1000 ~ 1400 °C, the viscous flow activation energy for stabilizing silica glass is 712 kJ/mol, much higher than 515 kJ/mol at high temperatures.

Table 1.6 Viscosity data of silica glass at 1200 ~ 2400 °C (viscosity unit is Poise) [32, 33]

Temperature/°C	lg(Viscosity)	Temperature/°C	lg(Viscosity)
2482	3.53	1652	7.79
2382	3.92	1599	8.08
2268	4.36	1438	9.48
2168	4.80	1375	9.97
2061	5.25	1306	10.95
1964	5.79	1250	11.40
1870	6.36	1192	12.15
1776	6.90		

Table 1.7 Viscosity data of silica glass measured by fiber stretching method in the range of 1000 ~ 1400 °C (viscosity unit is Poise) [33]

Temperature/°C	1400	1300	1200	1100	1000
lg(viscosity)	9.81	11.22	12.83	14.65	16.82

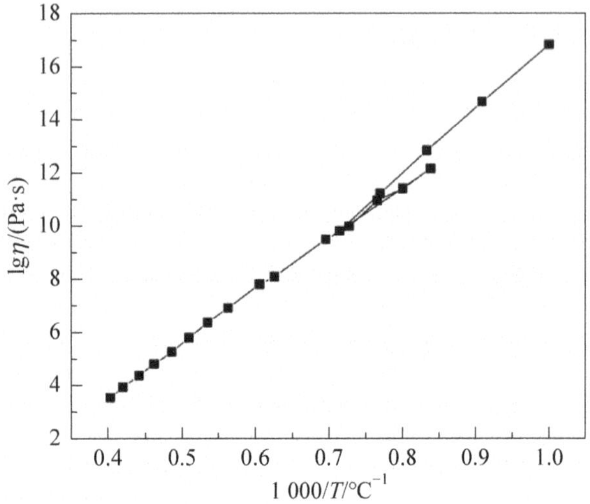

Fig. 1.20 Viscosity-temperature curve of silica glass

Figure 1.20 is the viscosity curve of silica glass in the range of 1000 ~ 2500 °C, derived from the data published by Urbain and Hetherington. It can be seen that there is a turning point in the relationship curve between viscosity and temperature around 1400 °C. In the temperature region below 1400 °C, the slope of the curve corresponding to the relationship between the logarithm of the viscosity of silica glass and the reciprocal of temperature (712 kJ/mol) is greater than the slope in the high temperature region above 1400 °C (515 kJ/mol). That is, different temperature ranges correspond to different flow activation energies.

Doremus [33] analyzed the change of the temperature characteristics of the viscosity of silica glass from the perspective of structural defects, and believed that the viscosity of silica glass depends on the movement of linear defects composed of SiO molecular defects in the glass structure. At high temperatures, the violent movement of silicon and oxygen atoms leads to the destruction of Si-O bonds in the silicon-oxygen tetrahedron, forming SiO and oxygen defects. The concentration of SiO defects decreases with decreasing temperature, and the amount of SiO defects below 1400 °C is related to time. Therefore, below 1400 °C, the measured viscosity of silica glass melt will change with time until it reaches the equilibrium viscosity value.

Impurities can cause changes in the viscosity and thermal diffusion coefficient of silica glass. For example, the room temperature thermal diffusion coefficient of silica glass can fluctuate in the range of 0.795 ~ 0.870 $mm^2 \cdot s^{-1}$. The high temperature viscosity of silica glass will also change due to different hydroxyl content. As

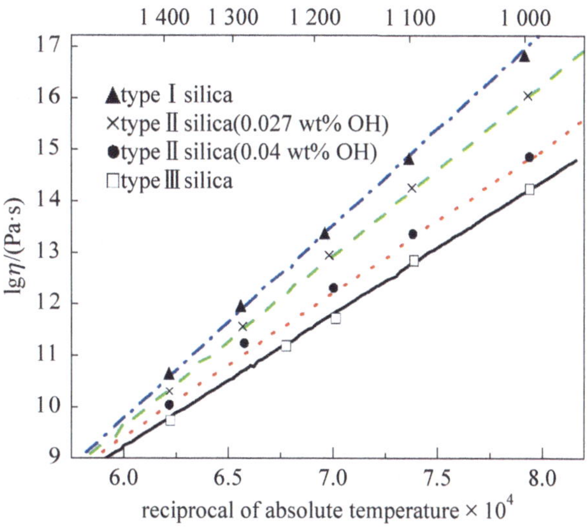

Fig. 1.21 Viscosity temperature curve of Type I, type II and type III silica glass [34]

shown in Fig. 1.21, the viscosity of Type I/II silica glass in the range of 1000 ~ 1400 °C is significantly higher than that of Type III silica glass with high hydroxyl content.

Tomozawa and others [35] used infrared spectroscopy to study the structural changes caused by the diffusion of water vapor into silica glass at a temperature range of 80 ~ 1150 °C and 0.467 atmospheric pressure. They found that a very small amount of water can significantly accelerate the structural relaxation of silica glass, and relaxation will further affect the water content in the glass. They believe that: the slower structural relaxation speed at lower temperatures will prevent the reaction between water and glass; the expansion caused by water entering the glass will accelerate the entry of water molecules into the glass; the relaxation process of bulk silica glass will reduce the solubility of hydroxyl in the glass.

1.3.4 Thermal Conductivity

Figure 1.22 shows the relationship between the thermal conductivity and temperature of several different crystals and glasses. It can be clearly seen that crystals and glasses have different thermal conductivity temperature dependencies. Similar to photons in optical energy transmission, phonons play an important role in thermal conduction. The longer the nominal free length l of the phonon, the higher the thermal conductivity:

$$\lambda_{\text{th}} = \frac{1}{3} cv \tag{1.1}$$

$$q = -\lambda_{\text{th}} \Delta T / \Delta x \tag{1.2}$$

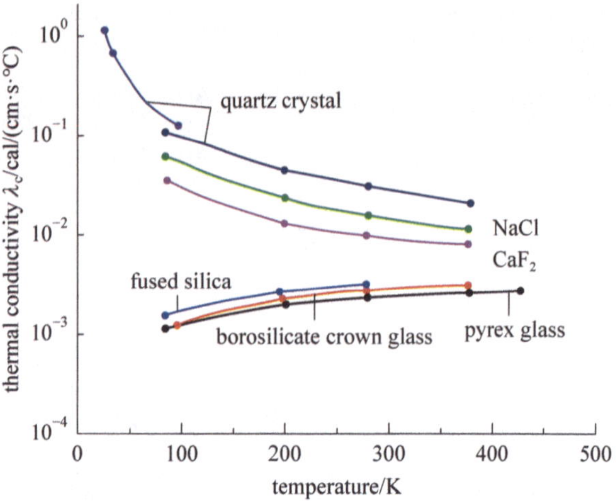

Fig. 1.22 Relationship between thermal conductivity and temperature of several different crystals and glasses [1]

where, λ_{th} is thermal conductivity; c is the heat capacity per unit volume; v is the speed of sound; l is the nominal free length; q is the heat conduction speed per unit area per unit time; Δx represents the distance with a temperature difference of ΔT. Crystals have a large free length l, so their thermal conductivity λ is large. Since glass lacks a periodic and symmetrical structure, its thermoelastic waves cross-transfer energy. Therefore, the nominal free length l is shorter, and its thermal conductivity is low. As can be seen from Fig. 1.22, the thermal conductivity of silica glass is much lower than that of quartz crystal, but the thermal conductivity of silica glass is the largest among several types of glass. This may be related to the more robust structure of silica glass.

Table 1.8 gives the thermal conductivity data of transparent silica glass at different temperatures from 90 to 500 K [36], from which it can be seen that its thermal conductivity increases with temperature.

1.3.5 Crystallization

Another important thermal property of silica glass is its crystallization performance. Although the composition of silica glass only contains one oxide, silicon dioxide, it is the simplest of all glass systems. However, silicon dioxide has complex polycrystallinity; under normal pressure and impurities, silicon dioxide has seven crystal phases, one liquid phase, and one gas phase. Under high pressure, there are new variants of silicon dioxide, which can transform into each other at certain temperatures and pressures. Therefore, the silicon dioxide system is a unit system with complex polycrystalline phase transitions.

Figure 1.23 is the phase diagram of SiO_2 system. Under normal pressure conditions, as the temperature gradually increases, one can observe α -quartz, β-quartz,

Table 1.8 Thermal conductivity data of transparent silica glass in the temperature range of 90 ~ 500 K

Temperature/K	Thermal conductivity/ (W·m $^{-1}$·K $^{-1}$)	Temperature/K	Thermal conductivity/ (W·m $^{-1}$·K $^{-1}$)
90	0.627	190	1.135
95	0.654	220	1.214
100	0.684	230	1.236
105	0.716	240	1.257
110	0.749	250	1.277
115	0.782	260	1.295
120	0.814	270	1.313
125	0.845	280	1.329
130	0.875	290	1.345
135	0.904	300	1.361
140	0.931	320	1.392
145	0.957	340	1.421
150	0.982	360	1.450
155	1.005	380	1.479
160	1.026	400	1.509
165	1.047	420	1.536
170	1.067	440	1.564
175	1.085	460	1.539
180	1.102	480	1.662
185	1.119	500	1.650

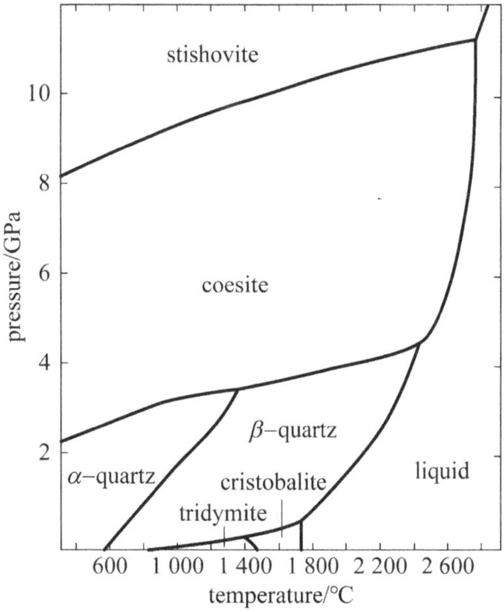

Fig. 1.23 Phase diagram of the SiO_2 system. (Adapted from reference [12])

tridymite, cristobalite, melt and other homogeneous polymorphs. Their transformation relationships are as follows:

$$\alpha-\text{quartz} \overset{573^{\circ}\text{C}}{\Leftrightarrow} \beta-\text{quartz} \overset{870^{\circ}\text{C}}{\Leftrightarrow} \text{tridymite} \overset{1470^{\circ}\text{C}}{\Leftrightarrow} \text{cristobalite} \overset{1713^{\circ}\text{C}}{\Leftrightarrow} \text{melt}$$

Below 573 °C, α -quartz is the thermodynamically most stable variant. When the temperature reaches 573 °C, α -quartz transforms into β -quartz. Since the transformation only involves changes in bond angles and does not involve breaking and restructuring of the crystal structure, the transformation process is rapid and reversible. As the temperature continues to rise, at 870 °C, β -quartz transforms into tridymite. Since the transformation involves breaking and restructuring of the crystal structure, this transformation process is very slow. Moreover, tridymite is less stable, and its structure usually needs to be stabilized by trace amounts of other metal oxides. Therefore, if the heating rate is fast, β -quartz overheats and directly melts at 1600 °C, no longer transforming into cristobalite. If the heating rate is slow, β-quartz first transforms into tridymite, and then transforms into cristobalite at 1470 °C. As the temperature continues to rise, cristobalite melts at 1713 °C.

As the pressure increases, the structurally unstable cristobalite and tridymite gradually disappear, and the more structurally stable α -quartz and β -quartz can continue to exist, and their melting points increase with increasing pressure. Under higher pressure conditions, α -quartz and β -quartz transform into coesite. Coesite can only exist stably under high pressure conditions of 4 ~ 10GPa (i.e., 40 ~ 100 kbar, 1 kbar = 0.1GPa). When the pressure is even greater, coesite will transform into denser stishovite. Coesite and stishovite are products of ultra-high pressure metamorphism, and can be found in meteorite craters and nuclear explosion craters.

A significant feature of silica glass crystallization is that the chemical composition of the crystalline phase is the same as that of the glass phase. Therefore, heterogeneous nucleation is the only way for its crystallization. The nucleus of heterogeneous nucleation can be internal impurities or the outer surface. If kept at a temperature above 1100 °C for a sufficient long time, crystallization can even occur on the surface of high-purity silica glass. Surface crystallization is a zero-order reaction, that is, the growth rate is linear. Different types of silica glass have different crystallization behaviors. The crystallization rate of Type III silica glass with higher OH^- is generally faster than that of Type I/II silica glass. The former often manifests as surface crystallization, while the latter, in the presence of a higher impurity content, may also undergo bulk crystallization. The higher the temperature, the faster the crystallization rate. The surface crystallization rates of dehydroxylated Type III silica glass at 1486 °C under different atmospheres are shown in Table 1.9. It can be seen that the atmosphere significantly affects the crystallization rate at the same reaction temperature. The fastest crystallization rate of the glass is under a steam atmosphere, followed by an oxygen atmosphere; the slowest crystallization rate is under vacuum conditions.

If the same type of glass at 1400 °C contains 0.32 wt% Na_2O, then its crystallization rate can reach 670 μm/min. This is mainly due to the impurity Na^+ ion

Table 1.9 Surface crystallization rate of Type III silica glass at 1486 °C under different atmospheres [1]

Atmosphere	Surface Crystallization Rate/(μm/min)	Atmosphere	Surface Crystallization Rate/(μm/min)
452 mm H_2O	1.02	Vacuum	0.12
Dry nitrogen or oxygen	0.42		

accelerating the ion migration rate at high temperatures, making structural reorganization easier.

An interesting phenomenon is that the measured surface crystallization rate of Type III silica glass under vacuum in Table 1.9 (0.12 μm/min) is far from the theoretical speed of 0.02 μm/min. Therefore, it cannot be ruled out that even in a vacuum environment, there are still a very small amount of impurities on the glass surface, making the actual crystallization rate much higher than the theoretical rate. Under high pressure conditions, the crystallization behavior of silica glass is closely related to the water content in the environment and the water content in the compressed silica glass. To a certain extent, water plays the role of a nucleating agent.

In two cases, there is preferential orientation of cristobalite crystallization: one is at 1200 °C for 6 h, where the (101) face of low-temperature cristobalite can be observed to have preferential orientation. The other is silica fiber crystallizes to oriented low-temperature and high-temperature cristobalite phases at 1200 °C or 1300 °C. However, if the silica fiber is heat-treated at 1000 °C for more than 10 h, no cristobalite with preferential orientation will be precipitated. This is because the annealing heat treatment intensifies the structural reorganization of the silica glass, reducing the possibility of preferential orientation of the crystalline phase.

A beneficial use of silica glass crystallization is the preparation of quartz ceramic crucibles or high-temperature refractory bricks. Since the melting point of the precipitated cristobalite crystal is 1723 °C, the quartz ceramic crucible can be used at high temperatures of 1550 ~ 1600 °C. At this temperature, transparent silica glass has already softened and deformed. It is worth noting that the quartz ceramic crucible will crack when cooled to below 272 °C, mainly due to the transformation of α -quartz to β -quartz, which causes changes in thermal expansion coefficient.

1.4 Mechanical Properties of Silica Glass

Silica glass is a brittle material with strong compressive strength. The mechanical properties of silica glass are closely related to its preparation method. The mechanical properties of silica glass can be represented by elastic modulus, including Young's modulus E, shear modulus G and Poisson's ratio μ. These parameters represent the ability of silica glass to resist deformation under external stress and temperature changes.

The following three equations express the relationship between Young's modulus(E), shear modulus (G) and Poisson's ratio(μ):

$$E = 2G(1+\mu) \tag{1.3}$$

$$G = \frac{E}{2(1+\mu)} \tag{1.4}$$

$$\mu = \frac{E - 2G}{2G} \tag{1.5}$$

Unlike common glass, the *E* and *G* of silica glass increase with temperature. The coefficient of temperature change for *E* between 0 ~ 900 °C is 10.198 ~ 13.600 MPa/°C, and the relationship between *G* and temperature is 3.098 MPa/°C. Figure 1.24 shows the relationship between Young's modulus E and temperature in temperature range of −200 ~ 1200 °C. It can be seen that the *E* value almost linearly increases with temperature in 0 ~ 900 °C. The increase in the *E* value slows down with temperature in 900 ~ 1200 °C.

Table 1.10 gives the main mechanical properties of different silica glasses. The density of opaque silica glass is lower than that of synthetic silica glass and transparent silica glass due to its higher porosity.

The compressive strength of silica glass is 23 times of its tensile strength. The effects of pressure on silica glass can be considered from the following three aspects: ① Below the fictive temperature, T_f, applying pressure to silica glass in an isostatic manner can achieve reversible changes in density (or volume) within a large pressure range; ② In solid glass, if sufficient pressure is applied, it will cause irreversible or permanent volume changes; ③ At or above the fictive temperature, T_f, the volume relaxation time of compressed glass is short. At such temperatures, applying pressure below 10 kbar can lead to a significant increase in density. Cooling with

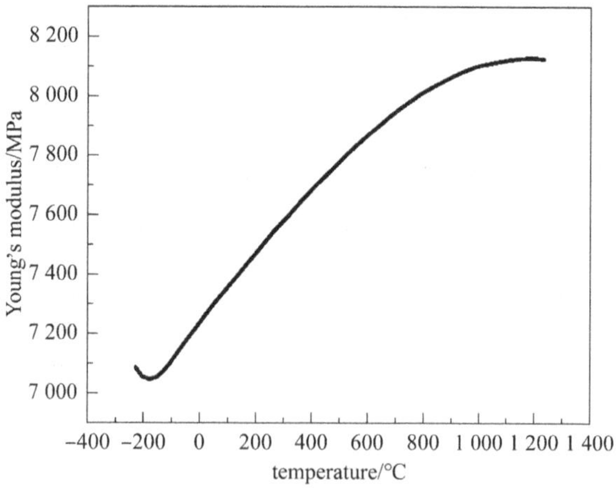

Fig. 1.24 Relationship between Young's modulus *E* of silica glass and temperature [36]

Table 1.10 Main mechanical properties of silica glass [5]

Properties	Transparent silica glass	Synthetic silica glass	Opaque silica glass
Density/(g/cm³)	2.203	2.201	2.07 ~ 2.12
Elastic modulus/MPa	7.25×10^4	7×10^4	
Poisson's ratio	0.17	0.17	
Compressive strength/MPa	1150	1150	About 500
Tensile strength/MPa	50	50	About 40
Bending strength/MPa	67	67	Approximately 67
Mohs hardness	5.5 ~ 6.5	5.5 ~ 6.5	
Microhardness	8000 ~ 9000	8000 ~ 9000	

pressure can permanently preserve such densification processes to a low temperature state.

Reversible volume changes can be represented by the compressibility coefficient (coefficient of compressibility) β_{m} as follows [1]:

$$\beta_{m} = a(T) + b(T)p \tag{1.6}$$

$$\beta_{m} = \frac{1}{V_0}\frac{V_0 - V_p}{p - p_0} \tag{1.7}$$

$$a(T) = (26.43 - 0.0025T) \times 10^{-7} \left(\mathrm{cm}^2/\mathrm{kg}\right) \tag{1.8}$$

$$b(T) = (43.6 - 0.080T) \times 10^{-12} \left(\mathrm{cm}^2/\mathrm{kg}\right)^2 \tag{1.9}$$

In the formula, the unit of T is °C. The above expression is applicable in the temperature range of 22 ~ 260 °C and the pressure range of 0 ~ 3000 kg/cm². In this temperature range, under a pressure of 12 kbar, there will be no permanent volume change.

Unlike common glass, the compressibility of silica glass increases with pressure, and as the temperature rises, its compressibility dependence on pressure decreases. This leads to an increase in thermal expansion coefficient of silica glass with pressures as shown in Fig. 1.25.

The refractive index of silica glass increases with the increase of isostatic pressure, but above 4 kbar pressure, its change with pressure deviates from linearity, as shown in Fig. 1.26. In addition, as shown in Fig. 1.27, the change in the refractive index of silica glass also shows a linear relationship with the Lagrangian strain below 7 kbar pressure [37]. This may be because under the action of isostatic pressure, there are vacancies composed of oxygen defects in the silica glass, and these oxygen defects form Si—O—Si bonds under pressure, making the vacancies smaller or even disappear. Therefore, the refractive index increases.

As the pressure continues to increase, permanent volume changes caused by shear stress will occur. Fig. 1.28 shows that, under the same room temperature

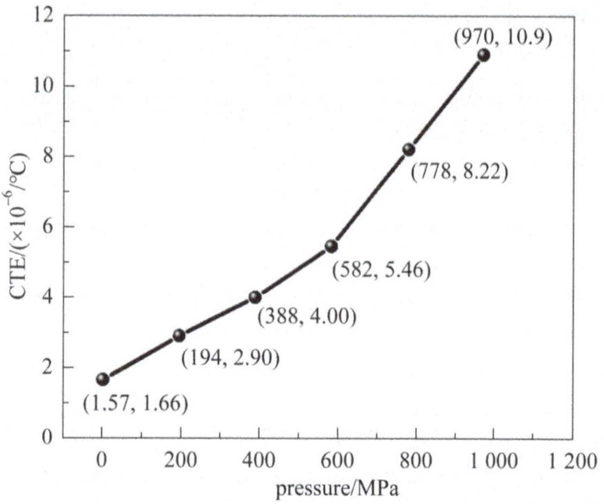

Fig. 1.25 Volume expansion coefficient of silica glass between 11 ~ 390 °C. Changes with pressure (adapted from reference [1])

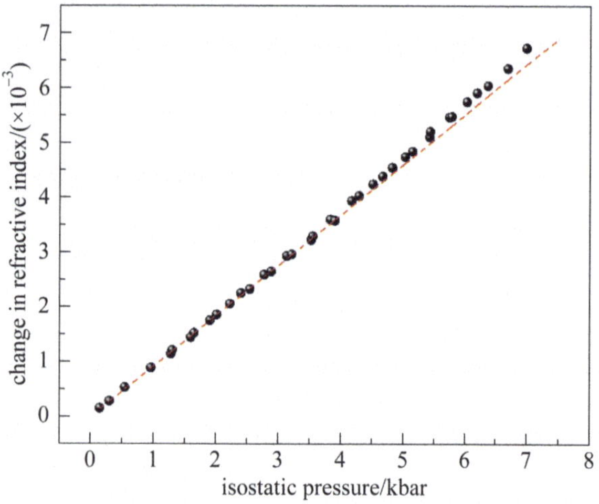

Fig. 1.26 Relationship between the refractive index of silica glass and isostatic pressure [37]

conditions, the greater the shear stress, the larger the proportion of permanent volume change. This phenomenon can be explained by the "flow" caused by shear stress. As shown in Fig. 1.29, in the absence of shear stress, silica glass exhibits elastic contraction. However, under the action of shear stress components and large stress, silicon-oxygen bonds break, and under the action of shear stress, the broken bond parts displace and form bonds, resulting in irreversible deformation [38].

Under the action of pressure, the densification degree of silica glass is closely related to its preparation process. Under the same pressure conditions, the densification degree increases from type IV silica glass, type III silica glass to type I silica

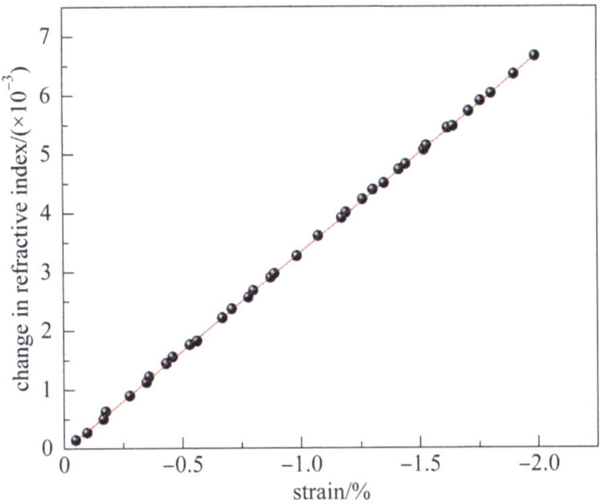

Fig. 1.27 Relationship between the change in the refractive index of silica glass and Lagrangian strain

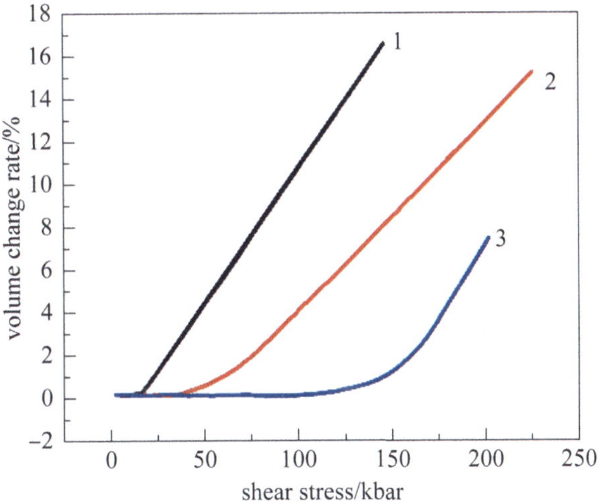

Fig. 1.28 Relationship of permanent volume change rate of silica glass at room temperature with shear stress [38]. 1—Large shear stress component; 2—Medium shear stress component; 3—Small shear stress component

glass. This may be related to the relatively high openness of the structure due to more metal impurities and oxygen defects in type I silica glass.

After annealing the compressed silica glass, the molar volume shows the pattern in Fig. 1.30 with annealing time and temperature. It can be seen that the higher the annealing temperature, the larger the molar volume after final stabilization; the first few minutes of annealing show a rapid increase in molar volume, and then gradually stabilize.

As shown in Fig. 1.31, near the glass transition temperature of 1080 °C, applying a pressure of 114 atm (1 atm = 101,325 Pa) can increase the density of silica glass by 5×10^{-4} g/cm^3 within 10 h. If cooled under pressurized conditions, this densified

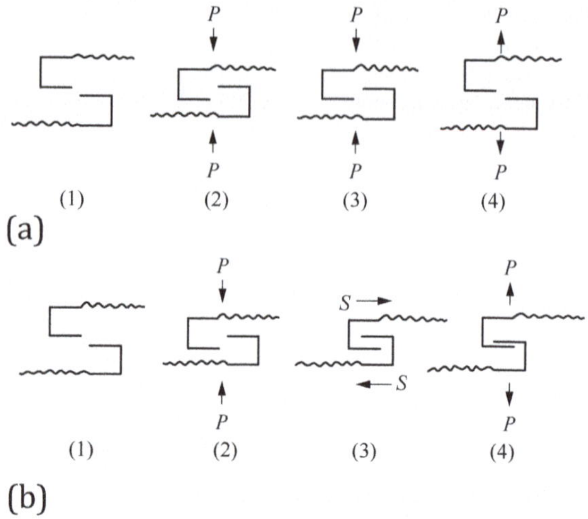

Fig. 1.29 Reversible and irreversible volume compression of glass. Under no shear stress (**a**) and shear stress (**b**) conditions

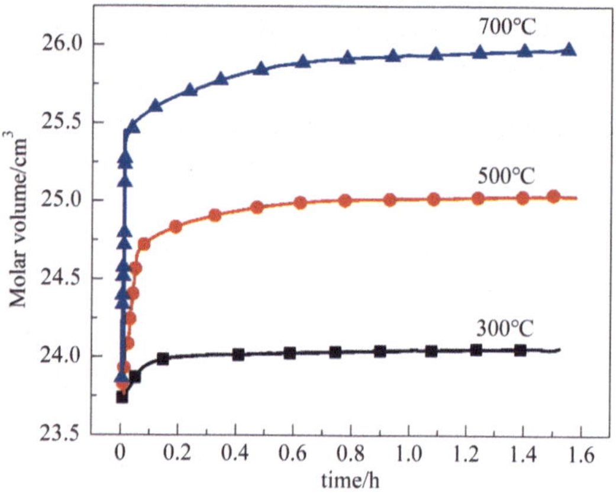

Fig. 1.30 Changes in molar volume during annealing at 300 °C, 500 °C, and 700 °C for silica glass with initial molar volume V_0 of 23.83 cm^3/mol [1]

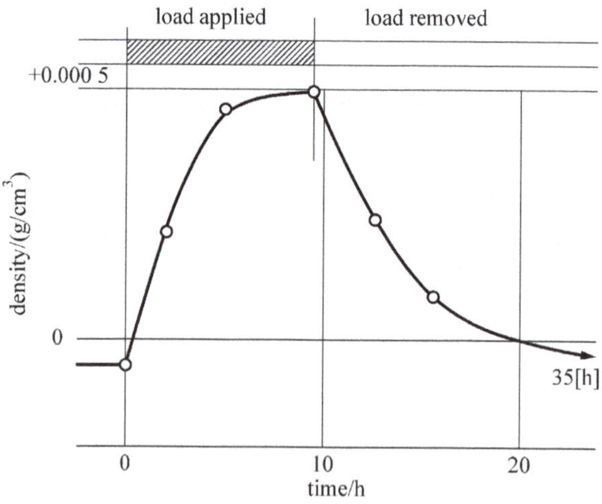

Fig. 1.31 densification of silica glass at 1080 °C at 114 atm and subsequent density changes without external pressure [1]

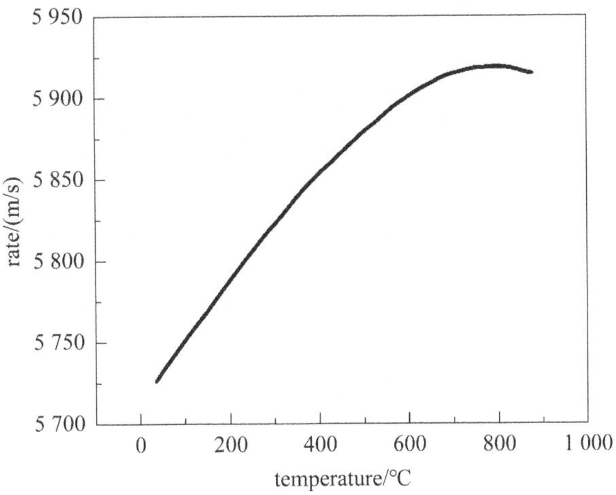

Fig. 1.32 Longitudinal wave propagation speed in synthesized silica glass and its relationship with temperature [36]

state can be preserved. If the pressure is released at this temperature, the volume contraction can be restored in about 10 h.

The propagation speed of sound waves in silica glass is affected by its optical homogeneity, and temperature also affects the propagation speed of sound waves in silica glass. The propagation speed of longitudinal waves in synthesized silica glass is 5720 m/s, and the propagation speed of transverse waves is 3769 m/s. As shown in Fig. 1.32, the propagation speed of longitudinal waves increases with temperature.

1.5 Optical Properties of Silica Glass

1.5.1 Refractive Index and Dispersion

Refractive index and dispersion are two important basic parameters of optical glass. The following introduces the refractive index and dispersion characteristics of silica glass.

As we all know, density and polarizability are two main factors affecting the refractive index. The refractive index of glass can be expressed by the Lorentz-Lorenz formula as follows:

$$\frac{n^2-1}{n^2+2}=\frac{4\pi}{3}\frac{N_A\rho}{M}\alpha \tag{1.10}$$

where, M is the molecular weight; N_A is Avogadro's constant; ρ is the density; α is for mean polarizability. The refractive index of glass depends on molecular weight, polarizability, and density. Since silica glass has lower molecular weight, density, and polarizability, it has the lowest refractive index among oxide glasses. The refractive index of high-purity silica glass at 589 nm at room temperature is 1.4584. Co-doping with germanium and fluorine can adjust the refractive index of silica glass at 633 nm wavelength from $+1.5 \times 10^{-3}$/mol% to -1.3×10^{-3}/mol%. This is mainly because the electronic polarizability of the dopant elements is different from that of silica glass. In addition, the dopant elements will also cause changes in the structure of silica glass, such as the reorganization of [SiO_2] tetrahedra, etc. This is also the reason for the change in the refractive index of silica glass.

Dispersion refers to the characteristic of the refractive index of glass changing with wavelength. Generally, the longer the wavelength, the smaller the refractive index. As early as 1965, Malitson of the National Bureau of Standards in the United States published the room temperature (20 °C) refractive index test data of silica glass at 60 wavelengths in the range of 210 ~ 3710 nm from Corning, Dynasil, and General Electric (GE), and fitted the Sellmier formula of eq. (1.11), where the wavelength unit is in micrometers:

$$n^2 - 1 = \frac{0.6961663\lambda^2}{\lambda^2 - 0.0684043^2} + \frac{0.4079426\lambda^2}{\lambda^2 - 0.1162414^2} + \frac{0.8974794\lambda^2}{\lambda^2 - 9.896161^2} \tag{1.11}$$

The refractive index of silica glass in the wavelength range from ultraviolet (0.213 μm) to infrared (3.7 μm) as shown in Table 1.11 is calculated from formula (1.11). It is found that as the wavelength increases, the refractive index of silica glass decreases from 1.534307 to 1.399389. The data in Table 1.11 shows that the Abbe number of silica glass in the visible band is 67.77 ~ 67.88.

Figure 1.33 shows the dependence of the refractive index of silica glass on wavelength. The experimental results are plotted according to the data in Table 1.11. Compared with the ultraviolet and infrared bands, the refractive index of silica glass changes relatively smoothly in the visible light band. Due to the influence of OH vibration absorption, no experimental data were measured in the wavelength range of 2.4 ~ 3.2 μm, and the calculation results were obtained from formula (1.11).

The property of the refractive index of a material changing with the frequency (or wavelength) of incident light is called the dispersion of the material. The relative dispersion D and the reciprocal of the relative dispersion D^{-1} can be obtained from formulas (1.12) and (1.13) respectively:

$$D = \frac{-dn/d\lambda}{n-1} \tag{1.12}$$

$$D^{-1} = \left(\frac{-dn/d\lambda}{n-1}\right)^{-1} \tag{1.13}$$

Figure 1.34 shows that relationship between the relative dispersion and its reciprocal with wavelength at room temperature. The larger the reciprocal of the relative dispersion, the smaller the chromatic aberration of the corresponding silica glass.

Table 1.11 Refractive index data of silica glass at room temperature [39]

Wavelength /μm	Spectrum Source	Refractive index obtained by calculation	C-D-GE [a] Deviation/ ($\times 10^{6}$)	Corning Deviation/ ($\times 10^{6}$)	Dynasil Deviation/ ($\times 10^{6}$)	GE Deviation/ ($\times 10^{6}$)
0.213856	Zn	1.534307	−27	−29	−42	−31
0.214438	Cd	1.533722	−2	−11	−21	−22
0.226747	Cd	1.522750	+70	+71	+68	+73
0.230209	Hg	1.520081	−21	−28	−31	−23
0.237833	Hg	1.514729	+1	+13	+23	+19
0.239938	Hg	1.513367	+3	+6	+2	+9
0.248272	Hg	1.508398	+2	+6	−1	+7
0.265204	Hg	1.500029	−29	−32	−25	−13
0.269885	Hg	1.498047	+3	+7	−4	+11
0.275278	Hg	1.495913	−3	+2	+8	+12
0.280347	Hg	1.494039	+1	−4	−9	−11
0.289360	Hg	1.490990	+20	+18	+22	+20
0.296728	Hg	1.488734	−14	−7	−12	−4
0.302150	Hg	1.487194	−4	−9	−2	+4
0.330259	Zn	1.480539	−9	+1	+10	+3
0.334148	Hg	1.479763	−3	−8	−1	+9
0.340365	Cd	1.478584	+6	+9	+2	−8
0.346620	Cd	1.477468	+2	−17	−12	−14
0.361051	Cd	1.475129	+1	+3	−9	−8
0.365015	Hg	1.474539	−19	−11	−15	−21
0.404656	Hg	1.469618	+2	+1	−1	+2
0.435835	Hg	1.466693	−3	+5	+1	+3
0.467816	Cd	1.464292	+8	+5	+3	+6
0.486133	H	1.463126	+4	+6	+5	+7
0.508582	Cd	1.461863	+7	+4	+1	+5
0.546074	Hg	1.460078	+2	+4	+1	−5
0.576959	Hg	1.458846	+4	+5	+3	+4
0.579065	Hg	1.458769	+1	+6	+6	+6
0.587561	He	1.458464	+6	+3	−2	+1
0.589262	Na	1.458404	−4	+6	+3	+7
0.643847	Cd	1.456704	+6	+9	+4	+7
0.656272	H	1.456367	+3	+7	+5	+7
0.667815	He	1.456067	+3	+8	+6	+3
0.706519	He	1.455145	+5	+10	+12	+7
0.852111	Cs	1.452465	+5	+8	+3	+5
0.894350	Cs	1.451835	+5	+11	+5	+10
1.01398	Hg	1.450242	+8	+6	+3	+6
1.08297	He	1.449405	−5	+8	+1	+9
1.12866	Hg	1.448869	+1	+7	+8	+9
1.3622	Hg	1.446212	−12	−6	−14	−12
1.39506	Hg	1.445836	+4	−1	+4	−3

(continued)

Table 1.11 (continued)

Wavelength /μm	Spectrum Source	Refractive index obtained by calculation	C-D-GE [a] Deviation/ (×10⁻⁶)	Corning Deviation/ (×10⁶)	Dynasil Deviation/ (×10⁶)	GE Deviation/ (×10⁶)
1.4695	Cs	1.444975	−5	+3	+9	+10
1.52952	Hg	1.444268	+2	+8	+6	0
1.6606	TCB[b]	1.442670	−20	−14	−19	−11
1.681	Poly[c]	1.442414	+6	−2	−10	+8
1.6932	Hg	1.442260	0	+7	−6	+1
1.70913	Hg	1.442057	+3	0	+3	−1
1.81307	Hg	1.440699	+21	−7	−7	+6
1.97009	Hg	1.438519	+1	+6	+12	+12
2.0581	He	1.437224	−4	−3	−9	−11
2.1526	TCB	1.435769	−29	−22	−25	−24
2.32542	Hg	1.432928	−18	−10	−2	−6
2.4374	TCB	1.430954	−24	−23	−21	−14
3.2439	Poly	1.413118	+32	+21	+29	+25
3.2668	Poly	1.412505	+25	+20	+30	+25
3.3026	Poly	1.411535	+25	+32	+30	+28
3.422	Poly	1.408180	+20	+40	+42	+37
3.5070	Poly	1.405676	−16	−26	−20	−10
3.5564	TCB	1.404174	−24	−27	−29	−18
3.7067	TCB	1.399389	−19	−22	−14	−9
Average absolute error			10.5	11.9	12.2	11.7

Note: [a] The deviation refers to the difference between the refractive index measurements of Corning, Dynasil, and GE and the calculated values
[b] TCB represents 1,2,4-trichlorobenzene
[c] Poly represents polystyrene

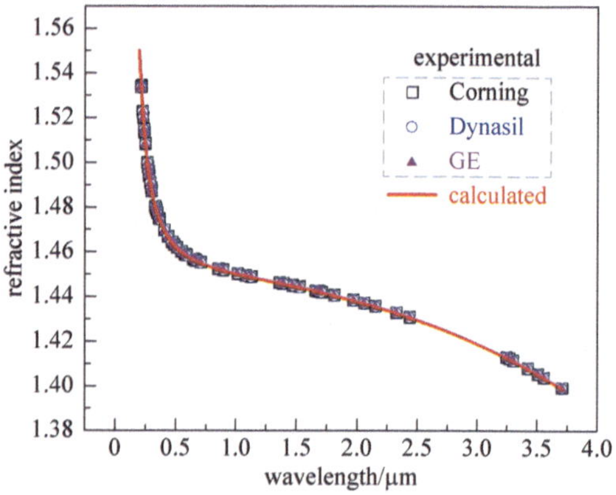

Fig. 1.33 Room temperature refractive index of silica glass in 210 ~ 3707 nm [39]

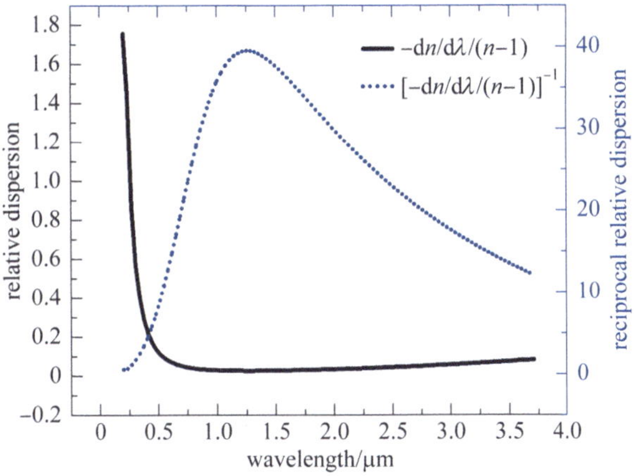

Fig. 1.34 Relationship between relative dispersion and its reciprocal of silica glass and wavelength [39]

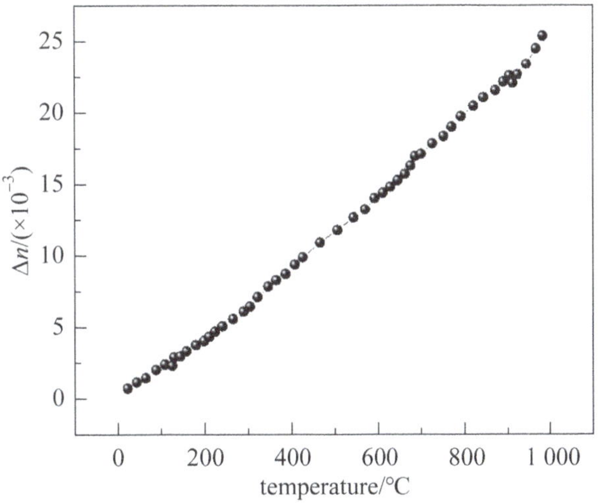

Fig. 1.35 Relationship between refractive index and temperature of silica glass [1]

In addition to being closely related to the wavelength, the refractive index of silica glass is also related to temperature, and its relationship with temperature can be expressed by the following formula: (Eq. 1.14).

$$\frac{dn}{dT} = \frac{\partial n}{\partial T} + \frac{\partial n}{\partial \rho}\frac{\partial \rho}{\partial T} = \frac{\partial n}{\partial T} - 3\alpha\rho\frac{\partial n}{\partial \rho} \tag{1.14}$$

In the formula, α is the expansion coefficient, ρ is the density. From the above formula, it can be seen that the refractive index of silica glass is affected by the thermal expansion coefficient. Figure 1.35 shows the relationship between the increase in the refractive index of Type II silica glass from room temperature to 1000 °C and temperature. It can be seen from the Figure that the refractive index of silica glass increases with the increase in temperature, that is, its d n/d T is positive. In 15 ~ 35 °C, d n/d T is from 95 × 10^{-7}/K to 103 × 10^{-7}/K .

Figure 1.36 shows the relationship between the dn/dT of silica glass and wavelength in 20 ~ 30 °C. It can be seen that the dn/dT of silica glass in the ultraviolet band significantly decreases with the increase of wavelength, and the dn/dT decreases from 15×10^{-6}/K at 210 nm to 10×10^{-6}/K at 400 nm. In the 600 nm ~ 2 μm band, the dn/dT of silica glass increases with the increase of wavelength.

The dn/dT value of silica glass can be adjusted by doping with other components. In order to develop silica glass or fibers with low or even negative thermo-optic coefficients, Dragic et al. [40] reported that negative dn/dT was obtained by doping high P into silica glass.

As shown in Fig. 1.37, in P_2O_5-SiO_2 glass with a P_2O_5 content of about 30 mol%, d*n*/d*T* begins to become negative. This is important for effectively controlling the thermal lensing and mode instability of silica fiber in high-power laser applications.

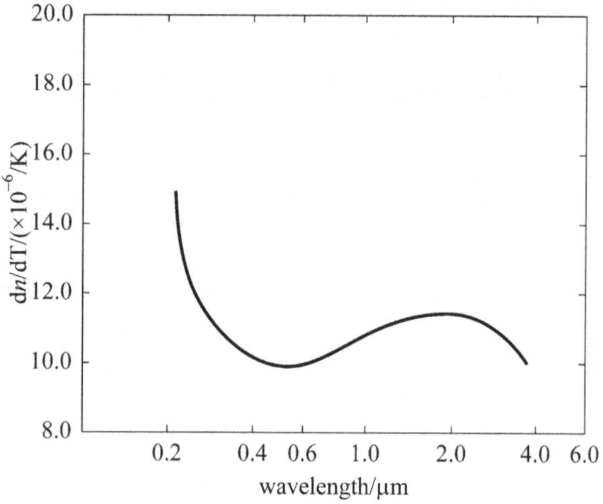

Fig. 1.36 Thermo-optic coefficient of silica glass between 20 ~ 30 °C (dn/dT) and its relationship with wavelength [39]

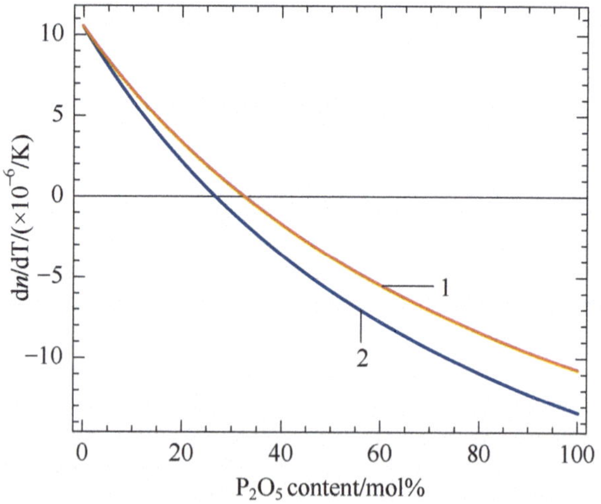

Fig. 1.37 Relationship between dn/dT of P_2O_5 $-SiO_2$ glass and its P_2O_5 content (Curve 1 is the testing result of fiber, Curve 2 is the testing result of planar waveguide) [40]

The refractive index of silica glass, in addition to being related to wavelength, doping element content, and temperature, is also increased by external pressure, ultraviolet light, electrons, and neutron beam irradiation. For example, electron radiation with an energy of 18 keV can increase the refractive index of silica glass by 10^{-2}, and applying a pressure of 2GPa can increase its refractive index by 3×10^{-2}.

Figure 1.38 shows that the refractive index of pure silica glass at different wavelengths of 473 nm, 633 nm, 974 nm, 1300 nm and 1544 nm increases with the increase of fictive temperature, T_f [41] .

As shown in Fig. 1.39, the density of silica glass also increases with the increase of fictive temperature, T_f.

Fig. 1.38 The relationship between the refractive index of silica glass and fictive temperature, T_f

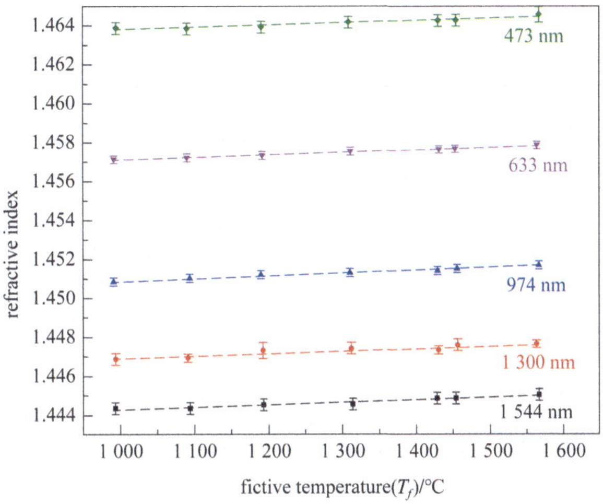

Fig. 1.39 The relationship between the room temperature density of silica glass and fictive temperature, T_f [41]

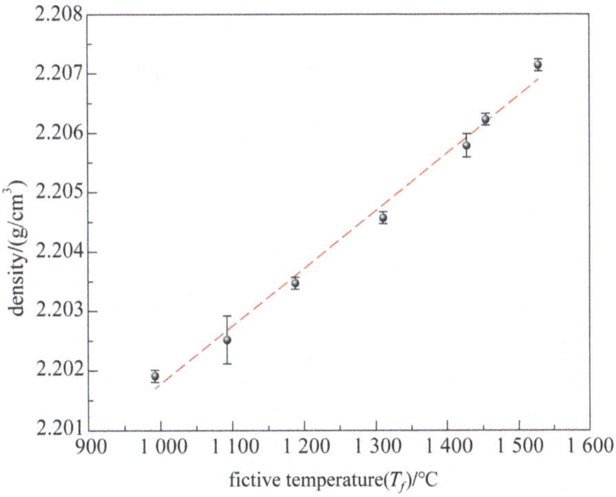

Table 1.12 Some properties of silica glass

Property	Value	Note	Reference
Density/(g.cm $^{-3}$)	2.2002 ~ 2.2060	Related to T_f	[1]
Molar volume/(cm^3 .mol $^{-1}$)	27.24 ~ 27.31	Related to T_f	[1]
Thermal diffusion coefficient/(m.m^2 s^{-1})	0.795 ~ 0.870	Room temperature	[42]
Refractive index	1.4583 ~ 1.4589	Related to T_f	[1]
Glass transition temperature T_g/K	1480	From calorimetry	[43]
Configurational entropy/(J.mol $^{-1}$ .K $^{-1}$)	5.1 ± 2	From calorimetry	[43]

Table 1.12 gives some properties of silica glass. As shown in the Table, the density, refractive index, and molar volume of silica glass are all related to the fictive temperature, T_f. A high fictive temperature corresponds to a fast cooling rate, so there are more silicon-oxygen tetrahedra composed of 3-membered and 4-membered rings in its structure [30], thus its density and refractive index are also higher.

1.5.2 *Absorption and Transmission Spectra*

The absorption spectrum of silica glass depends on the impurities and structural defects in the glass, as well as the migration of bridging oxygen to the conduction band affecting the absorption of the ultraviolet band, and the resonance of non-bridging oxygen affecting the absorption of the infrared band.

Silica glass is an amorphous material with the shortest ultraviolet cut-off wavelength, i.e., the best ultraviolet transmission characteristics. Its intrinsic vacuum ultraviolet cut-off wavelength is 153 nm [44]. The ultraviolet absorption edge of silica glass depends on the interaction of the internal electrons of the material with the high-frequency part of the electromagnetic wave. When the binding force of the internal electrons of the material is strong, this interaction occurs in a higher frequency area, i.e., the ultraviolet absorption cut-off wavelength of the material is shorter.

Impurities and structural defects have a significant impact on the absorption spectrum of silica glass. As shown in Fig. 1.40, the ultraviolet transmission performance of synthetic silica glass (Class III and Class IV silica glass) is greatly affected by the trace impurities in it. With higher OH^- and Cl^- the UV cut-off wavelength of synthetic silica glass significantly redshifts (curves A and B) because the absorption cross-sections of SiOH and SiCl at 157 nm are 16.8×10^{-20} cm^2 and 6.3×10^{-20} cm^2, respectively. The UV cut-off wavelength of synthetic silica glass with lower OH^- content blueshifts (curve C). Synthetic silica glass containing F^- has the shortest UV cut-off wavelength (curve D) because SiF does not introduce UV absorption. In addition, the synthetic glass containing F^- has fewer 3-membered and 4-membered ring structures. Therefore, the internal transmittance of synthetic silica glass containing F at 157 nm can be greater than 80% [45] .

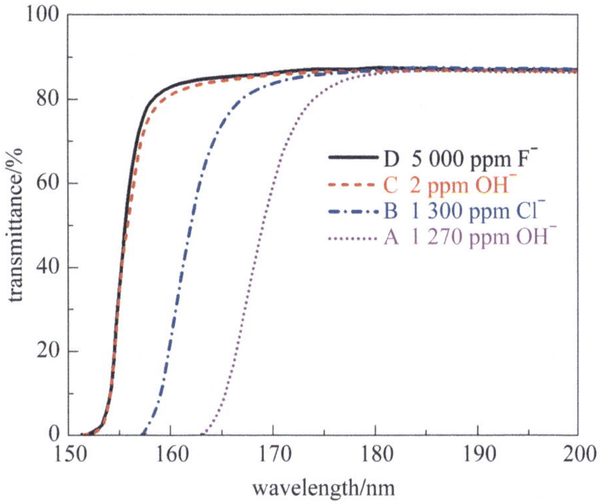

Fig. 1.40 The influence of trace impurities on the ultraviolet transmission of synthetic silica glass (10 mm sample thickness) [45]

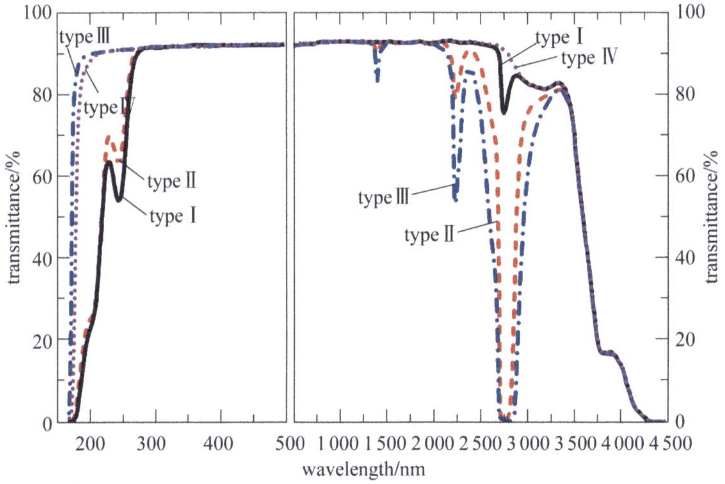

Fig. 1.41 Transmission spectra of four types of silica glass from UV to mid-infrared

Figure 1.41 shows the transmission spectra of I, II, III, and IV types of silica glass. It can be seen that impurities and defects in the glass severely affect the UV transmittance of I/II type silica glass, and the higher hydroxyl content in II/III type silica glass causes serious absorption in the infrared band.

In Fig. 1.41, the absorption peak near 240 nm in type I, II silica glass is related to oxygen vacancy defects. The impurity content in type III, IV silica glass is very low, and this absorption peak is not present. The absorption of silica glass in the long-wave direction mainly depends on the interaction between the vibration mode of the glass structure group and the incident light, and the hydroxyl content in the glass. Glass containing heavy ions and weakly bound glass has a longer infrared cut-off wavelength. Compared with heavy metal oxide and fluoride glass, the infrared cut-off wavelength of silica glass is shorter, about 3.5 μm. The fundamental vibration peak of the hydroxyl group is located at 2.7 μm, and its second and third harmonic peaks are located at 1.38 μm and 0.95 μm respectively. In Fig. 1.41, the absorption peak at 2.2 μm is caused by the overlap of OH^- vibration and Si—O—Si vibration. As can be seen from Fig. 1.41, type I silica glass has less hydroxyl than type II and III silica glass. Type II and III silica glasses have higher hydroxyl content because they are heated by hydrogen-oxygen flame during the preparation process. Type IV glass is heated by plasma, its hydroxyl content is very low, so the absorption at 2.7 μm is very low.

1.5.3 Scattering and Loss

So far, the 1.55 μm loss of silica fiber can be as low as 0.1419 dB/km [46], this loss mainly comes from scattering and multi-phonon absorption.

In a single-component glass like silica glass, the scattering coefficient can be given by the following formula [47] (Eq. 1.15):

$$\alpha_T = \frac{8\pi^3}{3\lambda^4}\frac{n^8 p^2}{\rho^2}\left\langle |\Delta\rho|^2 \right\rangle V = \begin{cases} \dfrac{8\pi^3}{3\lambda^4} n^8 p^2 \beta_T(T) k_B T \left(T > T_g\right) \\ \dfrac{8\pi^3}{3\lambda^4} n^8 p^2 \left[\beta_{T,\mathrm{rel}}\left(T_g\right) k_B T_g + \beta_{S,\infty}(T) k_B T\right] \end{cases} \left(T < T_g\right) \tag{1.15}$$

In the formula, $<|\Delta\rho|^2>$ is the nominal square amplitude of density fluctuations; λ is the wavelength of the incident light; p is the photoelastic coefficient; n is the refractive index; V is the scattering volume; ρ is the nominal density; k_B is the Boltzmann constant; T_g is the glass transition temperature; $\beta_T(T)$ is the isothermal compressibility; $\beta_{T,\mathrm{rel}}(T_g)$ is the relaxation unit of β_T at the glass transition temperature; $\beta_{S,\infty}(T)$ is the high-frequency limit of the adiabatic compressibility ratio when the temperature is T. Among them, in the formula where $T < T_g$, the first term represents Rayleigh scattering caused by static density inhomogeneity, and the second term represents the effect of Brillouin scattering. The static density inhomogeneity is closely related to the disorder of the glass transition temperature structure.

Saito et al. [48] tested the scattering of two types of silica glass, one without OH^- (A) and the other with 1200 ppm OH^- (G), in the temperature range of 300 ~ 1900 K. The heating and cooling rates during the test were both 100 K/h, and the polarized light from an argon-ion laser (10 MHz linewidth) was used. The test results are shown in Fig. 1.42. Although it is difficult to measure the glass transition temperature of silica glass using calorimetry and expansion curves, the T_g temperatures of the two types of silica glass are clearly visible from the results in Fig. 1.42.

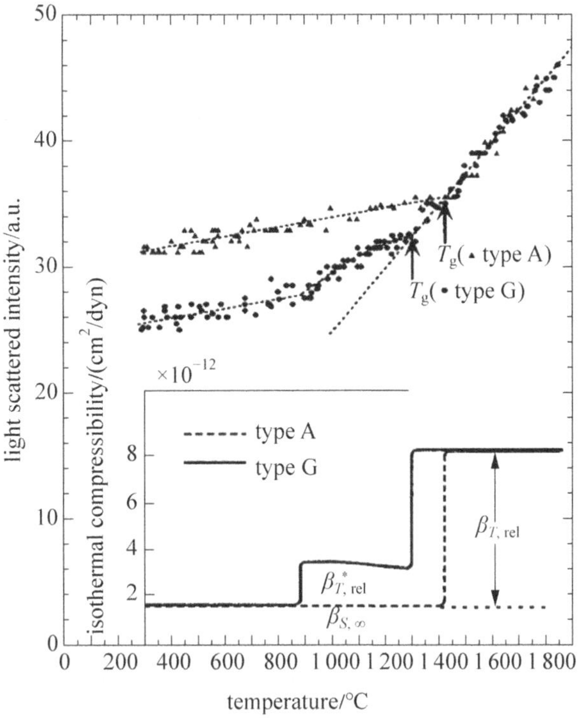

Fig. 1.42 Temperature and scattering intensity relationship in two types of silica glass [48]

The T_g temperature defines the viscosity of the glass at 10^{13} ~ $10^{14.5}$ Poise (strain point). In addition, below the glass transition temperature T_g, the scattering of G-type glass is significantly lower than that of A-type glass. The scattering of G-type silica glass in the T_g temperature ~ 900 K range decreases much faster than that of A-type glass. $\beta_{T, rel}$ parameter is the part related to structural relaxation in the isothermal compressibility coefficient. In A-type silica glass, the $\beta_{T, rel}$ parameter decreases to zero at the T_g temperature, but the $\beta_{T, rel}$ parameter of G-type silica glass slowly decreases to zero from the T_g temperature to 400 K below the T_g temperature. This indicates that the structural relaxation process of G-type silica glass containing hydroxyl groups continues over a large temperature range below the T_g temperature. Therefore, its scattering is also lower than that of A-type silica glass.

Saito et al. [49] systematically studied the effect of ppm-level Na_2O on the scattering loss of silica glass. Using a 10 MHz linewidth polarized light from an argon-ion laser to test the scattering loss of silica glass, it was found that it has the same effect as that of hydroxyl. That is, the higher the Na_2O content in the glass, the smaller the scattering loss. Compared with pure silica glass, adding 10 ~ 50 ppm Na_2O can reduce the loss of silica glass by 13% ~ 20%. And as shown in Fig. 1.43, at the same content, trace Na_2O has a greater reduction in scattering loss than hydroxyl. The reason is that the doping of trace Na_2O accelerates the relaxation speed of the silica glass structure, which can reduce the scattering loss caused by the uneven density in the silica glass.

Saito et al. [48] used the same 10 MHz linewidth polarized light from an argon-ion laser to test the scattering loss of silica glass and its supercooled liquid, and

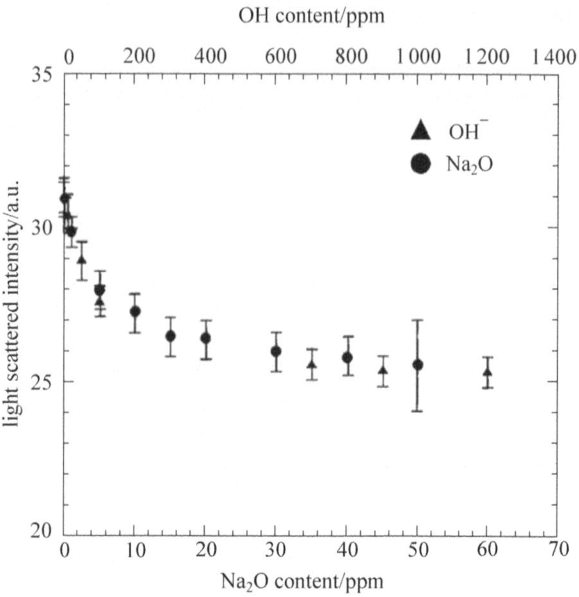

Fig. 1.43 Relationship between Na_2O and OH contents and scattering loss in silica glass [49]

found that both have depolarized scattering, thus speculating that there are micron-sized crystals in the silica glass. The existence of these micron-sized crystals is one of the reasons for the loss.

In summary, the sources of scattering in silica glass include micro-area density inhomogeneities and the presence of a small amount of microcrystals. Trace impurities such as hydroxyl and Na_2O help to extend the structural relaxation temperature range of silica glass, thereby reducing scattering and the loss caused by scattering.

1.5.4 Photoelastic Coefficient of Silica Glass

Homogeneous, annealed silica glass is an isotropic amorphous material. However, under the action of external mechanical stress, silica glass can show anisotropy and transient stress birefringence. The Photoelastic coefficient is used to quantifies the magnitude the size of the birefringence produced in the silica glass under the action of external stress. In the temperature range of 20 ~ 400 °C, the Photoelastic coefficient does not change much with temperature (about 2% ~ 8%), basically within the measurement error range. In the range of 20 ~ 100 °C, the Photoelastic coefficient can be considered a constant. The Photoelastic coefficient is measured by the difference in optical path when polarized light passes through the glass under stress, and its calculation formula is as follows:

$$p = \frac{R}{Bl} \tag{1.16}$$

In the formula, p is the compressive or tensile stress (MPa); R is the optical path difference (nm); l is the thickness of the glass sample (mm); B is the photoelastic coefficient $[(\times 10^{12}\ \mathrm{Pa})^{-1}]$. The photoelastic coefficient of transparent silica glass is $(3.5 \sim 3.7)\ (\times 10^{12}\ \mathrm{Pa})^{-1}$.

1.6 Radiation Resistance of silica Glass

Silica glass has excellent optical properties and radiation resistance. Rare earth-doped fibers with silica glass as the main component will be subjected to strong ionizing radiation in space applications. In addition, synthetic silica glass, which has the advantages of extremely low impurity content, good optical uniformity, and good transparency, is used as the window material and focusing lens for high-power laser systems. Even so, high-energy pulsed lasers can cause damage to silica glass. This section mainly introduces the effects of cosmic ray irradiation and pulsed laser on the optical properties of silica glass.

1.6.1 Radiation Effects on Silica Glass

Ionizing radiation will have a certain impact on the optical, mechanical, electrical, and thermal properties of silica glass [34, 50]. In terms of optics, the irradiation effect mainly manifests as a significant decrease in the transmittance of the ultraviolet-visible band, which is related to irradiation-induced color center absorption. In terms of mechanics, the irradiation effect mainly manifests as a decrease in volume and an increase in stress, which is related to the microcrystallization and densification of the silica glass structure. In terms of electricity, the irradiation effect mainly manifests as a decrease in resistivity and dielectric constant, which is related to the increase in color center carrier concentration induced by irradiation. In terms of thermodynamics, the irradiation effect mainly manifests as an increase in the glass transition temperature (T_g) and thermal conductivity, which is related to the decrease in the free enthalpy of the glass system and the increase in carrier concentration. It should be noted that the impact of ionizing radiation on the mechanical, electrical, and thermal properties of silica glass usually only becomes apparent under high-dose irradiation conditions, and is not obvious under low-dose irradiation conditions. Even under high-dose irradiation conditions, the slight changes in the mechanical, electrical, and thermal properties of silica glass will not seriously affect the normal use of silica glass. On the contrary, the impact of ionizing radiation on the optical properties of silica glass is very obvious even under low-dose irradiation conditions, and this irradiation effect will seriously restrict the normal application of silica glass. Therefore, the following focuses on the impact of ionizing radiation on the optical properties of silica glass.

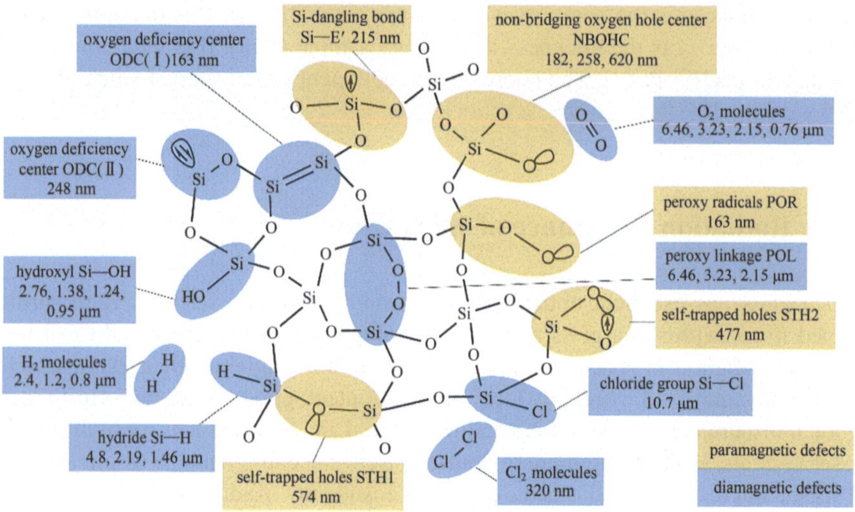

Fig. 1.44 Schematic diagram of common point defects in silica glass and the positions of absorption peaks

The impact of ionizing radiation on the optical properties of silica glass is mainly reflected in three aspects [5, 34, 51]: ① Radiation-induced absorption changes; ② Radiation-induced luminescence changes; ③ Radiation-induced refractive index changes. The fundamental reasons for these three changes are all related to radiation-induced color centers.

Figure 1.44 is a schematic diagram of common point defects in silica glass, and it also gives the positions of the absorption peaks corresponding to different point defects. Since the absorption of point defects is usually located in the ultraviolet-visible band, it will cause the glass to be colored, so these point defects are also commonly referred to as color centers. During the preparation of silica glass, different defects may be introduced due to different preparation conditions. For example, oxygen vacancy defects (ODC) may appear in silica glass prepared under vacuum or reducing atmosphere conditions. Oxygen-rich defects (POL) may appear in silica glass prepared under oxygen atmosphere conditions. Silica glass prepared with hydrogen-oxygen flame will contain a large amount of hydroxyl (Si—OH). In addition, when silica glass is exposed to high-energy radiation, it will produce dangling bond defects. Dangling bond defects usually have paramagnetism and can be characterized by electron paramagnetic resonance (EPR) spectra. Defects are divided into two categories according to whether they have paramagnetic signals, namely Frenkel defects (non-paramagnetic defects) and dangling bond defects (paramagnetic defects).

Griscom [52, 53] and Skuja [54] have conducted systematic research on the types, properties, precursors, and formation mechanisms of point defects in silica

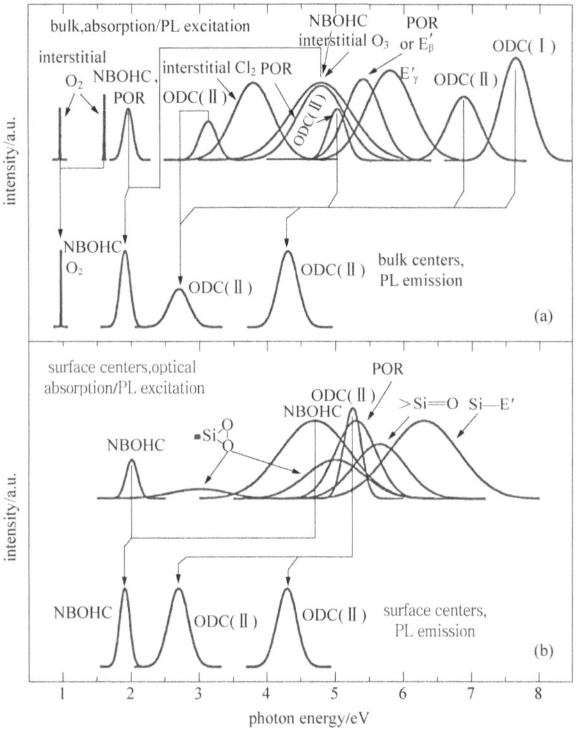

Fig. 1.45 Absorption and emission spectra of common point defects inside (**a**) and on the surface (**b**) of silica glass material

glass. Figure 1.45 shows the positions of the absorption peaks and luminescence peaks of common point defects in silica glass [54]. Since the specific surface energy of silica glass is larger and the structural defects are more complex, there are slight differences in the spectral characteristics of the defects on the surface and inside of silica glass.

Figure 1.46 shows the effect of neutron radiation on the density and refractive index of silica glass and quartz crystal [55]. As the dose of fast neutron radiation increases, the density and refractive index of silica glass gradually increase. In contrast, the density and refractive index of quartz crystal gradually decrease. When the radiation dose reaches 2×10^{20} n/cm^3, the density and refractive index of silica glass increase by less than 3%, while the density and refractive index of quartz crystal decrease by more than 10%. At this point, the density and refractive index of both silica glass and quartz crystal are very close and no longer change with the change in radiation dose. Studies have shown that when the radiation dose exceeds 2×10^{20} n/cm^3, the structural changes of silica glass and quartz crystal induced by radiation are almost saturated, and the structures of the two substances are similar at this time, deriving a new phase called the metamict phase. The structural units of silica glass and quartz crystal in the metamict phase are also silicon-oxygen tetrahedra, with a Si—O—Si bond angle distribution wider than quartz crystal but narrower than silica glass, with an average value of about 135°.

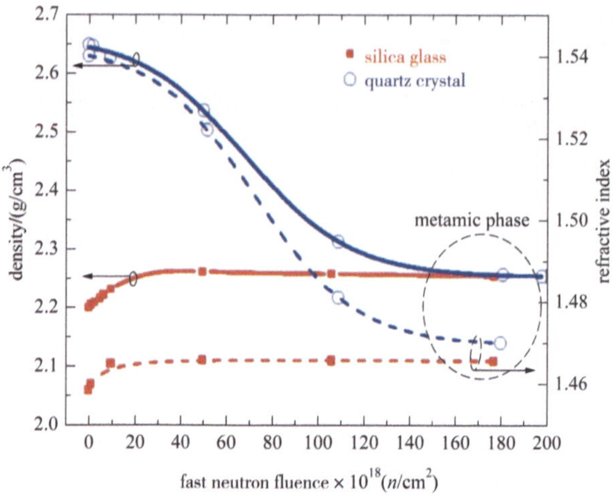

Fig. 1.46 The effect of neutron radiation on the density and refractive index of silica glass and quartz crystal

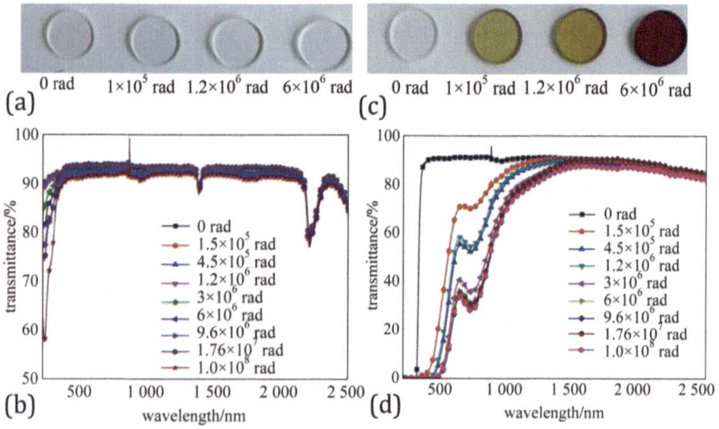

Fig. 1.47 Domestic high-purity silica glass (brand: JGS1) (**a**, **b**) and light flint glass (brand: QF) (**c**, **d**) after different doses of gamma-ray irradiation, sample photos (**a**, **c**) and transmittance curve (**b**, **d**)

1.6.2 Cosmic Ray Radiation Resistance

Studies have shown that under non-shielding conditions, the total cumulative irradiation dose of the International Space Station for 15 years is about 9.0×10^6 rad(Si), and the total cumulative irradiation dose of the geosynchronous orbit for 15 years is about 2.0×10^9 rad(Si).

Figure 1.47 compares the gamma-ray irradiation resistance of domestic silica glass (brand: JGS1) and light flint glass (brand: QF) [56]. The sample thickness is 3 mm, the goal is to ensure that gamma-rays can penetrate the glass homogeneously, and the change in sample transmittance is clear and testable. The gamma ray irradiation dose rate is 1.5×10^5 rad(Si)/h, the maximum total irradiation dose is 1×10^8 rad(Si), and the maximum single experiment total time is 28 days.

From the physical photos in Fig. 1.47a, c, it can be seen that as the total irradiation dose increases, the coloration of light flint glass gradually deepens, but no coloration phenomenon was observed in silica glass. From Fig. 1.47b, d, it can be seen that as the total irradiation dose increases, the transmittance of light flint glass in the ultraviolet to near-infrared band (200 ~ 1200 nm) significantly decreases, while the transmittance of silica glass only decreases in the ultraviolet band (<300 nm). From Fig. 1.47a–d, it can be seen that the gamma-ray irradiation resistance of silica glass is far superior to that of light flint glass.

The transmittance curves of four typical wide-spectrum transparent window materials, sapphire crystal, barium fluoride crystal, fused silica glass, and calcium fluoride crystal, after different doses of gamma-ray irradiation are shown in Fig. 1.48 [57]. From the Figure, it can be seen that after 10 Mrad gamma-ray irradiation, the transmittance of sapphire crystal almost does not change, and its radiation resistance is the best; the transmittance of fused silica glass only slightly decreases in the ultraviolet band (<5%), and its radiation resistance is second; the transmittance of calcium fluoride and barium fluoride crystals significantly decreases in the ultraviolet-visible band, and their radiation resistance is not as good as sapphire crystal and fused silica glass.

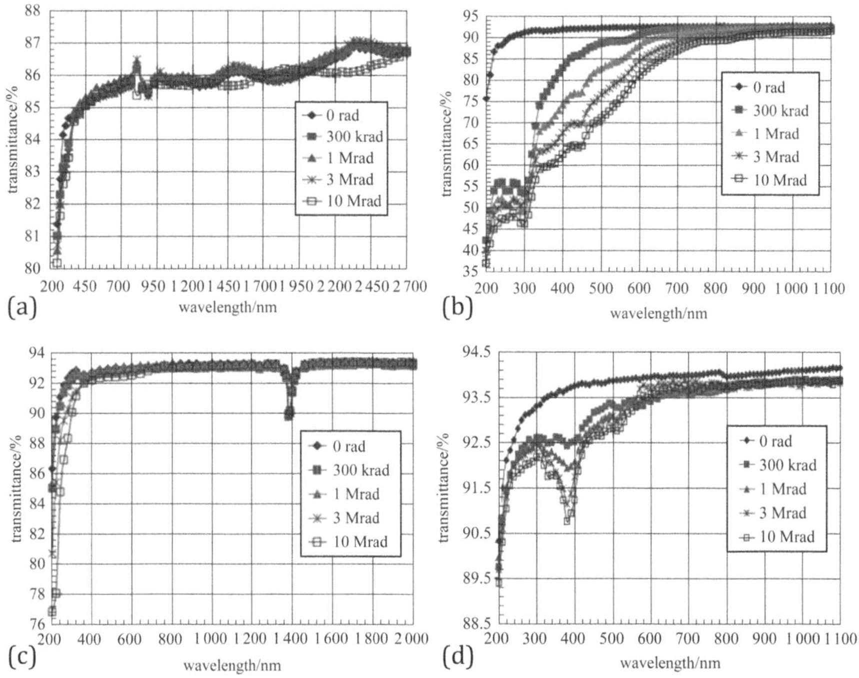

Fig. 1.48 Transmittance curves after different doses of gamma-ray irradiation of Sapphire crystal (**a**), barium fluoride crystal (**b**), fused silica glass (**c**), and calcium fluoride crystal (**d**)

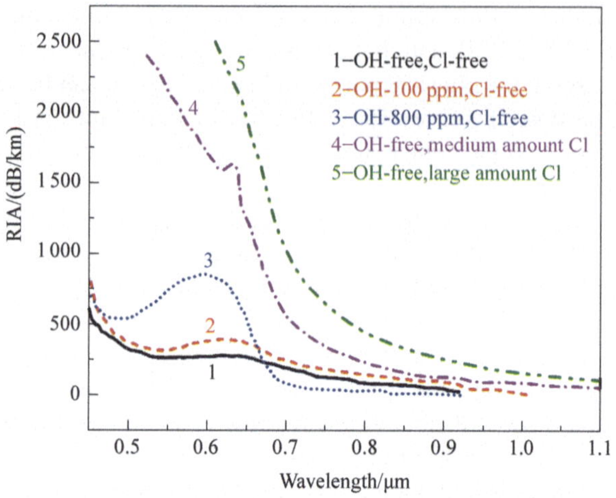

Fig. 1.49 The impact of non-metallic impurities (such as OH^-, Cl^-, etc.) on the radiation resistance of silica glass

The total irradiation dose of 10 Mrad is equivalent to the total irradiation dose accumulated by the International Space Station in 15 years. This shows that the space radiation tolerance of fused silica glass can meet the service requirements of the International Space Station for 15 years.

Studies have shown that the factors affecting the radiation resistance of silica glass mainly include three aspects: ① metal impurities; ② non-metal impurities, such as OH^- content, Cl^- content, etc.; ③ The fictive temperature, Tf of silica glass.

Ge Shi's research shows that the less metal impurities in silica glass, the stronger the radiation resistance, and vice versa. In addition, Ge Shi also compared the radiation resistance of silica glass doped with different Ce contents [58]. Compared with undoped silica glass, doping with Ce not only did not improve the radiation resistance of silica glass, but also tended to worsen it. This may be related to the destruction of the dense network structure of silica glass caused by Ce doping.

As shown in Fig. 1.49, non-metal impurities (such as OH^-, Cl^- etc.) have a great impact on the radiation resistance of silica glass [59] .

The higher the Cl content, the greater the radiation-induced absorption (RIA) intensity of silica glass in the ultraviolet and visible bands, which may be related to the absorption of silicon dangling bonds (Si–E′, absorption peak at 215 nm) and Cl_2 molecules (absorption peak at 330 nm). The chemical reaction equation is as follows:

$$\equiv \mathrm{Si}-\mathrm{Cl} \rightarrow \equiv \mathrm{Si}^{\cdot} + \mathrm{Cl}^{\bullet} \tag{1.17}$$

$$\mathrm{Cl}^{\bullet} + \mathrm{Cl}^{\bullet} \rightarrow \mathrm{Cl}_2 \tag{1.18}$$

The higher the hydroxyl content, the greater the RIA intensity of silica glass near 258 nm and 620 nm, which is related to the non-bridging oxygen hole center (NBOHC, absorption peak at 258 nm, 620 nm) produced by irradiation. At the same time, the higher the hydroxyl content, the smaller the RIA intensity of silica glass in the ultraviolet band, because the silicon dangling bond (Si–E′, absorption peak at 215 nm) is suppressed. The chemical reaction equation is as follows:

$$\equiv \mathrm{Si-OH} \rightarrow \equiv \mathrm{Si-O^{\circ}} + \mathrm{H^{\bullet}} + e- \tag{1.19}$$

$$\equiv \mathrm{Si^{\bullet}} + \mathrm{H^{\bullet}} \rightarrow \equiv \mathrm{Si-H} \tag{1.20}$$

The thermal history of silica glass can lead to differences in its fictive temperature, Tf. Generally speaking, the lower the fictive temperature of silica glass, the better its radiation resistance [60]. The fundamental reason is related to the suppression of color center precursors. A large number of studies have shown that 3-membered rings and 4-membered rings are precursors to the formation of silicon dangling bonds and non-bridging oxygen hole center.

The four methods to improve the radiation resistance of silica glass mainly include [51, 55, 60, 61]: ① Reduce Cl^- content and metal impurities; ② Lower the fictive temperature of silica glass; ③ Co-doping with fluorine; ④ Hydrogen or deuterium loading treatment for silica glass.

1.6.3 Resistance to Pulsed Laser Damage

Due to the extremely low impurity content, good optical uniformity, high ultraviolet transmittance, and high laser damage threshold of synthetic silica glass, it is the preferred material for main components such as windows, focusing lenses, and diffraction gratings in most high-power laser systems. For example, the National Ignition Facility (NIF) in the United States uses 2112 high-quality silica glass components [62], of which 1536 large-diameter high-performance silica glass is used as the window material for 1053 nm fundamental laser, and another 576 pieces of silica glass are used as the window and lens optical components for the third harmonic laser (351 nm wavelength). In practical applications, it is necessary to improve its resistance to third harmonic pulse laser damage. In addition, silica glass can also be used as the material for photolithography objective lenses and photomask substrates [63], and it is necessary to improve its resistance to excimer pulse laser damage in practical applications.

Table 1.13 compares the bulk damage thresholds of different types of optical glass. It can be seen that the bulk damage threshold of optical glass is related to the wavelength and pulse width of the pulse laser. The shorter the wavelength, the lower the bulk damage threshold; the smaller the pulse width, the lower the bulk damage threshold. The bulk damage threshold of fused silica glass is generally higher than other types of optical glass.

Studies have shown that in the case of third-harmonic ultraviolet lasers (351 nm, 3 ns) Under irradiation, the intrinsic damage threshold of fused silica glass exceeds 100 J/cm^2 [65], but the damage threshold of the fused silica surface is usually less

Table 1.13 Comparison of bulk damage thresholds of different types of optical glass

Glass classification	Glass Grade	Bulk damage threshold/(J/cm^2)			
		1064 nm @ 12 ns [a]	532 nm @ 10 ns [a]	1064 nm @ 74 ps [b]	532 nm @ 74 ps [b]
Boron crown	N-BK7	2017	74.4	31.8	8.2
Fluorine crown	N-FK5	1574	226	35.2	9.7
Flint	F2	690	7.7	16.7	3.5
Lanthanum dense Flint	N-LASF44	720	18.5	13.8	3.7
Lanthanum Flint	N-LAF21	933	15.0	12.6	4.7
Fused silica	Suprasil CG	1866	280	39.2	11

Note: [a] Data from https://www.edmundoptics.in/knowledge-center/application-notes/optics/bulk-laser-damage-in-glass/
[b] Data from reference [64]

Fig. 1.50 Main defects in silica glass

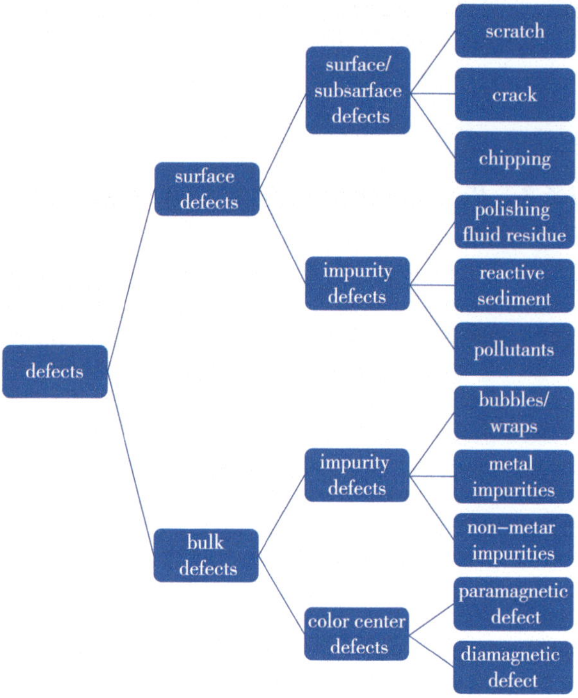

than 10 J/cm^2, much lower than its intrinsic damage threshold. The fundamental reason is related to the large number of defects on the surface of silica glass.

Figure 1.50 summarizes the main defects in silica glass. According to the location of the defects, the defects in silica glass can be divided into surface defects and bulk defects. During the cold processing of silica glass components, contact-based grinding and polishing processes will induce destructive defects such as cracks, scratches, and chipping on the surface and subsurface of silica glass; influenced by grinding tools and polishing fluid, impurity defects such as Al, Fe, Cu, Ce, Cr will remain on the surface of silica glass and between cracks. In addition, during the chemical wet etching process of silica glass, impurity defects such as SiF_6^{2-} and other reactants will deposit impurity defects; during the transportation and installation of silica glass components, contaminants such as fingerprints, water vapor, dust, etc. may be introduced. During the manufacturing process of silica glass, macro defects such as bubbles and inclusions will be generated inside it. With the improvement of manufacturing technology, the content of such defects has been greatly reduced. In addition, due to the influence of manufacturing process conditions, color center defects (such as ODC, POL, etc.), metal impurity defects (Al, Fe, etc.), and non-metal impurity defects (such as Cl, OH, etc.) will also be introduced. At present, the content of metal impurity defects in silica glass prepared by chemical vapor deposition has been reduced to the ppb level. During laser irradiation, more and more complex color center defects will be further induced, such as silicon dangling bonds, interstitial oxygen, oxygen vacancy centers, etc. The existence of these color center defects will cause the damage growth threshold of silica glass components to be far lower than their initial damage threshold.

Thanks to the innovation of silica glass surface processing technology, the damage threshold of the silica glass component surface has been greatly improved. Figure 1.51 gives the function relationship between the laser damage density on the surface of the silica glass used by the NIF of USA and the energy density of the third-harmonic laser (351 nm, 3 ns) [66]. In 1997, using typical optical precision

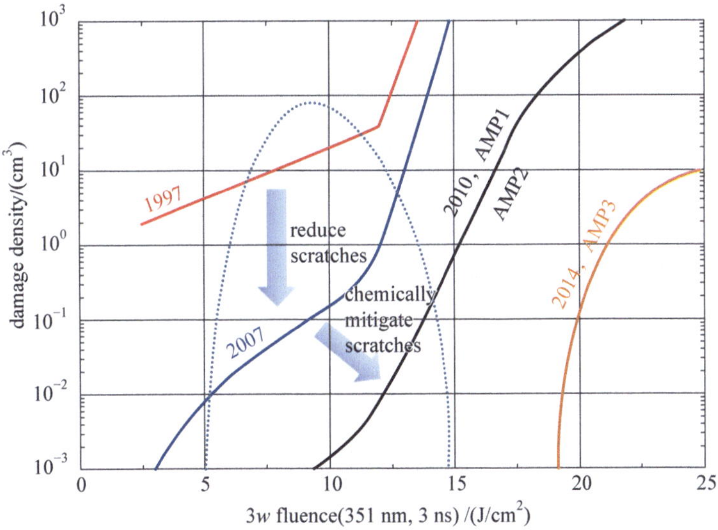

Fig. 1.51 The relationship between the laser damage density on the surface of the fused silica component and the function of energy density of third-harmonic laser. Solid line—Laser damage resistance level of fused silica component surface at different times; Dashed line—Relative energy distribution at 1.8 MJ NIF laser emission

machining technology, when exposed to a flux of 8 J/cm^2 third-harmonic laser, the initial damage point density on the surface of the silica glass was 15/cm^2. In 2007, by improving grinding and polishing processes, reducing surface microcracks or scratches on silica, under the irradiation of 8 J/cm^2 third-harmonic laser, the initial damage point density on the surface of the silica glass was reduced to 0.08/cm^2. From 1997 to 2007, the improvement of the damage threshold of the silica component surface mainly relied on reducing surface scratches. Despite significant progress, the zero damage threshold of the silica component surface is still less than 5 J/cm^2. In 2010, by treating the polished optical components with advanced mitigation process (AMP), the breakthrough of the zero damage threshold of the silica component surface from less than 5 J/cm^2 to close to 10 J/cm^2 was achieved. After 2014, by further optimizing AMP, under the irradiation of 8 J/cm^2 third-harmonic laser, the initial damage point density on the surface of the silica glass is close to zero.

As can be seen from Sects. 1.3–1.6 of this chapter, silica glass has excellent thermal, mechanical, optical, and radiation-resistant properties. In addition, silica glass also has excellent electrical properties and chemical stability. In terms of electrical properties, silica glass has the advantages of high resistivity, low dielectric loss, and high breakdown voltage. In terms of chemical stability, silica glass has excellent water resistance, acid resistance, and salt corrosion resistance, but like all glasses, silica glass is susceptible to alkali corrosion.

Figure 1.52 summarizes the six fundamental properties of silica glass. As can be seen, silica glass has a series of unparalleled properties of ordinary glass, such as high transparency from ultraviolet to near-infrared, extremely low thermal expansion coefficient, good temperature resistance and thermal shock resistance, high

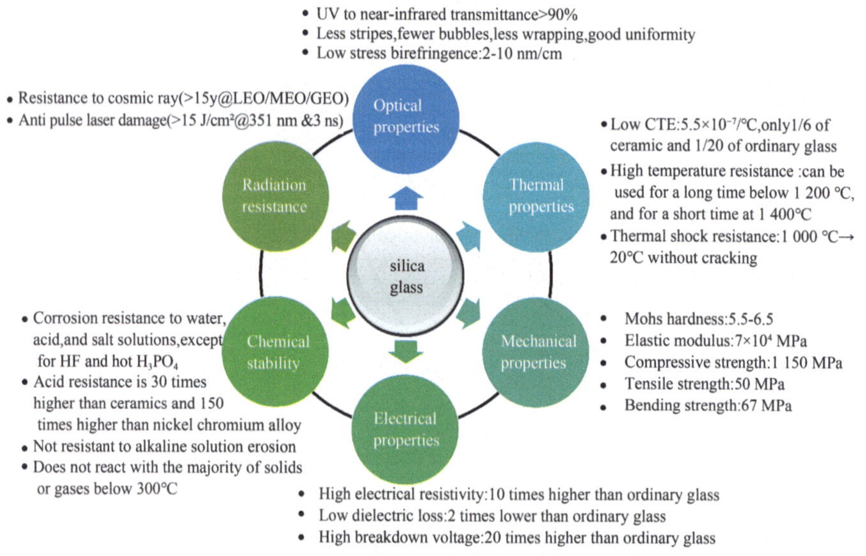

Fig. 1.52 Summary of silica glass properties

hardness and elastic modulus, good electrical insulation, excellent chemical stability, radiation resistance, laser damage resistance, etc., therefore silica glass is known as the "king of glass".

1.7 Rare Earth Doped Silica Fiber

Optical fiber is the abbreviation for optical guide fiber, which is basically composed of three parts: core, cladding, and coating. Optical fibers can transmit light signals, and their transmission principle is total reflection of light. This type of optical fiber is called passive optical fiber. In 1970, Maurer and others from Corning Company in the United States first prepared silica glass optical fibers with a background loss of 20 dB/km (at a wavelength of 632.8 nm) using the improved chemical vapor deposition (modified chemical vapor deposition, MCVD) process [67]. After more than half a century of development, nowadays, the loss of transmission passive optical fibers has been reduced to 0.142 dB/km (at a wavelength of 1560 nm) [46]. Passive optical fibers are widely used in optical communication, sensing, etc., providing important guarantees for modern communication, optical networks, and information society construction.

Optical fibers containing activated ions (rare earth ions and bismuth ions) in the core can amplify light signals or generate lasers. This type of optical fiber is called active optical fiber. Active optical fibers are the gain medium of fiber lasers or fiber amplifiers. Because fiber lasers and amplifiers have the advantages of high wall plug efficiency, small size, and maintenance-free during long-term operation, they have important applications in optical communication, sensing, industrial processing, defense application and other fields.

1.7.1 Luminescent Properties of Rare Earth Ions

The base material for active optical fibers can be silica glass, silicate glass, phosphate glass, fluoride glass, or other suitable glasses. Since Snitzer and colleagues first demonstrated lasing in neodymium-doped glass fibers in 1964 [68], rare earth ions have been the predominant dopants within the core of such fibers. The luminescent characteristics of these rare earth ions are primarily governed by their electronic structure. The radius of rare earth elements is large, and it is easy to lose the outermost two 6 s electrons and one 5d electron in the next layer to form trivalent ions. The electronic configuration of trivalent lanthanide rare earth ions is $1s^2\ 2s^2\ 2p^6\ 3s^2\ 3p^6\ 3d^{10}\ 4s^2\ 4p^6\ 4d^{10}\ 4f^n\ 5s^2\ 5p^6$. For La^{3+}, Ce^{3+}, Pr^{3+}, Nd^{3+}, Pm^{3+}, Sm^{3+}, Eu^{3+}, Gd^{3+}, Tb^{3+}, Dy^{3+}, Ho^{3+}, Er^{3+}, Tm^{3+} and Yb^{3+}, the main difference is the number of electrons in the 4f electron shell, which varies from 0 to 14. Among them, the 4f electrons of La and Lu are 0 and 14 respectively, showing optical inertia, they have no absorption and luminescence behavior. The 4f electrons of the remaining 13 rare earth ions vary from 1 to 13, showing spectral activity. The Coulomb interaction between electrons in the 4f configuration, the spin-orbit interaction, and the influence of the crystal field and magnetic field of the matrix in which they are located make the energy level structure of different rare earth ions unique. It is precise because of the transition of 4f electrons between different energy levels that absorption, luminescence, and lasers are produced. Due to the different number of 4f electrons of rare earth ions, their spectral properties are different. The 4f electrons of rare earth ions are shielded by the 5 s and 5p electron shells, and are less affected by external electromagnetic fields. The electric dipole transition between energy levels in the 4f configuration is forbidden, and the magnetic dipole transition is allowed. In solid-state materials, due to the effect of the crystal field, the state within the 4f configuration is no longer single, but a mixed state of two parities, which makes the electric dipole transition within the 4f configuration possible. Compared with transition metal ions, the spectral properties of rare earth ions are less affected by the outside crystal field. The absorption and emission spectra corresponding to the transition between 4f electron configurations are all narrow band peaks. In addition, the absorption and emission wavelengths of the f-f transition of rare earth ions are less affected by the matrix, and the absorption and emission wavelengths range from ultraviolet to mid-infrared.

In addition to the above-mentioned f-f transitions, rare earth ions can also observe 4f-5d transitions. The transitions between their $4f^{n-1}5d$ configuration and $4f^{n}$ configuration energy levels are parity-allowed transitions, and their luminescence intensity is much stronger than that of f-f transitions, and their transition probability is also much larger. Because 5d electrons are easily affected by the external crystal field, therefore, 4f-5d transitions show broadband luminescence from ultraviolet to visible light. Trivalent Ce^{3+}, Pr^{3+} and Tb^{3+}, as well as some divalent rare earth ions Eu^{2+}, Sm^{2+}, Yb^{2+}, Tm^{2+}, Dy^{2+}, Nd^{2+} etc. can observe 4f-5d transitions.

The f-f transitions and 4f-5d transitions of rare earth ions are affected in different degrees by the matrix in which they are located. The fine energy level structure of rare earth ions in glasses of different compositions will be different due to different coordination environments. The intensity and probability of the transition of rare earth ions from the ground state energy level to the excited state, and from the excited state to the lower energy level, will be affected by the crystal field effect of the matrix material. The spectral behavior of rare earth ions in different matrices can be studied through absorption and emission spectra. The stimulated emission cross-section and fluorescence lifetime of rare earth ions are important spectral parameters. The stimulated emission cross-section can be calculated by the Judd-Ofelt theory [69, 70] F-L formula [71] and McCumber formula [72] using the data of absorption spectrum and emission spectrum. The stimulated emission cross section of rare earth ions with four-level structure is usually calculated by Judd-Oflet theory, and the stimulated emission cross section of rare earth ions with three-level structure is usually calculated by F-L formula and McCumber formula. The ratio of the measured fluorescence lifetime to the theoretical fluorescence lifetime of the upper level of rare earth ions is its quantum efficiency.

Quantum efficiency is affected by the probability of non-radiative transition. The greater the probability of non-radiative transition, the lower the quantum efficiency. Non-radiative transition originates from the energy transfer between excited state of rare earth ions and the phonons, impurity ions or other rare earth ions in the matrix. This energy transfer eventually consumes the upper level ions in the form of heat.

The luminescent properties of rare earth ions are closely related to the composition and preparation process of the matrix glass. The composition of the matrix glass determines its phonon energy and structure, and will affect the solubility, absorption spectrum and emission spectrum, fluorescence lifetime, excited state absorption and other properties of rare earth ions. The glass preparation process affects its impurity content. Usually, hydroxyl and transition metal ions have a strong quenching effect on the upper level rare earth ions, and at the same time cause a sharp increase in the loss of glass and its optical fibers. The concentration of rare earth ions will affect their interaction degree. The higher the concentration of rare earth ions, the stronger the interaction, thus accelerating the energy transfer between the same kind of rare earth ions. The ways of interaction between rare earth ions include energy transfer, cross relaxation, cooperative upconversion and concentration quenching. To obtain good spectral properties, it is necessary to optimize the composition, structure and preparation process of the glass. As mentioned before, the structure of silica glass is a three-dimensional network structure connected by $[SiO_2]$ tetrahedra. Compared with other glass systems, its structure is more compact and robust. Because the radius of rare earth ions is large and they are usually trivalent, the solubility of rare earth ions in silica glass is relatively low, usually below several hundred ppm. Otherwise, it will cause clustering of rare earth ions and seriously affect their luminescent properties. In order to obtain ideal spectral and laser performance, it is necessary to add other ions, such as Al^{3+}, Ge^{4+}, P^{5+} while doping rare earth ions. The addition of these ions can to some extent disrupt the three-dimensional network structure of pure silica glass, allowing rare earth ions to be uniformly doped in silica glass. The introduction of rare earth ions and other co-doped ions will change the structure and optical properties of silica glass, causing changes in the spectral properties of rare earth ions. Therefore, in order to obtain the best laser performance of rare earth ion doped silica fiber, it is necessary to systematically study and optimize the spectral and optical properties of rare earth ion-doped silica glass. Chapters 2, 6, 7, and 8 of this book will detail the spectra and laser properties of Yb^{3+}, Er^{3+}, Tm^{3+}, Ho^{3+}, Nd^{3+} in silica glass.

Table 1.14 lists the rare earth ions and laser wavelengths that have achieved laser output in glass fibers so far. It can be seen that Pr^{3+}, Nd^{3+}, Sm^{3+}, Tb^{3+}, Dy^{3+}, Ho^{3+}, Er^{3+}, Tm^{3+}, Yb^{3+} and other rare earth ions have achieved laser output from the shortest 422 nm visible light to the longest 3950 nm mid-infrared band in glass fibers. Due to the higher phonon energy of silica glass, the longest laser wavelength achieved by rare earth ions in silica fiber is 2150 nm. Fluoride glass, due to its lower phonon energy, has achieved the longest 3.95 μm laser output in fluoride glass fibers with rare earth ions.

Table 1.14 Laser and amplification wavelength range of rare earth ions in glass fibers

Rare earth ions	Type of glass fiber	Laser wavelength/nm
Pr^{3+}	Fluoride glass	422, 442, 444, 476.5, 490, 520, 601 ~ 618, 631 ~ 641, 707 ~ 725, 880 ~ 886, 902 ~ 916, 1260 ~ 1350
	Oxide glass	1060 ~ 1110
Pr^{3+}/Yb^{3+}	Fluoride glass	849, 860, 850
Nd^{3+}	Oxide glass	890 ~ 950, 1000 ~ 1150, 1320 ~ 1466
	Fluoride glass	1000 ~ 1150, 1320 ~ 1400
Sm^{3+}	Oxide glass	651
Tb^{3+}	Fluoride glass	542.8
Dy^{3+}	Fluoride glass	478, 570, 575, 576, 3100
	Oxide glass	581, 582.5
Ho^{3+}	Fluoride glass	450, 550, 640, 643, 647, 753, 1380, 2040 ~ 2080, 3002, 3950
	Oxide glass	2040 ~ 2150
Ho^{3+}/Pr^{3+}	Fluoride glass	2900
Er^{3+}	Fluoride glass	543, 550, 800, 850, 980 ~ 1000, 1500 ~ 1620, 1660, 1720, 2700 ~ 2880, 3000, 3100, 3300 ~ 3800
	Oxide glass	1500 ~ 1620
Tm^{3+}	Fluoride glass	455, 480, 803 ~ 825, 1460 ~ 1510, 1700 ~ 2015, 2250 ~ 2400
	Oxide glass	480, 1700 ~ 2100
Yb^{3+}	Oxide glass	970 ~ 1180

1.7.2 Development of Rare Earth Doped Silica Fiber

Due to the excellent thermal properties, mechanical performance, laser damage resistance, radiation resistance of silica glass, and the mature application of chemical vapor deposition technology for ultra-low loss silica glass, the most of passive communication fibers and active fibers are currently based on silica glass. Rare earth doped silica fiber can be broadly categorized into the following two main development stages:

The first stage is from 1970 to 2000, mainly based on the invention of erbium-doped silica fiber and erbium fiber amplifiers. Erbium-doped silica fiber played an important role in the relay amplification of long-distance communication signals in the field of optical communication. During this period, the structure of rare earth doped silica fiber was improved, from single cladding to double cladding, the gradual reduction of fiber background loss, and the development of manufacturing process technology. In 1973, Stone et al. from Bell Labs in the United States [73] first reported neodymium-doped silica fiber. In 1985, Poole et al. from the University of Southampton in the UK [74] reported the preparation of low-loss rare earth doped neodymium and erbium silica fiber using the MCVD method, and achieved laser output. Mears et al. first reported a 1.55 μm laser and a near 1.54 μm optical amplifier based on erbium-doped fiber [75]. Because the laser wavelength of the erbium-doped fiber laser is exactly at the 1.55 μm low-loss window of the communication fiber, people began to realize that fiber amplifiers and fiber lasers will undoubtedly bring a revolution to fiber communication in terms of improving transmission rate and extending transmission distance. Erbium-doped fiber amplifier (EDFA) has developed rapidly and has become a matured technology. The output power of fiber lasers for fiber communication is generally at the milliwatt level, and has always been limited to fields such as optical communication. In 1988, Snitzer et al. [76] proposed double-cladding pumping technology, which changed people's perception that fiber lasers can only produce low power output, making it possible to produce high-power, high-brightness laser output using fiber lasers. Due to the four-level structure characteristics of neodymium ions, it has advantages such as low laser threshold power. The initial research mainly focused on neodymium ions and neodymium-doped silica fiber, as well as neodymium-doped cladding pumped fiber lasers. In 1992, Minelly et al. [77] reported a Nd-doped double-clad fiber laser with an output power larger than 1 W. In 1993, in the cladding pumped Nd^{3+} fiber laser experiment, Po et al. [78] achieved a laser with an output power of 5 W and a slope efficiency of 51%. In 1995, Zellmer et al. [79] reported a double-clad pumped Nd^{3+} fiber lasers, with a slope efficiency of only 25%. This is mainly due to the use of a circular cladding pump structure, which results in insufficient absorption of pump light by the single-mode core layer. However, as research progresses, it has been found that due to the very narrow absorption band of Nd^{3+} at 800 nm, it requires high stability and precision of the pump source wavelength. Yb^{3+} has a fairly wide absorption band at 900 ~ 1000 nm, and due to its simple energy level structure, there is no up-conversion luminescence caused by excited state absorption, high quantum efficiency, and high conversion efficiency and output power can be achieved. In addition, laser diodes suitable for pumping Yb^{3+} have also developed rapidly. The focus has shifted to the research of Yb^{3+} doped fiber lasers. In 1994, Pask et al. [80] first achieved cladding pumping in Yb^{3+} doped silica fiber, using 975 nm pump light to obtain 0.5 W laser output at 1040 nm wavelength, with a slope efficiency of up to 80%. In 1997, Muendel et al. from Polaroid Corporation [81] reported a 1100 nm, 35.5 W single-mode continuous laser output from a ytterbium-doped double-clad fiber laser. In 1999, Dominic et al. from SDL Corporation [82] successfully developed a 110 W single-mode continuous laser output ytterbium-doped double-clad fiber laser using four 45 W laser diodes pumping from both ends, with a light-to-light conversion efficiency of 58%. The emergence of ytterbium-doped silica fiber has continuously increased the output power of fiber lasers. Fiber lasers have gradually entered the field of industrial processing applications.

The second stage is from 2000 to the present. With the maturity and widespread application of high-power laser diode technology, fiber laser technology and rare earth doped silica fiber manufacturing technology have developed rapidly. Rare earth doped silica fiber has entered a stage dominated by high-power ytterbium-doped silica fiber, while also focusing on the development of high-power erbium, thulium, and holmium doped silica fiber. The important feature of this stage of rare earth doped silica fiber is the large mode area rare earth doped silica fiber to meet the development of high-power laser applications, and the fiber structure shows the characteristics of change from double-clad structure to multi-clad, and the cladding structure to photonic crystal fiber structure. The preparation technology of rare earth doped silica fiber preform has also diversified from the traditional MCVD combined with liquid and gas doping to nanoparticle doping, reactive powder sintering (REPUSIL) and sol-gel and other non-chemical vapor deposition methods, stacking methods. With the increase in output power of laser diodes and their widespread application, fiber lasers represented by ytterbium-doped fiber lasers have been developed rapidly, and the output power has grown rapidly. In 2004, IPG company achieved the commercialization of a continuous output of 10 kilowatts multimode fiber laser, and in 2009, it achieved the commercialization of a continuous output of 5 kW single-mode ytterbium-doped fiber laser. Using the same band pumping amplification technology, ytterbium-doped cladding fiber has achieved the highest continuous laser output power of single-mode 20 kilowatts and multimode 100 kilowatts [83]. The large mode area thulium-doped silica fiber with a core of 20 ~ 30 μm, an octagonal inner cladding, and a low numerical aperture (0.065) has become the mainstream laser fiber for high-power few-mode fiber lasers. In addition, since the birth of thulium-doped photonic crystal fiber in 2000 [84], the nonlinearity and dispersion of fiber lasers have been effectively controlled. With the emergence of new semiconductor mode-locking elements and mode-locking technology, the development, industrial processing, and scientific research applications of picosecond and femtosecond ultrafast fiber lasers have been greatly promoted. At present, large mode field thulium-doped photonic crystal fibers with core diameters of 40 ~ 135 μm have been successfully applied to fiber amplifiers in the range of nanoseconds to femtoseconds. Stutzki et al. [85] reported that using a large pitch thulium-doped photonic crystal fiber with a core diameter of 135 μm, they achieved a near-diffraction-limited beam quality pulse laser output with a 60 ns, 26 mJ, and 500 kW peak power. Eidam et al. [86] reported that using a large pitch thulium-doped photonic crystal fiber with a mode field diameter of 105 μm, they achieved a pulse laser output with an energy greater than 2 mJ and a peak power of up to 4GW.

At the same time, thanks to the development of fiber laser technology and the continuous increase in the output power of laser diodes used as pump sources, the output power of erbium-doped and thulium-doped silica fibers has also broken through 656 W [87] and 1050 W [88] respectively. Rare earth-doped silica fibers and fiber lasers have been promoted and applied in industrial processing, medical, space exploration, scientific research, military and other fields. Table 1.15 summarizes the laser wavelengths and output power levels of different rare earth ion-doped silica fibers. Figure 1.53 shows the output power and laser wavelength range distribution of different rare earth-doped fibers. This book will introduce the research progress and applications of thulium-doped, neodymium-doped, erbium-doped, thulium-doped, holmium-doped, Nd/Yb co-doped, and Tm/Ho co-doped silica fibers in subsequent chapters.

In summary, this chapter elaborates in detail the preparation, types, structure, and properties of silica glass, and briefly introduces the luminescent properties of rare earth ions and the development history of rare earth ion doped silica fibers. Since the twenty-first century, the research and development and mass production technology of rare earth doped silica fibers have greatly promoted the development and

Table 1.15 Rare earth ion-doped silica fiber laser output wavelength and power

Rare earth ions	Laser wavelength/nm	Maximum output power	Corresponding transition energy level
Sm^{3+}	651	~28 mW	$^4G_{5/2} \rightarrow {}^6H_{9/2}$
Dy^{3+}	582.5	18.4 mW	$^4F_{9/2} \rightarrow {}^6H_{13/2}$
Nd^{3+}	900 ~ 950	83 W	$^4F_{3/2} \rightarrow {}^4I_{9/2}$
Nd^{3+}	1000 ~ 1150	30 W	$^4F_{3/2} \rightarrow {}^4I_{11/2}$
Yb^{3+}	970 ~ 1040	20,000 W	$^2F_{5/2} \rightarrow {}^2F_{7/2}$
Nd^{3+}/Yb^{3+}	970 ~ 1150	1300 W	$^4F_{3/2} \rightarrow {}^4I_{11/2}$ & $^2F_{5/2} \rightarrow {}^2F_{7/2}$
Er^{3+}	1500 ~ 1620	656 W	$^4I_{13/2} \rightarrow {}^4I_{15/2}$
Er^{3+}/Yb^{3+}	1500 ~ 1600	338 W	$^4I_{13/2} \rightarrow {}^4I_{15/2}$
Tm^{3+}	1700 ~ 2015	1050 W	$^3F_4 \rightarrow {}^3H_6$
Ho^{3+}	2010 ~ 2100	407 W	$^5I_7 \rightarrow {}^5I_8$
Tm^{3+}/Ho^{3+}	2010 ~ 2100	38 W	$^5I_7 \rightarrow {}^5I_8$

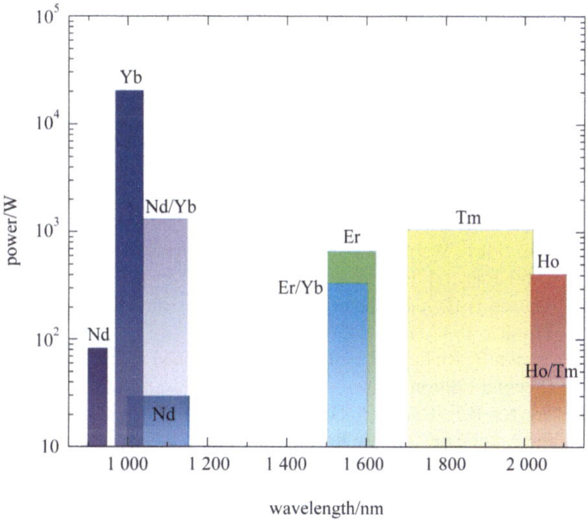

Fig. 1.53 Output power and laser wavelength distribution of different rare earth doped fibers

application of fiber lasers. Rare earth doped silica fibers will continue to improve laser output power while improving beam quality. The laser wavelength will also extend from the current 1 ~ 2 μm to visible light lasers. The structure of rare earth doped silica fibers will also continue to be improved and optimized. To further improve beam quality and output power, it is necessary to study in depth the mode instability and the impact of nonlinear effects such as stimulated Raman scattering and stimulated Brillouin scattering on beam quality and laser efficiency. In addition, in expanding the working wavelength range of fiber lasers and amplifiers, since 2005, the Dianov research group of the Russian Academy of Sciences [89] has successfully developed a bismuth-doped silica fiber with broadband amplification characteristics, providing a solution for further expanding the wavelength range of fiber lasers or amplifiers. Bismuth-doped fiber amplifiers are gradually moving towards practical applications. It is believed that with the deepening of research, special fibers based on silica glass will play an increasingly important role in technological progress and human social development.

References

1. Brückner R. Properties and structure of vitreous silica. I[J]. Journal of Non-Crystalline Solids, 1970,5(2):123-175.
2. Angell C A. Formation of glasses from liquids and biopolymers[J]. Science, 1995,267(5206):1924-1935.
3. Richet P, Bottinga Y. Thermochemical properties of silicate glasses and liquids: a review[J]. Reviews of Geophysics, 1986,24(1):1-25.
4. Yue Y. Anomalous enthalpy relaxation in vitreous silica[J]. Frontiers in Materials, 2015,2:54.
5. Wang Yufen, Liu Liancheng, Silica glass (in Chinese) [M] Chemical Industry Press, 2007
6. Warren B E. X-ray Determination of the Structure of Glass[J]. Journal of the American Ceramic Society, 1992,75(1):5-10.
7. Randall J T, Rooksby H P, Cooper B S. 13. X-ray Diffraction and the Structure of Vitreous Solids—I[J]. Zeitschrift für Kristallographie-Crystalline Materials, 1930,75(1):196-214.
8. Randall J T, Rooksby H P, Cooper B S. Atomic Physics and Related Subjects.: Communications to Nature.: The Diffraction of X-Rays by Vitreous Solids and its Bearing on their Constitution[J]. Nature, 1930,125(3151):458.
9. Zachariasen W H. The atomic arrangement in glass[J]. Journal of the American Chemical Society, 1932,54(10):3841-3851.
10. Mozzi R L, Warren B E. The structure of vitreous silica[J]. Journal of Applied Crystallography, 1969,2(4):164-172.
11. Warren B E. X-ray determination of the structure of liquids and glass[J]. Journal of Applied Physics, 1937,8(10):645-654.
12. Mysen B, Richet P. Chapter 5 - Silica[M]//MYSEN B, RICHET P. Silicate Glasses and Melts (Second Edition). Elsevier, 2019:143-183.
13. Warren B E, Biscce J. The structure of silica glass by X-ray diffraction studies[J]. Journal of the American Ceramic Society, 1938,21(2):49-54.
14. Wright A C. Neutron scattering from vitreous silica. V. The structure of vitreous silica: What have we learned from 60 years of diffraction studies?[J]. Journal of non-crystalline solids, 1994,179:84-115.
15. Malfait W J, Halter W E, Verel R. 29Si NMR spectroscopy of silica glass: T1 relaxation and constraints on the Si–O–Si bond angle distribution[J]. Chemical Geology, 2008,256(3-4):269-277.
16. Huang P Y, Kurasch S, Srivastava A, et al. Direct imaging of a two-dimensional silica glass on graphene[J]. Nano letters, 2012,12(2):1081-1086.
17. Dupree R, Farnan I, Capponi J J, et al. Pressure-induced bond-angle variation in amorphous SiO_2[J]. Physical Review B, 1987,35(5):2560-2562.
18. Salh R. Defect related luminescence in silicon dioxide network: a review[M]. INTECH Open Access Publisher, 2011.
19. Skuja L, Hosono H, Hirano M, et al. Advances in silica-based glasses for UV and vacuum UV laser optics[Z]. 2003: 5122.
20. Lo Piccolo G M, Cannas M, Agnello S. Intrinsic Point Defects in Silica for Fiber Optics Applications[J]. Materials, 2021,14(24):7682.
21. Galeener F L. Planar rings in vitreous silica[J]. Journal of Non-Crystalline Solids, 1982,49(1–3):53-62.

22. Newton M D, Gibbs G V. Ab initio calculated geometries and charge distributions for H4SiO4 and H6Si2O7 compared with experimental values for silicates and siloxanes[J]. Physics and Chemistry of Minerals, 1980,6(3):221-246.
23. Graetsch H, Mosset A, Gies H. XRD and 29Si MAS-NMR study on some non-crystalline silica minerals[J]. Journal of Non-Crystalline Solids, 1990,119(2):173-180.
24. Tool A Q. Relation between Inelastic Deformability and Thermal Expansion of Glass in Its Annealing Range[J]. Journal of the American Ceramic Society, 1946,29(9):240-253.
25. Agarwal A, Tomozawa M. Correlation of silica glass properties with the infrared spectra[J]. Journal of Non-Crystalline Solids, 1997,209(1):166-174.
26. Deng B, Shi Y, Yuan F. Investigation on the structural origin of low thermal expansion coefficient of fused silica[J]. Materialia, 2020,12:100752.
27. Vollmayr K, Kob W, Binder K. Cooling-rate effects in amorphous silica: A computer-simulation study[J]. Physical Review B, 1996,54(22):15808.
28. Ebrahem F, Bamer F, Markert B. The influence of the network topology on the deformation and fracture behaviour of silica glass: A molecular dynamics study[J]. Computational Materials Science, 2018,149:162-169.
29. Gross T M, Tomozawa M. Fictive temperature of GeO_2 glass: its determination by IR method and its effects on density and refractive index[J]. Journal of non-crystalline solids, 2007,353(52-54):4762-4766.
30. Heili M, Poumellec B, Burov E, et al. The dependence of Raman defect bands in silica glasses on densification revisited[J]. Journal of Materials Science, 2016,51(3):1659-1666.
31. Musgraves J D, Hu J, Calvez L. Springer Handbook of Glass[M]. Springer Nature, 2019.
32. Urbain G, Bottinga Y, Richet P. Viscosity of liquid silica, silicates and alumino-silicates[J]. Geochimica et cosmochimica acta, 1982,46(6):1061-1072.
33. Doremus R H. Viscosity of silica[J]. Journal of Applied Physics, 2002,92(12):7619-7629.
34. Brückner R. Properties and structure of vitreous silica. II[J]. Journal of Non-Crystalline Solids, 1971,5(3):177-216.
35. Davis K M, Tomozawa M. Water diffusion into silica glass: Structural changes in silica glass and their effect on water solubility and diffusivity[J]. 1995,185(3):203-220.
36. Fanderlik I. Silica glass and its application[M]. Elsevier, 2013.
37. Vedam K, Schmidt E, Roy R. Nonlinear variation of refractive index of vitreous silica with pressure to 7 Kbars[J]. Journal of the American Ceramic Society, 1966,49(10):531-535.
38. Mackenzie J D. High-pressure effects on oxide glasses: I, densification in rigid state[J]. Journal of the American Ceramic Society, 1963,46(10):461-470.
39. Malitson I H. Interspecimen Comparison of the Refractive Index of Fused Silica*,†[J]. Journal of the Optical Society of America, 1965,55(10):1205-1209.
40. Dragic P, Cavillon M, Ballato J. On the thermo-optic coefficient of P_2O_5 in SiO_2[J]. Optical Materials Express, 2017,7(10):3654-3661.
41. Kakiuchida H, Saito K, Ikushima A J. Refractive index, density and polarizability of silica glass with various fictive temperatures[J]. Japanese journal of applied physics, 2004,43(6A):L743.
42. Hofmeister A M, Whittington A G. Effects of hydration, annealing, and melting on heat transport properties of fused quartz and silica glass from laser-flash analysis [J]. Journal of Non-Crystalline Solids, 2012,358(8):1072-1082.
43. Richet P, Bottinga Y. Glass transitions and thermodynamic properties of amorphous SiO_2, $NaAlSi_n O_{2n+2}$ and $KAlSi_3 O_8$ [J]. Geochimica et Cosmochimica Acta, 1984,48(3):453-470.
44. Cheng S C, et al. Use of EELS to study the absorption edge of silica glass [J]. Journal of Non-Crystalline Solids, 2006,352(28-29):3140-3146.
45. Moore L A, Smith C M. Silica glass as an optical material [Invited] [J]. Optical Materials Express, 2022,12(8):3043.
46. Tamura Y, Sakuma H, Morita K, et al. Lowest-ever 0.1419 dB/km loss optical fiber: Optical Fiber Communications Conference and Exhibition, in Optical Fiber Communication Conference Postdeadline Papers[C].OSA Technical Digest, 2017, Th5D.1. https://doi.org/10.1364/OFC.2017.Th5D.1.

47. Saito K, Ikushima A J. Reduction of light-scattering loss in silica glass by the structural relaxation of "frozen-in" density fluctuations [J]. Applied Physics Letters, 1997,70(26):3504-3506.
48. Saito K, Kakiuchida H, Ikushima A J. Investigation of the origin of the Rayleigh scattering in SiO_2 glass [J]. Journal of Non-Crystalline Solids, 1997(222):329-334.
49. Saito K, et al. A new method of developing ultralow-loss glasses [J]. Journal of Applied Physics, 1997,81(11):7129-7134.
50. Wang Chengyu, Tao Ying. Handbook of Glass Properties and Processes (in Chinese) [M]. Beijing: Chemical Industry Press, 2008.
51. Shao Chongyun, Yu Chunlei, Hu Lili. Radiation-Resistant Active Fibers for Space Applications (in Chinese) [J]. Chinese Journal of Laser, 2020,47(5):233-261.
52. Griscom D L. Nature of defects and defect generation in optical glasses [C]//1985 Albuquerque Conferences on Optics. 1985: International Society for Optics and Photonics.
53. Griscom D L. Optical properties and structure of defects in silica glass [J]. Journal of the Ceramic Society of Japan, 1991,99(10):923-942.
54. Skuja L. Optically active oxygen-deficiency-related centers in amorphous silicon dioxide [J]. Journal of Non-Crystalline Solids, 1998,239(1-3):16-48.
55. Girard S, et al. Radiation effects on silica-based optical fibers: recent advances and future challenges [J]. IEEE Transactions on Nuclear Science, 2013,60(3):2015-2036.
56. Tian Hai. Research on the space radiation effects and test methods of quartz and flint optical glass[D]. Lanzhou: Lanzhou University, 2020:62.
57. Henson T D, Torrington G K. Space radiation testing of radiation-resistant glasses and crystals, in Inorganic Optical Materials III[C].SPIE, 2001(4452): 54-65.
58. Ge Shiming. Radiation resistance of silica glass[J]. Atomic Energy Science and Technology, 1983(2):170-173.
59. Nagasawa K, Tanabe M, Yahagi K. Gamma-ray-induced absorption bands in pure-silica-core fibers [J]. Japanese Journal of Applied Physics, 1984,23(12R):1608.
60. Lancry M, et al. Radiation hardening of silica glass through fictive temperature reduction [J]. International Journal of Applied Glass Science, 2017,8(3):285-290.
61. Hosono H, et al. Effects of fluorine dimer excimer laser radiation on the optical transmission and defect formation of various types of synthetic SiO_2 glasses [J]. Applied Physics Letters, 1999,74(19):2755-2757.
62. Baisden P A, et al. Large optics for the national ignition facility [J]. Fusion Science and Technology, 2016,69(1):295-351.
63. Chen Yali, et al. Research on low hydroxyl high purity silica glass for photomask [J]. Glass, 2021,48(3):6-10.
64. Jedamzik R, Dietrich V, Rossmeier T. Bulk laser damage threshold of optical glasses[C]. SCHOTT AG, Advanced Optics, 2012:55122.
65. Suratwala T I, et al. HF-Based etching processes for improving laser damage resistance of fused silica optical surfaces [J]. Journal of the American Ceramic Society, 2011,94(2):416-428.
66. Spaeth M L, et al. Optics recycle loop strategy for NIF operations above UV laser-induced damage threshold [J]. Fusion Science and Technology, 2016,69(1):265-294.
67. Kapron F P, Keck D B, Maurer R D. Radiation losses in glass optical waveguides [J]. Applied Physics Letters, 1970,17(10):423-425.
68. Koester C J, Snitzer E. Amplification in a fiber laser [J]. Applied Optics, 1964,3(10):1182-1186.
69. Ofelt G S. Intensities of crystal spectra of rare-earth ions [J]. The Journal of Chemical Physics, 1962,37(3):511-520.
70. Judd B R. Optical absorption intensities of rare-earth ions [J]. Physical Review, 1962,127(3):750-761.
71. Emission Cross Section, Füchtbauer-Ladenburg Equation, and Purcell Factor[C]. Dordrecht: Springer Netherlands, 2017:387-404. https://doi.org/10.1007/978-94-024-0850-8_19.
72. McCumber D E. Theory of phonon-terminated optical masers [J]. Physical Review, 1964,134(2A):A299.

73. Stone J, Burrus C A. Neodymium-doped silica lasers in end-pumped fiber geometry [J]. Applied Physics Letters, 1973,23(7):388-389.
74. Poole S, Payne D, Mears R, et al. Fabrication and characterization of low-loss optical fibers containing rare-earth ions [J]. Journal of Lightwave Technology, 1986,4(7):870-876.
75. Mears R J, Reekie L, Jauncey I M, et al. Low-noise erbium-doped fibre amplifier operating at 1.54μm [J]. Electronics Letters, 1987(23):1026-1028.
76. Snitzer E, Po H, Hakimi F, et al. Double clad, offset core Nd fiber laser. Optical Fiber Sensors [C]. New Orleans, Louisiana: Optica Publishing Group, 1988.
77. Minelly J D, Taylor E R, Jedrzejewski K P, et al. Laser-diode-pumped neodymium-doped fiber laser with output power>1W [C]. Conference on Lasers and Electro-Optics. Optical Society of America, 1992.
78. Po H, Cao J D, Laliberte B M, et al. High power neodymium-doped single transverse mode fibre laser [J]. Electronics Letters, 1993,17(29):1500-1501.
79. Zellmer H, Willamowski U, Tünnermann A, et al. High-power cw neodymium-doped fiber laser operating at 9.2W with high beam quality [J]. Optics Letters, 1995,20(6):578-580.
80. Pask H M, Archambault J L. Operation of cladding-pumped -doped silica fibre lasers in the 1μm region [J]. Electronics Letters, 1994,30(11):863-865.
81. Muendel M, Engstrom B, Kea D, et al. 35-watt cw single-mode ytterbium fiber laser at 1.1μm [C]. Conference on Lasers and Electro-optics, Optical Society of America,1997.
82. Dominic V A, MacCormack S, Waarts R, et al. 110W fibre laser [J]. Electronics Letters, 1999,35(14):1158-1160.
83. Shiner B. The impact of fiber laser technology on the World Wide Material Processing Market [C]. CLEO: Applications and Technology, Optica Publishing Group.
84. Wadsworth W J, Knight J C, Reeves W H, et al. Yb^{3+} -doped photonic crystal fibre laser [J]. Electronics Letters, 2000(36):1452-1454.
85. Stutzki F, Jansen F, Liem A, et al. 26mJ, 130W Q-switched fiber-laser system with near-diffraction-limited beam quality [J]. Optics Letters, 2012,37(6):1073-1075.
86. Eidam T, Rothhardt J, Stutzki F, et al. Fiber chirped-pulse amplification system emitting 3.8GW peak power [J]. Optics Express, 2011,19(1):255-260.
87. Lin Huaiqin, Feng Yujun, Feng Yutong, et al. 656W Er-doped, Yb-free large-core fiber laser [J]. Optics Letters, 2018,43(13):3080.
88. Ramírez-Martínez N J, Núñez-Velzquez M, Umnikov A A, et al. Highly efficient thulium-doped high-power laser fibers fabricated by MCVD [J]. Optics Express, 2019,27(1):196-201.
89. Dvoyrin V V, Mashinsky V M, Dianov E M, et al. Absorption, fluorescence and optical amplification in MCVD bismuth-doped silica glass optical fibres [C]. 2005, 31st European Conference on Optical Communication, https://doi.org/10.1049/cp:20050796.

Chapter 2
Properties and Structure of Ytterbium-Doped Silica Glass

Ytterbium-doped silica fiber (YDF) is the core component of high-power fiber lasers, the core material of which is ytterbium-doped silica glass (YDG). The performance of YDF is closely related to the properties of YDG. To achieve high power and high beam quality laser output, in addition to effectively controlling the loss of YDF, the following two prominent problems need to be solved. On the one hand, to suppress the nonlinear effects, it is necessary to increase the doping concentration of ytterbium ions (Yb^{3+}) in the YDG, thus to use as short as possible and larger mode-field-area YDF. On the other hand, to achieve high beam quality, it is necessary to reduce the numerical aperture of the inner cladding as much as possible, that is, to reduce the refractive index difference between the YDF core material and the cladding one. Therefore, in the core composition design of YDF, especially the high-power one, it is necessary to co-dope Al, P, Ge, F, and B elements to obtain ideal refractive index and spectral properties to meet the final use requirements. In addition, the preparation of YDF mainly includes two temperature-related steps: preform deposition and drawing process. During this process, YDF will undergo a thermal process of heating and cooling. The change in its thermal history will not only cause changes in Yb^{3+} absorption coefficient and the absorption peak position, but also will affect the refractive index of the cladding and core materials to varying degrees, leading to changes in the numerical aperture of YDF, thereby influencing the laser performance of YDF. Based on this, this chapter systematically elaborates the effects of co-doping elements and thermal history on the physical and optical properties of YDG. The corresponding mechanisms from the atomic microscale has been also analyzed by the Fourier transform infrared spectroscopy (FTIR), Raman, nuclear magnetic resonance (NMR), electron paramagnetic resonance (EPR) and other structural analysis techniques.

L. Hu, *Rare Earth Doped Silica Fiber and Its Applications*,
https://doi.org/10.1007/978-981-96-5306-5_2

2.1 Energy Level Structure and Spectral Properties of Ytterbium Ions

2.1.1 Energy Levels of Ytterbium Ions

Figure 2.1 shows the energy level diagram of Yb^{3+}. The electronic configuration of Yb^{3+} is Xe+$4f^{13}$. The Coulomb interaction between the electrons in the 4f shell causes the energy levels described by the electronic configuration $4f^{13}$ to split into a series of states with different energies. Each state is characterized by its total orbital angular momentum quantum number L (L = 3) and total spin angular momentum quantum number S (S = 1/2), referred to as ^{2S+1}L spectral term, represented as ^2F. Due to the spin-orbit interaction of 4f electrons, ^2F spectral term is split into ^2F_J multiplets, where $J = S + L, L + S - 1, \ldots, |L - S|$. Therefore, ^2F spectral term splits into $^2F_{7/2}$ and $^2F_{5/2}$ sub-energy levels with an interval of 10,000 cm^{-1}. In addition to the above effects, electrons will also be affected by the crystal field. The shielding effect of the 5s and 5p shells will weaken the effect of the crystal field on the $4f^{13}$ electrons, thus will not significantly change the energy level gap between $^2F_{7/2}$ and $^2F_{5/2}$ sublevels. But it will cause the $^2F_{7/2}$ and $^2F_{5/2}$ sublevels to split into four and three double degenerate Stark levels respectively. The double degenerate Stark levels will further split into Zeeman levels under the effect of a magnetic field. But their energy level differences are usually less than 1 cm^{-1}.

Due to the simple energy levels and small quantum defects of Yb^{3+}, it has advantages such as high energy conversion efficiency, low material thermal load, and long fluorescence lifetime (about 3 times that of Nd^{3+}), thus to be suitable for high-power laser applications.

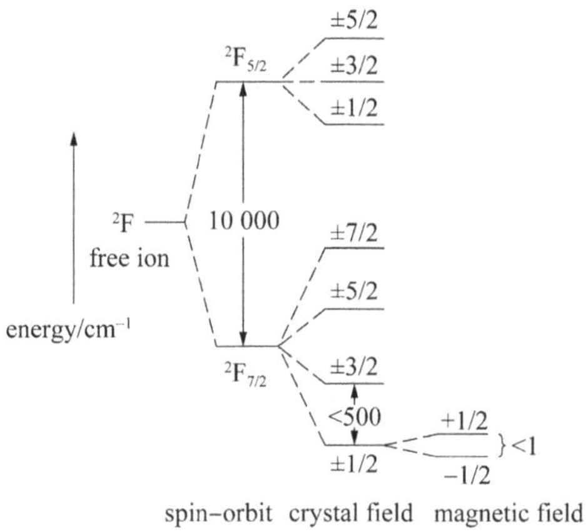

Fig. 2.1 Energy level diagram of Yb^{3+}

2.1.2 Spectral Properties of Ytterbium Ions

Figure 2.2a, b show the absorption and emission spectra of Yb^{3+} in pure silica glass (SY0.05), aluminate glass (Yb-AG), borate glass (Yb-BG), and phosphate glass (Yb-PG), respectively. There are significant differences in the absorption and emission spectra of Yb^{3+} in different glass systems. This is mainly due to the different crystal field strengths and asymmetry degrees of Yb^{3+} in different glass systems. By fitting the absorption and emission spectra of Yb^{3+} with Lorentzian peaks, the Stark split energy levels of Yb^{3+} can be obtained, as shown in Figs. 2.3.

Figure 2.3a, b are schematic diagrams of the Lorentzian peak fitting of the absorption and emission spectra of Yb^{3+} in single-doped silica glass (SY0.05). As shown in Fig. 2.3a, the absorption spectrum of Yb^{3+} consists of a sharp peak at 976 nm and a broad peak at 915 nm. Among them, the main absorption peak at

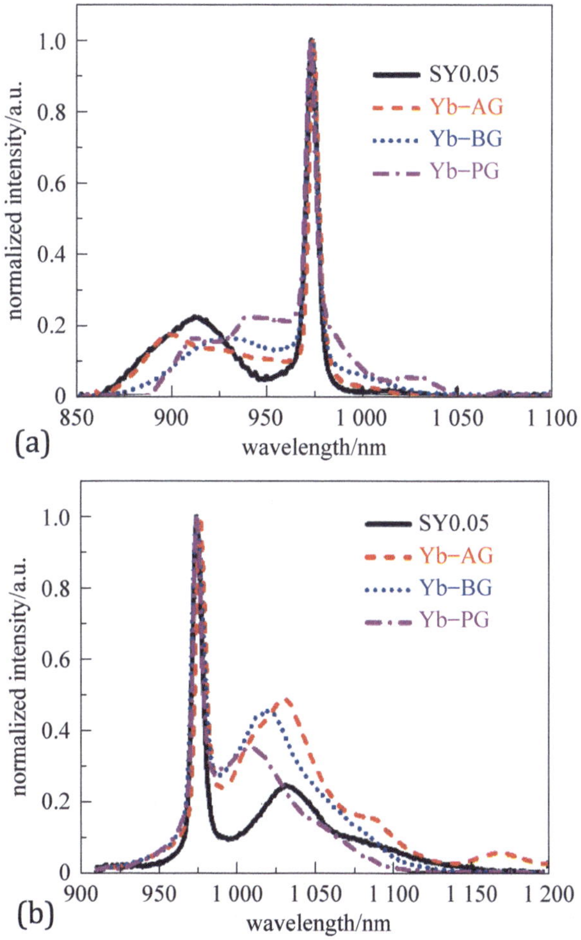

Fig. 2.2 The absorption (**a**) and emission (**b**) spectra in the Yb single-doped silica glass (SY0.05), aluminate (Yb-AG), borate (Yb-BG), phosphate (Yb-PG) glasses

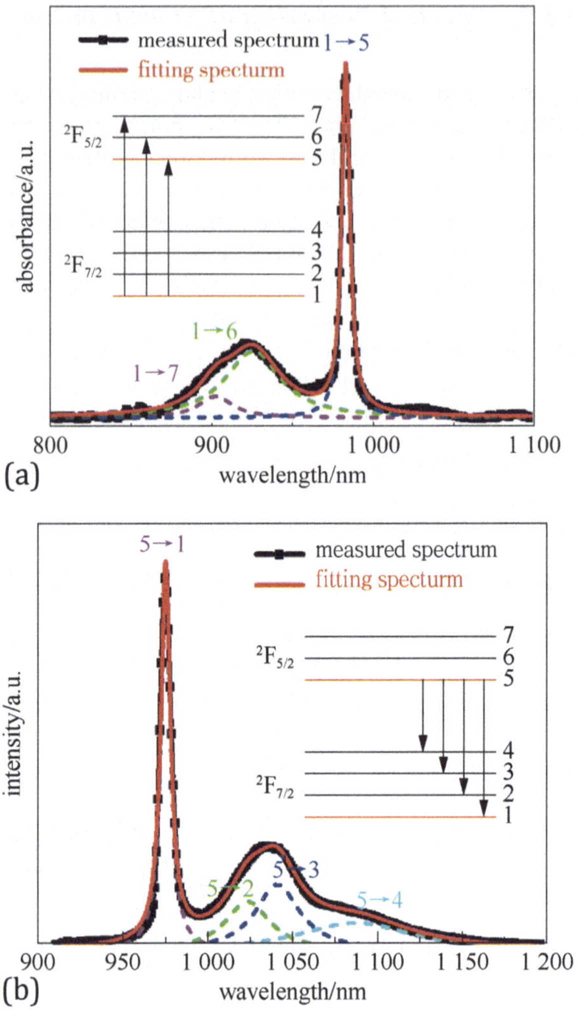

Fig. 2.3 Schematic diagram of Lorentz peak fitting of absorption spectrum (**a**) and emission spectrum (**b**) of Yb^{3+} in single-doped silica glass

976 nm corresponds to the zero-phonon transition 1 → 5 from the ground state $^2F_{7/2}$ to the excited state $^2F_{5/2}$. The broad absorption band at 915 nm covers the transition 1 → (6,7). As shown in Fig. 2.3b, the emission spectrum of Yb^{3+} consists of a sharp peak at 976 nm and a broad peak at 1030 nm. The main emission peak at 976 nm corresponds to the zero-phonon transition 5 → 1 from the excited state to the ground state. The broad emission band at 1030 nm covers the transitions 5 → (2,3,4).

Yang et al. [1] systematically studied the Stark splitting energy levels of Yb^{3+} in different glass systems, as shown in Figs. 2.4. For convenience, the main element symbols are used to represent different glass systems. For example, "Al" represents aluminate glass, "Ge" represents germanate glass, and so on. It can be seen that the Stark energy level differences of Yb^{3+} in different glass systems are significant. Among them, the energy level difference between two adjacent Stark levels of Yb^{3+}

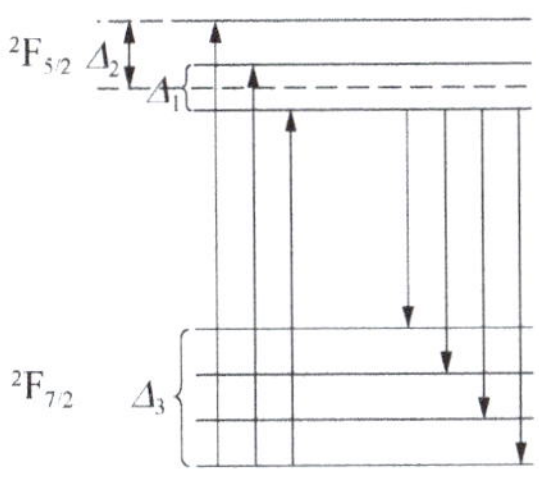

	Al	Ge	Si	B	Bi	FP	Te	F	P
7	11 127	11 038	10 977	10 917	10 846	10 846	10 834	10 753	10 776
6	10 720	10 684	10 593	10 537	10 493	10 504	10 504	10 449	10 460
5	10 261	10 246	10 246	10 246	10 246	10 267	10 246	10 267	10 267
4	998	901	874	830	758	734	722	624	596
3	550	509	488	452	509	511	509	453	307
2	312	316	265	246	396	277	286	257	207
1	0	0	0	0	0	0	0	0	0

Fig. 2.4 Stark splitting energy levels and energy values of each level E_i ($i = 1 \sim 7$, cm^{-1}) of Yb^{3+} in different glass systems

Fig. 2.5 The local crystal field intensity parameter (N_J) and degree of Yb^{3+} asymmetry (α_J) in different glass systems

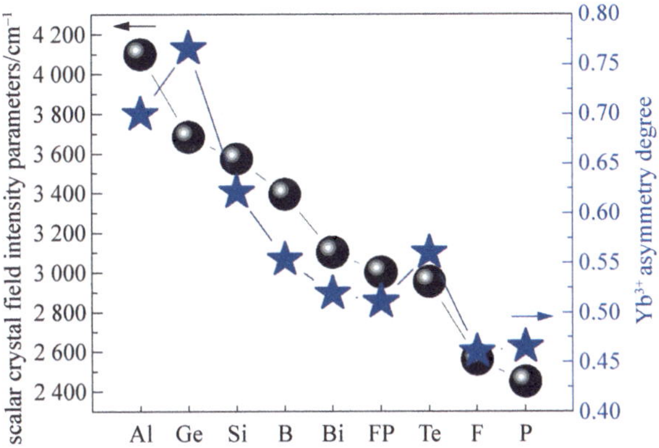

in aluminate glass is the largest, and the energy level difference between two adjacent Stark levels in phosphate glass is the smallest.

Studies have shown that the Stark energy levels of Yb^{3+} are closely related to the local crystal field strength and the degree of Yb^{3+} asymmetry. The scalar crystal field intensity parameter of Yb^{3+} has a linear relationship with the splitting value of the $^2F_{7/2}$ energy level of Yb^{3+}. Thus, it can be calculated based on the splitting of the $^2F_{7/2}$ energy level [2]. Robinson et al. [3] pointed out that the deviation from the regular octahedron of the crystal field of Yb^{3+}, i.e., the degree of asymmetry of the local crystal field around Yb^{3+}, could be calculated based on the splitting of the $^2F_{5/2}$ energy level of Yb^{3+}.

Based on the Stark energy level splitting in Fig. 2.4, Yang et al. [1] calculated the local crystal field intensity parameter (N_J) and degree of Yb^{3+} asymmetry (α_J) in different glass systems, as shown in Fig. 2.5.

According to the research of Auzel [2], N_J can be calculated using the following formula:

$$N_J = \Delta_3 / 0.245 \quad (2.1)$$

where, Δ_3 is the maximum Stark splitting energy level of the ground state $^2F_{7/2}$, as shown in Fig. 2.4 in details.

According to the research of Robinson et al. [3], α_J can be calculated through the following formulas:

$$\Delta_1 = E_6 - E_5 \tag{2.2}$$

$$\Delta_2 = E_7 - \Delta_1 / 2 \tag{2.3}$$

$$\alpha_J = \Delta_1 / \Delta_2 \tag{2.4}$$

where, E_5, E_6, E_7 are the energy values of Stark levels 5, 6, and 7, respectively, as shown in Fig. 2.4 in detail.

In addition to the transition spectrum and transition probability, the fluorescence lifetime of Yb^{3+} $F_{5/2} \rightarrow {}^2F_{7/2}$ transition is also closely related to the glass matrix, with its fluorescence lifetime ranging from several hundred microseconds to several milliseconds [4, 5]. For example, the fluorescence lifetime of Yb^{3+} in silica glass is about 600 ~ 1000 μs [6], more than 1.5 ms in fluorophosphate matrix [7], and can be more than 2 ms in some crystals [8]. Sect. 2.2 of this chapter will systematically introduce the influence of co-doped elements on the spectral properties of Yb^{3+} doped silica glass, and reveal its influence mechanism at the atomic microscale by NMR and EPR.

2.1.3 Spectral Theory of Ytterbium Ions

The absorption coefficient (α) and absorption cross section (σ_{abs}) of Yb^{3+} can be calculated by the following formulas [9]:

$$\alpha = \frac{1}{d} \times \ln\left(\frac{I_0}{I}\right) = \frac{1}{d \times \lg(e)} \times \lg\left(\frac{I_0}{I}\right) = \frac{OD(\lambda)}{d \times \lg(e)} \tag{2.5}$$

$$\sigma_{\text{abs}} = \frac{2.303 OD(\lambda)}{N_0 d} \tag{2.6}$$

$$N_0 = \frac{\rho}{M} \times \text{wt\%} \times N_{\text{A}} \tag{2.7}$$

In the formula, I and I_0 are the transmitted and incident intensities, respectively. d is the thickness of the glass. OD (λ) is the optical density. N_0 is the concentration of rare earth ions. ρ is the sample density. M is the molar mass of the rare earth ions. wt% is the mass fraction of the rare earth ions. N_A is Avogadro's constant.

The emission cross-section (σ_{em}) [10] of Yb^{3+} can be calculated using the reciprocity method and the Fuchbauer-Lademnurg (FL) formula. The calculation formula for the reciprocity method is as follows:

$$\sigma_{\text{em}} = \sigma_{\text{abs}} \frac{Z_l}{Z_u} \exp\left[\frac{hc}{kT}\left(\frac{1}{\lambda_0} - \frac{1}{\lambda}\right)\right] \tag{2.8}$$

where, h is Planck's constant. c is the speed of light in vacuum. k is Boltzmann's constant. T is the temperature in Kelvin. λ is the wavelength independent variable. λ_0 corresponds to the zero-line wavelength of the zero-line energy (the zero-line energy is the energy difference between the two lowest Stark split energy levels in the upper and lower main energy levels, and the zero-line wavelength is generally the peak absorption or emission wavelength). Z_l and Z_u are the partition functions of the lower and upper energy levels, respectively. At room temperature, the partition function is affected by the matrix (in silica glass, the ratio of the partition functions of Yb^{3+} at room temperature is approximately 1, i.e., $Z_l/Z_u \approx 1$).

The reciprocity method is more suitable for calculating the emission cross-section of rare earth ions with overlapping absorption and emission peaks at room temperature. This method is relatively simple because the emission cross-section can be calculated only based on the absorption spectrum. However, when the concentration of Yb^{3+} in the Yb^{3+} doped glass sample is low and the thickness of the glass is thin, there is a large measurement error in the absorption spectrum of Yb^{3+} at long wavelengths (>1000 nm). At this point, using the reciprocity method to calculate the emission cross-section of Yb^{3+} at long wavelengths could result in a significant error.

In addition, the cross-section of Yb^{3+} can be calculated using the FL formula based on its emission spectrum, and the FL formula is as follows [11]:

$$\sigma_{em} = \frac{4\lambda^4}{3\lambda_0^4}\Sigma_{abs} \times \frac{\lambda I(\lambda)}{\int \lambda I(\lambda)d(\lambda)} \tag{2.9}$$

$$\Sigma_{abs} = \int \sigma_{abs}(\lambda)d\lambda \tag{2.10}$$

In the formula, λ_0 is the average wavelength of the absorption band, generally the main peak wavelength. Σ_{abs} is the integral absorption of cross-section. I (λ) is the fluorescence intensity.

The calculation of emission cross-section using the FL formula requires simultaneous testing of absorption and emission spectra, which is complex. At the same time, it has large calculation error for samples with severe self-absorption effect due to the involvement of the emission spectrum function. Therefore, using FL formula to calculate the emission cross-section requires low doping concentration or thin sample thickness.

Figure 2.6 shows the absorption and emission cross-sections of Yb^{3+} in single-doped (SY0.1), Yb/Al co-doped (SYA4), and Yb/P co-doped (SYP4) silica glasses. Despite the same preparation methods and Yb_2O_3 content (0.1 mol%) of the three glass samples, their absorption and emission cross-sections differ greatly. It can be seen that the co-doping elements have a significant impact on the spectral properties of Yb^{3+}.

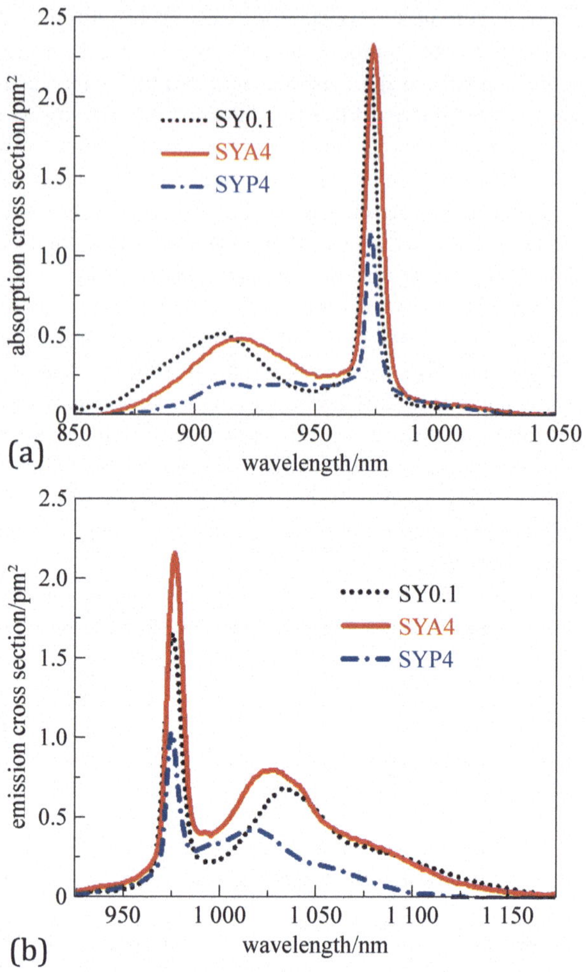

Fig. 2.6 The absorption (**a**) and emission cross-sections (**b**) of the Yb^{3+} single-doped (SY0.1), Yb/Al co-doped (SYA4), and Yb/P co-doped (SYP4) silica glasses

2.2 Relationship Among the Composition-Structure-Properties of Ytterbium-Doped Silica Glass

2.2.1 Preparation of Ytterbium-Doped Silica Glass

Ytterbium-doped silica fiber is the core component of high-power fiber lasers, and its core material is ytterbium-doped silica glass. The performance of ytterbium-doped silica fiber is closely related to the properties of ytterbium-doped silica glass. The existing studies usually use modified chemical vapor deposition (MCVD) method combined with gas or liquid phase doping process to prepare Yb-doped silica fiber preform, which is the most usual preparation technology for laser fiber. However, when the core rod glass prepared by this method is used as the research object to carry out basic research on spectral properties and structural defects, the following problems inevitably exist: ① The size of the core area is small, and the core diameter of the preform prepared by the solution doping method is only 1 ~ 2 mm. Therefore, when carrying out the absorption spectrum and fluorescence spectrum tests, the light transmission area is too small, resulting in a large signal-to-noise ratio of the spectrum, thus to affect the accuracy of the tests. ② The preparation of the core is expensive, and its process is relatively complex. So it is difficult to adjust the core composition and further carry out comparative research with property regulation as flexible as the traditional glass melting. Therefore, few studies have explored the core glass based on the optical fiber preform prepared by MCVD method. ③ The core rod glass is wrapped in a thick layer of silica skin. To avoid the interference of the cladding glass, the silica skin must be removed by mechanical processing combined with chemical etching before testing, which is a tedious process. ④ The refractive index distribution is uneven when the core is deposited by the traditional MCVD method, because the distribution of rare earth ions doped in the core area is uneven. However, the spectral properties reflect the average results of the entire core area, and cannot accurately reflect the effect of core composition on the spectrum, thereby affecting the conclusion.

To solve the above problems, the Hu Lili's research group have studied the preparation method of Yb-doped silica glass using the sol-gel method combined with high-temperature sintering process [12]. Compared with the core rod glass prepared by the traditional MCVD method (diameter ≤ 3 mm), the Yb-doped silica glass prepared by the sol-gel method has larger size (diameter > 25 mm), higher doping uniformity, higher component control precision and higher testing precision of performance, component and structure. Figure 2.7 is a flow chart of the preparation and testing of

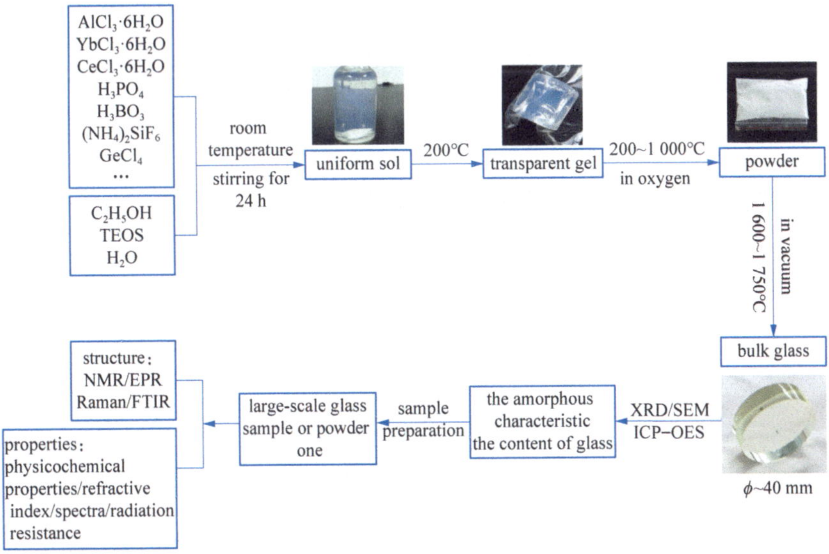

Fig. 2.7 Preparation and testing process of Yb-doped silica glass [13]

Yb-doped silica glass. The preparation process mainly includes the following parts: ① According to the glass formula, taking the corresponding raw materials and dissolving them in tetraethyl orthosilicate (TEOS), anhydrous ethanol and deionized water. Then, stirring at room temperature for 24 h to obtain a clear sol after catalyzing with the corresponding acid. ② The sol is heated at 200 °C for 15 h to obtain a uniform gel; ③ The glass gel is pre-treated at 140 °C for 5 h to remove most of the water and volatiles. Then, it is treated at 200 ~ 1000 °C for 20 h under an oxygen atmosphere to obtain white gel particles then to be made into white powder by ball milling; ④ The white powder is loaded into an alumina crucible and fired at 1600 ~ 1750 °C for 2 h under vacuum conditions to obtain a transparent glass; ⑤ The X-ray diffraction (XRD) technique and scanning electron microscope (SEM) are used to evaluate whether the prepared glass sample has crystallization or phase separation. The inductively coupled plasma optical emission spectrometer (ICP-OES) is used to test the actual composition of the glass sample; ⑥ The glass sample can be used for testing macroscopic performance (such as refractive index, absorption and emission spectra, etc.) as well as microscopic structure (such as glass network structure, Yb^{3+} local structure, etc.), after optical cold processing; ⑦ To establish the relationship among the composition, structure and performance of Yb-doped silica glass, as well as the relationship among thermal history, structure and performance.

2.2.2 The Effect of Co-Doping Elements on the Physical Properties of Ytterbium-Doped Silica Glass

The study by Liu et al. [14] shows that there is a turning point in the effect of Al_2O_3 on the refractive index of silica glass, as shown in Fig. 2.8. As the content of Al_2O_3 increases, the refractive index first decreases and then increases. When the content of Al_2O_3 is 0.55 mol%, the refractive index of the glass sample reaches the

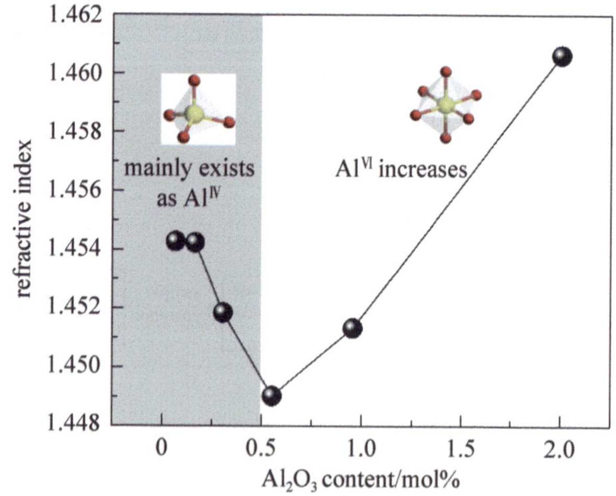

Fig. 2.8 The effect of Al_2O_3 content on the refractive index of silica glass

minimum. The density of the glass also shows a similar trend. This is related to the change in the coordination number of aluminum. Aluminum in silica glass has four, five, and six coordination numbers, which can be represented as Al^{IV}, Al^{V}, Al^{VI}.

Studies have shown that the coordination number of aluminum in aluminum-doped silica glass is closely related to the content of Al_2O_3. The study by Sen et al. [15] shows that in aluminum-doped silica glass, when the Al content is less than 0.59 mol%, aluminum mainly exists as Al^{IV}. Specifically, when the Al content does not exceed 0.24 mol%, aluminum almost entirely exists as Al^{IV}. When the Al content is greater than 0.59 mol%, the proportion of Al^{VI} significantly increases. Similarly, the study by Jiao et al. [16] shows that when the content of Al_2O_3 gradually increases from 0.5 mol% to 10 mol%, the proportion of Al^{IV} gradually decreases from 65% to 35%, while the proportions of Al^{V} and Al^{VI} gradually increase from 35% to 65%, as detailed in Fig. 2.22b.

Al ^{IV} mainly enters the glass network in the form of a body, its structural unit $[AlO_{4/2}]$ is similar to the silicon-oxygen tetrahedron structural unit $[SiO_{4/2}]$, and the molecular weight and volume of $[AlO_{4/2}]$ are smaller than those of $[SiO_{4/2}]$. Therefore, when Al^{IV} is dominant, the refractive index and density of the glass decrease with the increase of Al_2O_3 content. The structural units of Al^{V} and Al^{VI} are $[AlO_{5/2}]$ and $[AlO_{6/2}]$, respectively, which are quite different from $[SiO_{4/2}]$. They fill the network gaps in silica glass, making the glass structure more compact, and their polarization rate is larger than that of $[SiO_{4/2}]$. Therefore, when Al^{V} and Al^{VI} are dominant, the refractive index and density of the glass increase with the increase of Al_2O_3 content.

The study by Xu et al. [17] shows that there is also a turning point in the effect of the phosphorus-aluminum molar ratio (P/Al ratio) on the density and refractive index of silica glass, as shown in Fig. 2.9a. With the increase of the P/Al ratio, the density and refractive index of the glass first decrease and then increase. When P/Al = 1, the density and refractive index reach the minimum, which is mainly related to the spatial bonding structure between aluminum and phosphorus.

Figure 2.9b shows the schematic diagram of the effect of the P/Al ratio on the spatial bonding structure of silica glass. A large number of studies have shown that in alumina-phosphorus co-doped silica glass, aluminum and phosphorus atoms prefer to form the $[AlPO_4]$ structure. In $[AlPO_4]$, aluminum mainly exists as Al^{IV}, and phosphorus mainly exists as $P^{(4)}$. It should be noted that the superscript n in the $P^{(n)}$ group refers to the number of bridging oxygen atoms connected to the central atom P. This structural unit is equivalent to 2 $[SiO_{4/2}]$ structural units, contributing less to the density and refractive index of silica glass. When $P/Al < 1$, almost all P enters the $[AlPO_4]$ structure, and the excess aluminum exists as Al^{V} and Al^{VI}. As the P/Al ratio increases, the proportion of $[AlPO_4]$ gradually increases, and the proportion of Al^{V} and Al^{VI} gradually decreases, so the density and refractive index of the glass gradually decrease. When P/Al = 1, almost all Al and P enter the $[AlPO_4]$ structure, and the density and refractive index of the glass reach the minimum. It should be noted that at this time, the glass is particularly prone to phase separation, which leads to the glass becoming non-transparent and milky white. When $P/Al > 1$, almost all Al enters the $[AlPO_4]$ structure, and the excess phosphorus exists as $P^{(2)}$

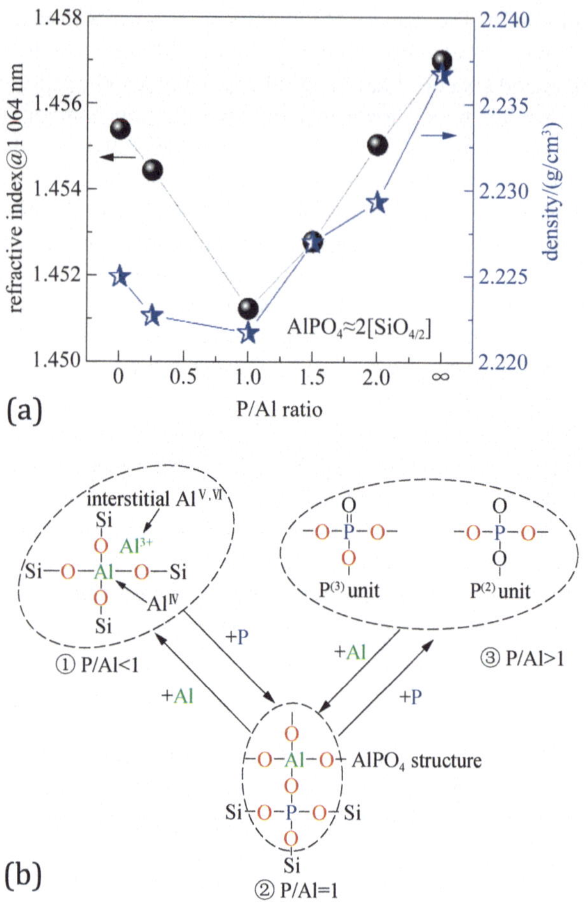

Fig. 2.9 The influence of P/Al ratio on the refractive index and density of silica glass (**a**) and the structure of the glass network (**b**)

and P [(3)]. As the P/Al ratio increases, P [(2)] and P [(3)] increase proportionally, leading to a gradual increase in the density and refractive index of the glass.

Figure 2.10 summarizes the effects of different compound doping on the refractive index of silica glass, where the vertical axis represents the refractive index difference between doped silica glass and pure silica glass at a wavelength of 1064 nm. The doping of Yb_2O_3, Al_2O_3 (>0.5 mol%), GeO_2 and P_2O_5 leads to an increase in the refractive index of silica glass. The $AlPO_4$, B_2O_3 and SiF_4 doping leads to a decrease in the refractive index of silica glass. The slope η represents the increase in the refractive index of silica glass by doping 1 mol% compound. For example, at 1064 nm, the increase in the refractive index of silica glass by doping 1 mol% of Yb_2O_3 is about 65×10^{-4}. That by doping 1 mol% of SiF_4 is about -50×10^{-4}.

Understanding the contribution of dopants to the refractive index of silica glass is of great significance for the design of core rod glass components. For example, at 1064 nm, the refractive index difference Δn between Yb/Al/P/F co-doped silica

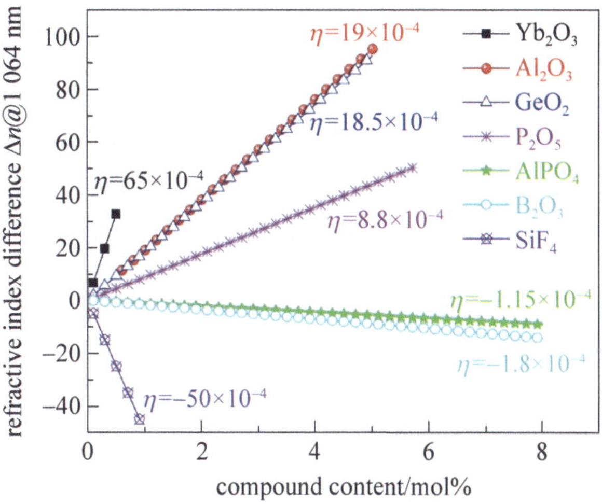

Fig. 2.10 The effect of different compound doping on the refractive index of silica glass [18–22]

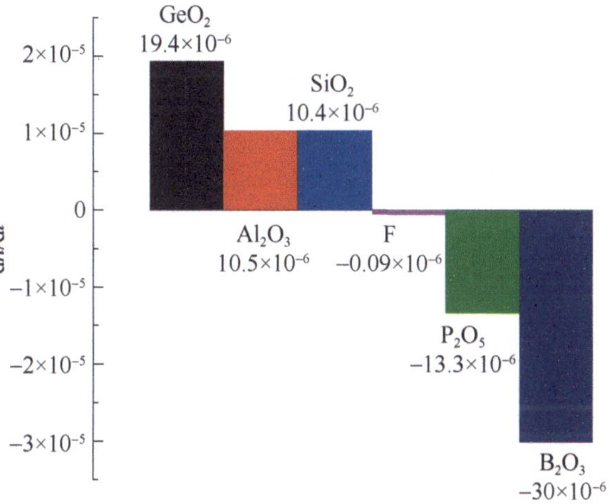

Fig. 2.11 The effect of different co-doping elements on the thermo-optic coefficient of silica glass [23, 24]

core rod glass and pure silica cladding glass can be estimated by the following formula:

$$\Delta n \times 10^4 = 65 \times n(Yb_2O_3) + 19 \times [n(Al_2O_3) - n(AlPO_4)] + 8.8 \times [n(P_2O_5) - n(AlPO_4)] - 1.15 \times n(AlPO_4) - 50 \times n(SiF_4) \quad (2.11)$$

In the formula, n (Yb_2O_3) is the molar ratio of Yb_2O_3, and the rest are similar.

Figure 2.11 summarizes the thermo-optic coefficients (TOC) of different compounds. TOC, also known as the temperature coefficient of refractive index, physically describes the rate of change of the refractive index of optical materials with temperature. Among them, the TOCs of GeO_2, Al_2O_3 and SiO_2 are positive, those of SiF_4, P_2O_5 and B_2O_3 are negative.

2.2.3 The Effect of Co-Doping Elements on the Spectral Properties of Ytterbium-Doped Silica Glass

Figure 2.12 summarizes the Yb^{3+} spectral properties of single-doped silica glass. The glass composition and main spectral parameters are shown in Table 2.1.

As can be seen from Fig. 2.12a, with the increase of Yb_2O_3 content, the absorption coefficient of the glass sample gradually increases, but the increase slows down. In addition, the absorption sub-peak at 915 nm shows a red shift with the increase of Yb_2O_3 content.

As can be seen from Fig. 2.12b, with the increase of Yb_2O_3 content, the luminescence intensity of the glass sample first increases and then decreases. When the Yb_2O_3 content is 0.05 mol%, the luminescence intensity reaches its maximum. In addition, the luminescence sub-peak at 1035 nm shows a blue shift with the increase of Yb_2O_3 content.

As can be seen from Fig. 2.12c and Table 2.1, when the Yb_2O_3 content is between 0.02 ~ 0.05 mol%, the absorption and emission cross-sections of the glass sample do not change. When the Yb_2O_3 content increases from 0.05 mol% to 0.18 mol%, the absorption cross-section decreases from 2.45 pm^2 to 1.98 pm^2, with a percent of 19.2%. The emission cross-section decreases from 0.85 pm^2 to 0.47 pm^2 with a percent of 44.7%. The excited state lifetime of Yb^{3+} decreases from 1.06 ms to 0.87 ms with a percent of 17.9%. The decrease in absorption and emission cross-sections with Yb is related to the decrease of the local crystal field strength and degree of asymmetry (Fig. 2.12e). In addition, it is also related to the increase in the degree of Yb^{3+} clustering (Fig. 2.13). The decrease in the fluorescence lifetime of Yb^{3+} is mainly related to the increase in the degree of Yb^{3+} clustering.

Figure 2.12d shows the Stark splitting energy levels of Yb^{3+}. According to the data in Fig. 2.12d, the local crystal field strength and degree of Yb^{3+} asymmetry can be calculated, as shown in Fig. 2.12e. When the content of Yb_2O_3 increases from 0.02 mol% to 0.18 mol%, the local crystal field strength of Yb^{3+} decreases from 4587 to 4303 with a percent of 6.2%. The degree of Yb^{3+} asymmetry both decreases from 1.17 to 0.9 with a percent of 23%. Figure 2.13 shows the inverse of the spin-spin relaxation time (T_2) of Yb^{3+} with temperature. T_2 time is usually tested by EPR at low temperature conditions. Since the Zeeman energy level difference of Yb^{3+} (<1 cm^{-1}) is smaller than the Stark energy level difference (<500 cm^{-1}), its EPR spectrum (corresponding to Zeeman energy levels) is more sensitive to changes in the external environment than its spectrum (corresponding to Stark energy levels). T_2 time can reflect the strength of the spatial interaction between the spin centers. The shorter the T_2 time, the stronger the spatial interaction between the spin centers, and the smaller the spatial distance between the spin centers. As can be seen from Fig. 2.13, as the content of Yb_2O_3 increases from 0.02 mol% to 0.15 mol%, the inverse of the T_2 time increases sharply, indicating that the spatial distance between Yb^{3+} decreases sharply, that is, the degree of Yb^{3+} clustering increases significantly. The research by Sen [25] shows that in pure silica glass, the introduction of Nd_2O_3 with a mass ratio of 700 ppm will produce Nd^{3+} clusters. But a small amount of

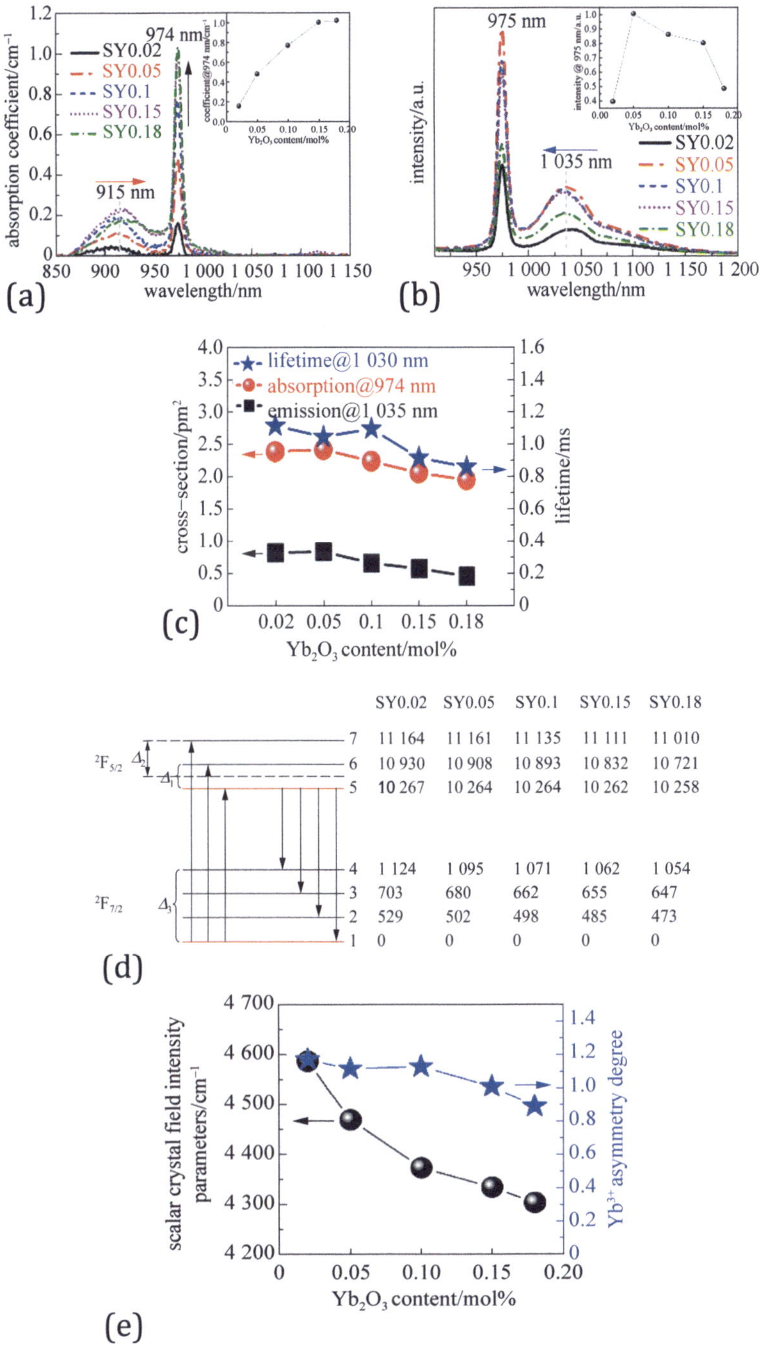

Fig. 2.12 Spectral properties of Yb^{3+} single-doped silica glass: absorption spectra (**a**), emission spectra (**b**), absorption/emission cross-sections and fluorescence lifetime (**c**), Stark split energy levels and energy values of each level E_i ($i = 1 \sim 7$,cm^{-1}) (**d**), scalar crystal field intensity parameters and degree of asymmetry (**e**)

Table 2.1 Theoretical composition and main spectral parameters of Yb^{3+} single-doped silica glass

Sample	Yb_2O_3 / mol%	SiO_2/ mol%	Fluorescence lifetime/ms	Absorption cross-section/pm^2	Emission cross-section/pm^2
SY0.02	0.02	99.98	1.13	2.42	0.84
SY0.05	0.05	99.95	1.06	2.45	0.85
SY0.1	0.1	99.9	1.11	2.27	0.67
SY0.15	0.15	99.85	0.93	2.09	0.59
SY0.18	0.18	99.82	0.87	1.98	0.47

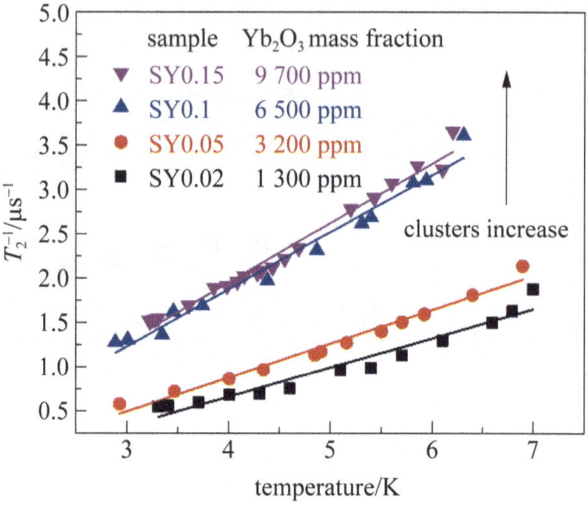

Fig. 2.13 The dependence of the inverse of Yb^{3+} Spin-spin relaxation time (T_2) on temperature

Al_2O_3 (~2000 ppm) is added, the Nd^{3+} clusters can be dissolved, as shown in Fig. 2.14.

Dong et al. [27] studied the spectral properties of Yb/Al co-doped silica glass, as shown in Fig. 2.15. The glass composition and main spectral parameters are detailed in Table 2.2.

As can be seen from Fig. 2.15a, b, as the content of Al_2O_3 increases, the absorption coefficient and luminescence intensity of Yb^{3+} in Yb/Al co-doped silica glass samples first increase and then decrease. When Al_2O_3 content is 1 mol%, the absorption coefficient and luminescence intensity of the glass sample reach the maximum. In addition, the absorption peak located at ~915 nm undergoes a red shift with the increase of Al_2O_3 content. While the secondary luminescence peak located at ~1030 nm undergoes a blue shift. This indicates that the degree of Stark splitting of Yb^{3+} decreases with the increase of Al_2O_3 content, as shown in Fig. 2.15d.

As can be seen from Fig. 2.15c and Table 2.2, as the content of Al_2O_3 increases, the absorption,emission cross-sections and fluorescence lifetime of the glass sample also first increase and then decrease. When the content of Al_2O_3 is 1 mol% (Al/Yb = 10), the absorption and emission cross-sections and fluorescence lifetime of

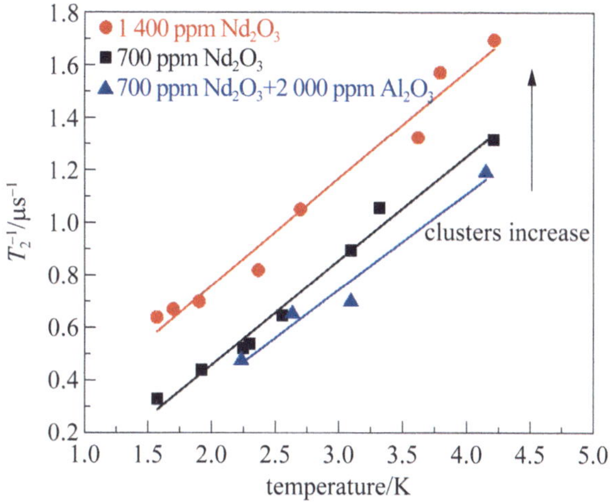

Fig. 2.14 The dependence of the inverse of Nd^{3+} Spin-spin relaxation time (T_2) on temperature [26]

the glass sample all reach their maximum values. When the content of Al_2O_3 increases from 0 to 1 mol%, the absorption cross-section increases from 2.27 pm^2 to 2.61 pm^2 with a percent of 15%. The emission cross-section increases from 0.67 pm^2 to 0.91 pm^2 with a percent of 35.8%. The fluorescence lifetime increases from 1.1 ms to 1.172 ms with a percent of 6.5%. When the content of Al_2O_3 increases from 1 mol% to 4 mol%, the absorption cross-section decreases from 2.61 pm^2 to 2.33 pm^2 with a percent of 10.7%. The emission cross-section decreases from 0.91 pm^2 to 0.82 pm^2 with a percent of 9.9%. The fluorescence lifetime decreases from 1.17 ms to 1.12 ms with a percent of 4.4%. The above results show that when Al/Yb = 10, the spectral properties of Yb^{3+} are optimal.

Figure 2.15d shows the Stark splitting energy levels of Yb^{3+}. According to the data in Fig. 2.15d, the local crystal field strength and degree of Yb^{3+} asymmetry can be calculated, as shown in Fig. 2.15e. When the content of Al_2O_3 increases from 0 to 4 mol%, the local crystal field strength decreases from 4388 to 3984 with a percent of 9.2%. The degree of Yb^{3+} asymmetry decreases from 1.21 to 0.96 with a percent of 21%.

Dong et al. [27] studied the spectral properties of Yb/P co-doped silica glass, as shown in Fig. 2.16. The glass composition and main spectral parameters are detailed in Table 2.3.

From Fig. 2.16a, it can be seen that the absorption and luminescence main peaks of Yb/P co-doped silica glass are located at 975 nm and 978 nm, respectively. The absorption sub-peak is relatively flat in the wavelength range of 915 ~ 965 nm. Therefore, the Yb/P co-doped silica fiber has lower requirements for the stability of the pump wavelength. The luminescence sub-peak of Yb^{3+} is located at 1018 nm. This means that Yb/P co-doped silica fiber is conducive to achieving 1018 nm laser output. As the content of P_2O_5 increases, the absorption coefficient and luminescence intensity of the glass samples both decrease. And the absorption sub-peak shows a slight red shift, and the luminescence sub-peak shows a slight blue shift.

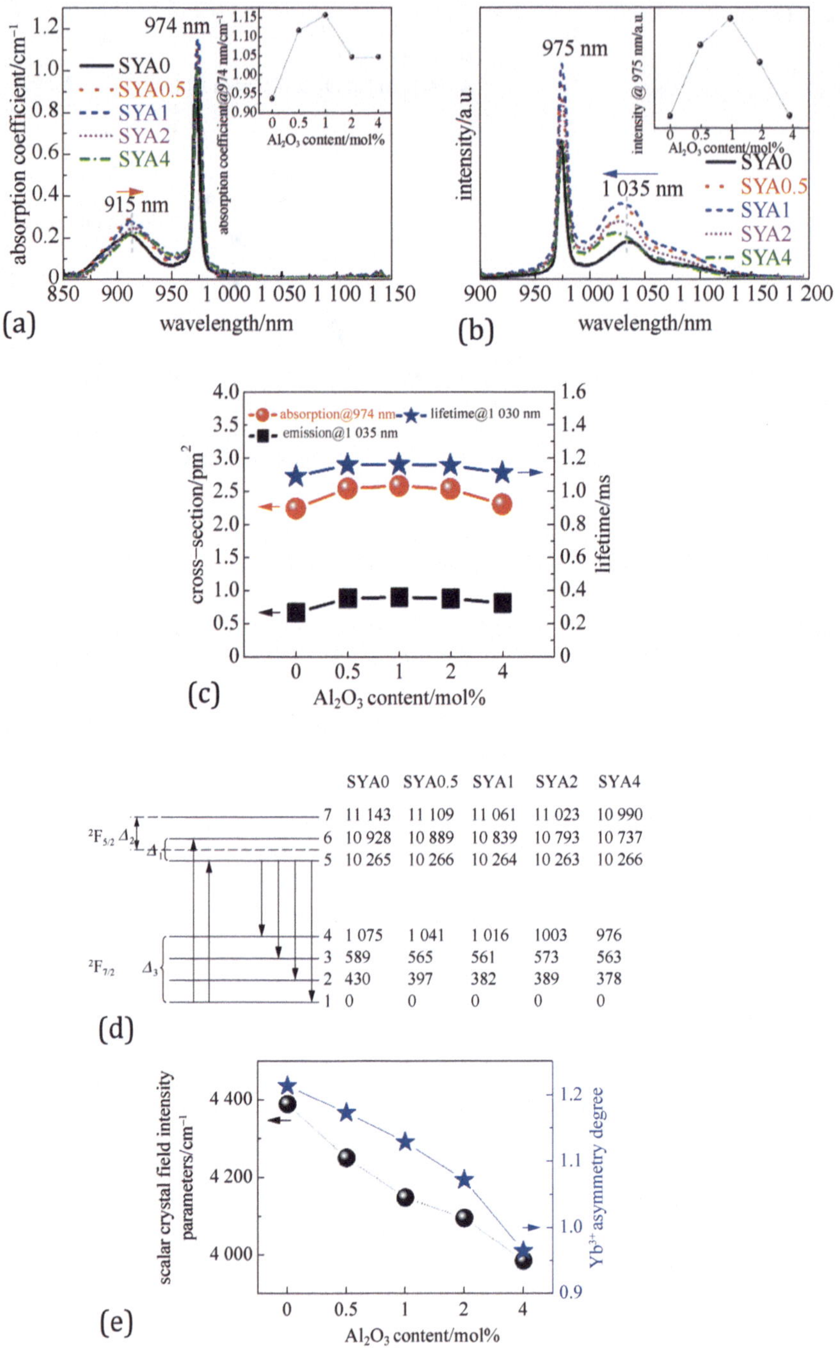

Fig. 2.15 Spectral properties of Yb/Al co-doped silica glass: absorption spectra (**a**), emission spectra (**b**), absorption/emission cross-sections and fluorescence lifetime (**c**), Stark Split energy levels and energy values of each level E_i ($i = 1 \sim 7$, cm^{-1}) (**d**), scalar crystal field intensity parameters and degree of asymmetry (**e**)

Table 2.2 Theoretical composition and main spectral parameters of Yb/Al co-doped silica glass

Sample	Yb_2O_3/ mol%	Al_2O_3/ mol%	SiO_2/ mol%	Fluorescence lifetime/ms	Absorption cross-section/pm^2	Emission cross-section/pm^2	Al/Yb ratio
SYA0	0.1	0	99.9	1.1	2.27	0.67	0
SYA0.5	0.1	0.5	99.4	1.169	2.57	0.89	5
SYA1	0.1	1	98.9	1.172	2.61	0.91	10
SYA2	0.1	2	97.9	1.166	2.56	0.89	20
SYA4	0.1	4	95.9	1.12	2.33	0.82	40

As can be seen from Fig. 2.16c and Table 2.3, as the content of P_2O_5 increases, the absorption and emission cross-sections of the glass samples slightly decrease, and the fluorescence lifetime slightly increases. When the content of P_2O_5 increases from 5 mol% to 12 mol%, the absorption cross-section decreases from 1.2 pm^2 to 1.12 pm^2 with a percent of 6.7%. The emission cross-section decreases from 0.53 pm^2 to 0.49 pm^2 with a percent of 7.5%. The fluorescence lifetime increases from 1.47 ms to 1.61 ms with a percent of 9.5%.

Figure 2.16d shows the Stark splitting energy levels of Yb^{3+}. According to the data in Fig. 2.16d, the local crystal field strength and degree of Yb^{3+} asymmetry can be calculated, as shown in Fig. 2.16e. When the content of P_2O_5 increases from 5 mol% to 12 mol%, the local crystal field strength decreases from 3193 to 3049 with a percent of 4.5%. degree of Yb^{3+} asymmetry decreases from 0.52 to 0.466 with a percent of 10.4%.

Xu et al. [17] studied the spectral properties of Yb/Al/P co-doped silica glass, as shown in Fig. 2.17. The glass composition and main spectral parameters are detailed in Table 2.4.

As can be seen from Fig. 2.17a, b, compared with those in the P/Al $\leq$ 1 samples, the absorption coefficient and luminescence intensity of Yb^{3+} in the P/Al > 1 samples significantly decreases.

As can be seen from Fig. 2.17c and Table 2.4, compared with those in the P/Al $\leq$ 1 samples, the absorption and emission cross-sections of Yb^{3+} in the P/Al > 1 samples significantly decreases, and the fluorescence lifetime significantly increases. When the P/Al ratio increases from 0 to 1, the absorption cross-section decreases from 2.82 pm^2 to 2.74 pm^2 with a percent of 2.8%. The emission cross-section decreases from 0.78 pm^2 to 0.73 pm^2 with a percent of 6.4%. The fluorescence lifetime increases from 0.893 ms to 0.908 ms with a percent of 1.7%. When the P/Al ratio is increased from 1 to 2, the absorption cross-section decreases from 2.74 pm^2 to 1.46 pm^2 with a percent of 46.7%. The emission cross-section decreases from 0.73 pm^2 to 0.41 pm^2 with a percent of 43.8%. The fluorescence lifetime increases from 0.908 ms to 1.209 ms with a percent of 33.1%.

Figure 2.17d shows the Stark splitting energy levels of Yb^{3+}. Based on the data in Fig. 2.17d, the local crystal field strength and degree of Yb^{3+} asymmetry can be calculated, as shown in Fig. 2.17e. When the P/Al ratio increases from 0 to 1, the local crystal field strength decreases from 4057 to 3947 with a percent of 2.7%. The degree of Yb^{3+} asymmetry decreases from 1.08 to 1 with a percent of 7.4%. When

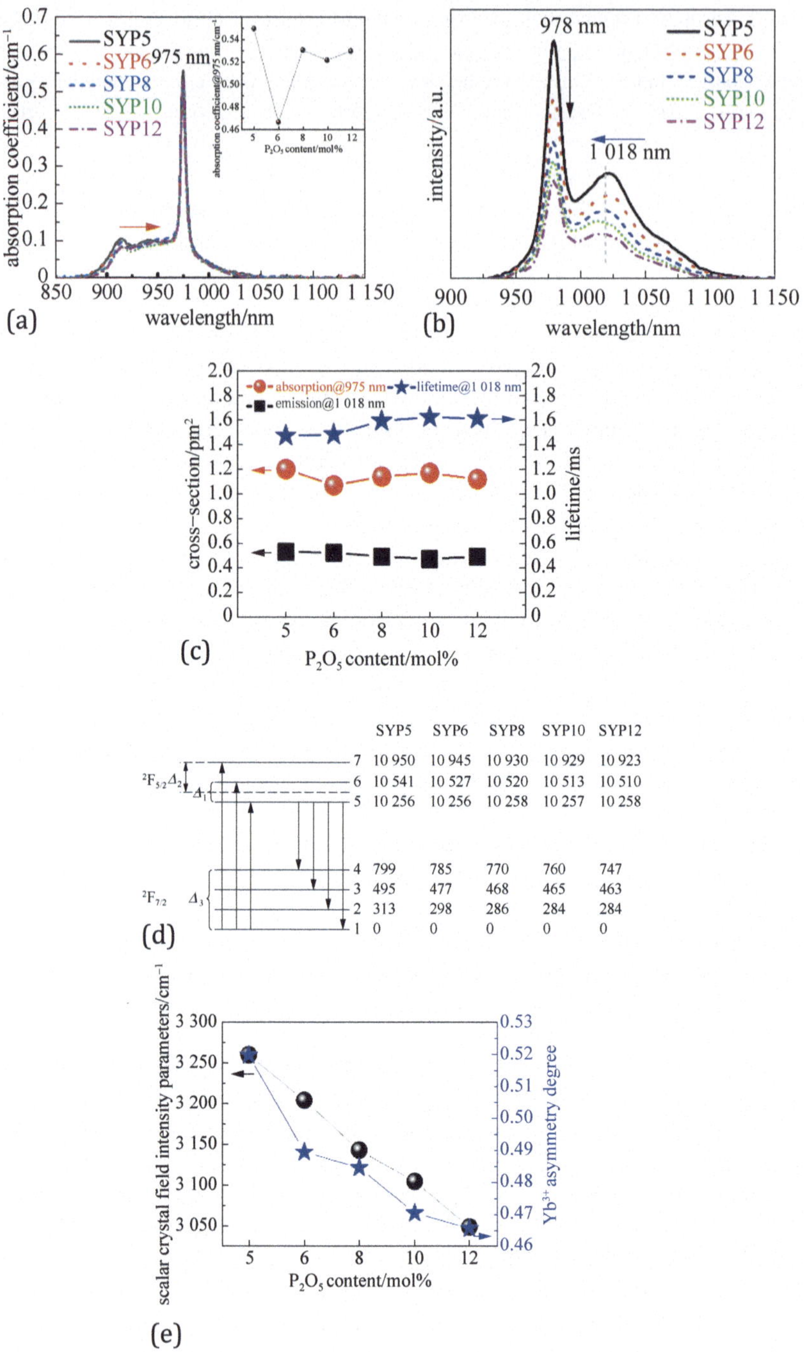

Fig. 2.16 Yb/P co-doped silica glass spectral properties: absorption spectra (**a**), emission spectra (**b**), absorption/emission cross-sections and fluorescence lifetime (**c**), Stark splitting energy levels and energy values of each level E_i ($i = 1 \sim 7$, cm^{-1}) (**d**), the scalar crystal field intensity parameters and degree of Yb^{3+} asymmetry (**e**)

Table 2.3 Theoretical composition and main spectral parameters of Yb/P co-doped silica glass

Sample	Yb_2O_3/ mol%	P_2O_5/ mol%	SiO_2/ mol%	Fluorescence lifetime/ms	Absorption cross-section/pm^2	Emission cross-section/pm^2	P/Yb ratio
SYP5	0.1	5	94.9	1.47	1.2	0.53	50
SYP6	0.1	6	93.9	1.48	1.07	0.52	60
SYP8	0.1	8	91.9	1.59	1.14	0.49	80
SYP10	0.1	10	89.9	1.62	1.17	0.47	100
SYP12	0.1	12	87.9	1.61	1.12	0.49	120

the P/Al ratio is increased from 1 to 2, the crystal field strength decreases from 3947 to 3482 with a percent of 11.8%. The degree of Yb^{3+} asymmetry decreases from 1 to 0.42 with a percent of 58%.

Wang et al. [18] studied the effect of $AlPO_4$ content on the spectral properties of Yb^{3+} in Yb/Al/P co-doped silica glass, as shown in Fig. 2.18. The glass composition and main spectral parameters are detailed in Table 2.5. Since Al and P prefer to interact with each other to form $AlPO_4$ structures in silica glass, the $AlPO_4$ content can be calculated by testing the Al_2O_3 and P_2O_5 content in the glass samples through ICP-OES. the calculation results are shown in Table 2.5.

As can be seen from Fig. 2.18a, as the content of $AlPO_4$ increases, the absorption coefficient of Yb^{3+} gradually decreases.

Figure 2.18b is the normalized (normalized peak position at 1025 nm) fluorescence spectrum of Yb^{3+}. The spectrum consists of a sharp peak at 975 nm and a secondary peak at 1000 ~ 1150 nm. The maximum peak is located at 1025 nm. As can be seen from the inset in Fig. 2.18b, as the concentration of $AlPO_4$ increases, the peak width of the secondary peak at ~1025 nm gradually decreases.

As can be seen from Fig. 2.18c and Table 2.5, as the $AlPO_4$ content increases from 3.28 mol% to 12.18 mol%, the absorption cross-section decreases from 2.2 pm^2 to 1.87 pm^2 with a percent of 15%. The fluorescence lifetime increases from 1.05 ms to 1.106 ms with a percent of 5.3%. Although the measured P/Al ratio in the above samples is close to 1, it can still be found that the emission cross-section is affected by the P/Al ratio, with a slightly larger emission cross-section in the P/Al < 1 samples and a slightly smaller one in the P/Al > 1 samples, as detailed in Table 2.5. The maximum change in the emission cross-section of the samples with different $AlPO_4$ contents does not exceed 7.3% (0.83 $pm^2 \rightarrow$ 0.89 pm^2), indicating that the $AlPO_4$ content will not significantly change the spectral properties of Yb^{3+}.

Figure 2.18d shows the Stark splitting energy levels of Yb^{3+}. According to the data in Fig. 2.18d, the local crystal field strength and degree of Yb^{3+} asymmetry can be calculated, as shown in Fig. 2.18e. When the $AlPO_4$ content increases from 3.28 mol% to 11.08 mol%, the crystal field strength decreases from 3997 to 3860 with a percent of 3.4%. The degree of asymmetry decreases from 1.00 to 0.99 with a percent of 1%. This result indicates that the $AlPO_4$ content will not significantly change the local crystal field strength and degree of Yb^{3+} asymmetry. This is consistent with the result that the $AlPO_4$ content will not significantly change the spectral properties of Yb^{3+}.

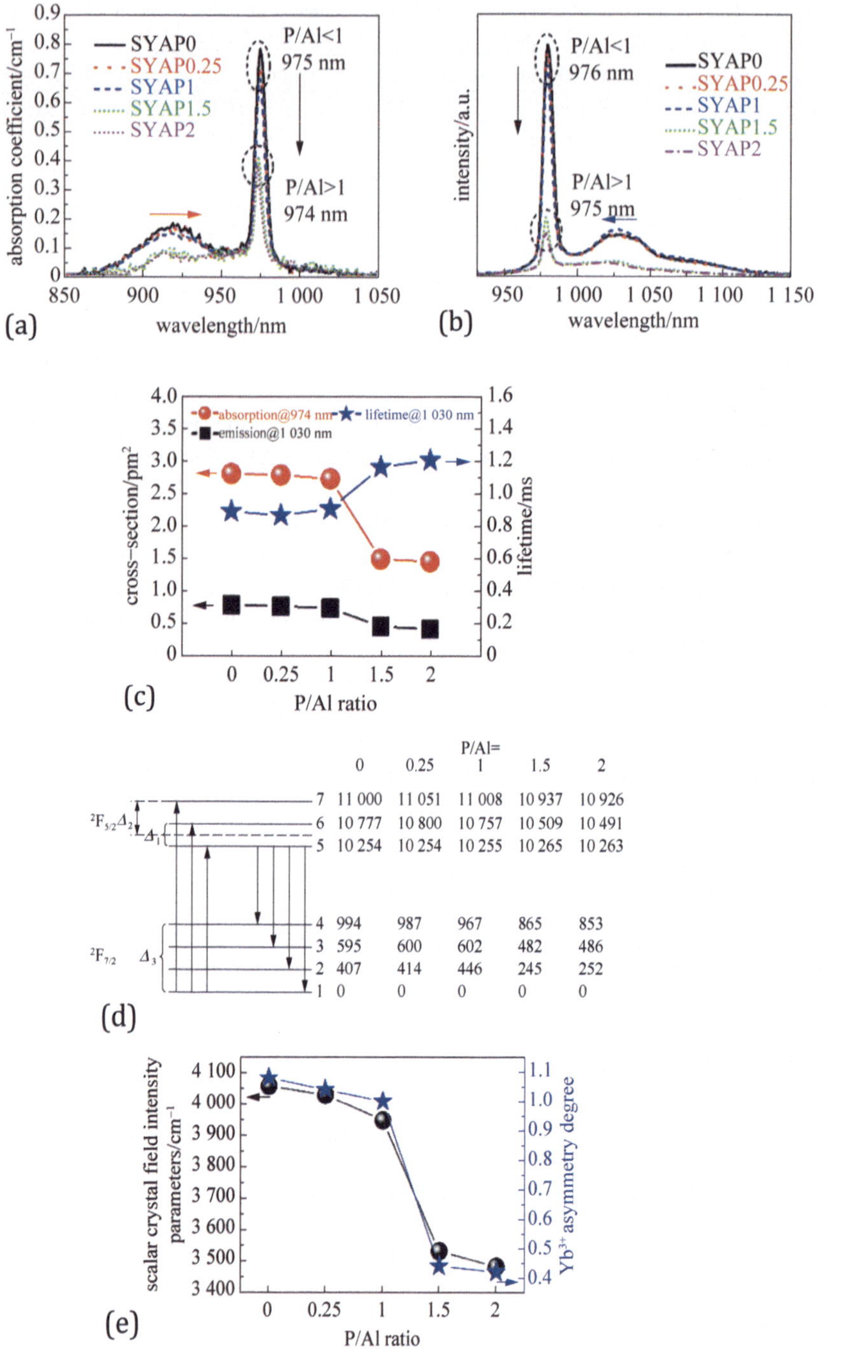

Fig. 2.17 Spectral properties of Yb/Al/P co-doped silica glass: absorption spectra (**a**), emission spectra (**b**), absorption/emission cross-sections and fluorescence lifetime (**c**), Stark split energy levels and energy values of each level E_i ($i = 1 \sim 7$, cm^{-1}) (**d**), scalar crystal field intensity parameters and degree of Yb^{3+} asymmetry (**e**)

Table 2.4 Theoretical composition and main spectral parameters of Yb/Al/P co-doped silica glass

Sample	Yb_2O_3/ mol%	Al_2O_3/ mol%	P_2O_5/ mol%	SiO_2/ mol%	Fluorescence lifetime/ ms	Absorption cross-section/ pm^2	Emission cross-section/ pm^2	P/Al Ratio
SYAP0	0.1	4	0	95.9	0.893	2.82	0.78	0
SYAP0.25	0.1	4	1	94.9	0.869	2.8	0.76	0.25
SYAP1	0.1	4	4	91.9	0.908	2.74	0.73	1
SYAP1.5	0.1	4	6	89.9	1.166	1.5	0.45	1.5
SYAP2	0.1	4	8	87.9	1.209	1.46	0.41	2

As can be seen from Fig. 2.18e and Table 2.5, although the $AlPO_4$ content of the YbAP-11 and YbAP-12 samples is not much different, the absorption/emission cross-sections, the crystal field strength as well as Yb^{3+} asymmetry degree in the YbAP-11 sample are all larger than those in the YbAP-12 sample, which is mainly affected by the P/Al ratio where the former is P/Al < 1, the latter is P/Al > 1.

Zhu et al. [28] studied the spectral properties of Yb/Al/Ge co-doped silica glass, as shown in Fig. 2.19. The glass composition and main spectral parameters are detailed in Table 2.6.

As can be seen from Fig. 2.19a, b, as the content of GeO_2 increases, the absorption coefficient and luminescence intensity of Yb^{3+} gradually increase. This may be related to the suppression of Yb^{2+} and the increase in the proportion of Yb^{3+}.

As can be seen from Fig. 2.19c and Table 2.6, as the content of GeO_2 increases from 0 to 12 mol%, the absorption cross-section of Yb^{3+} increases from 2.01 pm^2 to 2.04 pm^2 with a percent of 1.5%. The emission cross-section increases from 0.53 pm^2 to 0.61 pm^2 with a percent of 15.1%. The fluorescence lifetime increases from 1.073 ms to 1.183 ms with a percent of 10%.

Figure 2.19d shows the Stark splitting energy levels of Yb^{3+}. According to the data in Fig. 2.19d, the local crystal field strength and degree of Yb^{3+} asymmetry can be calculated, as shown in Fig. 2.19e. When the content of GeO_2 increases from 0 to 12 mol%, the local crystal field strength decreases from 3915 to 3794 with a percent of 3.1%. The degree of Yb^{3+} asymmetry decreases from 1.08 to 1.06 with a percent of 1.8%.

Doping of Ge leads to a slight decrement in the local crystal field strength and degree of Yb^{3+} asymmetry. Theoretically, The absorption/emission cross-sections of Yb^{3+} should decrease. However, experimental results show that the absorption/emission cross-sections and fluorescence lifetime of Yb^{3+} have a slight increase. There are two possible reasons: ① Ge doping suppresses Yb^{2+} [28], increasing the proportion of Yb^{3+}; ② A small amount of Ge coordinates around Yb^{3+}, reducing the local phonon energy of Yb^{3+}.

Shao et al. [29] studied the spectral and radiation-resistant properties of Yb/Al/Ce co-doped silica glass. The study shows that the more Ce_2O_3 content, the better the radiation-resistant properties of the glass. When the content of Ce_2O_3 does not exceed 0.125 mol%, the doping of Ce has little effect on the spectral properties (absorption, luminescence, fluorescence lifetime) of Yb^{3+}. However, Ce doping will cause a

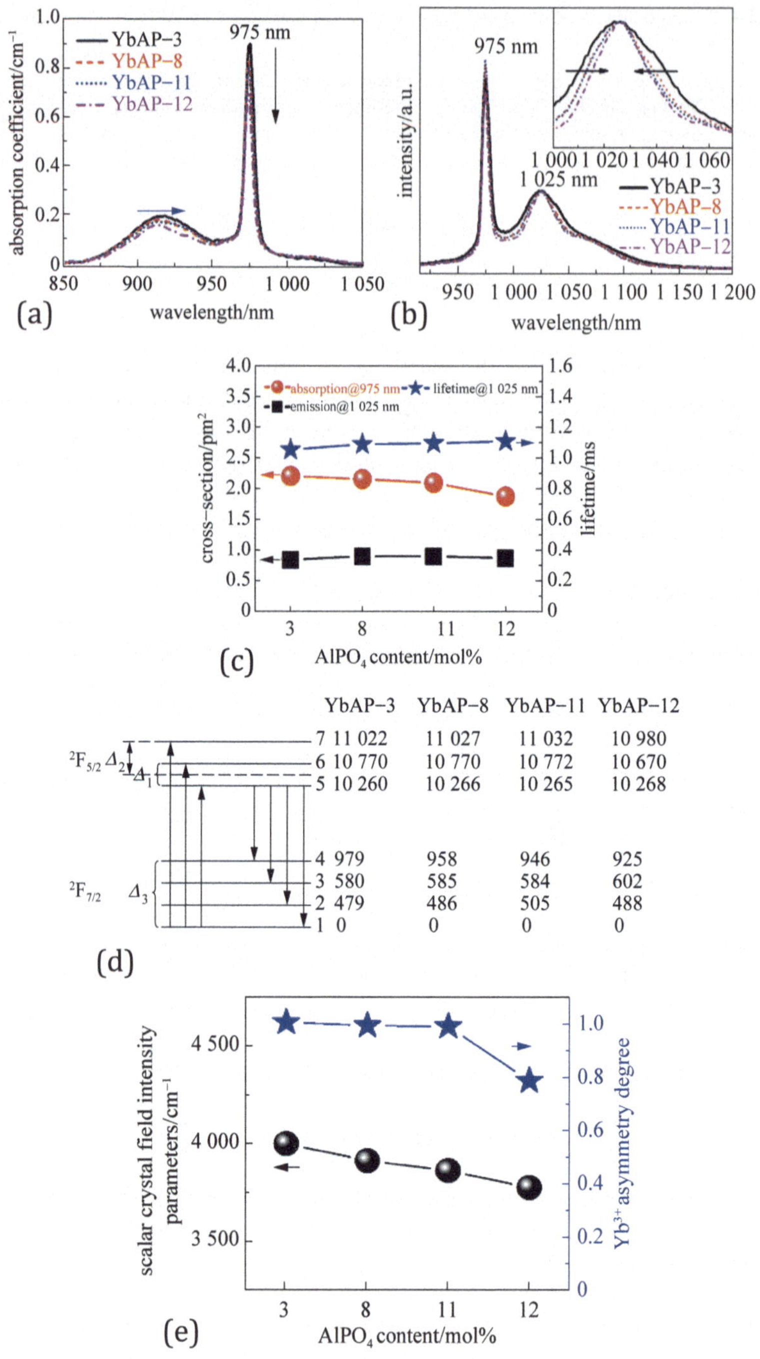

Fig. 2.18 Spectral properties of Yb-$AlPO_4$ co-doped silica glass: absorption spectra (**a**), emission spectra (**b**), absorption/emission cross-sections and fluorescence lifetime (**c**), Stark split energy levels and energy values of each level E_i ($i = 1 \sim 7$, cm^{-1}) (**d**), scalar crystal field intensity parameters and degree of Yb^{3+} asymmetry (**e**)

Table 2.5 Actual composition and main spectral parameters of Yb-$AlPO_4$ co-doped silica glass

Sample	Yb_2O_3/mol%	Al_2O_3/mol%	P_2O_5/mol%	SiO_2/mol%	Fluorescence lifetime/ms	Absorption cross-section/pm^2	Emission cross-section/pm^2	P/Al Ratio	$AlPO_4$ Content/mol%
YbAP-3	0.1	1.64	1.68	96.58	1.05	2.2	0.83	1.024	3.28
YbAP-8	0.09	4	3.92	91.99	1.086	2.15	0.89	0.98	7.84
YbAP-11	0.1	5.75	5.54	88.61	1.093	2.09	0.89	0.963	11.08
YbAP-12	0.1	6.1	6.59	87.21	1.106	1.87	0.86	1.08	12.18

significant increase in the refractive index of the glass, which is not conducive to the core composition design of large-mode-area fibers with low numerical aperture.

Shao et al. [30] further studied the optical, spectral, and radiation-resistant properties of Yb/Al/Ce/F co-doped silica glass. The glass composition and main spectral parameters are shown in Table 2.7. The results show that co-doping F can effectively reduce the refractive index of doped silica glass and improve the radiation-resistant properties of the glass. As the F content increases, the absorption and emission cross-sections of Yb^{3+} decreases, and the fluorescence lifetime increases. This conclusion is consistent with the research results of Xu et al. [31], as shown in Fig. 2.20.

From Fig. 2.20a, b, it can be seen that as the F content increases, the UV edge and the 1030 nm emission band of the glass appears blue-shifted, and the 915 nm absorption band appears red-shifted. In addition, the absorption coefficient and emission intensity of Yb^{3+} decrease. The blue shift of the UV edge is related to the glass structure relaxation [32].

From Fig. 2.20c, it can be seen that as the F content increases, the absorption and emission cross-sections of Yb^{3+} decrease, and the fluorescence lifetime increases. When the F/Si mass ratio increases from 0 to 8, the absorption cross-section of the glass sample decreases from 2.96 pm^2 to 2.4 pm^2 with a percent of 18.9%. The emission cross-section decreases from 0.85 pm^2 to 0.76 pm^2 with a percent of 10.6%. The excited state lifetime of Yb^{3+} increases from 1.153 ms to 1.26 ms with a percent of 9.3%.

Figure 2.20d shows the effect of F content changes on the Stark split energy levels of Yb^{3+}. It can be seen that the Stark energy difference of $1 \leftrightarrow 5$ does not change significantly, and the main peak (~976 nm) position of the corresponding absorption and emission spectra does not move. The transition energy differences of $1 \rightarrow 6,7$ and $5 \rightarrow 2,3,4$ all decrease with the decrease of F content, corresponding to the red shift of the ~915 nm absorption band ($1 \rightarrow 6,7$) and the blue shift of the ~1030 nm emission band ($5 \rightarrow 2,3,4$).

Figure 2.20e is the crystal field strength parameter of Yb^{3+} (N_J calculated based on the. Stark split energy levels in Fig. 2.20d.) and the degree of asymmetry. When

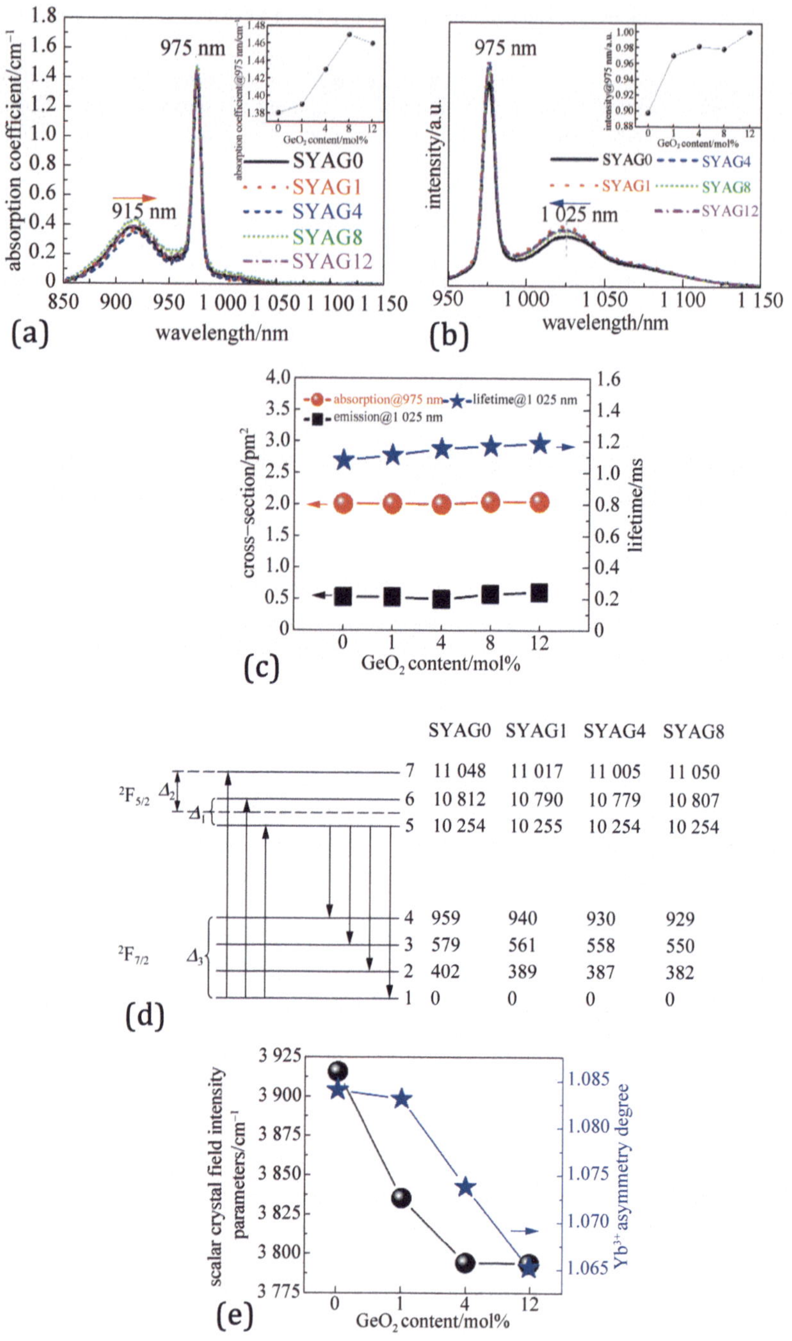

Fig. 2.19 Spectral properties of Yb/Al/Ge co-doped silica glass: absorption spectra (**a**), emission spectra (**b**), absorption/emission cross-sections and fluorescence lifetime (**c**), Stark split energy levels and energy values of each level E_i ($i = 1 \sim 7$, cm^{-1}) (**d**), scalar crystal field intensity parameters and degree of Yb^{3+} asymmetry (e) [28]

Table 2.6 Theoretical composition and main spectral parameters of Yb/Al/Ge co-doped silica glass [28]

Sample	Yb_2O_3/mol%	Al_2O_3/mol%	GeO_2/mol%	SiO_2/mol%	Fluorescence lifetime/ms	Absorption cross-section/pm^2	Emission cross-section/pm^2
SYAG0	0.15	1.5	0	98.35	1.073	2.01	0.53
SYAG1	0.15	1.5	1	97.35	1.107	2.01	0.52
SYAG4	0.15	1.5	4	94.35	1.148	2	0.49
SYAG8	0.15	1.5	8	90.35	1.164	2.04	0.57
SYAG12	0.15	1.5	12	86.35	1.183	2.04	0.61

Table 2.7 Theoretical composition and main spectral parameters of Yb/Al/Ce/F co-doped silica glass

Sample	Yb_2O_3/mol%	Al_2O_3/mol%	Ce_2O_3/mol%	F/Si Mass ratio	SiO_2/mol%	Fluorescence lifetime/ms	Absorption cross-section/pm^2	Emission cross-section/pm^2	Tested F content/wt%
YACF0	0.1	1	0.05	0	98.85	1.153	2.96	0.85	0
YACF2	0.1	1	0.05	2	98.85	1.182	2.62	0.84	0.44
YACF4	0.1	1	0.05	4	98.85	1.216	2.47	0.8	0.85
YACF8	0.1	1	0.05	8	98.85	1.26	2.4	0.76	1.1

the F/Si mass ratio increases from 0 to 8, the local crystal field strength parameter of Yb^{3+} decreases from 4184 to 4086 with a percent of 2.3%. The degree of asymmetry decreases from 1.13 to 0.9 with a percent of 20.4%.

Guo et al. [20] studied the spectral properties of Yb/Al/B co-doped silica glass, as shown in Fig. 2.21. The glass composition and main spectral parameters are detailed in Table 2.8.

From Fig. 2.21a, b, it can be seen that as the B content increases, the absorption coefficient of Yb^{3+} decreases, and the 915 nm absorption band shows a redshift, while the 1022 nm emission band shows a blueshift.

As can be seen from Fig. 2.21c, as the B content increases, the absorption and emission cross-sections of Yb^{3+} decrease, and the fluorescence lifetime increases. When the B_2O_3 content increases from 0 to 2 mol%, the absorption cross-section of the glass sample decreases from 2.450 pm^2 to 2.088 pm^2 with a percent of 14.8%. The emission cross-section decreases from 0.745 pm^2 to 0.624 pm^2 with a percent of 16.2%. The excited state lifetime of Yb^{3+} increases from 1.208 ms to 1.272 ms with a percent of 5.3%. When the B_2O_3 content increases from 2 mol% to 8 mol%, the absorption/emission cross-sections of the glass sample is basically unchange, and the fluorescence lifetime is increased by 9%. The increase in Yb^{3+} fluorescence lifetime may be related to the self-absorption effect.

Figure 2.21d shows the Stark splitting energy levels of Yb^{3+}. According to the data in Fig. 2.21d, the crystal field strength and the asymmetry degree of Yb^{3+} can be calculated, as shown in Fig. 2.21e. Those of Yb-doped borate glass (Yb-BG) are also added to Fig. 2.21d, e for comparison. The Yb-BG sample was prepared by the

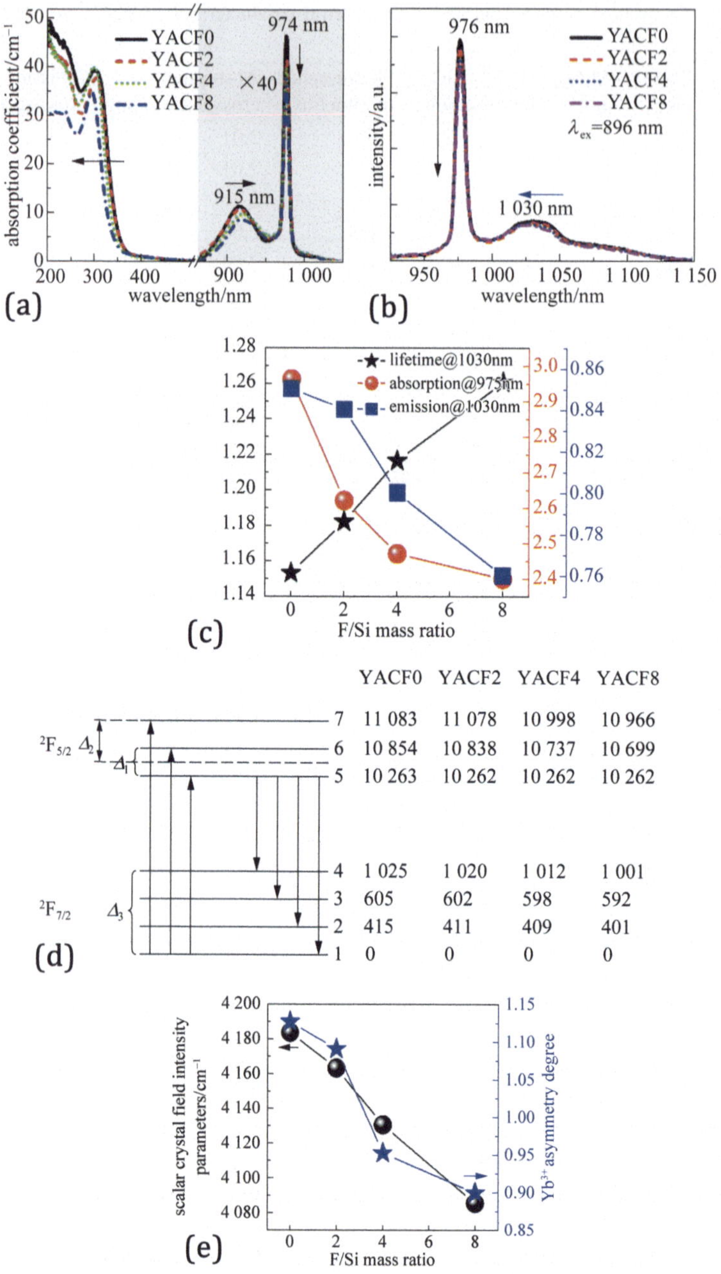

Fig. 2.20 Spectral properties of Yb/Al/Ce/F co-doped silica glass: absorption spectra (**a**), emission spectra (**b**), absorption/emission cross-sections and fluorescence lifetime (**c**), Stark split energy levels and energy values of each level E_i ($i = 1 \sim 7$, cm^{-1}) (**d**), scalar crystal field intensity parameters and degree of Yb^{3+} asymmetry (**e**) [30]

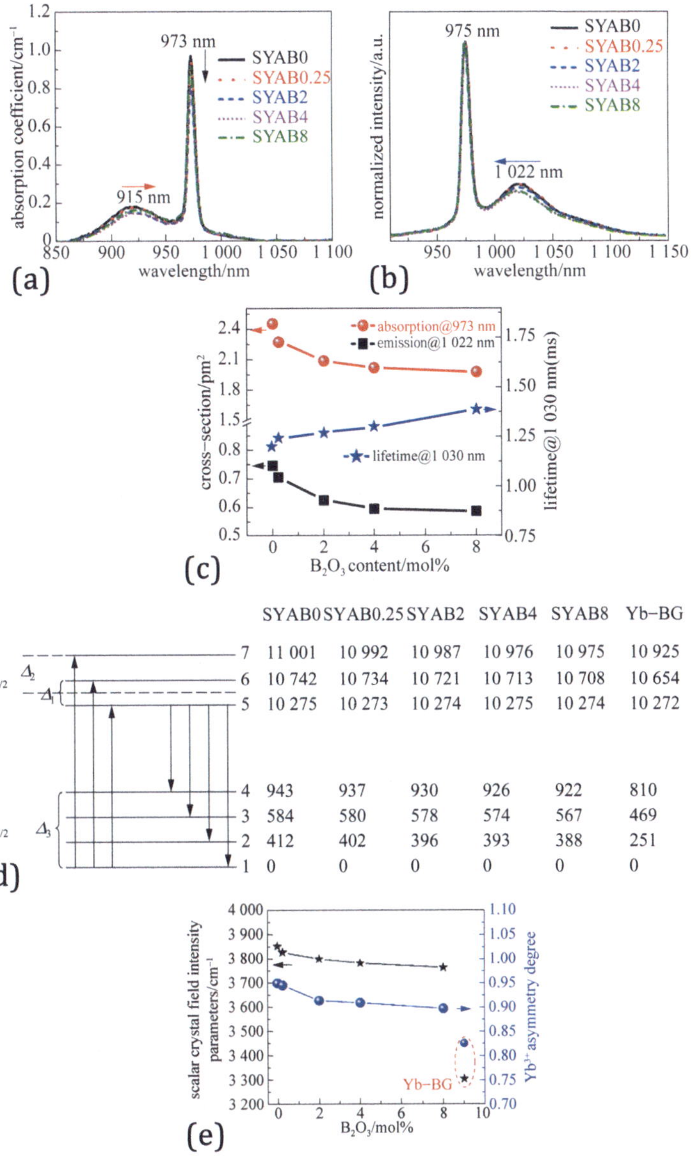

Fig. 2.21 Spectral properties of Yb/Al/B co-doped silica glass: absorption spectra (**a**); emission spectra (**b**); absorption/emission cross-sections and fluorescence lifetime (**c**); Stark splitting energy levels and energy values of each level E_i ($i = 1 \sim 7$, cm^{-1}) (**d**), scalar crystal field intensity parameters and degree of asymmetry (**e**) [20]

Table 2.8 Theoretical composition and main spectral parameters of Yb/Al/B co-doped silica glass

Sample	Yb_2O_3/mol%	Al_2O_3/mol%	B_2O_3/mol%	SiO_2/mol%	Fluorescence lifetime/ms	Absorption cross-section/pm^2	Emission cross-section/pm^2
SYAB0	0.1	6	0	93.9	1.208	2.45	0.745
SYAB0.25	0.1	6	0.25	93.65	1.241	2.276	0.704
SYAB2	0.1	6	2	91.9	1.272	2.088	0.624
SYAB4	0.1	6	4	89.9	1.3	2.019	0.595
SYAB8	0.1	6	8	85.9	1.388	1.976	0.586

Table 2.9 Influence of co-doping elements on the spectral properties of Yb-doped silica glass

Glass system	When variable increases	Special case	Absorption or emission cross-sections	Fluorescence lifetime	Local crystal field strength or degree of asymmetry
Yb-Si	Yb content	None	Decrease	Decreases	Decrease
Yb-Al-Si	Al content	Al/Yb = 10	Increase first, then decrease	Increases first, then decreases	Decrease
Yb-P-Si	P content	None	Decrease	Increases	Decrease
Yb-Al-P-Si	P/Al ratio	P/Al = 1	Decrease slowly first, then drop sharply	Increases slowly first, then rises sharply	Decrease slowly first, then drop sharply
Yb-Al-P-Si(P/Al = 1)	$AlPO_4$ content	None	Basically unchanged	Basically unchanged	Basically unchanged
Yb-Al-Ge-Si	Ge content	None	Increase	Increases	Decrease
Yb-Al-Ce-Si	Ce content	None	Basically unchanged	Basically unchanged	Basically unchanged
Yb-Al-Ce-Si-F	F content	None	Decrease	Increases	Decrease
Yb-Al-B-Si	B content	None	Decrease	Increases	Decrease

traditional high-temperature melting method, and the glass composition is $0.1Yb_2O_3$-$75B_2O_3$-25CaO(mol%). Compared with the SYAB glasses, the Stark splitting energy level difference of Yb^{3+}, the crystal field strength and the asymmetry degree of Yb^{3+} in Yb-BG glass are the smallest. This is consistent with the research results of Yang et al. [1]. When the B_2O_3 content increases from 0 to 2 mol%, the crystal field strength decreases from 3850 to 3797 with a percent of 1.4%. The degree of asymmetry decreases from 0.95 to 0.91 with a percent of 4.3%. When the B_2O_3 content increases from 2 mol% to 8 mol%, the crystal field strength decreases from 3797 to 3761 with a percent of 0.9%. The degree of asymmetry decreases from 0.91 to 0.90 with a percent of 1.1%.

In summary, the influence of co-doping elements on the spectral properties of Yb-doped silica glass is shown in Table 2.9.

2.2.4 *The Impact of Co-Doped Elements on the Network Structure of Silica Glass and Yb^{3+} Local Structure*

2.2.4.1 The Impact of Aluminum or Phosphorus Doping on the Network Structure of Silica Glass and Yb^{3+} Local Structure

The network structure of Al_2O_3-SiO_2 glass has been extensively studied. In x Al_2O_3–$(100-x)SiO_2$ glass, when the Al_2O_3 content $x \leq 10$ mol% or 60 mol% $\leq x \leq 67$ mol%, a uniformly distributed single-phase glass can be formed. When the Al_2O_3 content x exceeds the above range, the glass will phase separate. The main conclusions in single-phase Al_2O_3-SiO_2 glass are: ① the coordination number of silicon is always 4, and that of aluminum includes 4, 5, and 6. ② The coordination number of aluminum is closely related to the Al_2O_3 content. ③ There is an "Al $^{3+}$ avoidance" rule, that is, to avoid Al—O—Al connections. The following is a brief introduction.

In the magic angle spinning (MAS) NMR spectrum of ^{27}Al, Al^{IV}, Al^{V} and Al^{VI} units are located at ~60 ppm, ~30 ppm and ~ 0 ppm, respectively, as shown in Fig. 2.22. The percentage of the three coordination structures of aluminum is closely related to the Al_2O_3 content. Sen et al. [15] showed that when the Al content was less than 0.59 mol%, aluminum mainly existed as Al^{IV}. Specifically, when the Al content did not exceed 0.24 mol%, almost all aluminum existed as Al^{IV}. When the Al content was greater than 0.59 mol%, the proportion of Al^{VI} significantly increased. Similarly, Jiao et al. [16] showed that when the Al_2O_3 content gradually increased from 0.5 mol% to 10 mol%, the proportion of Al^{IV} gradually decreased from 65% to 35%, while the proportions of Al^{V} and Al^{VI} gradually increased from 35% to 65%, as shown in Fig. 2.22.

Sen et al. [15] used ^{17}O 3QMAS NMR experiments to confirm that there are no forms of Al—O—Al connections in single-phase Al_2O_3 -SiO_2 silica glass. Aluminum mainly enters the glass network in the form of four coordination, that is, there are Si—O—Al^{IV} connections. Al^{VI} mainly exists in the interstitial position to compensate for the charge of Al^{IV}, that is, there are no Si—O—Al^{VI} connections. When the Al content is less than 1 wt%, aluminum mainly exists as Al^{IV}, because Al^{VI} is not enough to compensate for the negative charge of Al^{IV}. Oxygen trimers will form at this time, that is, Si, Al = O − Si connections are formed.

Eckert et al. [33] systematically studied the network structure of x P_2O_5–$(100-x)$ SiO_2 silica glass. The results show as follows: when the P_2O_5 content x is less than 30 mol%, silicon mainly exists in the form of four coordination. When the P_2O_5 content x is not lower than 30 mol%, five and six coordinated silicon begin to form. Phosphorus mainly exists in the form of $P^{(3)}$ groups. The $P^{(3)}$ group contains three bridging oxygens and one non-bridging oxygen (i.e., P = O double bond). As the P_2O_5 content x gradually increases, the local structure of phosphorus is constantly changing, i.e. the number of P—O—P connections gradually increases, as shown in Fig. 2.23. $P^{(3)}_{3Si}$ represents 3 Si atoms connected to the central atom P through bridging oxygen. $P^{(3)}_{2Si,1P}$ represents 2 Si atoms and 1 P atom connected to the central atom P through bridging oxygen, and so on.

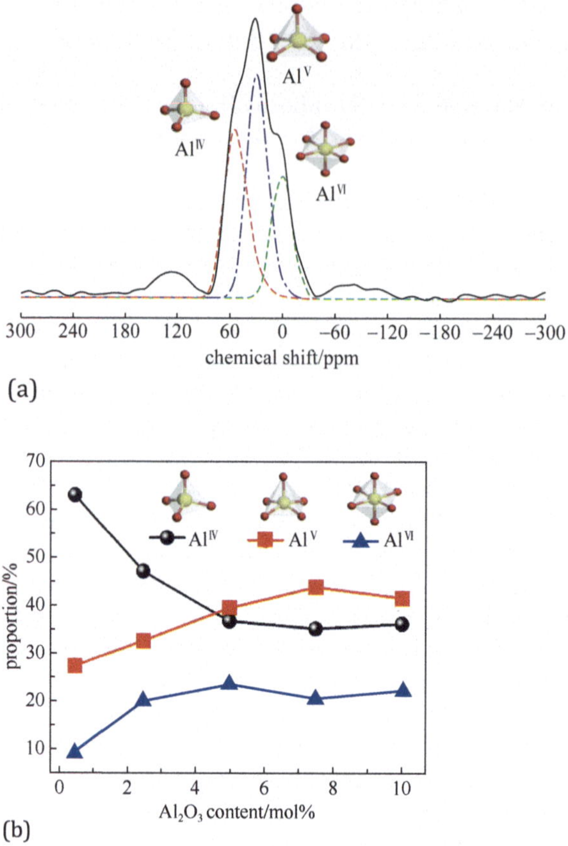

Fig. 2.22 Network structure of Al_2O_3 -SiO_2 silica glass. (**a**) ^{27}Al MAS NMR spectrum and its peak fitting. (**b**) The effect of Al_2O_3 content on the coordination number of aluminum

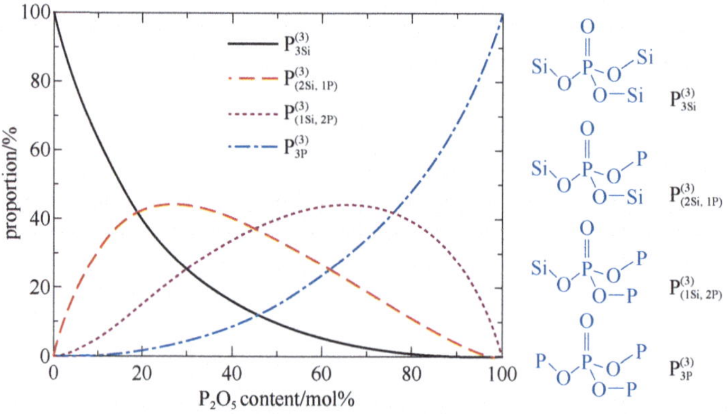

Fig. 2.23 The effect of P_2O_5 content on $P^{(3)}$ groups in the P_2O_5-SiO_2 glass [33]

Table 2.10 Theoretical composition of glass [34] Unit: mol%

Sample	SiO_2	Yb_2O_3	Al_2O_3	P_2O_5	CaO
SY	99.95	0.05	0	0	0
SYA	95.9	0.1	4	0	0
SYP	95.9	0.1	0	4	0
SYAP	87.9	0.1	4	8	0
Yb-AG	0	0.1	36	0	64
Yb-PG	0	0.1	0	50	50

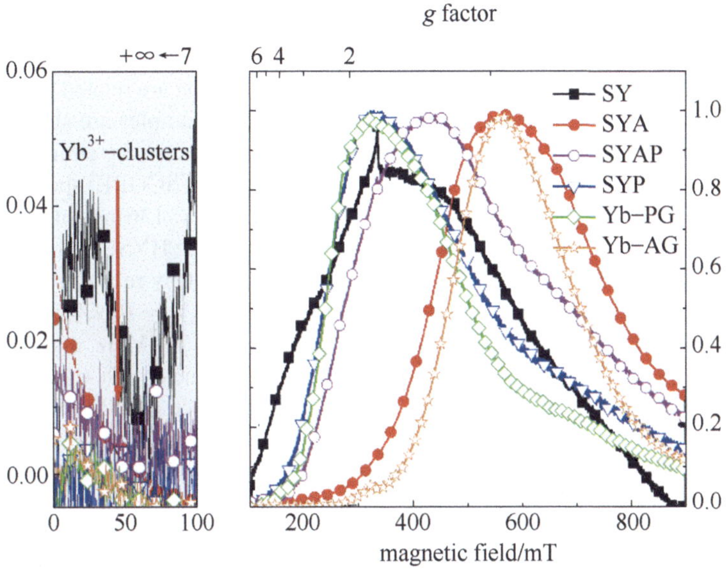

Fig. 2.24 Yb^{3+} echo detection field scanning spectrum (EDFS) in different glasses [34]

Shao et al. [34] used a pulsed EPR system to study the effect of aluminum/phosphorus doping on the local structure of Yb^{3+} in silica glass. Yb/Al/P doped silica glass was prepared by sol-gel method combined with high temperature melting method. Traditional high temperature melting method was used to prepare Yb^{3+} doped phosphate glass (Yb-PG) and Yb^{3+} doped aluminate glass (Yb-AG). The glass composition is detailed in Table 2.10.

Figure 2.24 shows the echo-detected field-swept (EDFS) spectra of Yb^{3+} singly doped (SY), Yb/Al co-doped (SYA), Yb/P co-doped (SYP), Yb/Al/P co-doped (SYAP) silica glasses. For comparison, those of the Yb^{3+} doped aluminate (Yb-AG) and Yb^{3+} doped phosphate (Yb-PG) glasses are also added to Fig. 2.24.

Theoretical studies show that the Landé factor g of a single Yb^{3+} in any matrix does not exceed 2 Λ*M, where the intrinsic Landé factor $\Lambda = 8/7$, the projection of the total angular momentum $M = 7/2$. Therefore, the g value of Yb^{3+} does not exceed 8 [35]. On the other hand, since Yb^{3+} is a Kramers ion with $S = 1/2$, there is no zero field splitting (must $S \geq 1$). Sen et al. [35] theoretically simulated the EPR spectrum of 2 ~ 4 Yb^{3+} cluster, and found that the more severe the Yb^{3+} cluster, the stronger the EPR signal ($g > 8$) at low magnetic field.

As can be seen from Fig. 2.24, the EDFS signal of the SY sample is strongest near zero magnetic field, while the EDFS signals of Yb-PG and Yb-AG are weakest. This indicates that the clustering degree of Yb^{3+} in SY glass is the most severe, while that in PG and AG glass is the weakest. Compared with the SY sample, the EDFS signals of SYA, SYAP and SYP samples near zero magnetic field have all weakened. This indicates that co-doping with Al^{3+} or P^{5+} can effectively reduce the clustering degree of Yb^{3+} in silica glass [36].

In addition, The EDFS spectra of all samples show very wide asymmetric peaks in the magnetic field range of 100 ~ 900 mT, which is related to the disordered structure of the glass. The EDFS spectrum and peak position are related to the local structure of Yb^{3+}. The EDFS spectra of Yb-PG and SYP samples are similar. This indicates that the local structure of Yb^{3+} in these two glasses is very similar.

Figure 2.25a–f are the hyperfine sublevel correlation (HYSCORE) spectra (magnetic field strength B_0 = 3500 G, pulse interval time t_{au} = 136 ns) of SY, SYA, Yb-AG, SYAP, SYP and Yb-PG samples, respectively. In the HYSCORE spectrum, the Larmor frequency ν_n of the magnetic nucleus is positively related to the gyromagnetic ratio γ_n and the external magnetic field B_0, that is, $\nu_n = B_0 \gamma_n/(2\pi)$. Under a 350 mT magnetic field, the resonance peaks at 3.0 MHz, 3.9 MHz, and 6.0 MHz correspond to the Larmor frequencies of the magnetic nuclei ^{29}Si (natural abundance = 4.7%, nuclear spin $I = 1/2$), ^{27}Al (natural abundance = 100%, $I = 5/2$) and ^{31}P (natural abundance = 100%, $I = 1/2$). All the resonance peaks of the nuclei only appear in the first or third quadrant (not shown), and all the resonance peaks are concentrated in the diagonal area, and no off-diagonal resonance peaks are observed. The fine coupling constant (A) is much smaller than the Larmor frequency of the nucleus. This result indicates that there is weak hyperfine coupling interactions among the magnetic nuclei ^{29}Si, ^{27}Al, ^{31}P and 4f orbital lone electron of Yb^{3+}. Therefore, it can be inferred that ^{29}Si, ^{27}Al, ^{31}P are mainly coordinated to the second or even further shells of Yb^{3+}. The oxygen atoms coordinated with Yb^{3+} in the first shell were not observed. This is because the natural abundance of the magnetic nucleus ^{17}O (~0.038%) is far below the detection limit of HYSCORE.

In the HYSCORE spectra of Yb-AG and Yb-PG glasses, only a strong and extended signal of ^{27}Al and ^{31}P is observed (Fig. 2.25c, f). Almost no cluster signals are detected in the EDFS spectra (Fig. 2.24), suggesting there is no Yb—O—Yb connection. This indicates that the Yb—O—Al connection is dominant in Yb-AG glass, and the Yb—O—P connection is dominant in Yb-PG glass. Due to the low natural abundance of the magnetic nucleus ^{43}Ca (~0.135%), its signal cannot be detected in the HYSCORE spectra of Yb-AG and Yb-PG glasses. Therefore, it is uncertain whether there is a Yb—O—Ca connection in Yb-AG and Yb-PG glasses.

In SY glass, the strong EDFS signal near zero magnetic field suggests the possible existence of the Yb—O—Yb connection (Fig. 2.24). However, in the HYSCORE spectra, only an extended and twisted ^{29}Si signal is observed, indicating the existence of a Yb—O—Si connection. The Yb—O—Yb connection is not

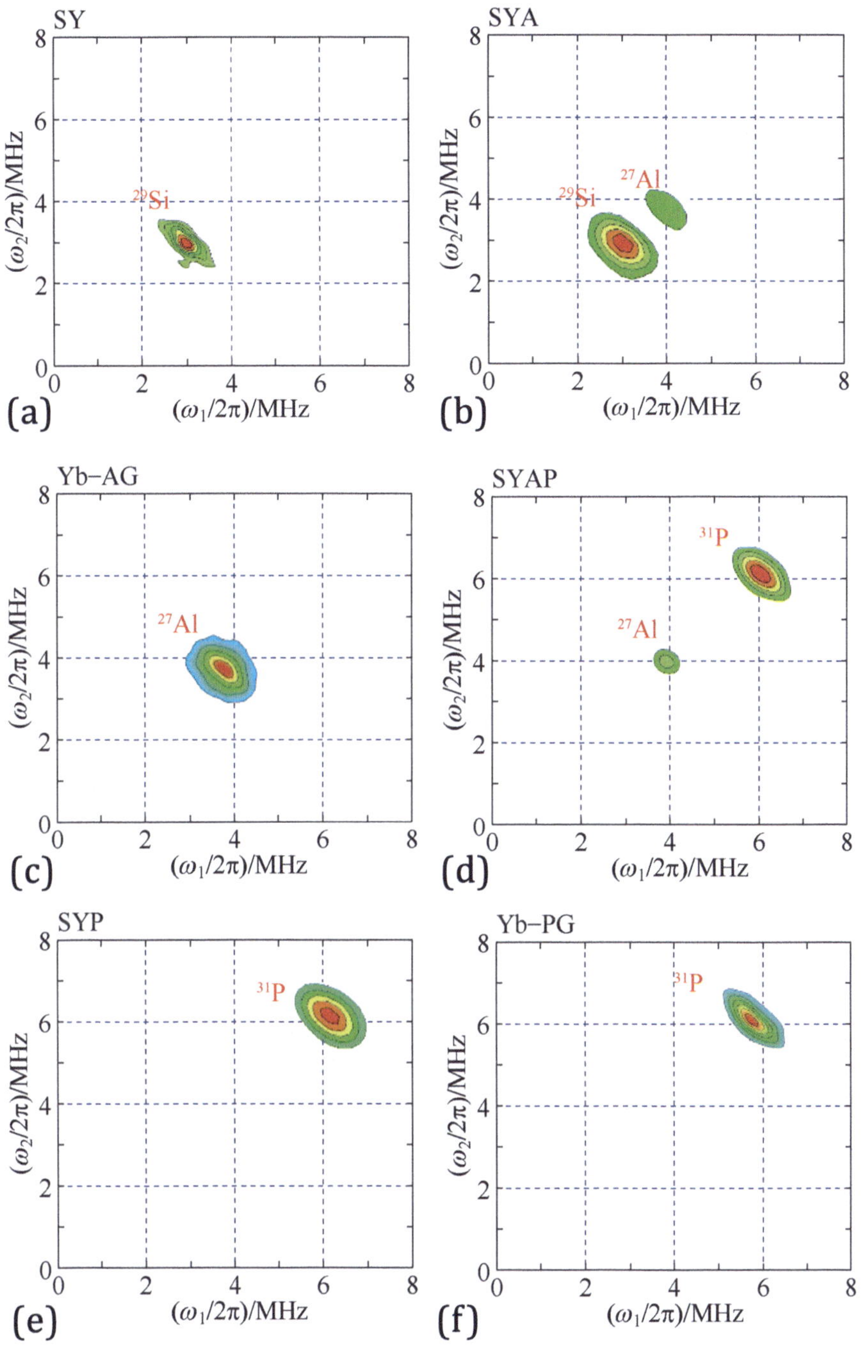

Fig. 2.25 The HYSCORE spectra of Yb^{3+} in different glasses SY (**a**), SYA (**b**), Yb-AG (**c**), SYAP (**d**), SYP (**e**) and Yb-PG (**f**) [34]

detected, which may be related to the low doping content of Yb^{3+} ([Yb] < 3000 ppm) (Fig. 2.25).

Compared with the SY sample, the Yb^{3+} cluster in SYA is reduced, and almost no cluster signal is detected in SYP glass (Fig. 2.24). In SYA glass, a strong and extended ^{29}Si signal and a weak signal in the diagonal area of ^{27}Al are detected (Figure 2.25b). In the SYP sample, only a strong and extended ^{31}P signal is detected (Fig. 2.25e). This indicates that when co-doping with 4 mol% Al_2O_3, not all Yb—O—Yb and Yb—O—Si are replaced by Yb—O—Al. While, when co-doping with 4 mol% P_2O_5, phosphorus can form a solvent shell structure around Yb^{3+}, that is, almost all Yb—O—Yb and Yb—O—Si connections are replaced by Yb—O—P connections. This shows that P^{5+} has a far superior ability to disperse rare earth ion than Al^{3+}.

In the SYAP sample, ^{27}Al and ^{31}P are both detected, and the signal of ^{31}P is much stronger than that of ^{27}Al. But the signal of ^{29}Si (Fig. 2.25d) and the cluster signal (Fig. 2.24) are not detected. This indicates that in the SYAP2 sample, there are a large number of Yb—O—P and a small number of Yb—O—Al, but there are no Yb—O—Yb and Yb—O—Si connections.

Currently, the opinion that co-doping phosphorus can form a solvent shell structure around rare earth ions (RE^{3+}) in silica glass has been widely accepted [37–39]. However, there is a controversy about whether co-doping aluminum can also form a solvent shell structure around RE^{3+} in silica glass. Sen et al. [38] showed that a strong aluminum signal could be observed in the HYSCORE spectrum of Al/Yb = 3 silica glass. Therefore, he speculates that when Al/RE > 10 in silica glass, Al can also form a solvent shell structure around RE^{3+}, that is, almost all RE—O—Yb and RE—O—Si connections in silica glass are replaced by RE—O—Al connections. Saitoh et al. [37] draw a conclusion as follows by testing and simulating the three-pulse electron spin echo envelope modulation (3P-ESEEM) of Er/Al co-doped silica glass. When Al/Er ≤ 20, Al^{3+} could not form a solvent shell around Er^{3+}. Funabiki et al. [40] used and simulated 3P-ESEEM to show that as the Al/Nd ratio increases, the coordination number of aluminum around Nd^{3+} changes from four coordination to six coordination, but Al^{3+} cannot completely replace Si^{4+} to form a solvent shell around Nd^{3+}.

Figure 2.26a–c show the 3P-ESEEM spectra of the SYA sample at 3500 G, 4500 G, and 5500 G magnetic fields, respectively. In 3P-ESEEM, the interval time (t_{au}) between the first two pulses and the external magnetic field (B_0) have a significant impact on the signal strength of ^{27}Al and ^{29}Si. In the 3P-ESEEM tested under the conditions of $B_0 = 3500$ G and $t_{au} = 176$ ns, almost no ^{27}Al signal can be observed. In the 3P-ESEEM tested under the conditions of $B_0 = 3500$ G and $t_{au} = 96$ ns, almost no ^{29}Si signal can be observed. Under the conditions of $B_0 = 4500/5500$ G and $t_{au} = 95$ ns, signals from ^{29}Si and ^{27}Al can be observed simultaneously in the 3P-ESEEM spectrum. This is due to the blind spot effect in the pulse EPR test process (see reference [41]). As can be seen from Fig. 2.26a–c, when $t_{au} = 136$ ns, signals from ^{29}Si and ^{27}Al can be detected simultaneously under three different magnetic field conditions, so $t_{au} = 136$ ns is selected for further HYSCORE testing.

Figure 2.26d–f are the HYSCORE spectra of the SYA sample under 350 mT, 450 mT, and 550 mT magnetic fields ($t_{au} = 136$ ns), from which it can be seen that

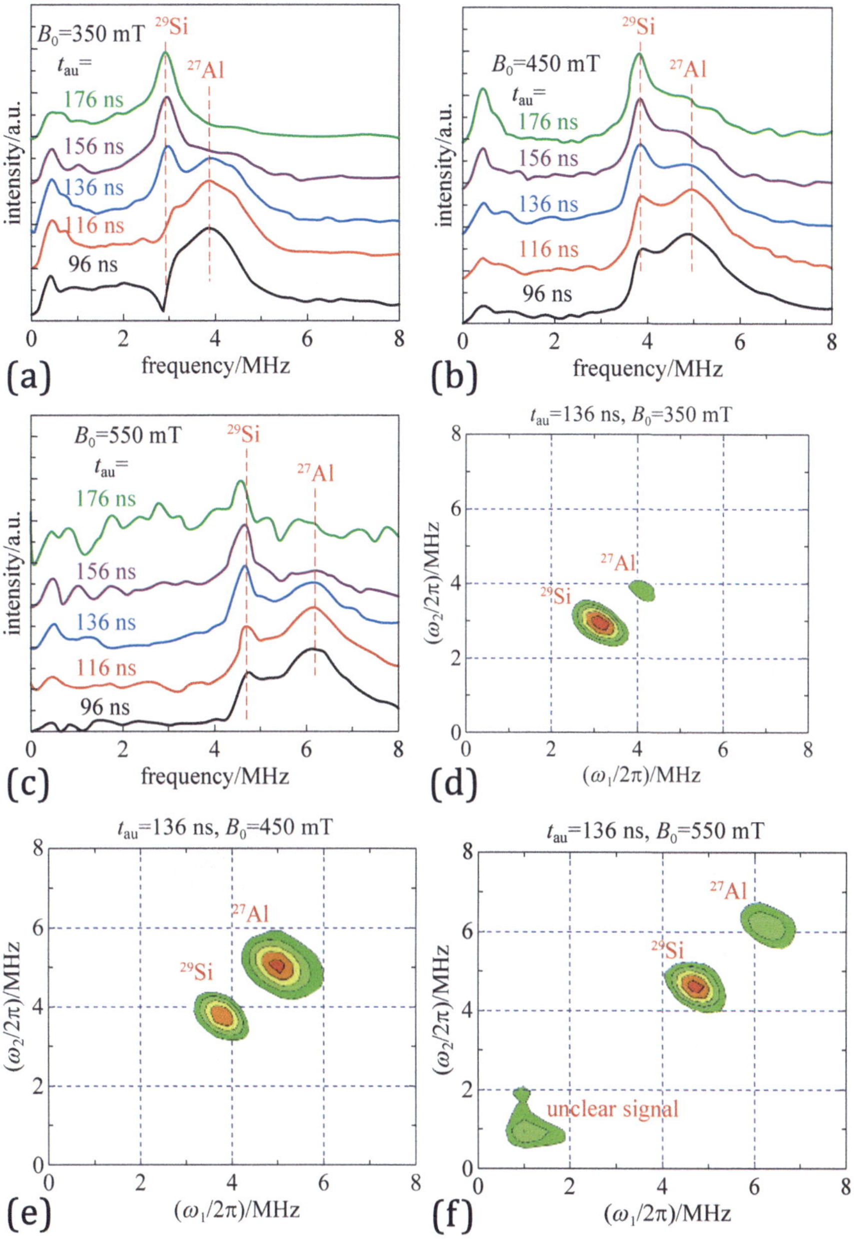

Fig. 2.26 3P-ESEEM (**a**–**c**) and HYSCORE spectra (**d**–**f**) under different magnetic fields B_0 in SYA sample [34]

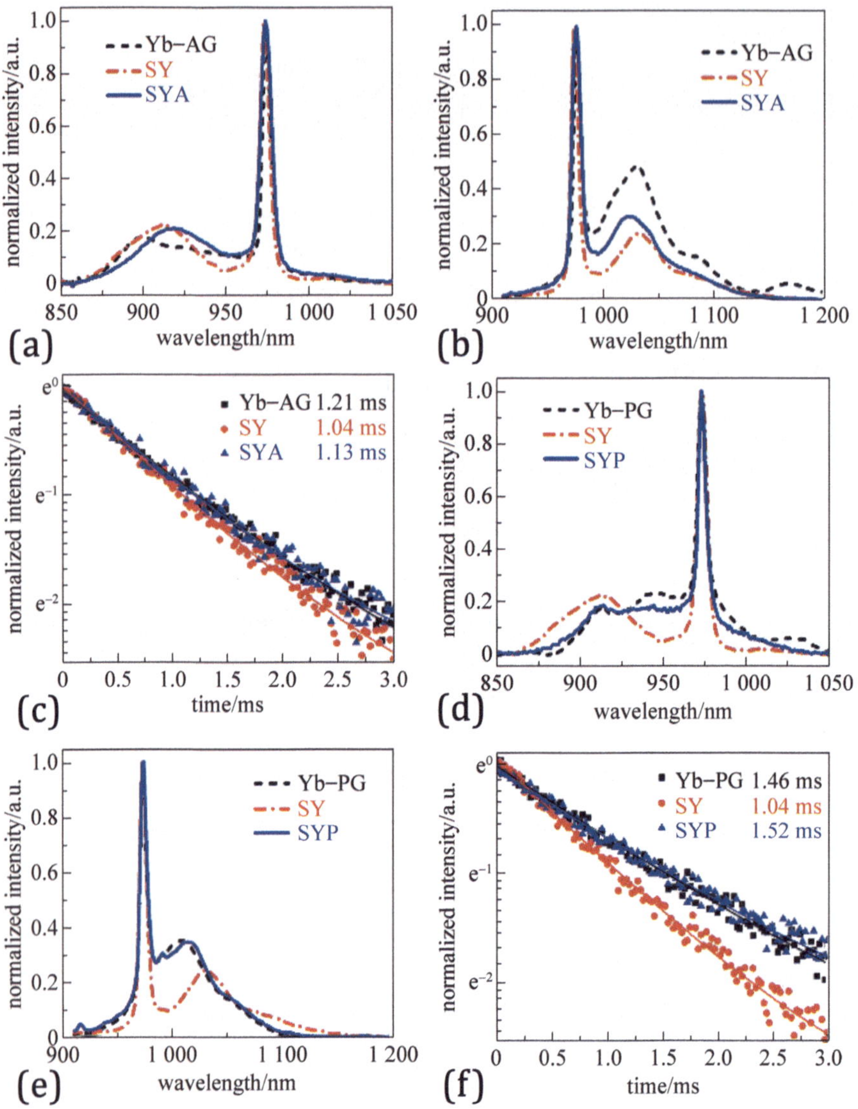

Fig. 2.27 Comparison of normalized absorption (**a**, **d**) and emission spectra (**b**, **e**), as well as fluorescence lifetime (**e**, **f**) of samples AG, SY, SYA, SYP, PG [34]

the signal strength of ^{29}Si and ^{27}Al is closely related to the magnetic field. Chiesa et al. [42] also observed this phenomenon in the Yb/Al/Ce co-doped silica glass. It is believed that this phenomenon is because rare earth ions (RE^{3+}) in silica glass may occupy different lattice sites: some RE^{3+} lattice sites are dominated by Al^{3+}, and some RE^{3+} lattice sites are dominated by Si^{4+}.

Figure 2.27 shows the normalized absorption (a, d) and emission spectra (b, e), and fluorescence lifetimes (e, f) of the Yb-AG, SY, SYA, SYP, and Yb-PG samples.

As can be seen from Fig. 2.27a–c, the absorption and emission spectra of the SYA sample show significant differences with those of SY and Yb-AG sample. While the fluorescence lifetime of the SYA sample is between that of the SY and that of the Yb-AG samples. It is well known that the spectral properties of rare earth ions mainly depend on the local environment of the rare earth ions. In the SY sample, there are many Yb—O—Si bonds and a very small amount of Yb—O—Yb ones. In the Yb-AG sample, it is mainly Yb—O—Al connections. In the SYA sample, there are a large number of Yb—O—Si and Yb—O—Al connections (Fig. 2.26). There are significant differences in the local environment of Yb^{3+} among above samples, leading to the variety in their spectral properties. In the Yb-PG and SYP samples, Yb^{3+} is wrapped by P^{5+} (Fig. 2.26), that is, there are only Yb—O—P connections. This results in the very similar absorption and emission spectra as well as fluorescence lifetime values (Fig. 2.27d–f).

Figure 2.28a shows the Stark split energy levels of Yb^{3+} in the Yb-AG, SYA, SY, SYP, and Yb-PG samples. It can be seen that the energy level difference between two adjacent Stark levels in the SY sample is the largest, followed by the Yb-AG sample. The Yb-PG sample has the smallest one.

Based on the Stark split energy levels of the five samples in Fig. 2.28a, the local crystal field strength parameter (N_J) and the degree of Yb^{3+}asymmetry in the five samples can be calculated, as shown in Fig. 2.28b. It can be seen that the N_J and the degree of Yb^{3+}asymmetry are the largest in the SY sample and the smallest in the Yb-PG and SYP samples.

The following discusses the dissolution mechanism of rare earth ion clusters in SiO_2 glass co-doped with Al or P. From the perspective of crystal chemistry, the

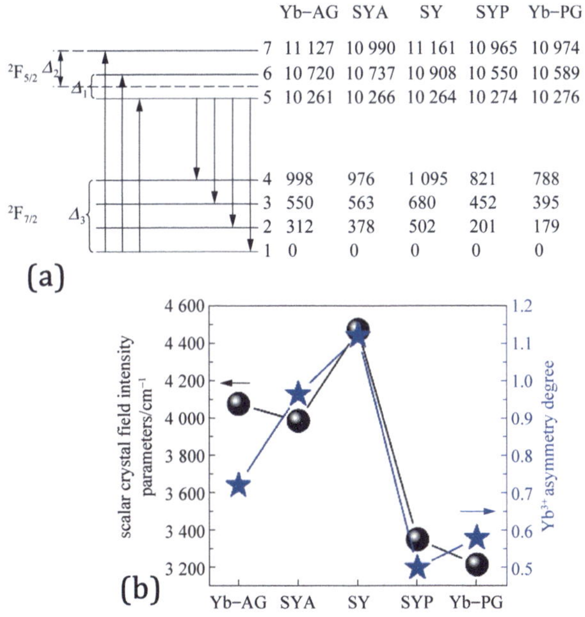

Fig. 2.28 Stark splitting levels and energy values of each level E_i ($i = 1 \sim 7$, cm^{-1}) (**a**), degree of asymmetry and crystal field strength parameter (**b**) [34]

essence of phase separation of oxide glass is the result of different cations competing for oxygen ions. In silicate melts, bridging oxygen ions are attracted to their surroundings by Si^{4+} in the form of $[SiO_{4/2}]$, so rare earth ions, as a kind of high field strength network external cation, tend to share non-bridging oxygen ions. From a thermodynamic perspective, under the constant temperature and pressure and under the condition without doing work externally, the premise for a process to proceed spontaneously is that the change in the Gibbs function is negative, that is,

$$\Delta G_{T,P} = \Delta H - T\Delta S < 0,\ \text{also}\ \Delta H < T\Delta S \tag{2.12}$$

Because the rare earth phase ($RE_2Si_2O_7$, $RE_2Ge_2O_7$, REB_3O_6) formed in single-component oxide glasses (such as SiO_2, GeO_2, B_2O_3) is very stable, the enthalpy change ΔH between the rare earth-rich phase and the pure oxide phase (SiO_2, GeO_2, B_2O_3, RE_2O_3) is very large. While the entropy change ΔS is very small, leading to $\Delta H > T\Delta S$. Therefore, the obtained glass often easily phase separates. Studies have shown that high-temperature melting combined with quenching can effectively suppress glass phase separation. However, the melting temperature of RE_2O_3-SiO_2 silica glass must be greater than 2300 K. Such a high melting temperature often needs to be carried out under vacuum conditions, which poses a challenge to the quenching process. There are two possible alternative solutions: ① reducing the enthalpy change ΔH value. ② Increasing the entropy change ΔS value. Assuming that the enthalpy change and entropy change of the system satisfy $\Delta H < 0 < T\Delta S$, there must be $\Delta G_{T,P} = \Delta H - T\Delta S < 0$, that is, the reaction process can proceed spontaneously. The glass phase separation can be suppressed.

Figure 2.29 is a schematic diagram of promoting the spontaneous progress of the system by changing the system enthalpy change ΔH and entropy change ΔS. Assuming that systems A and B are immiscible, this situation is similar to water and oil because of $\Delta G_{T,P} > 0$, as shown in Fig. 2.29a. Assuming that system C can

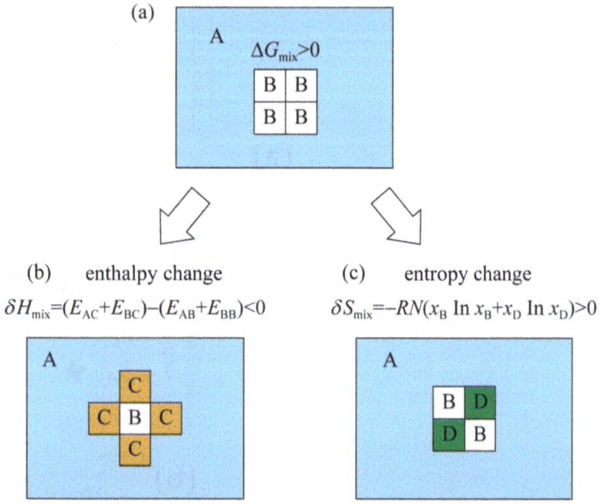

Fig. 2.29 The schematic diagram of enthalpy and entropy changes caused by the introduction of C and D in systems A and B [43]

wrap B and is miscible with system A, this situation is similar to adding a surfactant (such as detergent) to water and oil, because the large formation energy of the A–B mixture is replaced by the small formation energy of the A-C and B-C mixtures, resulting in enthalpy change $\Delta H = (E_{AC} + E_{BC})-(E_{AB} + E_{BB}) < 0$, as shown in Fig. 2.29b. Assuming that system D is similar to B, although they cannot be completely miscible with system A, the introduction of system D can dilute system B, this situation is similar to adding another oil to water and oil. This will inevitably lead to an increase in the disorder of the system (i.e., an increase in entropy change ΔS), but the change in mixing enthalpy ΔH is small, as shown in Fig. 2.29c.

Figure 2.30 is a schematic diagram of the dissolution of rare earth ion clusters co-doped with Al or P. Among them, Fig. 2.30a is a schematic diagram of the clustering of rare earth ions in silica glass. Studies have shown that P mainly exists as P $^{(3)}$ structural units in P_2O_5-SiO_2 glass. The P = O double bond in the $P^{(3)}$ structure attracts rare earth ions to gather around P. HYSCORE tests show that when P/ RE > 3, P can form a solvent shell structure to wrap rare earth ions. This is due to the enthalpy change effect caused by co-doping P, that is, the sum of the formation energy of P-RE and P-Si mixtures is lower than the sum of that of Si-RE and RE-RE clusters. This means that the system enthalpy change $\Delta H < 0$, the reaction process can proceed spontaneously, as shown in Fig. 2.30b. Studies have shown that Al exists in Al_2O_3-SiO_2 glass as Al^{IV}, Al^{V}, Al^{VI} structural units. Monteil et al. [44] showed by molecular dynamics methods that Al^{IV} could not effectively reduce the clustering of Er^{3+} in silica glass. Funabiki et al. [40] showed by 3P-ESEEM combined with computer simulation that as the Al/Nd co-doping ratio increased, the

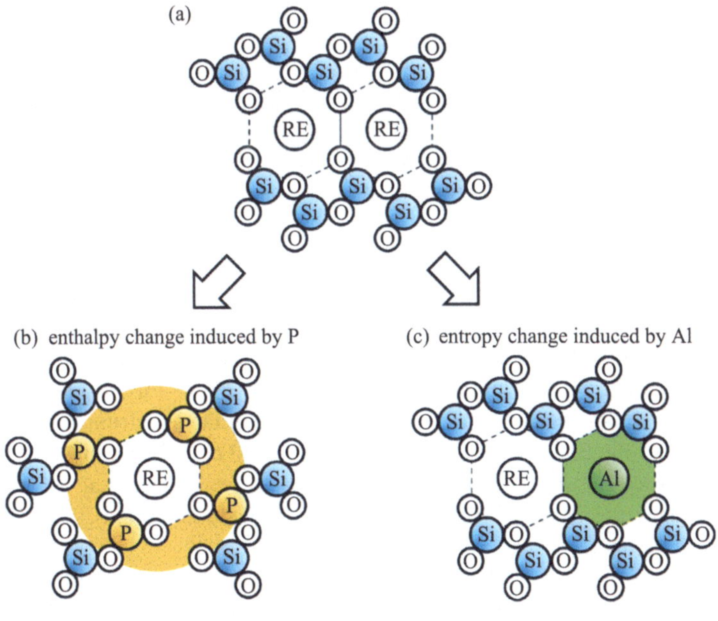

Fig. 2.30 The dissolution mechanism of RE ion clusters in silica glass co-doped with Al or P [43]

clustering of Nd^{3+} gradually decreased, and the coordination number of Al around Nd^{3+} gradually changed from four coordination to six coordination. Sen et al. [15] confirmed by NMR methods that in Al^{3+} single-doped silica glass, the coordination number of Al gradually changed from four coordination to six coordination as the content of Al increased. Lægsgaard [45] showed by density functional theory calculations that in silica glass, the formation energy of Er-Al mixture ($ErAl_3O_6$) was greater than the sum of that of Er-Er ($Er_2Si_2O_7$) Al-Al (Al_4O_6) clusters. This means that the system enthalpy change $\Delta H > 0$, so they believe that the entropy change effect caused by co-doping Al plays a key role in dispersing rare earth ion in silica glass. The local structure of six-coordinated Al and six-coordinated rare earth is similar, and both tend to enter the gap position of the glass network to compete for non-bridging oxygen ions. Six-coordinated Al dilutes the aggregation of rare earth ions to a certain extent, as shown in Fig. 2.30c. Similarly, Studies by Wang et al. [46] using spectroscopy, NMR and HYSCORE methods have shown that co-doping La^{3+} to a certain extent can improve the dispersion of Yb^{3+} in high Yb-doped silicate glass. The fundamental reason is that the introduction of La^{3+} to some extent dilutes the Yb^{3+} clusters.

2.2.4.2 The Effect of Co-Doping with Aluminum and Phosphorus on the Network Structure of Silica Glass and the Local Structure of Yb^{3+}

Shao et al. [36] further studied the effect of co-doping aluminum and phosphorus on the network structure of silica glass and the local structure of Yb^{3+}. The glass composition is shown in Table 2.4.

Figure 2.31a–c show the effects of different P/Al ratios on the ^{27}Al MAS NMR spectra, ^{31}P static NMR spectra, and Raman scattering spectra in the Yb/Al/P co-doped silica glass. From Fig. 2.31a, b, it can be seen that when P/Al = 1, only Al^{IV} and $P^{(4)}$ structures are detected, and their chemical shifts are located at 38.5 ppm and -30 ppm, respectively. In $AlPO_4$ glass, The chemical shifts of Al^{IV} and $P^{(4)}$ are located at 38 ppm and -28 ppm, respectively. This result indicates that in the SYAP1 sample, Al and P mainly exist as $AlPO_4$ structural units. When P/Al < 1, P preferentially connects with Al to form $AlPO_4$ structure, the excess Al connects with Si as Al^{IV} structural units, Al^{V}, Al^{VI} units are mainly located at the interstitial sites. When P/Al > 1, Al preferentially connects with P to form $AlPO_4$ structure, and the excess P exists as $P^{(2)}$ and $P^{(3)}$ structural units. Among them, $P^{(3)}$ mainly connects with Si, and $P^{(2)}$ mainly connects with Al^{V} and Al^{VI}. In the Raman spectrum of Fig. 2.31c, when P/Al = 1, the broad peak located at 1000 ~ 1250 cm^{-1} is mainly attributed to the $AlPO_4$ structural unit. When P/Al > 1, the sharp peak located at 1320 cm^{-1} is mainly attributed to the $P^{(3)}$ structural unit (i.e., P = O double bond). The structural schematic diagram is shown in Fig. 2.9b.

Figure 2.32a shows the effect of P/Al ratio on the EDFS spectra of Yb^{3+} doped silica glass. It can be seen that the EDFS spectra present two different line shapes when P/Al<1 and P/Al>1. This means that Yb^{3+} occupies two different lattice sites

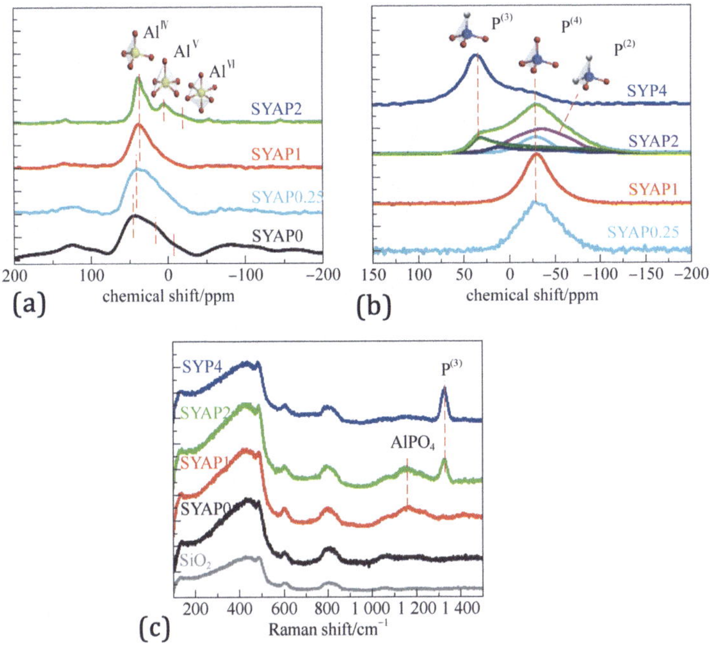

Fig. 2.31 Network structure of Yb/Al/P co-doped silica glass: The effect of different P/Al co-doping ratios on ^{27}Al MAS NMR spectra (**a**), ^{31}P static NMR spectra (**b**) and Raman scattering spectra (**c**) [36]

in these glasses. In addition, as the P/Al ratio increases, the EDFS signal at the near-zero magnetic field gradually weakens. This indicates that the Yb^{3+} cluster gradually decreases as the P/Al ratio increases.

Figure 2.32b shows the effect of P/Al ratio on the HYSCORE projection spectrum of Yb^{3+} doped silica glass. Those of Yb^{3+} singly doped (SY0.05) and Yb/P co-doped (SYP4) silica glasses are also included in Fig. 2.32b for comparison. At a magnetic field of 350 mT, the resonance peaks located at 3.0 MHz, 3.9 MHz, and 6.0 MHz correspond to the Larmor frequencies of magnetic nuclei ^{29}Si, ^{27}Al, and ^{31}P, respectively. It can be seen that in the SY0.05 sample, the nearest neighbor elements of Yb^{3+} are mainly Si. In the SYP4 sample, the nearest neighbor elements of Yb^{3+} are mainly P. In the SYAP series samples, as the P/Al ratio increases, Yb^{3+} gradually transfers from a silicon-rich environment to a phosphorus-rich environment. It is worth noting that when $P/Al \approx 1$, Al^{3+} and P^{5+} preferentially form [$AlPO_4$] units which cluster around Yb^{3+}. Therefore, in the HYSCORE spectrum projection, the signal intensities of Al and P in the SYAP1 sample are approximately equal. When $P/Al \leq 1$, Yb^{3+} is mainly in an aluminum-rich or silicon-rich environment. When $P/Al > 1$, Yb^{3+} is mainly in a phosphorus-rich environment.

According to NMR, Raman and pulse EPR results, Fig. 2.33 shows the schematic diagram of the network structure and Yb^{3+} local structure in Yb^{3+} doped silica glass with different P/Al ratio.

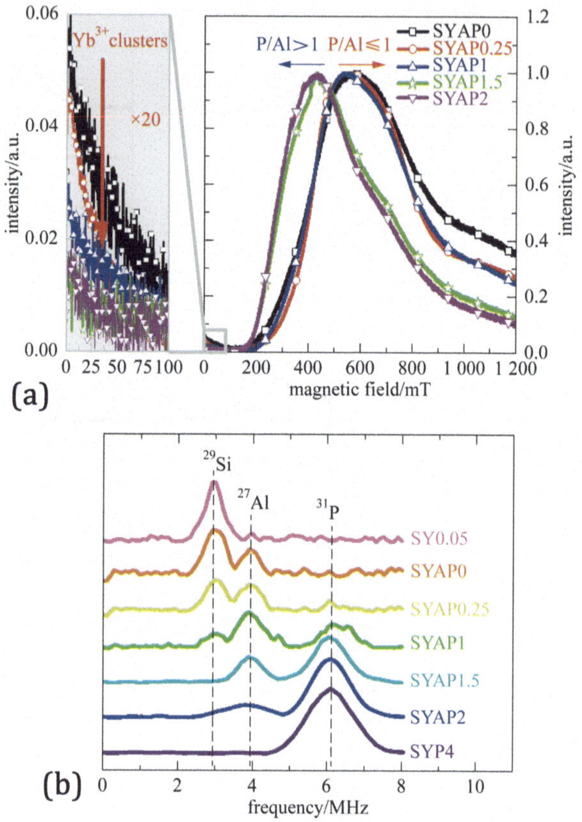

Fig. 2.32 Yb^{3+} local structure in Yb/Al/P co-doped silica glass: the effect of P/Al ratio on EDFS (**a**) and HYSCORE (**b**) (HYSCORE test magnetic field strength $B_0 = 350$ mT, pulse interval time $t_{au} = 136$ ns) [36, 47]

2.2.4.3 The Effect of Fluorine Doping on the Network Structure and Yb^{3+} Local Structure of Silica Glass

Shao et al. [30] systematically studied the effect of F/Si mass ratio on the glass structure, revealing the evolution process of the silica glass structure with the increase of F/Si mass ratio. Detailed glass composition is shown in Table 2.7, and the experimental results are shown in Fig. 2.34.

Figure 2.34a, b show the ^{19}F and ^{27}Al MAS NMR spectra, respectively. The MAS NMR spectrum of ^{19}F can be decomposed into two Gaussian peaks, located at −137 ppm and −146 ppm, respectively. According to the research of Youngman et al. [48, 49], fluorine enters the SiO_2 network in two ways, where the signal at −146 ppm corresponds to the [$SiO_{3/2}$ F] group, and that at −137 ppm corresponds to the [$SiO_{4/2}$ F] group. The ^{27}Al MAS NMR spectrum contains three asymmetric resonance peaks, located at 50 ppm, 25 ppm, and 4 ppm, respectively, which correspond to four, five, and six coordinated aluminum (represented as Al^{IV}, Al^{V}, and Al^{VI}). Al^{V} and Al^{VI} increase with the increase of F content, when the total Al content in the glass remains constant. This indicates that Al^{IV} decreases with the increase of F content. Zhang et al. [50] also observed the phenomenon of Al^{V} and Al^{VI} increasing

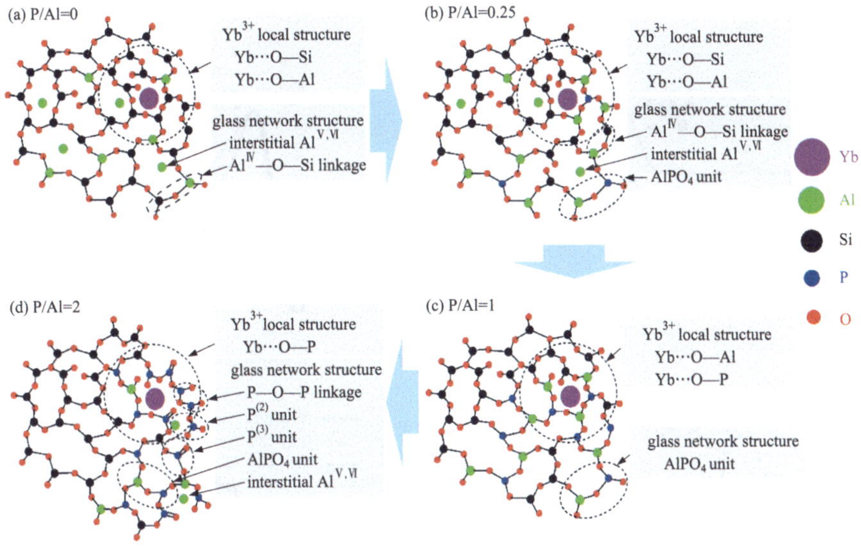

Fig. 2.33 Schematic diagram of the network structure and Yb^{3+} local structure as functions of P/Al ratio in the Yb/Al/P co-doped silica glass [36]

with the increase of F content in F-doped aluminum phosphate glass. The $^{27}Al\{^{19}F\}$ rotational echo double resonance (REDOR) results showed that this was due to the formation of F-Al.

Figure 2.34c shows the effect of F content on the Raman spectrum. As the F content increases, a weak vibration peak appears at 937 cm^{-1} in the Raman spectrum, and its intensity gradually increases. This peak is attributed to the stretching vibration of the F—Si bond. In addition, the vibration peaks at 490 cm^{-1} and 606 cm^{-1} in the Raman spectrum correspond to the planar 4-membered ring (D_1) and the planar 3-membered ring (D_2) structure. Their strength gradually decreases with the increase of F content.

Figure 2.34d shows the effect of F content on the 2260 cm^{-1} vibration band in FTIR spectrum. An exponential function is used to subtract the background of the FTIR curve in the YACF0 sample. The remaining part is decomposed into two Gaussian peaks II and III, among which Gaussian peak III is the target data. This peak corresponds to the asymmetric stretching vibration of Si—O—Si, the peak position is positively related to the Si—O—Si bond angle. Its full width at half maximum (FWHM) is positively related to the distribution of the Si—O—Si bond angle [32, 51].

Figure 2.34e shows the effect of F content on the peak and FWHM of Gaussian peak III in Fig. 2.34d. The fictive temperature of the glass (T_f) calculated based on the peak of Gaussian peak III is also included in Fig. 2.34e. T_f is negatively related to the peak position (ν) of Gaussian peak III, which can be calculated according to the following formula [51]:

$$\nu = 2228.64 + \left(43809.21 / T_f\right) \tag{2.13}$$

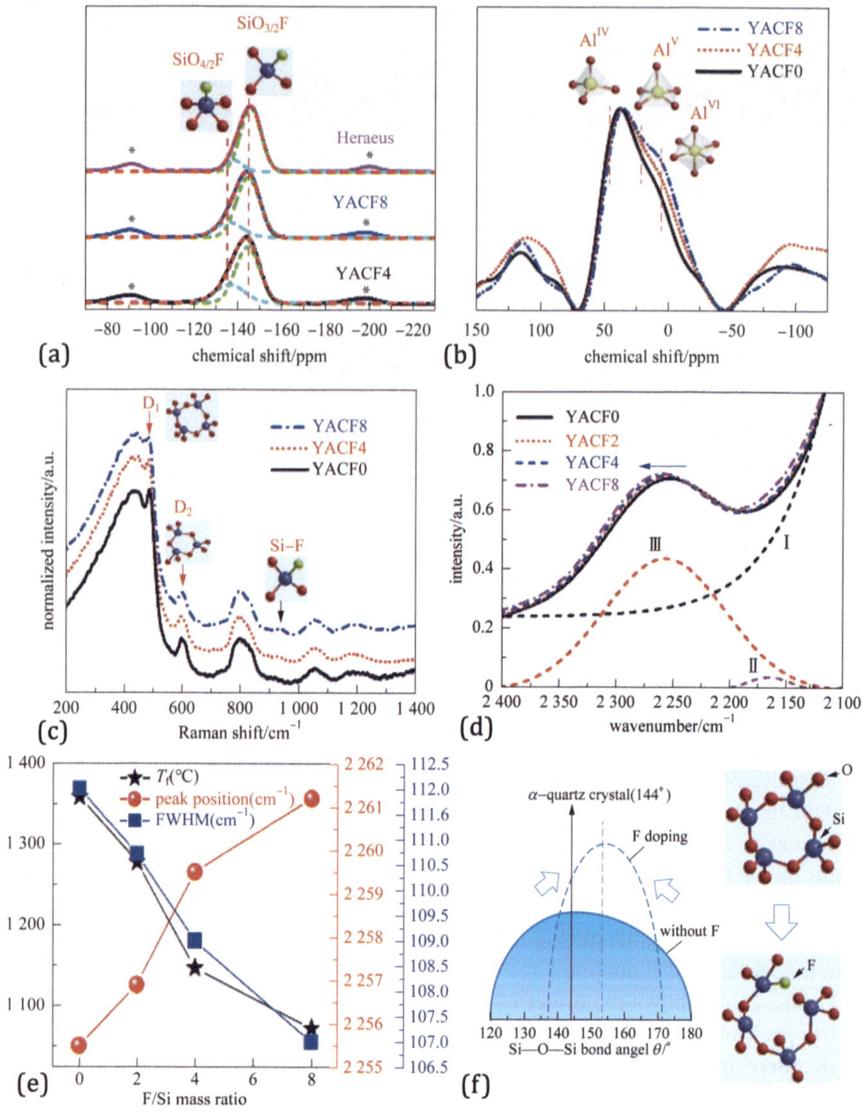

Fig. 2.34 Network structure of Yb/Al/Ce/F co-doped silica glass: ^{19}F (**a**) and ^{27}Al MAS NMR spectra (**b**), Raman (**c**) and FTIR absorption spectra (**d**), the peak position of the Gaussian peak III at 2260 cm^{-1} in the FTIR absorption spectra, full width at half maxima (FWHM), and fictive temperature (T_f) as a function of the F/Si ratio (**e**), and a schematic diagram of the distribution of Si—O—Si bond angles (**f**)

As can be seen from Fig. 2.34e, as the F content increases, the peak of the 2260 cm^{-1} band has a blue shift, but its FWHM gradually decreases. This indicates that as the F content increases, the average bond angle of Si—O—Si increases, but the distribution range of it becomes smaller, as shown in Fig. 2.34f.

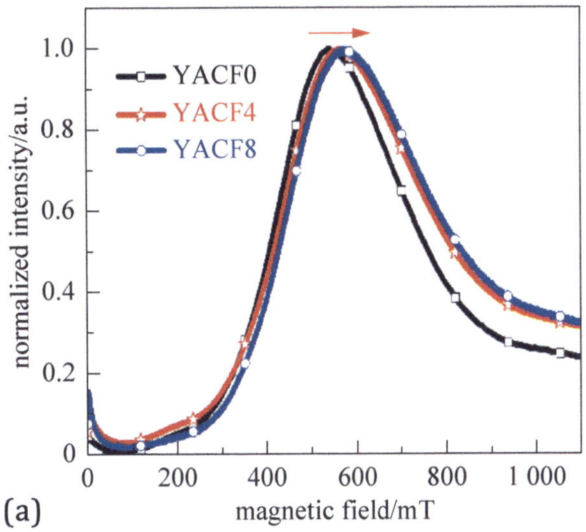

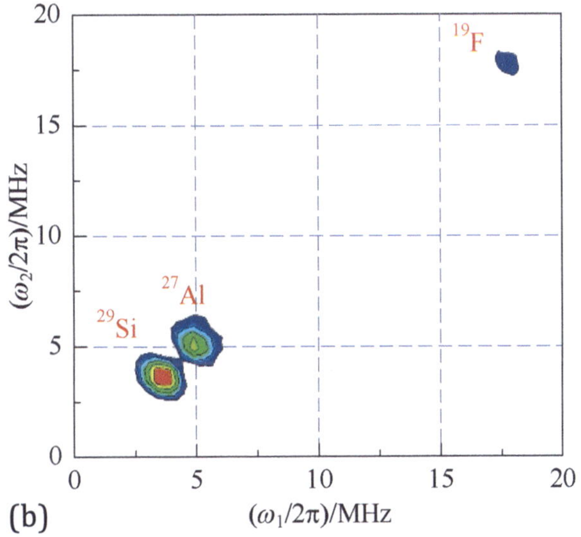

Fig. 2.35 The local structure of Yb^{3+} in the Yb/Al/Ce/F co-doped silica glass: EDFS (**a**), YACF8 (**b**) HYSCORE spectra [30]

Figure 2.35a shows the EDFS spectra of the YACF series of glasses. It can be seen that the EDFS spectra move towards the high magnetic field direction, as the F content increases. This phenomenon can also be observed in Yb^{3+} doped fluoroborate glass and fluorophosphate glass. This indicates that the addition of fluorine has a certain impact on the local environment of Yb^{3+}.

Figure 2.35b shows the two-dimensional HYSCORE spectrum of YACF8 glass. At B_0 = 450 mT magnetic field, the resonance peaks at 3.8 MHz, 5 MHz, and 18 MHz correspond to the Larmor frequencies of magnetic nuclei ^{29}Si, ^{27}Al, and ^{19}F (natural abundance = 100%, nuclear spin I = 1/2). All resonance peaks only appear in the (+ +) quadrant and are distributed along the diagonal. This indicates that the magnetic nuclei ^{29}Si, ^{27}Al, ^{19}F do not directly bond with Yb^{3+}. They should be located in the second shell of Yb^{3+} or even further away (4 ~ 8 Å).

In the HYSCORE spectra of Yb^{3+} doped fluoroborate glass and fluorophosphate glass, two types of ^{19}F signals can be observed. One is the resonance signal that only appears in the diagonal area (similar to Fig. 2.35b), corresponding to the weak coupling interaction between Yb^{3+} and F^-. The other is a pair of resonance peaks appearing in the non-diagonal area, their connecting line is perpendicular to the diagonal and the distance to the diagonal is equal. Easyspin simulation confirms that this signal comes from the strong coupling interaction between Yb^{3+} and F^-, that is, Yb-F bond. This signal was not observed in YACF8 glass, the main reason may be that the fluorine content in this glass sample is low (1.1 wt%). In Yb^{3+} doped fluoroborate glass and fluorophosphate glass with very low fluorine content (<2 wt%), the strong coupling interaction between Yb^{3+} and F^- was also not detected [52, 53]. Al^{3+} and P^{5+} ions perfer to coordinate around rare earth ones, playing a role in dispersing rare earth ion. Unlike the role of Al^{3+} or P^{5+}, F^- prefers to bond with Si^{4+} or Al^{3+}, so co-doping F^- has little effect on the dispersion of rare earth ions.

In silica glass, ring structures usually contain 3 ~ 9 rings. Among them, the 6 ~ 8 ring structure has the lowest energy, so it is the most stable. The energy of the ring structure deviating from the 6 ~ 8 ring is higher, thus they are unstable. Fluorine doping is often referred to as "chemical annealing", which promotes structural relaxation of silica glass. According to the above Raman, FTIR, NMR and pulse EPR research results, the structural diagrams of YACF0 and YACF8 glass samples are shown in Fig. 2.36a, b, respectively. In the YACF0 sample, the ring structure units around Yb^{3+} are relatively complex, possibly containing 3 ~ 9 rings. The local crystal field strength and the degree of Yb^{3+} asymmetry are high. In the YACF8 sample, the ring structure units around Yb^{3+} are relatively simple, possibly mainly 6 ~ 8 rings. The local crystal field strength and the degree of Yb^{3+} asymmetry are small.

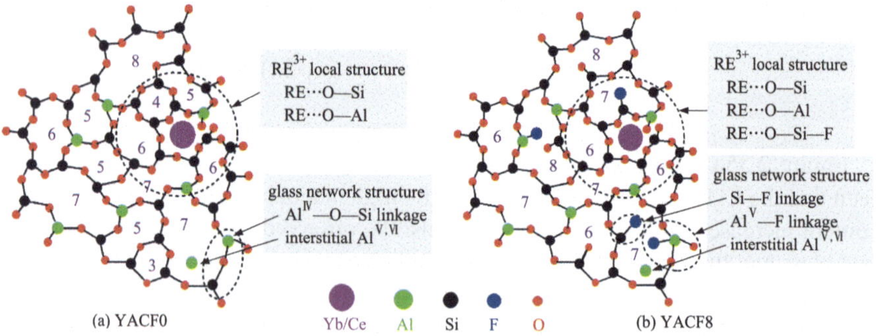

Fig. 2.36 Schematic diagram of the network structure and the local structure of Yb^{3+} in the Yb/Al/Ce/F co-doped silica glass as functions of fluorine content (The numbers in the figure represent the number of cations in the ring structure) [30]

2.2.4.4 The Influence of Boron Doping on the Network Structure of Silica Glass and the Local Structure of Yb^{3+}

Guo et al. [20] used NMR, FTIR and Raman methods to analyze the effect of boron oxide on the structure of Al/Yb co-doped silica glass. The intrinsic connection between glass composition, structure and properties were also pointed out. The glass composition is detailed in Table 2.8.

Figure 2.37a, b are the infrared and Raman spectra of the series SYAB glasses, respectively. All spectra are normalized to the vibration peak intensity at 800 cm^{-1}. For comparison, the infrared and Raman spectra of B_2O_3 doped silica glass (glass

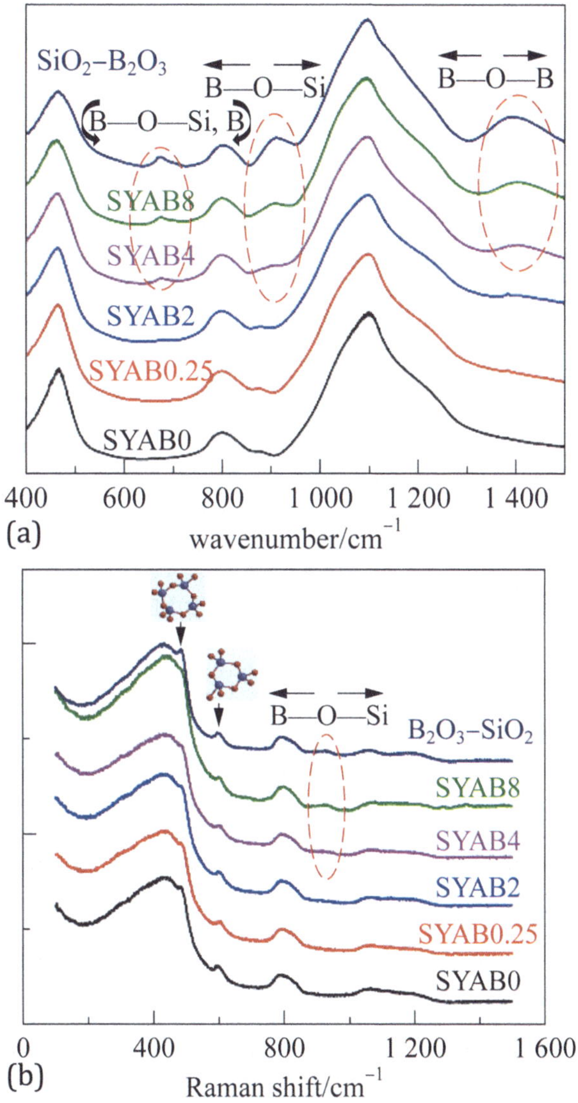

Fig. 2.37 Vibration spectrum characterization of the network structure of Yb/Al/B co-doped silica glass: FTIR (**a**) and Raman (**b**) spectra [20]

composition is $8B_2O_3$–$92SiO_2$, in mol%, marked as B_2O_3–SiO_2) are also added to Fig. 2.37a, b. When the B_2O_3 content is low, the infrared and Raman spectra of SYAB0.25 and SYAB2 glasses are similar to those of Yb/Al co-doped silica glass (SYAB0) without B doping. When the B_2O_3 content is higher than 2 mol%, three vibration peaks appear in the infrared spectra of SYAB4 and SYAB8, located at 677 cm^{-1}, 910 cm^{-1} and 1400 cm^{-1}. Among them, the infrared vibration peak at 677 cm^{-1} corresponds to the bending vibration of B—O—Si or B—O—B connection, and the infrared vibration peak at 1400 cm^{-1} corresponds to the antisymmetric stretching vibration of B—O—B connection [54, 55]. The infrared vibration peak at 910 cm^{-1} corresponds to the stretching vibration of the Si—O—B bond [56–58]. The Raman vibration peaks at 483 cm^{-1} and 603 cm^{-1} are attributed to the plane 4-ring Si—O—Si (D_1) and plane 3-ring (D_2) Si—O—Si [59]. As the B_2O_3 content increases, the peak intensities of D_1 and D_2 in the glass decrease, and the D_1 peak almost disappears when the content reaches 8 mol%. Compared with SYAB0.25 and SYAB2 glasses, a weak vibration peak appears at 930 cm^{-1} in SYAB4 and SYAB8 glasses, which is attributed to the stretching vibration of the Si—O—B bond [57, 58]. The FTIR and Raman results show that B_2O_3 leads to the formation of Si—O—B bonds in the glass.

Figure 2.38a, b show the ^{11}B and ^{27}Al MAS NMR spectra of the SYAB silica glass system, respectively. Deters et al. [60, 61] have shown that the interaction between paramagnetic rare earth ions and surrounding magnetic nuclei can broaden the MAS NMR spectra of the magnetic nuclei, thus to reduce their resolution. Therefore, Guo et al. [20] used Y_2O_3 without paramagnetism to replace Yb_2O_3 for comparison. The glass composition is $0.1Y_2O_3$–$6Al_2O_3$–$2B_2O_3$–$91.9SiO_2$ (in mol%), named as $SYAB_2$–Y_2O_3.

As can be seen from Fig. 2.38a, in the $SYAB_2$–Y_2O_3 silica glass, most B exists in the form of three-coordination (B^{III}), and only a small amount of B exists in the form of four-coordination (B^{IV}). With the increase of B_2O_3 content, the coordination number of B does not change significantly. Comparing the ^{11}B MAS NMR spectra of SYAB2 with that of SYAB4, it can be seen that when the B_2O_3 content is greater than 2 mol%, the resolution of the ^{11}B MAS NMR spectra significantly improves. This is because when the B_2O_3 content is greater than 2 mol%, most B atoms are far away from Yb^{3+}. Therefore, the broadening effect of the paramagnetism of Yb^{3+} on the ^{11}B MAS NMR spectra is weakened.

As can be seen from Fig. 2.38b, in the series of SYAB glass, Al mainly exists in the form of Al^{IV}. With the increase of B_2O_3 content, Al^{IV} gradually transforms to Al^{V} and Al^{VI}. The main reasons include two aspects: ① compared with Al^{IV}, B^{III} prefer to bond with Al^{V}, that is, it forms Al^{V}—O—B^{III} connection. ② Because the four-coordinated boron $[BO_{4/2}]^-$ has a negative charge, more Al^{VI} will form to compensate for its negative charge.

Figure 2.38c, d are the SYAB8 glass sample $^{27}Al\{^{11}B\}$ and $^{11}B\{^{27}Al\}$ REDOR spectra. To analyze the connection between Al and B, the REDOR spectrum of $Al_{18}B_4O_{33}$ crystal is also added to Fig. 2.38c, d. After comparison and calculation, there is the B—O—Al connection in the SYAB8 sample, and 0.5 Al is bonded in the

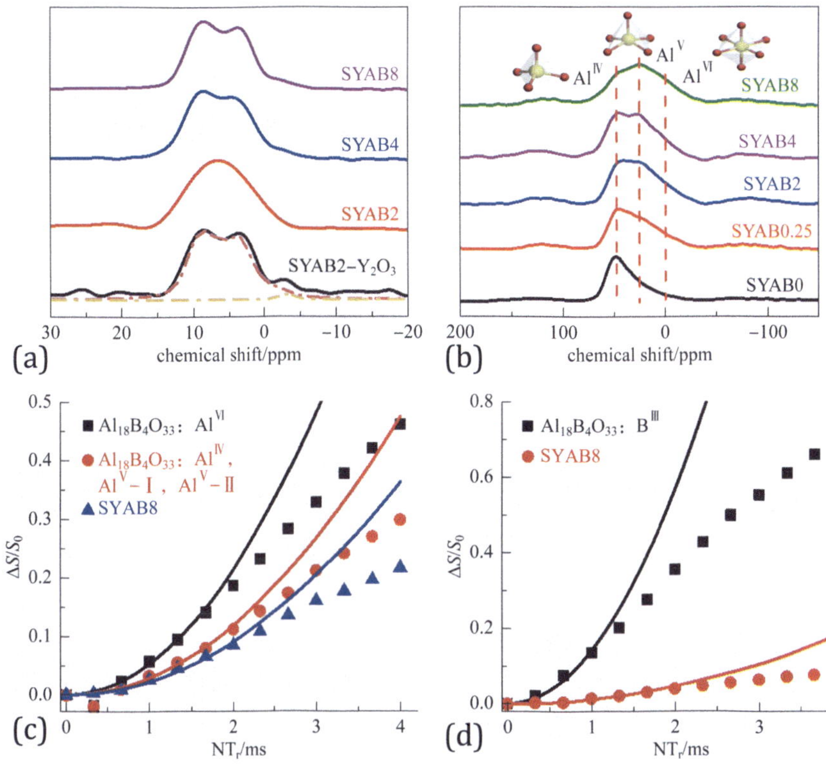

Fig. 2.38 NMR characterization of the network structure of Yb/Al/B co-doped silica glass: ^{11}B (**a**) and ^{27}Al (**b**) MAS NMR spectra, The $^{27}Al\{^{11}B\}$(**c**) and $^{11}B\{^{27}Al\}$(**d**) REDOR spectra of SYAB8 glass sample and $Al_{18}B_4O_{33}$ crystal [20]

second coordination layer of B, and vice versa. The detailed calculation process can be found in reference [20].

Figure 2.39a is the EDFS spectrum of Yb^{3+} -doped (SY0.05), Yb/Al -doped (SYAB0) and Yb/Al/B -doped (SYAB4) silica glasses. For easy comparison, The EDFS spectra of Yb^{3+} doped aluminate (Yb-AG) and Yb^{3+} -doped borosilicate glasses (Yb-BG) are also added in Fig. 2.39. It can be seen that the peak positions of the EDFS spectra in the SYAB0, SYAB4, and Yb-AG glasses are almost identical. This indicates that the local structure of Yb^{3+} in these samples is quite similar. Compared with SYAB0 and SYAB4, the EDFS spectrum of Yb-AG is narrower, indicating that the local structure of Yb^{3+} in Yb-AG glass is simpler. HYSCORE tests show that the next nearest neighbor of Yb^{3+} in Yb-AG glass is basically Al atoms, as shown in Fig. 2.25c. Figure 2.39b–d are the HYSCORE spectra of SYAB0, SYAB4, and Yb-BG samples. At a magnetic field of 450 mT, the resonance peaks at 2.06 MHz, 3.81 MHz, 5.0 MHz, and 6.15 MHz correspond to the Larmor frequencies of magnetic nuclei ^{10}B (natural abundance = 19.9%, nuclear spin I = 3), ^{29}Si, ^{27}Al, and ^{11}B (natural abundance = 80.1%, I = 3/2). It can be seen that in the Yb-BG

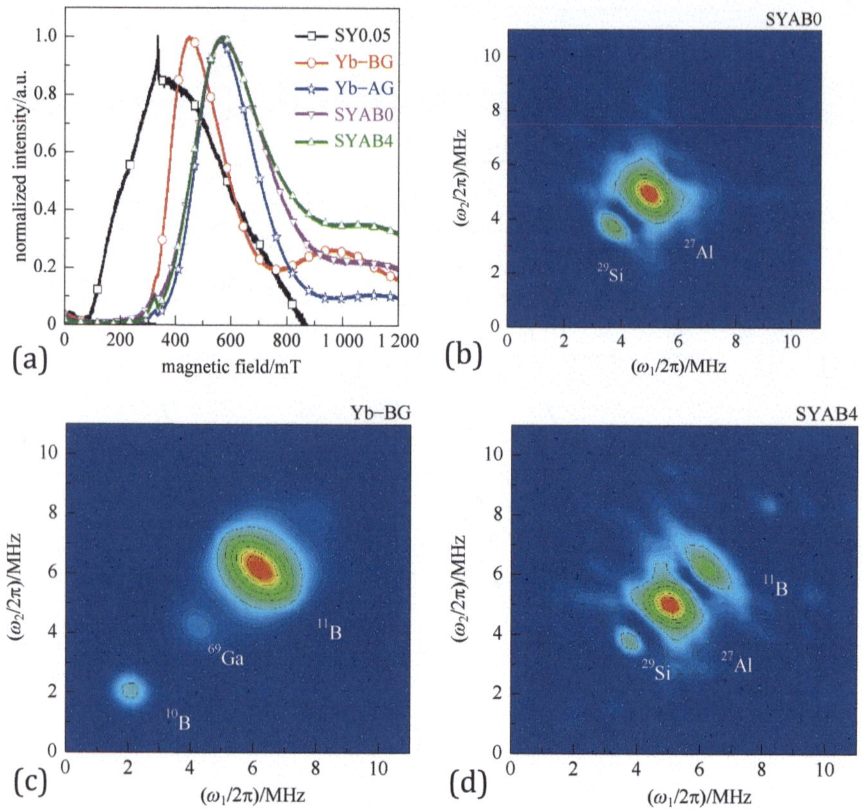

Fig. 2.39 Characterization of the local structure of Yb^{3+}: EDFS spectrum (**a**), HYSCORE spectrum (**b**–**d**)

sample, the next nearest neighbor of Yb^{3+} is mainly B atoms. In SYAB0 and SYAB4 glasses, the next nearest neighbor of Yb^{3+} is mainly Al atoms, which is indeed similar to the local structure of that in the Yb-AG glass. Compared with the SYAB0 sample, only a small amount of ^{11}B signal can be detected in the SYAB4 sample. This indicates that the addition of B_2O_3 does not form a a solvent shell structure around Yb^{3+}. In the SYAB4 sample, the ^{10}B signal was not detected, the reason is that the natural abundance of magnetic nucleus ^{10}B (~19.9%) is too low.

According to the above Raman, FTIR, NMR, and pulse EPR research results, Fig. 2.40a, b give the structural diagrams of the SYAB0 and SYAB8 glass samples, respectively. Similar to F doping, B doping will also have a "chemical annealing" effect. As the B_2O_3 content increases, the surrounding ring structure units of Yb^{3+} may change from 3 ~ 9 rings to 6 ~ 8 rings, causing the local crystal field strength and the degree of Yb^{3+} asymmetry decrease.

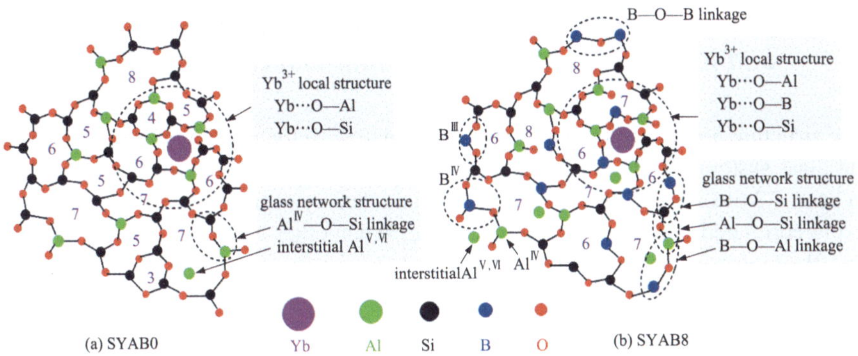

Fig. 2.40 Schematic diagram of the network structure and the local structure of Yb^{3+} of Yb/Al/B co-doped silica glass as functions of boron content (The numbers in the figure represent the number of cations in the ring structure)

2.3 Relationship Among the Thermal History, Structure, and Properties of Ytterbium-Doped Silica Glass

Glass is a metastable substance, the structure of which is between that of the melt and that of the crystal. When the melt is quenched into glass, some structural features of the melt are preserved, causing the refractive index, density, and thermal expansion coefficient of the glass to be closely related to the thermal history [62–74]. At the same time, the luminescent performance of rare earth ions mainly depends on their electric dipole transition probability, which is positively correlated with the asymmetry of the crystal field. Changes in thermal history cause changes in the network structure, which inevitably affects the asymmetry of the crystal field of rare earth ions and their luminescent performance [75]. Ytterbium-doped silica fiber is the core gain medium of fiber lasers, and the core material of which is ytterbium-doped silica glass. The refractive index and spectral properties of the core material directly affect the numerical aperture of the inner cladding and the spectral parameters of the fiber, thereby causing changes in the beam quality and laser performance of the fiber laser. Therefore, it is particularly important to realize the controllable design of the performance of the fiber laser by analyzing the influence mechanism of thermal history on the refractive index and spectral properties in ytterbium-doped silica glass. This section mainly introduces the influence of thermal history on the refractive index and spectral properties of Yb^{3+}-doped silica glass of different compositions. Then various structural analysis methods such as NMR and Raman are used to analyze its influence mechanism from the atomic microscale.

2.3.1 *The Influence of Thermal History on the Structure and Refractive Index of Ytterbium-Doped Silica Glass*

The structure and properties of glass will change with the thermal histories. different glass systems have various dependencies on thermal history [76–83]. As stated in Sect. 1.2 of Chap. 1, the refractive index and density of pure silica glass are positively correlated with the fictive temperature. When other elements such as Al, F, P, and B are doped into the silica glass, the dependency of the refractive index on thermal history will change. During the preparation process of silica fibers, a series of thermal processes are experienced, leading to changes in their thermal history, which in turn affects the refractive index (n) of the fiber core and cladding layer. As shown in Fig. 2.41a–c, the core-cladding refractive index difference (Δn) is reduced by 5×10^{-4} and 1×10^{-4} for Yb/Al and Yb/Al/F co-doped silica fibers after fiber drawing, respectively. However, for Yb/Al/P co-doped silica fiber (where P/Al = 1), the core-cladding refractive index difference is increased by 5×10^{-4} after fiber drawing. It can be seen that the dependency of the refractive index on thermal

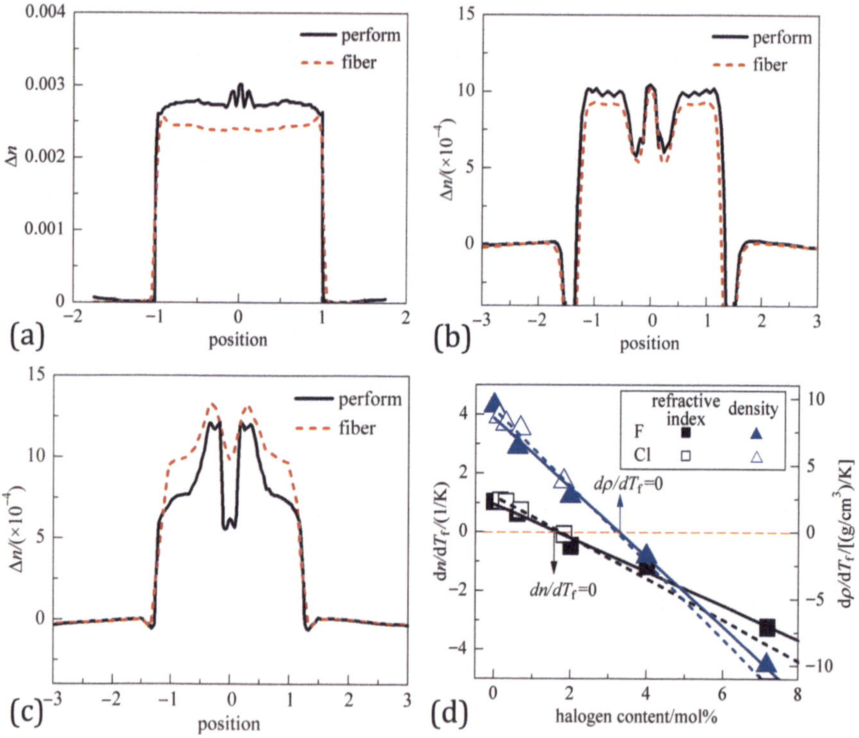

Fig. 2.41 The refractive index of Yb/Al [86] (**a**), Yb/Al/F (**b**) and Yb/Al/P (**c**) [87] co-doped silica fiber and preform, as well as the relationship among halogen doping amount in pure silica glass, dn/dT_f and dρ/dT_f (**d**) [85]

history is closely related to the composition in the doped silica glass. Therefore, the dependency of the refractive index and density (ρ) on the fictive temperature can be adjusted by altering the glass composition. Figure 2.41d illustrates the relationship among the halogen content, dn/dT_f and $d\rho/dT_f$, where T_f represents the fictive temperature. For pure silica glass, both dn/dT_f and $d\rho/dT_f$ are greater than 0. When a certain amount of halogen is doped, dn/dT_f and $d\rho/dT_f$ are equal to 0, which can eliminate the dependency of the refractive index and density of silica glass on thermal history [84, 85]. This section focuses on the impact of thermal history on the structure and refractive index of Yb^{3+} singly doped, Yb/Al, Yb/Al/B, Yb/Al/P, and Yb/Al/P/F co-doped silica glasses. It should be noted that when using the sol-gel method combined with high-temperature sintering technology to prepare Yb^{3+} single-doped, Yb/Al, Yb/Al/B and Yb/Al/P co-doped silica glasses, the T_f of the pristine glass is high (>1300 °C) due to the rapid cooling rate. When the glass is annealed at a certain temperature, the T_f of the glass will gradually change until it is equal to the annealing temperature.

2.3.1.1 The Influence Mechanism of Thermal History on the Refractive Index of Ytterbium Single-Doped Silica Glass

Figure 2.42 summarizes the effects of annealing temperature on the network structure, refractive index, and density of Yb^{3+} single-doped silica glass, with a glass composition of $0.05Yb_2O_3$–$99.95SiO_2$ (mol%).

Figure 2.42a shows the effect of annealing temperature on the FTIR spectrum of Yb^{3+} single-doped silica glass. In order to accurately obtain the infrared absorption peak (Gaussian peak III) at 2260 cm^{-1}, the infrared spectrum in Fig. 2.42a needs to be processed. The specific peak separation method is detailed in Figure 2.34d. The peak position and full width at half maxima (FWHM) of Gaussian peak III are shown in Table 2.11. As can be seen from Fig. 2.42b, the 2260 cm^{-1} absorption band moves to a higher wave number as the annealing temperature decreases. This indicates that the average Si—O—Si bond angle in the glass increases as the annealing temperature decreases. The FWHM of the 2260 cm^{-1} absorption band decreases as the annealing temperature decreases, indicating that the distribution range of Si—O—Si bond angles in the glass gradually narrows.

Figure 2.42c shows the Raman spectra of Yb^{3+}-single-doped silica glass annealed at 1200 °C and 900 °C. The Raman peaks at 483 cm^{-1} and 603 cm^{-1} correspond to the Si—O—Si planar 4-membered ring (D_1) and Si—O—Si planar 3-membered ring (D_2) [88]. As the annealing temperature decreases, the intensities of the Raman vibrational peaks of D_1 and D_2 decrease. This indicates that the content of planar 4-membered and 3-membered ring structures decreases as the annealing temperature decreases. It has been reported that the Si—O—Si bond angles in D_1 and D_2 deviate from the average bond angle of 145°, being 160.5° and 130.5°, respectively. Therefore, D_1 and D_2 are thermodynamically unstable, with their strain energies being 0.02 eV and 0.26 eV [89, 90]. The Si—O—Si bonds of the planar 3-membered ring will be broken and reorganize them to form larger rings, then distorts the

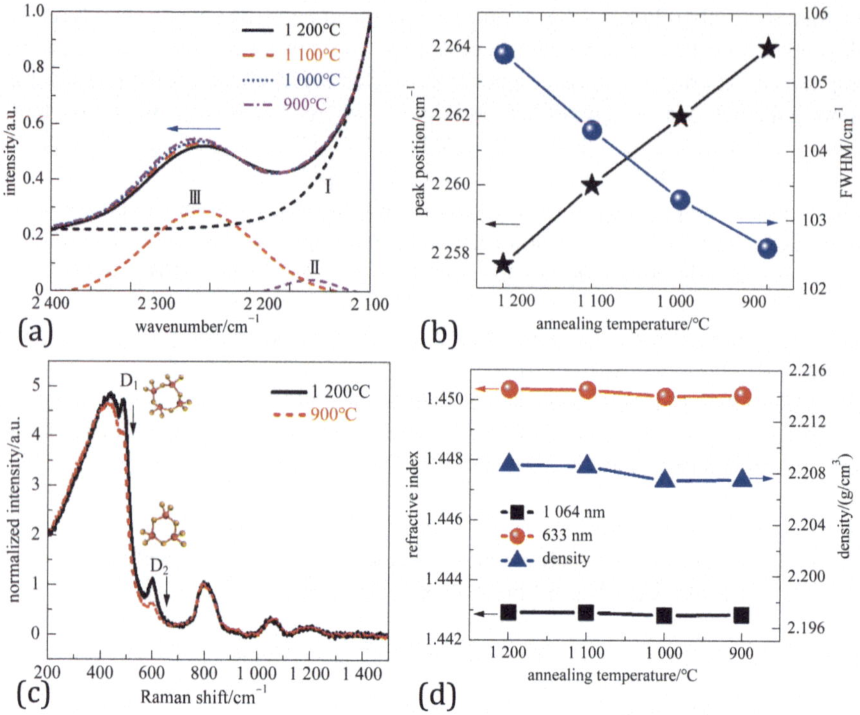

Fig. 2.42 The effect of annealing temperature on FTIR (**a**), the peak position and its full width at half maxima of the 2260 cm^{-1} Gaussian peak (**b**), Raman spectrum (**c**), refractive index and density of the Yb^{3+} single-doped silica glass (**d**) [34]

Table 2.11 The effect of annealing temperature on the peak position of the infrared absorption peak at 2260 cm^{-1}, its full width at half maxima and the refractive index and density at 633 nm and 1064 nm in Yb^{3+} single-doped silica glass [34]

Annealing Temperature /°C	2260 cm^{-1} Gaussian Peak Peak Position/ cm^{-1}	2260 cm^{-1} Gaussian Peak Full Width at Half Maxima, FWHM/ cm^{-1}	Refractive Index at 633 nm	Refractive Index at 1064 nm	Density/ (g/cm^3)
1200	2258	105.4	1.45035	1.44291	2.2086
1100	2260	104.3	1.45033	1.44291	2.2085
1000	2262	103.3	1.45012	1.44282	2.2074
900	2264	102.6	1.45018	1.44286	2.2075

ring plane by annealing. The planar 4-membered ring will directly distort the ring plane to reduce excess energy [91]. Compared with the planar 4-membered ring ($\theta_{<Si—O—Si>} = 160.5°$), the content of the planar 3-membered ring ($\theta_{<Si—O—Si>} = 130.5°$) decreases more. Therefore, under the annealing temperature of 900 °C, the average

Si—O—Si bond angle in the Yb^{3+} single-doped silica glass become larger, and the distribution of it becomes relatively narrowed.

Based on the impact of the fictive temperature on the structure of pure silica glass as described in Sect. 1.2 of Chap. 1, it can be inferred that when the annealing temperature decreases, in addition to the increase in the Si—O—Si bond angle of the two-bridge connected silicate tetrahedra, their O—O—O bond angles and the Si—Si distances in Yb^{3+} single-doped silica glass become larger. These will increase the looseness of the glass structure, thereby leading to a decrease in the density and refractive index of Yb^{3+} single-doped silica glass, as shown in Fig. 2.42d.

2.3.1.2 Mechanism of the Influence of Thermal History on the Refractive Index of Yb-Al and Yb-Al-B Co-Doped Silica Glasses

Guo et al. [92] used low-temperature annealing combined with air quenching to change the thermal history of the glass. Based on FTIR, Raman and NMR, they systematically studied the impact of thermal history on the network structure of Yb/Al(YbAP0), Yb/Al/B(SYAB8) and Y/Al/B(SYAB8-Y_2O_3) co-doped silica glasses, then analyzed the mechanism of refractive index changes. The detailed composition of the glass is shown in Table 2.12.

Figure 2.43a shows the effect of annealing time at 900 °C on the peak position of the infrared absorption peak at 2260 cm^{-1} and its full width at half maxima of YbAP0 glass. The specific peak division method of the infrared spectrum is detailed in Fig. 2.34d. Annealing causes a significant blue shift and a narrowing of the infrared absorption peak of YbAP0 glass, indicating that the average bond angle of Si—O—Si increases and the structural disorder decreases. Moreover, as the annealing time increases, the wavenumber and FWHM of the infrared absorption peak tend to be stable.

Figure 2.43b shows the effect of annealing time at 900 °C on the Raman spectrum of YbAP0 glass. The Raman peaks at 483 cm^{-1} and 603 cm^{-1} correspond to the Si—O—Si planar 4-membered ring (D_1) and Si—O—Si planar 3-membered ring (D_2), respectively [88]. Table 2.13 summarizes the relationship between the peak area of D_2 of YbAP0 glass and the annealing time. The peak areas of D_1 and D_2 decrease after thermal annealing. As the annealing time increases, the decreasing trend of the peak area of D_2 slows down, and the peak area gradually stabilizes.

Figure 2.43c shows the ^{27}Al MAS NMR spectra of the pristine and 900 °C annealed for 500 h YbAP0 glasses. The signals at 45 ppm, 20 ppm, and 0 ppm correspond to four-coordinated Al(Al^{IV}), five-coordinated Al(Al^{V}), and six-coordinated

Table 2.12 Composition of YbAP0, SYAB8 and SYAB8-Y_2O_3 glasses [92, 93] Unit: mol%

Sample	Yb_2O_3	Al_2O_3	SiO_2	B_2O_3	Y_2O_3
YbAP0	0.1	4	95.9		
SYAB8	0.1	6	85.9	8	
SYAB8-Y_2O_3		6	85.9	8	0.1

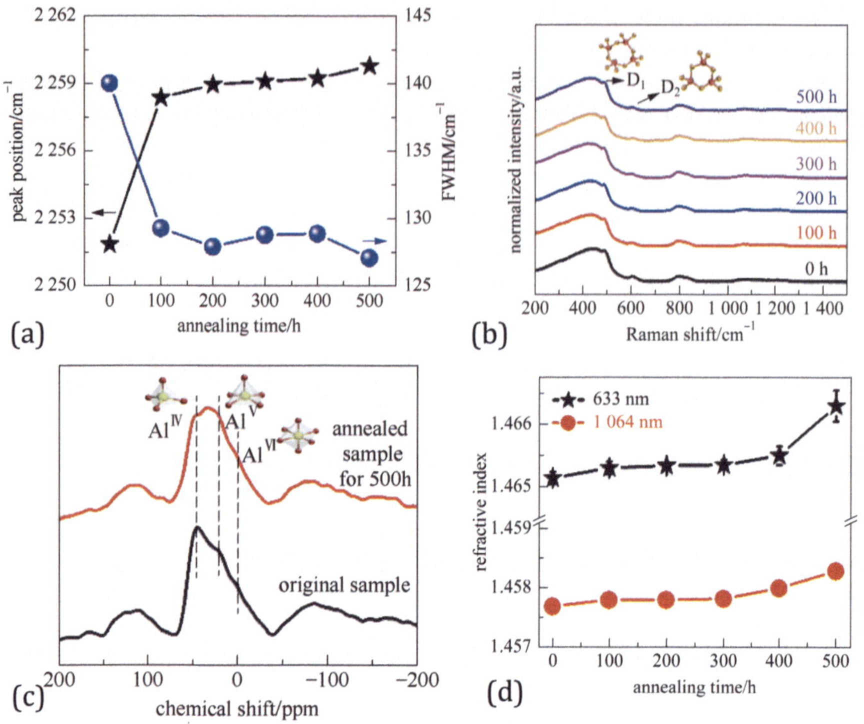

Fig. 2.43 The relationship among the annealing time at 900 °C, the peak position and full width at half maxima of 2260 cm^{-1} infrared absorption peak (**a**), Raman spectrum (**b**), ^{27}Al MAS NMR spectrum (**c**) and refractive index (**d**) of the YbAP0 glass [92]

Table 2.13 The relationship among the peak position and full width at half maxima of the 2260 cm^{-1} infrared absorption peak, the area of the 603 cm^{-1} Raman peak (D_2), the refractive index of the YbAP0 glass and the annealing time at 900 °C [92]

Annealing Time /h	2260 cm^{-1} Gaussian Peak Peak Position/cm^{-1}	2260 cm^{-1} Gaussian Peak Full Width at Half Maxima, FWHM/cm^{-1}	D_2 Peak Area	Refractive Index at 633 nm	1064 nm Refractive Index at 1064 nm
0	2251.9	140	16.7	1.46514	1.4577
100	2258.4	129.3	8.6	1.4653	1.45781
200	2258.9	127.9	8.1	1.46534	1.45781
300	2259.1	128.8	9.8	1.46535	1.45783
400	2259.2	127.1	7.1	1.4655	1.45801
500	2259.8	127.1	8.7	1.4663	1.4583

Al(Al^{VI}), respectively [36, 94]. In the pristine YbAP0 glass, Al mainly exists as Al^{IV}. After annealing, the Al spectrum becomes broad, and Al^{IV} transforms into Al^{VI}. High-temperature ^{27}Al MAS NMR results show that Al mainly exists as Al^{V} and Al^{VI} in the melt or rapidly cooled glass [95–98].

Figure 2.43d shows the effect of annealing time at 900 °C on the refractive index of YbAP0 glass. As the annealing time increases, the refractive index of YbAP0 glass increases. This phenomenon is contrary to that of Yb^{3+} single- doped and pure silica glasses [75]. In addition, when the annealing time is less than 400 h, the density of YbAP0 glass is about 2.24 g/cm^3. When the annealing time is more than 400 h, the density increases to 2.25 g/cm^3, which is consistent with the report by Pauli et al. [86]. As shown in Fig. 2.43a, b, as the annealing time increases, the Si—O—Si bond angle of the two-bridge connected silicate tetrahedron in the YbAP0 glass increases. That is, the looseness of the Si—O—Si glass network structure increases, which will in turn cause the decrease of density and refractive index of the glass. It is reported that Al^{VI} fills in the gaps in the silica glass network, increasing the density and refractive index of the glass [14]. During the annealing process of YbAP0 glass, the transformation from Al^{IV} to Al^{VI} is an important reason for the increase in the refractive index and density of YbAP0 glass.

Figure 2.44a, b are the ^{27}Al MAS NMR spectra of SYAB8 and SYAB8-Y_2O_3 glasses before and after annealing at 800 °C. The signal peaks at chemical shifts of 45 ppm, 25 ppm, and 0 ppm are attributed to Al^{IV}, Al^{V}, and Al^{VI}, respectively. Similar to YbAP0 glass, annealing leads to the transformation from Al^{IV} to Al^{VI}.

Figure 2.44c, d are the ^{11}B MAS NMR spectra of SYAB8 and SYAB8-Y_2O_3 glasses before and after annealing at 800 °C. As decribed in Sect. 2.2.4.4, in the pristine YbAB80 glass, most B exists in the form of B^{III}, and a small amount of B exists in the form of B^{IV}. Annealing at 800 °C has no significant effect on the ^{11}B MAS NMR spectrum, that is, the coordination number of B will not change significantly under this annealing condition. However, in borosilicate or borosilicate glass, the coordination number of B is affected by the fictive temperature. As the fictive temperature increases, B^{IV} transforms into B^{III} and non-bridging oxygen (NBO^-), that is, $BO_4^- \rightarrow BO_3 + NBO^-$ [99–101].

Figure 2.45a shows the effect of annealing time at 800 °C on the peak position and full width at half maxima of the infrared absorption peak at 2260 cm^{-1} in the SYAB8 glass, and the specific data is shown in Table 2.14. The specific peak division method of the infrared spectrum is detailed in Fig. 2.34d. With the increase of annealing time, the infrared absorption peak of SYAB8 glass shows a significant blue shift and narrowing. Annealing leads to the increase of the average bond angle of Si—O—Si and decrease of the structural disorder.

Figure 2.45b, c show the effects of annealing time at 800 °C on the FTIR and Raman spectra of SYAB8 glass, respectively. After annealing at 800 °C, there is no significant change in the FTIR and Raman spectra of SYAB8 glass.

Figure 2.45d shows the relationship among the refractive index, density of SYAB8 glass at 1064 nm and the annealing time at 800 °C, and the specific values are shown in Table 2.14. As the annealing time increases, the density and refractive index of SYAB8 glass increase. This result is consistent with those in the Yb/Al

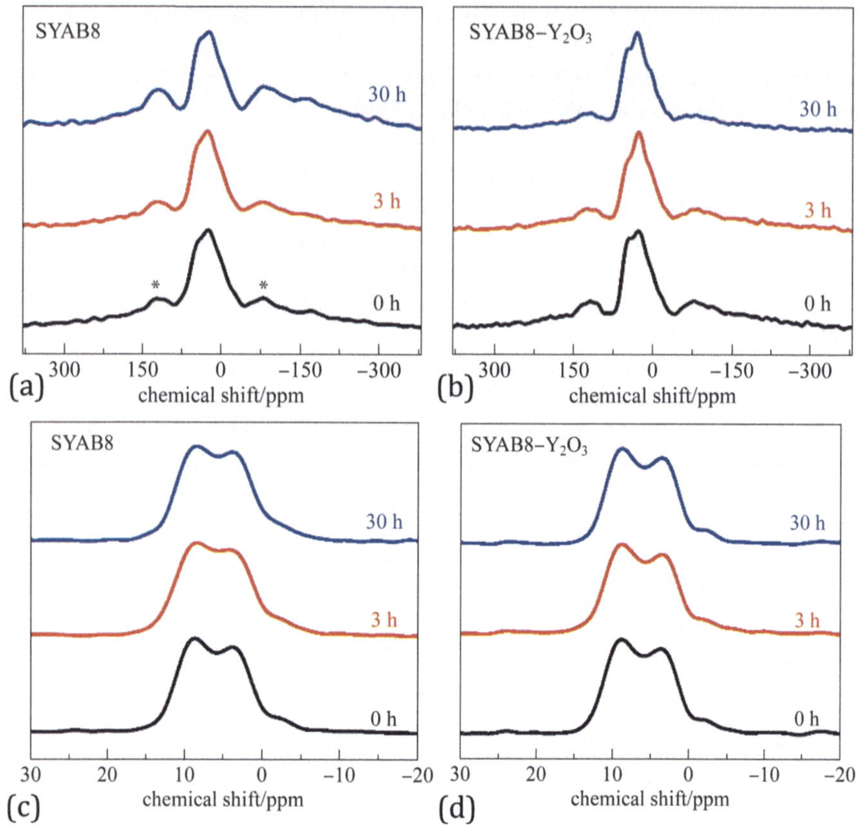

Fig. 2.44 The ^{27}Al (**a**) and ^{11}B MAS NMR(**c**) spectra of SYAB8 glass before and after annealing at 800 °C, ^{27}Al (**b**) and ^{11}B (**d**) MAS NMR spectra [93] of SYAB8-Y_2O_3 glass before and after annealing at 800 °C

co-doped silica glass. As shown in Fig. 2.45a–c, as the annealing time increases, the Si—O—Si bond angle of the two connected silica tetrahedron in SYAB8 glass increases. That is, the looseness of the Si—O—Si glass network structure increases, which will in turn lead to the decrease of the density and refractive index of the glass. It should be noted that the Raman vibration peaks intensities of the Si—O—Si planar 3-membered and 4-membered rings in the pristine SYAB8 glass are weak, so it is difficult to monitor the effect of annealing at 800 °C on them through Raman spectroscopy. As shown in Fig. 2.44, annealing at 800 °C will cause a transformation from Al^{IV} into Al^{VI}. Similar to that in the Yb/Al co-doped silica glass, the change in Al coordination number after annealing is an important reason for the increase in refractive index and density in the Yb/Al/B co-doped silica glass.

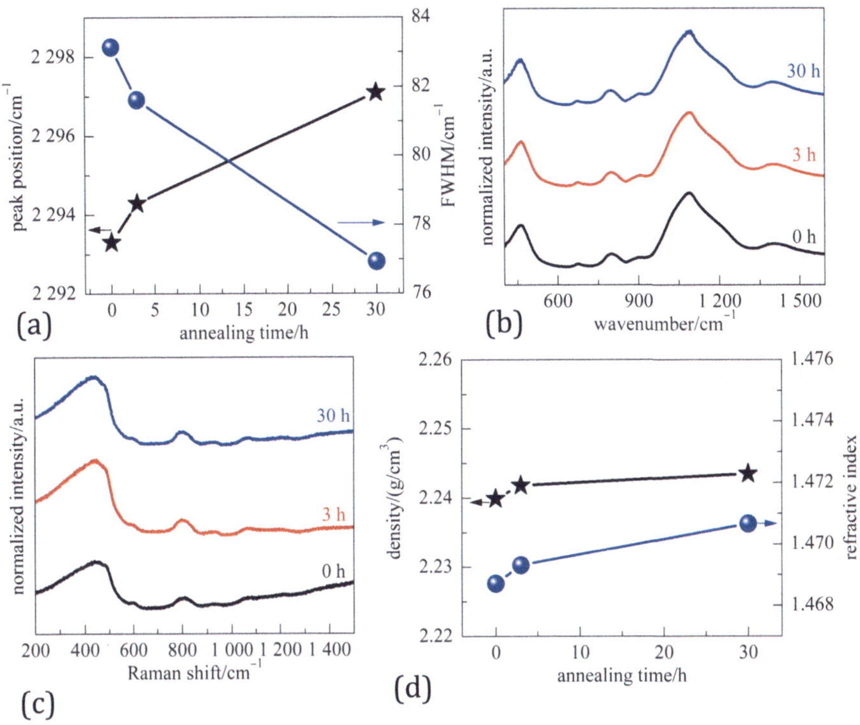

Fig. 2.45 The relationship among the annealing time at 800 °C, the peak position and full width at half maxima of 2260 cm^{-1} infrared absorption peak (**a**), infrared (**b**) and Raman (**c**) spectra, density and refractive index (**d**) of the SYAB8 glass [93]

Table 2.14 The relationship among the peak position and full width at half maxima of 2260 cm^{-1} infrared absorption peak, refractive index at 1064 nm, density and annealing time at 800 °C of the SYAB8 glass [93]

Annealing Time /h	2260 cm^{-1} Gaussian Peak Peak Position/cm^{-1}	2260 cm^{-1} Gaussian Peak Full Width at Half Maxima, FWHM/cm^{-1}	Refractive Index at 1064 nm	Density/(g/cm^3)
0	2293	83.1	1.46871	2.2399
3	2294	81.6	1.46931	2.2418
30	2297	76.9	1.47067	2.2435

2.3.1.3 The Mechanism of the Influence of Thermal History on the Refractive Index of Yb/Al/P Co-Doped Silica Glass

Guo et al. [102] used the annealing method in an argon atmosphere to change the thermal history of Yb/Al/P co-doped silica glass. They systematically studied the effect of annealing time at 850 °C on the glass network structure, and analyzed the mechanism of refractive index change caused by thermal history change. The glass composition is $0.1Yb_2O_3$-$4Al_2O_3$–$6P_2O_5$–$89.9SiO_2$ (mol%, YbAP1.5).

Figure 2.46a shows the relationship among the annealing time at 850 °C, the peak position and full width at half maxima at 2260 cm^{-1} infrared absorption peak of YbAP1.5 glass. The specific values are shown in Table 2.15. With the increase of annealing time at 850 °C, the infrared absorption peak near 2260 cm^{-1} of YbAP1.5 glass narrows and has a blue shift, which is consistent with those in the Yb^{3+} single-doped, Yb/Al and Yb/Al/B co-doped silica glasses. This indicates that annealing increases the average bond angle of Si—O—Si and reduces the structural disorder.

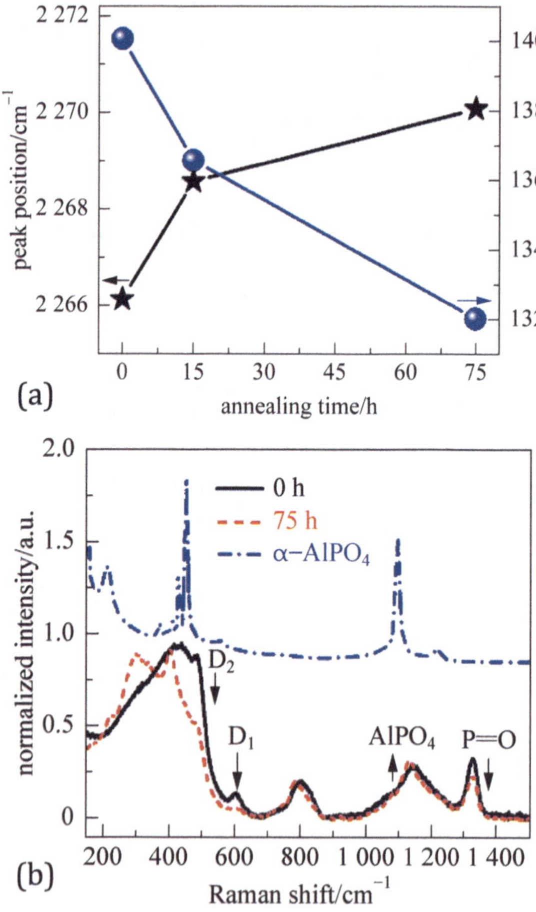

Fig. 2.46 The peak position and full width at half maxima of the infrared absorption peak at 2260 cm^{-1} (**a**) and Raman spectrum (**b**) of YbAP1.5 glass as functions of annealing time [102]

Table 2.15 The relationship between the peak position and full width at half maxima of the infrared absorption peak at 2260 cm^{-1}, the refractive index at 1064 nm, density of YbAP1.5 glass as function of the annealing time at 850 °C [102]

Annealing Time /h	Peak Position of the 2260 cm^{-1} Gaussian peak /cm^{-1}	FWHM of 2260 cm^{-1} Gaussian Peak /cm^{-1}	Refractive index at 633 nm	Density/ (g/cm^3)
0 h	2266	140.1	1.4568	2.2508
15 h	2269	136.6	1.4561	2.2459
75 h	2270	132	1.45575	2.2432

Figure 2.46b shows the Raman spectra of YbAP1.5 glass as function of annealing time. In the low-frequency range, the Raman peak of Yb/Al/P co-doped silica glass is similar to those of Yb^{3+} single-doped and Yb/Al co-doped silica glasses, but their Raman peaks in the range of 1000 ~ 1350 cm^{-1} are different. Among them, the wide peak at 1000 ~ 1200 cm^{-1} mainly originates from the stretching vibration of Al—O—P, and the Raman vibration peak at 1330 cm^{-1} originates from the stretching vibration of P = O [103, 104]. After annealing, the Raman peak intensity at 483 cm^{-1} and 603 cm^{-1} of YbAP1.5 glass significantly weakens, that is, D_1 and D_2 decrease. This phenomenon is consistent with those of Yb^{3+} single-doped and Yb/Al co-doped silica glasses. In addition, the intensity of the broad peak in the range of 1000 ~ 1200 cm^{-1} of YbAP1.5 glass increases, while that at 1330 cm^{-1} decreases. After annealing for 75 h, sharp peaks appear near 200 cm^{-1} and 300 cm^{-1}. Among them, the 200 cm^{-1} peak can be observed in the Raman spectrum of α-$AlPO_4$ crystal, and the 300 cm^{-1} peak is attributed to the bending vibration of the Al—O—P bond [105, 106]. In YbAP1.5 glass, annealing will lead to the decrement in the P = O content, the increase of Al—O—P connections, and even generating $AlPO_4$ crystals.

Figure 2.47a shows the ^{29}Si MAS NMR spectra of the pristine and annealed YbAP1.5 glasses for 75 h. The signal at −110 ppm is attributed to the Q^4 group [33]. In YbAP1.5 glass, Si mainly exists in the form of Q^4 groups, and its glass structure is similar to that of SiO_2 glass. After annealing, this signal peak becomes significantly narrower, indicating that the glass silicon-oxygen network structure becomes highly ordered.

Figure 2.47b shows the ^{27}Al MAS NMR spectra of the pristine and annealed YbAP1.5 glasses for 75 h. The signal peaks at 42.0 ppm, 7.4 ppm, and − 15.6 ppm belong to Al^{IV}, Al^{V}, and Al^{VI} [36, 107]. Annealing leads to a significant narrow of the ^{27}Al MAS spectrum, as well as a transformation from Al^{V} and Al^{VI} to Al^{IV}. This trend is opposite to those of Yb/Al and Yb/Al/B co-doped silica glasses. At the same time, the ^{27}Al MAS NMR spectrum of YbAP1.5 glass shows a spinning sideband at a chemical shift of -53 ppm after annealing, indicating that the paramagnetism around Al is enhanced after annealing.

Figure 2.47c shows the ^{31}P static NMR spectra of the pristine and annealed YbAP1.5 glass for 75 h. Annealing leads to a significant narrow of the ^{31}P NMR spectrum. Table 2.16 summarizes the changes in the $P^{(n)}$ unit. As can be seen from Fig. 2.47c, the ^{31}P NMR spectrum of the pristine YbAP1.5 glass can be divided into

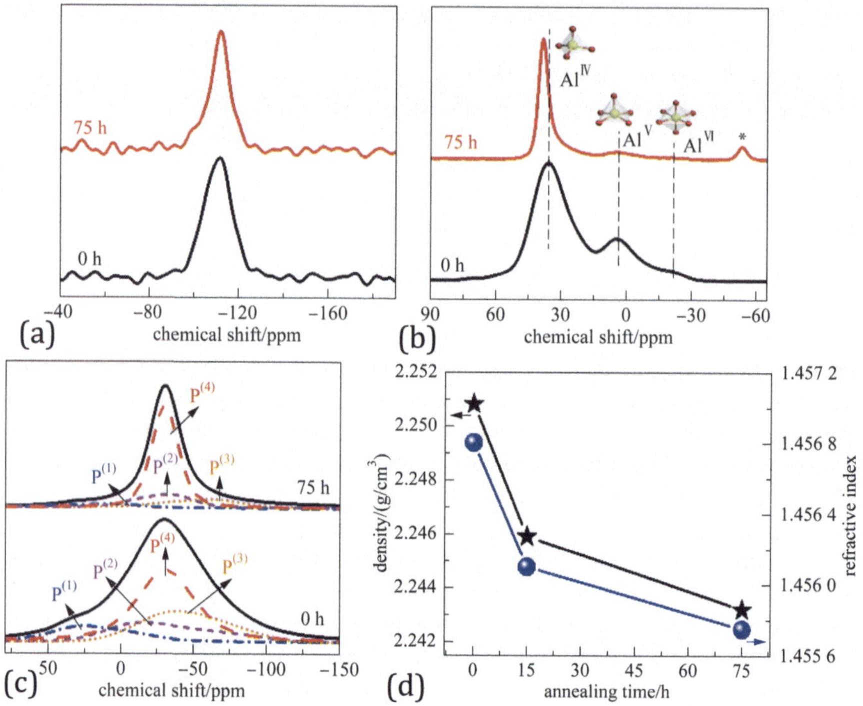

Fig. 2.47 The relationship among ^{29}Si (**a**), ^{27}Al (**b**) and ^{31}P (**c**) nuclear magnetic resonance, the density, the refractive index and annealing time in the pristine and 850 °C annealed YbAP1.5 glasses for 75 h (**d**) [102]

Table 2.16 Peak splitting parameters of ^{31}P static spectrum of YbAP1.5 glass before and after annealing [102]

Annealing time	Parameter	$P^{(1)}$	$P^{(2)}$	$P^{(3)}$	$P^{(4)}$
0 h	Chemical shift/(ppm, ±0.5)	18.0	−33.0	−45.8	−30.8
	Content/(%, ±3%)	12.0	17.8	24.9	45.4
75 h	Chemical shift/(ppm, ±0.5)	15.5	−28.4	−55.0	−30.2
	Content/(%, ±3%)	4.8	16.5	14.6	64.1

four peaks. The Gaussian peak at −30.8 ppm is attributed to the $P^{(4)}$ unit, which perfers to interact with Al^{IV} to form [$AlPO_4$] group [108, 109]. The axially symmetric signal at −45.8 ppm is attributed to the $P^{(3)}$ unit, such as O = $P(OSi)_3$ [110]. The remaining two non-axially symmetric signals are originated from $P^{(1)}$ and $P^{(2)}$ units, which tend to bond with the high-coordinated Al [107, 108]. As can be seen from Table 2.16, after annealing, the content of $P^{(4)}$ increases significantly, while the content of $P^{(1)}$ and $P^{(3)}$ decreases significantly. At the same time, annealing also leads to a increment in the content of Al^{IV}, and a decrement in the content of high-coordinated Al, as shown in Fig. 2.47b. The $P^{(4)}$ structure tends to bond with Al^{IV}, so it is speculated that annealing leads to a increment in the content of [$AlPO_4$] groups, and even form the nanoscale aggregation.

Table 2.17 Refractive index at 1064 nm of Yb/Al/P co-doped silica glass with a P/Al = 1 before and after annealing at 1000 °C [93]

			Refractive index at 1064 nm		Difference in refractive index
Sample	x value	P/Al	Pristine sample	Annealed sample	Annealed sample—Pristine sample
YbAP1–3	3	0.94	1.45142	1.45122	-2×10^{-4}
YbAP1–4	4	1.02	1.45125	1.45094	-3.1×10^{-4}
YbAP1–5	5	0.97	1.45124	1.45070	-5.4×10^{-4}

Figure 2.47d shows the effect of annealing time on the density and refractive index of YbAP1.5 glass. It is found that the density and refractive index of YbAP1.5 glass decrease after annealing. The change trend is opposite to those of Yb/Al and Yb/Al/B co-doped silica glasses, as shown in the Sect. 2.3.1.2 of this chapter. It can be seen that the doping of P in Yb/Al co-doped silica glass changes the dependence of the glass refractive index on the thermal history. On the one hand, based on the results of FTIR, Raman and ^{29}Si MAS NMR, annealing treatment improves the symmetry of the silicon-oxygen network structure, and increases the average bond angle of Si—O—Si. That is, the looseness of the Si—O—Si glass network structure increases, which will result in the decrement in the density and refractive index of the glass. On the other hand, according to the ^{27}Al MAS and ^{31}P static NMR (Table 2.16) results, annealing treatment leads to the transformation from high-coordinated Al, $P^{(1)}$ and $P^{(3)}$ to Al^{IV} and $P^{(4)}$ structure, which tend to form [$AlPO_4$] structure in the YbAP1.5 glass. The [$AlPO_4$] structure unit is equivalent to 2 [$SiO_{4/2}$] structure units, and the contribution of 1 mol% [$AlPO_4$] to the refractive index of silica glass is -1.1×10^{-4} [18]. After annealing, the increase in the average bond angle of Si—O—Si and the formation of the [$AlPO_4$] structure lead to the decrease of the density and refractive index of YbAP1.5 glass.

The higher the P/Al ratio, the lower the absorption and emission cross-sections of Yb^{3+} [17, 36]. When P/Al = 1, the refractive index of Yb/Al/P co-doped silica glass is the lowest [17]. Therefore, Yb/Al/P co-doped silica glass with a P/Al ratio close to 1 is usually chosen as the core glass of large-mode-field fibers with low numerical aperture. Table 2.17 summarizes the refractive indexes at 1064 nm of Yb/Al/P co-doped silica glass with a P/Al = 1 before and after high-temperature annealing, where the corresponding glass component is $0.1Yb_2O_3$- $x\,Al_2O_3$- $x\,P_2O_5$ -(99.9–2 x)SiO_2 (mol%, x = 3, 4, 5). According to the x value, the glass is abbreviated as YbAP1- x. The P/Al value in the table is obtained by ICP-OES.

The larger the x value in the pristine glass, the higher the absolute content of $AlPO_4$, and the lower the refractive index of the glass. This result is consistent with the report by Wang et al. [18]. After annealing, the refractive index of YbAP1- x glass decreases. As the absolute content of $AlPO_4$ increases, the degree of reduction in the refractive index increases. When P/Al = 1, Al and P basically exist as Al^{IV} and $P^{(4)}_{(mAl,hSi)}$ ($m + h = 4$, $m \le 4$), which tend to connect with each other to form $AlPO_4$ -like structures [36]. It should be noted that in the $P^{(n)}_{(mAl,hSi)}$ structure, the superscript n refers to the number of bridge oxygens connected to the central atom P, and the subscripts m and h refer to the number of Al and Si bonded to each P atom respectively. The above refractive index changes may be related to the change of $P^{(4)}_{(mAl,hSi)}$ structural units.

2.3.1.4 The Mechanism of the Influence of Thermal History on the Refractive Index of Yb/Al/P/F Co-Doped Silica Glass

Guo et al. [111] used annealing and fire polishing methods to change the thermal history of the Yb/Al/P/F co-doped silica core glass and commercial silica cladding glass. Then they analyzed the mechanism of refractive index changes caused by changes in thermal history based on Raman and NMR structural analysis results. The core glass composition is $0.12Yb_2O_3$–$3.2Al_2O_3$–$2.7P_2O_5$–$93.98SiO_2$–F(mol%, YbAPF0.85), where P/Al = 0.85 (molar ratio), and F/Si = 1% (mass ratio).

Figure 2.48a shows the treating process of YbAPF0.85 glass with different thermal histories in detail. First, the YbAPF0.85 glass is cut into four pieces, named as pristine, annealed (An), oxyhydrogen fire heated (FireP) and oxyhydrogen fire heated-annealed (FireP-An) YbAPF0.85 glass according to their thermal history. Among them, the pristine YbAPF0.85 glass refers to the glass whose thermal history has not been changed. Figure 2.48b is the differential scanning calorimetry (DSC) curve of the pristine YbAPF0.85 glass, with a heating rate of 15 K/min. The glass transition temperature (T_g), glass crystallization temperature (T_{x1} and T_{x2}) of the pristine YbAPF0.85 glass are 800, 1148 and 1362 °C, respectively. Annealing at 1000 °C temperature lower than T_x can effectively prevent the crystallization of the

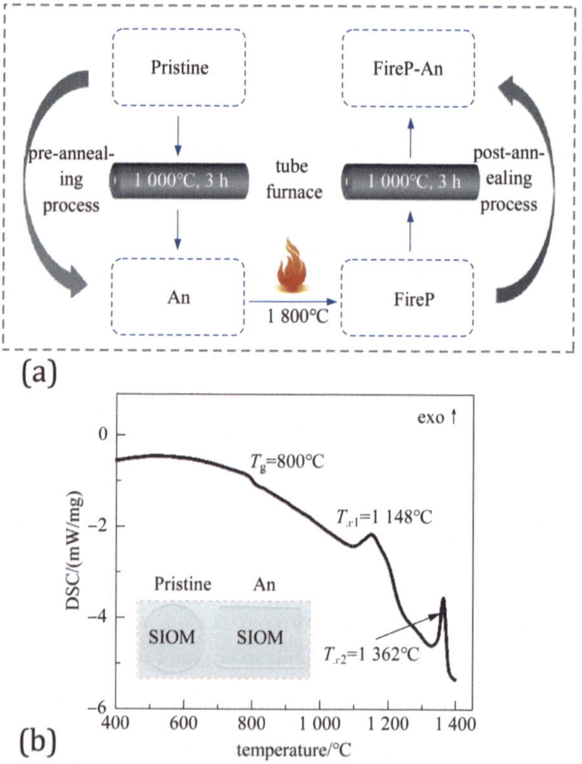

Fig. 2.48 The treating process of YbAPF0.85 and F300 glasses with different thermal histories. (**a**) The differential scanning calorimetry curve of pristine YbAPF0.85 glass (**b**). The inset in (**b**) shows pictures of the pristine and annealed YbAPF0.85 glasses [111]

glass. The pristine YbAPF0.85 glass was annealed for 3 h in a tube furnace heated to 1000 °C, with argon atmosphere protection. After annealing for a certain time, the sample was taken out at high temperature and quenched to room temperature to obtain the annealed YbAPF0.85 glass. The transparency of the annealed YbAPF0.85 glass is good, as shown in the inset of Fig. 2.48b. This annealing process is called the pre-annealing process. The annealed YbAPF0.85 glass was polished with a hydrogen-oxygen flame at about 1800 °C, then quenched to room temperature to simulate the fiber drawing process, and the fire-polished YbAPF0.85 glass was obtained. Finally, the fire-polished YbAPF0.85 glass was annealed, where the annealing condition is consistent with that in the pre-annealing process. This process is simply referred to as the post-annealing process. To monitor the changes in the numerical aperture (NA) of the fiber during different processes, Guo et al. [111] chose commercial pure silica glass (F300, Heraeus) as the cladding glass. Then, they used the same treatment to change the thermal history of F300 glass.

Figure 2.49a, b are the Raman spectra of YbAPF0.85 and F300 glasses with different thermal histories, respectively. All Raman spectra are normalized to the Raman peak intensity at 800 cm^{-1}. In F300 glass, the wide peaks at 440 cm^{-1} and 1000 ~ 1300 cm^{-1}, as well as the Raman peaks at 485 cm^{-1}, 604 cm^{-1} and 800 cm^{-1} both originate from the vibration of $[SiO_{4/2}]$ tetrahedra. Among them, the Raman peaks at 485 cm^{-1} and 604 cm^{-1} correspond to the Si-O-Si 4-membered ring (D_1)

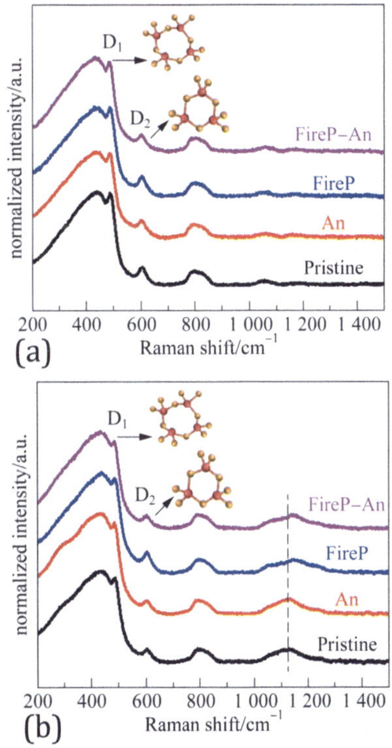

Fig. 2.49 Normalized Raman spectra of F300 (**a**) and YbAPF0.85 (**b**) glasses with different thermal histories [111]

Table 2.18 Peak area of D_2 in F300 and YbAPF0.85 glasses with different thermal histories, The Raman peak wavenumber in the range of 1000 ~ 1300 cm^{-1} in the YbAPF0.85 glass [111]

Sample	D_2 Peak Area		Wavenumber /cm^{-1}
	F300	YbAPF0.85	
Pristine	29.0	18.9	1128
An	24.9	17.3	1132
FireP	32.3	26.0	1145
FireP-an	22.6	17.0	1135

and plane 3-membered ring (D_2), respectively. The Raman spectrum of YbAPF0.85 glass is similar to that of F300 glass, but the Raman peaks at 300 cm^{-1} and 1000 ~ 1300 cm^{-1} are different from those of F300 glass. The Raman peak at 300 cm^{-1} is attributed to the bending vibration of the Al—O—P bond, and its peak intensity is only obvious when the content of this bond is high [105, 106]. The broad peak at 1000 ~ 1300 cm^{-1} originates from the stretching vibrations of Al—O—P, Al—O—Si, Si—O—Si and P—O—Si bonds [33, 103]. Compared with pure silica, the D_1 and D_2 peaks of YbAPF0.85 glass are weaker. The addition of Yb, Al, P and F leads to the breakage of Si—O—Si bonds and the formation of Si—O—Al, Si—O—P and Si—F bonds.

Table 2.18 summarizes the relationship between the peak area of the D_2 peak and the thermal history. After annealing, the peak areas of D_1 and D_2 in YbAPF0.85 and F300 glasses decrease, then increase after fire polishing, and decrease again after subsequent annealing. In addition to the D_1 and D_2 peaks, the Raman peaks at 300 cm^{-1} and the broad peaks at 1000 ~ 1300 cm^{-1} are also related to the thermal history. After annealing, the Raman peak at 300 cm^{-1} increases. This peak weakens after fire polishing, and even completely overlaps with the Si—O—Si bending vibration peak. The broad peak at 1000 ~ 1300 cm^{-1} show a blue shift. After fire polishing, the content of P—O—Si in the YbAPF0.85 glass is higher than that in the pristine and annealed samples.

Figure 2.50a shows the ^{19}F MAS NMR spectra of YbAPF0.85 glasses with different thermal histories, and that of the F-doped silica tube from Heraeus is also included for comparison. The ^{19}F MAS NMR spectra of all samples consist of a signal at about -145 ppm, which corresponds to the F—Si bond [49]. When the thermal history of the YbAPF0.85 glass is changed, the ^{19}F MAS NMR spectra of the samples show no obvious change. That is, when the F doping level is low, the coordination environment of F is independent of the thermal history.

Figure 2.50b shows the ^{27}Al MAS NMR spectra and peak splitting curves of YbAPF0.85 glasses with different thermal histories, where the fitting model is the Czjzek model. According to the isotropic chemical shift values of ^{27}Al with different coordination, the approximate contents of Al with different coordination in the glass can be obtained and listed in Table 2.19 [36, 107]. As can be seen from the table, the coordination number of Al is closely related to the thermal history. In the pristine YbAPF0.85 glass, Al mainly exists as Al^{IV}. After fire polishing, Al^{IV} transforms to Al^{V}. After annealing, Al^{V} transforms to Al^{IV}, and even the corresponding crystals are precipitated.

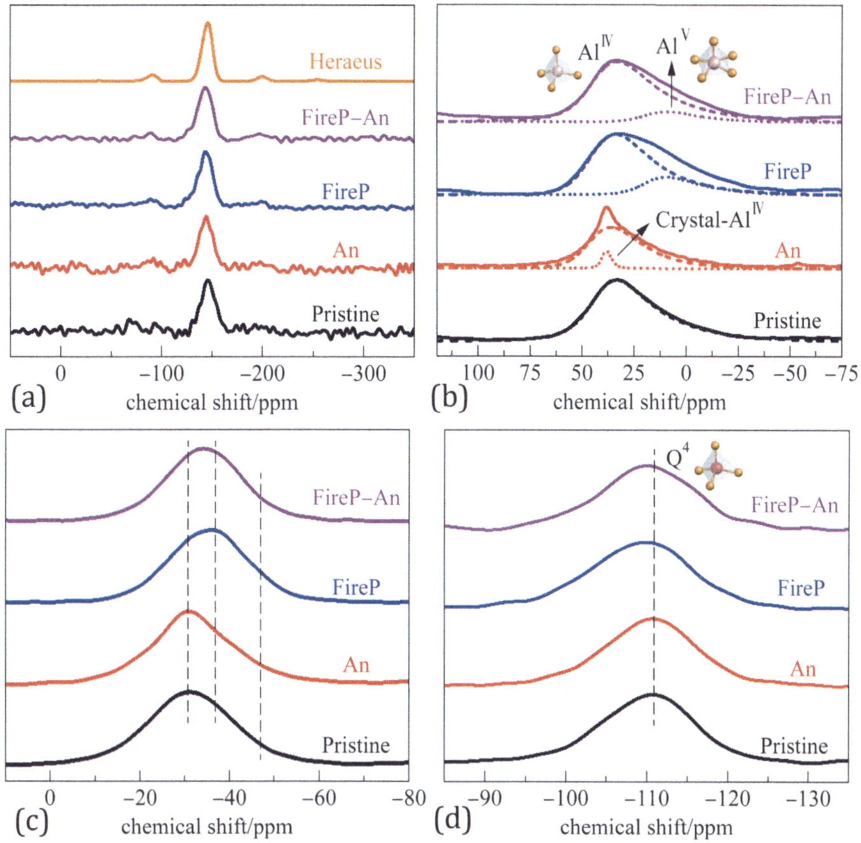

Fig. 2.50 ^{19}F (**a**), ^{27}Al (**b**), ^{31}P (**c**) and ^{29}Si (**d**) MAS NMR spectra of YbAPF0.85 glass with different thermal history [111]

Figure 2.50c shows the ^{31}P MAS NMR spectra of YbAPF0.85 glasses with different thermal histories. Changes in thermal history will significantly change the peak position and shape of the ^{31}P MAS NMR spectrum. ^{31}P MAS NMR spectrum of the pristine YbAPF0.85 glass is mainly composed of a relatively symmetrical signal located at about -30 ppm. The chemical shift of this signal is close to that of $P^{(4)}_{(4Al,0Si)}$ structure (−26 ppm), thus this signal can be attributed to the $P^{(4)}_{(4Al,0Si)}$ structure [108, 109]. It should be noted that the superscript n in $P^{(n)}_{(mAl,hSi)}$ represents the number of bridging oxygens, and the subscripts m and h represent the number of Al and Si bonded to each P atom, respectively. After annealing, the ^{31}P MAS NMR spectrum of YbAPF0.85 glass is significantly narrowed, which is consistent with that of ^{27}Al MAS NMR spectrum. Guo et al. [111] speculated that $AlPO_4$ nanocrystals precipitated or highly symmetrical $AlPO_4$ clusters formed in the glass after annealing. After fire polishing, the signal peak at -30 ppm in the sample decreases, and two signal peaks appear at about -36 ppm and -47 ppm, which are attributed to

$P^{(4)}_{(3Al,1Si)}$ and $P^{(4)}_{(2Al,2Si)}$ structures. After fire polishing, the P—O—Al bonds in the glass are broken to form P—O—Si and Al—O—Si bonds. After annealing the fire-polished YbAPF0.85 glass, its ^{31}P MAS NMR spectrum is transformed towards that of the pristine YbAPF0.85 glass.

Figure 2.50d shows the ^{29}Si MAS NMR spectra of YbAPF0.85 glass with different thermal histories. The ^{29}Si MAS NMR spectrum of the pristine YbAPF0.85 glass consists of a signal located at about -110 ppm, which is attributed to the Q^4 structure [33]. The ^{29}Si MAS NMR spectrum of the annealed sample shows no significant changes, proving that the precipitated nanocrystals are $AlPO_4$ crystals. After fire polishing, the signal moves towards the high frequency direction, possibly forming a small amount of Q^3_{1Al} structure, as can be seen in Table 2.19. That is, the Si—O—Si bonds in the fire-polished sample are destroyed and Si—O—Al bonds are formed, so that the fire-polished YbAPF0.85 glass contains more Al^V.

Figure 2.51 is the deconvolution result of ^{31}P MAS NMR spectra, and the specific parameters are shown in Table 2.20. In the pristine YbAPF0.85 glass, the signals at -14 ppm, −30 ppm, −39 ppm, and −47 ppm are respectively attributed to $P^{(3)}_{(2Al,1Si)}$, $P^{(4)}_{(4Al,0Si)}$, $P^{(4)}_{(3Al,1Si)}$ and $P^{(4)}_{(2Al,2Si)}$ structures. The relative content of the $P^{(4)}_{(4Al,0Si)}$ structure is closely related to the thermal history. After annealing, the relative content of the $P^{(4)}_{(4Al,0Si)}$ structure in the glass increases, even forming the corresponding crystals. At the same time, the relative content of $P^{(4)}_{(3Al,1Si)}$ and $P^{(4)}_{(2Al,2Si)}$ structures decreases. After fire polishing, the relative content of $P^{(4)}_{(3Al,1Si)}$ and $P^{(4)}_{(2Al,2Si)}$ structures increases, while that of $P^{(4)}_{(4Al,0Si)}$ structure decreases. After annealing the fire-polished samples, it is found that the relative content of the $P^{(4)}_{(4Al,0Si)}$ structure increases, and it evolves towards the structure of the pristine YbAPF0.85 glass.

Table 2.19 ^{27}Al average chemical shift, the relative content of different coordinated Al, quadrupole coupling constant (C_q) and ^{29}Si average chemical shift (centroid) of YbAPF0.85 glass with different thermal history [111]

Sample	Type	Chemical Shift/ (ppm, ±0.5)	Content/ (%, ±3%)	C_q/ (MHz, ±0.3)
Pristine	^{29}Si	−110.1		
	Al^{IV}	40.4	100	6.6
An	^{29}Si	−110.5		
	Al^{IV}	43.5	92.8	5.8
	Crystal-Al^{IV}	41.3	7.2	2.9
FireP	^{29}Si	−109.0		
	Al^{IV}	43.0	78.9	6.6
	Al^{V}	17.7	21.1	6.8
FireP-An	^{29}Si	−109.7		
	Al^{IV}	41.0	88.5	6.5
	Al^{V}	17.4	11.5	6.5

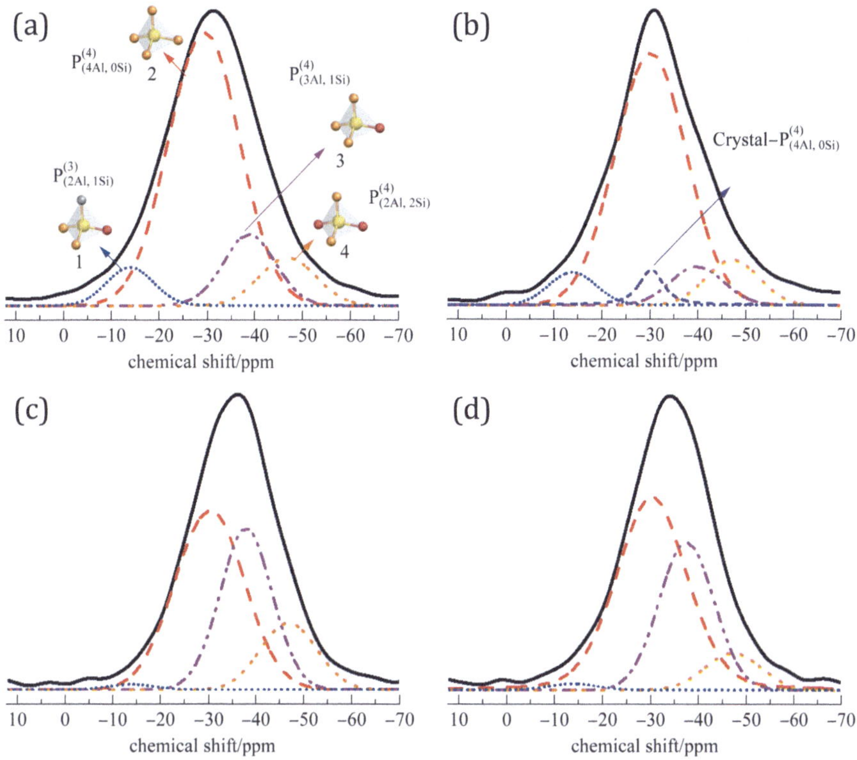

Fig. 2.51 Deconvolution results of ^{31}P MAS NMR spectra of pristine (**a**), annealed (**b**), fire polished (**c**) and fire polished-annealed (**d**) YbAPF0.85 glasses. The dashed lines labeled as 1, 2, 3, and 4 represent $P^{(3)}_{(2Al,1Si)}$, $P^{(4)}_{(4Al,0Si)}$, $P^{(4)}_{(3Al,1Si)}$ and $P^{(4)}_{(2Al,2Si)}$ structures [111]

In order to verify the rationality of peak splitting of the ^{31}P MAS NMR spectrum, Guo et al. further tested the $^{27}Al\{^{31}P\}$ REDOR, $^{31}P\{^{27}Al\}$ rotational echo adiabatic passage double resonance (REAPDOR), and ^{31}P dipolar recoupling effects nuclear alignment reduction (DQ-DRENAR) nuclear magnetic resonance spectra. Fig. 2.52a shows the $^{27}Al\{^{31}P\}$ REDOR results of YbAPF0.85 glass with different thermal histories. The solid line in the figure represents the parabolic fitting results in the range of $\Delta S/S_0 \leq 0.2$. Table 2.21 summarizes the number of Al nuclei connected to a single P (N_{Al-P}) in YbAPF0.85 glass with different thermal histories. The average N_{Al-P} of the pristine YbAPF0.85 glass is 2.9, which is lower than that of $AlPO_4$ glass ($N_{Al-P} = 4$) [109, 112]. That is, in the pristine YbAPF0.85 glass, most Al is connected to four P to form $P^{(4)}_{(4Al,0Si)}$ structure, and a small part of Al bonds with Si. These results are consistent with those of ^{27}Al MAS NMR. After annealing, the N_{Al-P} value of the sample increases from 2.9 to 3.0, then drops to 2.3 after fire polishing. This indicates that annealing leads to an increase of the Al—O—P connections, but fire polishing results in an increase of the Al—O—Si and P—O—Si bonds. In addition, the presence of Al^{V} in the fire-polished YbAPF0.85 glass also implies the existence of

Table 2.20 Deconvolution results of ^{31}P MAS NMR spectra [111]

Sample	Type	Chemical Shift/(ppm, ±0.5)	Content/(%, ±3%)
Pristine	$P^{(3)}_{(2Al,1Si)}$	−14.0	6.5
	$P^{(4)}_{(4Al,0Si)}$	−29.5	70.0
	$P^{(4)}_{(3Al,1Si)}$	−38.9	13.6
	$P^{(4)}_{(2Al,2Si)}$	−47.0	9.9
An	$P^{(3)}_{(2Al,1Si)}$	−14.0	6.1
	$P^{(4)}_{(4Al,0Si)}$	−30.4	71.6
	Crystal-$P^{(4)}_{(4Al,0Si)}$	−30.4	4.6
	$P^{(4)}_{(3Al,1Si)}$	−40.0	8.1
	$P^{(4)}_{(2Al,2Si)}$	−47.4	9.7
FireP	$P^{(3)}_{(2Al,1Si)}$	−14.0	1.1
	$P^{(4)}_{(4Al,0Si)}$	−30.4	49.8
	$P^{(4)}_{(3Al,1Si)}$	−38.0	34.0
	$P^{(4)}_{(2Al,2Si)}$	−47.0	15.1
FireP-An	$P^{(3)}_{(2Al,1Si)}$	−14.0	1.3
	$P^{(4)}_{(4Al,0Si)}$	−30.4	57.9
	$P^{(4)}_{(3Al,1Si)}$	−37.7	32.3
	$P^{(4)}_{(2Al,2Si)}$	−47.4	8.5

Al—O—Si bonds. After annealing the fire-polished YbAPF0.85 glass, it is found that its N_{Al-P} value increases. $^{27}Al\{^{31}P\}$ REDOR results confirm that the interaction between Al and P is closely related to the thermal history. In the fire-polished YbAPF0.85 glass, the interaction between Al and P is relatively weak, and the content of Al—O—Si is higher.

Figure 2.52b shows the $^{31}P\{^{27}Al\}$ REAPDOR results of YbAPF0.85 glass with different thermal histories. Under the condition that the CQ value of ^{27}Al is 6.6 MHz, the $^{31}P\{^{27}Al\}$ REAPDOR spectrum is fitted using the SIMPSON model [113]. The simulation results where ^{31}P bonds with 1 or 2 ^{27}Al are shown in Fig. 2.52b. From the deconvolution results of ^{31}P MAS NMR of the pristine and fire-polished YbAPF0.85 glass, it can be known that ^{31}P is averagely bonded with nearly 4 and more than 3 ^{27}Al respectively. But the simulation results shows a slight deviation from the above results. This deviation also exists in the similar glass systems where

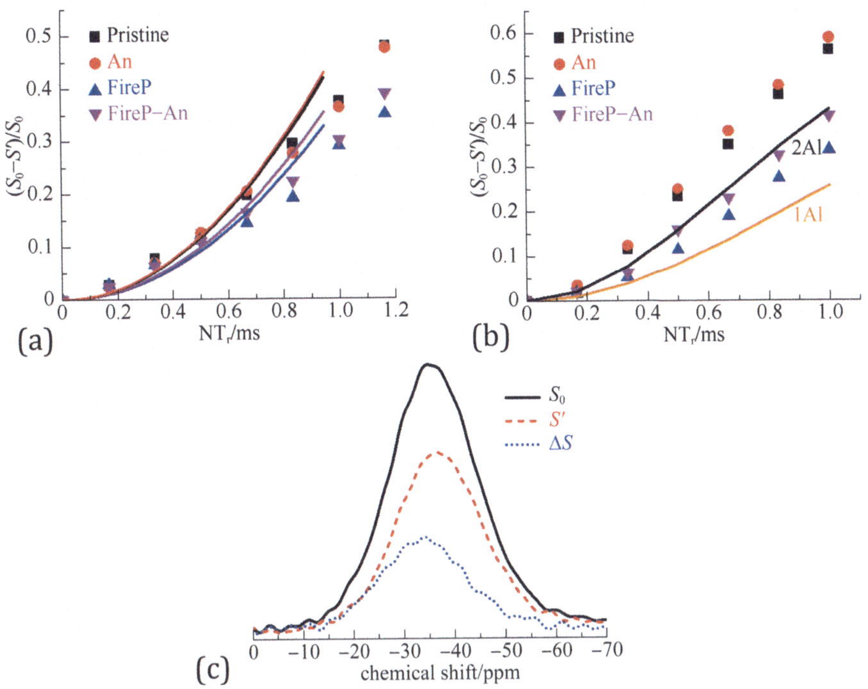

Fig. 2.52 $^{27}Al\{^{31}P\}$ REDOR (**a**) and $^{31}P\{^{27}Al\}$ REAPDOR (**b**) for YbAPF0.85 glass with different thermal histories. Under the condition that the dipole evolution time is fixed at 1 ms, the ^{31}P spectra of the fire-polished YbAPF0.85 glass loaded (S') and unloaded $(S_0)^{31}$ P $-^{27}$ Al dipole interaction and their difference spectrum (ΔS) (**c**) [111]

Table 2.21 Values of M_2^{Al-P}, number of Al nuclei connected with a single P N_{Al-P} for YbAPF0.85 glass with different thermal histories [111]

Sample	M_2^{Al-P} /10 ^{6} (rad ^{2}·s $^{-2}$, ±10%)	N_{Al-P} (±10%)
Pristine	3.5 [a]	2.9 [a]
An	3.6 [a]	3.0 [a]
FireP	2.7 [a]	2.3 [a]
FireP-An	2.9 [a]	2.5 [a]
$AlPO_4$	4.7	4

Note: The superscript a indicates the average value

the corresponding glass components are x Al_2O_3 -(30- x)P_2O_5 -70SiO_2. This deviation may originate from the deviation of the C_q value used in the simulation from the actual value. In YbAPF0.85 glass, most Al atoms are connected to P to form $AlPO_4$ structure, only a small part of Al bonds with Si atoms. In addition, different connections of Al correspond to different C_q values, so the C_q distribution of ^{27}Al in YbAPF0.85 glass is wide. But when simulating $^{31}P\{^{27}Al\}$ REAPDOR spectrum, only a certain C_q value can be used, so the simulation results will have a certain deviation. Although it is difficult to accurately obtain the number of ^{27}Al atoms using SIMPSON simulation, it can be obtained the trend as functions of thermal history from the decay curve of $^{31}P-^{27}Al$ dipole interaction. Similar to the $^{27}Al\{^{31}P\}$ REDOR spectrum, the ^{31}P-^{27}Al dipole interaction is enhanced after annealing, and weakened after fire polishing. This result shows that the number of P-O-Al in YbAPF0.85 glass with different thermal histories follows the rule: An>Pristine>FireP-An>FireP.

Figure 2.52c is the ^{31}P spectrum of the fire-polished YbAPF0.85 glass obtained when the ^{31}P-^{27}Al dipole interaction is loaded (S′) and not loaded (S_0) at a dipole evolution time of 1 ms. The entire wide spectrum decreases sharply under the influence of ^{27}Al, and the decay of the high-frequency signal is greater. On the one hand, this result proves that all P in the fire-polished YbAPF0.85 glass are bonded with Al atoms. On the other hand, the signal in the high-frequency direction originates from the $P^{(4)}_{(4Al,0Si)}$ structure, and the signal in the low-frequency direction originates from the $P^{(4)}_{(mAl,hSi)}$ structure ($m + h = 4$, $m < 4$).

Figure 2.53a shows the CT-DRENAR-POST-C7 decay curves of YbAPF0.85 glass with different thermal histories, and the corresponding $\sum_j b_{jk}^2$ values are shown in Fig. 2.53a. The $\sum_j b_{jk}^2$ value of the pristine and annealed YbAPF0.85 glass is about half of that of $AlPO_4$ glass of 3.3×10^5 Hz, where each P is bonded with 4 Al [114]. On the one hand, this result confirms that there is no P—O—P connection in all YbAPF0.85 glasses, indicating that P only exists in the form of P $^{(4)}$. The diversity of P $^{(4)}$ structure only originates from its second nearest neighbor atoms, further verifying the rationality of $P^{(4)}_{(mAl,hSi)}$ structure attribution. On the other hand, the dipole-dipole interaction of $^{31}P-^{31}P$ in the annealed YbAPF0.85 glass is comparable to that of $AlPO_4$ glass, suggesting that $AlPO_4$ nanocrystals may precipitate in the annealed YbAPF0.85 sample. Although the annealing temperature used (1000 °C) is lower than the glass crystallization temperature (1148 °C), long-term annealing may lead to the precipitation of crystals. After fire polishing, the $\sum_j b_{jk}^2$ value of the sample decreases, indicating that the degree of P aggregation decreases. It can be inferred that fire polishing will disintegrate the $AlPO_4$ clusters, and increase the interaction between the [$AlPO_4$] structure and the silica network. Therefore, The relative content of $P^{(4)}_{(3Al,1Si)}$ and $P^{(4)}_{(2Al,2Si)}$ structures is higher in the fire-polished YbAPF0.85 glass. The subsequent annealing treatment can make the $\sum_j b_{jk}^2$ value of the YbAPF0.85 glass transform into its pristine value. Therefore, the relative content of $P^{(4)}_{(3Al,1Si)}$ and $P^{(4)}_{(2Al,2Si)}$ structures in the FireP-An YbAPF0.85 glass decreases.

Figure 2.53b shows the ^{31}P spectra of the fire-polished YbAPF0.85 glass loaded (S′) and unloaded (S_0) the ^{31}P-^{31} P dipole interaction. Under the influence of ^{31}P, the decay of the ^{31}P spectrum in the high-frequency direction is higher than that in the

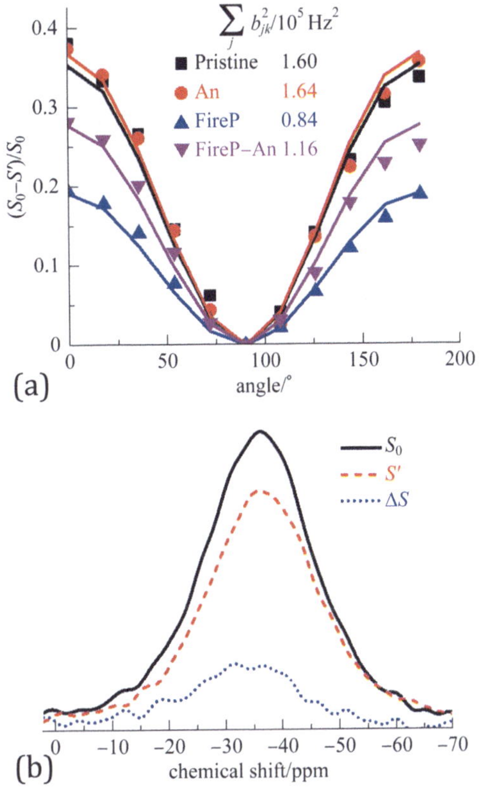

Fig. 2.53 CT-DRENAR-POST-C7 decay curves of YbAPF0.85 glass with different thermal histories (**a**), the solid line in (**a**) represents the simulation result. The ^{31}P spectra of the fire-polished YbAPF0.85 glass loaded (S′) and not loaded (S_0) ^{31}P –^{31}P dipole interactions and their difference spectra (ΔS) (**b**) [111]

low-frequency direction. This is because the $P^{(4)}_{(4Al,0Si)}$ structure exists in the $AlPO_4$ clusters, where the ^{31}P-^{31}P dipole interaction is strong. While in the low-frequency direction, the $P^{(4)}_{(3Al,1Si)}$ and $P^{(4)}_{(2Al,2Si)}$ structures are mainly located at the junction of $AlPO_4$ structure and the silicon-oxygen tetrahedron. Therefore, their ^{31}P –^{31}P dipole interactions are relatively weak. This result further proves the validity of the ^{31}P structure attribution.

The NMR results of all samples show that Al tends to connect with PO_4 structure to form $AlPO_4$ agglomerates wrapped by silica network. The $AlPO_4$ agglomerates are connected to the silica network through Al—O—Si and P—O—Si bonds. In the $AlPO_4$ agglomerates, each Al bonds with four P, and Al exists in a four-coordinated form. At the boundary of $AlPO_4$ agglomerates and silica network, Al bonds with Si, and Al exists in a five-coordinated form. Annealing treatment mainly causes the transformation of $AlPO_4$ agglomerates into the ordered one in the long range. Fire polishing will disintegrate $AlPO_4$, then forming Al—O—Si and P—O—Si bonds. Thus the connection between $AlPO_4$ agglomerates and the silica network improves. After annealing treatment of the fire-polished YbAPF0.85 sample, it is found that Al—O—Si and P—O—Si bonds are broken to form $AlPO_4$, i.e., the structure of the glass transforms into that of the pristine glass.

Figure 2.54a shows the refractive index at 1064 nm for core glass YbAPF0.85 and cladding glass F300 with different thermal histories. The error bars in the figure represent the standard deviation, and the specific values can be found in Table 2.22. Annealing at 1000 °C leads to a decrease in the refractive index ofYbAPF0.85 and F300 glasses, and the difference in refractive index (Δn) also decreases. However, after fire polishing, their refractive index and the difference in refractive index significantly increase. When the fire-polished samples are annealed, their refractive index and Δn significantly decrease.

Figure 2.54b shows the NA of YbAPF0.85 fiber under different thermal histories, where the fiber core is YbAPF0.85 glass and the cladding is F300 glass. The NA is calculated as follows:

$$\mathrm{NA} = \sqrt{n_{YbAPF0.85}^{2} - n_{F300}^{2}} \tag{2.14}$$

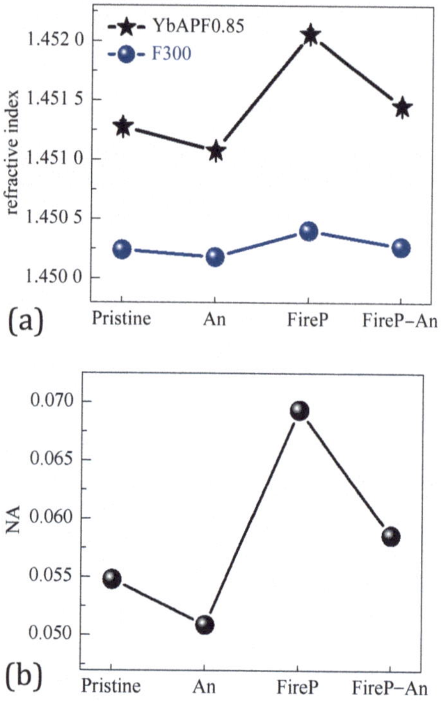

Fig. 2.54 Refractive index (**a**) and NA (**b**) at 1064 nm of core glass YbAPF0.85 and cladding glass F300 with different thermal histories

Table 2.22 Refractive index, the difference in refractive index (Δn) and numerical aperture (NA) at 1064 nm of core glass YbAPF0.85 and cladding glass F300 with different thermal histories [111]

	1064 nm Refractive Index			
Sample	F300	YbAPF0.85	Δn (YbAPF0.85-F300)	NA
Pristine	1.45024	1.45127	10×10^{-4}	0.055
An	1.45018	1.45107	8.9×10^{-4}	0.051
FireP	1.4504	1.45206	17×10^{-4}	0.069
FireP-An	1.45027	1.45145	12×10^{-4}	0.059

After annealing at 1000 °C for 3 h, the NA of the YbAPF0.85 fiber is decreased by 7%. As can be seen from Fig. 2.54a, the decrease in NA is due to the fact that under the same treatment conditions, the decrease in the refractive index of the core glass YbAPF0.85 is greater than that of the cladding glass F300. When the preform rod composed of YbAPF0.85 and F300 glasses is drawn into a fiber, the NA value of the fiber sharply increases. Subsequent annealing further reduces the NA value of the fire-polished samples. Under the same annealing conditions, the ability of the post-annealing process to reduce the refractive index is much greater than that of the pre-annealing process. The structure and refractive index of YbAPF0.85 and F300 glasses are closely related to their thermal history. The changes in the refractive index of glasses with different thermal histories are mainly related to the changes in the structure and content of SiO_4, AlO_x, PO_4, and $AlPO_4$.

As can be seen from Table 2.17, the peak areas of D_1 and D_2 of YbAPF0.85 and F300 glasses decrease after annealing, and that increases after fire polishing, then decreases after subsequent annealing. The D_2 peak area of F300 glass follows the same trend as that of the refractive index at 1064 nm with thermal history. This result is consistent with the reported results in the literature [115–117]. The change in the refractive index of pure silica glass with thermal history is mainly related to the existence form of the Si—O—Si ring structure (mainly planar 3-membered rings and planar 4-membered rings), see Sect. 1.2, Chap. 1 of this book.

The relationship between the refractive index of YbAPF0.85 glass at 1064 nm and thermal history is consistent with that of F300 glass, but the dependence of the refractive index of YbAPF0.85 glass on thermal history is stronger, thus the change in refractive index under the same treatment conditions is greater. In addition to the D_1 and D_2 ring structures, the changes in the AlO_x and PO_4 structures are also closely related to thermal history. After fire polishing, the amount of Al^{V} in the glass increases sharply which can fill the gaps in the glass structure network. The higher its content, the higher the density and refractive index of the glass.

The content of $P^{(4)}_{(4Al,0Si)}$ structure is negatively correlated with the refractive index. $P^{(4)}_{(4Al,0Si)}$ tends to bond with Al^{IV} to form the [$AlPO_4$] structure. This structure is similar to the [$SiO_{4/2}$] tetrahedral structure, which can increase the looseness of the glass network. It is reported that for every 1 mol% increase in the [$AlPO_4$] structure, the refractive index of the glass will decrease by 1.1005×10^{-4} [18]. After annealing, the content of $P^{(4)}_{(4Al,0Si)}$ in YbAPF0.85 glass increases, and the decrease in refractive index caused by post-annealing treatment (16.3%) is higher than that caused by pre-annealing treatment (8.9%). That is, compared with pre-annealing treatment, the glass refractive index decreases more during the post-annealing treatment process.

Thermal history can cause reversible structural changes in pure silica and Yb/Al/P/F co-doped silica glasses, thus leading to changes in the refractive index. For F300 glass, its refractive index dependence on thermal history is mainly related to the Si—O—Si ring structure. The refractive index dependence of YbAPF0.85 glass on thermal history originates from the Si—O—Si ring structure, the coordination number of Al, and $P^{(4)}_{(4Al,0Si)}$ structure, where the coordination number of Al and $P^{(4)}_{(4Al,0Si)}$ contents are more improtant. Therefore, under the same treatment conditions, the refractive index change of YbAPF0.85 glass is greater than that of F300 glass, which leads to the change of NA with thermal history.

Based on the mechanism of refractive index change of YbAPF core and F300 cladding glasses with different thermal histories, Guo et al. [111] used post-annealing treatment process to decrease the NA of Yb/Al/P/F co-doped photonic crystal fiber (PCF), and then control the beam quality of the fiber. The pristine YbAPF PCF has a NA of 0.065. Annealing at 1000 °C reduces the NA to 0.035, as shown in Fig. 2.55a, b. Figure 2.55c, d are the far-field beam spots of the pristine and annealed YbAPF PCF. It is found that annealing can significantly improve the beam quality of the fiber. The beam spot of the pristine fiber is non-Gaussian symmetric, containing higher-order modes. But the beam quality of the fiber after annealing is significantly improved, and its corresponding beam quality factor M^2 is 1.9.

Figure 2.56a–c respectively summarize the influence mechanism of refractive index changes in pure silica and Yb^{3+} single-doped, Yb/Al and Yb/Al/B co-doped, Yb/Al/P and Yb/Al/P/F (low F content) co-doped silica glasses with different thermal histories. When the fictive temperature decreases, the refractive index of pure silica and Yb^{3+} single-doped silica glasses decreases. This change is mainly related to the Si-O-Si ring structure, that is, the ring structure deviating from 6 ~ 8 rings will

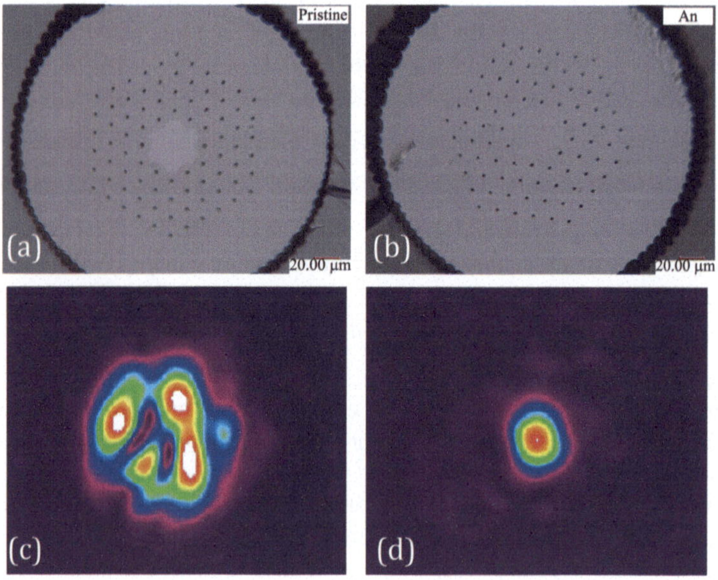

Fig. 2.55 The cross-section (**a**) and far-field light spot (**c**) of the pristine YbAPF PCF. The cross-section (**b**) and far-field light spot (**d**) of the annealed YbAPF PCF [111]

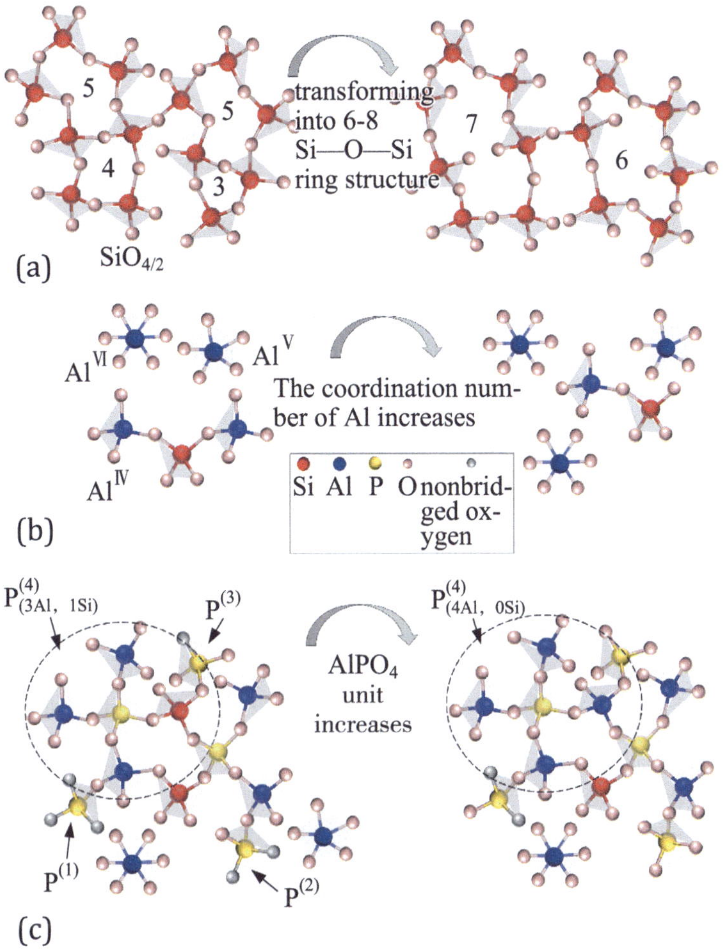

Fig. 2.56 The effect of decreasing fictive temperature on the structure of silica glass. (**a**) Pure silica and Yb^{3+} single-doped silica glasses. (**b**) Yb/Al and Yb/Al/B co-doped silica glasses. (**c**) Yb/Al/P and Yb/Al/P/F (low F content) co-doped silica glasses

be broken and rearranged as 6 ~ 8 rings, thus the Si—Si distance increases, as shown in Fig. 2.56a. However, when the fictive temperature decreases, the refractive index of Yb/Al and Yb/Al/B co-doped silica glasses increases, which is mainly related to the increase in the coordination number of Al, as shown in Fig. 2.56b. After doping P_2O_5 into Yb/Al co-doped silica glass, the refractive index of Yb/Al/P or Yb/Al/P/F (low F doping amount) co-doped silica glasses decreases with the decrease of fictive temperature, which is mainly related to the formation of $AlPO_4$, as shown in Fig. 2.56c.

2.3.2 The Effect of Thermal History on the Spectral Properties of Ytterbium-Doped Silica Glass

Yb^{3+} doped silica fiber is the gain medium of high-power fiber lasers, where Yb^{3+} doped silica glass is the core material. So the spectral properties of Yb^{3+} have an important impact on the laser performance of Yb-doped silica fiber. Studies have shown that thermal history can affect the absorption coefficient and absorption peak position of Yb^{3+} [75, 92, 102]. To reveal the effect of thermal history on the spectral properties of Yb^{3+}, Guo et al. [92, 102] chose the Yb^{3+} single-doped, Yb/Al, Yb/Al/B and Yb/Al/P co-doped silica glasses as the research objects. The local environment of Yb^{3+} is analyzed by EPR technology, and the relationship between the local environment of Yb^{3+} and spectral properties during the annealing process is established.

Figure 2.57 summarizes the effect of annealing temperature on the spectral properties of Yb^{3+} single- doped silica glass, with a glass composition of $0.05Yb_2O_3$–$99.95SiO_2$ (mol%). The main spectral parameters of the glass are shown in Table 2.23.

As can be seen from Fig. 2.57a, as the annealing temperature decreases, the UV absorption edge of Yb^{3+} single-doped silica glass shows a blue shift, the absorption coefficient of Yb^{3+} decreases and the shoulder peak at 915 nm shows a red shift, but the main absorption peak position remains basically unchanged.

As can be seen from Fig. 2.57b, as the annealing temperature decreases, the luminescence intensity of Yb^{3+} decreases and the shoulder peak at 1035 nm shows a blue shift, but the main emission peak position remains basically unchanged.

As can be seen from Fig. 2.57c and Table 2.23, as the annealing temperature decreases, the absorption and emission cross-sections of Yb^{3+} decrease. The decrease in these parameters is related to the decrease in the the local crystal field and degree of Yb^{3+} asymmetry(Fig. 2.57e).

Figure 2.57d shows the Stark splitting energy levels of Yb^{3+}. According to the data in Fig. 2.57d, the local crystal field and degree of Yb^{3+} asymmetry can be calculated, as shown in Fig. 2.57e. They are positively correlated with the annealing temperature.

Guo et al. [92] studied the effect of annealing time at 900 °C on the spectral properties of Yb/Al co-doped silica glass, as shown in Fig. 2.58. The glass composition is $0.1Yb_2O_3$-$4Al_2O_3$–$95.9SiO_2$ (mol%), and the main spectral parameters of the glass are shown in Table 2.24.

Similar to Yb^{3+} single-doped silica glass, as the annealing time increases, the UV absorption edge, secondary absorption peak, and secondary emission peak of Yb/Al co-doped silica glass all show a blue or red shift. In addition, the absorption and emission cross-sections decrease, as shown in Fig. 2.58a–c.

As can be seen from Fig. 2.58c and Table 2.2, annealing at 900 °C increases the excited state lifetime of Yb^{3+}. The lifetime decay curve of the pristine YbAP0 glass is fitted by a single exponential component, while that is fitted by two exponential components after annealing for 200 h, corresponding to the fast decay process of 0.13 ~ 0.16 ms and the slow decay process of 1.15 ~ 1.18 ms. It is speculated that there are two types of lattice sites in YbAP0 glass after annealing.

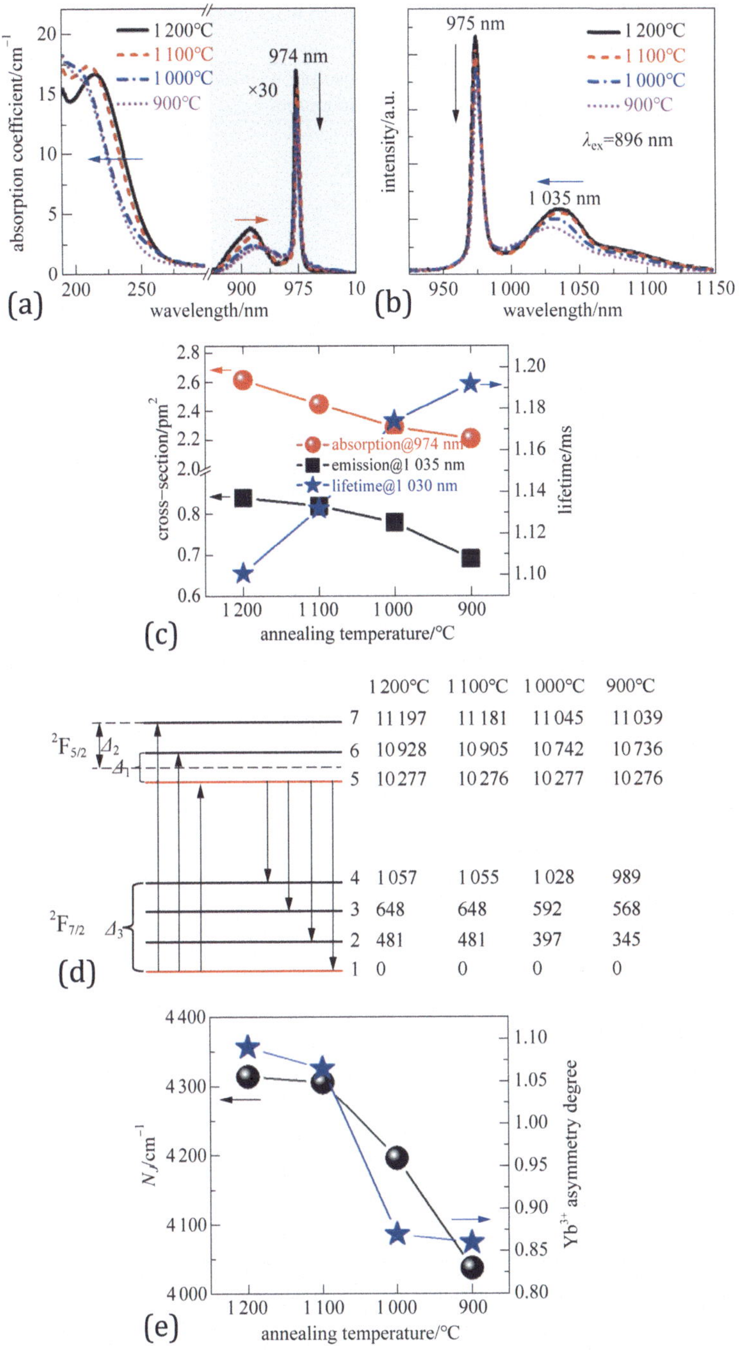

Fig. 2.57 The effect of annealing temperature on the absorption (**a**) and emission (**b**) spectra, absorption cross-section at 974 nm, emission cross-section and fluorescence lifetime at 1035 nm (**c**), Stark splitting energy levels and energy values of each level E_i ($i = 1 \sim 7$, cm^{-1}) (**d**), local crystal field strength parameters and degree of asymmetry (**e**) of Yb^{3+} single-doped silica glass [34]

Table 2.23 The effect of annealing temperature on the main spectral parameters of Yb^{3+} single-doped silica glass [34]

Annealing Temperature/°C	Fluorescence Lifetime/ms	Absorption Cross-Section/pm ^{2}	Emission Cross-Section/pm ^{2}
1200	1.1	2.62	0.84
1100	1.13	2.45	0.82
1000	1.17	2.29	0.78
900	1.19	2.21	0.69

Figure 2.58d shows the Stark split energy levels of Yb^{3+}. The local crystal field strength parameters and degree of Yb^{3+} asymmetry are shown in Figure 2.58e. After annealing, the local crystal field strength parameters and degree of Yb^{3+} asymmetry decrease. When the annealing time is not more than 200 h, N_J and the degree of asymmetry decrease significantly with the increase of annealing time, but further increasing the annealing time leads to a low decrease extent of N_J and the degree of asymmetry. When the annealing time is 500 h, the crystal field strength parameters and the degree of Yb^{3+} asymmetry in the YbAP0 glass are decreased by 8.6% and 15%, respectively.

Figure 2.59 summarizes the effect of annealing time at 800 °C on the spectral properties of Yb/Al/B co-doped silica glass, with the glass composition being $0.1Yb_2O_3$-$6Al_2O_3$ -$8B_2O_3$–$85.9SiO_2$ (mol%, SYAB8). The main spectral parameters of the glass are shown in Table 2.25.

Similar to Yb^{3+} single-doped silica glass, with the increase of 800 °C annealing time, the UV absorption edge, secondary absorption peak and secondary emission peak of SYAB8 glass all show a blue or red shift. In addition, the cross-section, local crystal field strength parameters and degree of Yb^{3+} asymmetry linearly decrease, and the excited state lifetime of Yb^{3+} linearly increases, as shown in Fig. 2.59.

Guo et al. [102] studied the effect of 850 °C annealing time on the spectral properties of Yb/Al/P co-doped silica glass, as shown in Fig. 2.60. The glass composition is $0.1Yb_2O_3$-$4Al_2O_3$ -$6P_2O_5$–$89.9SiO_2$ (mol%, YbAP1.5). The main spectral parameters of the glass are shown in Table 2.26.

Similar to Yb^{3+} single-doped and Yb/Al/B co-doped silica glasses, as the annealing time at 850 °C increases, the UV absorption edge, secondary absorption peak, and secondary emission peak of YbAP1.5 glass all show a blue or red shift. At the same time, the Yb^{3+} cross-section, local crystal field strength parameters, and degree of Yb^{3+} asymmetry linearly decrease, as shown in Fig. 2.59. Similar to Yb/Al co-doped silica glass, annealing increases the excited state lifetime of Yb^{3+}. The lifetime decay curve of the pristine YbAP1.5 glass is fitted by a single exponential component, and after annealing for 15 h, it is be fitted by two exponential component, i.e., $I(t) = \sum A_i e^{-t/t_i}$ + constant ($i = 1, 2$), corresponding to the fast decay process of 0.41 ~ 0.51 ms and the slow decay process of 1.42 ~ 1.59 ms.

Figure 2.61 summarizes the absorption spectra of Yb/Al/P co-doped silica glass with different P/Al ratio before and after annealing at 1000 °C. The glass composition and main spectral parameters are shown in Table 2.27.

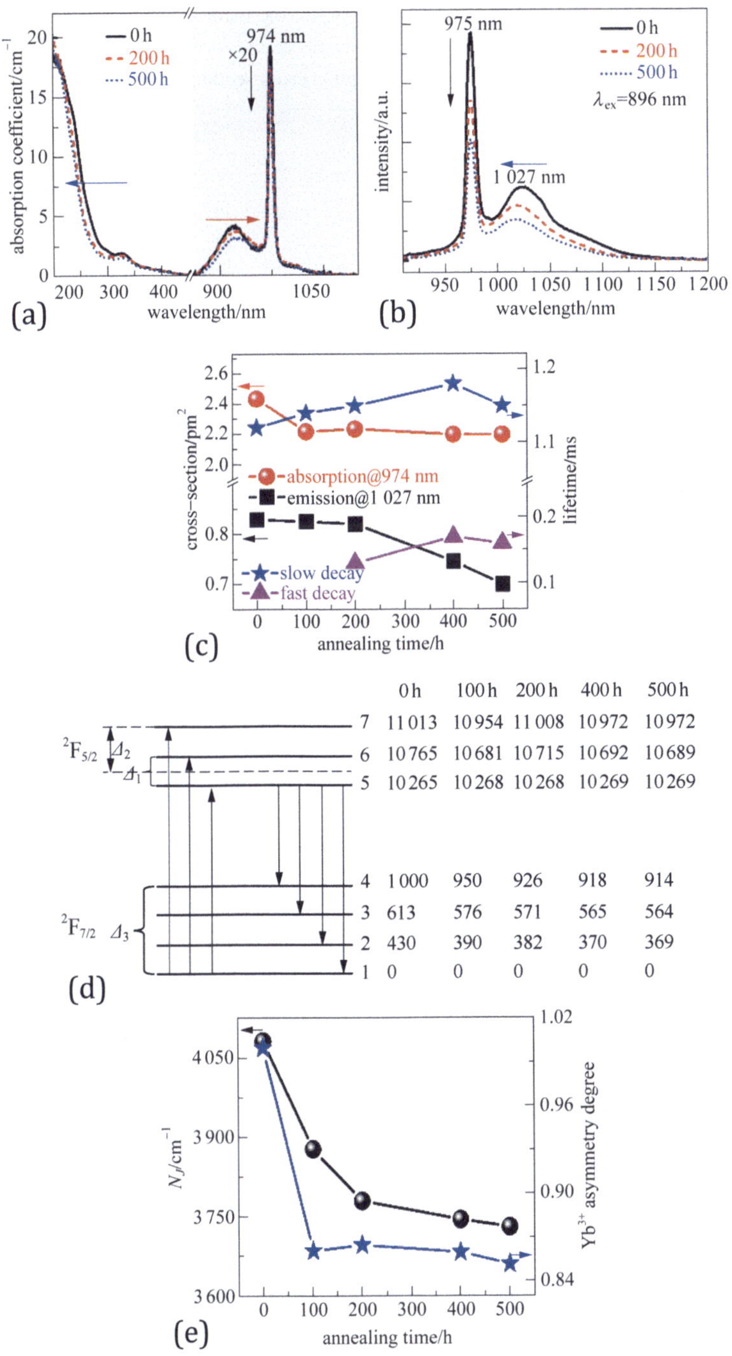

Fig. 2.58 The effect of annealing time at 900 °C on the absorption (**a**) and emission (**b**) spectra of Yb/Al co-doped silica glass, the absorption cross-section at 974 nm, the emission cross-section at 1027 nm, and the fluorescence lifetime at 1030 nm (**c**), Stark split energy levels and energy values of each level E_i ($i = 1 \sim 7$, cm^{-1}) (**d**), local crystal field strength parameters and degree of Yb^{3+} asymmetry (**e**) [92]

Table 2.24 The effect of annealing time at 900 °C on the main spectral parameters of Yb/Al co-doped silica glass [92]

Annealing time/h	Fluorescence lifetime/ms	Absorption cross-section/pm^2	Emission cross-section/pm^2
0	1.12	2.436	0.829
100	1.14	2.217	0.825
200	1.15/0.13	2.234	0.82
400	1.18/0.17	2.196	0.744
500	1.15/0.16	2.196	0.699

As can be seen from Fig. 2.61a–d, annealing at 1000 °C for 4 h does not change the absorption spectrum shape of 0.25Yb, SYbAP0, SYbAP1, and SYbAP2 glasses, but it leads to a reduction in the absorption coefficient and a red shift in the absorption shoulder peak at 915 nm with different decrease extent of the glass.

Figure 2.61e, f respectively show the peak positions of the absorption peaks corresponding to the (1 → 7), (1 → 6), and (1 → 5) level transitions of Yb^{3+} in 0.25Yb, SYbAP0, SYbAP1, and SYbAP2 glasses before and after annealing at 1000 °C. As can be seen from Fig. 2.61e, annealing causes a red shift in the absorption peaks corresponding to the (1 → 7) and (1 → 6) transitions of Yb^{3+}. As the P/Al ratio increases, the amount of redshift decreases. When P/Al > 1, the sub-peak position of SYbAP2 glass remains basically unchanged. However, as can be seen from Fig. 2.61f, the peak position of the main absorption peak corresponding to the (1 → 5) transition of Yb^{3+} does not change with annealing time.

Figure 2.61g shows the absorption coefficients of the main and secondary absorption peaks of 0.25Yb, SYbAP0, SYbAP1, and SYbAP2 glasses before and after annealing at 1000 °C. It should be noted that the peak positions of the main absorption peaks are 975 nm (0.25Yb), 975 nm (SYbAP0), 975 nm (SYbAP1), and 974 nm (SYbAP2), respectively, and the peak positions of the secondary absorption peaks are 916 nm (0.25Yb), 924 nm (SYbAP0), 919 nm (SYbAP1), and 914 nm (SYbAP2), respectively. Annealing at 1000 °C for 4 h significantly reduces the absorption coefficients of the main absorption peaks of 0.25Yb, SYbAP0, SYbAP1, and SYbAP2 glasses, and the reduction in the absorption coefficients of the secondary absorption peaks is relatively small, which is related to the redshift of the secondary absorption peaks.

The electronic configuration of Yb^{3+} is Xe+$4f^{13}$. Under the influence of the crystal field, the two sub-energy levels $^2F_{7/2}$ and $^2F_{5/2}$ split into several Stark levels with small intervals, generally less than 500 cm^{-1}, as shown in Fig. 2.1. The redshift of the secondary absorption peak of Yb^{3+} indicates that the degree of Stark level splitting of Yb^{3+} decreases, and the influence of the crystal field on Yb^{3+} weakens. After annealing, the redshift of the secondary absorption peak of Yb^{3+} is related to the P/Al ratio, which may be related to the local coordination environment of Yb^{3+}. When the doping concentration of Yb is high, the nearest neighbor atoms of Yb^{3+} in single-doped silica glass are mainly Si and Yb. That in the Yb/Al co-doped and Yb/Al/P co-doped (P/Al ≤ 1) silica glasses are mainly Si and Al. But in the Yb/Al/P co-doped (P/Al > 1) silica glass, Yb^{3+} is wrapped by P, which can shield the influence

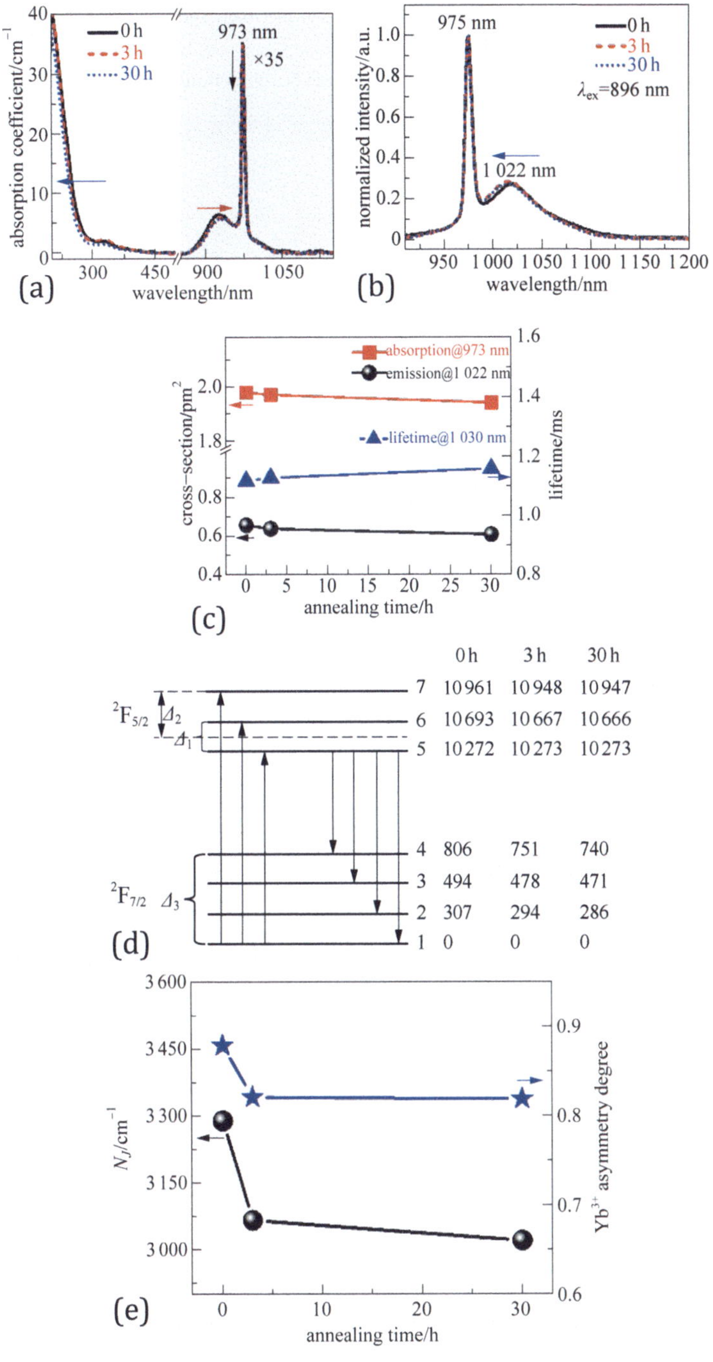

Fig. 2.59 The effect of annealing time at 800 °C on the absorption (**a**) and emission (**b**) spectra of Yb/Al/B co-doped silica glass, absorption cross-section at 973 nm, emission cross-section at 1022 nm and fluorescence lifetime at 1030 nm (**c**), Stark split energy levels and energy values of each level E_i (i = 1 ~ 7, cm^{-1}) (**d**), the influence of local crystal field strength parameters and degree of Yb^{3+} asymmetry (**e**)

Table 2.25 The effect of annealing time at 800 °C on the main spectral parameters of Yb/Al/B co-doped silica glass

Annealing time/h	Fluorescence lifetime/ms	Absorption cross-section/pm^2	Emission cross-section/pm^2
0	1.12	1.98	0.66
3	1.13	1.97	0.64
30	1.16	1.94	0.61

of crystal field changes on the spectral properties of Yb^{3+} with a certain extent. That is, its crystal field strength parameter is much lower than that of Yb/Al/P co-doped silica glass with P/Al ≤ 1 [17, 36]. Therefore, under the same treatment conditions, annealing basically does not change the Stark splitting parameters of Yb^{3+} in the Yb/Al/P co-doped silica glass with P/Al > 1, and the shift of the secondary absorption peak of Yb^{3+} is relatively small.

The luminescence of Yb^{3+} -doped silica glass mainly originates from the electric dipole transition of $4f^n$ configurations in the Yb^{3+} ion. In the static conditions, its electric dipole transition is forbidden, there is only weak magnetic dipole and electric quadrupole interaction transition. But when Yb^{3+} is in a non-centrosymmetric crystal field, the electronic state does not have a definite parity. Therefore, it will not follow the parity selection rule, and the electric dipole transition is no longer strictly forbidden [92]. According to the parity selection rule, the probability of electric dipole transition is positively correlated with the asymmetry of the coordination field. It can be seen that annealing will affect the degree of Yb^{3+}asymmetry, thereby affecting its absorption and emission cross-sections.

Figure 2.62 is the energy level diagram of Eu^{3+}. The transitions of Eu^{3+} $^7F_0 \rightarrow {}^5D_2$ and $^5D_0 \rightarrow {}^7F_2$ correspond to electric dipole transitions, $^7F_0 \rightarrow {}^5L_6$ and $^5D_0 \rightarrow {}^7F_1$ correspond to magnetic dipole transitions. Therefore, the asymmetry degree of the Eu^{3+} coordination field can be evaluated according to the transition intensity ratio between $^7F_0 \rightarrow {}^5D_2$ and $^7F_0 \rightarrow {}^5L_6$ and the transition intensity ratio between $^5D_0 \rightarrow {}^7F_2$ and $^5D_0 \rightarrow {}^7F_1$. The larger the transition intensity ratio, the stronger the asymmetry [52, 118–120]. The valence state of Eu^{3+} is closely related to the glass components. In silica glass, when P/Al > 1, Eu ions are trivalent. To characterize the effect of annealing on the degree of Yb^{3+} asymmetry, Guo et al. [102] used Eu^{3+} as a probe, and treated Yb/Al/P and Eu/Al/P co-doped silica glasses (P/Al > 1) under the same annealing conditions. The specific glass components are $0.1Eu_2O_3/Yb_2O_3$-$4Al_2O_3$-$6P_2O_3$–$89.9SiO_2$ (mol%, EuAP1.5/YbAP1.5), respectively. They revealed the effect of annealing on the degree of Yb^{3+} asymmetry, then characterized the effect of annealing on the local coordination atoms of Yb^{3+} using EPR structural analysis technology.

Figure 2.63a shows the normalized emission spectrum of Eu^{3+}, and all emission spectra are normalized at 591 nm. The emission peak at 591 nm corresponds to the magnetic dipole transition $^5D_0 \rightarrow {}^7F_1$, the emission peak at 611 nm corresponds to the electric dipole transition $^5D_0 \rightarrow {}^7F_2$. The inset in Fig. 2.63a shows the relationship between the transition intensity ratio of $^5D_0 \rightarrow {}^7F_2$ and $^5D_0 \rightarrow {}^7F_1(\alpha_1)$ and the annealing time. After annealing, α_1 decreases. Figure 2.63b shows the normalized

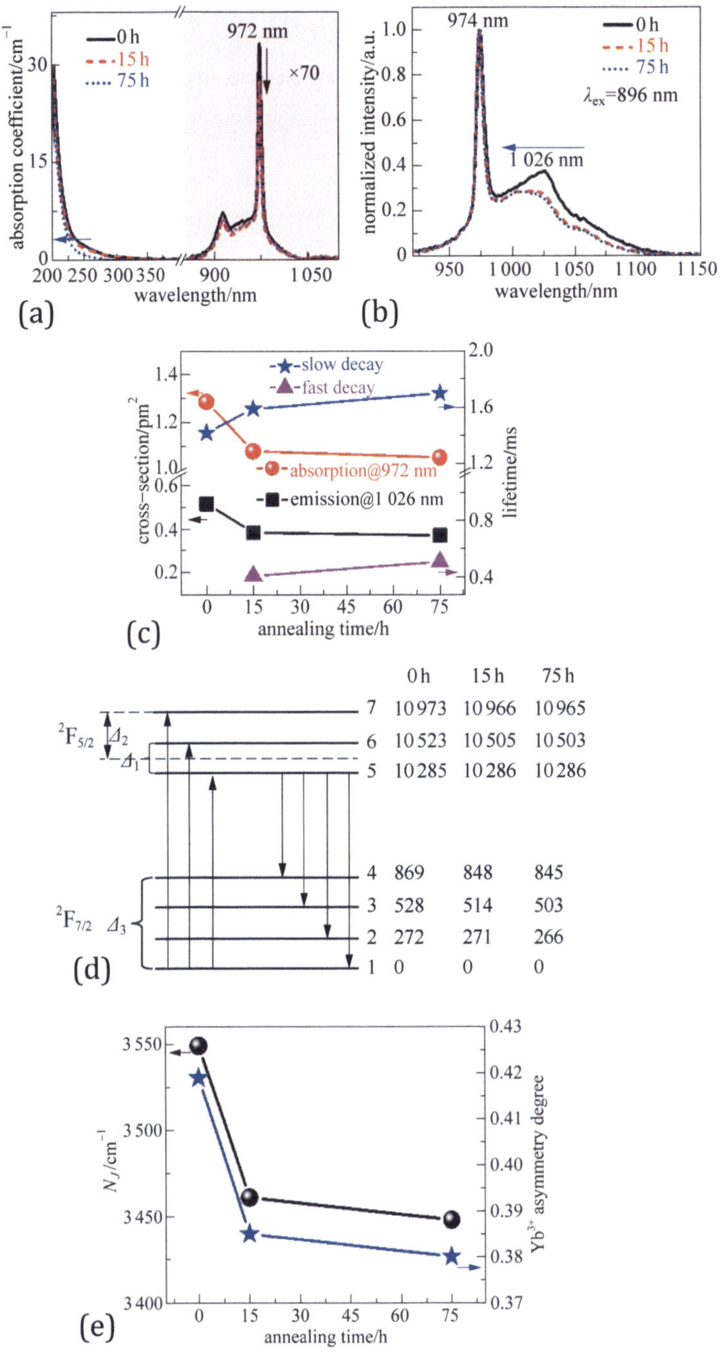

Fig. 2.60 The effect of annealing time at 850 °C on the absorption (**a**) and emission (**b**) spectra, absorption cross-section at 972 nm, emission cross-section at 1026 nm, and fluorescence lifetime at 1030 nm (**c**), Stark splitting energy levels and energy values of each level E_i ($i = 1 \sim 7$, cm^{-1}) (**d**), local crystal field strength parameters and degree of asymmetry (**e**) of Yb/Al/P co-doped silica glass [102]

Table 2.26 The effect of annealing time at 850 °C on the main spectral parameters of Yb/Al/P co-doped silica glass

Annealing time/h	Fluorescence lifetime/ms	Absorption cross-section/ pm^2	Emission cross-section/ pm^2
0	1.42	1.29	0.52
15	1.59/0.41	1.08	0.38
75	1.70/0.51	1.05	0.37

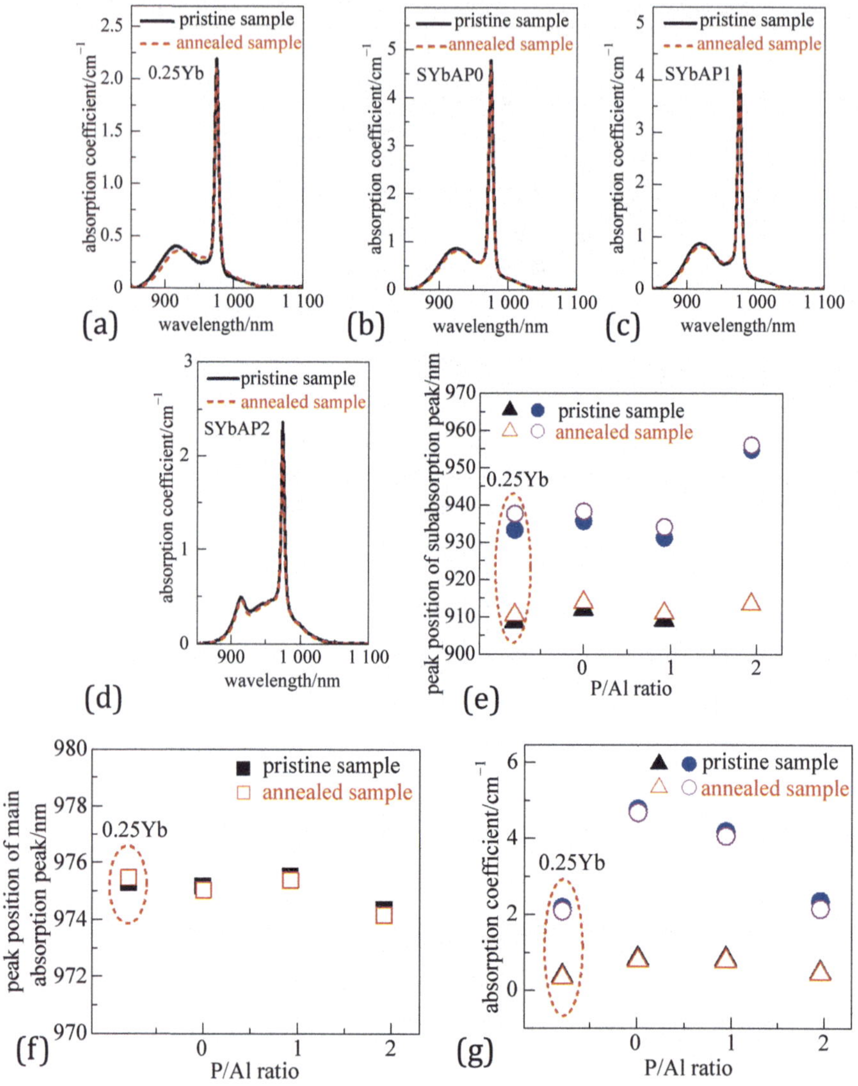

Fig. 2.61 Absorption spectra of 0.25Yb (**a**), SYbAP0 (**b**), SYbAP1 (**c**) and SYbAP2 (**d**) glasses before and after annealing at 1000 °C for 4 h, peak positions of the main absorption peaks (**e**), secondary absorption peaks (**f**) and absorption coefficients corresponding to the main and secondary absorption peaks (**g**) of 0.25Yb, SYbAP0, SYbAP1 and SYbAP2 glasses before and after annealing at 1000 °C for 4 h

Table 2.27 High Yb-doped silica glass composition, wavelengths corresponding to Yb^{3+} (1 → 7), (1 → 6) and (1 → 5) transitions before and after annealing, absorption coefficients of the main and secondary absorption peaks

	0.25Yb		SYbAP0		SYbAP1		SYbAP2	
Parameter	Pristine	Annealed	Pristine	Annealed	Pristine	Annealed	Pristine	Annealed
Yb_2O_3/mol%	0.25		0.50		0.50		0.50	
Al_2O_3/mol%	0.00		10.00		5.00		5.00	
P_2O_5/mol%	0.00		0.00		5.00		10.00	
Measured P/Al	–		0.00		0.93		1.94	
(1 → 7) wavelength/nm	908.7	910.7	912.0	914.0	909.2	911.1	913.6	913.7
(1 → 6) wavelength/nm	933.5	937.8	935.8	938.4	931.3	934.2	954.9	956.2
(1 → 5) wavelength/nm	975.3	975.5	975.2	975.0	975.5	975.4	974.3	974.2
Main peak absorption coefficient/ cm^{-1}	2.19	2.10	4.78	4.69	4.19	4.07	2.36	2.16
Sub-peak absorption coefficient/ cm^{-1}	0.40	0.34	0.85	0.80	0.84	0.80	0.49	0.47

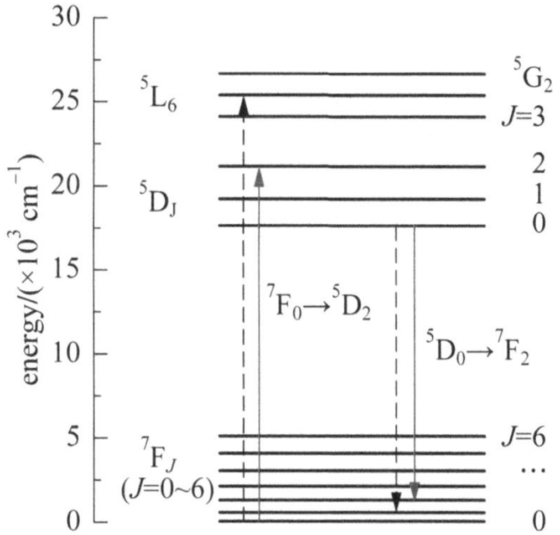

Fig. 2.62 Energy level diagram of Eu^{3+} ion

excitation spectrum of Eu^{3+}, all spectra are normalized at 393 nm. The absorption peak at 393 nm corresponds to the magnetic dipole transition $^5D_0 \rightarrow {}^7F_1$. The absorption peak at 464 nm corresponds to the electric dipole transition $^5D_0 \rightarrow {}^7F_2$. The inset in Fig. 2.63b shows the relationship between the transition intensity ratio (α_2) of $^5D_0 \rightarrow {}^7F_2$ and $^5D_0 \rightarrow {}^7F_1$ and the annealing time. It is found that the α_2 value decreases after annealing. This indicates that the asymmetry degree of the local coordination field of Eu^{3+} decreases after annealing. That is, the asymmetry of the coordination field of Yb^{3+} decreases.

Figure 2.64a, b are the HYSCORE spectra of the pristine and 850 °C annealed YbAP1.5 glasses under a 400 mT magnetic field, respectively. The HYSCORE spectra of the samples before and after annealing were obtained under the same conditions, so the relative intensity of their resonance signals is comparable. The signal at 4.5 MHz is weak and concentrated in the diagonal area. The signal at 6.9 MHz is strong and shows an extended trend. They originate from ^{27}Al nuclei and

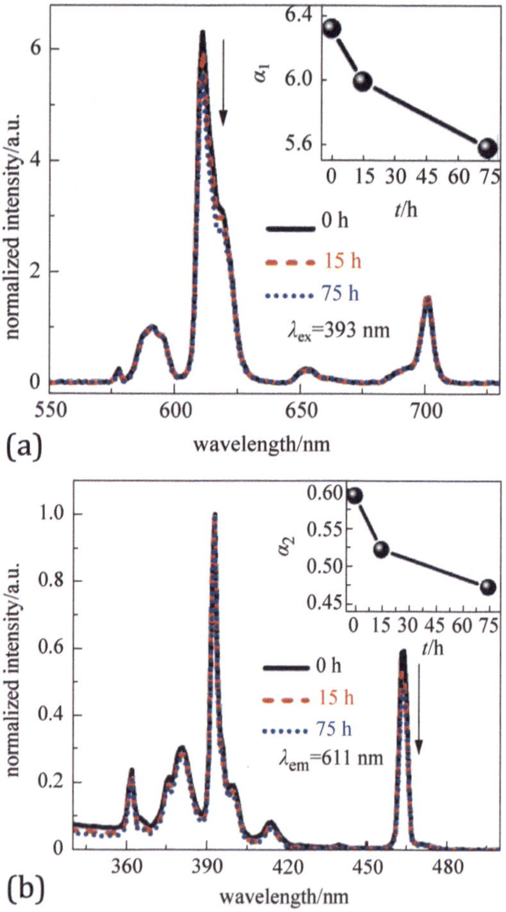

Fig. 2.63 The normalized emission spectrum of Eu^{3+} (**a**), the inset shows the relationship between the transition intensity ratio of $^5D_0 \rightarrow {}^7F_2$ and $^5D_0 \rightarrow {}^7F_1$ (α_1) and the annealing time. The normalized excitation spectrum of Eu^{3+} (**b**), the inset shows the relationship between the transition intensity ratio (α_2) of $^5D_0 \rightarrow {}^7F_2$ and $^5D_0 \rightarrow {}^7F_1$ and the annealing time

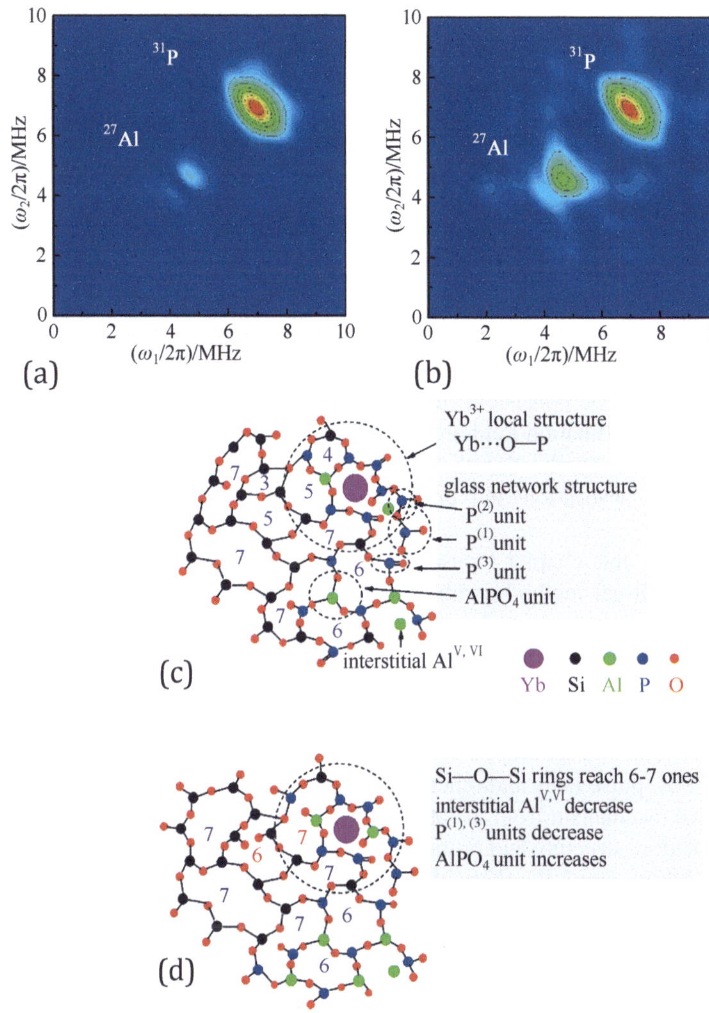

Fig. 2.64 HYSCORE spectra of pristine (**a**) and 850 °C annealed for 75 h (**b**) YbAP1.5 glasses, and schematic diagrams of the structure of YbAP1.5 glass before (**c**) and after annealing for 75 h (**d**) [102]

^{31}P nuclei, respectively. After annealing at 850 °C for 75 h, the intensity ratio of ^{31}P nuclei and ^{27}Al nuclei decreases. That is, the relative content of ^{27}Al nuclei around Yb^{3+} increases. As can be seen from Fig. 2.47, after annealing, the Al^{IV} and $P^{(4)}$ content increase simultaneously, and they tend to interact with each other to form [$AlPO_4$] structures. It is inferred that the number of [$AlPO_4$] structures around Yb^{3+} increases after annealing. Because there are still a small amount of $P^{(3)}$ structures in YbAP1.5 glass after annealing, it is inferred that some Yb^{3+} are still mainly surrounded by ^{31}P nuclei. The solubility of RE^{3+} in silica glass is low, and it is easy to form RE^{3+} clusters [121]. The [$AlPO_4$] structure is similar to that of pure silica glass, so the fluorescence lifetime of Yb^{3+} located around the [$AlPO_4$] structure is shorter. It has been reported that P can completely occupy the second coordination layer of Yb^{3+}, avoiding the formation of Yb—O—Yb and Yb—O—Si bonds, so P can greatly increase the solubility of Yb^{3+} in silica matrix [39]. It is inferred that the decay of Yb^{3+} located around the $P^{(3)}$ structure is slower, and the fluorescence lifetime is relatively longer.

According to the results of FTIR, Raman and NMR (see Figs. 2.46 and 2.47), Figure 2.64c, d show the schematic diagram of the structure of YbAP1.5 glass before and after annealing at 850 °C for 75 h. Annealing leads to the increment in the asymmetry degree of the silicon oxygen, aluminum oxygen, and phosphorus oxygen around Yb^{3+} in YbAP1.5 glass. In addition, the local glass network ring structure tends to be consistent, thereby causing a decrement in the asymmetry degree of the coordination field around Yb^{3+}. Because the transition probability of Yb^{3+} is positively correlated with the asymmetry degree of the coordination field. Therefore, the absorption and emission cross-sections of Yb^{3+} decrease after annealing.

In summary, this chapter focuses on the properties and structure of Yb-doped silica glass. Sol-gel method combined with vacuum sintering process is used to prepare the Yb-doped silica glass containing different co-doping elements (Al, P, Ge, Ce, F, B). The thermal annealing or fire polishing methods are used to change the thermal history (or fictive temperature, T_f) of the Yb-doped silica glass. The effects of glass composition and thermal history on the refractive index and Yb^{3+} spectral properties of Yb-doped silica glass are systematically studied, and NMR, Raman, FTIR, pulse EPR and other structural research methods are used to reveal its influence mechanism from the atomic level microscale. The relationship among the composition, structure and properties of ytterbium-doped silica glass, as well as the relationship among the thermal history, structure and properties of ytterbium-doped silica glass, have been established. The following main conclusions are drawn:

1. The influence rules and mechanisms of different elements on the refractive index of silica glass are revealed. Among them, doping with Yb, P, Ge and Ce will increase the refractive index of silica glass, but doping with $AlPO_4$, F and B will reduce the refractive index of silica glass. The effect of Al doping on the refractive index of silica glass has a turning point. When the content of Al_2O_3 is less than 0.59 mol%, aluminum mainly exists in the form of four coordination, thus to reduce the refractive index of silica glass. When the content of Al_2O_3 is greater than 0.59 mol%, the proportion of four-coordinated aluminum gradually decreases, and the proportion of five and six-coordinated aluminum gradually increases, thus to increase the refractive index of silica glass. Al and P co-doping also has a turning point on the refractive index of silica glass. As the P/Al ratio increases, the refractive index of silica glass first decreases and then increases. When P/Al = 1, the refractive index reaches the minimum value. The fundamental reason is related to the coordination number of aluminum and the change of phosphorus $P^{(n)}$ group.

2. The influence rules and mechanisms of different elements on the spectral properties of Yb^{3+} are revealed. In Yb^{3+} single- doped silica glass, when the content of Yb_2O_3 exceeds 0.05 mol%, Yb^{3+} clusters will appear, leading to the deterioration of Yb^{3+} spectral properties (absorption and emission cross-sections, and fluorescence lifetime all decrease). Co-doping with Al or P can effectively suppress Yb^{3+} clusters. The difference is that, because there is no Al—O—Al connection in silica glass (Al avoidance rule), Al cannot form a solvent shell around Yb^{3+}, thus it will not significantly reduce the local crystal field and the degree of Yb^{3+} asymmetry. Therefore, co-doping with Al can effectively increase the absorption and emission cross-sections of Yb^{3+}. There is a P—O—P connection in silica glass, thus P tends to form a solvent shell around Yb^{3+} based on the enthalpy change effect. it will significantly reduce the local crystal field and asymmetry degree of Yb^{3+}. Therefore, co-doping with P usually reduces the absorption and emission cross-sections of Yb^{3+}. In the Yb/Al co-doped silica glass, the introduction of Ge, F and B has weak effect on the degree of Yb^{3+} clustering and the spectral properties of Yb^{3+}. Among them, co-doping with Ge can slightly increase the absorption and emission cross-sections of Yb^{3+}, which is related to the suppression of Yb^{2+} and the reduction of phonon energy around Yb^{3+}. Co-doping with F or B will slightly reduce the absorption and emission cross-sections of Yb^{3+}, which is related to the structure relaxation of the silica glass promoted by F or B doping, and the reduction of the local crystal field strength and asymmetry degree of Yb^{3+}.
3. The influence of the fictive temperature on the refractive index of Yb-doped silica glass with different components is revealed. For Yb^{3+} single-doped silica glass, the refractive index of the glass is positively correlated with the fictive temperature. The fundamental reason is related to the ring structure. Specifically, as the fictive temperature decreases, the ring structure transforms from 3 ~ 9 to 6 ~ 8 rings. For Yb/Al and Yb/Al/B co-doped silica glasses, the refractive index of the glass is negatively correlated with the fictive temperature. The fundamental reason is related to the change in the coordination number of aluminum. Specifically, as the fictive temperature decreases, the coordination number of aluminum transforms from four-coordination to five, six-coordination. For Yb/Al/P and Yb/Al/P/F co-doped silica glasses, the refractive index of the glass is positively correlated with the fictive temperature. The fundamental reason is related to the change in the coordination number of aluminum and the change in the $P^{(n)}$ group of phosphorus. Specifically, as the fictive temperature decreases, the coordination number of aluminum transforms from five, six-coordination to four-coordination, and the $P^{(n)}$ group of phosphorus transforms from $P^{(2)}$ and $P^{(3)}$ groups to $P^{(4)}$ group, that is, the absolute content of $AlPO_4$ increases.
4. The influence of the fictive temperature on the spectral properties of Yb-doped silica glass with different components is revealed. For Yb^{3+} single-doped, Yb/Al, Yb/Al/B, Yb/Al/P and Yb/Al/P/F co-doped silica glasses, the influence of the fictive temperature on the spectral parameters of Yb^{3+} is basically consistent. As the fictive temperature decreases, the symmetry degree of the silicon-oxygen, aluminum-oxygen, and phosphorus-oxygen tetrahedral structures increases, leading to a decrease in the absorption and emission cross-sections of Yb^{3+}, and an increase in fluorescence lifetime.

Co-doping elements will affect the numerical aperture, thermo-optic coefficient, and laser performance of Yb-doped silica fiber. From the preform rod to the fiber, a series of heating and cooling processes are required, that is, there is a difference between the fictive temperature of the fiber and that of the preform rod. Yb-doped silica glass is the core material of Yb-doped silica fiber, and the refractive index and spectral parameters of Yb-doped silica glass are closely related to the beam quality and laser performance of Yb-doped silica fiber.

References

1. Yang Binhua, Liu Xueqiang, Wang Xin, et al. Compositional dependence of room-temperature Stark splitting of Yb^{3+} in several popular glass systems [J]. Optics Letters, 2014,39(7):1772-1774.
2. Auzel F. On the maximum splitting of the ($^2F_{7/2}$) ground state in Yb^{3+} -doped solid state laser materials [J]. Journal of Luminescence, 2001,93(2):129-135.
3. Robinson C C, Fournier J T. Co-ordination of Yb^{3+} in phosphate, silicate, and germanate glasses [J]. Journal of Physics and Chemistry of Solids, 1970,31(5):895-904.
4. Dai Nengli, Hu Lili, Yang Jianhu, et al. Spectroscopic properties of Yb^{3+} -doped silicate glasses [J]. Journal of Alloys and Compounds, 2004,363(1-2):1-5.
5. Ferhi M, Hassen N B, Bouzidi C, et al. Near-infrared luminescence properties of Yb^{3+} doped $LiLa(PO_3)_4$ powders [J]. Journal of Luminescence, 2016(170):174-179.
6. Barua P, Sekiya E H, Saito K, et al. Influences of Yb^{3+} ion concentration on the spectroscopic properties of silica glass [J]. Journal of Non-Crystalline Solids, 2008,354(42-44):4760-4764.
7. Zhang Liyan, Leng Yuxin, Zhang Junjie, et al. Yb^{3+} -doped fluorophosphate glass with high cross section and lifetime [J]. Journal of Materials Science and Technology, 2010,26(10):921-924.
8. Camy P, Doualan J L, Benayad A, et al. Comparative spectroscopic and laser properties of Yb^{3+} -doped CaF_2, SrF_2 and BaF_2 single crystals [J]. Applied Physics B, 2007,89(4):539-542.
9. Abitan H, Bohr H, Buchhave P. Correction to the beer-lambert-bouguer law for optical absorption [J]. Applied Optics, 2008,47(29):5354-5357.
10. Gao Guojun, Peng Mingying, Wondraczek L. Temperature dependence and quantum efficiency of ultrabroad NIR photoluminescence from Ni^{2+} centers in nanocrystalline Ba-Al titanate glass ceramics [J]. Optics Letters, 2012,37(7):1166-1168.
11. Pollnau M, Eichhorn M. Emission cross section, füchtbauer-ladenburg equation, and purcell factor [C]. Nano-Optics: Principles Enabling Basic Research and Applications, 2017:387-404.
12. Hu Lili, Wang Shikai, Lou Fengguang, et al. Method for preparing Yb-doped silica fiber preform core rod: China, CN103373811B [P]. 2015-05-13.
13. Wang Fan, Hu Lili, Xu Wenbin, et al. Manipulating refractive index, homogeneity and spectroscopy of Yb^{3+} -doped silica-core glass towards high-power large mode area photonic crystal fiber lasers [J]. Optics Express, 2017,25(21):25960-25969.
14. Liu Xuefeng, Gu Zhen'an, Ouli H. Study of the aluminium anomaly in Al-doped silica glasses [J]. Journal of Non-Crystalline Solids, 1989,112(1-3):169-172.
15. Sen S, Youngman R E. High-resolution multinuclear NMR structural study of binary aluminosilicate and other related glasses [J]. The Journal of Physical Chemistry B, 2004,108(23):7557-7564.
16. Jiao Yan, Guo Mengting, Wang Renle, et al. Influence of Al/Er ratio on the optical properties andstructures of Er^{3+}/Al^{3+} co-doped silica glasses [I]. Journal of Applied Physics, 2021,129(5):053104.

17. Xu Wenbin, Ren Jinjun, Shao Chongyun, et al. Effect of P^{5+} on spectroscopy and structure of $Yb^{3+}/A1^{3+}/P^{5+}$ co-doped silica glass [J]. Journal of Luminescence, 2015(167):8-15.
18. Wang Fan, Shao Changyun, Yu Chunlei, et al. Effect of $AlPO_4$ join concentration on optical properties and radiation hardening performance of Yb-doped Al_2O_3-P_2O_5-SiO_2 glass [J]. Journal of AppliedPhysics, 2019,125(17):173104.
19. Kirchhof J, Unger S, Schwuchow A. Fiber lasers: materials, structures and technologies [C]. Optica Fibers and Sensors for Medical Applications III, 2003(4957):1-15.
20. Guo Mengting, Shao Chongyun, Zhang Yang, et al. Effect of B_2O_3 addition on structure and propertiesof $Yb^{3+}/A1^{3+}/B^{3+}$-co-doped silica glasses [J]. Journal of the American Ceramic Society, 2020,103(8):4275-4285.
21. Bubnov M M, Vechkanov V N, Gur'Yanov A N, et al. Fabrication and optical properties of fibers with an Al_2O_3-P_2O_5-SiO_2 glass core [J]. Inorganic Materials, 2009,45(4):444-449.
22. Wang Fan, Dong Hehe, Guo Mengting , et al. Optical, spectroscopic and structural properties of Yb^{3+}-doped silica glasses [J]. Journal of the Chinese Ceramic Society, 2022, 50(4): 991 (in chinese).
23. Hawkins T W. The materials science and engineering of advanced YB-doped glasses and fibers for high-power lasers [D]. Clemson: Clemson University, 2020.
24. Dragic P D, Cavillon M, Ballato A, et al, A unified materials approach to mitigating optical nonlinearities in optical fiber. II. B, The optical fiber, material additivity and the nonlinear coefficients [J]. International Journal of Applied Glass Science, 2018,9(3):307- 318.257
25. Sen S. Atomic environment of high-field strength Nd and Al cations as dopants and major components in silicate glasses: a Nd LIII-edge and Al K-edge X-ray absorption spectroscopic study [J]. Journal of Non-Crystalline Solids, 2000,261(1-3)226-236.
26. Sen S, Orlinskii S B, Rakhmatullin R M. Spatial distribution of Nd^{3+} dopant ions in vitreous silica: a pulsed electron paramagnetic resonance spectroscopic study [J]. Journal of Applied Physics, 2001, 89(4):2304-2308.
27. Dong Hehe, Wang Zhongyue, Shao Chongyun, et al. Effect of co-dopants on the spectral property of Yb^{3+} doped silica glasses at 1018 nm [J]. Optical Materials, 2021(122):111761.
28. Zhu Yiming, Jiao Yan, Cheng Yue, et al. Influence of GeO_2 content on the spectral and radiation-resistant properties of Yb/Al/Ge co-doped silica fiber core glasses [J]. Materials, 2022,15(6):2235.
29. Shao Chongyun, Xu Wenbin, Ollier N, et al. Suppression mechanism of radiation-induced darkening by Ce doping in Al/Yb/Ce-doped silica glasses: Evidence from optical spectroscopy, EPR and XPS analyses [J]. Journal of Applied Physics, 2016,120(15):153101.
30. Shao Chongyun, Wang Fan, Guo Mengting, et al. Structure-property relations in $Yb^{3+}/Al^{3+}/Ce^{3+}/F^-$-doped silica glasses [J]. Journal of the Chinese Ceramic Society, 2019,47(1):120-131.
31. Xu Wenbin, Yu Chunlei, Wang Shikai, et al. Effects of F^- on the optical and spectroscopic properties of Yb^{3+}/Al^{3+}-co-doped silica glass [J]. Optical Materials, 2015(42):245-250.
32. Saito K, Ikushima A J. Effects of fluorine on structure, structural relaxation, and absorption edge in silica glass [J]. Journal of Applied Physics, 2002,91(8):4886-4890.
33. De Oliveira Jr M, Aitken B, Eckert H. Structure of P_2O_5-SiO_2 pure network former glasses studied by solid state NMR spectroscopy [J]. The Journal of Physical Chemistry C, 2018,122(34):19807-19815.
34. Shao Chongyun. Study on the structure, spectrum, radiation resistance and radiation-induced darkening mechanism of Yb^{3+}-doped silica glasses [D]. Beijing: University of Chinese Academy of Sciences, 2019.
35. Sen S, Rakhmatullin R, Gubaydullin R, et al. A pulsed EPR study of clustering of Yb^{3+} ions incorporated in GeO_2 glass [J]. Journal of Non-Crystalline Solids, 2004,333(1):22-27.
36. Shao Chongyun, Ren Jinjun, Wang Fan, et al. Origin of radiation-induced darkening in $Yb^{3+}/Al^{3+}/P^{5+}$-doped silica glasses: effect of the P/Al ratio [J]. The Journal of Physical Chemistry B, 2018,122(10):2809-2820.

37. Saitoh A, Matsuishi S, Se-Weon C, et al. Elucidation of codoping effects on the solubility enhancement of Er $^{3+}$ in SiO $_2$ glass: striking difference between Al and P codoping [J]. The Journal of Physical Chemistry B, 2006,110(15):7617-7620.
38. Sen S, Rakhmatullin R, Gubaidullin R, et al. Direct spectroscopic observation of the atomic-scale mechanisms of clustering and homogenization of rare-earth dopant ions in vitreous silica [J]. Physical Review B, 2006,74(10):100201.
39. Deschamps T, Ollier N, Vezin H, et al. Clusters dissolution of Yb $^{3+}$ in codoped SiO_2 - Al_2O_3 -P_2O_5 silica fiber and its relevance to photodarkening [J]. The Journal of Chemical Physics, 2012,136(1):014503.
40. Funabiki F, Kajihara K, Kaneko K, et al. Characteristic coordination structure around Nd ions in sol-gel-derived Nd-Al-codoped silica glasses [J]. Journal of Physical Chemistry B, 2014,118(29):8792-8797.
41. Schweiger A, Jeschke G. Principles of pulse electron paramagnetic resonance [M]. New York: Oxford University Press, 2001.
42. Chiesa M, Mattsson K, Taccheo S, et al. Defects induced in Yb $^{3+}$ /Ce $^{3+}$ co-doped alumino-silicate silica fiber preforms under UV and γ-ray irradiation [J]. Journal of Non-Crystalline Solids, 2014,403:97-101.
43. Funabiki F, Kamiya T, Hosono H. Doping effects in amorphous oxides [J]. Journal of the Ceramic Society of Japan, 2012,120(1407):447-457.
44. Monteil A, Chaussedent S, Alombert-Goget G, et al. Clustering of rare earth in glasses, aluminum effect: experiments and modeling [J]. Journal of Non-Crystalline Solids, 2004(348):44-50.
45. Lægsgaard J. Dissolution of rare-earth clusters in SiO_2 by Al codoping: a microscopic model [J]. Physical Review B, 2002,65(17):174114.
46. Wang Xue, Zhang Ruili, Ren Jinjun, et al. Mechanism of cluster dissolution of Yb-doped high-silica lanthanum aluminosilicate glass: Investigation by spectroscopic and structural characterization [J]. Journal of Alloys and Compounds, 2017(695):2339-2346.
47. Shao Chongyun, Yu Chunlei, Jiao Yan, et al. Radiation-induced darkening and its suppression methods in Yb $^{3+}$ -doped silica fiber core glasses [J]. International Journal of Applied Glass Science, 2022,13(3):457-475.
48. Youngman R E, Sen S. Structural role of fluorine in amorphous silica [J]. Journal of Non-Crystalline Solids, 2004,349(1):10-15.
49. Youngman R E, Sen S. The nature of fluorine in amorphous silica [J]. Journal of Non-Crystalline Solids, 2004,337(2):182-186.
50. Zhang Long, de Araujo C C, Eckert H. Structural role of fluoride in aluminophosphate sol-gel glasses: high-resolution double-resonance NMR studies [J]. The Journal of Physical Chemistry B, 2007,111(35):10402-10412.
51. Agarwal A, Davis K M, Tomozawa M. A simple IR spectroscopic method for determining fictive temperature, T_f of silica glasses [J]. Journal of Non-Crystalline Solids, 1995,185(1-2):191-198.
52. De Oliveira Jr M, Uesbeck T, Gonçalves T S, et al. Network structure and rare-earth ion local environments in fluoride phosphate photonic glasses studied by solid-state NMR and electron paramagnetic resonance spectroscopies [J]. The Journal of Physical Chemistry C, 2015,119(43):24574-24587.
53. Zhang Ruili, de Oliveira M, Wang Zaiyang, et al. Structural studies of fluoroborate laser glasses by solid state NMR and EPR spectroscopies [J]. The Journal of Physical Chemistry C, 2017,121(1):741-752.
54. Handke M, Sitarz M, Rokita M, et al. Vibrational spectra of phosphate-silicate biomaterials [J]. Journal of Molecular Structure, 2003(651):39-54.
55. Gaafar M S, Marzouk S Y. Structural investigation of Na_2O - B_2O_3 - SiO_2 glasses doped with NdF $_3$ [J]. International Journal of Materials and Metallurgical Engineering, 2016,9(10):1249-1257.

56. Möncke D, Kamitsos E I, Palles D, et al. Transition and post-transition metal ions in borate glasses: Borate ligand speciation, cluster formation, and their effect on glass transition and mechanical properties [J]. The Journal of Chemical Physics, 2016,145(12):124501.
57. Tenney A S, Wong J. Vibrational spectra of vapor-deposited binary borosilicate glasses [J]. The Journal of Chemical Physics, 1972,56(11):5516-5523.
58. Bell R J, Carnevale A, Kurkjian C R, et al. Structure and phonon spectra of SiO_2 , B_2O_3 and mixed SiO_2 -B_2O_3 glasses [J]. Journal of Non-Crystalline Solids, 1980(35):1185-1190.
59. Pasquarello A, Car R. Identification of Raman defect lines as signatures of ring structures in vitreous silica [J]. Physical Review Letters, 1998,80(23):5145-5147.
60. Deters H, de Camargo A S S, Santos C N, et al. Structural characterization of rare-earth doped yttrium aluminoborate laser glasses using solid state NMR [J]. The Journal of Physical Chemistry C, 2009,113(36):16216-16225.
61. Deters H, de Camargo A S S, Santos C N, et al. Glass-to-vitroceramic transition in the yttrium aluminoborate system: structural studies by solid-state NMR [J]. The Journal of Physical Chemistry C, 2010,114(34):14618-14626.
62. Lancry M, Régnier E, Poumellec B. Fictive temperature in silica-based glasses and its application to optical fiber manufacturing [J]. Progress in Materials Science, 2012,57(1):63-94.
63. Lancry M, Régnier E, Poumellec B. Fictive temperature measurements in silicabased optical fiber and its application to Rayleigh loss reduction [M]. InTechopen, 2009:125-159.
64. Tool A Q. Relation between inelastic deformability and thermal expansion of glass in its annealing range [J]. Journal of the American Ceramic Society, 1946,29(9):240-253.
65. Hetherington G. The viscosity of silica glass [J]. Physics and Chemistry of Glasses, 1964,123(5):130-136.
66. Fraser D B. Factors influencing the acoustic properties of silica glass [J]. Journal of Applied Physics, 1968,39(13):5868-5878.
67. Saito K, Ikushima A J, Kotani T, et al. Effects of preirradiation and thermal annealing on photoinduced defects creation in synthetic silica glass [J]. Journal of Applied Physics, 1999,86(7):3497-3501.
68. Babu B H, Lancry M, Ollier N, et al. Radiation hardening of sol gel-derived silica fiber preforms through fictive temperature reduction [J]. Applied Optics, 2016,55(27):7455-7461.
69. Lancry M, Babu B H, Ollier N, et al. Improving optical fiber preform radiation resistance through fictive temperature reduction [C]. Bragg Gratings, Photosensitivity, and Poling in Glass Waveguides, 2016: BW5B.2.
70. Agarwal A, Tomozawa M. Correlation of silica glass properties with the infrared spectra [J]. Journal of Non-Crystalline Solids, 1997,209(1-2):166-174.
71. Saito K, Ikushima A J. Structural relaxation of the frozen-in density fluctuations in silica glass [J]. Progress of Theoretical Physics Supplement, 1997(126):277-280.
72. Watanabe T, Saito K, Ikushima A J. Fictive temperature dependence of density fluctuation in SiO_2 silica glass [J]. Journal of Applied Physics, 2003,94(8):4824-4827.
73. Watanabe T, Saito K, Ikushima A J. Density and concentration fluctuations in F-doped SiO_2 glass [J]. Journal of Applied Physics, 2004,95(5):2432-2435.
74. Hong J. FTIR investigation of amorphous silica fibers and nanosize particles [M]. New York: Rensselaer Polytechnic Institute, 2003.
75. Saito K, Yamamoto R, Kamiya N, et al. Fictive temperature dependences of optical properties in Yb-doped silica glass [C]. Solid State Lasers and Amplifiers III, 2008(6998):368-375.
76. Spinner S, Napolitano A. Further studies in the annealing of a borosilicate glass [J]. Journal of Research of the National Bureau of Standards Section A: Physics and Chemistry, 1966,70(2):147-152.
77. Haken U, Humbach O, Ortner S, et al. Refractive index of silica glass: influence of fictive temperature [J]. Journal of Non-Crystalline Solids, 2000,265(1-2):9-18.
78. Brueckner. R. Properties and structure of vitreous silica: I [J]. Journal of Non-Crystalline Solids, 1970,5(2):123-175.

79. Brueckner. R. Properties and structure of vitreous silica: II [J]. Journal of Non-Crystalline Solids, 1971,5(3):177-216.
80. Hsich H S Y. A non-linear structural relaxation model for the refractive index of glass during annealing [J]. Journal of Materials Science, 1978,13(4):750-758.
81. Haken U, Humbach O, Ortner S, et al. Refractive index of silica glass: Influence of fictive temperature [J]. Journal of Non-Crystalline Solids, 2000,265(1-2):9-18.
82. Gross T M, Tomozawa M. Fictive temperature of GeO_2 glass: its determination by IR method and its effects on density and refractive index [J]. Journal of Non-Crystalline Solids, 2007,353(52-54):4762-4766.
83. Heili M, Poumellec B, Burov E, et al. The dependence of Raman defect bands in silica glasses on densification revisited [J]. Journal of Materials Science, 2015,51(3):1659-1666.
84. Kakiuchida H, Shimodaira N, Sekiya E H, et al. Refractive index and density in F- and Cl-doped silica glasses [J]. Applied Physics Letters, 2005,86(16):161907.
85. Kakiuchida H, Sekiya E H, Shimodaira N, et al. Refractive index and density changes in silica glass by halogen doping [J]. Journal of Non-Crystalline Solids, 2007,353(5-7):568-572.
86. Kiiveri P, Koponen J, Harra J, et al. Stress-induced refractive index changes in laser fibers and preforms [J]. IEEE Photonics Journal, 2019,11(6):1-10.
87. Kuhn S, Hein S, Hupel C, et al. High-power fiber laser materials: influence of fabrication methods and codopants on optical properties [C]. Optical Components and Materials XVI, 2019(10914):27-39.
88. Geissberger A E, Galeener F L. Raman studies of vitreous SiO_2 versus fictive temperature [J]. Physical Review B, 1983,28(6):3266-3271.
89. Uchino T, Kitagawa Y, Yoko T. Structure, energies, and vibrational properties of silica rings in SiO_2 glass [J]. Physical Review B, 2000,61(1):234-240.
90. Awazu K, Kawazoe H. Strained Si—O—Si bonds in amorphous SiO_2 materials: A family member of active centers in radio, photo, and chemical responses [J]. Journal of Applied Physics, 2003,94(10):6243-6262.
91. Galeener F L. Planar rings in vitreous silica [J]. Journal of Non-Crystalline Solids, 1982,49(1-3):53-62.
92. Guo Mengting, Shao Chongyun, Wang Fan, et al. Effect of structural relaxation on the structure and properties of Al^{3+} /Yb^{3+} co-doped silica glasses [J]. Journal of the Chinese ceramic society, 2018,46(11):1499-1506.
93. Guo Mengting. Study on the influence mechanism of thermal history on the performance of Yb^{3+} doped silica glass and fiber [D]. Beijing: University of Chinese Academy of Sciences, 2021.
94. Poe B T, McMillan P F, Angell C A, et al. Al and Si coordination in SiO_2 - Al_2O_3 glasses and liquids: A study by NMR and IR spectroscopy and MD simulations [J]. Chemical Geology, 1992,96(3-4):333-349.
95. Kanehashi K, Stebbins J F. In situ high temperature ^{27}Al NMR study of structure and dynamics in a calcium aluminosilicate glass and melt [J]. Journal of Non-Crystalline Solids, 2007,353(44-46):4001-4010.
96. Wang Min, You Jinglin, Sobol A, et al. In-situ studies of structure transformation and Al coordination of $KAl(MoO_4)_2$ during heating by high temperature Raman and ^{27}Al NMR spectroscopies [J]. Materials, 2017,10(3):310.
97. Poe B T, McMillan P F, Coté B, et al. Magnesium and calcium aluminate liquids in situ high-temperature ^{27}AI NMR spectroscopy [J]. Science, 1993(259): 786-788.
98. Poe B T, McMillan P F, Coté B, et al. Silica-alumina liquids: in-situ study by high-temperature ^{27}Al NMR spectroscopy and molecular dynamics simulation [J]. Journal of Physical Chemistry, 2002,96(21):8220-8224.
99. Greaves G N, Sen S. Inorganic glasses, glass-forming liquids and amorphizing solids [J]. Advances in Physics, 2007,56(1):1-166.
100. Angeli F, Villain O, Schuller S, et al. Effect of temperature and thermal history on borosilicate glass structure [J]. Physical Review B, 2012,85(5):054110.

101. Stebbins J F, Ellsworth S E. Temperature effects on structure and dynamics in borate and borosilicate liquids: high-resolution and high-temperature NMR results [J]. Journal of the American Ceramic Society, 1996,79(9):2247-2256.
102. Guo Mengting, Shao Chongyun, Shi Feng, et al. Effect of thermal annealing on structures and properties of $Yb^{3+}/Al^{3+}/P^{5+}$ co-doped silica glasses [J]. Journal of Non-Crystalline Solids, 2019(522): 119563.
103. Deschamps T, Vezin H, Gonnet C, et al. Evidence of AlOHC responsible for the radiation-induced darkening in Yb doped fiber [J]. Optics Express, 2013,21(7):8382-8392.
104. Ding Jia, Chen Youkuo, Chen Wei, et al. Effect of P_2O_5 addition on the structural and spectroscopic properties of sodium aluminosilicate glass [J]. Chinese Optics Letters, 2012,10(7):071602.
105. Kosinski S G, Krol D M, Duncan T M, et al. Raman and NMR spectroscopy of SiO_2 glasses co-doped with Al_2O_3 and P_2O_5 [J]. Journal of Non-Crystalline Solids, 1988,105(1-2):45-52.
106. Handke M, Mozgawa W, Rokita M. Vibrational spectra of $AlPO_4$ as a structure model for silica glass [C]. Progress in Fourier Transform Spectroscopy, 1997:511-513.
107. Aitken B G, Youngman R E, Deshpande R R, et al. Structure-property relations in mixed-network glasses: multinuclear solid state NMR investigations of the system $x$$Al_2O_3$:(30-x) P_2O_5:70SiO_2 [J]. The Journal of Physical Chemistry C, 2009,113(8):3322-3331.
108. Zhang Long, Bögershausen A, Eckert H. Mesoporous $AlPO_4$ glass from a simple aqueous sol-gel route [J]. Journal of the American Ceramic Society, 2005,88(4):897-902.
109. De Araujo C C, Zhang Long, Eckert H. Sol-gel preparation of $AlPO_4$ - SiO_2 glasses with high surface mesoporous structure [J]. Journal of Materials Chemistry, 2006,16(14):1323-1331.
110. Douglass D C, Duncan T M, Walker K L, et al. A study of phosphorus in silicate glass with ^{31} P nuclear magnetic resonance spectroscopy [J]. Journal of Applied Physics, 1985,58(1):197-203.
111. Guo Mengting, Wang Shikai, Zhao Tongyao, et al. Structural origin of thermally-induced refractive index changes in $Yb^{3+}/Al^{3+}/P^{5+}/F^{-}$ -co-doped silica fiber [J]. Journal of the American Ceramic Society, 2021,104(10):5016-5029.
112. Zhang Long, Eckert H. Short- and medium-range order in sodium aluminophosphate glasses: new insights from high-resolution dipolar solid-state NMR spectroscopy [J]. The Journal of Physical Chemistry B, 2006,110(18):8946-8958.
113. Bak M, Rasmussen J T, Nielsen N C. SIMPSON: a general simulation program for solid-state NMR spectroscopy [J]. Journal of Magnetic Resonance, 2011,213(2):366-400.
114. Ren Jinjun, Eckert H. Applications of DQ-DRENAR for the structural analysis of phosphate glasses [J]. Solid State Nuclear Magnetic Resonance, 2015(72):140-147.
115. Micoulaut M, Cormier L, Henderson G S. The structure of amorphous, crystalline and liquid GeO_2 [J]. Journal of Physics: Condensed Matter, 2006,18(45):R753-R784.
116. Shimodaira N, Saito K, Sekiya E H, et al. In-situ observation of relaxation process in F-doped silica glass by Raman spectroscopy [J]. Advances in Glass and Optical Materials, 2006(173):79-86.
117. Krol D M. Femtosecond laser modification of glass [J]. Journal of Non-Crystalline Solids, 2008,354(2-9):416-424.
118. Ebendorff-Heidepriem H, Ehrt D. Spectroscopic properties of Eu^{3+} and Tb^{3+} ions for local structure investigations of fluoride phosphate and phosphate glasses [J]. Journal of Non-Crystalline Solids, 1996,208(3):205-216.
119. Zhang Qiang, Qiao Yanbo, Qian bin, et al. Luminescence properties of the Eu-doped porous glass and spontaneous reduction of Eu^{3+} to Eu^{2+} [J]. Journal of Luminescence, 2009,129(11):1393-1397.
120. Zhang Qiang, Liu Xiaofeng, Qiao Yanbo, et al. Reduction of Eu^{3+} to Eu^{2+} in Eu-doped high silica glass prepared in air atmosphere [J]. Optical Materials, 2010,32(3):427-431.
121. Galant E I, Kondrat'ev Y N, Przhevuskii A K, et al. Stimulated emission of neodymium ions in quartz glass [J]. Soviet Journal of Experimental and Theoretical Physics Letters, 1973(18):372.

Chapter 3
Preparation and Characterization of Rare Earth Doped Silica Fiber

The first step in the preparation of rare earth doped silica fiber is the preparation of the preform. The preform consists of rare-earth doped core glass rod and silica cladding. The preform is then heated and softened in a high-temperature furnace, and drawn into a fiber of a certain diameter under the action of traction, a process known as fiber drawing completed by drawing tower. In order to ensure the quality of the fiber material, appropriate detection methods and analysis methods must be selected to characterize the prepared fiber. This chapter will elaborate on the preparation of rare earth doped silica fiber preforms by chemical vapor deposition, non-chemical vapor deposition, and the drawing and characterization of rare earth doped silica fiber.

3.1 Preparation of Rare Earth Doped Silica Fiber Preform by Chemical Vapor Deposition

The performance of the fiber depends on the internal structure and the quality of the core glass rod, so the preparation of the preform is the most important part of the fiber preparation process. The laser fiber preform consists of a rare earth doped core glass rod and a cladding structure used to confine light. The commonly used method for preparing preforms is the modified chemical vapor deposition (MCVD) combined with liquid doping technology. This technology is mainly used to prepare double-clad fibers, which are widely used in the preparation of fiber laser. With the development of high-power laser technology, the increase in laser power of traditional small-core-diameter double-clad fiber lasers is limited by laser damage of the fiber end-face and nonlinear effects, and it is necessary to improve or develop new preform preparation technologies. In addition, different companies have developed

L. Hu, *Rare Earth Doped Silica Fiber and Its Applications*,
https://doi.org/10.1007/978-981-96-5306-5_3

their own preform preparation technologies to overcome technical barriers. This section mainly introduces the preparation technologies of rare earth doped fiber preforms based on chemical vapor deposition.

3.1.1 *Modified Chemical Vapor Deposition (MCVD)*

The Modified Chemical Vapor Deposition (MCVD) process [1] was invented by Bell Labs in 1973. The process is shown in Fig. 3.1: A deposition tube is installed in the middle of a glass lathe, with the left end of the deposition tube connected to the feed system and the right end connected to the waste gas treatment system. The feed system introduces $SiCl_4$, $GeCl_4$, and $POCl_3$ vapors into the deposition tube through oxygen bubbling. The hydrogen-oxygen flame torch below the deposition tube heats the deposition tube at a controllable speed and flame temperature, causing high-temperature chemical reactions in the heating area, forming the required dopant silica glass. The torch moves to the exhaust end of the deposition tube and quickly returns to the intake end, starting the second deposition. By repeating the above mentioned deposition steps, multiple depositions can be performed. After the deposition is completed, the deposited high-purity materials and the deposition tube are melted together at a high temperature (above 2000 °C) to form a solid glass rod, which is the core glass rod. Then, the fiber preforms can be made using various cladding technologies.

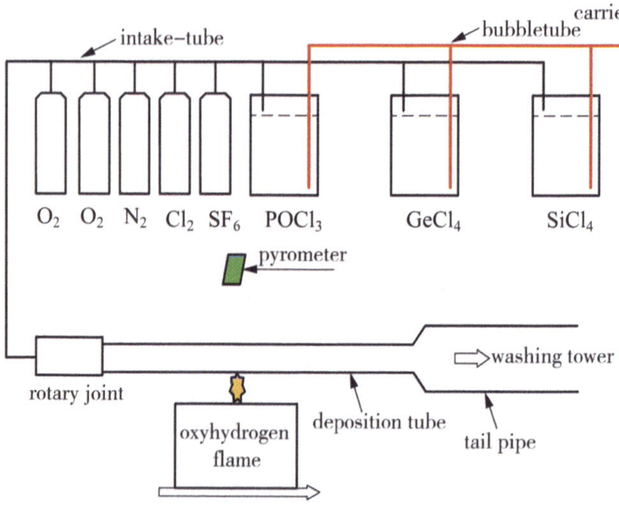

Fig. 3.1 Schematic of MCVD equipment

The reaction mechanism of the MCVD process is shown as follows:

$$SiCl_4 + O_2 \xrightarrow{\Delta} SiO_2 + 2Cl_2 \tag{3.1}$$

$$GeCl_4 + O_2 \xrightarrow{\Delta} GeO_2 + 2Cl_2 \tag{3.2}$$

$$4POCl_3 + 3O_2 \xrightarrow{\Delta} 2P_2O_5 + 6Cl_2 \tag{3.3}$$

$$4BCl_3 + 3O_2 \xrightarrow{\Delta} 2B_2O_3 + 6Cl_2 \tag{3.4}$$

$$9SiO_2 + SF_6 \xrightarrow{\Delta} 6Si_{1.5}F + SO_2 + 8O_2 \tag{3.5}$$

$$H_2O + Cl_2 \xrightarrow{\Delta} 2HCl + HClO \tag{3.6}$$

In these formulas, $SiCl_4$ is the base material for fiber deposition, $GeCl_4$ and $POCl_3$ can increase the refractive index, while BCl_3 and SF_6 are used to reduce the refractive index, and Cl_2 is used to remove hydroxyl group in the fiber. All the above reactions need to be carried out at high temperature. There is a competitive relationship among these chemical reactions. The degree of the reaction, the rate of chemical reaction, the direction of the reaction, and the equilibrium constant are all affected by the reaction temperature, gas pressure, and the composition of the gas in the deposition tube.

The MCVD technology deposits SiO_2 inside the tube, it is not easily affected by the environment, so it can produce low optical loss, high optical quality fiber preforms. At the same time, benefit from its simplicity and flexibility, it is widely used to prepare active fiber preforms. Unlike communication fibers, the preparation of active fiber preforms requires doping with rare earth and other elements, which puts higher requirements on the doping technology of MCVD equipment.

The glass network structure of silica is dense, and the rare earth ions have large ionic radii, so it is difficult for rare earth ions to directly enter the gaps in the silica structure. Silica glass is a refractory oxide with a high viscosity at high temperatures, and the diffusion distance of rare earth ions is limited, so it is difficult to obtain rare earth doped silica glass with high homogeneity. To overcome these difficulties, it is necessary to make the rare earth ions and silica well-mixed before melting. At the same time, co-dopants such as Al_2O_3 and P_2O_5 need to be added to disrupt the dense structure of silica, improve the solubility, homogeneity, and the spectral performance of rare earth ions in silica glass. The main methods for preparing rare earth-doped fiber preforms are MCVD combined with solution soaking method and MCVD combined with gas phase doping method. They will be introduced in detail as follows.

3.1.1.1 MCVD Combined with Solution Soaking Method

The MCVD combined with solution soaking method is a classic method for preparing rare-earth doped fiber preforms [2–5], with Nufern in the United States as a representative of rare earth doped laser fiber manufacturers using this method to prepare commercial fibers. The main process flow is shown in Fig. 3.2. Before starting the deposition, SF_6 is usually used to clean the inner wall of the silica tube. The preparation of rare earth doped preform core glass rod will be elaborated from the four process steps of loose body deposition in the tube, rare earth solution soaking and post-treatment, MCVD lathe sintering, and MCVD lathe rod shrinking.

1) Loose Body Deposition in the Tube

When depositing loose body in the silica casing, $SiCl_4$, $GeCl_4$, $POCl_3$ react with O_2 at about 1300 ~ 1650 °C to form submicron particles, which are deposited on the colder tube wall through thermal migration. However, due to the insufficient sintering at this temperature range, the sintering is not complete, thus forming a porous structure, which is called loose body. The thickness of the loose body is about tens of micrometers, it is mainly determined by the flow rate of each reactant and the deposition temperature. Although loose body can be formed at as low as 1200 °C, the obtained loose body is more likely to fall off from the tube wall. In the process, according to the volatility of the deposit, loose body deposition can be divided into forward deposition and reverse deposition, and the reverse deposition process is suitable for the deposition of highly volatile loose body.

Forward deposition [2, 3] refers to the same direction of the reaction gas flow and the hydrogen-oxygen flame movement. The reactants undergo oxidation reactions in the high-temperature area to generate SiO_2, GeO_2, P_2O_5 and other oxides, which migrate to the tube wall through the thermophoresis process, and then are sintered and fixed by the moving hydrogen-oxygen flame. The deposition and sintering of the loose body are completed in the same hydrogen-oxygen flame movement process, as shown in Figure 3.3a.

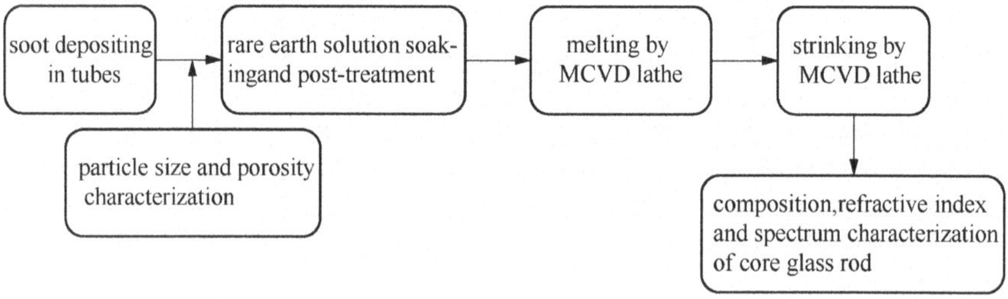

Fig. 3.2 Process flow chart of preparing rare earth doped fiber preform core glass rod by MCVD combined with solution soaking method

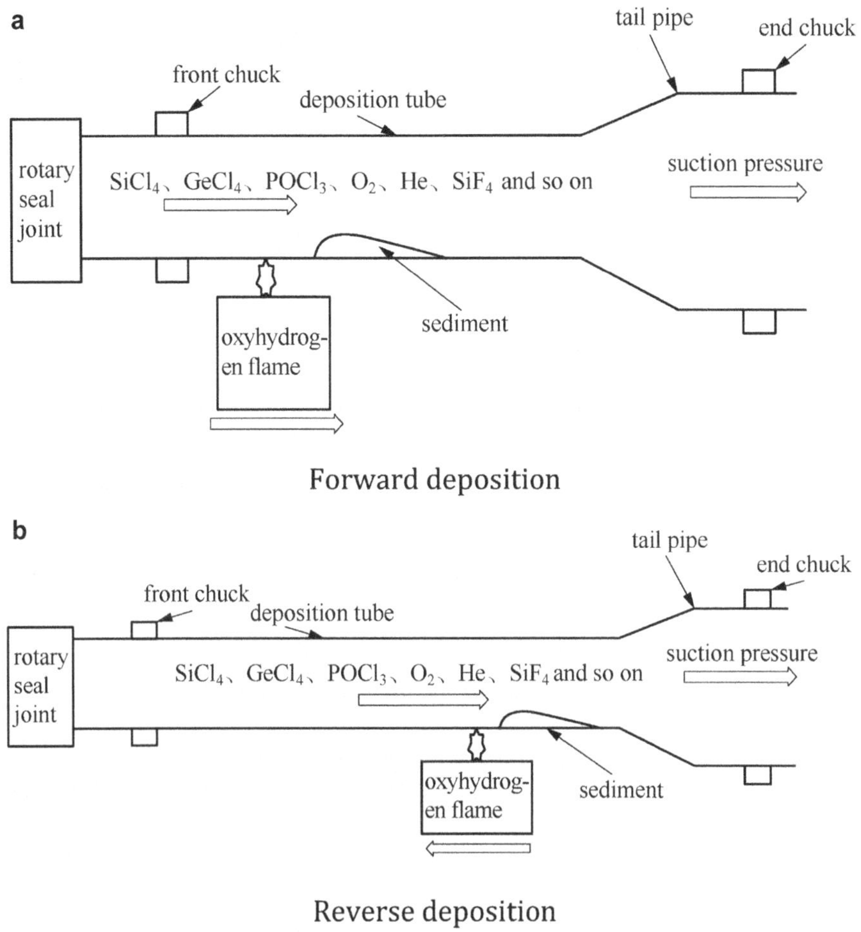

Fig. 3.3 Schematic diagram of MCVD tube internal deposition process (drawn according to references [2, 3, 6, 7]). (**a**) Forward deposition. (**b**) Reverse deposition

Reverse deposition [6, 7] refers to the reaction gas flow and the hydrogen-oxygen flame moving in opposite directions. Because the hydrogen-oxygen flame moves in the opposite direction, the oxide does not experience high temperature after it is formed, but simply adheres to the tube wall. After the deposition is finished, the hydrogen-oxygen flame is moved again to fix the loose body generated in the first step on the tube wall at a temperature 200 ~ 300 °C lower than the deposition temperature, to prevent it from falling off during the soaking process, as shown in Figure 3.3b. Because the deposition temperature of the reverse deposition process is higher, it is conducive to the efficient reaction of chlorides to form oxides, but the sintering temperature is lower, which is conducive to the retention of volatile substances in the loose body, suitable for the preparation of loose bodies with high doping of P, Ge and other volatile substances.

The porosity of the loose body is related to the sintering temperature. For forward deposition, the deposition temperature is the sintering temperature, while the porosity of the loose body in the reverse deposition process mainly depends on the secondary sintering temperature. If the sintering temperature is too low, the loose body is prone to fall off during the soaking process. If the sintering temperature is too high, the loose body will be too dense, which is not conducive to the entry of doping ions from the solution. Figure 3.4 is a comparison of scanning electron microscope photos of the SiO_2 -P_2O_5 loose body before and after secondary sintering in reverse deposition. After secondary sintering, the particles stick to each other and are denser than before sintering.

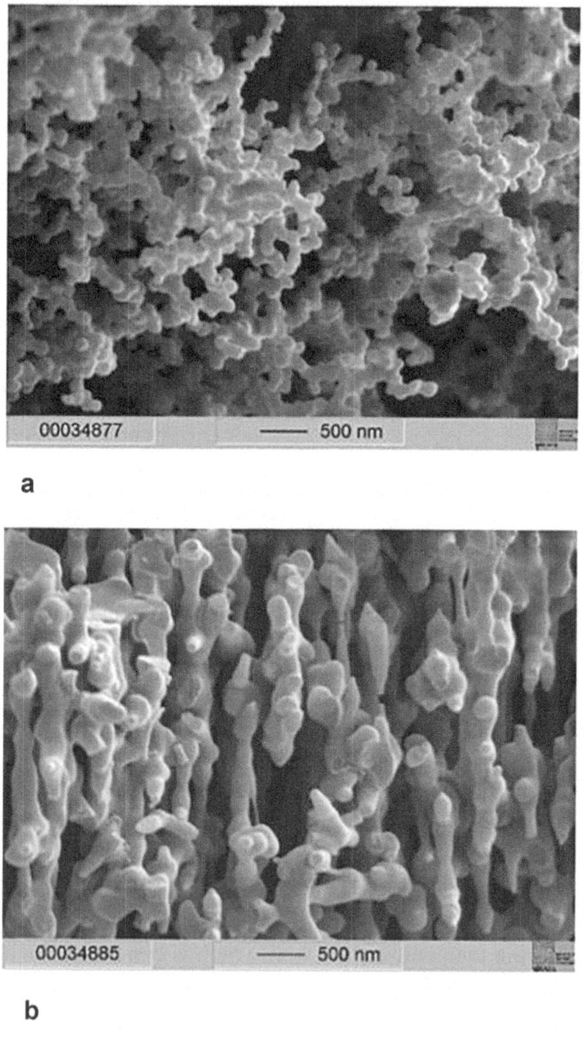

Fig. 3.4 Comparison of scanning electron microscope photos of SiO_2 -P_2O_5 loose body before (**a**) and after (**b**) secondary sintering in reverse deposition [6]

Table 3.1 lists the parameters of the reverse deposition loose body at different sintering temperatures, where φ_{rel} is the density ratio before and after the glassification of the loose body, used to characterize the compactness of the loose body. The smaller the φ_{rel}, the higher the porosity of the loose body, the more conducive to the entry of rare earth ions. It can be seen that the loose body prepared at a lower secondary sintering temperature has a higher porosity and the content of volatile matter P_2O_5 is higher. This is because the higher sintering temperature causes the higher degree of sintering densification of the porous body and more serious volatilization of P_2O_5.

In the forward deposition process, the smaller the porous body, the closer its deposition position to the tube wall, as shown in Fig. 3.5. However, because smaller particles have a larger specific surface area than larger particles, they are sintered faster, making the density of the porous body near the tube wall even more higher.

For pure silica or lightly doped GeO_2 /P_2O_5 /F porous body, the sintering temperature needs to be 1450 ~ 1600 °C. If the temperature is too low, the porous body is likely to fall off from the tube wall. For porous body with a large amount of GeO_2 /P_2O_5 /F co-deposited during the deposition process, the deposition temperature of the porous body can be significantly reduced depending on the content of the doped elements. For example, Fig. 3.6 shows the scanning electron microscope photos of the porous bodies obtained by forward deposition at deposition temperatures of 1220 °C and 1295 °C with a $GeCl_4$ /$SiCl_4$ flow ratio of 0.86. Figure 3.7 shows the scanning electron microscope photo of the porous body obtained by forward deposition at a deposition temperature of 1100 °C with a $POCl_3$ /$SiCl_4$ flow ratio of 0.48.

Table 3.1 Parameters of reverse deposition porous body at different sintering temperatures [6]

Sample Number	Secondary Sintering Temperature/°C	φ_{rel}	P_2O_5/mol%
A	Unsintered	0.05	9.9
B	950	0.06	8.1
C	1050	0.12	7.8
D	1100	0.23	7.2
E	1135	0.36	7.1
F	1175	0.75	6.0

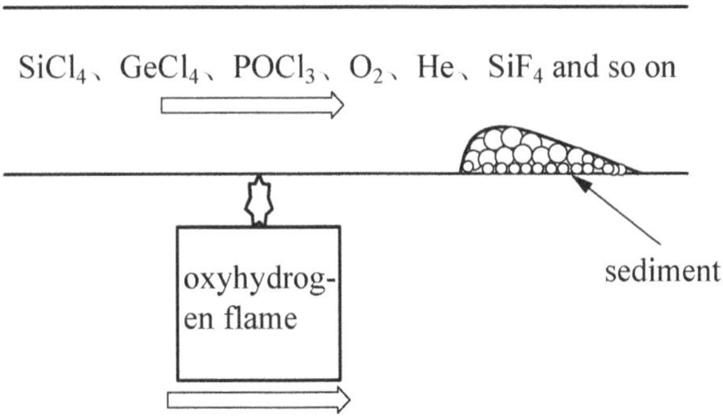

Fig. 3.5 Schematic diagram of the distribution of porous particles of different sizes (Drawn according to literature [2])

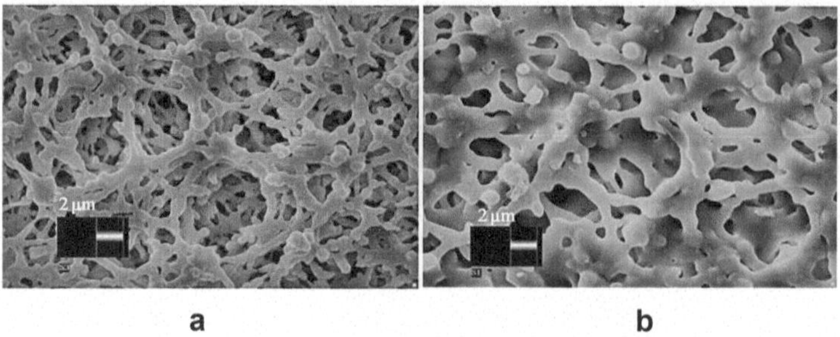

Fig. 3.6 SiO_2-GeO_2 porous bodies at different deposition temperatures [8]. (**a**) 1220 °C and (**b**) 1295 °C

Fig. 3.7 SiO_2 -P_2O_5 porous body deposited at 1100 °C [8]

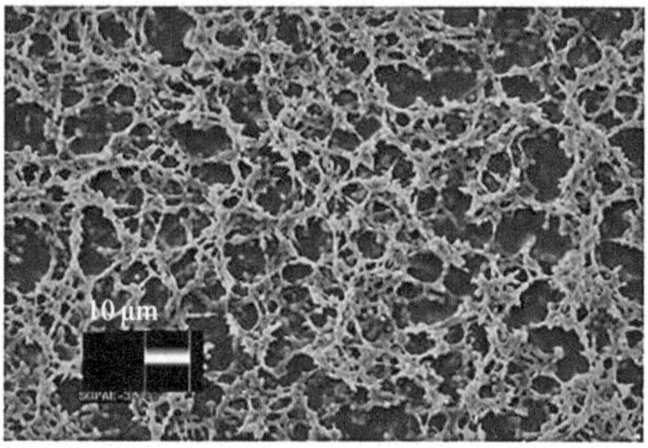

As can be seen from Figs. 3.6 and 3.7, under conditions below 1300 °C, the particles of the porous body have already adhered to each other, which is due to the fact that GeO_2 /P_2O_5 /F can significantly reduce the melting point of SiO_2.

2) Solution Doping

The quartz tube with the deposited porous body is usually vertically soaked in a prepared solution for about 60 min, allowing the doped ions to be adsorbed on the surface or in the pores of the porous body. The device is shown in Fig. 3.8, and the operation steps are as follows:

(1) Install the deposition tube on the Teflon seal plug.
(2) Open the valve and the solution is pressed into the deposition tube using N_2.
(3) Close the valve, let it stand for 60 min, and let the porous body fully absorb the solution.
(4) Open the valve and let the solution in the deposition tube flow back into the conical flask.
(5) Close the valve and reinstall the deposition tube on the lathe.

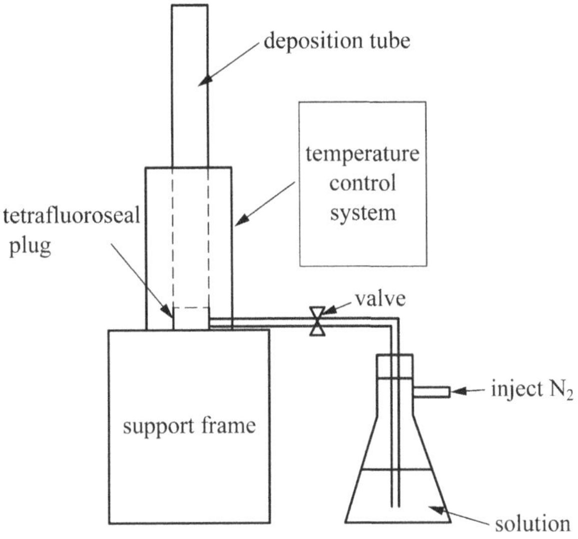

Fig. 3.8 Schematic diagram of the solution soaking device [9]

The effect of solution doping is mainly determined by two factors: the porosity of the porous body and the adsorption efficiency of the porous body. The porosity depends on the deposition process of the porous body, and the adsorption efficiency of the porous body can be described by the capillary model [9]. As shown in Fig. 3.9, the pores of the porous body are abstracted as capillaries, and the adsorption height conforms to the relationship of Eqs. (3.7) and (3.8):

$$h = \frac{F}{\pi a^2 g \rho} \tag{3.7}$$

$$F = 2\pi a \gamma \cos\theta \tag{3.8}$$

where, h is the adsorption height; F is the capillary force; γ is the surface tension of the liquid; θ is the angle between the tangent of the concave liquid surface and the tube wall; g is the gravitational constant; ρ is the density of the liquid; a is the radius of the capillary. The larger the h, the more solution the porous body absorbs per unit area, and the more rare earth ions will be doped. From Eqs. (3.7) and (3.8), h is proportional to the capillary force, and the capillary force is proportional to the surface tension. The surface tension of the liquid is related to the properties of the liquid itself, and it is also a function related to temperature. As the temperature rises, the surface tension decreases. Therefore, the larger the surface tension of the solvent and the lower the soaking temperature, the easier to obtain a highly doped porous body, theoretically.

The selection of solvent considers the following aspects. Firstly, consider the solubility of the doped ions in the solvent. Secondly, consider the adsorption efficiency of the solvent. And thirdly, consider the volatility of the solvent. Since Al^{3+}

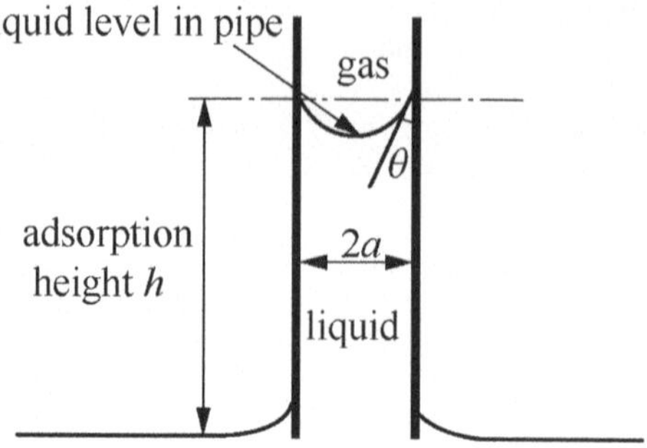

Fig. 3.9 Schematic diagram of the capillary model [9]

Table 3.2 Surface tension of each solvent at room temperature (20 °C). Surface tension [9]

Solvent	Surface tension/(mN/m)
Water	72
Methanol	22.7
Ethanol	22.4
Propanol	21.3

has the effect of increasing the doping concentration of rare earth ions in silica glass, reducing the clustering of rare earth ions, and improving the luminescence performance of rare earth ions. Therefore, the choice of solvent needs to consider the solubility of Al^{3+} and rare earth ions in it. Common solvents are water, ethanol or methanol, etc. Table 3.2 is the surface tension of four solvents at room temperature, and it can be seen that the surface tension of the water solution is the largest. According to the aforementioned capillary model, it can adsorb more solvent, and the solubility of Al^{3+} and rare earth ions in water is greater, which is conducive to achieving high doping. Compared with water solvents, the advantages of solvents such as ethanol or methanol are that they are more volatile and cause less hydroxyl loss.

Table 3.3 shows the doping concentrations of Al^{3+} and Er^{3+} in the preform core glass rods prepared by soaking in solutions at different temperatures. Experiments show that under the same conditions, low-temperature soaking is more likely to achieve high doping. This is because the surface tension of the solution at low temperatures is larger, the adsorption force of the porous body is stronger, the adsorption height is larger, which is consistent with the expectations of capillary theory.

Table 3.4 shows the fiber parameters of the porous body prepared under different deposition conditions. The absorption of Er^{3+} at 980 nm can directly reflect the concentration of Er^{3+} in the fiber. As can be seen from the data in the table, as the deposition temperature increases, the porosity of the porous body decreases, and under the same soaking solution conditions, the concentration of Er^{3+} is approximately proportional to the porosity of the porous body.

In addition, the doping concentration of ions can be further increased by repeating the solution soaking, drying, and dehydrating process multiple times. The results obtained by soaking the porous body multiple times in $AlCl_3$ solution of different

Table 3.3 Doping concentration of Al^{3+} and Er^{3+} in preform core glass rods prepared by soaking in solution at different temperatures

Solution composition	Soaking temperature/°C	Er^{3+} concentration/wt%	Al^{3+} concentration/wt%
0.6 mol/L $AlCl_3$ + 0.06 mol/L $ErCl_3$	65	0.33	0.32
	13	0.34	0.4
	−20	0.58	0.78

Table 3.4 Porous body preparation process parameters and fiber performance [8]

Reactant composition	Soaking solution concentration	Deposition temperature/°C	Porosity of the porous body	Er^{3+} Absorption @980 nm
$GeCl_4$ / $SiCl_4$ = 0.86	0.3 mol/L $AlCl_3$ + 0.01 mol/L $ErCl_3$	1255	32%	3.12
$GeCl_4$ / $SiCl_4$ = 0.86	0.3 mol/L $AlCl_3$ + 0.01 mol/L $ErCl_3$	1295	24%	2.265
$POCl_3$ / $SiCl_4$ = 0.48	0.3 mol/L $AlCl_3$ + 0.01 mol/L $ErCl_3$	1100	50%	4.990

Table 3.5 Changes in the refractive index of core rod glass soaked in $AlCl_3$ solution multiple times [10]

Soaking solution concentration	Refractive index difference between core rod glass and pure silica/($\times 10^{-3}$)			Percentage increase in refractive index	
	First soaking	Second soaking	Third soaking	From first to second	From second to third
0.3 mol/L $AlCl_3$	1.44	1.97	2.2	37	17
0.7 mol/L $AlCl_3$	2.7	4.5	6.0	67	33
1.2 mol/L $AlCl_3$	5.4	8.1	10.0	50	23.5

concentrations are shown in Table 3.5. Since Al_2O_3 can increase the refractive index in silica glass, the refractive index difference can be used to indirectly represent the content of Al_2O_3 in the glass. As the number of soakings increases, the doping concentration in the core rod glass gradually increases, but the percentage of refractive index increase for each soaking shows a decreasing trend, indicating that as the number of soakings increases, the concentration of Al_2O_3 in the glass will gradually tend to saturate, that is, the method of multiple soakings to increase the doping concentration is also limited.

3) Sintering and Rod Collapse

The dehydrated deposited quartz tube is sintered at 1800 ~ 2100 °C to form transparent silica glass, and the rod collapse is completed at about 2200 ~ 2300 °C. Due to the high temperature of the rod collapse process, it is easy to cause volatilization and form a refractive index center depression. For Yb-doped fibers, the formation of the center depression can easily generate more laser modes, which is not conducive to single mode maintaining or quasi-single mode operation at high power, especially for Al-P-Yb co-doped silica fiber. Due to the strong volatility of phosphorus pentoxide, the problem of center depression is often very serious. The literature [11] avoids the center depression of Yb-doped fiber by the combined use of two process means: ① A certain flow of $POCl_3$ is introduced during the collapse process; ② A low-temperature collapse process of -500 Pa is adopted. Figure 3.10 is the refractive index distribution diagram of the Yb-doped fiber preform of the aluminum-phosphorus co-doped system under different process conditions. Figure (a) is the refractive index diagram obtained by the normal collapse process, with a clear refractive index depression in the middle; figure (b) is the refractive index depression during the collapse process with a certain flow of $POCl_3$, the refractive index depression has been improved to some extent; figure (c) is the refractive index depression during the collapse process with a certain flow of $POCl_3$, and in the last step of the collapse process, a super high negative pressure of -500 Pa is used, and a collapse temperature lower than the conventional collapse temperature by 400 ~ 500 °C is used, which can completely avoid the refractive index depression.

3.1.1.2 MCVD Combined with In-Situ Solution Soaking Doping Method

The main problem of the above mentioned MCVD combined with solution soaking process is that it requires multiple removals and installations of the deposited quartz tube from the lathe, which will result in the loss of usable length of the deposited quartz tube and may introduce contaminants. These problems are more serious when depositing multiple layers of loose body such as preparing large-core fiber preforms. The in-situ solution soaking technology has been developed by University of Southampton, UK [12], as shown in Fig. 3.11. A thin glass tube is inserted into the deposited silica tube through the tailstock of the lathe. The tip of the thin glass tube has a specific structure to prevent damage to the porous layer. The thin glass tube is connected to a peristaltic pump, and the methanol solution of chloride is fed into the deposited silica tube at a speed of 7 ~ 15 ml/min. Once a thin layer of solution immerses the porous layer in the length direction, the thin glass tube can be pulled out, rather than completely filling the deposited silica tube with solution as in vertical solution soaking. The solvent is then evaporated, leaving the doped ions in the porous body. The deposited silica tube is kept rotating throughout the entire process. After drying, dehydrating, and sintering the porous body, the next layer of porous body is deposited.

Using the in-situ solution soaking technology, a large mode field fiber preform with 10 layers of Yb^{3+} /Al^{3+} co-doped is prepared (porous body preparation parameters: $SiCl_4$ flow rate 100 ml/min, deposition temperature 1475 °C). The process takes about 8 h, i.e., a single layer of porous body deposition, solution soaking, drying, dehydration, and vitrification process takes about 50 min. The refractive index distribution (measured by PK2600) of the fiber preform is shown in Fig. 3.12, with

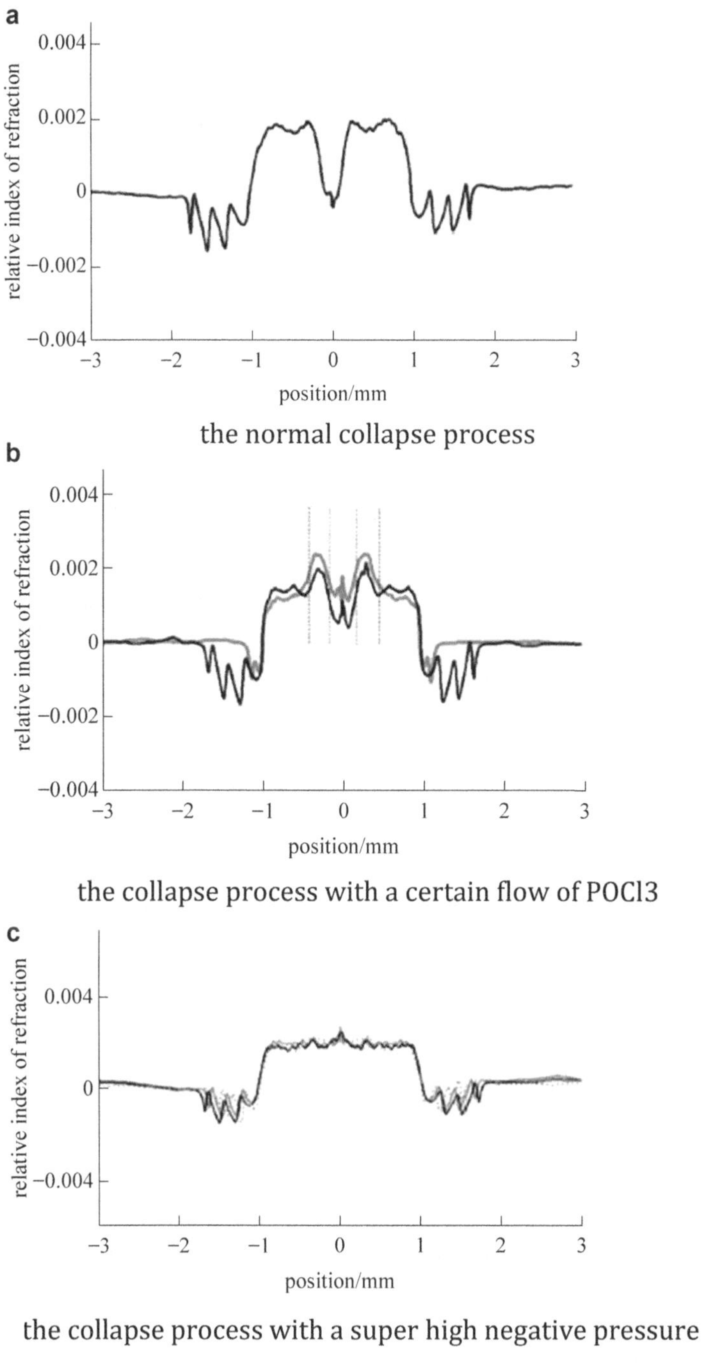

Fig. 3.10 Refractive index distribution diagram of the Yb-doped fiber preform of the aluminum-phosphorus co-doped system under different process conditions. (**a**) the normal collapse process, (**b**) the collapse process with a certain flow of POCl3. (**c**) the collapse process with a super high negative pressure

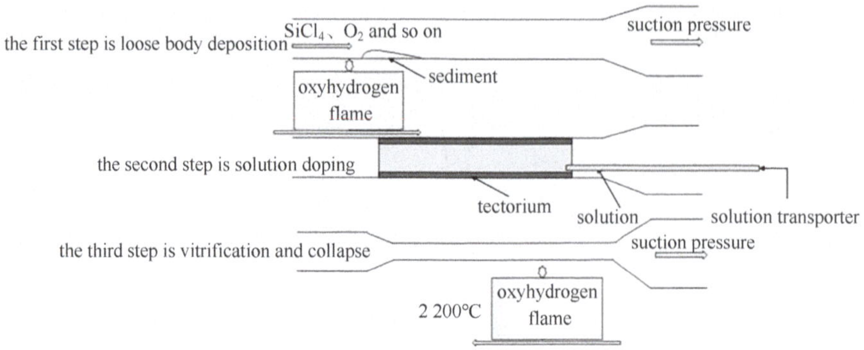

Fig. 3.11 Schematic diagram of the in-situ solution soaking technology process [12]

Fig. 3.12 Refractive index distribution of large mode field fiber preform with 10 layers of Yb^{3+} /Al^{3+} co-doped layers [12]

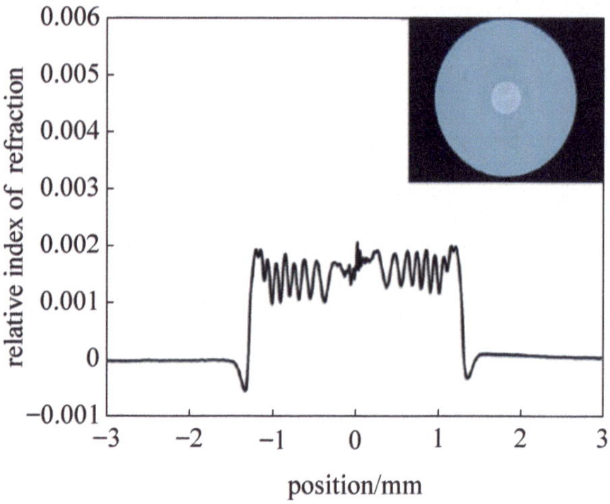

a refractive index fluctuation of about 0.001, and the 10 fluctuation peaks also distinguish each deposition layer. The average NA of the preform is 0.07, and the core diameter is 2.6 mm. After being drawn into a fiber, the refractive index fluctuation is reduced to less than 0.0001 (measured by S14).

In order to compare the ion doping effects of in-situ solution soaking technology and conventional vertical soaking technology, they prepared a single porous body layer (porous body deposition temperature 1550 °C) using the same deposition process, and the same Yb^{3+} /Al^{3+} co-doped solution was subjected to in-situ solution soaking (A) and conventional vertical soaking (B) respectively. The refractive index distribution of the fiber preform rod (measured by PK2600) is shown in Fig. 3.13. It can be seen that the refractive index of the preform rod prepared by in-situ solution soaking is higher. This is because the "dripping dry" process of vertical soaking causes many doped ions to be lost from the porous body, while the in-situ solution soaking method uses the "evaporation" process to remove the solvent from the porous body, thus retaining more doped ions in the porous body.

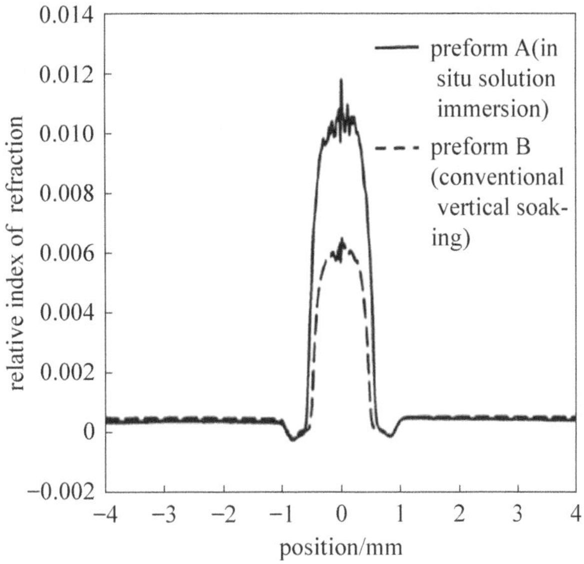

Fig. 3.13 Refractive index distribution of fiber preforms prepared by in-situ doping and conventional solution soaking technology [12]

3.1.1.3 MCVD Combined with Sol-Gel Solution Soaking Doping Method

In addition to the above two solution soaking doping methods, Hu Lili and other researchers from the Shanghai Institute of Optics and Mechanics, Chinese Academy of Sciences invented a new method for preparing rare-earth doped silica fiber preform rods: sol-gel solution soaking doping [13]. Unlike in-situ solution soaking and conventional vertical soaking using rare earth chloride dissolved in organic alcohol solution, sol-gel solution soaking doping can configure Yb, Al and other dopants' chlorides and other raw materials with a suitable amount of tetraethyl orthosilicate, ethanol, water, etc. in a certain proportion into a sol-gel solution, so that Yb^{3+} and co-dopant Al^{3+} can be pre-dispersed uniformly in the silica suspension, and then the sol-gel solution soaking doping is carried out by vertical soaking. This method is a combination of MCVD technology and sol-gel technology, which not only retains the molecular-level uniform doping characteristics of the sol-gel solution, but also has the low-loss advantage of MCVD technology, improves the homogeneity of the preform rod core doping, greatly reduces the Yb^{3+} clustering degree, and effectively alleviates the photodarkening effect of Yb-doped fibers. The refractive index distribution and fiber loss of the preform rod prepared by this method are shown in Fig. 3.14. The refractive index depression in the middle of the preform rod is effectively filled, the doping homogeneity is improved, and the low-loss characteristic of MCVD fiber is maintained. The photodarkening of Yb-doped silica fiber drawn from the preform rod prepared by this method was tested, and the results showed that the fiber could work under 1000 W pump conditions for 500 h, with a power reduction within 5%, meeting the requirements for long-term stable operation of Yb-doped fibers [13]. Yang [14] prepared a high-concentration Er^{3+} doped silica fiber using this method, with an Er^{3+} doping concentration of up to 15,000 ppm, and

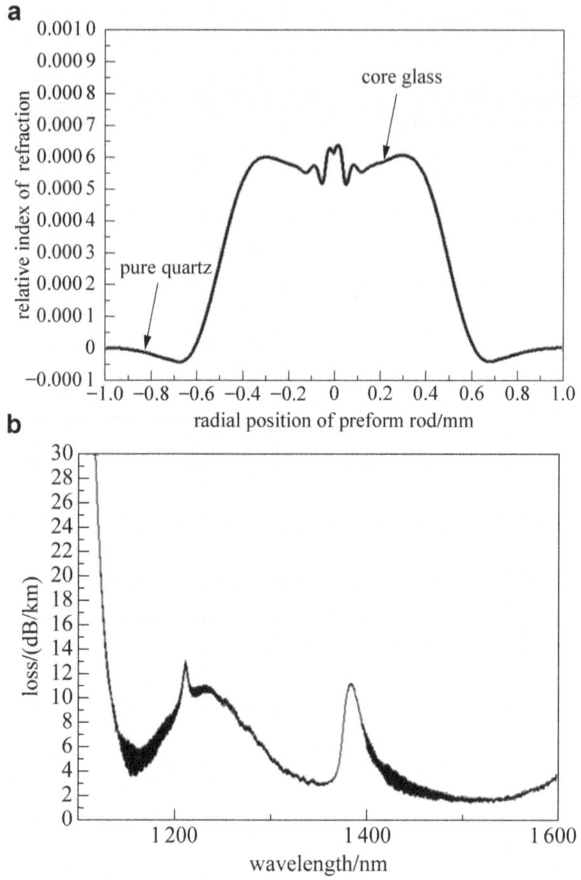

Fig. 3.14 Refractive index distribution of the core rod of the preform rod prepared by MCVD combined with sol-gel solution soaking doping (**a**) and optical loss of the drawn fiber (**b**)

uniformly dispersed in silica glass. A cladding fiber with a core of 8 μm was drawn from this preform rod, and the broadband gain coefficient and noise index of this fiber at different lengths were tested, as shown in Fig. 3.15. It can be seen that, at a length of 25 cm, the fiber has a maximum gain of 23 dB at a wavelength of 1530 nm, and the 75 cm long fiber has the highest gain at the 1560 nm band, greater than 23 dB. The noise figure is controlled at around 5 dB. The greatest advantage of this fiber is the high doping concentration of Er^{3+}, which can greatly shorten the length of the fiber to achieve high gain amplification performance per unit length.

3.1.1.4 MCVD Gas Phase Doping Method

To overcome the size limitation of the preform rod prepared by the solution doping method, gas phase doping technology has developed rapidly. In 2009, Lenardic et al. reported a vapor doping technology for rare earth and aluminum chelates [15]. Rare earth chelates have a high vapor pressure at around 200 °C, which can be

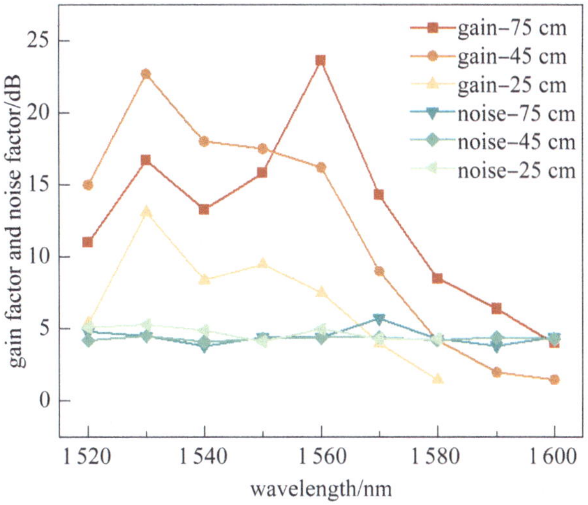

Fig. 3.15 Broadband gain coefficient and noise figure of Erbium-doped silica fiber prepared by MCVD combined with sol-gel immersion doping at different lengths

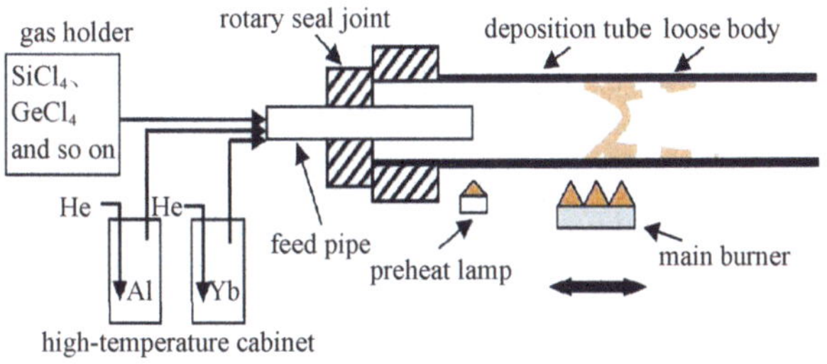

Fig. 3.16 MCVD device with gas phase delivery unit [16]

co-deposited with $SiCl_4$. This process has good repeatability and can effectively control the doping concentration, but the entire feeding system needs to be maintained at about 200 °C, which demands on the equipment. Because the organics in the chelates are difficult to be completely removed, the resulting material has a high hydroxyl content and is prone to impurities.

Saha and others from the Central Glass and Ceramic Research Institute of India used the MCVD process equipment with a gas phase doping delivery unit from the Finnish company Nextrom to prepare doped silica fiber preforms [16], which is shown in Fig. 3.16. This equipment uses helium as the delivery unit, bringing the heated gaseous doping chelate to the reaction area, where it reacts with the $SiCl_4$ gas delivered from another pipeline and deposits on the inner wall of the base tube, achieving doping. The most critical part of this process is the control of the temperature of the band burner at the head of the deposition tube, to avoid overheating causing the rare earth chelate to decompose before reaching the reaction area, and to avoid too low temperature causing the rare earth chelate to condense and recrystallize before reaching the burner. Through optimization, this process can deposit more than 30 core layers, used to prepare a uniformly doped core glass rod with a diameter of 10.5 ~ 11.1 mm, a length of 400 mm, and a Yb^{3+} concentration of 1.5 mol%.

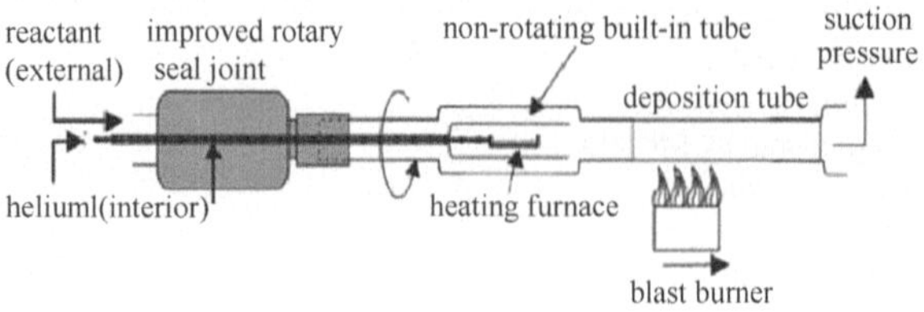

Fig. 3.17 Device for preparing preforms by MCVD combined with chemical crucible method

The University of Southampton in the UK proposed another gas phase doping process-the chemical crucible deposition method [17], the process flow is shown in Fig. 3.17. The MCVD deposition tube is equipped with a crucible connected to the heat source, which can heat the chelate in the crucible in the deposition tube, with a temperature control error of ±1 °C, ensuring the uniform and stable volatilization of the chelate. The gaseous chelate is transported to the reaction zone with nitrogen as the carrier gas, and is deposited on the inner wall of the silica glass tube with $SiCl_4$ at the same time, realizing doping. The entire doping process keeps the deposition tube rotating. Finally, it undergoes drying, sintering, rod collapsing, and other treatments to obtain a rare-earth doped preform rod. The chelate is directly heated inside the deposition tube, which is conducive to improving the longitudinal uniformity of the preform rod. The Yb^{3+} /Al^{3+} /P^{5+} co-doped silica fiber prepared by this method has a loss of 30 ~ 70 dB/km at 1285 nm.

MCVD technology has been widely used in the preparation of optical fiber preform rods, and well-known users include IPG, Nufern, and SPI. China's self-made rare-earth doped optical fiber preform rods are almostly obtained by MCVD technology, especially the commercialized MCVD technology combined with liquid doping. However, for the field of high-power large mode area fibers, a large core-cladding ratio, complex fiber structures are usually required, so higher requirements are put forward for the preform core glass rod, such as large size, high uniformity core rod, high concentration rare-earth ion doping, and low core NA. Restricted by the size, uniformity, and numerical aperture of the core glass rod, MCVD liquid doping technology has encountered a bottleneck in the preparation of large mode area fiber preform rods. Relatively speaking, thanks to the characteristics of large size and high optical uniformity, MCVD technology combined with gas phase doping is more favored in the preparation of large mode area fiber preform rods. But at present, this technology is not stable enough in process control, and there are still some difficulties in achieving low NA preform rod preparation.

3.1.2 Outside Vapor Deposition (OVD) and Vapor Axial deposition (VAD)

OVD is a simple process developed by Corning in the United States in 1970, as shown in Figure 3.18a. This process uses oxygen as the carrier, transports the gaseous precursors containing Si^{4+}, Yb^{3+}, Al^{3+} to the hydrogen-oxygen flame or

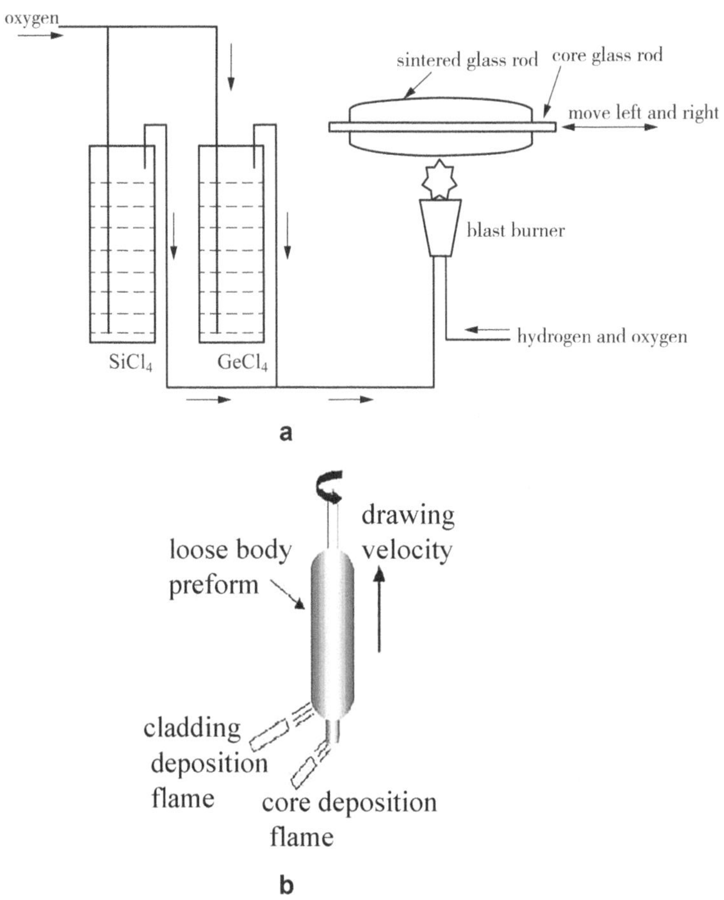

Fig. 3.18 OVD(**a**) and VAD(**b**) core rod preparation process (drawn according to references [18, 19])

methane flame torch, causing them to undergo hydrolysis to form dust particles, which are deposited on the outer surface of the base rod made of silica, graphite, or alumina material; after multiple depositions, the base rod is removed, and the loose preform rod with a central hole is put into a high-temperature furnace for purification and sintering, finally obtaining a bubble-free core rod. The diameter of the base rod, moving speed of the base rod, and hydrogen-oxygen flow rate are the three main parameters of the OVD process: the base rod with a large diameter increases the probability of dust capture; increasing the moving speed of the base rod increases the area of the thermal boundary layer per unit time, but too fast moving speed will lead to low density; the larger the flow rate of hydrogen and oxygen, the higher the dust temperature, which is conducive to dust movement, but too high temperature will make the dust particles too small to capture, and at the same time, the loose body will become dense due to excessive heating. The preform rod produced by this method is efficient, and its size is not limited by the base rod, which can be used to prepare large-size preform rod. To avoid patent disputes with Corning's OVD, Nippon Telegraph and Telephone Corporation (NTT) proposed the vapor axial deposition (VAD) in 1977 [18], as shown in Figure 3.18b. The similarity between this process and the OVD process is the use of flame hydrolysis; the difference is that the preform rod prepared by the VAD method grows from bottom to top along the vertical axis, and sintering and deposition are completed simultaneously in different spaces of the same equipment. Currently, there are no reports on the preparation of Yb^{3+} doped fiber preform rods by the VAD method.

All processes of the OVD method do not involve non-synthetic materials, and the preform rods produced have low impurity content and low loss. The "all-glass" double-clad fiber with a high numerical aperture (≥ 0.3) prepared by the OVD process achieved a laser output of 110 W, with a slope efficiency of 80%, and the fiber loss at 1310 nm was about 3 dB/km [17]. However, using graphite or alumina as the base rod, the removal process will cause the refractive index distribution of the center layer of the preform rod to be disturbed, affecting the transmission performance of the fiber.

3.1.3 Direct Nanoparticles Deposition Method

Tammela and others from the Finnish company Liekki (now merged with the American company nLight) developed a nanoparticles direct doping (DND) technology similar to the OVD technology [20–22] the process flow is shown in Fig. 3.19. The solution with doped ions and $SiCl_4$ materials are transported to the hydrogen-oxygen flame nozzle under computer-controlled flow rate, forming doped SiO_2 particles of 10 ~ 100 nm. Then it is directly deposited on the base rod. After the deposition is completed, the base rod is removed, dried at high temperature, sintered, and the rod is shrunk to obtain a doped silica core rod. Compared with MCVD technology, DND technology completes the doping process in one step, and the doping at the nanoscale has higher uniformity. The factors that affect the quality of the core glass rod are mainly the flow rate of the material and the size of the doped particles. A narrow particle distribution helps to achieve higher uniformity. At present, the Yb1200–20/400 DC fiber prepared by DND technology achieves 450 W laser output under 920 nm laser pumping. The coupling pump light-laser slope efficiency reaches 69%.

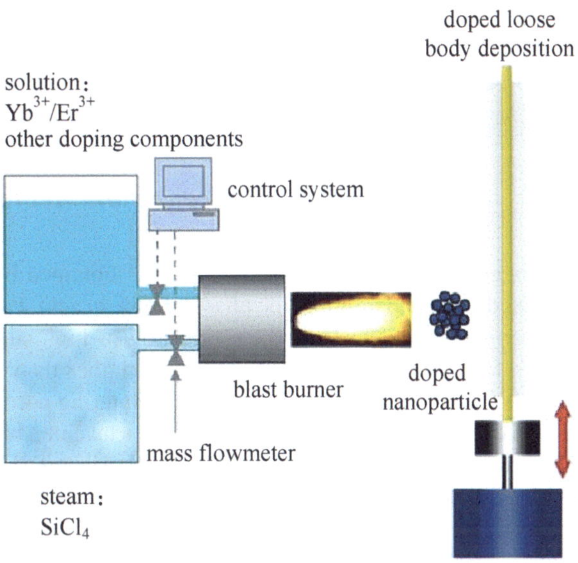

Fig. 3.19 DND core rod preparation process (drawn according to references [20–22])

3.2 Preparation of Rare Earth Doped Silica Fiber Preform by Non-chemical Vapor Deposition Method

For rare earth-doped fiber lasers, the fiber core is the key part that determines the performance of the fiber laser. Therefore, the preparation of the core preform has always been the most important technology for fiber lasers. As mentioned earlier, most commercial doped fibers are mainly produced using improved chemical vapor deposition (MCVD) combined with solution doping method [23]. Technically, MCVD combined with solution doping method has obvious advantages in preparing conventional cladding structure doped fibers with low optical loss. However, in the preparation of ultra-large mode field cladding fiber and rare earth doped large mode field photonic crystal fiber (large mode area photonic crystal fiber, LMA PCF), this method has disadvantages such as uneven refractive index distribution, difficulty in preparing large core diameter, and less doping elements [24]. At present, the core diameter of doped fibers prepared by MCVD combined with solution doping method is generally 10 ~ 40 μm, and the maximum doping concentration is generally below 20,000 ppm. Based on the demand for large size and high doping uniformity cores of Yb^{3+} doped large mode field silica PCF and ultra-high power fiber lasers, in recent years, a series of non-MCVD rare earth doped core and preform preparation technologies have been developed at home and abroad, mainly including the porous glass phase separation method of Shanghai Institute of Optics and Mechanics and Huazhong University of Science and Technology, the active powder sintering method of Heraeus Company in Germany, the sol-gel method combined with high temperature sintering technology of Shanghai Institute of Optics and Mechanics, and the laser sintering rare earth doped silica powder technology of South China Normal University. The following will give a detailed introduction to these non-vapor deposition methods for preparing rare earth doped silica fiber preforms and ytterbium doped fibers.

3.2.1 Porous Glass Phase Separation Method

Porous glass is a porous structure material obtained by acid leaching the phase-separated glass and draining the phase soluble in acid. Porous glass was initially just an intermediate product for the production of high silica glass [25] . High silica glass refers to glass with a SiO_2 content higher than 96%. Different from silica glass, high silica glass is usually made by sintering porous glass. In 1934, researchers at Corning Company in the United States invented high silica glass while studying borosilicate glass phase separation, and in 1939 they produced and sold products under the trade name Vycor [26] . Porous glass has now developed into an indispensable functional material in the fields of rare earth luminescent materials, biomedicine, and chemical and industrial catalysis.

Porous glass has characteristics such as large specific surface area, controllable micropore diameter, regular shape, and stable chemical properties. The pore size of this material can be modulated by the preparation process. According to the professional terminology recommended by the International Union of Pure and Applied Chemistry (IUPAC), porous materials with pore diameters <2 nm are called microporous materials (micropore), such as activated carbon; porous materials with pore diameters of 2 ~ 50 nm are called mesoporous materials (mescopore), such as mesoporous glass; porous materials with pore diameters >50 nm are called macroporous materials (macropore), such as aerogels and porous ceramics [25] . The pore size of porous glass used in luminescent materials is usually between 1 ~ 100 nm; in comparison with the definition of nanomaterials, this type of porous glass can be referred to as nanoporous glass.

For luminescent materials, solving the problem of concentration quenching of doped ions in the matrix material is very critical. Thanks to the high specific surface area of nanoporous glass, researchers have proposed using nanoporous glass as the matrix material for doped ions, thereby achieving the purpose of dispersing doped ions in the glass [27, 28].

In 2005, Chen Danping and others [28] used the high specific surface area of porous glass to disperse luminescent active ions, inhibit the spontaneous clustering of rare earth ions, and thereby enhance the luminescence intensity of active ions. The specific process is: porous glass with connected nanoscale pores (pore diameter of 2 ~ 10 nm) and high specific surface area is immersed in a solution of luminescent active ions such as rare earth or transition metal ions at a certain concentration. After drying, the luminescent active ions are evenly dispersed and adsorbed in the nanoholes. After high-temperature sintering, a pore-free transparent luminescent high silica glass can be obtained. Figure 3.20 is a process flow chart for preparing porous glass [29] . High-temperature melting to prepare alkali borosilicate glass,

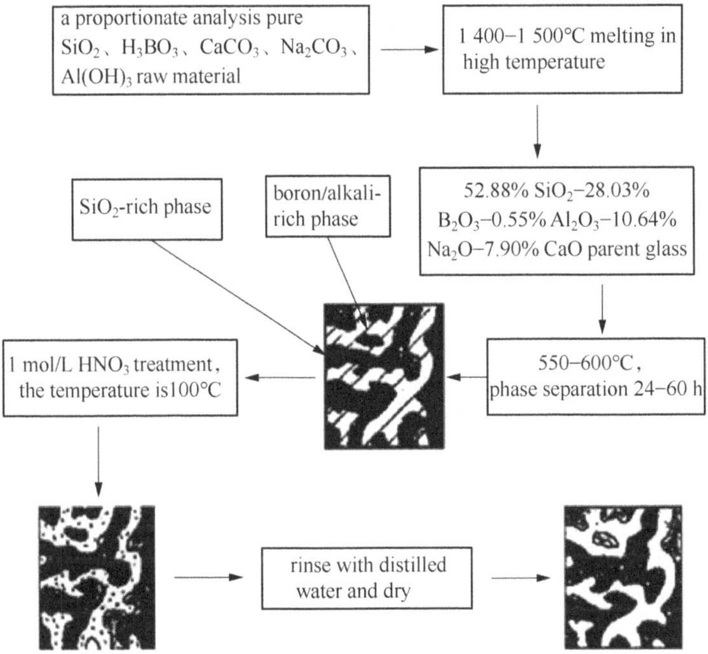

Fig. 3.20 Process flow of porous glass preparation

and then heat treatment to achieve phase separation. Then the glass is crushed for acid treatment and deionized water cleaning to obtain glass particles with nano-holes. This porous glass can be prepared into a rare earth-doped core rod through rare earth solution soaking and high-temperature sintering. The size of the core rod prepared by this process is not limited by the size of the bulk porous glass itself, and a large-size rare earth-doped core rod can be realized. Chen Danping and others used this method to prepare a core rod to draw a 38 μm core doped with Yb^{3+} photonic crystal fiber, achieving a laser output of 34.8 W, and the slope efficiency of pump light absorption-laser output is 71.3%. The results are shown in Fig. 3.21 [30].

Yang Luyun, Chu Yingbo and others from Huazhong University of Science and Technology used the phase separation method to prepare rare earth-doped glass core rods. The process flow is shown in Fig. 3.22. The specific process is: first, the parent glass is processed into cylindrical thin rods, which are corroded after phase separation, then the thin rods are soaked in a rare earth ion solution for doping, and finally dried and impurities removed, and high-temperature collapse is used to prepare into preforms. The difference from the process of Chen Danping and others is that: directly use cylindrical thin rods for doping, instead of grinding bulk glass into powder, avoiding the secondary sintering of the core rod, thus significantly reducing the risk of secondary impurity introduction. In addition, the addition of Cl_2 during the removal of impurities can effectively reduce the background loss of the fiber. It is reported that the background loss of the fiber prepared by this method is

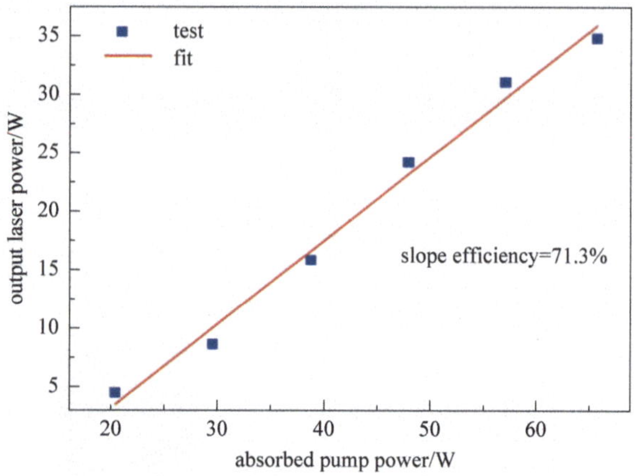

Fig. 3.21 Output power-absorbed pump power curve of the prepared photonic crystal fiber

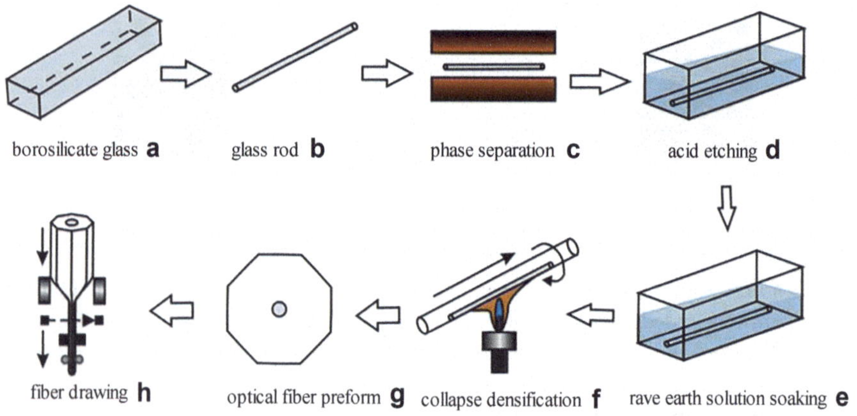

Fig. 3.22 Process flow of rare earth-doped fiber preparation by phase separation method of Huazhong University of Science and Technology [31]

20 dB/km. Chu and others used this method to obtain 18 W laser output in a 2.3 m cladding fiber, with a slope efficiency of 72.8% relative to the absorbed pump light [31]. Using the same Yb-doped core rod preparation method, a laser output of 12.8 W was obtained in a 1 m long photonic crystal fiber (photonic crystal fiber, PCF), with a slope efficiency of 64.5% relative to the absorbed pump light [30]. Because the doping is carried out by the solution soaking method, it is affected by ion diffusion, so only when the size of the core rod is less than 3 mm can the doping concentration and uniformity be guaranteed. In addition, the fiber prepared by this method also has problems such as high hydroxyl content and difficulty in removing impurities.

3.2.2 Active Powder Sintering Method

Since 2008, Langner and others from Heraeus Company in Germany, in collaboration with Jena University [32–35] developed the active powder sintering (REPUSIL) method. This technology dopes rare earth and other elements into nano-porous high-purity SiO_2 particles to form doped powder, and sinter this powder to prepare large-size rare earth-doped silica glass core rods. The preparation process is shown in Fig. 3.23. The specific process is: first, the high-purity porous SiO_2 obtained by hydrolyzing $SiCl_4$ Nanoparticles are dispersed in a solution to obtain a thixotropic suspension. Then, a solution containing rare earth ions and co-doped ions is added to the suspension. After dehydration, purification, and some subsequent treatments, SiO_2 particles doped with ions are obtained from the doped SiO_2 suspension. The size of the porous SiO_2 particles in this step is crucial. Langner and others believe that the size of the initial particles should be 10 ~ 50 nm, and the size of the particles with open pores before sintering should be 10 ~ 50 μm. Then, the doped particles are formed into a dense rod-shaped blank by isostatic pressing, and impurities such as hydroxyl groups are removed by heating in a Cl_2 atmosphere. Finally, the rod-shaped blank is sintered and vitrified at high temperature to obtain a uniform, bubble-free Yb-doped silica glass. The maximum amount of the glass prepared by the REPUSIL process can reach 50 g [36] .

Figure 3.24 shows SiO_2 particles doped with ions and the final Yb-doped silica core rod. Heraeus Company used this core rod to prepare large mode field and ultra-large mode field cladding fibers with core diameters of 45 ~ 150 μm, and conducted continuous research and improvement on the laser performance and thermal treatment of the fiber pump end face in cooperation with Jena University, and published

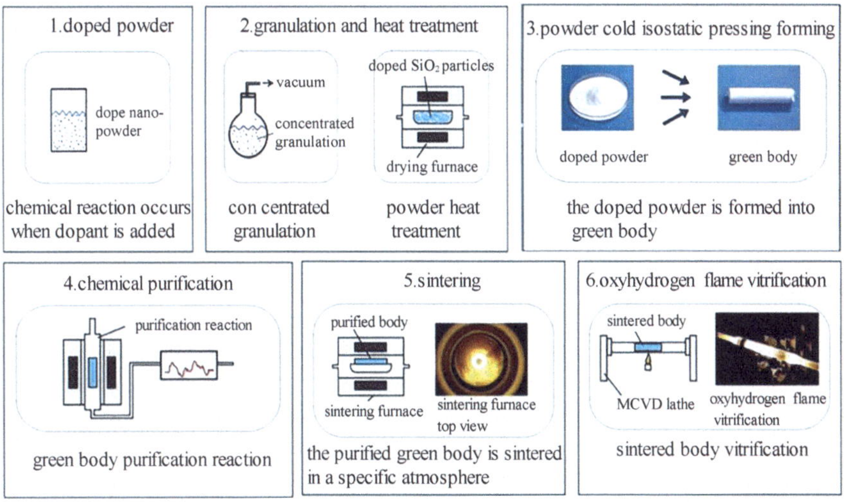

Fig. 3.23 Schematic diagram of the REPUSIL process [36]

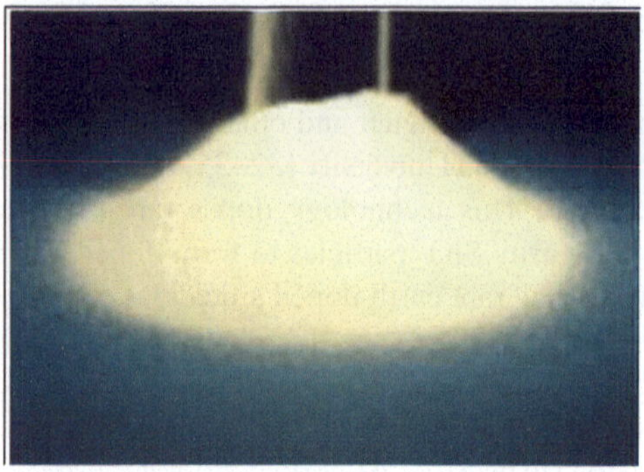

a

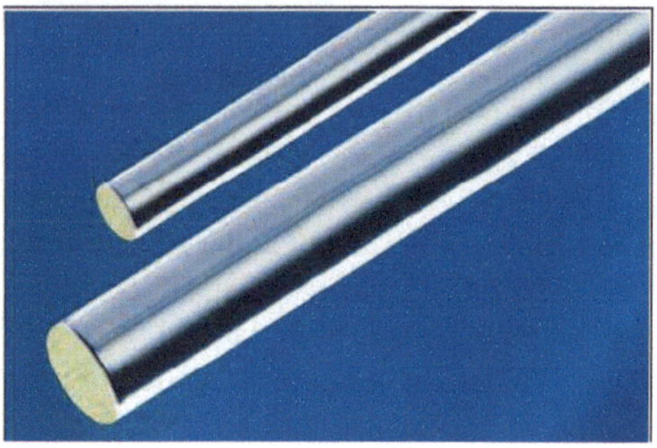

b

Fig. 3.24 Doped SiO_2 particles (**a**) and the final doped silica core rod (**b**)

a series of reports [32–35]. In 2008, it was reported [32] that an ultra-large mode field circular cladding fiber with a core of 150 μm was prepared, and a continuous laser output of 490 W was obtained through side pumping, with a light-to-light efficiency of 30%. In 2010, it was reported [33] that a 4D-type cladding fiber was prepared, which reduced the spiral light loss of the circular cladding fiber, and the core size was 50 μm. This fiber achieved a laser power output of 650 W, but researchers pointed out that further power increase was limited by thermal damage to the fiber end face, and then began to study fiber end caps. In 2012, researchers welded fiber end caps to the end faces of large mode area 4D-shaped cladding fibers with a 50 μm core to reduce thermal effects, achieving a 4 kW laser output with an optical-to-optical efficiency of 56% [34].

Through further optimization, in 2013 [35] a maximum output power of over 5 kW was obtained from the same type of single fiber. Figure 3.25 shows the power input-output curve of this fiber and a photo of the fiber end face (inset). In the photo, (a) represents the Yb-doped fiber core, (b) the Heraeus pure silica glass (model F300), which is the well-processed 4D-shaped inner cladding, (c) the high-fluorine, low-refractive-index circular silica outer cladding, and (d) the fiber coating layer. . This fiber, combined with the fiber end cap welding and a fiber cooling device, achieved a maximum power of over 5 kW, with a plug-in efficiency of 26%.

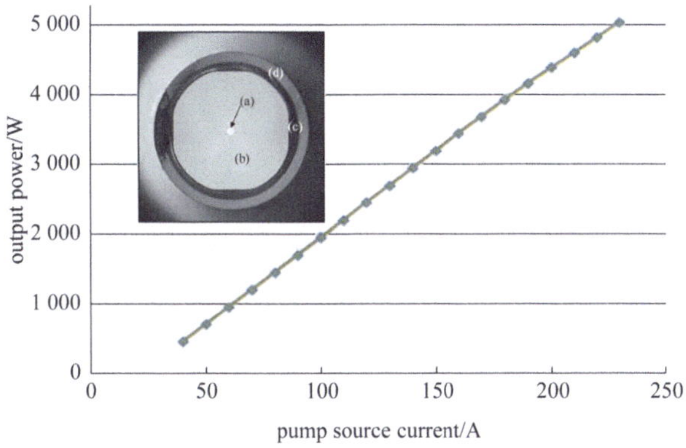

Fig. 3.25 Ytterbium-doped large mode field 4D type cladding fiber laser input-output curve and fiber end face photo (inset) [36]

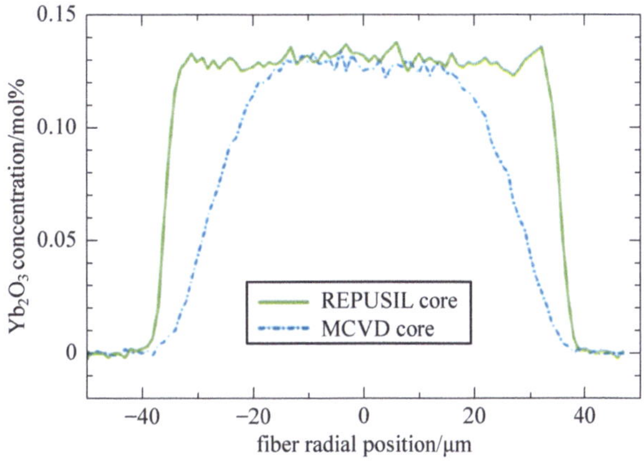

Fig. 3.26 Powder sintering process and MCVD process preparation fiber Yb_2O_3 concentration distribution diagram

Unlike direct deposition processes such as MCVD, the SiO_2 particles prepared by the REPUSIL process are uniformly dispersed in the suspension and are not directly deposited in the tube, so the final doped silica rod is not limited by the external cladding tube and can be directly used as an active fiber core. It can also be easily processed into various shapes of core rods, such as square, hexagonal, octagonal, etc., to meet the preparation requirements of single-, double-, or multiple-cladding fibers, multicore fibers, photonic crystal fibers, and many other complex fiber designs.

The REPUSIL process can produce Yb_2O_3 core rods with a concentration of up to 0.25 mol% (~15,000 ppm Yb), and can achieve precise control of the doping concentration (Fig. 3.26), therefore the doped silica glass rod produced has very good repeatability. The highly uniform doping ensures the uniformity of the core rod refractive index, making it easy to achieve a strict step-index distribution. All materials in the REPUSIL process are artificially chemically synthesized, strictly controlling the sources of pollution during the process, and the doped silica rod produced has a high purity, which is conducive to obtaining a lower background loss. As shown in Fig. 3.27, the loss at 1200 nm of the ytterbium-doped silica fiber prepared by this method is less than 20 dB/km.

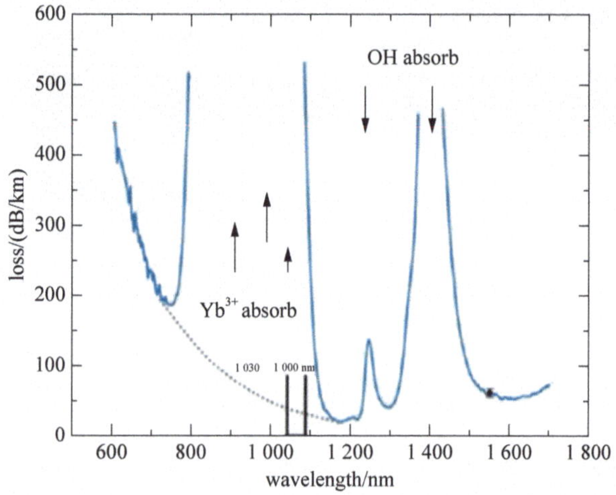

Fig. 3.27 Loss spectrum of fiber core prepared by powder sintering method [34]

Through the optimization of the sintering process, the REPUSIL process can obtain large-size core rods (diameter ≥ 15 mm, length ≥ 150 mm) with low Yb^{2+} content, low defect rate, and few bubbles. This makes the design and preparation of extra-large mode area (XLMA) Yb-doped silica fibers possible. For XLMA design, the larger the doped core diameter under a certain number of activated ions, the lower the required doping concentration, thus the photodarkening effect related to the doping concentration can be significantly improved, which is also beneficial for thermal management under high-power laser output. This is very beneficial for fiber lasers to achieve stable, high-quality, high-power laser output. In addition, by combining the powder sintering method with plasma outside deposition (POD) technology, a layer of silica glass is first deposited on the Yb-doped silica core rod, followed by a layer of low-refractive-index fluorine-doped silica glass, forming an all-glass double-clad fiber preform. This method overcomes the disadvantage that the traditional polymer outer cladding can hardly withstand the thermal effect of the fiber during high-power laser operation, and has unique advantages in the preparation of high-power laser fiber preforms, especially some special-shaped preforms. The disadvantage of the REPUSIL process is that it is difficult to co-dope with volatile elements such as P and F.

3.2.3 Sol-Gel Method Combined with High-Temperature Sintering Technology

The sol-gel method refers to the process in which metal organic or inorganic compounds undergo hydrolysis and subsequent condensation, gradually forming a gel and, after necessary heat treatments, obtaining an oxide or another solid compound. The traditional sol-gel method for preparing rare earth-doped preforms generally first uses the impregnation drawing method to form a sol film containing rare earth-doped ions on the inner wall of the glass tube, and finally obtains the rare earth-doped preform through subsequent heat treatment steps [37] . This process was initially used more for the preparation of Nd^{3+} doped glass [38] .

In 2007, Pedrazza et al. from Bern University, Switzerland [39] published the results of using the sol-gel method to prepare Yb/Al co-doped silica fibers and achieve laser output. The specific preparation process is to first use tetraethyl orthosilicate (TEOS), $Al(NO_3)_3$, and $Yb(NO_3)_3 \cdot 5H_2O$, a doped sol is obtained by mixing deionized water and anhydrous ethanol. The doped sol is poured into a quartz tube that has been cleaned with hydrofluoric acid and alcohol and dried with nitrogen. The sol in the quartz tube is poured out at a speed of 200 cm/min, and then nitrogen is blown into the quartz tube for 20 min to form a gel. Optimizing the viscosity of the sol and the speed of pouring can ensure the formation of a transparent, crack-free, thick gel on the inner wall of the quartz tube. The quartz tube is placed on an automatic lathe, and the heating furnace moves at a speed of 5 cm/min to heat the quartz tube twice. The heating temperature of (1800 ± 100)°C vitrifies the gel, completing one coating process, and a glass layer with a thickness of 2.7 μm is obtained from one coating. After three such coating processes, this quartz tube is drawn into a Yb/Al-doped silica glass rod. In the process of preparing rare earth-doped silica preform rods by the sol-gel method, the ions are doped in the liquid phase, so multiple doping ions, high concentration, and molecular-level uniform doping can be achieved. In addition, the sol-gel method makes it possible to prepare doped silica glass rods at a lower temperature (far lower than the 2000 °C or even higher required by general processes), thereby reducing the preparation cost. In 2012, researchers at Bern University proposed another method for preparing rare earth-doped silica preform rods by the sol-gel method. The first step of this process still uses a process similar to the traditional sol-gel method to prepare SiO_2 sol containing rare earth ions. Unlike the traditional sol-gel process, the prepared sol directly gels at 70 ~ 150 °C, and after several heat treatment steps such as drying, removing residual organic matter, and sintering, porous doped gel material is obtained. The prepared doped gel material is ball milled, and a CO_2 laser is used to melt it into transparent doped small balls (particle size about 250 μm). Finally, the prepared doped small balls are placed in a quartz tube with one end closed, and sintered to obtain a rare earth-doped silica preform rod. Compared with the traditional sol-gel method, this process has a higher background loss (350 dB/km) of the doped silica preform rod finally obtained because the porous doped gel material easily absorbs impurities such as moisture in the air. In addition, the organic matter introduced in the sol-gel process is not easy to remove completely. If carbon elements remain in the prepared preform rod, it may cause some Yb elements to be reduced to Yb^{2+}, and Yb^{2+} will significantly increase the photodarkening effect and affect the long-term stability of the final product laser operation. In 2013, Baz et al. at the University of Lyon in France soaked a transparent porous silica rod obtained by the sol-gel method in an ethanol solution doped with Yb to achieve rare earth doping, and made a Yb-doped silica glass rod with a diameter of 1.5 cm and a length of 7 cm. In the center of the doped glass rod with a diameter of 14 mm, the Yb concentration was measured at (0.25 ± 0.03)wt%, and the refractive index fluctuation was less than 1.1×10^{-4}, the core loss measured at 1200 nm was about 40 dB/km. The maximum output power of 1034 nm laser obtained from the photonic crystal fiber prepared by the stack and draw method is 230 mW, and the corresponding slope efficiency is 73.5%. This method can prepare large size silica glass rods with uniform Yb doping, but the output laser power is low. This may be due to the fact that this process has not been able to effectively remove the residual carbon in the bulk glass. Therefore, this method is mature for the preparation of rare earth doped silica glass rods and photonic crystal fiber preforms, and needs further optimization.

In 2013, Hu Lili et al. from the Shanghai Institute of Optics and Fine Mechanics developed a method for preparing Yb-doped silica glass core rods using the sol-gel method combined with high-temperature sintering technology to meet the demand for large-sized core rods in large mode field doped silica fibers. The core rod preparation process of this method is shown in Fig. 3.28. Specifically, the preparation process of large mode field photonic crystal fiber with $0.05Yb_2O_3$-$2Al_2O_3$-P_2O_5-$96.95SiO_2$ (mol%) doped silica glass core rod is as follows:

In the preparation stage of the sol, according to the designed glass composition $0.05Yb_2O_3$-$2Al_2O_3$-P_2O_5-$96.95SiO_2$ (mol%), different proportions of tetraethyl orthosilicate (TEOS), anhydrous ethanol (C_2H_5OH), H_3PO_4, $AlCl_3 \cdot 6H_2O$, $YbCl_3 \cdot 6H_2O$ are mixed as precursors, and a certain amount of deionized water is added to maintain the hydrolysis reaction. In silica glass, Al_2O_3 is an effective rare earth ion dispersant, which can significantly increase the doping concentration of Yb^{3+} in silica glass, prevent the concentration quenching caused by Yb^{3+} aggregation and even more serious glass crystallization. P_2O_5 not only has the function of dispersing rare earth ions similar to Al_2O_3, but also can form [$AlPO_4$] structure, reducing the refractive index of doped silica glass. More importantly, the introduction of P_2O_5 can prevent or alleviate the photo-darkening phenomenon (photo darkening, PD) that occurs in Yb^{3+} doped silica fiber during laser operation. The above mixed liquid is fully mixed and stirred for 20 h to obtain a uniform transparent sol. Then the transparent sol is subjected to a polymerization reaction and dehydration and decarbonization heat treatment, the heat treatment temperature is from 20 °C to 800 °C, to obtain dry gel powder. The gel powder is melted into transparent bulk silica glass at 1750 °C for 3 h in a high-temperature vacuum furnace (vacuum degree 10^{-4} Torr, 1 Torr ≈ 133 Pa). The bulk silica glass is processed into a ϕ 2.5 mm × 50 mm core rod by high-temperature hydrogen-oxygen flame rod making and subsequent processing and polishing, as the core rod of the fiber preform. The tube-rod method is used to prepare the large mode field cladding fiber preform, and the stacking method is used to prepare the large mode field photonic crystal fiber (LMA PCF) preform, and then drawn into a large mode field cladding fibers or LMA PCF at 2000 °C.

The advantage of this method is that it can achieve uniform doping of rare earth ions in the centimeter size range, and at the same time, it can achieve co-doping of Al, P, F and other elements, which can effectively suppress the photo-darkening of ytterbium ions and regulate the refractive index of the core rod. In addition, this method can more fully remove hydroxyl and residual organic carbon. The

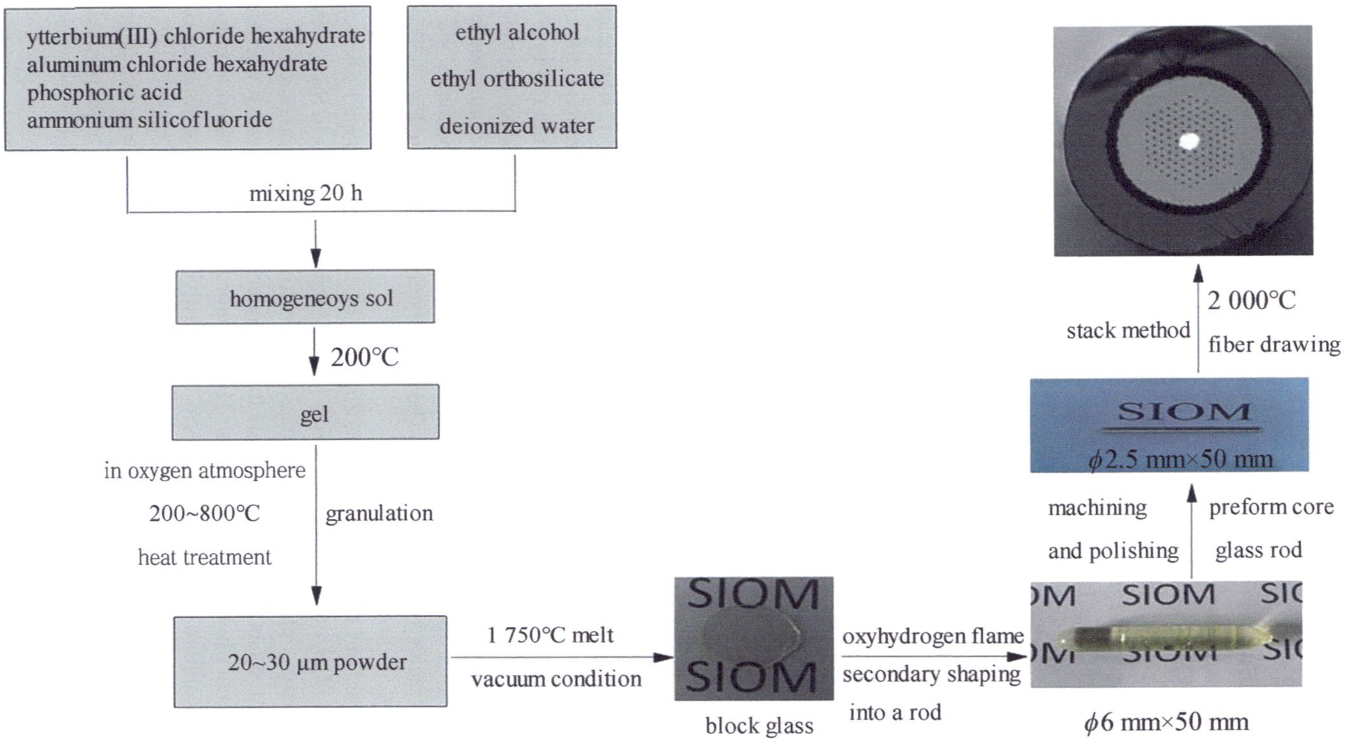

Fig. 3.28 Sol-gel method combined with high-temperature sintering technology to prepare rare earth ion (Yb, Nd, etc.) doped silica glass and fiber process flow

disadvantage is that the preparation process is time-consuming, which generally takes 7 to 10 days to complete. In addition, compared with MCVD technology, this method operates in an open atmospheric environment, resulting in higher fiber losses, usually 300 ~ 1000 dB/km at the 1200 nm band. To address this problem, a chlorine gas purification process is introduced to reduce the optical loss to 30 ~ 50 dB/km. This method enables the preparation of low-loss, highly uniform ytterbium-doped silica core rods and fibers, erbium-doped silica core rods and fibers, thulium-doped silica glass and fibers, and neodymium-doped silica glass and fibers.

In the research and development of ytterbium-doped silica glass core rods and large mode field fibers, In 2013, Wang et al. [40] prepared Al/Yb co-doped silica core rods, evaluated their optical and spectral properties, and used these core rods to prepare a 30 μm photonic crystal fiber (PCF), the refractive index fluctuation uniformity Δn of the fiber core is about 1×10^{-3}, the fiber loss at the 1200 nm band is 1200 dB/km. An initial laser power of 6.8 W was obtained with a slope efficiency of 48%. In the same year, the team further prepared Al/P/Yb co-doped silica glass and a 35 μm large mode field PCF [41], the refractive index fluctuation uniformity Δn of the fiber core is about 8×10^{-4}, the doping uniformity has been improved. This fiber achieved a continuous laser output of 34.6 W at the 1040 nm band, with a slope efficiency of 62%. In the same year, Wang and others also prepared a large mode field PCF with an Al/Yb co-doped core of 35 μm [42], the core refractive index fluctuation is less than 2×10^{-4}, the fiber loss at the 1200 nm band is 410 dB/km. A continuous laser output of 81 W was obtained from this fiber, with a slope efficiency of 70.8%. The laser wavelength is 1040 ~ 1070 nm, it is a multimode fiber, as shown in Fig. 3.29.

In 2016, Xu and others [43] prepared Al/F/Yb co-doped silica core rods and large mode field PCF with a core size of 50 μm. The introduction of F significantly reduced the refractive index of the Al/Yb-doped silica core rod, reducing the core NA to 0.02. However, F is unstable at high temperatures, and the core rod is prone to F volatilization during high-temperature drawing, thereby reducing the uniformity of the fiber in the radial and length directions. Finally, a maximum power of 8 W and a slope efficiency of 43.3% were obtained through fiber bending, and the beam quality factor M^2 is less than 1.1. The laser results are shown in Fig. 3.30, and the laser mode is shown in Fig. 3.31.

As described in Chap. 2 of this book, doping P_2O_5 in Al/Yb co-doped silica core rod glass and fiber, on the one hand, forms [$AlPO_4$] structure similar to SiO_2, with a similar refractive index, which can reduce the refractive index of the core rod. On the other hand, it can effectively alleviate the photodarkening problem of ytterbium doped fiber under high power operation. The traditional MCVD method to prepare co-doped phosphorus silica fiber preform rods easily causes a large amount of P volatilization during the collapse stage, resulting in a "pit" in the middle of the core, and cannot achieve uniform doping of the volatile element phosphorus. Wang et al. used the sol-gel method combined with high-temperature sintering to prepare Al/P/Yb-doped silica core rod glass and fiber [44]. The uniform distribution of various

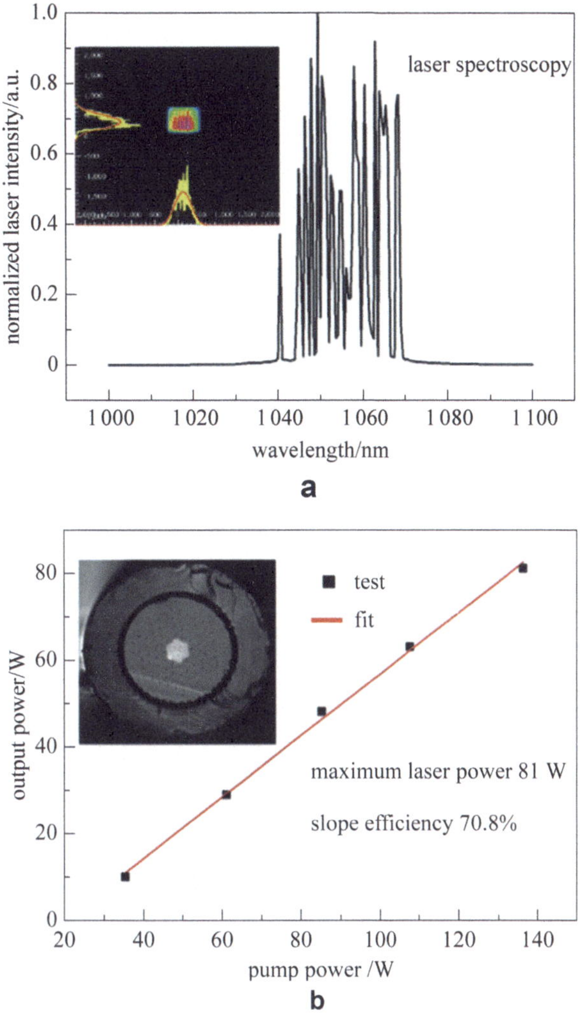

Fig. 3.29 Al/Yb co-doped large mode field PCF laser spectrum, laser mode (**a**) and laser input-output curve, PCF fiber end face photo (**b**)

doping elements in the doped silica core rod was characterized by an electron probe microanalyzer (EPMA), as shown in Fig. 3.32. From this, it can be observed that the distribution of each doping element is uniform, which overcomes the issue of significant P evaporation caused by the conventional MCVD method used for preparing preforms.

The corresponding large mode field cladding fiber and PCF fiber are prepared from this core rod, the fiber loss at 1200 nm is reduced to 240 dB/km, and a quasi-single mode laser output with M^2 1.3 is obtained from the large mode field (core 35 μm) cladding fiber, and a continuous laser output with the highest power of 46 W

Fig. 3.30 Al/F/Yb-doped silica PCF laser output-output curve (**a**) and laser spectrum (**b**)

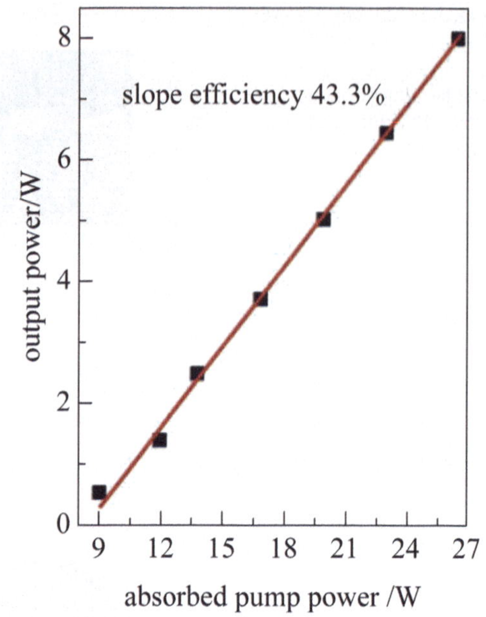

a

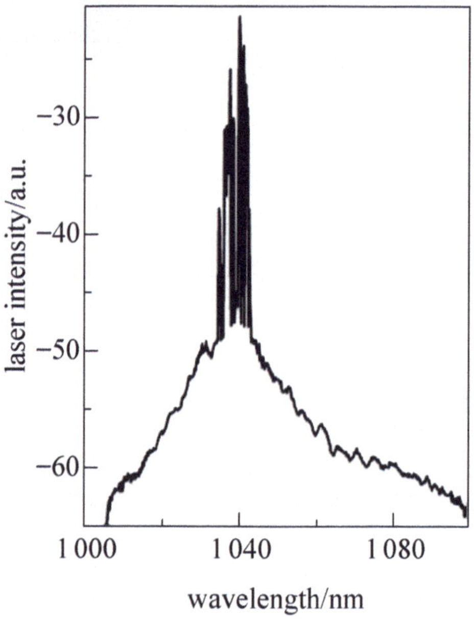

b

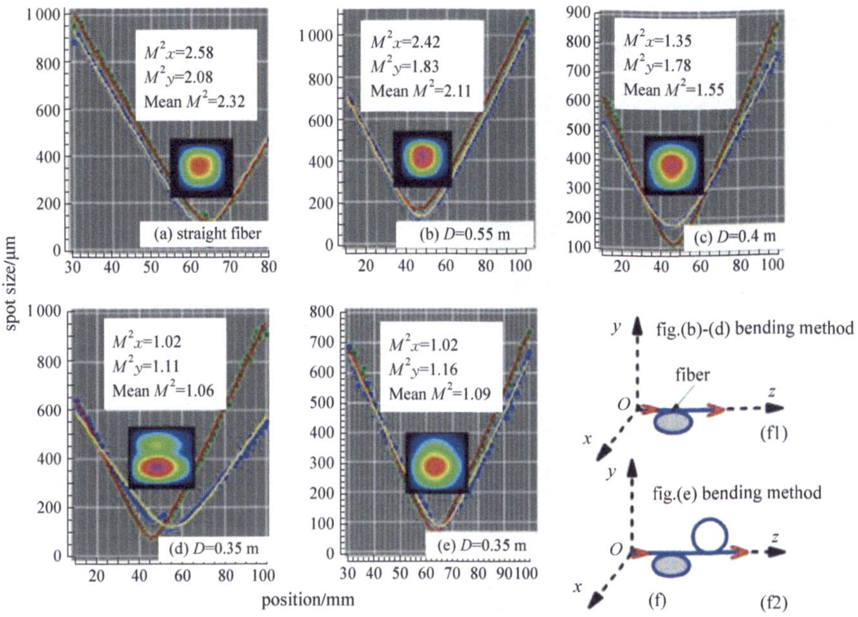

Fig. 3.31 Al/F/Yb-doped silica PCF tested beam quality and laser spot (inset) under different bending diameters

Fig. 3.32 EPMA line scan of the end face of Al/P/Yb-doped silica core rod and comparison with the uniformity of P element doping in the early stage

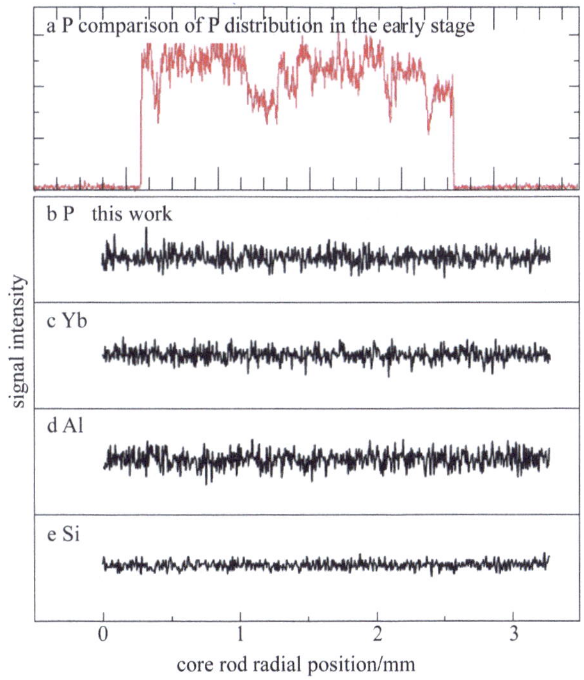

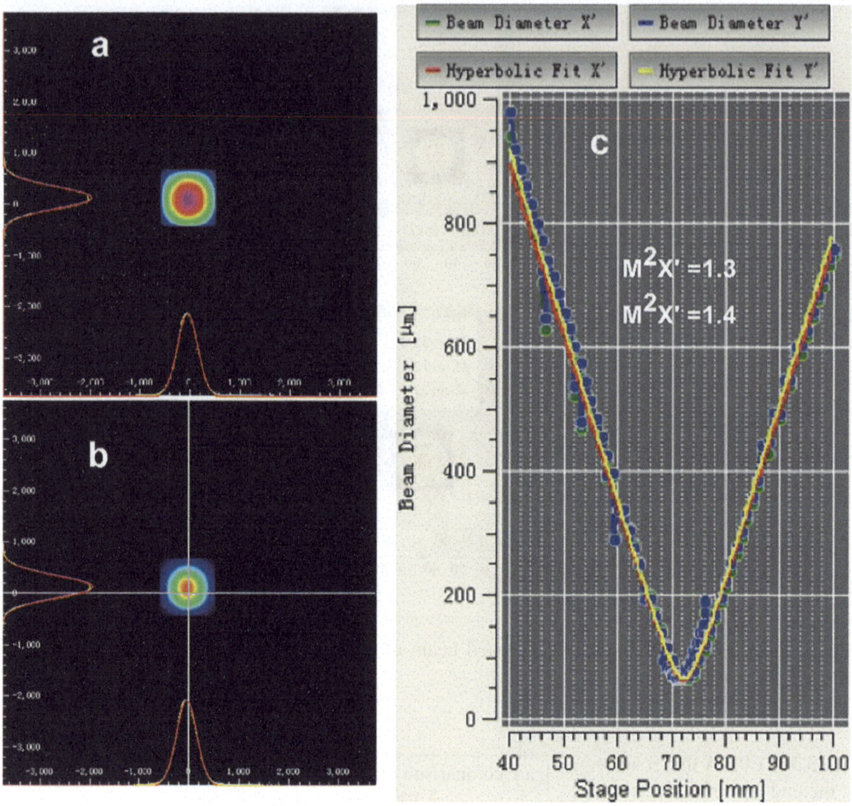

Fig. 3.33 Laser mode of Al/P/Yb-doped large mode field cladding fiber

and slope efficiency of 61% is obtained from the large mode field (core 50 μm) PCF. The laser mode of the large mode field (core 35 μm) cladding fiber is shown in Fig. 3.33.

In 2017, Wang et al. prepared Al/P/F/Yb-doped silica core rod using sol-gel method combined with high temperature sintering method [45], which greatly reduced the refractive index of the core rod, the core NA was 0.027, and the refractive index distribution of the core rod is shown in Fig. 3.34. It can be seen from the figure that the refractive index fluctuations in the radial and axial length directions of the core rod are both less than 2×10^{-4}. Compared with the Al/F/Yb-doped core rod, its axial length uniformity and repeatability of the core rod were greatly improved. Using this core rod to prepare a large mode field doped with thulium PCF with a core of 50 μm, a pulsed laser output with an average power of 97 W was obtained, the amplified light-to-light efficiency was 54%, and the beam quality factor M^2 was 1.4, which was a quasi-single mode laser, as shown in Fig. 3.35.

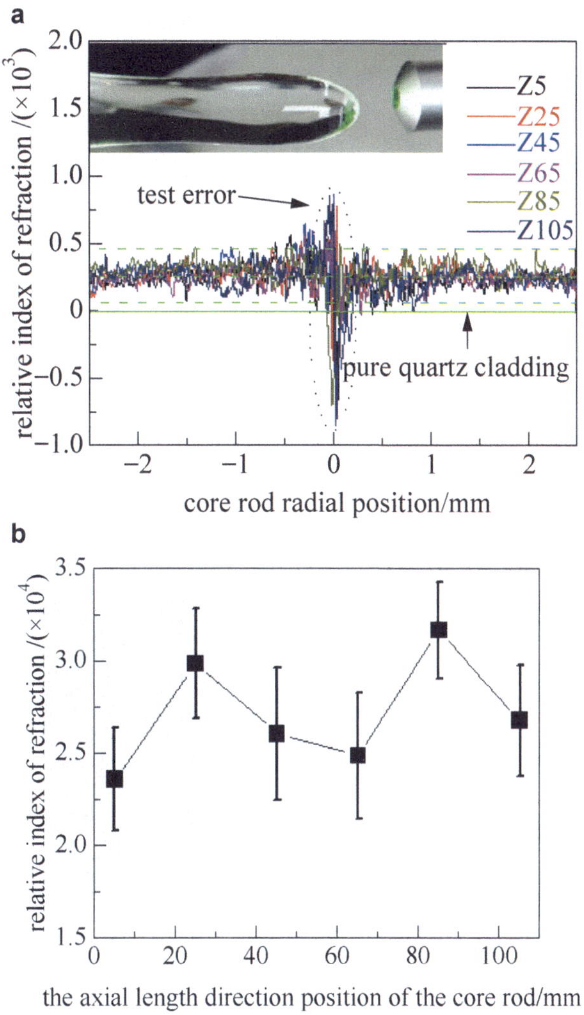

Fig. 3.34 Radial refractive index distribution of Al/P/F/Yb-doped silica core rod at different lengths (**a**) and the refractive index change distribution in different length directions of the core rod (**b**)

In 2017, Wang et al. reported [45, 46] that through purification technology, they greatly improved the purity of the Yb-doped silica core rod, reducing the fiber loss to 50 dB/km, as shown in Fig. 3.36. Using this core rod to prepare an Al/Yb co-doped ultra-large mode field (core 100 μm) PCF, a continuous laser output with a slope efficiency of up to 83.3% was achieved. In the pulse amplification experiment, a peak power of 1.5 MW and an average power of 310 W were obtained. The laser results are shown in Fig. 3.37.

By using the sol-gel method combined with high-temperature sintering technology, different rare earth ions can be doped into silica preforms and fibers. In

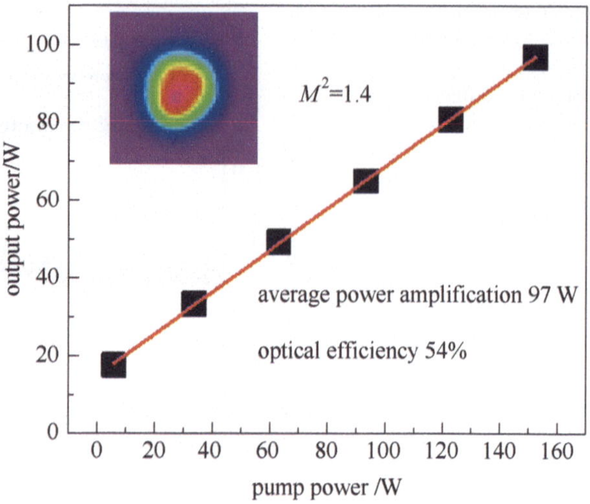

Fig. 3.35 Laser amplification curve and beam quality (inset) of Al/P/F/Yb-doped large mode field PCF (core 50 μm)

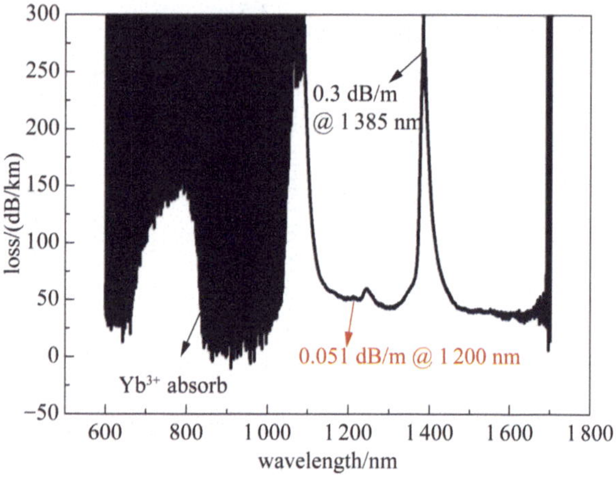

Fig. 3.36 Loss spectrum of Yb-doped silica fiber

addition to Yb-doped silica glass and fibers, the research group also developed thulium-doped, erbium-doped, erbium-ytterbium co-doped, neodymium-doped silica glass and various fibers. Figure 3.38 shows the silica glass doped with various rare earth and other elements prepared using this technology. Detailed results are introduced in Chaps. 6–8 of this book.

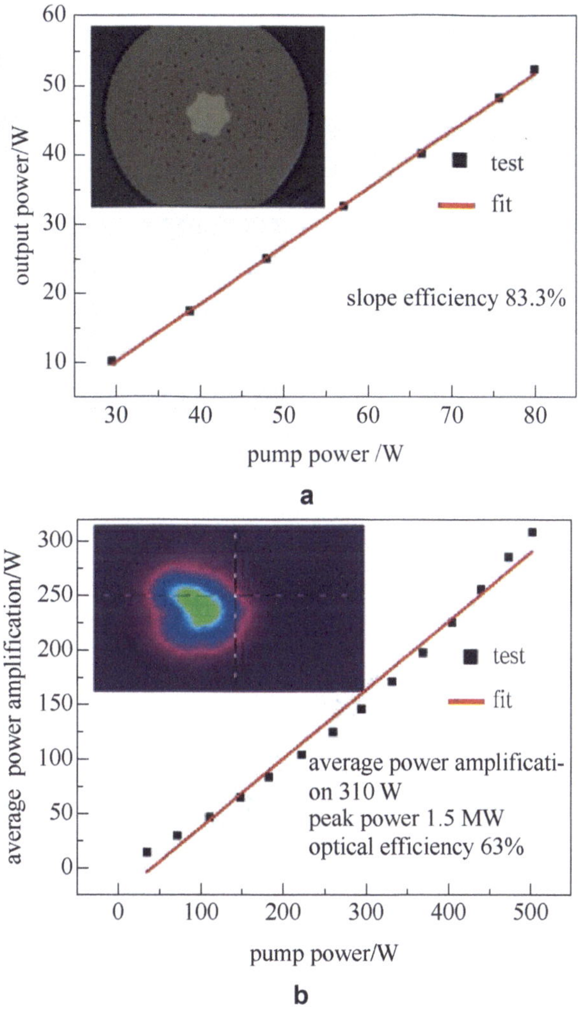

Fig. 3.37 Continuous laser (**a**) and pulse amplified laser performance (**b**) of Al/Yb co-doped ultra-large mode field (core 100 μm) PCF

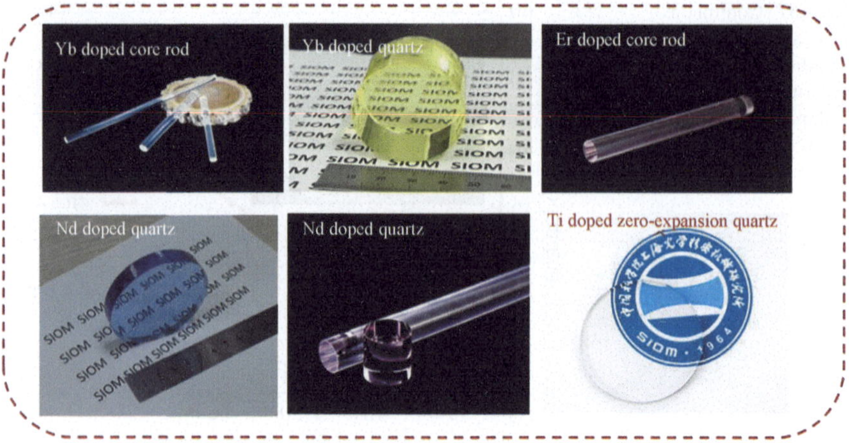

Fig. 3.38 Silica glass doped with different rare earth ions prepared by the sol-gel method combined with high-temperature sintering technology

3.2.4 Other Non-CVD Preparation Techniques

Zhou Guiyao from South China Normal University et al. reported a method for laser sintering rare earth-doped silica powder [47]. The specific process is: first, the precursor of the rare earth ion and the co-dopant is dissolved in water to form a doped solution. $SiCl_4$ The solution is introduced for hydrolysis, mixing, and doping. The rare earth-doped silica powder is obtained through subsequent dehydration and heat treatment. The silica powder is melted into rare earth-doped silica glass at high temperature in the built laser sintering device. The preparation process is shown in Fig. 3.39. Using O_2, the doped silica powder is blown to the top of the mother rod, and it is melted into a glass state by a carbon dioxide laser. In this process, as the temperature rises, the viscosity of the glass liquid decreases, the diameter of the bubbles in the glass liquid gradually increases and continues to rise, and they are finally removed from the glass liquid. At the same time, the mother rod rotates and descends, the glass liquid gradually leaves the melting area, the temperature decreases and the glass liquid gradually cools into a glass rod, finally obtaining a uniformly doped silica glass rod. The doped silica glass rod is cold-processed and polished in different sizes according to the needs of drawing optical fibers, and used to prepare optical fiber preforms. The optical fiber prepared by this technology has a minimum optical loss of 0.25 dB/m, and the prepared fiber core 105 μm photonic crystal fiber has a slope efficiency of 70.2%, and a laser power of 6 W, as shown in Fig. 3.40 [47].

In 2009, Devautour et al. form France [48] developed the granulated silica method for preparing rare earth-doped silica preforms, as shown in Fig. 3.41. The specific process is: first, Yb/Al co-doped silica sand (silica sand size ~0.125 μm^3) obtained by the solution doping method is filled in a pure silica glass tube (Figure 3.41a). This silica tube (inner tube) is placed in a larger silica tube (outer

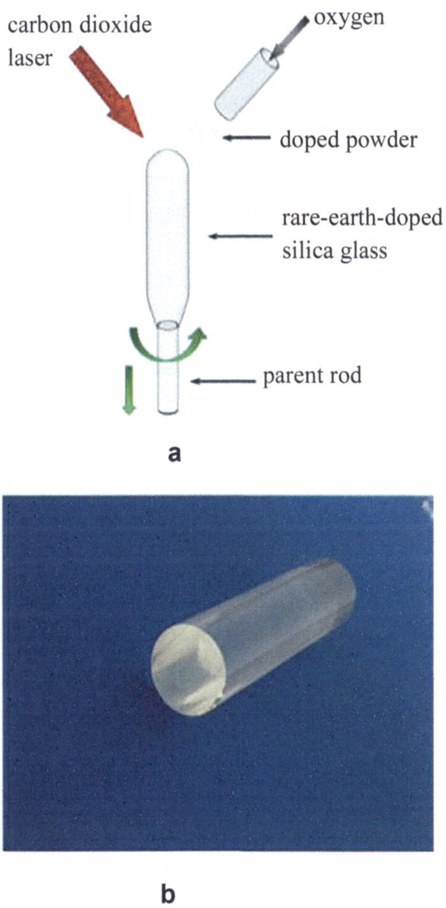

Fig. 3.39 Schematic diagram of laser sintering of rare earth-doped silica powder (**a**) and prepared Yb-doped silica glass (**b**)

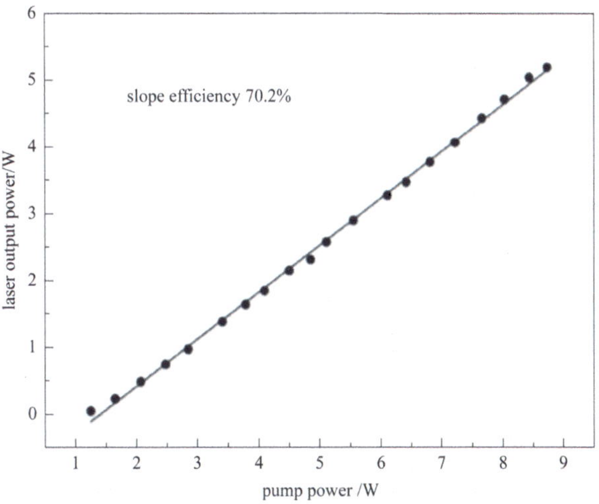

Fig. 3.40 Input-output curve of Yb-doped large mode field PCF

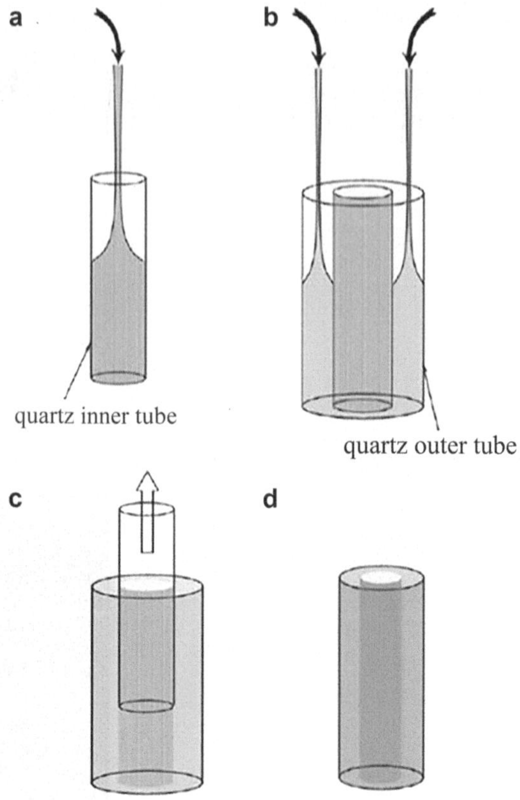

Fig. 3.41 Silica particle vitrification process

tube), and the gap between the two is about 3.4×10^6 μm^3 of large-grain silica sand filling (Figure 3.41b). After removing the inner tube, the entire structure is heated to 1950 °C in a clean air flow to vitrify the silica sand (Figure 3.41c), finally obtaining a dense preform rod (Figure 3.41d). The vitrification process of silica particles can conveniently adjust the chemical composition of various parts of the material to obtain the desired refractive index difference between the core and the cladding. Coupled with the flexibility of the structure arrangement, this process is particularly suitable for preparing complex structure fibers, such as multicore fibers, microstructure fibers, etc. Unlike the MCVD process, which requires multi-layer deposition to prepare large-core preform rods, the simplicity of the silica sand sintering method can overcome the problem of poor refractive index uniformity caused by multi-layer deposition of the MCVD process. In addition, the time required to prepare large-core preform rods using this process is independent of the size of the doped core finally obtained, so large-core preform rods can be quickly prepared. However, since the silica sand sintering method cannot effectively eliminate impurities, especially in the case of high-concentration rare earth doping, these impurities may cause crystallization in the core of the preform rod during the fiber drawing process, resulting in a higher background loss of the final fiber (about 800 dB/km).

Table 3.6 Comparison of the technical level of non-MCVD technology in preparing Yb-doped large-mode-field silica fibers, both domestically and internationally

Performance parameters	REPUSIL Technology of Heraeus, Germany	Sol-Gel Technology of the University of Bern, Switzerland	Phase Separation Method of Porous Glass by Shanghai Institute of Optics and Fine Mechanics (SIOM), CAS, and Huazhong University of Science and Technology	Sol-Gel Combined with High-Temperature Sintering Technology by Shanghai Institute of Optics and Fine Mechanics (SIOM), CAS
Refractive index Uniformity Fluctuation Δn	2×10^{-4}	4×10^{-4}	$(2 \sim 4) \times 10^{-4}$	4.5×10^{-5}
Doped with P	No	Yes	Yes	Yes
Doped with F	No	No	Yes	Yes
Core rod with low refractive index (close to pure silica)	No	No	Yes	Yes
Loss @1200 nm (dB/m)	0.02 ~ 0.05	0.35	0.02 ~ 0.35	0.03 ~ 0.05
Fiber laser performance	Ultra-large mode field single fiber continuous power over 4 kW	Watt level	38 μm PCF 34.8 W, 71.3%	100 μm PCF Pulse: 310 W, 1.5 MW Continuous: 83.3% slope efficiency
Application	Laser processing, laser cleaning	None	None	Ultrafast laser, laser cleaning

The current comparison of several mainstream non-MCVD technologies for preparing rare earth-doped silica glass and fibers, both domestically and internationally, is shown in Table 3.6.

3.3 Drawing and Performance Characterization of Rare Earth-Doped Silica Fiber

3.3.1 Drawing of Rare Earth-Doped Silica Fiber

After preparing by Chemical Vapor Deposition or non-Chemical Vapor Deposition, the preform is heated to the softening state of silica glass in a high-temperature furnace, and the silica preform is drawn into the required specification of fiber under controlled traction.

Although the fiber core contains rare earth dopants, its primary composition remains silica glass, endowing it with excellent physicochemical properties: such as the low loss, low expansion, high strength and high chemical stability.

Similar to the drawing process of communication fibers, the drawing of rare earth doped silica fibers is to heat the preform glass to the softened state through a high-temperature drawing furnace. Within the appropriate temperature range, the traction force overcomes the internal friction of the glass (i.e., viscosity) and surface tension, then the preform is drawn to the fiber.

When the temperature fluctuations are minor, the relationship between viscosity and temperature follows the Arrhenius equation:

$$\eta = A_f \exp\left(E_a / RT\right) \tag{3.9}$$

In the formula, A_f is a constant; R is the gas constant; T is the absolute temperature; E_a is the activation energy. E_a reflects the sensitivity of melt viscosity to temperature, the larger E_a is, the more sensitive the viscosity is to temperature changes.

Taking the logarithm of the Arrhenius equation, it becomes

$$\lg \eta = \lg A_f + \frac{E_a}{2.303RT} \tag{3.10}$$

Let $A = \lg A_f$, $B = \dfrac{E_a}{2.303R}$, Eq. (3.10) transforms into

$$\lg\eta = A + \frac{B}{T} \tag{3.11}$$

In the formula, η represents the internal friction force between two layers of liquid under the unit contact area and unit velocity gradient. The unit is Pa·s.

The empirical formula for the relationship between silica glass viscosity and temperature can be expressed as

$$\lg\eta = -2.4 + \frac{19541}{T} \tag{3.12}$$

Table 3.7 shows some characteristic temperatures during the change of silica glass viscosity with temperature, which are critical during the drawing process of rare earth doped silica fiber.

Table 3.7 Silica glass viscosity and corresponding characteristic temperatures [49]

η /(Pa·s)	Characteristic point	Corresponding temperature/°C
10	Melting and clarifying temperature	2000
$<10^3$	Melting temperature	1713 ~ 1756
$3 \times 10^6 \sim 1.5 \times 10^7$	Softening temperature	1580 ~ 1650
10^{12}	Annealing temperature	1084
$10^{13.5}$	Strain point	956

Rare earth doped silica fiber and communication fiber have significant structural differences. Figure 3.42 compares the structures of traditional communication fiber and rare earth doped fiber.

Communication fiber is mainly used for signal transmission, utilizing the core of the fiber. Therefore, as shown in Fig. 3.42a, the structure of communication fiber generally includes a core doped with germanium ions, a silica cladding, and high refractive index organic coating layers. Rare earth doped silica fiber utilizes the core to absorb pump light and emit or amplify laser, mainly used in fiber lasers or amplifiers. Therefore, the structure of rare earth doped fiber generally includes a core doped with rare earth ions, an octagonal-shaped inner cladding, a low refractive index organic coating layer, and a high refractive index organic coating layer, known as double cladding fiber, as shown in Fig. 3.42b. The core of the rare earth doped fiber absorbs pump light through the doped rare earth ions, and then emits lasers or amplifies signal light. To increase the core's absorption of pump light, the inner cladding is designed with an octagonal structure, which can effectively reduce the formation of spiral light by the pump light. The silica glass inner cladding is covered by a low refractive index coating layer, serving as the second inner cladding of the rare earth doped fiber. Through the large refractive index difference between the organic coating and the silica glass, a large numerical aperture waveguide structure is formed, which is beneficial for improving the coupling efficiency of the pump light. The outermost layer of the rare earth doped fiber is a high refractive index coating layer, which serves to protect the fiber and enhance its strength. Figure 3.43

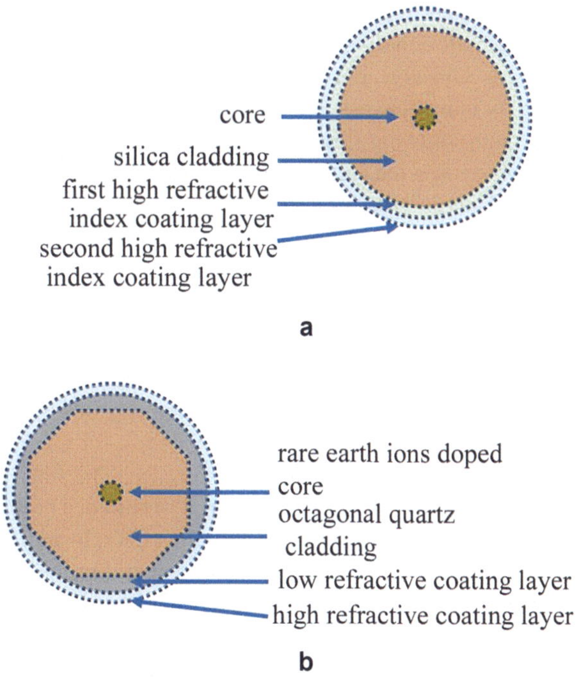

Fig. 3.42 Comparison of the structures of traditional communication fiber (**a**) and ytterbium-doped laser fiber (**b**)

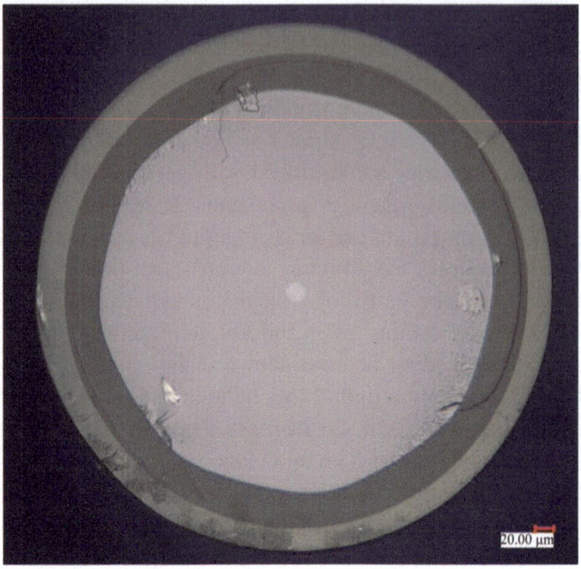

Fig. 3.43 Typical 20/400 ytterbium-doped double-clad microscopic structure of the fiber end face

shows the microscopic structure of a typical 20/400 ytterbium-doped double-clad fiber end face.

The drawing process of rare earth doped silica fiber has its uniqueness. During the drawing process of rare earth doped silica fiber, it is necessary to maintain the octagonal inner cladding structure and the second cladding structure with a low refractive index coating. It generally requires the use of a special fiber drawing tower.

Figure 3.44 is a schematic diagram of the structure of a special fiber drawing tower. Its main functional components include a preform feeding system, a high-temperature drawing furnace, a fiber diameter measurement system, a coating and curing system, a tension detection system, a traction system, and a fiber collection system. The functions of each part are introduced below.

1. Preform feeding system. The rare earth-doped fiber preform, which is processed into an octagonal shape, is clamped on the feeding system with a three-jaw chuck. According to the set fiber drawing process parameters, the preform is slowly fed into the high-temperature fiber drawing furnace. Due to the small diameter and short length of the rare earth-doped fiber preform, the feeding speed is generally less than 10 mm/min due to the limitation of the drawing speed.
2. High-temperature fiber drawing furnace. The fiber preform is heated to above 2000 °C, causing the silica material to soften. Through the action of the traction wheel, the preform rod is drawn into a fiber. In addition to providing a high temperature above 2000 °C, the high-temperature fiber drawing furnace also needs to maintain a stable temperature field control in the drawing area. Within the softening range, the accuracy of the fiber is affected by the stability of the temperature field, and any fluctuation in the temperature gradient will cause fluctua-

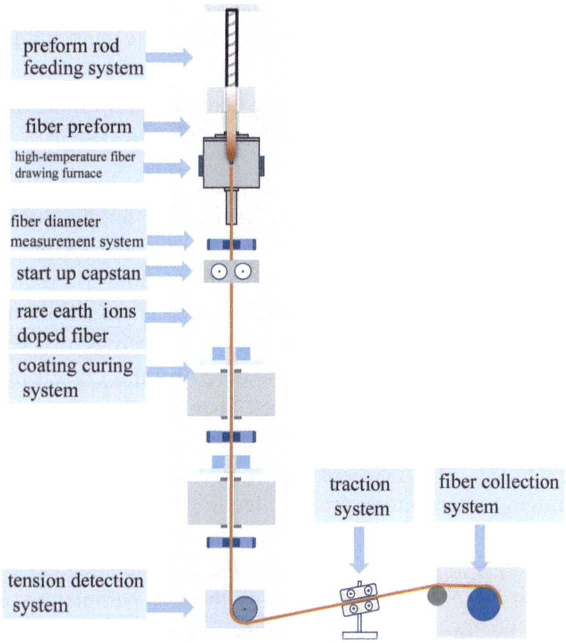

Fig. 3.44 Schematic diagram of the structure of a typical special fiber drawing tower

tions in the fiber diameter. The more common high-temperature fiber drawing furnaces in special fiber drawing towers are graphite resistance furnaces and graphite induction furnaces, both of which can maintain a temperature fluctuation of less than 1 °C.

3. Fiber diameter measurement system. It is used to measure the outer diameter of the fiber, and based on this, the fiber drawing parameters are fine-tuned to obtain the required rare earth-doped fiber. Because the rare earth-doped fiber has an octagonal and other irregular structure, as well as a low refractive index inner coating and a high-refractive-index outer coating structure, multiple sets of fiber diameter measurement systems are needed to measure the bare fiber diameter, inner coating diameter, and outer coating diameter, etc.
4. Coating and curing system. The rare earth-doped fiber needs to use a fluorine-doped acrylate coating as the inner coating, providing a divergence angle of not less than 0.46 numerical aperture, thereby improving the coupling efficiency of the pump light; a high-modulus high-strength acrylate coating is used as the outer coating to protect the fiber. Therefore, the coating system of the special fiber drawing tower generally adopts a wet-to-dry two-layer coating and two sets of UV curing systems.
5. Tension detection system. During the fiber drawing process, in order to judge the appropriateness of the drawing temperature, drawing speed, and other process conditions, it is necessary to monitor the tension borne by the fiber in real time. By installing a tension meter at the bottom of the fiber drawing tower, the tension of the fiber can be detected in real time.

6. Traction system. Through the rotation of the traction wheel, the fiber is drawn from the preform. The drawing speed is controlled by the rotation speed of the traction wheel.
7. Fiber collection system. The fiber drawn from the traction system is collected on the fiber reel through the fiber collection system.

The specific process of preparing rare earth-doped fiber includes: clamping the preform, heating the preform to the drawing temperature, and drawing the fiber preform into a fiber through traction. During the drawing process, the geometric size of the fiber is controlled by adjusting the feeding speed of the preform and the drawing speed. The fiber of the appropriate size is passed through the coating mold of the drawing tower, and the inner and outer coatings are evenly coated on the surface of the fiber. The liquid coating is cured by the UV curing system to form a double-layer fiber structure. The coated fiber is collected on the fiber reel through the fiber collection system to obtain the fiber product.

During the process of fiber preparation, the process parameters mainly include drawing temperature, preform feeding speed, drawing speed, drawing tension, coating adhesive heating temperature and pressure, and UV curing furnace power. The drawing temperature determines the "melting degree" of the fiber preform, which is one of the main factors affecting the size parameters of the fiber. At a certain preform feeding speed and drawing speed, the higher the drawing temperature, the smaller the drawing tension, and the smaller the fiber stress, which is beneficial to improve the fiber strength. However, too high a drawing temperature will cause serious rounding of the octagonal structure of the fiber cladding, thereby affecting the absorption coefficient of the rare earth doped fiber. In addition, too small drawing tension will exacerbate the fluctuation of fiber size. The lower the drawing temperature, the larger the drawing tension, which is beneficial to control the stability of the fiber size and maintain the octagonal structure of the cladding, but an overly pronounced octagonal structure is easy to introduce coating defects. In addition, excessive drawing tension will produce additional surface micro-cracks, affecting the strength of the fiber. Therefore, the drawing process of rare earth doped fiber must use an appropriate drawing temperature. The preform feeding speed determines the speed at which the fiber preform enters the drawing furnace, and the drawing speed determines the speed at which the fiber leaves the drawing furnace. Only when the preform feeding speed and the drawing speed match can the target size fiber be prepared. Assuming the preform feeding speed is V_r, the drawing speed is V_f, the outer diameter of the fiber preform is D, and the outer diameter of the target fiber is d, according to the principle that the volume of the preform before melting is equal to the volume of the fiber drawn after melting, then we have

$$V_r \times D^2 = V_f \times d^2 \tag{3.13}$$

For a fiber preform with a certain outer diameter, if the diameter of the target fiber is determined, the faster the rod feeding speed, the faster the drawing speed, and the greater the drawing tension of the fiber under a certain drawing temperature.

Excessive drawing tension will produce coating defects and additional micro-cracks, affecting the strength of the fiber. The slower the preform feeding speed, the slower the drawing speed, and the smaller the drawing tension. Too small drawing tension will affect the structure of the cladding and the stability of the fiber size. Therefore, an appropriate preform feeding speed and drawing speed must be used in the fiber preparation process. The drawing tension is affected by factors such as the viscosity of the preform forming area, the cross-sectional area of the preform, and the gradient of the drawing speed. Its relationship can be expressed as

$$F = 2\eta S \frac{\partial_{v_z}}{\partial_z} \tag{3.14}$$

In the formula, η is the viscosity in the forming area; S is the cross-sectional area of the fiber preform; $\frac{\partial_{v_z}}{\partial_z}$ is the gradient of the drawing speed.

During the drawing process, the diameter of the preform is determined, that is, the cross-sectional area is determined, so the drawing tension of the fiber mainly depends on the viscosity and speed gradient. If the drawing temperature is kept constant, the viscosity is constant, and the drawing tension is affected by the speed gradient. The faster the drawing speed, the greater the speed gradient, and the greater the tension. Similarly, if the drawing speed is constant, the lower the drawing temperature, the greater the viscosity, and the greater the drawing tension. Too large or too small drawing tension is not conducive to the preparation of high-quality fibers, so the appropriate drawing tension must be determined during the drawing process. The heating temperature and pressure of the coating glue determine the thickness and defects of the coating layer. The higher the heating temperature of the coating glue, the smaller the viscosity of the coating glue, under a certain pressure, the smoother the coating surface, the thinner and more uniform the coating thickness. But the smaller the viscosity, the easier it is for bubbles to form in the coating glue, causing coating defects. The lower the heating temperature, the greater the viscosity of the coating glue, under a certain pressure, the thicker the coating, but it is easy to have uneven coating thickness. Therefore, the appropriate heating temperature and pressure of the coating glue must be determined during the drawing process.

The power of the UV curing furnace determines the degree of curing of the coating layer. If the power is too low, it is easy to cause insufficient curing of the coating layer, affecting the environmental tolerance of the fiber. If the power is too high, it is easy to cause over-curing of the coating layer, causing aging of the coating, affecting the strength of the fiber.

For the influencing factors in the preparation process of rare earth doped fibers, several typical rare earth doped fiber drawing process parameters are listed in Table 3.8.

Table 3.8 Several typical rare earth doped fiber drawing process parameters

Fiber Type	Drawing Temperature(°C)	Feed Preform Speed/(mm/min)	Drawing Speed/(m/min)	Drawing Tension(g)	Coating adhesive heating Temperature/°C	Coating adhesive pressure/bar [a]	UV curing furnace Power/W
20/400 ytterbium-doped double-clad fiber [b]	2120	4.96	15	40	36	0.3	240
8/125 erbium-doped double-clad fiber [c]	2140	3.05	50	80	40	0.4	350
10/130 thulium-doped double-clad fiber [d]	2130	1.69	40	60	40	0.4	350

Note: [a] 1 bar = 0.1 MPa
[b] The diameter of the Ytterbium-doped fiber preform is 30 mm
[c] The diameter of the Erbium-doped fiber preform is 16 mm
[d] The diameter of the Thulium-doped fiber preform is 20 mm

3.3.2 *Performance Characterization of Rare Earth-Doped Silica Fiber*

The performance parameters of rare earth-doped silica fiber mainly include fiber geometric dimensions, cladding loss, absorption coefficient, fiber strength, and environmental resistance, etc. These performance parameters will directly affect the power, efficiency, and service life of fiber lasers and amplifiers using rare earth-doped silica fiber. The following will be introduced one by one.

3.3.2.1 Fiber Geometric Dimension Parameters

There are many specifications of rare earth-doped silica fiber, most of which are named by the doped ions and the core diameter and cladding diameter of the fiber. For example, YDF-20/400, that is, ytterbium-ion-doped 20/400 double-clad fiber, with a core diameter of 20 μm and a face-to-face diameter of 400 μm for the octagonal inner cladding. From this, we can see the importance of fiber geometric dimension parameters. The geometric dimension parameters of the fiber include core diameter, inner cladding diameter, coating diameter, core/cladding concentricity, and core ellipticity, etc. When rare earth-doped silica fiber is used in fiber lasers and amplifiers, these fiber geometric dimension parameters will greatly affect the matching degree of active fiber with passive devices in the system. If the matching degree is poor, it will cause the fusion point loss between the active fiber and the passive device to increase, causing abnormal temperature rise and making the active fiber scrapped. Therefore, the geometric size of rare earth-doped silica fiber is the first parameter to be controlled in the fiber drawing process.

Usually, the same specification of active fiber has an allowable tolerance range for geometric dimensions. For example, for YDF-20/400 active fiber, the core diameter is 20 μm, and the tolerance range is 2 μm; the inner cladding is 400 μm, and the tolerance range is 10 μm. However, if the fiber size is controlled ideally, the size can be controlled within (20 ± 0.5) μm and (400 ± 2) μm. Through precise control of drawing speed, feeding speed, drawing temperature, and feedback from the fiber diameter measurement system, rare earth-doped fiber with small geometric size fluctuations can be prepared.

The prepared fiber can be tested for its geometric dimensions using a high-depth-of-field microscope.

The test process includes using a large-core fiber cutter to cut the fiber sample to obtain a complete measurement end face. Place the fiber sample on the test platform, obtain a clear fiber end face, and use measurement software to obtain the fiber geometric dimensions. Repeat the measurement more than three times, and average the multiple measurement values to reduce the human testing error.

Figures 3.45 and 3.46 show the bare fiber diameter and coating diameter measurements of the typical 20/400 Ytterbium-doped double-clad fiber, respectively.

In addition, some equipment manufacturers have also developed special fiber geometric dimension testing equipment for rare earth-doped silica fiber, which can quickly determine the geometric dimension parameters of active fiber.

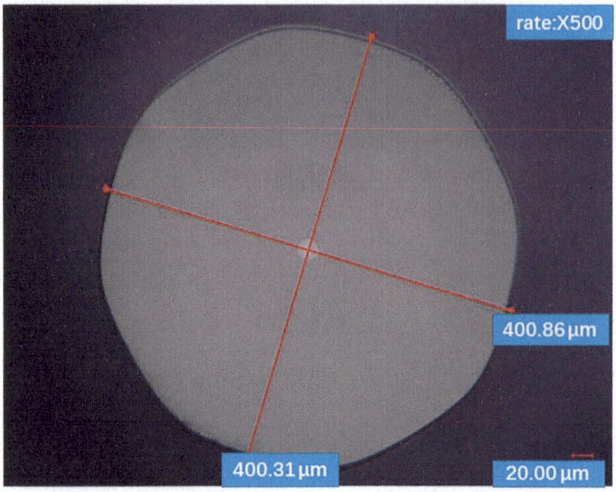

Fig. 3.45 Diameter of bare fiber of 20/400 Ytterbium-doped double-clad fiber

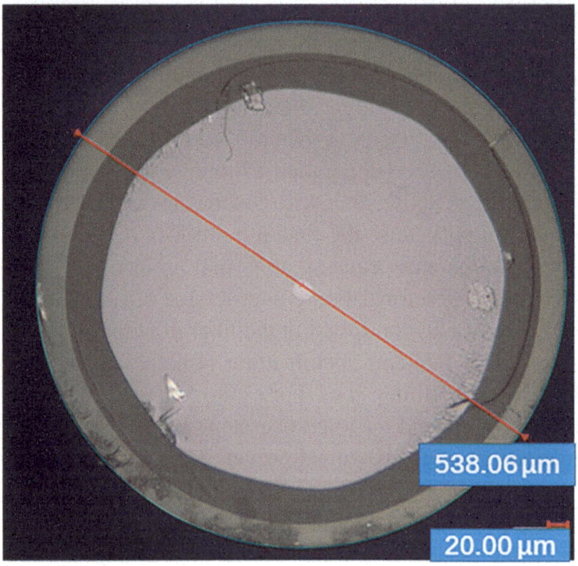

Fig. 3.46 Diameter of coating of 20/400 Ytterbium-doped double-clad fiber

3.3.2.2 Cladding Loss of Rare Earth-Doped Silica fiber

During the process of obtaining lasers through the application of rare earth-doped silica fiber, the pump light will be injected into the cladding and absorbed by the rare earth ions in the core during its transmission in the cladding. Therefore, the cladding loss of the rare earth-doped silica fiber will affect the transmission of the pump light, thereby affecting the laser efficiency. During the drawing process of the rare earth-doped silica fiber, improper drawing tension will increase defects such as micro-cracks on the surface of the bare fiber, which will lead to an increase in the cladding loss of the fiber. Therefore, it is necessary to reasonably design and optimize parameters such as drawing speed and drawing temperature to obtain high-quality rare earth-doped fibers with low cladding loss.

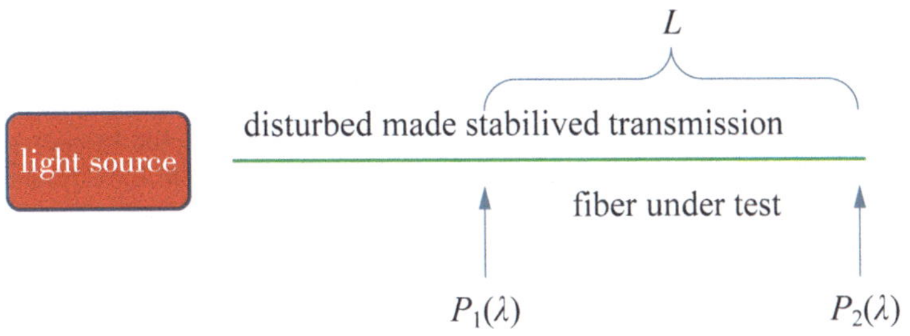

Fig. 3.47 Principle diagram of the cut-off method test

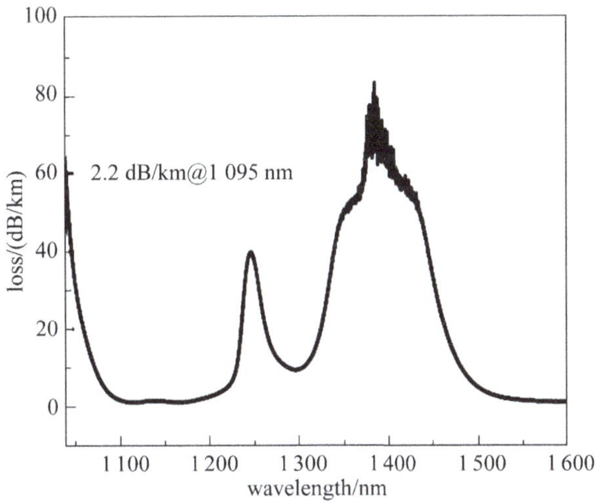

Fig. 3.48 Cladding loss spectrum of 20/400 Ytterbium -doped double-clad fiber

The cladding loss of the drawn rare earth-doped fiber can be tested by the cut-off method. The test principle is shown in Fig. 3.47.

The test process is as follows: Use a broadband incoherent light source to fill the entire cladding of the fiber to be tested, forming stable transmission in the fiber. Repeatedly prepare the end face by cleaving to obtain stable and reliable P1(λ) and P2(λ). The measured end face output powers are P1(λ) and P2(λ) for a fiber length difference L. The fiber cladding loss is calculated according to formula (3.15). After repeated measurements, take the average value.

$$\alpha(\lambda) = \frac{10}{L} \lg \frac{P_1(\lambda)}{P_2(\lambda)} \tag{3.15}$$

Figure 3.48 shows the typical cladding loss spectrum of a 20/400 Ytterbium-doped double-clad silica fiber, with a loss of 2.2 dB/km at a wavelength of 1095 nm.

3.3.2.3 Absorption Coefficient of Rare Earth Doped Silica fiber

The absorption coefficient of rare earth doped silica fiber refers to the absorption of pump light by the rare earth ions doped in the fiber core during the transmission of the pump light in the inner cladding in the laser system. It determines the length of the fiber used in the fiber laser. The larger the absorption coefficient, the shorter the fiber length; the smaller the absorption coefficient, the longer the fiber. However, the total absorption is equivalent. For the conventional 20/400 Ytterbium-doped double-clad silica fiber, the normal nominal absorption coefficient at the 915 nm band is (0.4 ± 0.05) dB/m, so the fiber length used in the laser system is between 32 ~ 40 m, the specific length is determined according to the needs of the laser. The absorption coefficient of rare earth doped silica fiber is mainly determined by the concentration of rare earth ions doped during the preform rod manufacturing process, and the core composition will also have a certain impact on the absorption coefficient. During the fiber drawing process, the shape of the fiber cladding structure will also have a certain impact on the absorption coefficient, which is largely affected by the drawing temperature and drawing tension.

The absorption coefficient of rare earth doped silica fiber can be tested by the cut-off method. The test principle is shown in the schematic diagram of the cut-off method in Fig. 3.47. The test process is as follows: first, select an appropriate fiber length, and control the cladding pump absorption of the cut fiber within 3 ~ 7 dB. Use a pump light source with the wavelength to be tested, fill the entire cladding to be tested with the pump light, and form stable transmission in the corresponding cladding of the fiber. Repeatedly prepare the end face by cleaving to obtain stable and reliable $P_1(\lambda)$ and $P_2(\lambda)$. The measured output power at the end face is for fiber length L of $P_1(\lambda)$and $P_2(\lambda)$. The pump absorption coefficient of the fiber cladding is calculated according to formula (3.15). Repeat the measurement several times and take the average value.

Figure 3.49 is a typical absorption coefficient test diagram of a 20/400 Ytterbium-doped double-clad silica fiber. The red line in the figure represents the test light transmission spectrum before the cut-off, the yellow line represents the test light transmission spectrum after the cut-off, and the green line represents the difference between the red and yellow curves.

3.3.2.4 Strength of Rare Earth Doped Silica fiber

The strength of rare earth-doped optical fibers refers to the amount of tension the fiber can withstand without breaking, reflecting the state of surface defects in the fiber and, to some extent, the lifespan of the fiber. During the fiber drawing process, inappropriate parameters such as drawing temperature, drawing speed, and drawing tension can introduce additional surface micro-cracks in the fiber, thereby affecting the strength of the fiber. In addition, improper coating pressure and temperature can

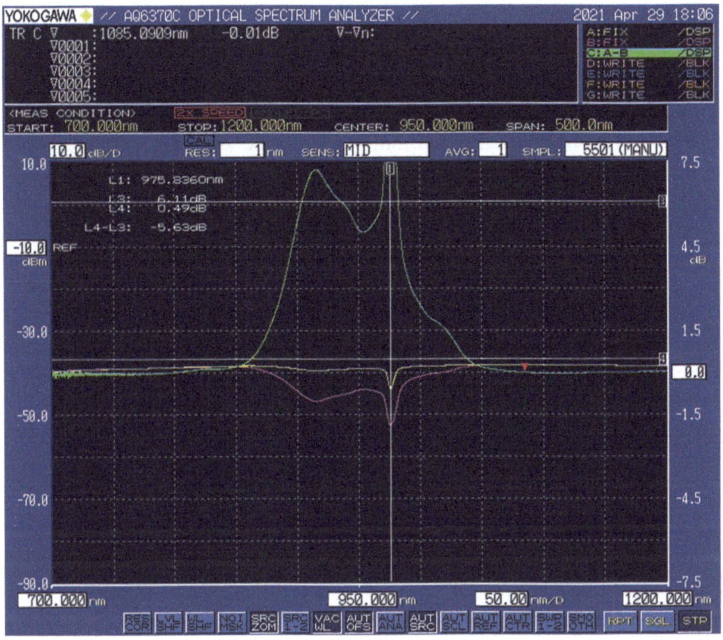

Fig. 3.49 20/400 Ytterbium-doped double-clad silica fiber absorption coefficient test diagram

easily cause bubbles to remain in the coating layer, which also affects the strength of the fiber. Therefore, to ensure that the fiber has sufficient strength, it is necessary to optimize parameters such as drawing temperature, drawing speed, drawing tension, and the pressure and temperature of the coating adhesive during the drawing process.

To test the strength of rare earth-doped optical fibers, the general method is to apply a certain tension to the fiber; if the fiber does not break, it indicates that the fiber has reached the required strength. For rare earth-doped optical fibers, the current common standard requires that the fiber can meet a screening tension strength of 100kpsi (kilopounds per square inch), and for conventional 20/400 Ytterbium-doped double-clad fibers, the fiber is required to withstand approximately 87 N of tension. The tension screening of the fiber is generally carried out on a fiber screening machine. The specific screening process is as follows: use a full-length fiber as a sample, remove a small section of the end, about 20 ~ 50 m. Wind the fiber on the fiber screening machine in the prescribed manner, and set the screening tension to 100kpsi (or 690 MPa). Start the equipment, and if the fiber passes through the fiber screening machine at a certain speed without breaking, it indicates that the sample fiber meets the strength requirement of 100kpsi. If a break occurs at a certain position, it indicates that there are defects at this position that affect the strength. Remove the fiber from this point and continue to screen the remaining fiber until the end.

3.3.2.5 Coating Environmental Resistance

During long-term use of rare earth-doped optical fibers, the organic coating will gradually age until it loses the ability to confine pump light, leading to the scrapping of the fiber. Therefore, the environmental resistance of the coating of rare earth-doped optical fibers largely determines the lifespan of the fiber. During the fiber drawing process, inappropriate parameters such as drawing speed can cause micro-cracks on the surface of the fiber, improper pressure and temperature of the coating adhesive can cause bubbles in the coating, and improper curing process can lead to poor curing effect of the coating. These factors will affect the quality of the coating of the rare earth-doped optical fiber, compromise the environmental resistance of the coating, and thereby shorten the lifespan of the fiber. The assessment of coating environmental resistance is generally conducted by accelerating the aging of the fiber in extreme environments such as high temperature and high humidity to reflect the quality of the coating, thereby providing a reference for optimizing the drawing process parameters.

The test method generally uses an environmental accelerated aging device to provide a high temperature and high humidity assessment environment. After a certain aging time, the red light transmission power through the fiber cladding before and after aging is compared to judge the change in the constraint ability of the coating on the red light after the rare earth-doped optical fiber undergoes environmental aging assessment, thereby reflecting the environmental resistance of the fiber coating.

The test process is as follows: a certain length of rare earth doped fiber is wound into a circle of a certain diameter. Before the environmental resistance treatment, a red light with a power of 100 mW and a wavelength of 633 nm is coupled into the inner cladding of the fiber to be tested by fusion splicing, and the output power at the end of the fiber after passing through the fiber is tested. Then, the sample fiber is subjected to accelerated aging treatment under high temperature and high humidity. The high temperature and high humidity conditions can be a temperature of 120 °C, a relative humidity of 100%, and a treatment time of 24 h. After the aging treatment, the fiber is reconnected to the above-mentioned red light test system, and the red light output power at the end of the fiber after treatment is tested under the same conditions. The result requires that there are no obvious light leakage points on the surface of the fiber, and the power attenuation of the red light passing through meets the requirements.

In summary, this chapter mainly discusses the three main links in the preparation of rare earth-doped silica fiber, including the preparation of fiber preforms (core rods), the drawing of preforms into fibers, and the screening and testing of basic performance parameters of fibers. Among them, the preform preparation technology is the key to determine the laser performance of the fiber. This chapter introduces the current conventional commercial rare earth-doped silica fiber preform preparation technology based on improved chemical vapor deposition, and the non-MCVD rare earth-doped preform preparation technology developed to adapt to various large mode field laser fibers. The fiber drawing technology determines the size, mechanical strength, environmental resistance and other performance parameters of the final fiber product. Whether the laser fiber meets the product requirements requires strict screening and testing.

References

1. Nagel S R, Macchesney J B, Walker K L. An overview of the modified chemical vapor-deposition (MCVD) process and performance [J]. Ieee Journal of Quantum Electronics, 1982,18(4):305-322.
2. Aljamimi S M, Anuar K M S, et al. On the fabrication of aluminum doped silica preform using MCVD and solution doping technique [C]. 2012 IEEE 3rd International Conference on Photonics, 2012.
3. Aljamimi S M, Yusoff Z, Abdul-Rashid H A, et al. Aluminum doped silica preform fabrication using MCVD and solution doping technique: effects of various aluminum solution concentrations [M]//ABDULLAH F. 2013 Ieee 4th International Conference on Photonics, 2013:268-271.
4. Halder A, Lin D, et al. Yb-doped large-mode-area Al-P-silicate laser fiber fabricated by MCVD [M]. 2018 Conference on Lasers and Electro-Optics, 2018.
5. Yi Yongqing, Pang Lu, Pan Rong, et al. Optimization of dehydration conditions for active optical fiber [J]. Optical Communication Technology, 2014,38(7):58-59.
6. Yi Yongqing, Tian Haisheng, Ning Ding. Study of the influence of the deposition temperature by using the MCVD technique on the rare earth ion concentration of active fiber [J]. Optical Communication Technology, 2007,31(1):60-61.
7. Lindner F, Kriltz A, et al. Influence of process parameters on the incorporation of phosphorus into silica soot material during MCVD process [J]. Optical Materials Express, 2020,10(3):763-773.
8. Lindner F, Unger S, et al. Phosphorus incorporation into silica during modified chemical vapour deposition combined with solution doping [J]. Physics and Chemistry of Glasses-European Journal of Glass Science and Technology: Part B, 2015,56(6):278-284.
9. Dhar A, Paul M C, et al. Characterization of porous core layer for controlling rare earth incorporation in optical fiber [J]. Optics Express, 2006,14(20):9006-9015.
10. Kim Y H, Paek U C, et al. Effect of soaking temperature on concentrations of rare-earth ions in optical fiber core in solution doping process [M]//Jiang S. Rare-Earth-Doped Materials and Devices V, 2001:123-132.
11. Aljamimi S M, Anuar M S K, et al. Multiple soaking with different solution concentration in doped silica preform fabrication using modified chemical vapor deposition and solution doping [J]. Fiber and Integrated Optics, 2014,33(1-2):105-119.
12. Webb A S, Boyland A J, et al. MCVD in-situ solution doping process for the fabrication of complex design large core rare-earth doped fibers [J]. Journal of Non-Crystalline Solids, 2010,356(18-19):848-851.
13. Lou Fengguang, Hu Lili, Yu Chunlei, et al. A kind of ytterbium-doped silica fiber resistant to photodarkening and its preparation method: China, ZL201810587237.2 [P/OL]. 2018.
14. Yang Q, Jiao Y, et al. Gain and laser performance of heavily Er-doped silica fiber fabricated by MCVD combined with the sol-gel method [J]. Chinese Optics Letters, 2021,19(11):110603-(1-5).
15. Lenardic B, Kveder M, IEEE. Advanced vapor-phase doping method using chelate precursor for fabrication of rare earth-doped fibers [C]. 2009 Optical Fiber Communication Conference and National Fiber Optic Engineers Conference, 2009.
16. Saha M, Pal A, et al. Vapor phase chelate delivery technique for fabrication of rare earth doped optical fiber [C]. 2012 International Conference on Fibre Optics and Photonics, 2012.

17. Boyland A J, Webb A S, et al. Rare-earth doped optical fiber fabrication using novel gas phase deposition technique [C]. 2010 Conference on Lasers and Electro-Optics, 2010.
18. Izawa T M T, Hanwa F. Continuous optical fiber preform fabrication method: United States, US4062665 [P]. 1977.
19. Wang J, Walton D T, et al. All-glass high NA Yb-doped double-clad laser fibres made by outside-vapour deposition [J]. Electronics Letters, 2004,40(10):590-592.
20. Tammela S, Soderfund M, et al. The potential of direct nanoparticle deposition for the next generation of optical fibers [M]//Digonnet M J F, Jiang S. Optical Components and Materials III, 2006.
21. Koponen J J, Petit L, et al. Progress in direct nanoparticle deposition for the development of the next generation fiber lasers [J]. Optical Engineering, 2011,50(11):111605-(1-11).
22. Ye C, Koponen J, et al. Characterization of chirally-coupled-core (3C (R)) fibers fabricated with direct nanoparticle deposition (DND) [M]//Ballato J. Fiber Lasers: Technology, Systems, and Applications, 2016.
23. Poole S B, Payne D N, Mears R J, et al. Fabrication and characterization of low-loss optical fibers containing rare-earth ions [J]. Lightwave Technol, 1986,4(7):870-876.
24. David M, Victor K, Upendra M, et al. Large-mode-area double-clad fibers for pulsed and CW lasers and amplifiers [J]. Proc. SPIE, 2004(5335):140-150.
25. Gong Fanhan. Research on nanoporous glass-based composite luminescent materials [D]. Guangzhou: South China University of Technology, 2011.
26. Northwest Light Industry College. Glass Technology [M]. Beijing: China Building Materials Industry Publishing House, 1994.
27. Chen Danping, Masui H, Miyoshi H, et al. Extraction of heavy metal ions from waste colored glass through phase separation [J]. Waste Management, 2006,26(9):1017-1023.
28. Chen Danping, Miyoshi H, Akai T, et al. Colorless transparent fluorescence material: sintered porous glass containing rare-earth and transition-metal ions [J]. Applied Physics Letters, 2005,86(23):231903-231908.
29. Liu Shuang, Wangmeng, Zhou Qinling, et al. Ytterbium-doped silica photonic crystal fiber laser fabricated by the nanoporous glass sintering technique [J]. Laser Physics, 2014,24(6):065801-065801-5.
30. Liu Shuang. Preparation and performance study of Yb-doped silica glass and large mode field fiber [D]. Shanghai: Shanghai Institute of Optics and Fine Mechanics, Chinese Academy of Sciences, 2015.
31. Chu Y, Ma Y, Yang Y, et al. Yb3+-doped large core silica fiber for fiber laser prepared by glass phase-separation technology [J]. Optics Letters, 2016, 41:1225-1228.
32. Langner A, Schötz G, Kayser T, et al. A New Material for High Power Laser Fibers [J]. Proc. of SPIE, 2008, 6873:687311(1-9).
33. Langner A, Such M, Schötz G, et al. Development, Manufacturing and lasing behavior of Yb-doped ultra large mode area fibers based on Yb-doped fused bulk silica [J]. Proc. of SPIE, 2010, 7580:75802X-(1-12).
34. Langner A, Such M, Schötz G, et al. Multi-kW single fiber laser based on an extra large mode area fiber design [J]. Proc. of SPIE, 2012, 8237:82370F-(1-12).
35. Langner A, Such M, Schötz G, et al. Design evolution, Long term performance and application tests of extra large mode area (XLMA) fiber lasers [J]. Proc. of SPIE, 2013, 8601:86010G(1-13).
36. Schuster K, Unger S, Aichele C, et al. Material and Technology Trends in Fiber Optics [J]. Advanced Optical Technologies, 2014, 3(4): 447-468.
37. D. Etissa, M. Neff, S. Pilz, et al. Rare earth doped optical fiber fabrication by standard and sol-gel derived granulated oxides [J]. SPIE, 2012, 8426: 842611(1-6).
38. Thomas. I M, Payne. S A, Wilke. G D, et al. Optical properties and laser demonstration of Nd-doped sol-gel silica glasses [J]. Journal of Non-Crystalline Solids, 1992, 151:183-194.
39. U. Pedrazza, V. Romano, W. Luthy, et al. Yb3+: Al3+: sol–gel silica glass fiber laser [J]. Optical. Materials, 2007, 29(7): 905–907.

40. Wang, S, Li Z, Yu C, et al. Fabrication and laser behaviors of Yb3+doped silica large mode area photonic crystal fiber prepared by sol–gel method [J]. Optical Materials, 2013. 35(9): 1752-1755.
41. Wang, S., et al. Optical and laser properties of Yb3+-doped Al2O3–P2O5–SiO2 large-mode-area photonic crystal fiber prepared by the sol–gel method. Laser Physics Letters, 2013. 10(11):115802.
42. Wang, S., et al. Characteristics and laser performance of Yb3+-doped silica large mode area fibers prepared by Sol-Gel method. Fibers, 2013. 1(3): 93-100.
43. Xu, W, Lin Z, Wang M, et al. 50 μm core diameter Yb3+/Al3+/F– codoped silica fiber with M2< 1.1 beam quality [J]. Optics letters, 2016. 41(3):504-507.
44. Wang, S, Xu W, Lou F, et al. Spectroscopic and laser properties of Al-P co-doped Yb silica fiber core-glass rod and large mode area fiber prepared by sol-gel method [J]. Optical Materials Express, 2016. 6(1): 69-78.
45. Wang, F, et al. Manipulating refractive index, homogeneity and spectroscopy of Yb3+-doped silica-core glass towards high-power large mode area photonic crystal fiber lasers. Optics Express, 2017. 25(21):25960-25969.
46. Wang, S, Xu W, Wang F, et al. Yb3+-doped silica glass rod with high optical quality and low optical attenuation prepared by modified sol-gel technology for large mode area fiber. Optical Materials Express, 2017. 7(6): 2012-2022.
47. Zhang W, Wu J, Zhou G, et al. Yb-doped silica glass and photonic crystal fiber based on laser sintering technology [J]. Laser Physics, 2016, 26:035801(1-5).
48. Devautour M, Roy P, Férier S, et al. Nonchemical-vapor-deposition process for fabrication of highly efficient Yb-doped large core fibers [J]. Applied Optics, 2009, 48(31): G139-G142.
49. Wei Zhongcheng, et al. Preparation Technology of Optical Fiber Materials [M]. Beijing: Beijing University of Posts and Telecommunications Press, 2016.

Chapter 4
High-Power Ytterbium-Doped Cladding-Pumped Silica Fiber and Its Applications

High-power fiber lasers, compared to traditional solid-state lasers, have obvious advantages such as compact structure, high conversion efficiency, good beam quality, and ease of heat dissipation. They have significant application prospects in many fields, including advanced manufacturing, military defense, scientific research, and medicine, and are a research hotspot in the international laser technology field. Thanks to the continuous development and improvement of double-clad fiber drawing technology and high-power semiconductor laser manufacturing processes, the conversion efficiency and output power of fiber lasers have been greatly enhanced. Breakthroughs have been made in various key technologies, such as new structural designs for doped fibers, various coupling technologies including end and side pumping, and laser beam combining techniques. Especially driven by the application demands in industrial processing, directed energy weapons, long-range telemetry, and lidar, foreign research institutions and companies represented by IPG Photonics Corporation, Nufern, nLight Corporation, SPI Lasers UK, and TRUMPF Group in Germany have actively developed high-power fiber lasers. They have successively launched commercial fiber laser products with continuous wave powers ranging from tens of watts to tens of kilowatts, and some of these products are subject to restrictive export or embargo to China. Although domestic research on high-power fiber lasers started later than abroad, rapid development has been achieved in recent years. Many universities, research institutes, and companies have participated in the research and development of fiber lasers and have successively launched a variety of mature products.

Active fiber is the gain medium in fiber lasers, which converts high numerical aperture pump light (e.g., a core numerical aperture of 0.22) into low numerical aperture (e.g., a core numerical aperture of 0.06), high-brightness laser light. The performance of active fiber directly affects the quality of the output laser, especially in terms of laser power, slope efficiency, nonlinear effects, modal instability, and

L. Hu, *Rare Earth Doped Silica Fiber and Its Applications*,
https://doi.org/10.1007/978-981-96-5306-5_4

photodarkening effects. Active fiber has evolved from early single-clad designs to double-clad, and from early small cores (with core diameters typically below 10 μm) to large cores (i.e., large mode area fibers). With the improvement in rare-earth doping techniques, large mode area active fibers have been realized. Early rare-earth doping systems were relatively simple, with lower purity of raw materials, and a lack of in-depth research on the mechanisms of co-doping with elements like aluminum and phosphorus resulted in lower doping concentrations of Yb^{3+}. In recent years, with the continuous maturation of processes such as vapor phase method, liquid phase method, sol-gel method, and nanoparticle doping, the concentration of rare-earth doping has been increased. Even with Yb^{3+} concentrations as high as 0.5 wt%, there is no occurrence of rare-earth ion clustering or opacification, and the core refractive index distribution is more reasonable. This results in a fiber mode field that is nearly top-hat in distribution, further increasing the fiber's damage threshold. Currently, imported 20/400 μm active fibers have broken through power levels of 3 kW, and some ytterbium-doped fibers with core diameters of 30 μm or more can achieve output powers exceeding 10 kW.

4.1 Structure and Properties of Ytterbium-Doped Large Mode Area Fiber

Fiber lasers have developed from zero to hundreds of watts of power over nearly four decades. As early as 1961, American scientist Snitzer proposed that using rare-earth-doped fibers within the laser cavity could yield stable single-mode laser output [1], but due to limitations in fiber fabrication and pumping sources, it did not experience rapid development. The 1970s to 1980s were 20 years of rapid development for semiconductor lasers and fiber-drawing techniques. Because it was difficult to efficiently couple pump light into a fiber core with a diameter of just a few micrometers, fiber lasers could only produce laser output in the milliwatt range for a long time.

In the early 1960s, laser oscillation in Yb^{3+}-doped glass was achieved [2]. However, the activation ion of general interest at that time was Nd^{3+}, and Yb^{3+} was more commonly used as a sensitizing ion. It wasn't until the late 1980s that attention was refocused on Yb^{3+}-doped single-mode silica fiber lasers [3–6].

In China, Shanghai Institute of Optics and Fine Mechanics, Chinese Academy of Sciences (SIOM) has carried out early research work on Yb^{3+}-doped fiber lasers. By pumping Yb^{3+} fibers of varying lengths with tunable titanium-sapphire lasers ranging from 860 to 980 nm, researchers achieved laser emissions at three and four energy levels, respectively [7]. The simplest toroidal cavity structure was achieved through the use of wavelength division multiplexing (WDM) [8]. An experimental and theoretical investigation into the variation of the emission wavelength of Yb-doped fiber lasers with respect to fiber length and threshold was conducted [8, 9]. Leveraging the benefits of the simple Yb^{3+} energy level structure, the absence of excited state absorption at higher doping concentrations, a wide pump wavelength range, and a broad tuning range for emission wavelengths, Yb^{3+} fiber has emerged as the preferred material for high-power fiber lasers. Particularly in 1988, the advent of double cladding fiber and advancements in semiconductor lasers propelled fiber lasers into a rapid phase of development.

4.1.1 Double Cladding Fiber

In the late 1980s, Polaroid Corporation in Massachusetts, USA. [9] first proposed the application of cladding pumping technology on the basis of double-cladding fiber, Fig. 4.1 is the schematic diagram of the principle. The difference between double cladding fiber and traditional single-mode fiber is that by designing the fiber structure and selecting the appropriate material, an inner cladding layer is formed outside the core of the fiber doped with rare-earth ions that can be used for multi-mode pumping light transmission. Therefore, a high-power multimode laser can be used as the pump source to couple the multimode pump light into the inner cladding through the cladding pumping technique. The pumped light will cross the core several times as it propagates axially along the inner cladding of the fiber and is gradually absorbed by the rare earth ions in the core, resulting in laser gain in the core.

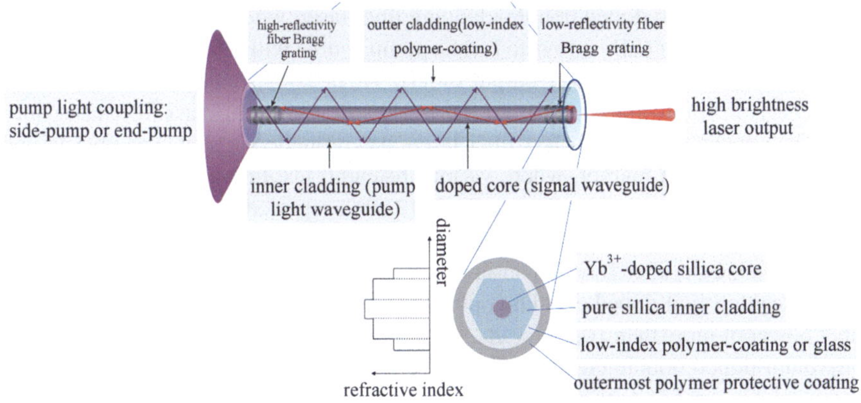

Fig. 4.1 Working principle of rare earth doped double cladding fiber

For double cladding fiber, the core is composed of silicon dioxide doped with rare earth elements, which acts as a laser medium as a channel for laser oscillation; the inner cladding is composed of pure quartz glass with radial dimensions and numerical aperture (NA) much larger than the core and smaller refractive index than the core, with dimensions in the order of hundreds of microns, which can withstand high pumping power. The outer cladding layer consists of an organic coating layer with a smaller refractive index than the inner cladding layer, usually based on UV-curable acrylate material. The protective layer is also a layer of organic material, usually of the same type as the low refractive index outer cladding layer, which is an acrylate material, thus allowing better bonding properties between the two coatings. The difference between the two organic coatings is that the refractive index and modulus of the protective layer is much higher than that of the inner coating (outer cladding). The organic material has good bending properties, which protects the fiber and prevents breakage. The common kilowatt-class 20/400 μm ytterbium-doped double cladding fiber in the market today has a core NA of ~0.06 and an inner cladding size of 400 μm, representing a distance of 400 μm between the octagonal inner cladding structure facing each other, with a cladding NA of 0.46.

At the early stage of cladding pumping technology development, attention was mainly focused on the research of Nd^{3+}-doped double-cladding fiber lasers. Until the 1990s, with the popularization of semiconductor lasers, significant progress was made in the development of ytterbium-doped silica fiber lasers and amplifiers. In 1994, the cladding pump was first implemented in the doped Yb^{3+} silica fiber by H.M. Pssk et al. [10], and a maximum laser output of 0.5 W at the wavelength of 1040 nm was obtained, and the slope efficiency reached 80%. Because of the higher slope efficiency of Yb^{3+} doped double cladding fiber laser than Nd^{3+} doped fiber laser, and the Yb^{3+} simple energy level structure, wide absorption band and high quantum efficiency, people's attention gradually turns to the study of ytterbium-doped double-coated fiber laser. In 1999, V. Dominic et al. [11] reported the results of a 100 W class Yb-doped double-cladding fiber laser, where they used four 45 W semiconductor laser (LD) arrays to form a pump source with a total power of 180 W and obtained a 1120 nm laser output of 110 W. In 2004, Jeong et al. [12] at the University of Southampton achieved the world's first kilowatt-class fiber laser output. The advent of kilowatt-class fiber lasers has propelled high-power fiber lasers into the application market, where various research institutions and startups have emerged, presenting a vibrant scene. In 2012, IPG reported a 20 kW quasi-single-mode and a 500 kW multi-mode fiber laser, which represents the current maximum power of a single fiber for fiber lasers. In China, around 2018, several units used domestic ytterbium-doped cladding optical fiber to achieve the laser output of single fiber power of more than 10,000 watts. Chen Xiaolong of SIOM used the 30/600 μm ytterbium-doped double cladding fiber developed by Hu Lili team of the institute, and used 976 nm LD dual-end pump to achieve the amplification efficiency of 89% [13]. In 2021, Lin Aoxiang's group at the Chinese Academy of Engineering Physics, in collaboration with Tsinghua University [14], successfully achieved a single-fiber 20 kW laser output based on a 48/400 μm domestic double cladding fiber using the

1018 nm pumping scheme, which became a new milestone for single-fiber output power in China.

4.1.2 Influence of the Shape of the Inner Cladding on the Performance of Optical Fiber

In the development history of double coating optical fiber material, the initial inner cladding structure is circular and symmetrical, and part of the pump light will reflect many times without propagating to the fiber core. This light is commonly known as "spiral light", which leads to the very low absorption efficiency of the fiber core.

In the study of the performance of double-clad fiber lasers, A. Liu and K. Ueda investigated the shape of the fiber inner cladding cross-section and the absorption efficiency of the core for pumping light [15, 16]. The results show that the pumping efficiency of double cladding fiber increases with the increase of core diameter and decreases with the increase of inner cladding diameter. In order to improve the pumping efficiency, they proposed two schemes: (1) designing an eccentric double cladding fiber (2) break the circular symmetry of the inner envelope. The centric double cladding fiber helps the core to absorb incident pump light in a larger angle range and improve the pump efficiency. With the increase of core deviation displacement, the pump absorption coefficient gradually increases, and the absorption coefficient of eccentric fibers with different core diameters is nearly equal. The principle of improving the absorption coefficient of the cladding pump is to break the circular symmetry structure of the inner cladding.

In order to improve the absorption efficiency of pump light and optimize the boundary conditions of the inner cladding, in addition to circular structures, D-shaped, rectangular, square, plum-shaped, single-cut rectangular, double-cut rectangular, double-D-shaped, 4D-shaped and positive octagonal inner cladding shapes have been proposed, and double cladding fibers of these inner cladding shapes have been drawn. Fig. 4.2 shows the end faces of typical inner cladding structure fibers. Experiments show that these cladding shapes have substantially higher absorption efficiency of pumped light compared to circular cladding fibers. However, D-shaped, rectangular and square also have several localized modes where light is steadily reflected in the cladding and does not enter the core. Therefore, designing double cladding fibers with novel inner cladding shape and feasible process to remove as much as possible the local stable modes of the fiber can make the absorption efficiency of pump light higher and the required fiber shorter with constant core doping concentration. The analysis of the propagation of light in the new inner cladding shape fiber in double-clad fibers is carried out in the book "High Power Fiber Lasers and Their Applications" by Qihong Lou, who used the two-dimensional ray tracing method to carry out the calculation and analysis of pump absorption [17]. The effect of the structure on the cladding pump absorption is mathematically illustrated.

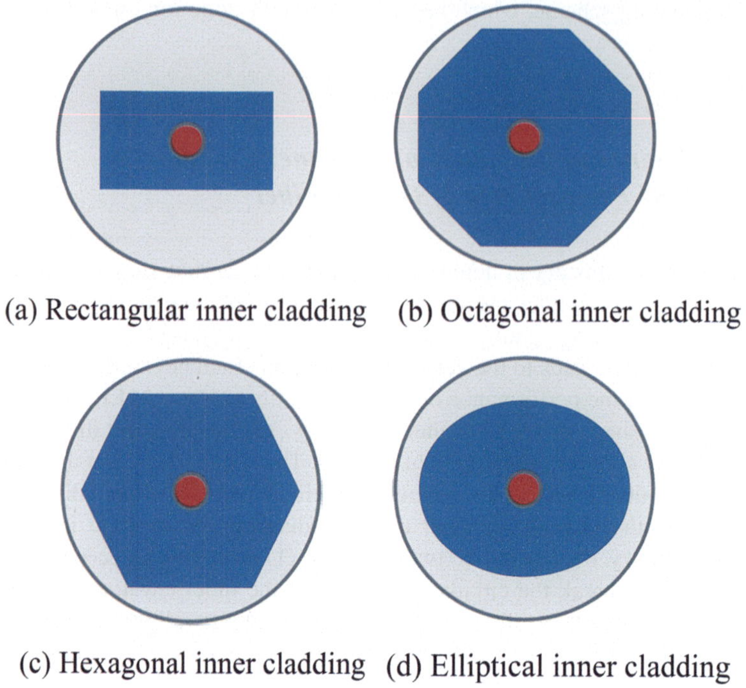

Fig. 4.2 Schematic diagram of typical inner cladding structure. (**a**) Rectangular inner cladding. (**b**) Octagonal inner cladding. (**c**) Hexagonal inner cladding. (**d**) Elliptical inner cladding

In general, after more than thirty years of development of double cladding optical fiber, its structure design gradually tends to standardize, that is, in the current practical engineering applications, from the processing process difficulties and consideration of low-loss fusion with circular passive fiber perspective, the most common inner cladding structure for the positive octagonal structure, which is due to the large inner cladding cross-section can effectively collect pump light, while the small core can keep the fiber laser good. This is due to the large inner cladding cross-section for efficient pump light collection and the small core for good beam quality. Similarly, the requirements for fiber structure are the same for pulsed and continuous fiber lasers.

4.1.3 Multi-Cladding Fiber Structure

With the development of double cladding fiber preparation technology, additional layers of fiber cladding are being considered to more effectively control the mode field diameter, bending loss, dispersion, pump power tolerance, and reliability of the fiber for sustained use.

Recent advances in fiber laser technology will benefit many laser applications. Thanks to advances in semiconductor lasers and fiber optics, compact and reliable fiber lasers can now deliver continuous wave (CW) outputs in the kilowatt power range near the diffraction limit in both unpolarized and polarized cases. Generating such high power requires a fiber effective mode area of about 500–1000 μm^2 to prevent undesirable nonlinear effects such as stimulated Brillouin scattering (SBS) and stimulated Raman scattering (SRS). This can be achieved with double-cladding large-mode area (LMA) fiber designs, such as those utilizing fibers with core diameters ranging from 20 to 30 μm and numerical apertures (NA) of approximately 0.06. However, this fiber waveguide structure, which is essentially a multimode optical waveguide, conflicts with the high beam quality necessary to maintain near diffraction-limited output. Moreover, a low numerical aperture (NA) fiber would further diminish the core's capacity to confine the signal light. To prevent this, practical fiber amplifiers typically employ bend filtering. This involves suppressing the gain of higher-order modes by exploiting the difference in bend loss between the fundamental mode and the higher-order modes. The current fiber coiling diameter in the fiber water-cooled tray of high-power fiber laser amplifiers is typically in excess of ten centimeters. It is important to note that this method of achieving higher power single-mode laser output also encounters several bottlenecks: to minimize nonlinear effects, the fiber diameter must be increased, thereby reducing the core laser power density. However, an increase in core diameter will reduce the difference in bending loss between the fundamental mode and higher-order modes, resulting in a smaller allowable window for bending filtering. In other words, achieving higher-order mode filtering without a significant increase in fundamental mode loss becomes more challenging. Achieving higher-order mode filtering without a substantial increase in fundamental mode loss is highly dependent on the coil diameter, which greatly complicates engineering applications. For this reason, in 2006, André Croteau et al. of the National Optics Institute (INO) in Canada [18] proposed the concept of a bend-insensitive ytterbium-doped large-mode field tripple cladding fiber, shown in Fig. 4.3, which achieves a high rare-earth ion concentration doping (~4 wt%) and therefore has a high absorption coefficient. To reduce the core NA in highly doped conditions, Ge-doped quartz glass is employed as the first cladding to achieve a lower core NA. Additionally, to decrease the sensitivity of bending to mode filtering, the core is partially doped with rare earth ions [19]. In this study, a polarization-preserving 22 μm core Yb-doped tripple cladding fiber was tested for

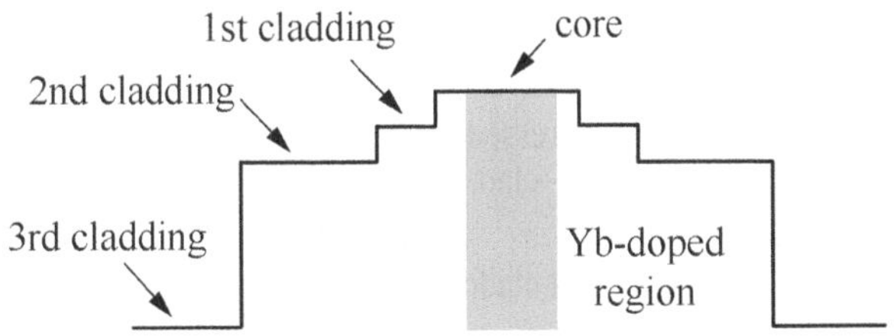

Fig. 4.3 Refractive index distribution of three-clad fiber design

the first time. It exhibited a laser slope efficiency of 86%, a polarization extinction ratio of more than 24 dB, and an output beam quality factor M^2 of less than 1.1, making it suitable for lasers and amplifiers. A special note: The beam quality and output power remain largely unaffected when the fiber is wound to a diameter of 12 cm. This technology offers a solution for the production of compact, high-power fiber amplifiers and laser sources.

In 2007, INO Canada reported an optimized design for a Yb-doped tripple cladding fiber structure, as depicted in Fig. 4.4. This design features a refractive index depression within the first cladding, as illustrated in Figure (b) compared to Figure (a) [20]. Theoretical simulations suggest that by establishing a lower limit for the refractive index in the first cladding, the loss difference between the fundamental mode and higher-order modes under bending filter conditions can be enhanced, thereby promoting its engineering application.

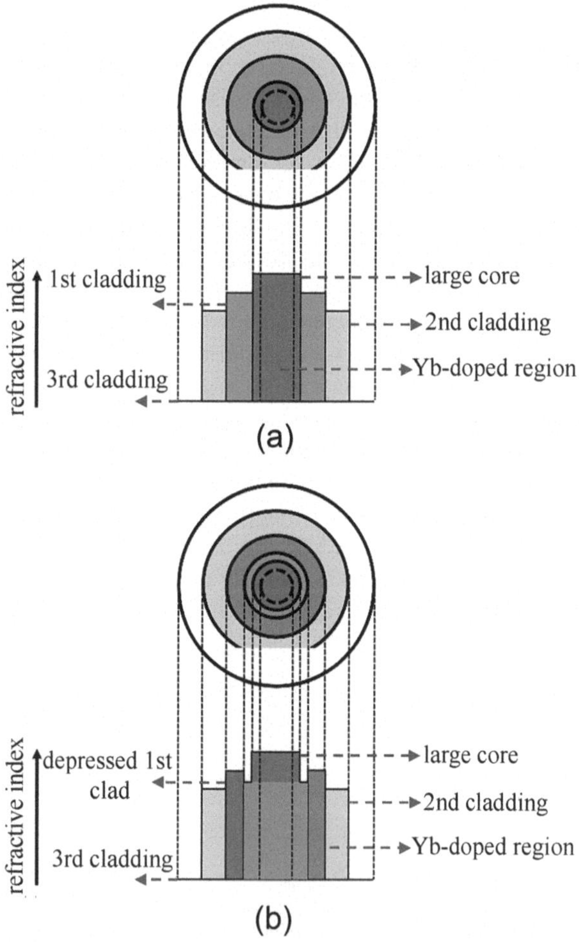

Fig. 4.4 Refractive index distribution of ytterbium-doped three-clad fiber design with the addition of the lower refractive index limit of the first cladding [20]

Examining the fiber structure above, we observe an extra cladding layer compared to the standard double cladding fiber, leading researchers to refer to it as a triple-cladding structure to differentiate it from the conventional double-cladding structure (additional layers can indeed be incorporated). However, due to the minimal numerical aperture (NA) difference between this additional cladding layer and the doped core as well as the second cladding layer, it diverges from the cladding concept defined in traditional double-cladding fibers. Consequently, it lacks the necessary conditions to function as an independent cladding layer when subjected to semiconductor pumping. Therefore, it essentially remains within the category of double-cladding fiber structures.

Currently, the active fiber predominantly used in kilowatt-class fiber lasers is primarily ytterbium-doped, double-cladding structured silica fiber. For instance, Nufern has commercialized various double-cladding Yb-doped silica fibers, and is one of the leading suppliers of high-power fiber lasers in China. Fig. 4.5a depicts a micrograph of the end face of a 20/400 μm Yb-doped double-cladding fiber, featuring the core, inner cladding, low refractive index organic outer cladding, and external organic protective coating, arranged from the center outward. As the output power of the double cladding fiber laser continues to increase, the inner cladding requires higher power injection. Consequently, the interface between the outer and inner cladding interacts more intensely with the pump light. The temperature rise due to interfacial defects will inevitably lead to the interfacial organic coating experiencing a higher thermal load. For conventional UV-curable coating resins, the glass transition temperature is approximately 80 °C, and the tolerable operating temperature is only a few tens of degrees Celsius. This characteristic results in a significant conflict between the existing low refractive index organic coatings and the requirement for increased power pumping. Although most fiber lasers are water-cooled during high-power operation, the persistent impact of upconversion luminescence and frequency doubling light on UV-curable organic coatings within Yb-doped fibers under high-power pumping will inevitably hasten the aging of these organic

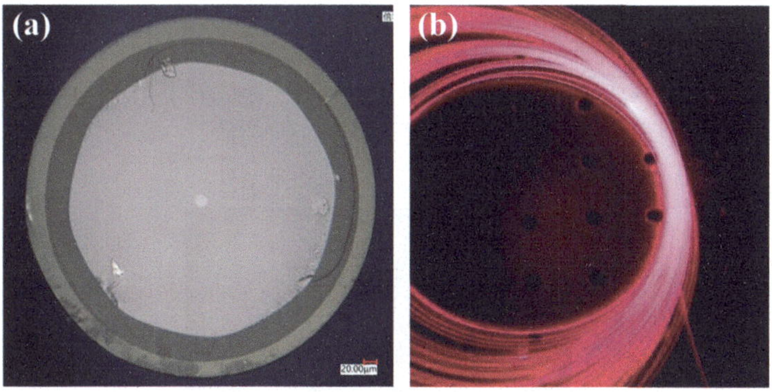

Fig. 4.5 Photomicrographs of the end face of 20/400 μm Yb-doped double-clad fiber (**a**) and organic coating defects after 500 hours of operation (**b**)

coatings. After an extended period of high-power operation, the ability of outer cladding to confine the pump light diminishes, and defects may even emerge, as illustrated in Fig. 4.5b by the bright spot. This will inevitably result in a decline in the performance of the Yb-doped fiber laser. This indicates that the current double-cladding structure of the fiber struggles to achieve long-term stable operation under 10,000-watt high-power conditions. Therefore, there is an urgent need to develop a tripple cladding structure Yb-doped fiber with a higher numerical aperture for the inner cladding and improved thermal performance specifically for single-fiber 10,000-watt Yb-doped large mode area fiber lasers, starting from the fiber structure itself.

In 2017, Han Shuai and Chen Danping from the Shanghai Institute of Optics were the first to apply for an invention patent for a multi cladding optical fiber. They proposed a structure that utilized highly fluorine-doped quartz glass as the outer cladding [21]. In 2019, Lili Hu and colleagues from the same group applied for a patent titled "Triple Cladding Ytterbium-Doped Quartz Optical Fiber and Method of High Concentration Fluorine Layer Quartz Tube Sleeve" [22]. The patent discloses a process technology for preparing high-efficiency, low-loss Yb-doped triple-cladding optical fiber using highly fluorine-doped quartz glass tubes. Fig. 4.6 illustrates the structure of the end face of the Yb-doped triple cladding fiber and the Yb-doped double cladding fiber fabricated by the Shanghai Institute of Optics and Fine Mechanics. The fiber features a highly fluorine-doped quartz glass as the second cladding layer, and the numerical aperture (NA) of the pure quartz glass inner cladding layer is approximately 0.22, which is suitable for existing laser diode (LD) pumping injection conditions. Since 2019, the preparation and application of ytterbium-doped triple-cladding fibers have also been reported by research institutions and companies such as CMI and Wuhan Changfei [23, 24]. Currently, Yb-doped triple cladding fiber has been experimentally verified and industrially applied in several fiber laser companies in China. In 2021, the Shanghai Institute of

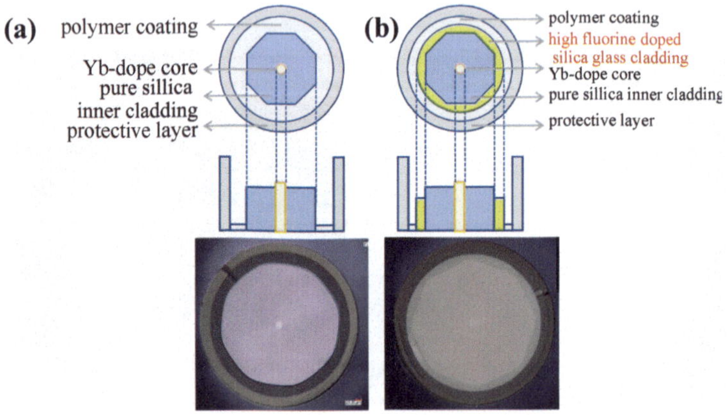

Fig. 4.6 End face structures of optical fibers with ytterbium-doped cladding structure: double-clad fiber (**a**) and triple-clad fiber (**b**)

Optics enhanced the performance of Yb-doped tri-clad fiber to a level capable of meeting the power output of more than 12 kW for a single module in the domestic industrial laser field, significantly improving the reliability of high-power fiber lasers for long-term industrial applications.

4.1.4 Novel Structured Ytterbium-Doped Large Mode Area Cladding Fiber

Despite the rapid advancement of high-power fiber laser technologies, several technical challenges have surfaced as output power continues to escalate. The constraints on fiber laser output power primarily encompass pump power, nonlinear effects such as Stimulated Raman Scattering (SRS) and Stimulated Brillouin Scattering (SBS), mode instability—including transverse mode instability (TMI)—and photodarkening effects. Nonlinear effects can cause part of the laser energy to be converted into longer wavelength, useless light, which can even exceed the laser power, thereby reducing the effective output power of the laser. These effects can typically be suppressed by reducing the fiber length and increasing the core size. TMI, the Transverse Mode Instability, is the interconversion between the fundamental and higher-order modes of the laser at high power operation with millisecond periods, severely degrading the beam quality. The currently more widely accepted model is the so-called thermogenic mode instability. The emergence of mode instability effects alters the power enhancement level of high-power fiber lasers. Energy coupling from fundamental to higher modes occurs in large mode field fibers, ultimately degrading the output spot and limiting the power enhancement of near-diffraction-limited lasers. Additionally, the photon darkening effect, a common concern, is also an important influence factor, which is greatly related to the core doping. Currently, by optimizing the core composition, the photon darkening effect has been largely improved.

To address the aforementioned technical challenges of fiber lasers, a range of new structures for ytterbium-doped large-mode field cladding fibers have been introduced internationally. The aim is to achieve higher output power, improved beam quality, and a higher mode instability threshold. These types primarily include low-NA ytterbium-doped large-mode field cladding fiber, GTwave fiber, 3C fiber, tapered fiber, and pitted fiber, which are detailed below.

4.1.4.1 Low NA Ytterbium-Doped Large Mode Area Cladding Fiber

Since 2014, institutions such as Jena (Germany) and MIT (USA) have begun reporting the results of low numerical aperture (NA) double-clad fiber preparation using modified chemical vapor deposition (MCVD) [25, 26]. The University of Massachusetts has published the results of the MCVD process, revealing the preparation of very low NA Yb-doped double-clad fibers with an NA as low as 0.04 or even 0.025. These fibers were created using gas-phase and liquid-phase doping processes, respectively, to achieve a single-mode laser output of 3000 W under direct LD pumping. Since then, the development of low-NA Yb-doped large-mode field fibers and lasers has become a research hotspot. The preparation of these low-NA fibers often involves a trade-off in doping concentration, typically resulting in a Yb^{3+} doping level of only 5000 ppm, significantly lower than that of conventional 20/400 μm ytterbium-doped double-clad fibers (over 8000 ppm). This necessitates the use of longer fibers for the main amplifier, which can lead to nonlinear effects and is not conducive to the amplified output of high-power lasers. The preparation of low-NA single-mode fibers with larger doping concentrations is a current research focus and challenge in the field of high-power fiber lasers.

Meanwhile, as the power level of narrow linewidth fiber lasers, particularly those required for single-mode fiber lasers and fiber grouping technology, increases, the nonlinear SBS effect of the fiber (which mainly affects the laser linewidth) and the instability of the laser mode (affecting the laser beam quality) emerge as two key factors limiting the development of high brightness narrow linewidth fiber lasers. Xu Xiaojun and Zhou Pu's group at the National Defense University of Science and Technology (NDU) have studied the intrinsic Stimulated Brillouin Scattering (SBS) threshold properties of conventional doped quartz fiber materials. Their theoretical analysis concluded that the traditional quartz fiber can achieve single-frequency laser power, but it is more difficult to surpass 2 kW [27].

As previously mentioned, the mode instability threshold of fiber lasers, a significant scientific issue, limits the brightness of high-power fiber lasers. The prevailing understanding is that this issue is directly related to thermo-optical properties. The understanding in materials science is that the thermal-optical coefficient of the rare earth doped silica glass core is greater than that of the pure quartz glass cladding. As laser power increases and fiber temperature rises, the core's numerical aperture (NA) becomes larger, and the mode quality deteriorates. By incorporating phosphorus and boron, which possess negative thermo-optical coefficients, into the ytterbium-doped core glass to counterbalance the positive thermo-optical coefficients introduced by the rare earth and aluminum doping, the mode instability can be improved from a material perspective. Therefore, maintaining a low refractive index of the Yb-doped core, while also achieving a lower thermo-optical coefficient, is extremely advantageous for the preparation of Yb-doped large-mode field fibers with a high mode instability threshold.

As we all know, the traditional MCVD rare earth doping process encompasses both liquid phase and gas phase doping processes. The liquid phase process involves preparing a solution—typically an alcohol solution—of rare earth chloride and aluminum chloride, including co-doping elements in the appropriate proportions. This solution is then used to soak the MCVD-prepared silica preform to complete the doping of rare earth ions. Subsequently, the prefabricated rods are produced through a vitrification and rod shrinking process. The vapor phase doping process entails depositing rare earth chelate or rare earth chloride vapor onto the interior of silica fibers, which then reacts with oxygen at high temperatures. This is followed by high-temperature vitrification and rod shrinkage to produce fiber preform. It is important to note that when employing gas phase doping with various metal element chelates or chlorides, the significantly different saturated vapor pressures restrict the co-doping of multiple elements. In the case of the traditional liquid phase doping process, while multiple co-doping elements can be combined into alcohol solutions, the disparities in ionic radius and electric field strength result in substantial variations in the uniformity of adsorption of the co-doping elements within the silica spar, often leading to ion clustering and uneven gain.

In 2016, the Shanghai Institute of Optics and Mechanics developed the sol-gel process to fabricate fibers with an extremely low numerical aperture (NA) of approximately 0.02 and a Yb ion concentration exceeding 5000 ppm. This was achieved by co-doping fluorine into the sol-gel and carefully controlling the isobaric phosphorus-aluminum doping, resulting in single-mode laser output at a core diameter of 50 μm [28]. This indicates that cores with higher fluorine content can be prepared using the sol-gel process. Additionally, it is very convenient to introduce other modifier components, such as Y, Mg, Ca, etc., during the sol-gel formulation. It is particularly significant that through molecular-level mixing of the sol, multiple coordination modifier elements can be dispersed more uniformly around the rare earth ions in the appropriate proportion. This results in the generation of more stable rare earth modifier colloids through hydrolytic condensation reactions. Furthermore, by utilizing the MCVD deposited silica sparse body immersion technique, the proportional adsorption and doping of rare earth ligand groups can be more effectively accomplished. This facilitates the adsorption of rare earth ions and modifier elements, making it easier to prepare Yb-doped large-mode field fibers with a low refractive index. This is particularly advantageous for the preparation of Yb-doped large-mode area fibers with complex compositions through deposition.

4.1.4.2 GTwave Side-Pumping Structure Fiber

The GTwave fiber is a general name for a side-pumped active fiber. In 2021, Huang Zhanhe et al. from the National University of Defense Technology introduced the structure of GTwave fiber and reviewed the research results of domestic and international high power GTwave fiber lasers [29]. GTwave fiber typically comprises an active signal fiber and at least one passive pump fiber, both enveloped by a shared outer cladding, as depicted in Fig. 4.7. In this figure, 1 represents the pump fiber, 2 is the signal fiber, 2a is the signal fiber core, 2b is the signal fiber inner cladding, and 3 is the outer cladding. The pump light is confined within the fiber by the outer cladding, which has a lower refractive index. It is coupled between the pump fiber and the signal fiber via evanescent waves and subsequently absorbed by the active core, resulting in laser generation.

Internationally, the University of Southampton was the first to fabricate the GTwave fiber, initially demonstrating the powerful pump injection and power scaling capabilities of the GTwave fiber. In 2004, Norman et al. [30] reported on the structure and principle of the GTwave fiber laser, which was utilized in a 100-watt

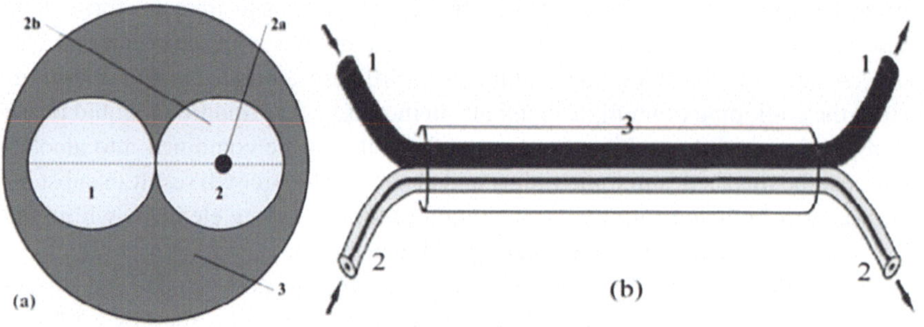

Fig. 4.7 Schematic diagram of GTwave fiber structure [29] (a) Cross-section; (b) Side view

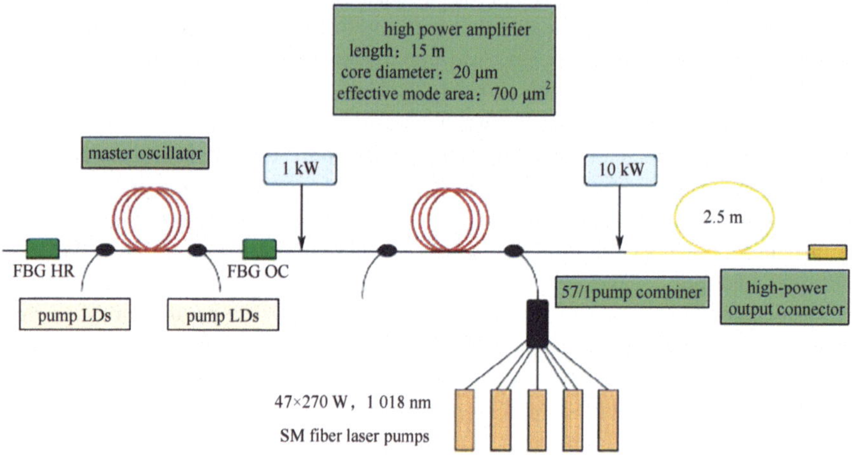

Fig. 4.8 Schematic diagram of IPG's 10 kW fiber laser structure [32]

fiber laser product. In 2011, Zimer et al. [31] from the University of Jena, Germany, in collaboration with SPI, UK, reported a 1.1 kW GTwave fiber amplifier. In 2009, IPG Photonics announced via its official website the groundbreaking results of its research on high-power, all-fiber, single-mode lasers, achieving an output of 10 kW from a single fiber [32]. As depicted in Fig. 4.8, this single-mode 10,000-watt laser system utilizes GTwave fiber backward pumping technology and a 1018 nm fiber laser cascade pumping arrangement. This setup enables the injection of high-power, high-brightness pump light into the active fiber while also isolating the influence of signal light on the pump laser. The cascade pumping wavelength is closer to the laser wavelength, resulting in reduced quantum loss and a lower thermal load on the GTwave fiber.

The main domestic institutions conducting research on GTwave fiber laser technology are the National University of Defense Technology, the 23rd Research Institute of China Electronics Technology Group Corporation, and the China Academy of Engineering Physics. In 2013, the National University of Defense

Technology collaborated with the 23rd Research Institute to successfully developed usable GTwave fiber and achieved a hundred-watt fiber laser output [33]. In the following years, further kilowatt-level power output was achieved [34–37]. The National University of Defense Technology was the first in China to conduct theoretical and experimental research on GTwave fiber. It verified the domestic kilowatt-class GTwave fiber laser and undertook research on several kilowatt bidirectional pumping multi-stage cascade amplifiers and multi-stage cascade pumping oscillators using domestically produced GTwave fiber, achieving superior results.

In 2016, Aoxiang Lin's team at CMI [38] reported the first instance of a (2 + 1) GTwave fiber laser with a power output of 2 kW, as depicted in Fig. 4.9. This achievement highlights the development of multiple pumped fiber GTwave technology.

Building on this, the team has continued to enhance the level of fiber preparation technology and has achieved further power breakthroughs [39–42]. In 2018, they reported an 8.72 kW (8 + 1) GTWave fiber amplifier, as depicted in Fig. 4.10. They utilized a 79.8 W oscillator as the seed source, and each of the six backward pumping ports injected a total of 10.66 kW of pump light through a 976 nm LD with 1.7 kW power to achieve an amplified power output of 8.72 kW with a slope efficiency of 81%. The fiber structure depicted in Fig. 4.11 demonstrates that the eight pump fibers are nearly fully integrated with the central signal fiber, closely resembling a pumped fiber combiner. By refining the scheme and boosting the pumping power, they achieved an additional laser output of 10.45 kW using the same fiber in the same year [41]. According to the literature, the team possesses the flexibility to design the structure of the GTWave fiber in accordance with the parameters of

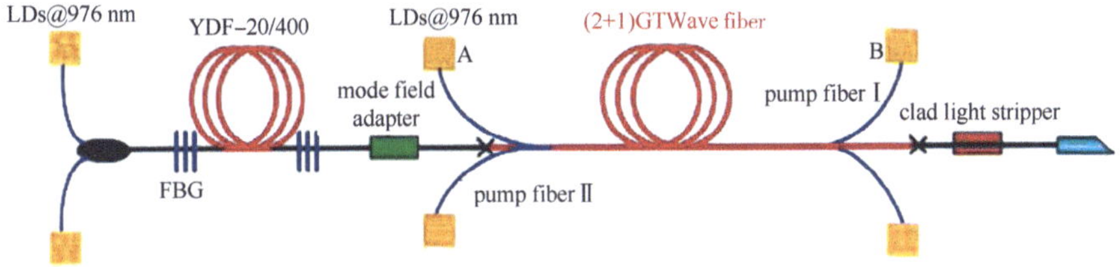

Fig. 4.9 (2 + 1) Schematic diagram of GTwave fiber laser structure [38]

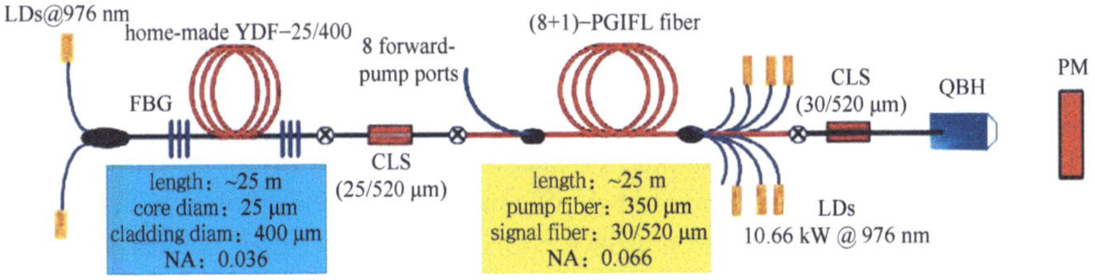

Fig. 4.10 (8 + 1) Schematic diagram of GTWave fiber amplifier structure [41]

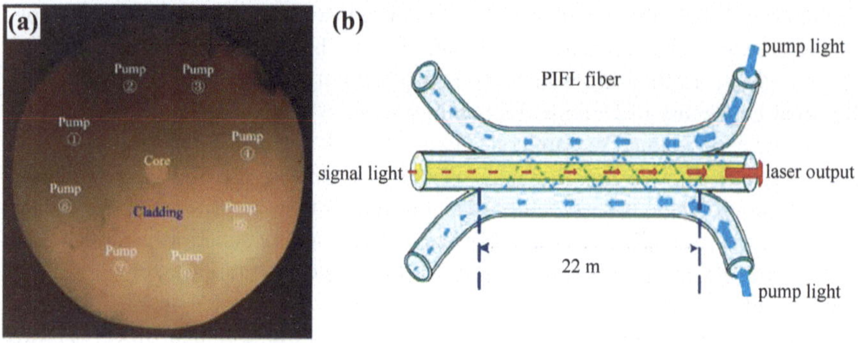

Fig. 4.11 (8 + 1) GTWave fiber structure diagram [41]: Cross-sectional photo (**a**), side view (**b**)

available pump sources, without being constrained by the parameters of commercial products.

According to the analysis of Huang Zhenhe et al. [29] From the perspective of the research process of GTWave fiber laser, its development ideas mainly focus on two aspects:First, through the power expansion in the longitudinal cascade amplification of the fiber laser. It is easier to achieve the power expansion, but the increase of the length of the fiber makes it vulnerable to the stimulated Raman effect. The second objective is to enhance the structure of GTWave fiber by incorporating additional pump fibers horizontally, thereby achieving power amplification. This approach can increase the laser's pump injection capacity, leading to power expansion. However, due to the increased complexity of the structure, the fiber becomes thicker, and the process requirements are more stringent, the scalability in the number of pump fibers will be constrained. As fiber laser technology rapidly advances, the GTwave fiber preparation technology is certain to progress as well. With the growing interest from practitioners, related fundamental and applied research is certain to achieve new breakthroughs.

4.1.4.3 Chiral Waveguide Structure, Chirally-Coupled-Core (3C) Fiber

In recent years, 3C chiral coupling core fibers have increasingly become the focus of interest within the family of fiber laser devices. The concept of a chiral waveguide was first proposed by N. Engheta and P. Pelet in 1989 [43]. It is composed by filling a general columnar waveguide with an isotropic chiral medium, meaning the core layer is chiral while the cladding layer consists of conventional materials. Chiral optical waveguides cannot independently support TEM, TE, and TM modes; instead, their modes exist as coupled TE and TM pairs. This differs from traditional optical waveguides and has garnered significant interest from scholars both domestically and internationally. The reason for this attention is that it is anticipated to surmount the drawbacks associated with increased fundamental mode loss, which are typically encountered with mode control techniques such as high-order mode

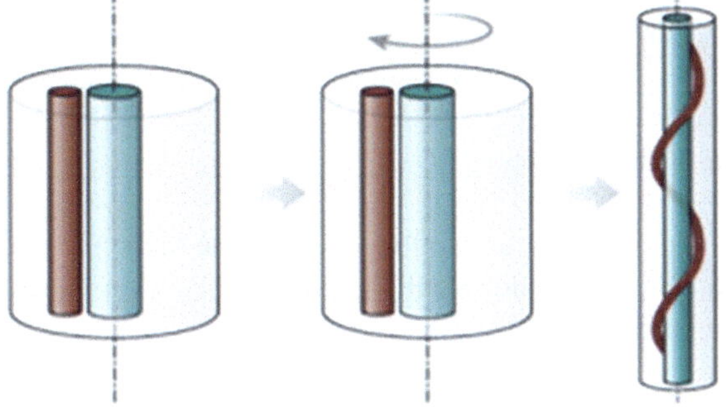

Fig. 4.12 3C fiber pulling principle

transmission, special excitation, or bending coiling in conventional large mode field area fiber or photonic crystal fiber (PCF) lasers at high power outputs. In 2007, the Ultrafast Optics Research Center at the University of Michigan, USA [44], proposed a new fiber structure known as the chiral coupled core 3C fiber. This innovation overcomes the limitations of the normalized cutoff frequency(V = 2.405) of traditional single-mode fibers, enabling stable single-mode operation at larger core sizes (over 30 μm) without the need for any mode control techniques. Thus, the output power of the fiber laser can be increased, and the fiber can be easily integrated into complex systems. The 3C fiber pulling principle, as illustrated in Fig. 4.12, offers a novel approach to achieving high peak power and high energy in fiber laser systems.

The structure of the 3C chiral coupled fiber is depicted in Fig. 4.13: it features two cores within the quartz cladding. One is the central core, which runs along the axial direction and has a larger diameter, typically exceeding 30 μm, designed for signal light transmission. The side core, spirally distributed around the central core, has a diameter significantly smaller at approximately 10 microns. Its primary function is to manage the central core's mode, coupling higher-order modes into the side core and inducing high loss (exceeding 100 dB/m), thereby ensuring the fundamental mode in the central core is transmitted with minimal loss (less than 0.1 dB/m). A suitable combination of R and Λ enables the side core to control and select the central core's mode [45].

3C optical fiber achieves stable single-mode transmission at large core diameters due to the special spiral structure of its side cores. The composite structure, where the side core spirals around the central core in this new fiber, enables three functions:

1. To achieve phase velocity matching of the modes in the central fiber core base mode and the side core, and to enable mode coupling, it is necessary for the coupling between two waveguides to satisfy the phase velocity matching condition ($\beta^{(1)} = \beta^{(2)}$) [46] However, in the 3C structure, the transmission constants of

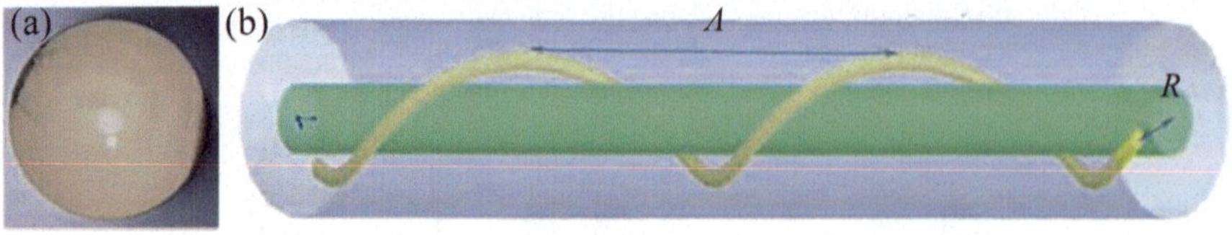

Fig. 4.13 Structure of 3C optical fiber

the modes in the two cores are no longer equal due to the presence of the spiral factor, which leads to an additional phase difference. Consequently, the matching condition becomes [47]:

$$\beta^{(side\ mode)} + \Delta\beta_{helix} = \beta^{(central\ mode)} \tag{4.1}$$

$$\Delta\beta_{helix} = \frac{2\pi n}{\lambda\sqrt{(2\pi R/\Lambda)^2 + 1} - 1} \tag{4.2}$$

in the form $\beta^{(central\ mode)}$ and $\beta^{(side\ mode)}$ are the propagation constants of the mode in the central core and the side core, respectively, and $\Delta\beta_{helix}$ is the additional phase velocity of the side core due to the helix, which can be controlled by R and Λ to achieve the matching condition.

2. By satisfying the quasi-phase matching (QPM) condition, an effective symmetric selective coupling of higher-order modes between the central core and the side core can be achieved. This QPM condition is [48]:

$$\beta_{l_1m_1} - \beta_{l_2m_2} \cdot \sqrt{1 + K^2R^2} - \Delta m \cdot K = 0 \tag{4.3}$$

where $\beta_{l_1m_1}$ is the main core in the LP_{11m_1} the propagation constant of the mode, and $\beta_{l_2m_2}$ is the propagation constant in the side core LP_{12m_2} is the propagation constant of the mode, and $\sqrt{1+K^2R^2}$ is the spiral correction factor of the side core. Since the QPM condition requires the side core transmission constant to be the value projected to the central axis, it needs to be corrected. $K = 2\pi/\Lambda$, denotes the side core helix rate. $\Delta m = \Delta l + \Delta s$, where Δl consists of four possible combinations $\Delta l = \pm l_1 \pm l_2$, and Δs The possible values of are −2, −1, 0, +1, and + 2. This equation shows that the two modes interact when the phase difference between them due to the helix factor is compensated by the helix rate K. Only the modes in the two cores that satisfy this QPM condition can be coupled so that the higher-order modes of the central fiber core are coupled into the side core.

3. Rational selection of side core size, offset R and spiral period Λ to achieve high loss characteristics of high order modes in the side core.

By satisfying the QPM condition and the high loss characteristic of the side core, the higher-order mode of the central core can be coupled to the side core and thus be lost, leaving only the fundamental mode in the central core for stable

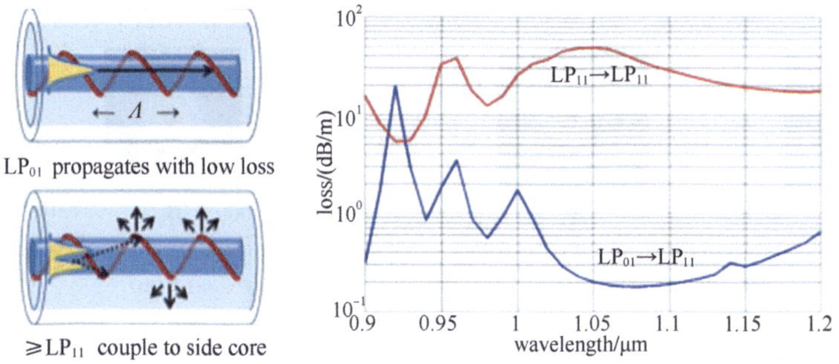

Fig. 4.14 33 μm nLight 3C fiber and numerical simulation mode loss

transmission. And by using characteristic 1), the base mode of the central core is partially coupled with the side core mode, and the phase velocity and dispersion characteristics of the base mode can be easily controlled, as shown in Fig. 4.14.

The research team at the University of Michigan's Center for Ultrafast Optics has conducted a series of experiments that effectively validated the single-mode transmission characteristics of 3C fibers and discovered that the fibers also possess excellent bias-preserving properties [48]. In 2009, the team constructed an amplification system featuring a double cladding Yb-doped 3C fiber to investigate its amplification characteristics [48]. A continuous wave power output of 250 W and a pulsed output of 150 W (with a pulse width of 10 ns, pulse energy of 0.6 mJ, and peak power of 60 kW) were achieved with an amplification slope efficiency of 74%. Additionally, the output spot of the system remained single-mode at all power levels. In 2010, the team applied the 3C fiber to the master oscillator power-amplifier (MOPA) structure to enhance the system output power [49]. In 2010, the team utilized a 2.7 m-long air-clad ytterbium-doped fiber as the gain medium for the power amplifier and achieved a power output of 511 W with an amplifier slope efficiency of 70% when excited with 2.2 W of signal light. et al. [50] In 2012, Thomas Sosnowski et al. achieved a heavy-frequency pulsed laser output of 257 W at 200 kHz, with a pulse duration of 8.5 ns and an energy of 1.3 mJ using a 33/250 μm 3C fiber. Additionally, they demonstrated a high-energy pulsed laser output of 41 W average power, 8.3 mJ pulse energy, and a peak power of 640 kW with a 55 μm 3C fiber. The superiority of the 3C fiber has been further confirmed. It has been demonstrated that this fiber can serve as a gain medium for lasers, similar to ordinary fibers. The resulting fiber laser boasts high slope efficiency and a low threshold power, and the output beam quality has been significantly enhanced. The 3C chiral coupling core fiber from n-Light, for instance, features a central core of 33 μm, a side core of 3 μm, a cladding diameter of 250 μm, and a pump absorption rate of 1.8 dB/m at 920 nm. This allows for the generation of a pulsed laser output with an energy of 2 mJ, a peak power of 300 kW, and a beam quality characterized by an M^2 value of less than 1.15. The system can operate continuously at an average power of 25 W and a pulse energy of 15 μJ for over 1.75 years, with the fiber mode variation depicted in Fig. 4.15 [51].

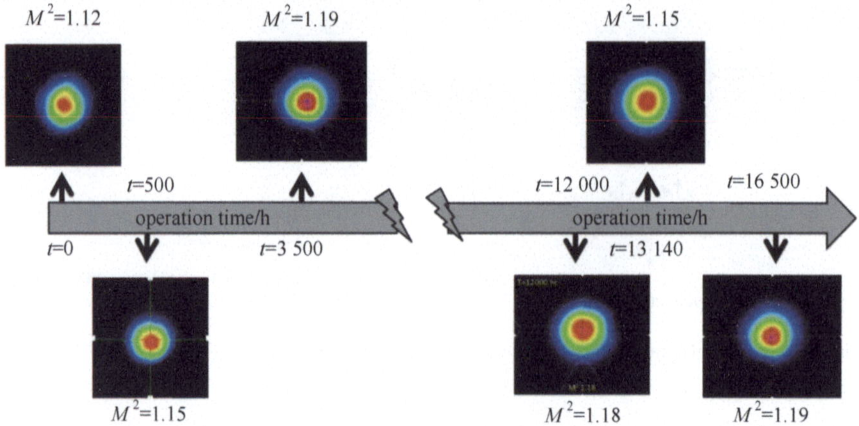

Fig. 4.15 Beam quality of the fiber laser obtained with nLight's 3C gain fiber [51]

In addition to achieving stable single-mode transmission, we predict that the 3C fiber is also capable of suppressing certain nonlinear effects due to its special structure. For instance, through selective coupling between the base mode and the side mode of the central core, the polarization state of the base mode is transferred into the side core. Consequently, the polarization state reflected back is opposite to the initial polarization state, thereby effectively suppressing Stimulated Brillouin Scattering (SBS). Additionally, the transmission spectrum of the 3C fiber, after a special structural design, exhibits a wavelength suppression region. This suppression region aligns with the peak region of the Stokes Stimulated Raman Scattering (SRS) gain spectrum, effectively suppressing the SRS effect. The effect of SRS can be effectively suppressed by aligning this suppression region with the peak region of the Stokes SRS gain spectrum [52]. Simultaneously, adjusting the wavelength suppression region allows for wavelength selection in Yb-doped fiber lasers and amplifiers. The theoretical analysis of 3C fiber indicates that its output beam possesses angular momentum. Consequently, it is anticipated that 3C fiber could enable new applications, including particle capture and manipulation, quantum communication, quantum computing, and information encoding in multidimensional quantum space [53].

In summary, 3C fiber exhibits characteristics of good fundamental mode and polarization state output, without bending loss, effectively suppresses nonlinear effects during pulse power amplification, and enables high-energy, high-peak-power pulsed laser output. Considering these characteristics, the optical specifications of pulsed fiber lasers can be significantly enhanced to meet the demands of contemporary research and industrial applications for high-quality light sources [48].

Although 3C optical fiber offers numerous advantages for the aforementioned applications, its preparation is extremely challenging. The small size of its side core presents significant difficulties during actual processing, and the processing errors that arise from this directly determine the success or failure of fiber preparation. In order to reduce preparation difficulty, Professor Haichao Zhang from Tsinghua University proposed a new fiber structure, defined as an Annular Chiral Coupled Core (ACCC) fiber, with a larger side ring instead of a small side core [54]. The feasibility of this new structure, along with its single-mode characteristics similar to those of conventional 3C fiber, has been demonstrated. The effect of fiber parameters on mode loss was investigated using the finite element method, which incorporates a spiral coordinate system to analyze the fiber mode field, and the beam propagation method. The following results were obtained:To achieve efficient single-mode transmission, the fiber structure was optimized through theoretical simulation. With a fiber core diameter of 35 μm, a beam quality M^2 value of 1.04, and a light-optical conversion efficiency of 84%, this fiber allows the fundamental mode (FM) to propagate with acceptable loss, while the high-order modes (HOMs) attenuate rapidly.

4.1.4.4 Tapered Structure Fiber

Tapered fibers possess unique geometric characteristics that give rise to numerous optical properties, including normalized frequency, effective cross-sectional area, optical intensity density, group velocity dispersion, and nonlinear coefficients. The tapered structure has found numerous applications in fiber laser technology. In recent years, as high peak power ultrafast fiber lasers have become an essential tool in industry, stringent requirements have been placed on ytterbium-doped laser fibers. Conventional cladding structured Yb-doped fibers possess a low nonlinear threshold, limiting the maximum peak power achievable. Consequently, for potential applications of pulsed fiber lasers, various types of large mode area fibers have been developed to surmount this constraint. These include photonic crystal fibers, leaky channel fibers, step refractive index fibers with microstructured cladding that suppresses higher order modes, helical core fibers, chiral coupling core fibers, photonic band gap fibers, and others. Photonic crystal fibers, as representatives, have been utilized in numerous applications for ultrafast processing lasers in industrial settings. The commercial photonic crystal fibers introduced by NKT have also found extensive use. Nevertheless, this type of fiber is challenging to prepare, costly, and difficult to fully fiberize due to its air-hole microstructure, which demands high operator proficiency for practical engineering applications. Consequently, the Fiber Optic Research Center of the Russian Academy of Sciences has persisted in developing long tapered fibers, achieving significant progress in their application [55–58]. The tapered fiber features a triple cladding structure and a panda-type, boron-doped stress zone, which ensures the fiber's polarization maintaining properties. The parameters of this fiber are detailed in Table 4.1, corresponding to the core and cladding geometries, as well as the spectral parameters for the signal input (fine diameter) and signal output (coarse diameter), respectively (Fig. 4.16).

The Fiber Optic Research Center of the Russian Academy of Sciences has introduced a tapered fiber, primarily of the YDF-DC-40/400-PM-TPR series, designed for use in very high peak power cladding pumping amplifiers. This tapered fiber

Table 4.1 Information of tapered optical fiber products of Russian Optical Fiber Research Center

	YDF-DC-40/400-PM-TPR	
Optical fiber specifications	Signal input	Signal output/pump input
Fiber core diameter (μm)	9 ± 1	>40
Cladding diameter (μm)	90 ± 10	400 ± 50
Cut-off wavelength (μm)	<1.0	
MFD	10.0 ± 2.0	>20 (25 ÷ 30 special)
Package layer type	Panda type with F-doped quartz as the second cladding	
Fiber Core NA	0.085 ± 0.01	
Cladding NA	>0.26	
915 nm cladding absorption coefficient (dB/m)	>4	
976 nm cladding absorption coefficient (dB/m)	>15	
1150 nm core loss (dB/km)	<50	
Anti-light darkening performance	More than 20 times that of Yb^{3+} doped Al_2O_3 -SiO_2 fiber	

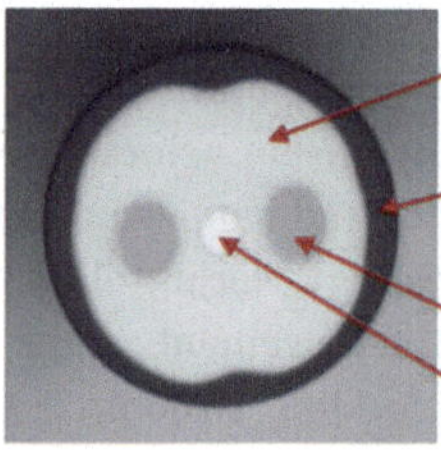

Fig. 4.16 End face of tapered fiber introduced by Russian Fiber Research Center [55]

features a single-mode end for signal input (with a typical size of 8/80 μm) and a very large mode-area end (with a typical size of 40/400 μm) for signal output and pump input. The relationship of the fiber geometry variation is depicted in the Fig. 4.17a. As illustrated in the figure, the geometry of this fiber does not merely follow the linear gradient of the fiber cladding geometry. Instead, there are several areas with relatively gentle variations. Specifically, the taper variation is more pronounced at the signal input end (fine diameter), whereas at the signal output end, the variation is more subdued. The fiber product is a modular product with a fixed length and geometry, and the absorption coefficient of this module is shown in Fig. 4.17b. The absorption coefficients for the 915 nm and 976 nm wavelengths are 6.5 dB/m and 24 dB/m, respectively. It is important to note that the absorption coefficients for the fiber cladding provided in this figure are not the same as those in the product data sheet. However, they still fall within the range of values given in the table. This indicates that there is some fluctuation in the performance parameters of the fiber product during actual preparation. Therefore, the specific product information. This indicates that there is some fluctuation in the performance parameters during the actual preparation of the fiber, hence the specific product information allows for a certain degree of variation. The fiber product features an all-glass

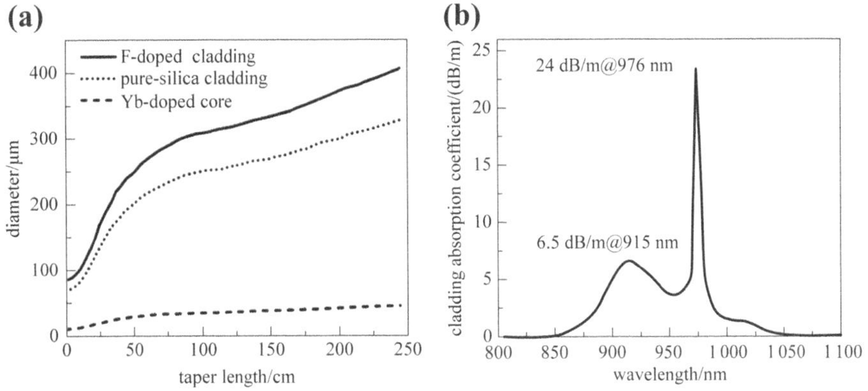

Fig. 4.17 Schematic diagram of the variation of the axial geometry of the Russian tapered fiber (**a**) and the absorption coefficient of the fiber module (**b**) [55]

double-clad fiber design (based on a highly F-doped second cladding with a typical NA of 0.26), allowing for simple polishing of the thick 400 μm fiber ends. Owing to the high Yb concentration and short tapered fiber length, this fiber-based amplifier has the highest nonlinearity threshold in the market (up to 0.5 MW), while maintaining diffraction-limited beam quality at the output.

When conducting fiber laser amplification applications, it is essential to properly adjust the parameters of the tapered amplifier to achieve maximum peak power and average output power. Firstly, it is natural to employ a reverse pumping scheme, where the fiber's cladding diameter is largest on this side. Furthermore, the threshold for nonlinear effects is typically higher in this configuration compared to forward pumping. Research conducted at this fiber research center [56] indicates that the selection of signal and pumping wavelengths is also crucial. The maximum threshold for nonlinear effects can be achieved with a signal near 1064 nm and a stable pumping at 976 nm. It is important to note that in this case, the threshold for nonlinear effects increases as the tapered fiber length increases [56, 58]. Compared to short tapered fibers (~1.5 m), a tapered fiber length of 2–2.5 m is more preferable. Based on available fiber product specifications and experimental data, the standard tapered length ranges from 200 to 300 cm. The reason for maintaining this length is that it has been tested extensively and is deemed the most appropriate for achieving maximum peak power and average power laser output. To understand the aforementioned results, researchers specifically indicatedthat the above case does not apply to signal amplification near 1030 nm [56]. In this case, the first-order Raman gain (with the Raman peak corresponding to the 1030 nm signal located at approximately 1080 nm) falls within the amplified wavelength range of the Yb-doped fiber. Here, the amplified spontaneous emission (ASE) signal serves as the seed for Raman amplification. Consequently, the longer the tapered fiber amplifier, the greater the seed optical power generated and the lower the stimulated Raman scattering (SRS) threshold.

It is important to note that long, tapered fibers can enable operation at very high average power levels. The main factor affecting power enhancement in tapered optical fibers is the optical halo effect that occurs as the fiber narrows from thick to thin. The halo originates from the light leakage caused by the decrease in waveguide confinement as it propagates from the thick end to the thin end along a tapered optical fiber. Leakage of light at certain points in tapered optical fibers can cause localized heating, leading to the degradation of the polymer coating and resulting in damage. For longer conical optical fibers, this adverse effect can be greatly suppressed. In 2017, Russian researchers have achieved a peak power of 5 MW and perfect beam quality in a picosecond MOPA power amplifier using a double cladding ytterbium-doped tapered fiber (4 m long) [56]. However, the practical application may still face two serious problems: significant ASE and Raman nonlinear effects.

Researchers from the University of Jena, Germany [59], indicated that in the aforementioned study conducted by the Russian group on tapered fibers, the tapered region spans several meters. This extended taper is accomplished by rapidly altering the parameters of the fiber drawing process during fabrication. Typically, precise control over the taper stretching conditions and taper ratio is challenging when producing tapered fibers. Consequently, they proposed the localized short taper method, which concentrates on optimizing the taper and minimizing diffusion during splicing. Fig. 4.18 illustrates the primary characteristics of a tapered LMA fiber amplifier featuring a highly fluorine-doped quartz glass cladding structure: (1) The fusion splice between the seed transmission fiber and the tapered waist is crucial, as it determines the power coupling efficiency. The waist of the tapered active fiber must be strictly maintained in a single mode at the fusion splice to prevent the excitation of higher-order modes. (2) Adiabatic tapering is a crucial element for sustaining single-mode operation in large-mode-area (LMA) fibers. It refers to the smooth transition in the geometry of the taper, ensuring that there is no power coupling between the fundamental and higher-order modes. The essence of the method involves utilizing a local adiabatic taper to create a transmission path for selectively excited fundamental modes within multimode fibers.(3) The uniform refractive index distribution of the LMA fiber core is an essential foundation for achieving a near-Gaussian mode field distribution, as any perturbation in the refractive index can result in mode distortion and consequently impact beam quality.

Based on the aforementioned analysis, researchers at Jena University employed a newly designed silica fiber with a high concentration of Yb and Al doping [59]. This was accomplished by lowering the Al^{3+} concentration in both the fiber core and cladding, and by applying a highly fluorine-doped quartz glass cladding tube to decrease the fiber's viscosity. Consequently, this reduced the drawing temperature and the rate of diffusion. A micrograph of the fiber's end face is presented in Fig. 4.19a, which is specifically labeled to indicate a reduction in the Al^{3+} doping concentration, thereby inhibiting high-temperature diffusion. Fig. 4.19b provides a refractive index distribution diagram of the optical fiber, from which the change in refractive index between the core and the inner and outer cladding can be observed. By comparing Figs. 4.19a, b, the change in refractive index at the interface produced during the preparation of the optical fiber preform can be clearly seen. The

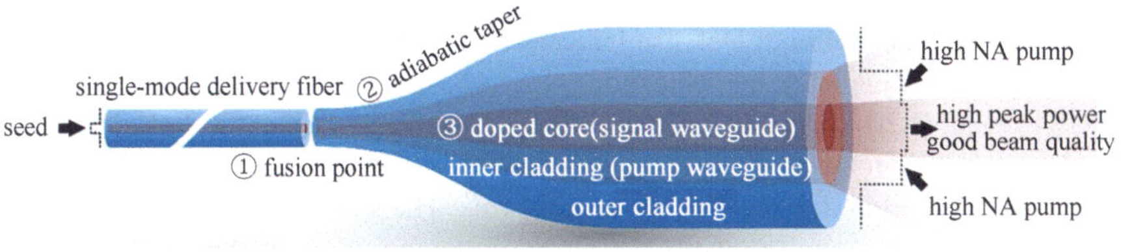

Fig. 4.18 Schematic diagram of a tapered fiber drawn using a double-clad fiber [59]

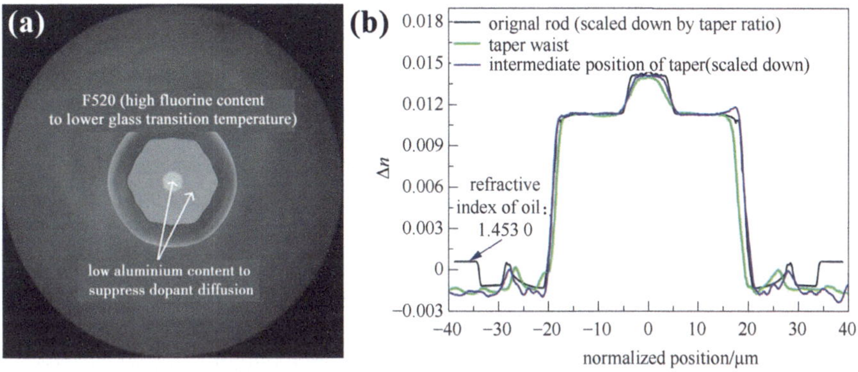

Fig. 4.19 LMA fiber cross-section (**a**) (Yb/Al/Ce doped core, aluminum doped inner and outer layers, highly fluorine doped quartz glass tube and Refractive index distribution of rod-shaped optical fibers (**b**) [59]

design of the fiber must also accommodate the requirements for adiabatic tapering and splicing to the MOPA system. Adiabatic tapering means that the geometric shape change of the taper is sufficiently smooth, thus there is no power coupling between the fundamental mode and higher-order modes.

Moreover, INO Canada offers mature tapered fiber products for sale. The input core diameter of a tapered fiber commonly used in the domestic market is 35 μm, whereas the Russian product's diameter is only 8 ~ 9 μm. The larger signal input core diameter may present a greater challenge in achieving good beam quality and could lead to mode instability.

In China, represented by the National Defense University, a series of new tapered fibers have been developed in recent years for application in high-power continuous laser research. These include the traditional saddle-shaped structure with a constant core-to-cladding ratio, spindle-shaped structures, and new fibers featuring a constant cladding geometry with a gradual change in the core-to-cladding ratio [60, 61]. The power amplification based on conventional ytterbium-doped fibers is limited by optical nonlinear effects and mode instability effects. They propose new long tapered fibers with a constant cladding diameter and a tapered core (CCTC) along its axial direction. The tapered core region of the fiber is designed to enhance the excited Raman scattering threshold and suppress the higher-order mode resonance in the laser cavity. This structure was prepared through a collaborative effort between the National Defense University of Science and Technology (NDU) and Jinyan Li′s team at the Huazhong University of Science and Technology (HUST). Fig. 4.20 illustrates the process of fiber preparation: the basic method involves partially refining the prepared Yb-doped fiber preforms, followed by optical processing tailored to the size of the preforms. The geometry of the core can be altered, while the geometry of the outer cladding can remain unchanged. This has the advantage of allowing it to be more easily fused to passive fiber devices. According to the study's findings, CCTC has significant potential to mitigate both Stimulated Raman Scattering (SRS) and Transverse Mode Instability (TMI) effects. Furthermore, additional power enhancement can be realized by optimizing the Ytterbium-Doped Fiber (YDF) structure.

In summary, for the future development of higher power lasers, from the perspective of engineering applications, the suppression of fiber nonlinear effects, prevention of end-face damage, and the increase in output power levels are crucial. The development of large core diameter tapered all-glass cladding fibers represents an important direction for progress. Simultaneously, to enhance the inner cladding pumping power injection level, utilizing a laser pump of the same band, such as 1018 nm, is currently a more viable solution. However, the absorption coefficient of conventional Yb-doped double cladding fiber at 1018 nm is extremely low, making it challenging to achieve sufficient gain without inducing nonlinear Raman effects. In other words, it is difficult to attain sufficient pump absorption using a shorter fiber. In theory, the absorption coefficient of the fiber cladding can be enhanced by increasing the doping concentration of rare-earth ions and enlarging the core-to-cladding ratio. However, excessively high doping concentrations can cause

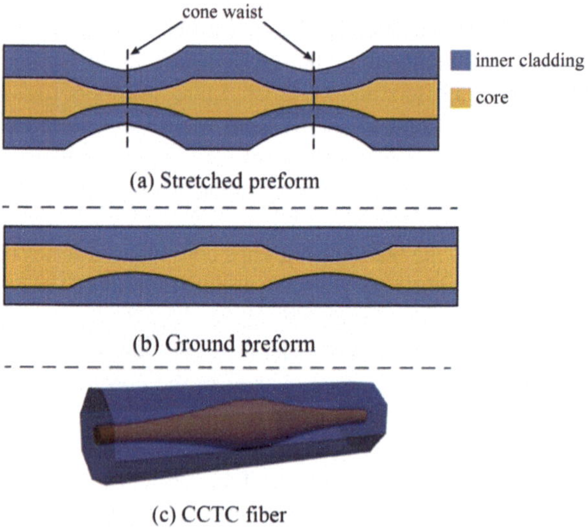

Fig. 4.20 Schematic diagram of the preparation process of CCTC-type optical fiber [61]

phase-separation and crystallization of the core glass, significantly increasing the background loss of the fiber. This also leads to a reduction in the dispersion of doping ions, resulting in clusters that are prone to photodarkening effects, which is detrimental to the stable output of tens of thousands of watts of power lasers. It is evident that the development of ytterbium-doped cladding fibers with a larger core-to-cladding ratio has become a significant choice. Consequently, overcoming the challenges of laser mode degradation due to increased core diameter and the heightened difficulty of fiber fusion coupling with the front-end pre-amplification module has emerged as a key technical issue. Simultaneously, the power and temperature tolerance of the organic coating on conventional double cladding fibers greatly increase the engineering difficulty of the laser system under laser cladding pumping conditions with tens of thousands of watts of power. The development of a cladding fiber with a gradual change in the core-to-cladding ratio, where the active fiber end face with a smaller core-to-cladding ratio is fused with existing passive fiber components, can reduce the difficulty of full fiber-based fusion. By gradually increasing the core-to-cladding ratio and enhancing the mode adaptation characteristics, as well as progressively increasing the pump absorption coefficient of the cladding, one can achieve a balanced temperature field distribution between the fiber pump end and the middle section of the fiber. This also aids in controlling the laser mode.

To address the aforementioned issues, high fluorine-doped low refractive index quartz glass can be used as the outer cladding to fabricate fully glass-clad, core-cladding graded multi-cladding ytterbium-doped fibers with unchanged cladding geometrical dimensions. This approach not only ensures compatibility with existing passive fiber devices but also allows for the control of the cladding absorption coefficient through the internal core-to-cladding ratio, suppressing the nonlinear Raman effect. It is beneficial for balancing the temperature field distribution of the fiber and can appropriately overcome the degradation of the laser mode and mode instability caused by the increase in core diameter, thereby enhancing the overall stability of the fiber laser system.

4.1.4.5 Trench Structure Type Fiber

The structural characteristics of trench fiber: A low refractive index trench surrounds the core, along with an annular layer structure that is close to the core's refractive index. According to the difference in refractive index between the core and the cladding, they can be divided into two types (which can be considered as leaky and confined waveguides from the perspective of light guiding principles, or as gain fibers and energy transmission fibers from the perspective of applications). The one with a higher refractive index of the core compared to the cladding is more suitable for the case of rare earth-doped fibers, as shown in Fig. 4.21a. Similar to the introduction of a resonant ring structure in ordinary rare-earth-doped fibers, which assists higher-order modes to resonate into the ring, thereby increasing their loss. The second type, due to the refractive index of the core being the same as that of the cladding, results in all core modes being leaky modes (meaning that a thin layer of

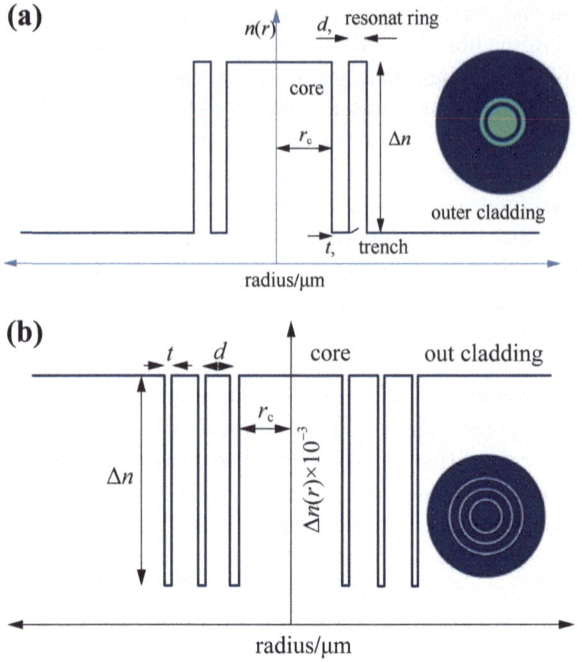

Fig. 4.21 (Schematic diagram of (**a**) single trench fiber (restricted trench fiber, gain fiber) and (**b**) multi-trench fiber (leaking trench fiber, energy transfer fiber) structures

the channel has difficulty in confining light within the core). Therefore, to reduce the loss of the fundamental mode and increase the loss ratio between the fundamental mode and higher-order modes, leaky trench fibers typically have a multi-trench structure [62, 63],

The disadvantages of microstructured fibers, such as photonic crystal fiber (PCF) and leakage channel fiber (LCF), include their complex fabrication process, which makes preparation more challenging. Additionally, the air hole structure complicates fusion, potentially affecting the stability of the laser system. Compared to other methods, the preparation process of trench fiber is simpler, utilizing the MCVD combined with the rod-in-tube method, and it features a fully solid-state structure. Additionally, as illustrated in Fig. 4.22, the resonant ring structure within the trench fiber enhances the loss ratio of both the fundamental mode and higher-order modes under bending conditions, thereby improving the suppression of higher-order modes [64]. Trench fibers primarily utilize the resonant ring structure to achieve higher-order mode filtering, enabling single-mode operation.

The preparation of multi-trench optical fiber preforms can be achieved by combining the MCVD process with the rod-in-tube method [65, 66]. The specific preparation process is as follows: first, the MCVD method is used to deposit low-refractive index trench structures and high-refractive index resonant ring structures inside a Heraeus F300 quartz tube. Then, the quartz tube is partially collapsed to form a

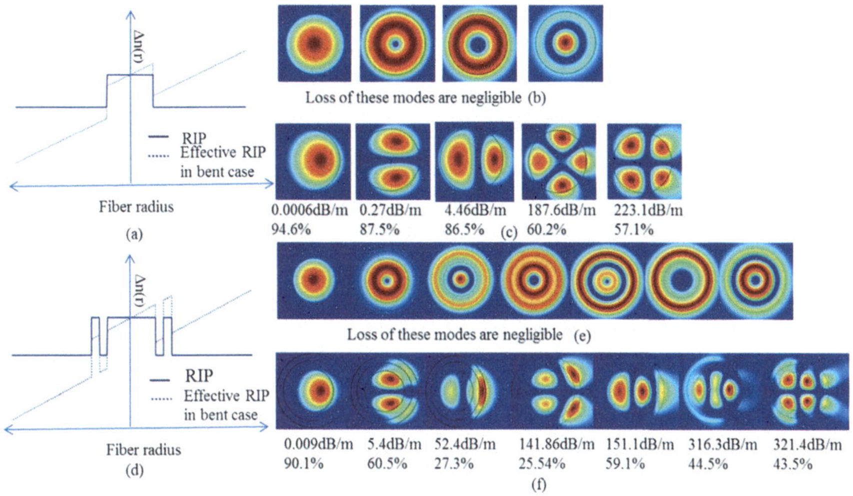

Fig. 4.22 Simulation of higher-order mode filtering effect of pitted fiber and conventional step-type fiber [64]

hollow, non-enclosed structure. Next, a size-matched F300 quartz rod is inserted into the preform rod, and finally, the two are collapsed into one. It is noteworthy that in the literature [65], the core rod and the groove cladding tube are drawn together, fusing during the process, which results in higher losses. In the literature [66], an improvement plan is mentioned where the double-barrel structure is collapsed on the MCVD lathe after completion, followed by the drawing of the fiber. For single-mode fibers, the MCVD process is directly combined with liquid-phase doping, where rare earth elements are doped into the core and germanium into the cladding to achieve this.

From the application effects, in 2015 Jain et al. reported on a single-mode fiber with a core diameter of 30 μm and an ultra-low NA of 0.038. They utilized this fiber for picosecond pulse amplification, ultimately achieving an average power of 52 W, a peak power exceeding 160 kW, and an output beam M^2 less than 1.15 [67]. It is important to note that although the other fibers within the amplifier are polarization-maintaining, the single-mode fiber is not. Nonetheless, the final output laser preserves a polarization extinction ratio of 15 dB, suggesting that the larger bending diameter (~32 cm) has a reduced effect on the polarization state of the signal light.

In summary, the fiber laser system can be significantly enhanced by making adjustments such as changing the pumping method, pumping wavelength, pumping power distribution, improving pumping brightness, and optimizing the gain fiber coiling method. Fundamentally, the limiting factor for further increasing the output power of fiber lasers is the interaction between the laser and the material. Every improvement in the performance of fiber lasers is the result of optimizing the core material properties and fiber structure design of ytterbium-doped fibers. It is also an effective way to further enhance the output power of high-power fiber laser systems.

4.2 Main Factors Affecting the Performance of Ytterbium-Doped Large Mode Area Fiber

4.2.1 Fiber Core Components

The quartz glass matrix of Yb-doped fiber ensures the high damage threshold characteristics of the fiber however, the solubility of Yb ions in quartz glass is low, and clustering and precipitation are likely to occur at high doping concentrations. The introduction of co-doping ions such as Al^{3+}, P^{5+}, Ge^{4+}, F^{-}, B^{3+}, Ce^{3+}, etc., significantly enhances the solubility of Yb ions, but also alters the absorption and emission spectra of Yb ions and the refractive index of the core.

4.2.1.1 Effect of Co-Doped Elements on the Absorption and Emission Cross Sections of Ytterbium Ions

Figure 4.23 reveals the effects of the common co-doping elements Al^{3+}, P^{5+}, Ge^{4+}, etc. on the absorption cross section and emission fluorescence spectrum of Yb^{3+} ions. It can be seen that: the absorption and emission cross sections of Yb^{3+} ions vary significantly with the introduction of different co-dopants in quartz glass, the addition of P^{5+} and Ge^{4+} significantly reduces the absorption cross section of Yb^{3+} ions compared with Al^{3+}.

In the Yb/Al/P co-doped fiber, keeping Yb^{3+}, Al^{3+} unchanged and increasing the content of P^{5+}, the absorption cross section of Yb^{3+} ions will gradually decrease. Especially, when the P/Al ratio exceeds 1, the absorption cross section of Yb^{3+} ions decreases significantly, as shown in Fig. 4.24.

4.2.1.2 Influence of Components on Refractive Index of Fiber Core

The influence of core components on the optical and spectral properties of rare-earth doped glasses has been introduced systematically in the previous sections. In this section, we briefly summarize the representative results and describe the variation of core refractive index with component. Fig. 4.25a shows the contribution of common co-dopants to the refractive index of quartz fiber cores at 633 nm, and Fig. 4.25b shows the contribution of common rare earth oxides to the refractive index of quartz fiber cores at 633 nm. The slope of the graph ($\eta \times 10^{-4}$) represents the contribution of each 1 mol% co-doped compound to the refractive index of the core. As can be seen from the figure, there are three common co-dopants that can reduce the refractive index of quartz glass, among which SiF_4 can significantly reduce the refractive index of quartz glass, followed by B_2O_3 and $AlPO_4$. among several rare earth oxides, Ce_2O_3 makes the largest contribution to the increase of refractive index of quartz glass.

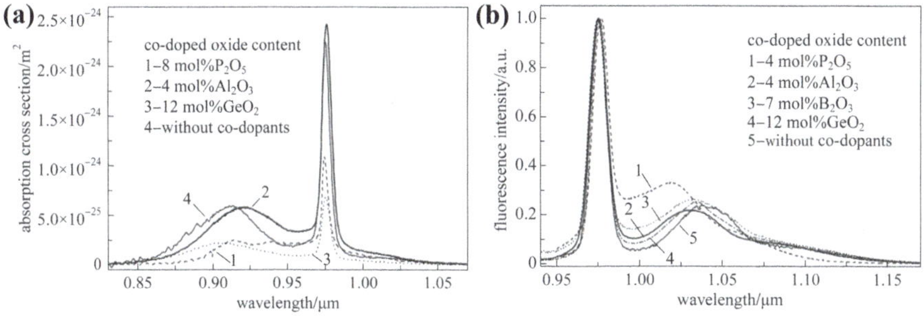

Fig. 4.23 Effect of common core co-doping elements on the absorption cross section (**a**) and emission cross section (**b**) of ytterbium ions in quartz fiber [68]

Fig. 4.24 Effect of variation of aluminum and phosphorus content on the absorption cross section of Yb3+ ions in quartz glass [69]

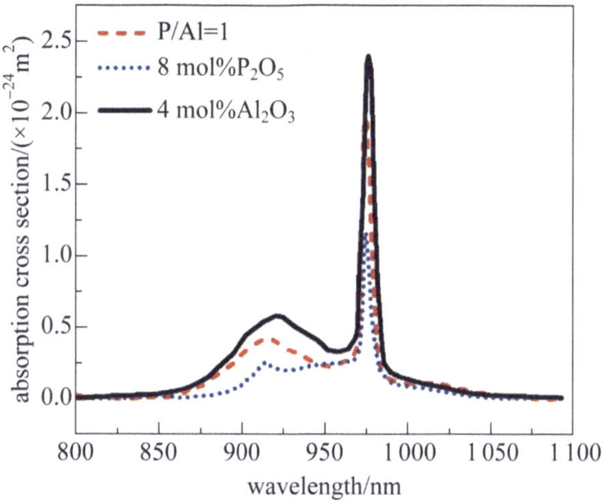

Fig. 4.25 Contribution of co-dopant and rare earth oxide per mole to the refractive index difference (Δn) of quartz fiber core and cladding

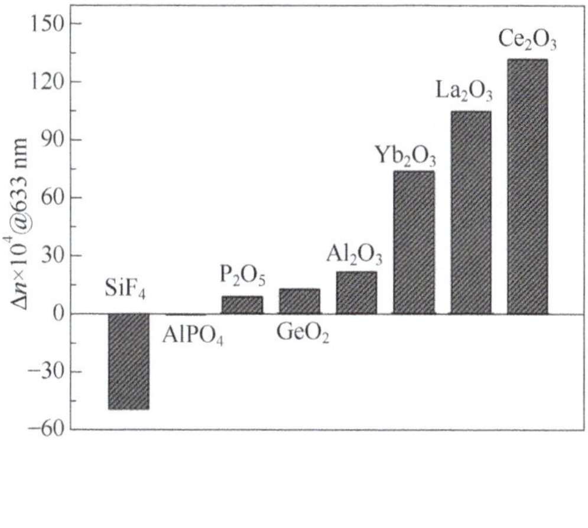

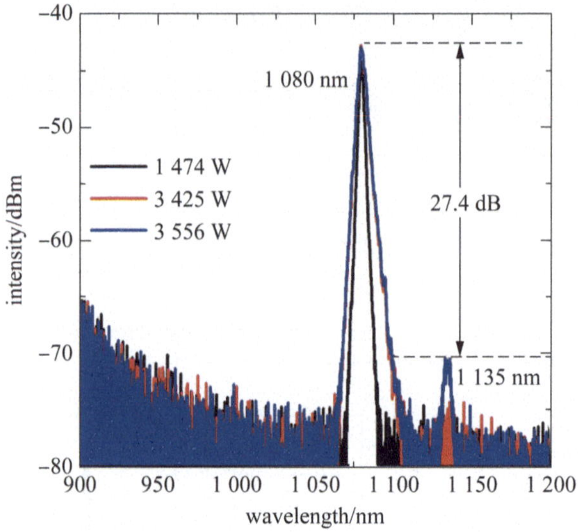

Fig. 4.26 Excited Raman scattering effect in a fiber laser

4.2.2 *Nonlinear Effects*

The special waveguide structure of optical fibers confines the laser energy within the fiber core, which measures in the micron level. As the power of fiber lasers continues to increase, extremely high power densities are formed within the core, causing various harmful nonlinear effects that impede further enhancement of laser power. Nonlinear effects in fiber lasers primarily encompass stimulated Raman scattering, stimulated Brillouin scattering, self-phase modulation, and four-wave mixing. As depicted in Fig. 4.26, this represents a typical example of stimulated Raman scattering within a ytterbium-doped fiber laser.

Increasing the fiber core size to reduce the laser power density within the core, and enhancing the absorption coefficient of the Yb-doped fiber to decrease the fiber's length, are two effective methods to prevent nonlinear effects.

4.2.3 *Photodarkening Effect*

The photodarkening effect is a phenomenon in which high-power ytterbium-doped fiber lasers undergo a reduction in power output, an increase in threshold power, and an increase in fiber loss over extended periods of operation. As depicted in Fig. 4.27a, the output power of Yb/P, Yb/Al, and Yb/Al/P co-doped quartz fibers fluctuates over time. As illustrated in Fig. 4.27a, the laser output power of the Yb/P co-doped fiber diminishes at the slowest rate over time, exhibiting the least degree of

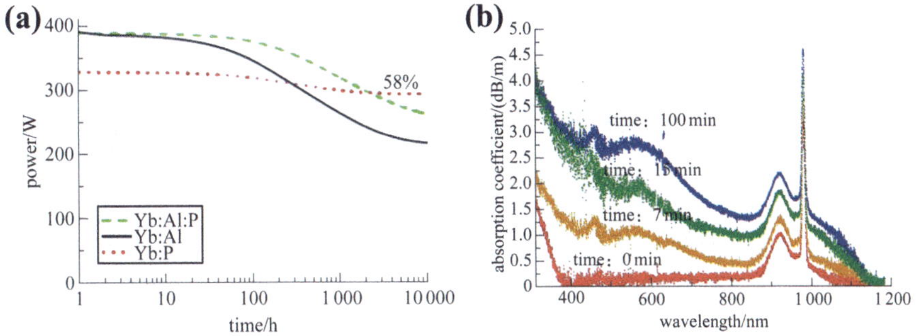

Fig. 4.27 Photon darkening effect of Yb-doped quartz fibers with different core compositions (**a**) and additional loss due to Yb-doped fiber color centers with pumping time (**b**) [69]

photodarkening. The photodarkening effect in Yb/Al co-doped fibers is the most pronounced. The photobleaching effect of Yb/Al/P co-doped fibers falls between the two. The widely accepted explanation for the photodarkening effect is that it essentially involves pump photons inducing the formation of color centers with absorption peaks in the ultraviolet and visible wavelength bands in ytterbium-doped fiber lasers. The absorption band of the color center is very wide, extending to the near-infrared 1 μm working band of ytterbium-doped fiber lasers, as depicted in Fig. 4.27b. Color centers result in an increase in loss, leading to a decrease in the output power of the laser and degrading the performance of the fiber laser output.

It is now widely believed that in ytterbium-doped fiber lasers, as the pump photons in the near-infrared 1 μm band, and pump-induced additional loss mainly in the ultraviolet and visible bands, the occurrence of photon darkening process must be the effect of the joint action of multiple pump photons. Therefore, doping with co-dopants to improve the dispersion of Yb ions and reduce clustering can enhance the fiber's anti-darkening performance.

In the discussion of the formation mechanism of color center defects in the Yb/Al/P co-doped quartz glass system, aluminum-oxygen defects are the color center defects that have the greatest impact on photodarkening. P co-doping can shift the charge transfer absorption band of ytterbium ions to higher energy levels, thereby reducing the density of alumina defects and improving photodarkening resistance. As the phosphorus-to-aluminum ratio increases, when the P/Al ratio reaches 1, alumina defects are significantly reduced, and the photodarkening resistance of the optical fiber is significantly enhanced.

The variable valence property of the Ce ion allows its two valence states, Ce^{3+} and Ce^{4+}, to coexist and interconvert under certain conditions. Additionally, the introduction of Ce^{3+} ions into Yb^{3+}-doped fiber can significantly enhance the fiber's resistance to photodarkening. On one hand, Ce^{3+} can suppress the creation of hole-type color centers by capturing holes to form Ce^{4+}.On the other hand, Ce^{4+} can effectively inhibit the formation of electron-type color centers by capturing electrons to form Ce^{3+}. Therefore, Ce-doping can effectively improve the photodarkening resistance of ytterbium-doped fibers. However, during the experiment, we observed that in the Ce^{3+}-doped fiber, the fiber's absorption coefficient increased slightly. Nonetheless, the optical-to-optical efficiency of the output laser did not markedly improve compared to the fiber without Ce^{3+}, likely due to the added Ce^{3+} ions increasing the fiber's thermal load.

4.2.4 Mode Instability

Increasing the core size of the optical fiber and reducing the power density within the core is one of the effective methods to avoid the generation of nonlinear effects. However, increasing the size of the fiber core inevitably leads to the fiber supporting multiple modes. If the gain fiber supports multiple modes, another phenomenon occurs—mode instability. As illustrated in Fig. 4.28, once the output power exceeds a certain threshold, the output mode of the fiber laser shifts from a stable fundamental mode to an unstable mode. In this state, the relative components of the fundamental mode and higher-order modes fluctuate rapidly and randomly over time. Mode instability can lead to a significant deterioration in laser beam quality, severely restricting the application of fiber lasers and becoming the greatest limiting factor for power and beam quality enhancement in large-mode-area ytterbium-doped fiber lasers.

Currently, researchers generally agree that mode instabilities are related to thermal effects caused by quantum losses in optical fibers. To suppress the generation of nonlinear effects such as stimulated Brillouin scattering (SBS) and stimulated Raman scattering (SRS), high-power fiber lasers use ytterbium-doped large-mode area fibers, and fibers with large mode field areas typically support multiple laser modes themselves. When a signal light is injected into the gain fiber, the main energy is concentrated in the fundamental mode. However, a small number of higher-order modes will inevitably be excited. These fundamental and higher-order modes interfere with each other, resulting in a periodic light intensity distribution

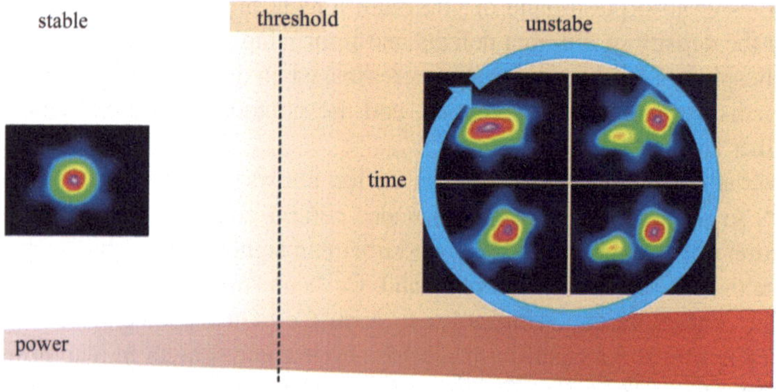

Fig. 4.28 Mode instability in a fiber laser [70]

within the fiber core. The period of this distribution is inversely proportional to the effective refractive index difference between the two modes. When the pump light is injected and the signal light begins to be amplified, periodic extraction of the pump light occurs within the doped region of the fiber core. The heat generated from quantum loss is related to the absorption of the pump light, resulting in a quasi-periodic oscillation of the heat load distribution. This is primarily characterized by the periodic light intensity distribution. The temperature rise is not immediate; rather, a periodic temperature distribution in the fiber core is established only after a certain duration of heat load generation. Due to the thermo-optical effect, the periodic signal light distribution within the core and the periodic temperature distribution resulting from quantum loss eventually create a long-period refractive index grating. This grating is capable of coupling energy between low-order and high-order modes, and it is the cause of the mode instability phenomenon observed in high-power fiber lasers.

4.2.5 *Fiber Coating*

The advent of double-clad fiber structures has facilitated rapid increases in the output power of fiber lasers. A schematic of a typical double-clad fiber structure is depicted in Fig. 4.29, where the pump light is primarily confined to the transmission within the low refractive index coating. The performance of the coating, the coating process, and other factors significantly influence the confinement of the pump light under high-power pumping conditions, the surface temperature of the fiber, and the long-term reliability of the coating material.

To fabricate a conventional double-clad fiber, a secondary coating is necessary. This includes a layer of soft, low refractive index inner coating that is cured by ultraviolet light, and a layer of hard, high refractive index outer coating also cured by ultraviolet light. Coatings with a low refractive index are primarily based on polyacrylates and typically have an operating temperature that does not exceed 80 °C. As the output power of single fibers continues to increase, high-power pump lasers will inevitably exert tremendous pressure on low refractive index coatings,

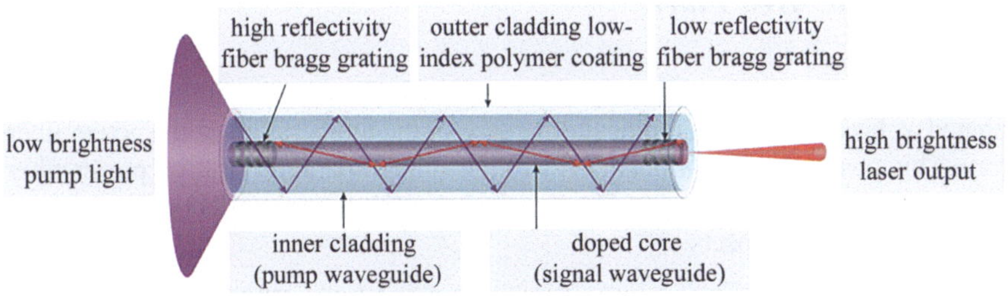

Fig. 4.29 Schematic diagram of common double-clad fiber structure

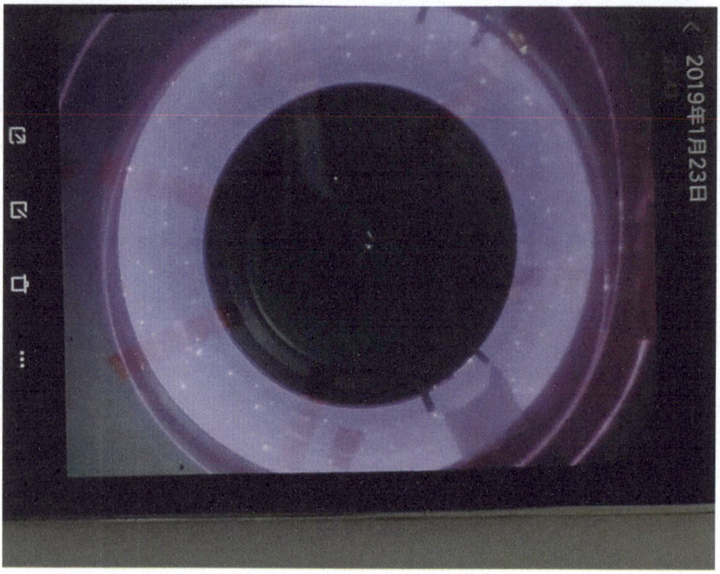

Fig. 4.30 Aging leakage of the coating layer after a long period of double-clad fiber

making them a significant factor affecting the reliability of high-power ytterbium-doped double-clad fibers in engineering applications. Fig. 4.30 illustrates the phenomenon of coating aging and light leakage observed during the long-term aging assessment of a double-clad fiber laser.

Triple-cladding fiber features a pure quartz cladding with an additional low refractive index fluorinated quartz glass layer interposed between the double-cladding fiber layers. This design ensures that the majority of the pump laser beam is confined within the fluorinated quartz glass layer for transmission, significantly reducing the impact of high-power pump light on the low refractive index coating. Consequently, it enhances the long-term operational reliability of high-power fiber lasers.

4.2.6 *Effect of Temperature Rise on Ytterbium-Doped Large Mode Area Fiber*

When operating at high power, a portion of the pump energy in fiber lasers is dissipated as heat, causing a significant increase in the temperature of the active fiber core. This poses significant challenges to the stable operation of high-power fiber lasers and amplifiers. In active fibers, the inevitable thermal effects mainly stem from the following aspects:

1. Quantum defect between pump light and signal light. When the ground state electrons of rare earth ions absorb pump energy, they non-radiatively relax to

metastable states or the ground state with the assistance of multiple phonons. The heat generated during this process is deposited into the fiber, causing a temperature rise in the fiber.
2. Some of the pump energy is lost during the upconversion luminescence process, being deposited into the fiber as heat.
3. The ultraviolet and mid-infrared components of the pump light can be absorbed by the matrix, generating heat.
4. The active fiber absorbs the pump energy, which is then converted into heat.
5. The process losses (such as those at splicing points) generated during the preparation and processing of active optical fibers result in a temperature rise.

Compared to solid-state lasers, fiber lasers have a larger surface area to volume ratio, thus possessing superior heat dissipation capabilities. However, as the pump power gradually increases, the thermal load on the fiber also becomes heavier, and natural cooling alone is no longer sufficient to suppress the rise in fiber temperature. High-power fiber lasers typically require air cooling or water cooling for the fiber. Nonetheless, irrespective of the cooling method employed, heat must be conveyed from the core to the outer layer. The heat transfer coefficient between the core and the inner layer, as well as between the glass cladding and the organic coating layer, significantly influences this process, making it challenging to ensure rapid heat dissipation. For high-power YDF, the temperature rise of the fiber has always been a key factor limiting the increase in its power. As illustrated in Fig. 4.31, under LD pumping, the heat generated in the active fiber core of the Yb^{3+} doped fiber laser is first gradually transferred to the cladding through heat conduction. Subsequently, the heat in the cladding is transferred to the air via convection and radiation. Ultimately, the heat generation and dissipation within the fiber reach equilibrium, establishing a steady-state temperature field.

Currently, there have been many studies on the internal temperature field distribution of fiber lasers and amplifiers. For instance, Y. Fan et al. [71] considered the fiber core as a heat source with uniform heat generation. When pumped

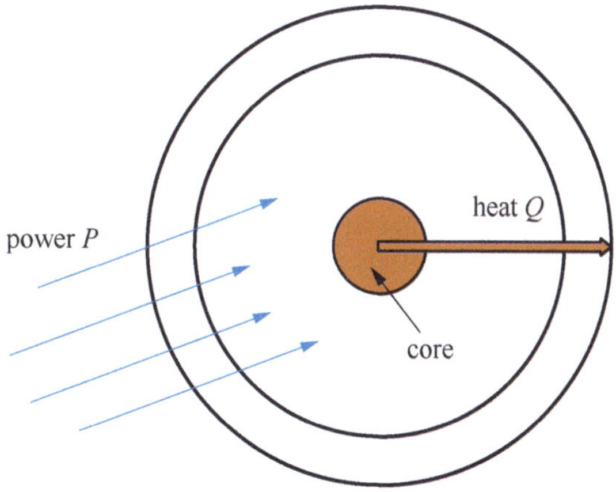

Fig. 4.31 Schematic diagram of active fiber radial heat transfer

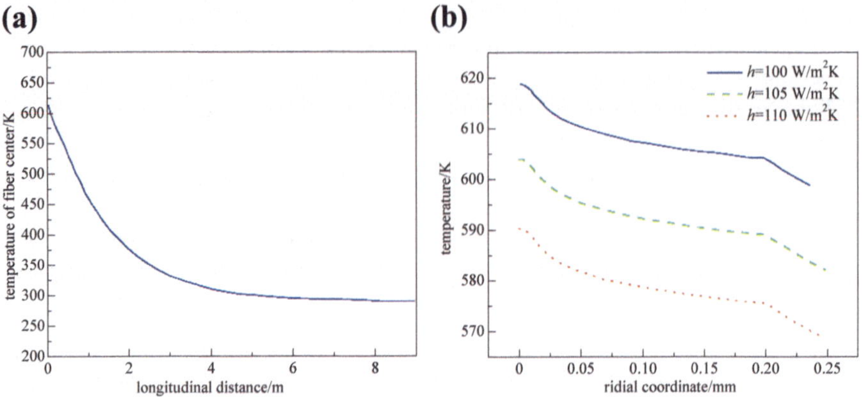

Fig. 4.32 Distribution of core temperature in the fiber axis (**a**) and in the radial direction for different h values (**b**) [71]

unidirectionally, the temperature distribution in the radial and axial directions of the active fiber can be determined using a thermal conduction theory model. As depicted in Fig. 4.32, when the pump power is set to 1.25 kW, the thermal load at the front end of the fiber is severe, and the temperature decreases approximately exponentially along the axial direction. For fibers with identical core diameters, a higher convective heat transfer coefficient (h) yields a lower temperature value. Radially, the temperature peak is located at the center of the fiber core, while the coating layer experiences the lowest temperature.

Thermal effects have a variety of impacts on high-power fiber lasers, influencing aspects such as fiber temperature, refractive index, pump absorption, output wavelength, laser efficiency, output power, beam quality, nonlinear effects, and mode instability.

In 2004, D. A. Grukh et al. [72] indicated that an increase in temperature affects the population distribution of energy levels and the spectral characteristics of Yb^{3+} ions, resulting in a decrease in laser output efficiency, which facilitates the generation of long-wavelength lasers. In 2005, the experimental results of X. S. Gao et al. were published [73]. They indicated that with the increase of thermal load, the temperature of the active fiber may rise to 150–200 °C. Typically, double-clad fibers are sheathed in low refractive index organic polymers. However, exposure to high temperatures can damage the coating layer, compromising the stability of the output laser. Although high-power lasers often employ circulating cooling water to maintain the fiber's temperature, the rise in temperature within the ytterbium-doped fiber core cannot be entirely prevented. In 2011, K. R. Hansen et al. [74] utilized a model to investigate the thermal effects in a large-mode-area (LMA) single-mode fiber amplifier, indicating that a larger radial temperature gradient results in thermal lensing effects and a reduction in the mode field area, negatively impacting beam quality. In 2015, J. M. Daniel et al. [75] investigated the impact of temperature elevation on amplified spontaneous emission and laser emission wavelength in Yb^{3+}-doped fiber lasers (Yb^{3+}-doped fiber lasers, YDFLs). Their experiments, conducted within the temperature range of 25–400 °C, revealed that as temperature rises, there is a notable decrease in emission at shorter wavelengths (<1050 nm) and a corresponding increase in emission at longer wavelengths (>1100 nm).

Through comparative analysis of the spectral properties of Yb/Al/P co-doped quartz glass cores in the temperature range of 25–200 °C, we discovered that the spectral properties of Yb^{3+} ions in ytterbium-doped core glasses are influenced by both the P/Al molar ratio and temperature [76]. The main characteristics are as follows:

1. The absorption cross sections at ~915 nm and 975 nm are highly dependent on temperature and P/Al molar ratio, and the temperature dependence of the ~940 nm absorption cross section is much lower than the above two wavelengths. Due to the rearrangement of Yb^{3+} sub-level particles and the broadening of absorption peaks, the absorption cross-section significantly decreases. As the temperature rises from 25 to 200 °C, the highest absorption peak at 975 nm undergoes a blue shift, and the absorption coefficient at the peak wavelength decreases.
2. As the temperature and P/Al molar ratio increase, the ratio of the absorption cross-section at 975 nm to that at 915 nm decreases, and the advantage of the high absorption coefficient at 975 nm diminishes.

The aforementioned characteristics suggest that during the operation of high-power lasers in ytterbium-doped fibers, the temperature increase within the fiber core results in a decrease in the pump absorption coefficient. This is directly reflected as a reduction in the efficiency of the fiber laser, with variations in the extent of this decrease depending on the wavelength of the lasers.

4.3 Composition Design of Ytterbium-Doped Large Mode Area Fiber

Considering the various limiting factors and their mechanisms during the power enhancement of high-power fiber lasers, the compositional design of large-mode-area, high-power Yb-doped laser fibers needs to be comprehensively optimized from four main aspects: core spectroscopy and optical properties, nonlinear suppression, photodarkening suppression, and mode instability threshold enhancement. Since the requirements for suppressing nonlinear effects in fibers and preventing mode instability are mutually exclusive, it is essential to consider the balance between the two when designing the fiber composition. The photodarkening effect necessitates, while adhering to the basic optical parameters of the fiber, minimizing the doping concentration of elements such as Yb^{3+}, and simultaneously, enhancing the dispersion of Yb^{3+} ions to diminish the formation of color centers and clusters.

4.3.1 Optimization of Spectral Properties and Refractive Index

According to the optical fiber optical basic parameters such as absorption coefficient, numerical aperture and other requirements, combined with the co-doping elements Al^{3+}, P^{5+}, F^{-}, etc. on the Yb^{3+} ion absorption cross section, emission cross section and core refractive index, the following core component design principles are developed:

1. The Yb^{3+} content is as low as possible under the premise of meeting the requirements of fiber absorption coefficient.
2. The content of Al^{3+}, P^{5+}, etc. as the main dispersant elements meets the requirement of reducing or even eliminating Yb^{3+} ion clusters.
3. According to the requirement of refractive index of fiber core, add appropriate content of F element.
4. Other co-doping elements are added in small amounts or not according to demand.

4.3.2 Suppression of Nonlinear Effects

Considering the causes of nonlinear effects, which include high laser power density and a sufficient nonlinear gain coefficient, measures to fundamentally suppress these effects in fiber lasers can involve: (1) Utilizing large mode area fibers to reduce the laser power density within the fiber core; (2) enhancing the fiber's absorption coefficient, shortening the fiber laser's cavity length, and thereby decreasing the nonlinear gain coefficient.

Therefore, the concentration of rare earth ions should be increased in component design, and the mode field area should be expanded in optical design.

4.3.3 Suppression of Photodarkening Effect

Although the mechanism of photodarkening in ytterbium-doped fibers remains unclear, there is a common understanding that, compared to passive germanium-doped fibers and pure silica fibers, the photodarkening-induced additional loss or radiation-induced loss in rare-earth-doped fibers is significantly higher. The reason is that both the rare-earth elements and the co-dopants in rare-earth-doped fibers have valence states that differ from those of silicon and germanium elements. This disruption affects the silica glass network structure, resulting in the generation of numerous induced color center defects and additional absorption losses during use, thereby degrading the performance of laser fibers.

While increasing the concentration of P^{5+} can enhance the photodarkening effect of the fiber compared to Al/Yb co-doped fibers, to fundamentally improve the photodarkening resistance of ytterbium-doped fibers, it is essential to reduce the doping levels of Yb^{3+}, Al^{3+}, and P^{5+} to the lowest possible extent while still meeting the design requirements for fiber parameters.

4.3.4 Elevation of the Mode Instability Threshold

Mode instability occurs because the fiber supports multiple modes and the quantum loss of the ytterbium-doped fiber generates thermal effects. To fundamentally suppress mode instability in fiber lasers, measures can be taken such as (1) utilizing absolute single-mode fiber; (2) decreasing the doping concentration of rare-earth ions to reduce the thermal load per unit length of the core.

Both methods will lead to an appropriate reduction in the core area of the ytterbium-doped cladding fiber. Furthermore, the actual length of the fiber employed may extend due to the diminished concentration of ytterbium-doped ions, both outcomes being contrary to the desired effect of suppressing nonlinear effects in the fiber.

4.4 Optical Design of Ytterbium -Doped Large Mode Area Fiber

It is well known that the normalized frequency in a step refractive index fiber is.

$$V = \frac{2\pi a_{core}}{\lambda} NA \approx \frac{2\pi a_{core}}{\lambda}\sqrt{2n_{core}\Delta n} \tag{4.4}$$

where a_{core} is the core radius, λ is the operating wavelength, n_{core} is the refractive index of the core, and Δn denotes the refractive index difference between core and cladding.

Figure 4.33 gives the effect of V the effect of the value on the dispersion effect and beam quality in different modes. When the fiber in $V = 2.405$ time, the second-order mode cuts off and only the fundamental mode transmission, i.e., LP_{01} mode, can be accommodated in the fiber, as shown in Fig. 4.34. From the normalized frequency V expression, it can be found that to obtain single-mode output in a large mode field step refractive index fiber, it is necessary to decrease the core NA while increasing the core diameter. But the core NA cannot be decreased indefinitely, and an increase in the fiber core diameter inevitably increases the modes in which the fiber supports propagating light.

Figure 4.34 shows the field intensity distribution of different orders of linearly polarized modes (LP) in a step fiber, which is usually referred to as LP_{mn} mode. Where m is the number of polarizations in the circumferential direction and n is the number of polarizations in the radial direction.

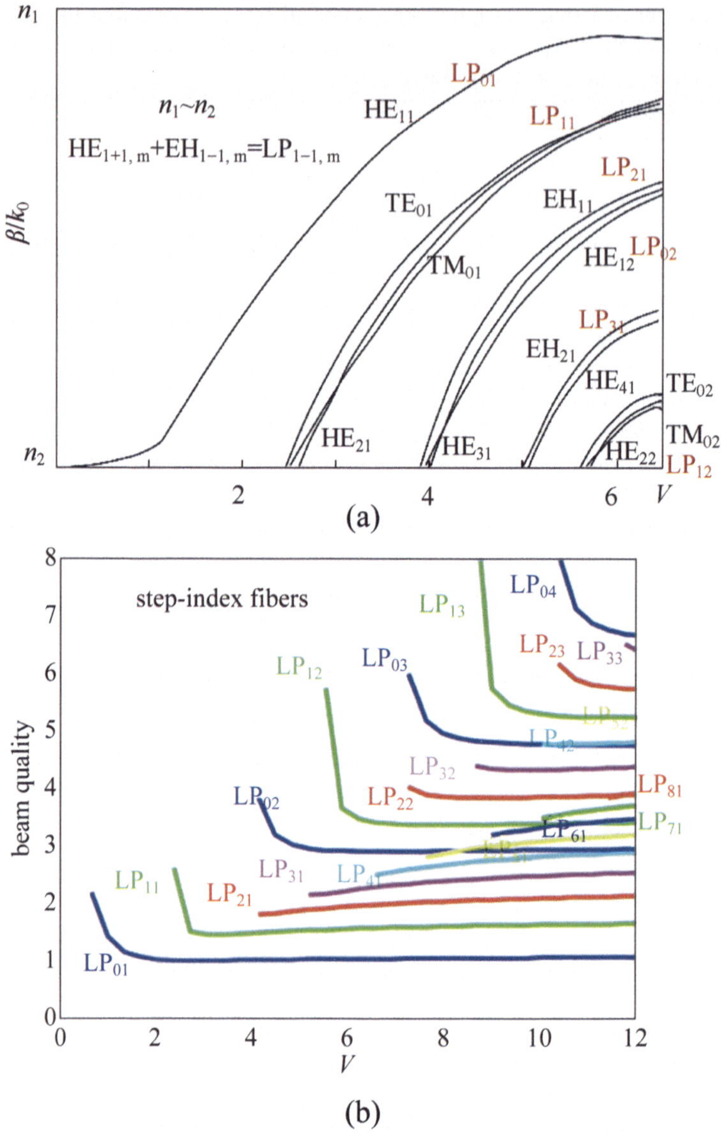

Fig. 4.33 Step fiber normalized frequency dispersion curve

Traditional optical fiber waveguide theory typically analyzes step-index fiber structures; however, in the preparation of actual active fibers, due to various process conditions, it is nearly impossible to fabricate an absolute step-index fiber. What is typically achieved is a geometric structure with a certain graded refractive index distribution. It is evident that variations in the shape of the refractive index distribution within ytterbium-doped fiber cores will inevitably result in different effective

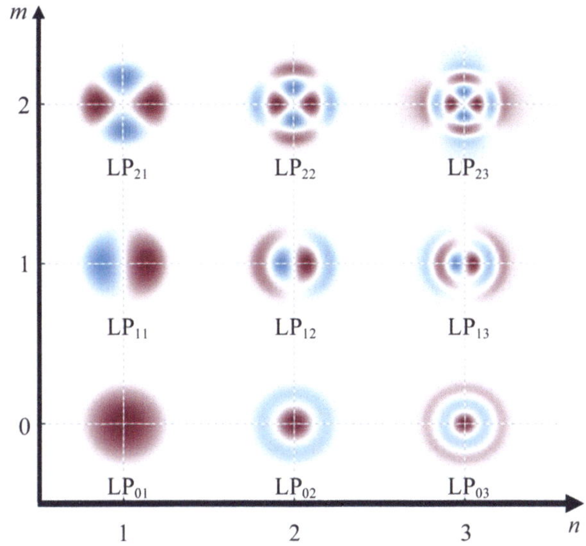

Fig. 4.34 Mode field distribution of LP

refractive indices for various modes. This crucially affects the distribution and proportion of different order modes within the core. Commonly available large-mode-area ytterbium-doped fibers in the current market, based on the calculation of the cutoff frequency, are generally single-mode fibers. Their waveguide structure still permits the existence and transmission of certain higher-order modes. To achieve quasi-single-mode laser amplification using this type of few-mode fiber, it is necessary to increase the loss of higher-order modes through bending or other methods while keeping the loss of the fundamental mode nearly unchanged. The suppression of higher-order modes enhances the beam quality of fiber lasers. However, as laser power escalates, these modes reemerge, causing dynamic coupling between the fundamental mode and the higher-order modes, which in turn leads to mode instability. Fibers introduced earlier, such as those with a low numerical aperture and other novel structures, are designed to increase the output power of single-mode operation.

It is particularly important to note that the fiber waveguide structure should be designed to meet the needs of practical applications. The current industrial laser market, particularly in laser cutting and other applications, has relatively low requirements for fiber laser modes. This is mainly to achieve higher output power and laser efficiency, and to permit the existence of a certain amount of higher-order modes, which are more favorable for practical applications. This necessitates communication with users during the actual design process of large-mode-area ytterbium-doped fibers, and the proposal of waveguide structures that can meet the application requirements to achieve efficient laser output. The following illustrates the design of the core refractive index distribution for the most commonly used Yb-doped large mode field fibers in the industrial fiber laser market, specifically those in the kilowatt and 10,000-watt classes.

4.4.1 Design of Refractive Index Distribution of Kilowatt-Class Ytterbium-Doped Large Mode Area Fiber

Here, a commonly used 20/400 m Yb-doped fiber with a numerical aperture (NA) of 0.06 serves as an example to analyze how to increase the mode instability threshold. This is achieved by modulating the core refractive index distribution, increasing the effective refractive index difference between higher-order modes and fundamental modes, and suppressing higher-order modes.

In the preparation of Yb-doped large-mode field fibers using the MCVD method, various core refractive index distributions can be achieved through process adjustments. Fig. 4.35 provides an example of the refractive index distribution for two fibers.

The maximum NA (Numerical Aperture) value of the fiber is determined by the difference between the highest point of the refractive index distribution and the refractive index of quartz glass (with pure quartz glass having a refractive index of approximately 1.44692). Typically, only the fiber's transmission performance is considered, neglecting its gain performance, dispersion, and nonlinear effects. When evaluating various refractive index distributions, the focus is usually on the effective refractive index of each mode and its transmission performance at 1080 nm, which includes the bending loss of each mode and the mode field diameter.

In Fig. 4.35, the numerical aperture (NA) corresponding to the highest point of the refractive index for both core distribution 1 and distribution 2 is 0.065.The computational analysis indicates that both refractive index distributions exhibit two modes: the fundamental mode LP(0,1) and the higher-order mode LP(1,1).

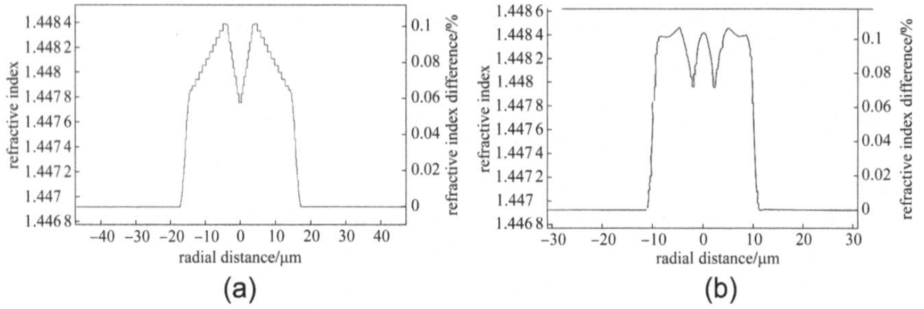

Fig. 4.35 Fiber core refractive index distribution1 (**a**) and refractive index distribution2 (**b**)

4.4.1.1 Effective Refractive Index

For the refractive index distribution 1, the effective refractive index of $LP_{(0,\ 1)}$ is 1.44781 and that of $LP_{(1,\ 1)}$ is 1.44727, with a difference of 0.00054; for the refractive index distribution 2, the effective refractive index of $LP_{(0,\ 1)}$ is 1.44797 and that of $LP_{(1,\ 1)}$ is 1.44749, with a difference of 0.00048. It can be seen that the difference between the effective refractive index of the fundamental mode and the higher-order mode is larger for the refractive index distribution 1, which is easier to distinguish the fundamental mode from the higher-order mode, which is beneficial to improve the mode instability threshold of the fiber.

4.4.1.2 Mode Field Diameter

For refractive index distribution 1, the LP(0, 1) mode field diameter measures 17.35 μm, while the LP(1, 1) mode field diameter is 24.06 μm, resulting in a difference of 6.71 μm. For refractive index distribution 2, the LP(0, 1) mode field diameter is 18.75 μm, and the LP(1, 1) mode field diameter is 22.51 μm, with a difference of 3.76 μm. In comparison, the spatial distribution of the two modes for refractive index distribution 1 is more distant, making them more difficult to couple with each other. This characteristic is beneficial for improving the mode instability threshold of the fiber.

4.4.1.3 Bending Loss

As indicated in Table 4.2, the bending loss for refractive index distribution 1 is greater than that of distribution 2 for both fundamental and higher-order modes. In particular, for higher-order modes, the bending loss of refractive index distribution 1 is an order of magnitude greater than that of distribution 2, which facilitates an increase in the mode instability threshold of the fiber.

Table 4.2 Calculation analysis of refractive index distribution of fiber core and results of fiber laser TMI threshold test

Refractive index of fiber core	Mode	Effective refractive index	Mold field diameter μm	Bending loss dB/ km@10 cm	Bending loss dB/ km@14 cm	TMI Threshold (W)
Refractive index distribution1	LP(0,1)	1.44781	17.35	155	0.98	>3000
	LP(1,1)	1.44727	24.06	366,438	97,195	
Refractive index distribution2	LP(0,1)	1.44797	18.75	6.47	0	~900–1100
	LP(1,1)	1.44749	22.51	24,233	1739	

4.4.1.4 Fiber Laser Mode Instability Threshold Test Case

The absorption coefficient is approximately 1.2 dB/m at 976 nm, and the loss coefficients are comparable. Subsequently, 25 meters of each fiber were placed in the same fiber laser oscillation system for testing. It was found that for the fiber with core refractive index distribution 1, mode instability does not occur until the output power exceeds 3 kW. For the fiber with a core refractive index distribution 2, mode instability occurred when the output power reached approximately 900–1100 W. The fundamental mode coupled with the higher-order mode, preventing further increases in output power.

The aforementioned design analysis and experimental results indicate that the core refractive index of ytterbium-doped large-mode area fiber with distribution 1 can significantly enhance the mode instability threshold of the fiber.

4.4.2 Design of Refractive Index Distribution for 10,000-Watt High-Brightness Fiber

Based on the results of the aforementioned study on the impact of the core's refractive index distribution on the mode instability threshold of single-mode or quasi-single-mode fibers, a 30/600 μm ytterbium-doped large-mode field fiber was designed for 10,000-watt single-mode lasers. This design utilized a specific refractive index distribution, as depicted in Fig. 4.36, and yielded a maximum theoretical numerical aperture (NA) of approximately 0.065 for the core.

The V value of this fiber is about 5.6 at 1080 nm. The computational analysis shows that the refractive index distribution contains four modes, including the fundamental mode $LP_{(0,1)}$ and the higher modes $LP_{(1,\ 1)}$, $LP_{(0,\ 2)}$ and $LP_{(2,\ 1)}$. The calculation results of effective refractive index, mode field diameter and bending

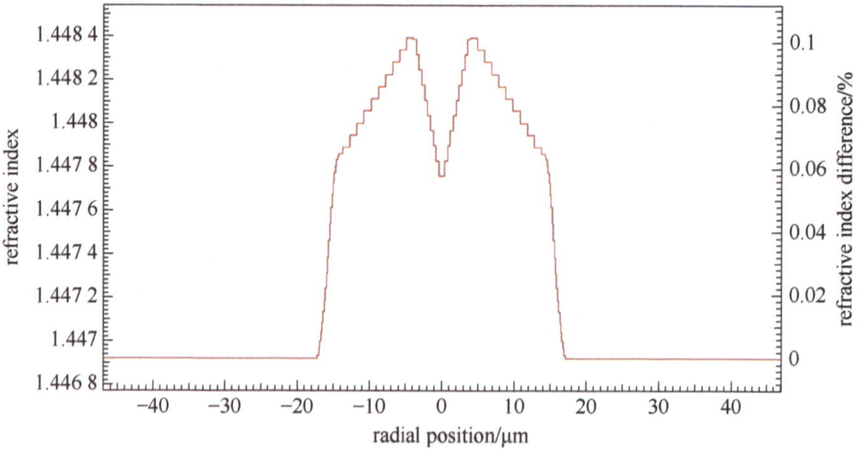

Fig. 4.36 30/600 μ m fiber core design for 10,000 watt single-mode lasers (NA ~ 0.065)

loss for each mode are shown in Table 4.3. It can be seen that the effective refractive index difference between the base mode $LP_{(0,\ 1)}$ and the higher-order modes $LP_{(0,\ 2)}$ and $LP_{(2,\ 1)}$ is larger, and the bending loss of $LP_{(0,\ 2)}$ and $LP_{(2,\ 1)}$ is larger at 10 cm and 15 cm, so these two higher-order modes can be disregarded in actual use.

Furthermore, when considering the data presented in Table 4.2 and Table 4.3, it becomes evident that the effective refractive index difference between the LP(1,1) and LP(0,1) modes diminishes significantly as the core diameter increases, relative to the 20/400 μm fiber. As illustrated by the bending loss of these modes, the LP(1,1) mode experiences a substantial reduction in bending loss compared to the 20/400 μm fiber, and at this juncture, the LP(1, 1) mode is most susceptible to exhibiting a lower mode instability threshold.

To further control the $LP_{(1,\ 1)}$ mode of the fiber, the NA is further reduced based on the above refractive index distribution. as shown in Fig. 4.37, the maximum theoretical NA of the fiber core is about 0.060 and the fiber V value at 1080 nm is about 5.2. The calculated results of the effective refractive index, mode field diameter and bending loss for each mode are shown in Table 4.4. It can be seen that: compared with the refractive index distribution in Fig. 4.36, the bending loss of the $LP_{(1,\ 1)}$ mode in Fig. 4.37 increases substantially with the effective refractive index difference (Δn) between $LP_{(0,\ 1)}$ and $LP_{(1,\ 1)}$ being almost equal (Δn about 34E^{-5} in Fig. 4.36 and Δn about 33E^{-5} in Fig. 4.37), while the bending loss of the fundamental mode

Table 4.3 Computational analysis of refractive index distribution of 30/600 μm fiber cores NA~0.065

Mode	Effective refractive index	Mold field diameter μm	Bending loss dB/km@10 cm	Bending loss dB/km@15 cm
$LP_{(0,1)}$	1.44800	22.36	4.21	0.001
$LP_{(1,1)}$	1.44766	28.9	3675	30.26
$LP_{(0,2)}$	1.44722	29.13	838,717	220,810
$LP_{(2,1)}$	1.44727	32.59	102,070	18,724

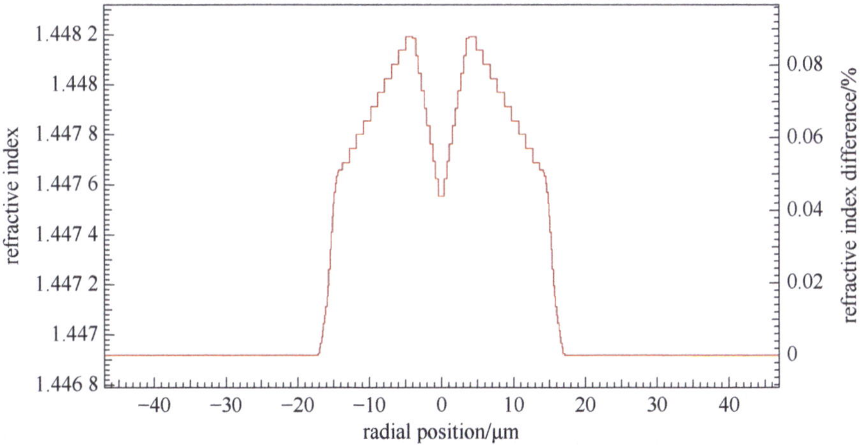

Fig. 4.37 30/600 μ m fiber core design for 10,000-watt single-mode lasers (NA~0.06)

Table 4.4 Computational analysis of refractive index distribution of 30/600 μ m fiber cores NA~0.06

Mode	Effective refractive index	Mold field diameter μm	Bending loss dB/km@10 cm	Bending loss dB/km@15 cm
$LP_{(0,1)}$	1.44780	22.5	201	0.42
$LP_{(1,1)}$	1.44747	28.9	62,701	2743
$LP_{(0,2)}$	1.44706	32.82	2,025,710	1,195,730
$LP_{(2,1)}$	1.44710	34.58	208,580	102,810

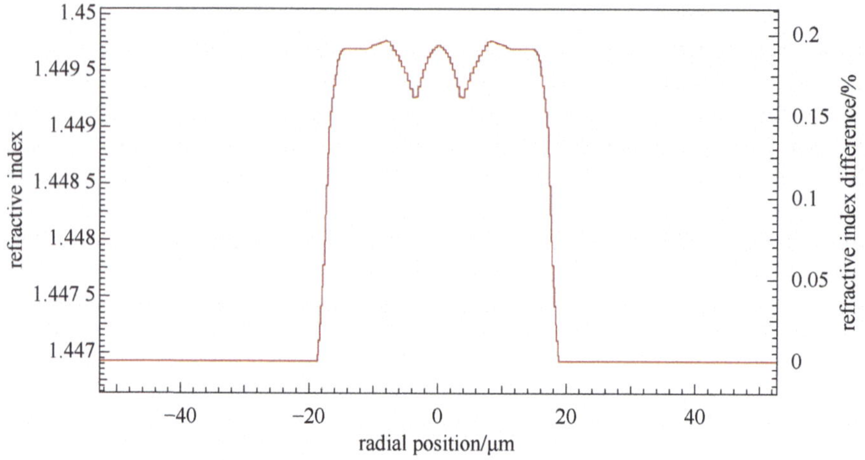

Fig. 4.38 35 μm core design for 10,000-watt multimode laser output (NA ~ 0.09)

is still within the acceptable range, which will make the mode instability threshold of the fiber in actual use The core refractive index distribution design is therefore suitable for 10,000-watt high brightness large mode field fibers.

4.4.3 Design of Refractive Index Distribution for Multimode Fiber Cores in 10,000-Watt Class

The design focus for multi-mode fibers doped with ytterbium at the 10 kW level must consider avoiding excessive bending loss of the few higher-order modes adjacent to the fundamental mode, which could result in low efficiency of fiber lasers.

Therefore, a fiber with a core ~35 μm for multimode 10,000-watt laser output was designed using a refractive index distribution 2 as shown in Fig. 4.38, with a maximum theoretical NA of about 0.09 for the core.

The V value of the fiber at 1080 nm is about 8.9. The calculation analysis shows that the fiber supports 13 modes such as $LP_{(0,1)}$, $LP_{(0,2)}$, $LP_{(0,3)}$, $LP_{(1,1)}$, $LP_{(1,2)}$, $LP_{(1,3)}$, $LP_{(2,1)}$, $LP_{(2,2)}$, $LP_{(3,1)}$, $LP_{(3,2)}$, $LP_{(4,1)}$, $LP_{(5,1)}$, $LP_{(6,1)}$, etc. The effective refractive

Table 4.5 Calculated analysis of refractive index distribution of fiber core ~35 μm NA~0.09

Mode	Effective refractive index	Mode	Effective refractive index	Mode	Effective refractive index
$LP_{(0,1)}$	1.44947	$LP_{(1,2)}$	1.44827	$LP_{(3,2)}$	1.44718
$LP_{(1,1)}$	1.44927	$LP_{(4,1)}$	1.44810	$LP_{(6,1)}$	1.44701
$LP_{(2,1)}$	1.44896	$LP_{(2,2)}$	1.44776	$LP_{(1,3)}$	1.44698
$LP_{(0,2)}$	1.44874	$LP_{(0,3)}$	1.44764		
$LP_{(3,1)}$	1.44856	$LP_{(5,1)}$	1.44758		

indices of each mode are shown in Table 4.5. Calculate the bending loss for the two modes LP(5,1) and LP(3,2) as examples. The effective refractive index of LP(5,1) is 1.44764, significantly different from that of pure silica, which is 1.44692. The difference is substantial, and its bending loss at 1080 nm is 0.94 dB/km at a bend radius of 10 cm. In practical applications, it is nearly impervious to bending and can be effectively excited for transmission. Conversely, the effective refractive index of LP(3,2) is 1.44718, very close to the refractive index of pure silica. However, its bending loss at 1080 nm is a staggering 15,217 dB/km at a bend radius of 10 cm, rendering this mode unsuitable for excitation and transmission in practice.

The above analysis indicates that this ten-kilowatt-level ytterbium-doped multimode fiber design supports a total of 10 modes in transmission, including LP(0,1), LP(1,1), LP(2,1), LP(0,2), LP(3,1), LP(1,2), LP(4,1), LP(2,2), LP(0,3), and LP(5,1).

4.5 Important Applications of Ytterbium-Doped Large Mode Area Fiber

Currently, the advancement of industry and technology has led to the adoption of high-power fiber lasers across various sectors, including laser marking, laser cutting, laser medicine, and military defense. Ytterbium-doped large-mode area fiber serves as the primary gain medium for high-power fiber laser main amplifiers. The performance requirements for Yb-doped large-mode area fibers also differ depending on the application of lasers. Generally, in industrial market applications such as laser marking, cutting, and cleaning, the quality index requirements for fiber laser beams are typically low. As long as the power, efficiency, and long-term reliability indicators are met, these requirements are satisfied. Special scenarios in defense and military applications require ytterbium-doped large-mode area fiber lasers to possess both high power and excellent beam quality. Consequently, the thresholds for fiber nonlinear effects and mode instability are set extremely high. Ytterbium-doped large-mode area cladding fiber has been utilized in various high-power fiber lasers, achieving significant commercial success. Recently, several applications have been introduced, including narrow linewidth fiber lasers, 1018 nm fiber lasers, 980 nm fiber lasers, and industrial fiber lasers, each representing a certain degree of significance.

4.5.1 Application of Ytterbium-Doped Large Mode Area Fiber in High Power Narrow Linewidth Fiber Lasers

In recent years, narrow linewidth fiber lasers with a spectral width of less than or equal to 0.1 nm have garnered widespread attention from researchers both domestically and internationally due to their significant application value in beam synthesis, gravitational wave detection, and other fields. However, stimulated Brillouin scattering and mode instability severely limit further power enhancement of narrow linewidth fiber lasers. Currently, methods to enhance the output power of narrow-linewidth fiber lasers mainly focus on improvements and optimizations at the system level of the fiber laser. The following provides a brief introduction to the recent research achievements in ytterbium-doped silica glass fibers for narrow-linewidth fiber lasers and amplifiers, focusing on the perspectives of optical fiber materials and structures. It also looks forward to the future development trends of narrow-linewidth fiber laser technology.

Research indicates that the increase in output power of narrow linewidth fiber lasers is primarily constrained by stimulated Brillouin scattering and mode instability effects [77]. Once the laser power within the fiber core attains the threshold for stimulated Brillouin scattering, the resultant backward propagating Stokes light can severely damage the fiber laser system. Consequently, it is regarded as the primary factor limiting the output power of narrow linewidth fiber lasers [78].The mode instability effect causes the output beam of the fiber laser system to transition from a stable fundamental mode to a dynamic coupling between the fundamental mode and higher-order modes. This transition limits the power scaling of single-mode lasers and increases the thermal load on devices such as cladding light strippers [79]. Current researchers primarily suppress Stimulated Brillouin Scattering (SBS) and Transverse Modal Instability (TMI) by increasing the effective mode field area of optical fibers [80, 81], enhancing the absorption coefficient, broadening the seed light bandwidth, limiting bending modes, and optimizing the pumping method. Currently, the single-mode output power of fiber amplifiers with a linewidth of the order of hundreds of GHz has been increased to over 5 kW [82–84].Increasing the core diameter and enhancing the absorption coefficient can indeed raise the Stimulated Brillouin Scattering (SBS) threshold. However, this also intensifies the fiber's thermal effect and increases the presence of higher-order mode components within the core, consequently lowering the Transverse Mode Instability (TMI) threshold. It is precisely because of this "contradiction" between the two that the key to achieving higher output power from narrow-linewidth fiber lasers lies in how to balance Stimulated Brillouin Scattering (SBS) and Transverse Mode Instability (TMI) [85].

From a physical mechanism perspective, Stimulated Brillouin Scattering arises from the interaction between the optical field and the acoustic field within the fiber, while mode instability effects originate from the thermo-optic effect caused by waste heat during the conversion of laser energy. Both phenomena can be attributed to the interaction between light and matter. Therefore, studying and controlling the internal acoustic and thermo-optic properties of the optical fiber material itself is the top priority for fundamentally overcoming the power enhancement bottleneck of narrow linewidth fiber lasers. This section will provide a brief introduction to the research progress on ytterbium-doped silica fibers for high-power, narrow-linewidth laser amplifiers. It will begin with the regulation of the glass optical fiber material composition and fiber structure design, with a focus on suppressing Stimulated Brillouin Scattering (SBS) and Transverse Mode Instability (TMI) effects.

4.5.1.1 Fiber Core Composition Regulation

The composition of the glass core in gain fibers is crucial to the properties of the fiber. Rare-earth-doped silica fibers generally necessitate the incorporation of dopants, such as aluminum, phosphorus, fluorine, boron, etc., into the silica matrix to alter the glass, rendering it more appropriate for high-power fiber laser applications. The introduction of these elements will not only affect the optical refractive index of the core glass but also alter acoustic parameters, such as the speed of sound and phonon lifetime within the glass, and impact its thermo-optic properties. Thus, the rational control of the core doping ion composition can effectively mitigate stimulated Brillouin scattering and mode instability effects in optical fibers. Aluminum stands out as the most significant doping element in large-mode-area ytterbium-doped silica fibers, aside from the rare-earth elements themselves [86, 87]. Al ions can reduce the clustering of rare earth ions, thereby increasing the doping concentration of rare earth ions in quartz glass. Moreover, Al_2O_3 possesses the unique property of enhancing the optical refractive index of SiO_2 glass while simultaneously decreasing its acoustic refractive index, as depicted in Fig. 4.39 [88]. Consequently, the speed of sound in the aluminum-doped core will be greater than that in the pure silica cladding, resulting in acoustic waves that are leaky modes with higher loss within the core, thereby reducing the interaction between the optical field and the acoustic field. Researchers have utilized the melt-cladding method to prepare crystal-derived optical fibers with a high aluminum content [89–91]. In 2012, Dragic et al. [92] prepared alumina crystal derivative optical fibers featuring a silica cladding and a core composed of aluminosilicate glass, which was produced through the diffusion reaction between alumina crystals and a silica glass cladding. The concentration of aluminum ions in the core can reach up to 55 mol%. Experimental measurements indicated that the Brillouin gain coefficient was merely 3.1×10^{-13} mW^{-1}, which is 1/100th of that found in commercial communication fibers. This result indicates that glass optical fibers with high aluminum doping are significantly effective in suppressing Brillouin gain. However, fibers prepared using the core-melting method typically have a higher numerical aperture and background loss, making them mainly suitable for mechanism research. There is still a considerable distance to cover before they can be applied in high-power fiber laser practical applications.

The TMI effect is inextricably linked to the thermo-optical effect of the fiber under high power operation, and the TMI threshold can be increased by reducing

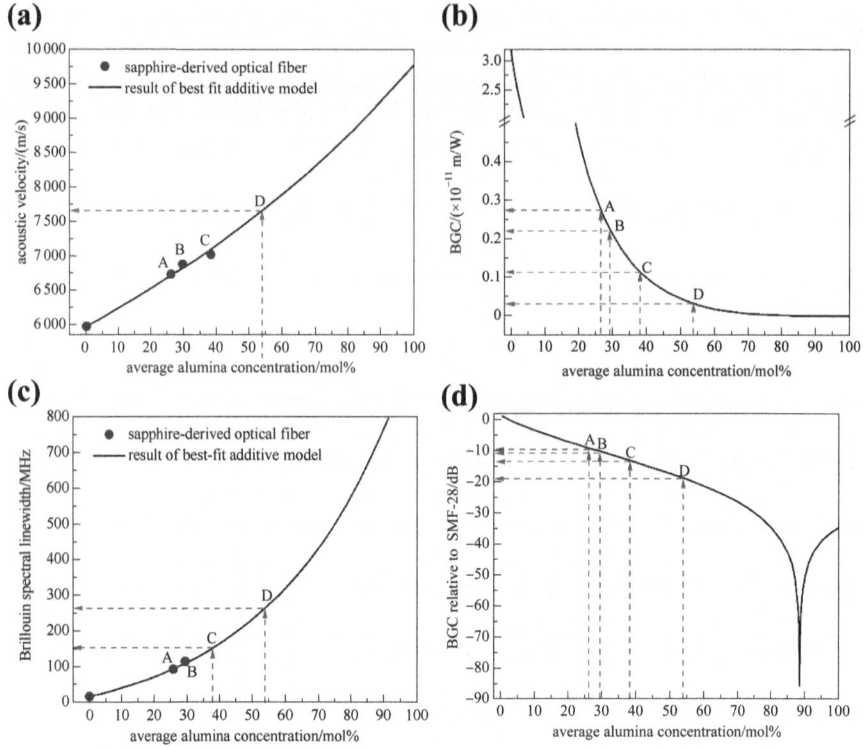

Fig. 4.39 Effect of aluminum element content in Al O_{23} crystal derivative fiber on sound velocity (**a**), Brillouin linewidth (**b**), absolute Brillouin gain coefficient (**c**), and relative Brillouin gain coefficient (**d**) with SMF-28 fiber [92]

the thermo-optical coefficient of the core material [93]. Among the common silica fiber dopants, SiO_2, GeO_2, and Al_2O_3 have positive thermo-optic coefficients, while P_2O_5 and B_2O_3 have negative thermo-optic coefficients, so the ratio of P and B elements in the core can be controlled to achieve a low or even "zero" thermo-optic effect fiber [94]. Since P_2O_5 has a negative thermo-optical coefficient, the introduction of phosphorus can reduce the thermo-optical coefficient of quartz glass fibers and help maintain the beam quality at high power operation. In conclusion, the Al-P-Si ternary system glass is characterized by low NA, low darkening effect, and low thermal-optical coefficient, all of which can improve the TMI threshold to some extent. 2019, Yu N. et al. also systematically investigated the effect of the introduction of [$AlPO_4$] bond on the optical and acoustic properties of quartz optical fibers by preparing Al-P co-doped fibers [95]. Their results showed that the [$AlPO_4$] had a significant impact on the optical and acoustic properties of the quartz fiber. Their results show that the introduction of [$AlPO_4$] increases the Brillouin gain bandwidth by increasing the internal phonon lifetime of the quartz glass, although it slightly decreases the acoustic velocity. In addition, [$AlPO_4$] reduces the degree of Brillouin scattering response to temperature changes and decreases the interaction between the acoustic and optical fields by reducing the electrostriction constants. It is because of these characteristics that most of the current high-power narrow linewidth fiber lasers are based on glass fibers in the AlPS ternary system.

John Ballato's group at Clemson University has done a lot of research in the field of developing negative thermo-optic effect fibers. 2020, they measured the thermo-optic coefficients of B_2O_3 silica fibers using the temperature characteristics of the free spectral range of the ring cavity laser and determined the thermo-optic coefficients of B_2O_3 and SiO_2 to be $-24.4\ K^{-1}$ and $9.96 \times 10^{-6}\ K^{-1}$, respectively. In 2021, they reported a low Brillouin gain coefficient and low thermo-optical coefficient ytterbium-doped fiber based on Al_2O_3-P_2O_5 -B_2O_3-SiO_2 system, and experimentally characterized the thermo-optical coefficient, Brillouin gain spectrum (BGS) and additional loss of optical darkening [96].Yb-doped fiber preforms doped with 14.6 mol% and 6.8 mol% B_2O_3 were prepared by MCVD and drawn into 21/400 μm low-NA fibers for comparison with commercial fibers. The thermal-optical coefficient of $6.8 \times 10^{-6}\ K^{-1}$ is lower than that of $9.0 \times 10^{-6}\ K^{-1}$ for commercial Yb-doped fibers. The Yb/Al/P/B fiber's Brillouin gain spectrum (BGS), as shown in Fig. 4.40a, is compared with that of the Nufern YDF-15/130 fiber. The commercial fiber exhibits a single-peak structure that aligns with the Lorentzian line shape, with a small peak around 11 GHz resulting from the use of SMF-28 single-mode fiber in testing. In contrast, B-doped fibers with BSG (Boron Silicate Glass) display multiple acoustic modes due to the reduced core glass sound velocity from high levels of doping with elements such as P (Phosphorus) and B, which endows them with a broader gain bandwidth and a lower gain coefficient. Additionally, the high doping of P elements imparts exceptional resistance to photodarkening to the fiber, as illustrated in Fig. 4.40b. When researchers employed this fiber to amplify signals with a 10 GHz linewidth, the amplifier achieved the TMI threshold at an output power of 1099 W. However, the proportion of backscattered light power at this point remained below 0.01%, and no Stimulated Brillouin Scattering (SBS) phenomenon was detected.

Although the Al/P/B/Si quaternary glass system offers significant advantages in customizing the thermal and acoustic properties of quartz optical fibers, the preparation of fiber preforms for this composition demands high technical standards. It is essential to carefully control the specific components to prevent issues such as excessively low refractive index and glass crystallization phase separation. Consequently, the practical engineering application of this type of fiber is currently quite rare.

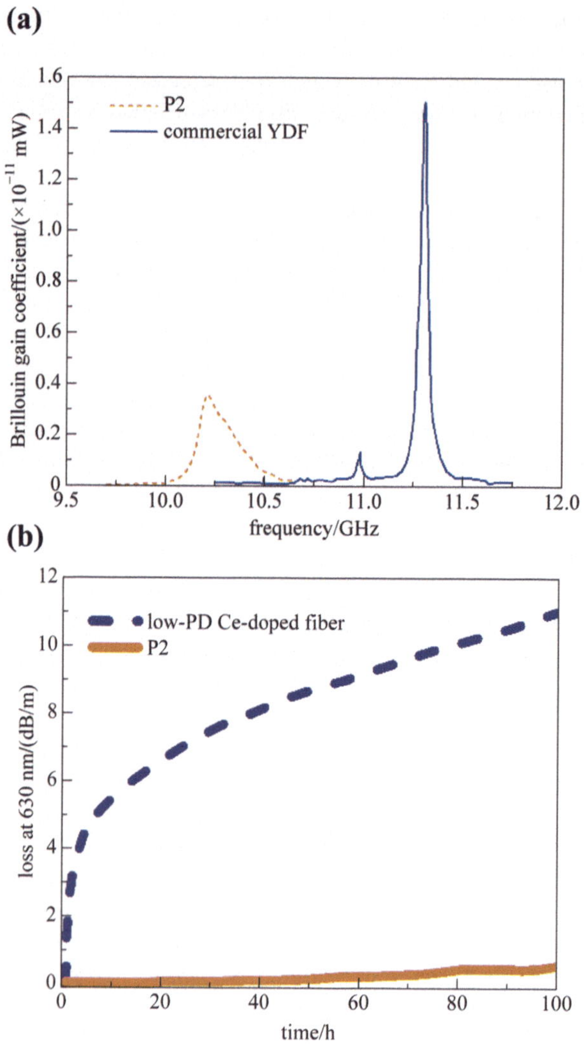

Fig. 4.40 Brillouin gain spectrum (**a**) and additional loss of optical darkening (**b**) for Al-phosphorus-Boron-Si quaternary glass system fiber and commercial fiber [96]

4.5.1.2 Improvement of Narrow Linewidth Laser Performance Using Different Fiber Structures

The purpose of core component modulation is to enhance the fiber's Stimulated Brillouin Scattering (SBS) and Transverse Mode Instability (TMI) thresholds by optimizing the intrinsic acousto-optical and thermo-optical properties of the material. Meanwhile, fiber structure design aims to modulate the distribution of optical, acoustic, and even thermal fields within the fiber. Several specialized fiber design options to suppress SBS or TMI effects are detailed below.

To address the contradiction between the increasing core diameter and mode degradation in conventional large-mode field cladding structured fibers, several international schemes have been proposed. These aim to enhance the output beam quality of large-mode area fibers by fabricating microstructured fibers, including photonic crystal fibers, photonic bandgap fibers, and chiral coupled fiber core fibers. Several of these microstructured fibers have achieved a series of landmark results in single-frequency and narrow-linewidth laser amplification, respectively. In 2019, NKT Photonics reported a 30/250 μm all-solid-state deflection-preserving PCF with a single-frequency laser amplification power of tens of watts [97]. In 2015, the Liang Dong group at Clemson University, in collaboration with the U.S. Air Force Research Laboratory, achieved a 400 W single-frequency laser output using a 50 μm core diameter all-solid-state photonic bandgap fiber [98]. In 2021, Pulford et al. achieved a 1.37 kW, 8 GHz narrow linewidth fiber amplification output using a 25/400 μm all-solid photonic bandgap fiber (AS-PBF) [99]. In 2022, Matniyaz et al. utilized an optimized all-solid photonic bandgap fiber with a 500 W single-frequency laser amplification output structure to determine a laser linewidth of 6.2 kHz at the highest power output using outlier measurements. Additionally, a smaller core diameter AS-PBF was utilized as a gain medium for narrow-linewidth laser amplifiers on the GHz scale to enhance the TMI threshold, achieving kilowatt-scale average power output [100]. In recent years, both Hannover University and n-Light have been working on the application of 3C fibers in all-fiber high-power, narrow-linewidth laser systems. In 2022, Hochheim and colleagues at Hannover University integrated various devices, including mode field adapters, cladding optical strippers, isolators, and beam combiners, into a single ytterbium-doped 3C fiber, which was side pumped. The pump light was coupled into the cladding of the gain fiber, which ultimately increased the single-frequency laser amplification output power to 336 W [101].

While the three types of microstructured large-mode field fibers possess unique advantages in suppressing higher-order modes, their complex structures make them challenging to fabricate. Additionally, the complete fiberization of the laser system necessitates specialized passive components (such as beam expanders, mode field adaptors, cladding optical strippers, etc.) that are compatible with the active fiber. These are the key factors that limit the practical application of microstructured large mode field fibers on a large scale.

The effective Brillouin gain coefficient along the entire fiber is the integral of the Brillouin gain coefficients at various positions along the fiber's length. Therefore, by introducing different Brillouin frequency shifts along the fiber's longitudinal direction, the Brillouin gain spectrum can be broadened, thereby reducing the Brillouin gain coefficient [102]. Since the frequency shift of the Stokes light is related to the speed of sound within the optical fiber material, different temperature gradients [103, 104] and stress gradients [105–107] are typically applied along the fiber's longitudinal direction to enhance the Stimulated Brillouin Scattering (SBS) threshold. Simultaneously, acoustic wave fields within optical fibers of varying core diameters exhibit different frequencies, which correspond to distinct Brillouin frequency shifts [108]. Thus, long tapered fibers exhibit a broader Brillouin gain spectrum than traditional uniform fibers. It is precisely due to the advantages of long tapered optical fibers in suppressing stimulated Brillouin scattering that significant progress has been made in the field of single-frequency fiber amplifiers in recent years. The National University of Defense Technology in China has conducted a series of studies on the application of high-power lasers in conical optical fibers [60, 109, 110]. In 2020, a team led by Lai from the National University of Defense Technology constructed a fully fiber single-frequency fiber amplifier utilizing polarization-maintaining conical ytterbium-doped fibers, ultimately achieving a single-frequency laser output of 550 W, which is currently the highest single-frequency laser output power record based on a fully fiber structure [111]. In this experiment, the input and output core-cladding sizes of the tapered gain fiber used were 36.1/249.3 μm and 57.8/397.3 μm, respectively. The larger core-cladding ratio ensures a sufficient absorption coefficient, reducing the fiber's usable length to 1.27 meters, with the tapered section being 0.74 meters long.

4.5.1.3 Composite Modulation of Components and Structures to Improve Narrow Linewidth Laser Performance

The SBS effect is ultimately the result of the interaction between the optical and acoustic fields. Therefore, special design of the acoustic waveguide structure can be used to reduce the overlap between the acoustic and optical fields and increase the SBS threshold. For instance, a region with lower acoustic velocity, doped with elements such as Ge, F, and P, can be designated as the acoustic guide layer in the outer ring of the core. This design ensures that most of the acoustic energy is concentrated in the outer ring of the core, while the fundamental mode optical energy remains focused in the core. Consequently, this achieves the separation of the acoustic and optical wave fields. In 2007, Li et al. from Corning proposed an Al/Ge-doped acoustic field tailoring fiber, as depicted in Fig. 4.41a. The doping concentration gradient of the two elements within the fiber core is carefully controlled to ensure the consistency of the core's refractive index [112]. This structure significantly reduces the overlap between the acoustic and optical fields, resulting in a 6 dB increase in the fiber's SBS threshold compared to conventional fibers. Using this specially designed acoustic field trimming fiber, researchers achieved single-frequency optical laser amplification results of 502 W [113]. In 2014, Robin et al. at the U.S. Air Force Research Laboratory achieved an 811 W single-frequency fiber laser output using a photonic crystal fiber cropped by an acoustic field. This result represents the highest reported single-frequency fiber laser power output to date [114]. The core of this photonic crystal fiber comprises seven distinct doped regions, as depicted in Fig. 4.41b. Regions υ1 and υ2 are doped with Ytterbium (Yb), while the doping of Germanium (Ge), Aluminum (Al), and Fluorine (F) elements modifies them to exhibit different sound velocities. Region υ3 is an undoped pure quartz region, which together with the others, results in three distinct sound velocity regions that serve to broaden the Brillouin gain spectrum.

Furthermore, Marciante et al. from the University of Rochester have shown both theoretically and experimentally that the gain coefficient of higher-order modes in optical fibers can be diminished through area doping. Additionally, they have utilized the gain competition between the fundamental and higher-order modes to

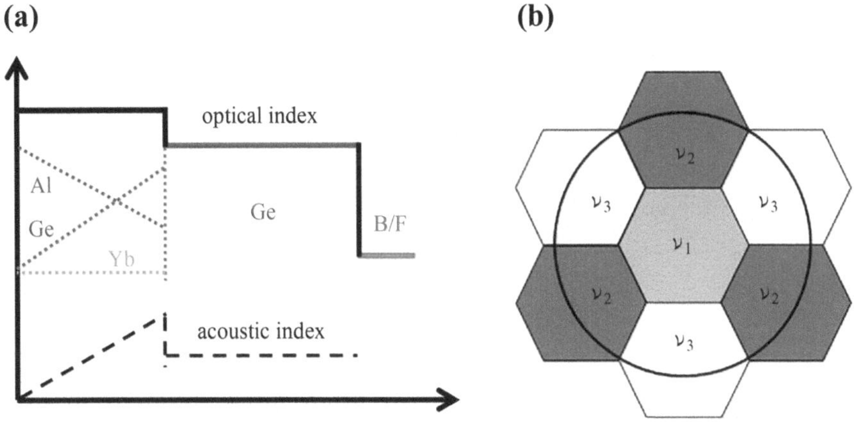

Fig. 4.41 Al/Ge co-doped Yb-doped fiber design (**a**) [112] and acoustic field cropped photonic crystal fiber core structure (**b**) [114]

"clean up" the output laser mode [115, 116]. In 2016, the Japanese company Fujikura reported continuous output results of 3 kW [117, 118], 5 kW [119], and 8 kW [120] for oscillators by optimizing fiber design and laser systems to suppress nonlinear effects. In 2019, Zhang et al. from Huazhong University of Science and Technology prepared a ytterbium-doped silica fiber with a 70% area doping ratio and increased the TMI threshold from 717 W to 1.25 kW compared with the conventional fiber [120]. The specific parameters of this fiber are shown in Fig. 4.42, where 70% of the core center is co-doped with Yb/Al/Ce/F, while the outer 30% is only doped with Al/F elements. In 2021, Huang et al. achieved laser amplification with a power output of 3.5 kW and a narrow linewidth of 0.32 nm, along with a beam quality of $M^2 = 1.86$, utilizing domestically produced area-doped fiber at the Center for Laser Fusion Research, CIMB. No TMI phenomenon was observed at the highest power output [121]. The doped area ratio of the fiber used in this work is 60%, and the mode instability threshold is increased by 2.2 times compared to traditional fibers under 976 nm pumping. In the same year, Hanshuo Wu et al. from the National University of Defense Technology reported a 1018 nm cascade pumping fiber amplifier based on a 30/40/250 μm region-doped fiber with a maximum output power of 6.2 kW [122].The mode instability threshold for the doped fiber in this region under 1018 nm pumping conditions is 4.74 kW, which represents a 170% increase compared to fully doped fiber. In 2022, Wu Hanshuo and colleagues utilized a single-frequency laser with phase-modulated white noise as the seed source. This was pre-amplified and subsequently subjected to main power amplification through cascaded pumped regionally doped fiber, ultimately resulting in a narrow linewidth laser output with a maximum power of 5.96 kW and a 3 dB linewidth of 0.42 nm. Given that the TMI threshold was approximately 5.01 kW, the beam quality degraded to $M^2 = 2.0$ at the highest power output. Prior to reaching the TMI threshold, at an output power of 4.97 kW, the beam quality was $M^2 = 1.64$ [123].

Non-rare-earth doped regions in regionally doped optical fibers typically incorporate elements such as Ge, Al, and P to enhance the refractive index, ensuring a

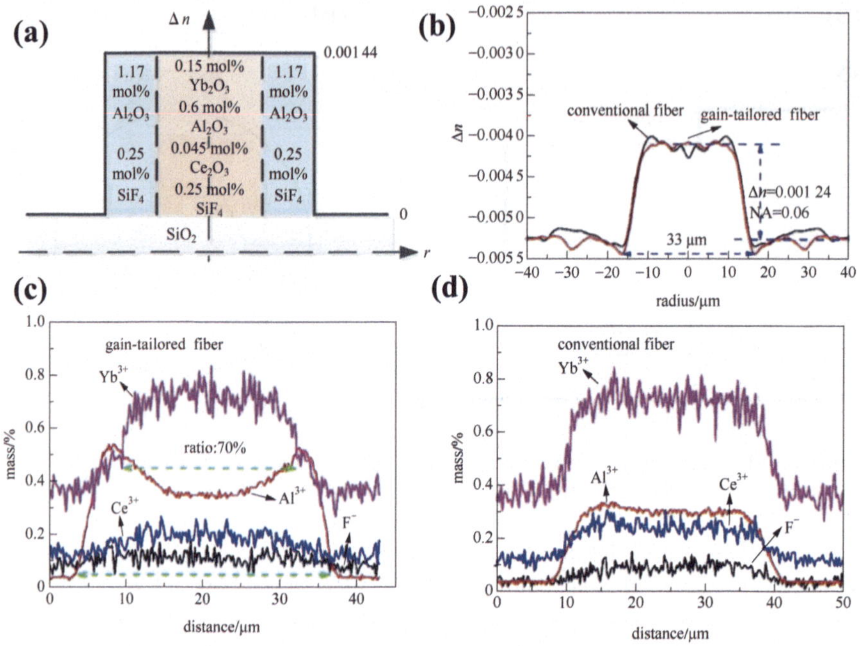

Fig. 4.42 Area doped fiber design (**a**), refractive index distribution (**b**), area doped fiber element distribution (**c**) and conventional fiber optic element distribution (**d**) [121]

uniform refractive index profile across the entire core region. However, the addition of these elements impacts the material's acoustic properties, thereby modifying the acoustic field distribution within the core. In 2022, Cooper et al. from the University of Central Florida theoretically investigated the properties of Stimulated Brillouin Scattering (SBS) in Ge-doped fiber with ring-shaped doping regions and experimentally achieved a single-frequency laser amplification output of 123 W based on ring-doped fibers, with a beam quality factor of $M^2 = 1.11$ [124]. Theoretical studies have shown that regional doping structures effectively broaden the Brillouin gain bandwidth of optical fibers and increase the loss of acoustic wave modes, thereby suppressing the Stimulated Brillouin Scattering (SBS) effect. Furthermore, compared to traditional fully doped fibers, regional doped fibers can enhance the single-frequency laser output power by 2.4 times by raising the SBS threshold.

While regional doped fibers hold significant promise for high-power, narrow linewidth fiber lasers, it is crucial to recognize that their absorption coefficient is lower than that of traditional fibers post-doping. Consequently, longer fibers are typically needed to achieve adequate pump absorption, which can hinder the suppression of nonlinear effects. Additionally, the presence of interfaces formed by various doping elements within regionally doped optical fibers can lead to stress and diffusion, potentially altering the refractive index profile. Therefore, it is essential to meticulously design parameters such as the rare-earth doping ratio, doping elements, and concentration during the preparation of optical fibers, and the preparation process of preforms must meet higher requirements.

Additionally, conventional double-clad fibers typically feature a low-refractive-index fluoropolymer as an outer cladding layer over a pure quartz inner cladding to ensure the transmission of pump light within the inner cladding. However, the low thermal conductivity of the organic coating hampers heat dissipation. Consequently, employing metal-coated ytterbium-doped cladding fibers can also enhance the Transverse Mode Instability (TMI) threshold.

The suppression of stimulated Brillouin scattering and mode instability effects will remain a research focus for future high-power, narrow-linewidth fiber lasers. Utilizing glass fiber materials and structures to modulate the intrinsic properties of fibers is expected to achieve breakthroughs in the field of narrow-linewidth fiber lasers.

Moreover, the passive delivery fiber at the amplifier's terminus is also a crucial component in high-power fiber laser systems. However, at equal lengths of usage, the delivery fiber has a longer nonlinear interaction distance compared to the gain fiber. Therefore, in high-power, narrow-linewidth fiber amplifiers, the delivery fiber may require more specialized acoustic field tailoring designs to suppress SBS (stimulated Brillouin scattering) effects than the gain fiber. In 2021, Tsvetkov and colleagues from the Russian Academy of Sciences prepared a passive fiber featuring a special multimode acoustic field waveguide structure. This was achieved by non-uniformly doping the core with phosphorus and fluorine elements, which resulted in an 8 dB increase in the SBS threshold power [125].This result offers valuable insights for the development of solid-state high-power energy transmission fibers.

4.5.2 Application of Ytterbium-Doped Large Mode Area Fiber in ~1018 nm Lasers

Thermal effects are critical for lasers with average power above the kilowatt level. Among them, TMI driven by quantum defect heating is one of the main challenges. The power of domestic laser diode-pumped ytterbium-doped fiber lasers (YDFL) is limited to the 10 kW level due to the low brightness of the laser diodes and the thermal effects resulting from quantum defects. Another method to mitigate thermal management issues (TMI) is by reducing quantum defect heating. The quantum defect is defined as $\eta=1-(h\nu_1)/(\mathrm{h}\nu_2)$, where ν_1 and ν_2 represent the laser frequency and pump frequency, respectively, and h is Planck's constant. A 1018 nm fiber laser is utilized for pumping within the same band to achieve high brightness and a low quantum defect in the laser output. To achieve a 1070 nm laser output, for instance, the quantum defect is approximately 9% for the 976 nm LD pumped YDFL, whereas it is only 5% for the 1018 nm fiber laser pumped YDFL. Moreover, the fiber laser is significantly brighter than the LD by more than an order of magnitude, allowing for easier coupling into the gain fiber. Utilizing high-efficiency, high-power fiber lasers in tandem with output wavelengths below 1020 nm presents a solution that can enhance pump brightness and diminish thermal load, which is crucial for achieving fiber lasers with outputs exceeding 10 kW. It is well known that the single-mode YDFL, reported by IPG in 2009, has reached 10 kW, and its implementation relies on the co-band pumping of a 1018 nm fiber laser. However, the emission cross section of Yb^{3+} at 1018 nm is significantly smaller than that at 1030 nm (the emission peak of Yb^{3+}). Operating at such a short wavelength as 1018 nm, for instance, results in gain competition between the laser and the amplified spontaneous emission (ASE), which diminishes the laser power and complicates the achievement of efficient output at 1018 nm. Consequently, there has been limited reporting in this area.

In 2014, Wang Yanshan et al. of CIMB analyzed theoretically the effect of gain fiber parameters on the design of short wavelength YDFL [126]. Building upon this foundation, they developed a 1018 nm YDFL system and conducted experimental studies on the output characteristics of 1018 nm lasers with varying lengths of gain fiber and core-to-cladding ratios. The findings indicated that reducing the length of the gain fiber, lowering the Yb^{3+} ion doping concentration, and increasing the core-to-cladding area ratio are effective strategies to suppress amplified spontaneous emission (ASE) gain and ensure stable laser operation at 1018 nm. The effective way to ensure the operation of the laser at 1018 nm is crucial. Additionally, the composition of the ytterbium-doped core also significantly impacts its performance. Guancheng Gu and colleagues at Clemson University, USA [127] demonstrated that the phosphosilicate glass system is more advantageous than the aluminosilicate glass system for laser output in the 1000 ~ 1030 nm range, as illustrated in Fig. 4.43.

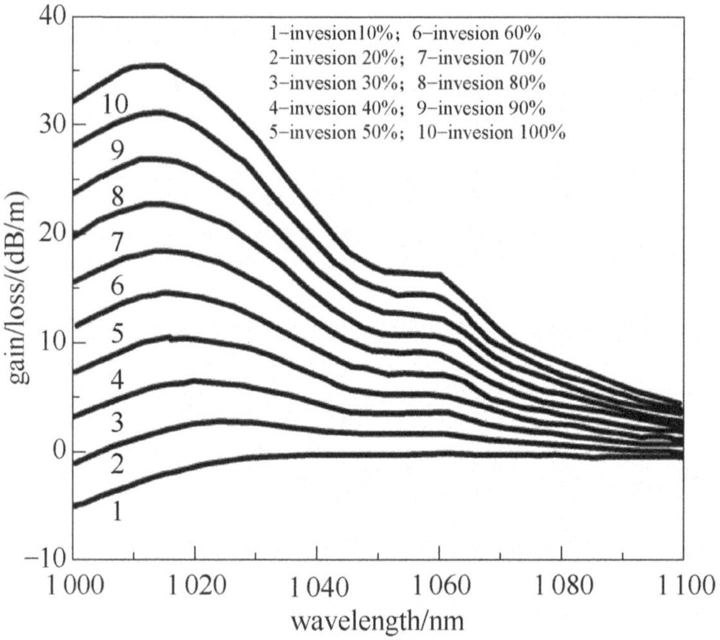

Fig. 4.43 Gain variation characteristics of ytterbium-doped phosphorosilicate glass with wavelength

It can achieve a net gain with merely a 20% ion number inversion. They proposed a method to further suppress amplified spontaneous emission (ASE) at wavelengths above 1030 nm by combining a leakage channel large core fiber (LCF) structure design to prepare a 50/420 μm core-to-package. They proposed a phosphor-silica glass fiber with a core-to-package ratio of 50/420 μm and conducted a comparative study on the performance of short-wavelength lasers, such as those with a 1018 nm wavelength, which can achieve an efficiency of 73% for 1018 nm laser oscillation.

4.5.3 Application of Ytterbium-Doped Large Mode Area Fiber in ~980 nm Fiber Lasers

The absorption cross section of Yb^{3+} ions near 1018 nm is 1 ~ 2 orders of magnitude lower than near 976 nm, which places high demands on the design of Yb-doped fiber and the selection of pumping scheme. The pumping of Yb-doped fiber with 980 nm wavelength (974 ~ 980 nm, hereinafter referred to as 980 nm) fiber laser can improve the brightness of the pumped light while facilitating the absorption of Yb-doped fiber, which has potential application value for achieving high power laser output. In addition, the 980 nm fiber laser can be used to obtain new wavelength laser output by frequency conversion through nonlinear crystals, etc. The 480–490 nm laser generated by frequency doubling can replace the argon ion laser as a blue laser source. It can be seen that the 980 nm fiber laser has high research value.

In 2012, Ying Liu et al. from the National University of Defense Technology [128] indicated that a 980 nm laser output generates gain competition with ASE in a four-level energy system (1020 ~ 1100 nm). To achieve a 980 nm signal light output in an Yb-doped fiber laser, it is necessary to invert over 50% of the lower energy level particles in the gain medium to the upper energy level. In contrast, ASE (Amplified Spontaneous Emission) generation requires only about 5% inversion. Furthermore, the emission cross-section at 980 nm is significantly larger than that at the ASE band, resulting in more dominant emission at 980 nm when the particle number inversion condition is fulfilled. Therefore, to achieve 980 nm laser output, it is necessary to enhance the pump absorption of the gain medium and increase the number of inverted particles to suppress ASE self-excited oscillation. It has been demonstrated that the key to achieving 980 nm laser output and suppressing ASE self-excited oscillation lies in the rational design of the gain fiber structure and parameters. While ASE can be suppressed through various microstructure modulations, including the use of photonic crystal fibers and single-frequency laser output at the watt level with phosphate glass fibers, both microstructured fibers and multi-component glass fibers present challenges when attempting to fuse them with conventional passive fibers at low loss, particularly from the perspective of high-brightness fiber laser pumping sources. Consequently, it is evidently more practical to utilize conventional double-clad fibers to achieve 980 nm laser output for applications such as all-fiberization systems and pump beam combining.

Generally, the use of high-phosphorus, low-doping concentration, large core-to-cladding ratio, W-type ytterbium-doped double-clad fiber as the research subject, combined with the pulling cone treatment for filtering, etc., could be the future direction for the development of 980 nm high-efficiency output fiber lasers.

4.5.4 Industrial Laser Applications

Since 2010, the rapid development of kilowatt-level fiber lasers has been the most representative in the industrial field, primarily used for metal cutting, welding, and other applications. The domestically produced fiber, with Nufern's 20/400 μm ytterbium-doped double-clad fiber being the most representative, has dominated the majority of the domestic laser market. Since 2015, domestically produced 20/400 μm ytterbium-doped double-clad fibers have been progressively applied and promoted, achieving laser applications with power levels ranging from 1 to 2 kW. Laser manufacturers have specific requirements for 20/400 μm ytterbium-doped double-clad fiber products, which encompass basic fiber parameters such as dimensional, optical performance, and mechanical performance parameters. Additionally, it includes fundamental performance parameters for fiber usage, such as laser oscillation efficiency, coating environmental testing, and photonic darkening performance. Taking the 20/400 μm ytterbium-doped double-cladding fiber produced by the Shanghai Institute of Optics and Fine Mechanics as an example, the primary performance parameters of the fiber are listed in Table 4.6. Regarding the numerical aperture of the fiber core, due to the refractive index distribution of the ytterbium-doped core, as shown in Fig. 4.44, not being step-like, it can only be considered as an equivalent numerical aperture. To enable fiber lasers to be applied in all-fiber systems, they must be spliced with compatible passive fiber devices and coiled into bends of a specific diameter. This process can easily result in cladding mode leakage, significantly impacting the efficiency of the fiber lasers. Given that the diameter of the fiber optic water-cooling disc and the passive components were previously designed and customized to meet the fiber standards of foreign companies, domestic active fibers must concentrate on the distribution of the core's refractive index during development to attain the desired laser efficiency after cladding light filtering.

In order to further improve the nonlinear threshold of the fiber, Shanghai Institute of Optics further introduced 25/400 μm ytterbium-doped double-clad fiber (YDF) for applications in fiber lasers with powers above 3 kW. Currently, according to the actual results of fiber laser applications, 25/400 μm YDF has higher requirements for refractive index distribution control due to larger core diameter and smaller effective refractive index difference between different modes. From a practical application perspective, 20/400 μm YDF has been widely utilized in the production of 3 kW continuous fiber lasers, following the replacement of the 915 nm LD with a 976 nm LD as the pump source. Although 25/400 μm YDF can endure higher power levels, its double-cladding structure imposes limitations. The increased laser power presents a significant challenge to the long-term reliability of the low

Table 4.6 Basic parameters and performance indicators of ytterbium-doped fiber with large mode field in 20/400 μm double cladding

YDF-20/400-DC 20/400 double cladding large mode field ytterbium doped fiber				
Serial number	Technical specifications	Unit	Indicator requirements	
			Customer needs	Internal control
1	Envelope pump absorption (@915 nm)	dB/m	0.40 ± 0.05	0.39 ± 0.04
2	Core numerical aperture NA	/	0.065 ± 0.005	0.063 ± 0.003
3	Cladding numerical aperture (5%)	/	≥0.46	≥0.46
4	Core loss (@1200 nm)	dB/km	≤15.0	≤10.0
	Core loss (@1300 nm)	dB/km	≤30.0	≤20.0
5	Cladding loss (@1095 nm)	dB/km	≤15.0	≤10.0
6	Fiber core diameter	Mm	20.0 ± 1.5	20.0 ± 1.2
	Cladding diameter (face to face)	Mm	400.0 ± 10.0	400.0 ± 5.0
	Concentricity of fiber core cladding	Mm	≤2.0	≤1.5
7	Diameter of coating layer	Mm	550.0 ± 15.0	545.0 ± 10.0
8	Tension screening test	Kpsi	≥100	≥100
9	Laser oscillation efficiency @ 915 nm pump absorption	/	≥70%	≥70%
10	Coating accelerated aging red light power decay	/	≤5%	≤5%
11	Photon darkening saturation additive loss @ 633 nm	dB/m	/	<80

refractive index organic cladding. Consequently, 25/400 μm YDF is currently employed in relatively few applications within the domestic industrial laser market.

Since 2018, institutions such as the Shanghai Institute of Optics and Fine Mechanics and Changfei have begun introducing ytterbium-doped triple-clad fibers to meet the needs of industrial lasers operating at power levels above 3 kW with long-term reliability. To date, the Shanghai Institute of Optics and Fine Mechanics has successfully developed ytterbium-doped triple-clad fibers for single-module industrial use at power levels of 6 kW and above, achieving ideal application results. The specific parameters of this type of fiber have been detailed in the previous section and are not reiterated here. Only the fundamental parameters of a 34 μm core diameter ytterbium-doped triple-clad fiber are listed in Table 4.7 for reference. This fiber is capable of meeting the application requirements for industrial single-module fiber lasers with power exceeding 5 kW.

According to information released by Laser Focus World, the market size of China's fiber laser in 2020 was approximately $1.38 billion, with domestic products accounting for about 56%, reaching $773 million. China's fiber laser industry began around 2006, and over the past decade plus, it has given rise to fiber laser companies

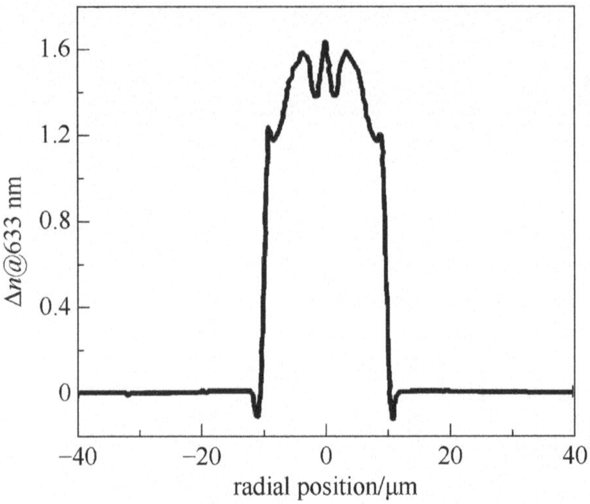

Fig. 4.44 Refractive index distribution of YDF-20/400 double-clad Yb-doped fiber core with large mode field in Shanghai Institute of Optical Machinery

such as Raycus Laser, JPT, and Chuangxin Laser. These companies have gradually achieved domestic substitution in the medium and low-power fields, breaking the monopoly of foreign companies in the high-power field, and greatly alleviating the "chip shortage" dilemma in the industrial laser sector. From a technical development perspective, the domestic fiber laser industry has essentially followed the mature experiences of foreign countries. Through advancements in pump sources, doping fibers, and beam synthesis technology, among other methods, the power and processing capabilities of lasers have been continuously enhanced to meet market demands in the fields of marking and cutting. As domestic manufacturers have completed the independent research and development of some high-power fiber lasers and introduced them into the market, the localization rate has steadily increased. In 2018, the market share of domestically produced high-power fiber lasers reached 34.5%, and it is anticipated that medium and high-power fiber lasers will swiftly achieve import substitution in the near future. With the innovation and enhancement of manufacturing technology by domestic laser producers, the energy efficiency of domestic high-power lasers has markedly improved, progressively competing with foreign products in the high-power fiber laser market.

According to the "2019-2025 China Industrial Laser Industry Market Supply and Demand Forecast and Development Prospects Report" published by Wise Research Consulting, as the application of fiber lasers in the field of industrial processing continues to expand, the development of the fiber laser industry in the next few years will exhibit five major trends:

Table 4.7 Basic parameters and performance indicators of ytterbium-doped fibers with large mode fields for 34/460/530 μ m triple cladding

TYDF-34/460/530-TC 34/460/530 μm triple cladding large mode field ytterbium doped fiber				
Serial number	Technical specifications	Unit	Indicator requirements	
			Customer needs	Internal control
1	Envelope pump absorption (@915 nm)	dB/m	0.9 ± 0.2	0.9 ± 0.2
	Internal cladding pump absorption (@915 nm)	dB/m	1.2 ± 0.2	1.2 ± 0.2
2	Core numerical aperture NA	/	0.09 ± 0.005	0.09 ± 0.005
3	Numerical aperture of the inner cladding (50%)	/	0.21 ± 0.02	0.21 ± 0.02
	Cladding numerical aperture (5%)	/	≥0.46	≥0.46
4	Core loss (@1200 nm)	dB/km	≤15.0	≤15.0
	Core loss (@1300 nm)	dB/km	≤30.0	≤30.0
5	Cladding loss (@1095 nm)	dB/km	≤20.0	≤20.0
6	Fiber core diameter	Mm	34.0 ± 2	34.0 ± 2
	Diameter of inner cladding (face to face)	Mm	460.0 ± 10.0	460.0 ± 10.0
	Outer cladding diameter	Mm	530 ± 10.0	530 ± 10.0
	Concentricity of fiber core cladding	Mm	≤2.0	≤2.0
7	Diameter of coating layer	Mm	635.0 ± 15.0	635.0 ± 5.0
8	Screening test	Kpsi	≥100	≥100
9	Laser oscillation efficiency @ 971 nm pump absorption	/	/	≥80%
10	Coating accelerated aging red light power decay	/	≤2%	≤2%

4.5.4.1 Development Towards Higher Power

Driven by the demand in high-tech fields such as maritime and aerospace, as well as the widespread application of additive manufacturing technology, one of the main trends in fiber lasers is towards higher output power. The output power of fiber lasers is evolving from the hundred-watt and kilowatt levels towards the ten-thousand-watt level. Industrial equipment equipped with kilowatt to tens of kilowatt high-power fiber lasers will become the mainstream equipment in high-end manufacturing industries.

4.5.4.2 Development Towards High Average Power, High Peak Power Pulsed Fiber Lasers

In many applications, there is a need for pulsed fiber lasers that offer high average and peak power. By utilizing a laser with high beam quality and low power as the seed source, and employing a double-clad fiber as the amplifier, it becomes straightforward to achieve pulsed laser output with high average power and high pulse energy. This represents a current focus and challenge in the industry's development.

4.5.4.3 Developing Towards Ultra-Short Pulse Fiber Lasers

In the field of laser precision machining, which includes drilling brittle materials and cutting sapphire glass, ultrafast and ultrashort pulse fiber lasers are essential. Currently, medium to high power ultra-short pulse fiber lasers are a focal point of research and development.

4.5.4.4 Development Towards Higher Brightness

High-power fiber lasers with high beam quality are in strong demand in scientific research and military fields. Currently, developed countries abroad consider high beam quality high-power fiber lasers as the preferred light source for tactical laser weapons. Special military needs will drive fiber lasers to develop towards higher brightness while increasing power, that is, enhancing the output power while maintaining the beam quality of the fiber lasers.

4.5.4.5 Developing Towards Modularization and Intelligence

To adapt to the diverse needs of lasers in the market, fiber lasers will gradually move towards serialization, modularization, standardization, and generalization. By utilizing a limited range of specifications and varieties, and through the combination and pairing of different modules, the development cycle for new products can be shortened, and the stability and reliability of the products can be improved. At the same time, by adopting advanced communication technologies and design concepts, remote diagnosis, remote maintenance, remote control, and data statistics of fiber lasers can be realized. Through real-time monitoring of the operating status of fiber lasers, potential faults can be detected and addressed in advance.

Comprehensively, this chapter takes the main line of Yb-doped large-mode field cladding fibers and their applications. Firstly, it introduces the basic characteristics of double-cladding structure fiber, multi-cladding structure fiber, and other new structured Yb-doped large-mode field fibers, focusing on the structure and characteristics of Yb-doped large-mode field cladding fibers. Secondly, the primary factors influencing the performance of Yb-doped large-mode field fiber cladding are examined, including the core composition, nonlinear effects, photodarkening effect, mode instability effect, the reliability of the fiber coating layer, and the temperature increase in Yb-doped large-mode field fibers. The principles of optimized core and optical design for Yb-doped large-mode field cladding fibers are proposed. Examples include the waveguide design of a 20/400 μm Yb-doped double cladding fiber suitable for kilowatt-class fiber lasers, a 30/600 μm double cladding structure for 10,000-watt-class single-mode and multimode fiber lasers, and a three-cladding structure for Yb-doped large-mode field fibers. Finally, the significant applications of Yb-doped large mode area cladding fibers, such as narrow linewidth lasers, 1018 nm fiber lasers for co-band pumping, 980 nm fiber lasers, and fiber lasers for industrial applications, are briefly introduced.

In summary, ytterbium-doped large-mode-area (LMA) fibers constitute the core gain materials for high-power fiber lasers. In recent years, China has achieved rapid advancements in the field of high-power fiber laser technology, making significant breakthroughs in key areas. These advancements are inseparable from the breakthroughs in the material preparation technology of ytterbium-doped large-mode-area fibers. China has rapidly transitioned from heavily relying on imported ytterbium-doped fibers during the "12th Five-Year Plan" period to now having domestically produced ytterbium-doped fibers dominating over half of the market share. Regarding the power level of ytterbium-doped large-mode-area fiber applications, we have entered the era of ten-thousand-watt lasers. In particular, the ytterbium-doped large-mode-area triple-clad fiber developed by the Shanghai Institute of Optics and Fine Mechanics, which is utilized in single-module ten-thousand-watt industrial lasers, has achieved performance indicators that meet or even exceed international standards. This demonstrates that domestic understanding of high-power ytterbium-doped laser fibers has reached a new level. At the same time, we must also recognize that there is still a certain gap in the preparation technology of ytterbium-doped large-mode-area fibers for ten-thousand-watt single-mode output compared to international standards, and more effort is needed in the design of new waveguide structure fibers.

References

1. Snitzer E. Proposed fiber cavities for optical masers [J]. Journal of Applied Physics, 1961, 32(1): 36-9.
2. Etzel H, Gandy H, Ginther R. Stimulated emission of infrared radiation from ytterbium activated silicate glass [J]. Applied Optics, 1962, 1(4): 534-6.
3. Hanna D, Percival R, Perry I, et al. Continuous-wave oscillation of a monomode ytterbium-doped fibre laser [J]. Electronics Letters, 1988, 24(17): 1111-3.
4. Hua RZ, Yin HB. Yb-doped fiber laser pumped by titanium gemstone laser [J]. China Laser, 1999, 26(9): 781-4.
5. Paschotta R, Hanna D, De Natale P, et al. Power amplifier for 1083 nm using ytterbium doped fibre [J]. Optics communications, 1997, 136(3-4): 243-6.

6. Armitage J, Wyatt R, Ainslie B, et al. Efficient 980-nm Operation of a Yb3+-Doped Silica Fiber Laser; proceedings of the Advanced Solid State Lasers, F , 1989 [C]. Optical Society of America.
7. Chen B, Chen L R, Lin Zun Q, et al. LD pumping of a Yb-doped ring-cavity fiber laser operating at 1041 nm [J]. China Laser, 1999, 26(11): 965-8.
8. Chen B, Lin Zunqi. Study on the relationship between excitation wavelength and threshold value of Yb-doped fiber laser [J]. Journal of Optics, 2000, 20(6): 750-4.
9. Snitzer E, Po H, Hakimi F, et al. Double clad, offset core Nd fiber laser; proceedings of the Optical fiber sensors, F, 1988 [C]. Optical Society of America.
10. Paskh M, Archambault J, Hanna D. Operating of cladding pumped Yb3+ doped silica fibre lasers in 1∃ m region [J]. Electron Lett, 1994, 30(11): 863-5.
11. Dominic V A, Maccormack S, Waarts R, et al. 110 W fibre laser [J]. Electronics Letters, 1999, 35(14): 1158-60.
12. Jeong Y, Sahu J, Payne D, et al. Ytterbium-doped large-core fiber laser with 1 kW continuous-wave output power; proceedings of the Advanced Solid- State Photonics, F, 2004 [C]. Optical Society of America.
13. Chen, X. L., Lou, G., He, Y., et al. High efficiency fully localized 10 kW fiber laser [J]. Journal of Optics, 2019,(3): 423-5.
14. Lin Aoxiang, Ni L, Peng Kun, et al. Domestic YDF active fiber achieves single-fiber 20 kW laser output [J]. China Laser, 2021, 48(9): 1.
15. Ueda K-I. Optical cavity and future style of high-power fiber lasers; proceedings of the Laser Resonators, F, 1998 [C]. SPIE.
16. Liu A, Song J, Kamatani K, et al. Effective absorption and pump loss of double-clad fiber lasers, Solid State Lasers VI, Photonics West '97, 1997. Proceedings Volume 2986 [C].
17. Lou Q. Hong. High-power fiber lasers and their applications [M]. High Power Fiber Lasers and Their Applications, 2010.
18. Croteau A, Paré C, Zheng H, et al. Bending insensitive, highly Yb-Doped LMA triple-clad fiber for nearly diffraction-limited laser output [C]. Proceedings of SPIE - The International Society for Optical Engineering, 2006, 6101(61010G-G-10.
19. Sousa J, Okhotnikov O. Multimode Er-doped fiber for single-transverse-mode amplification [J]. Applied physics letters, 1999, 74(11): 1528-30.
20. Laperle P, Paré C, Zheng H, et al. Yb-doped LMA triple-clad fiber for power amplifiers, SPIE. Proceedings Volume 6453, Fiber Lasers IV: Technology. Systems, and Applications; 645308 (2007)[C].
21. Han Shuai, Chen Danping. A structure of multi-clad quartz optical fiber:China, CN201710359327.1 [P].
22. Hu L, Wang M, Zhang L, et al. Triple cladding ytterbium-doped quartz fiber and high concentration fluorine layer quartz tube sleeve rod method:China,CN201911382514.7 [P].
23. Liu S, Zhan H, Peng K, et al. Yb-doped aluminophosphosilicate triple-clad laser fiber with high efficiency and excellent laser stability [J]. IEEE Photonics Journal, 2019, 11(2): 1-10.
24. Meng Y, Tong W J, Zheng W, et al. Special active fiber for high power lasers, Proceedings Volume 11763, Seventh Symposium on Novel Photoelectronic Detection Technology and Applications; 117639R (2021) [C].
25. Beier F, Hupel C, Nold J, et al. Narrow linewidth, single mode 3 kW average power from a directly diode pumped ytterbium-doped low NA fiber amplifier [J]. Optics express, 2016, 24(6): 6011-20.
26. Petit V, Tumminelli R P, Minelly J D, et al. Extremely low NA Yb doped preforms (< 0.03) fabricated by MCVD; proceedings of the Fiber Lasers XIII. Technology, Systems, and Applications, F, 2016 [C]. International Society for Optics and Photonics.
27. Zhang HW, Zhou P, Wang S, et al. Single-frequency limiting output power of Yb3+ doped fiber with different substrates [J]. Journal of Optics, 2014, 1): 140-5.
28. Xu W, Yu C, Wang S, et al. Effects of F- on the optical and spectroscopic properties of Yb3+/Al3+-co-doped silica glass [J]. Optical Materials, 2015, 42(245-50.

29. Huang Z. H., Cao J. Q., Chen J. B. Advances in high power GTWave fiber lasers [J]. China Laser, 2021, 48(4): 0401010.
30. Norman S, Zervas M N, Appleyard A, et al. Latest development of high-power fiber lasers in SPI; proceedings of the Fiber Lasers: Technology, Systems. and Applications, F, 2004 [C]. SPIE.
31. Zimer H, Kozak M, Liem A, et al. Fibers and fiber-optic components for high-power fiber lasers [J]. Fiber Lasers VIII: Technology, Systems, and Applications, 2011, 7914(236-52.
32. Stiles E. New developments in IPG fiber laser technology; proceedings of the Proceedings of the 5th International Workshop on Fiber Lasers, F, 2009 [C].
33. Yu Y, Huang Z. He, Cao J. Q., et al. Experimental study of bi-directional distributed laterally pumped single-mode fiber amplifier [J]. Advances in Lasers and Optoelectronics, 2016, 53(7): 108-12.
34. Huang Z, Cao J, An Y, et al. A kilowatt all-fiber cascaded amplifier [J]. IEEE Photonics Technology Letters, 2015, 27(16): 1683-6.
35. Chen, J. B., Cao, J. Q., Pan, C. Y. Multi-stage cascaded distributed lateral pumping fiber oscillator based on domestic fiber to achieve 2 kW power output [J]. China Laser, 2017, 44(4): 264-.
36. Chen, J.B., Cao, J.Q., Huang, Z.H., et al. Multi-stage cascaded distributed lateral pumping fiber oscillator based on domestic fiber for strong Raman suppression with 3kW-scale power output [J]. China Laser, 2018, 45(3): 330-.
37. Chen H, Cao J, Huang Z, et al. Experimental investigations on multi-kilowatt all-fiber distributed side-pumped oscillators [J]. Laser Physics, 2019, 29(7): 075103.
38. Zhan H, Wang Y, Peng K, et al. 2 kW (2+ 1) GT-wave fiber amplifier [J]. Laser Physics Letters, 2016, 13(4): 045103.
39. Zhan H, Liu Q, Wang Y, et al. 5kW GTWave fiber amplifier directly pumped by commercial 976nm laser diodes [J]. Optics express, 2016, 24(24): 27087-95.
40. Zhan H, Peng K, Wang Y, et al. 6kW GTWave fiber amplifier; proceedings of the Asia Communications and Photonics Conference, F, 2017 [C]. Optical Society of America.
41. Zhan H, Peng K, Liu S, et al. Pump-gain integrated functional laser fiber towards 10 kW-level high-power applications [J]. Laser Physics Letters, 2018, 15(9): 095107.
42. Lin Ao-Xiang, Zhan Huan, Peng Kun, et al. Domestic composite functional fiber achieves 10,000 watts laser output [J]. Intense Lasers and Particle Beams, 2018, v.30; No.252(06): 7.
43. Engheta N, Pelet P. Modes in chirowaveguides [J]. Optics Letters, 1989, 14(11): 593-5.
44. Swan M C, Liu C-H, Guertin D, et al. 33μm Core Effectively Single-Mode Chirally-Coupled-Core Fiber Laser at 1064-nm; proceedings of the Optical Fiber Communication Conference, F, 2008 [C]. Optical Society of America.
45. Zhao N, Li JY. Introduction and research progress of chiral coupled fiber cores [J]. Advances in Laser and Optoelectronics, 2014, 51(4): 16-9.
46. Liu C-H, Chang G, Litchinitser N, et al. Effectively single-mode chirally-coupled core fiber; proceedings of the Advanced Solid-State Photonics. F, 2007 [C]. Optical Society of America.
47. Liu C-H, Chang G, Litchinitser N, et al. Chirally coupled core fibers at 1550-nm and 1064-nm for effectively single-mode core size scaling. proceedings of the 2007 Conference on Lasers and Electro-Optics (CLEO), F, 2007 [C]. IEEE.
48. Ma X, Liu C-H, Chang G, et al. Angular-momentum coupled optical waves in chirally-coupled-core fibers [J]. Optics express, 2011, 19(27): 26515-28.
49. Zhu C, Hu I-N, Ma X, et al. Single-frequency and single-transverse mode Yb-doped CCC fiber MOPA with robust polarization SBS-free 511W output. Proceedings of the Advanced Solid-State Photonics, F, 2011 [C]. Optical Society of America.
50. Sosnowski T, Kuznetsov A, Maynard R, et al. 3C Yb-doped fiber based high energy and power pulsed fiber lasers; proceedings of the Fiber Lasers X. Technology, Systems, and Applications, F, 2013 [C]. International Society for Optics and Photonics.

51. Želudevičius J, Danilevičius R, Viskontas K, et al. Femtosecond fiber CPA system based on picosecond master oscillator and power amplifier with CCC fiber [J]. Optics Express, 2013, 21(5): 5338-45.
52. Mccomb T S, Mccal D, Farrow R, et al. High-peak power, flexible-pulse parameter, chirally coupled core (3C) fiber-based picosecond MOPA systems. Proceedings of the Fiber Lasers XI: Technology, Systems, and Applications, F, 2014 [C]. International Society for Optics and Photonics.
53. Hu I-N, Ma X, Zhu C, et al. Experimental demonstration of SRS suppression in chirally-coupled-core fibers; proceedings of the Advanced Solid-State Photonics, F, 2012 [C]. Optical Society of America.
54. Zhang H, Hao H, He L, et al. Single-mode annular chirally-coupled core fibers for fiber lasers [J]. Optics Communications, 2018, 410(297-304.
55. Filippov V, Chamorovskii Y, Kerttula J, et al. Double clad tapered fiber for high power applications [J]. Optics express, 2008, 16(3): 1929-44.
56. Bobkov K, Andrianov A, Koptev M, et al. Sub-MW peak power diffraction-limited chirped-pulse monolithic Yb-doped tapered fiber amplifier [J]. Optics Express, 2017, 25(22): 26958-72.
57. Bobkov K K, Aleshkina S S, Khudyakov M M, et al. Active tapered fibers for high peak power fiber lasers; proceedings of the Micro-structured and Specialty Optical Fibres VII, F, 2021 [C]. International Society for Optics and Photonics.
58. Bobkov K, Levchenko A, Kashaykina T, et al. Scaling of average power in sub-MW peak power Yb-doped tapered fiber picosecond pulse amplifiers [J]. Optics Express, 2021, 29(2): 1722-35.
59. Zhu Y, Leich M, Lorenz M, et al. Yb-doped large mode area fiber for beam quality improvement using local adiabatic tapers with reduced dopant diffusion [J]. Optics Express, 2018, 26(13): 17034-43.
60. Ye Y, Lin X, Xi X, et al. Novel constant-cladding tapered-core ytterbium-doped fiber for high-power fiber laser oscillator [J]. High Power Laser Science and Engineering, 2021, 9(
61. Lin X, Zhang Z, Chu Y, et al. Fabrication and laser performance of cladding uniform core tapered fiber [J]. Optical Fiber Technology, 2021, 64(102561.
62. Jain D, Baskiotis C, Sahu J K. Mode area scaling with multi-trench rod-type fibers [J]. Optics express, 2013, 21(2): 1448-55.
63. Jain D, Baskiotis C, Sahu J K. Bending performance of large mode area multi-trench fibers [J]. Optics express, 2013, 21(22): 26663-70.
64. Jain D, Jung Y, Nunez-Velazquez M, et al. Extending single mode performance of all-solid large-mode-area single trench fiber [J]. Optics Express, 2014, 22(25): 31078-91.
65. Jain D, Baskiotis C, May-Smith T C, et al. Large Mode Area Multi-Trench Fiber With Delocalization of Higher Order Modes [J]. IEEE Journal of Selected Topics in Quantum Electronics, 2014, 20(5): 242-50.
66. Jain D, Jung Y, Kim J, et al. Robust single-mode all-solid multi-trench fiber with large effective mode area [J]. Opt Lett, 2014, 39(17): 5200-3.
67. Jain D, Alam S, Codemard C, et al. High power, compact, picosecond MOPA based on single trench fiber with single polarized diffraction-limited output [J]. Opt Lett, 2015, 40(17): 4150-3.
68. Kirchhof J, Unger S, Schwuchow A, et al. Dopant interactions in high-power laser fibers; proceedings of the Optical Components and Materials II, F. 2005 [C]. International Society for Optics and Photonics.
69. Manek-Hönninger I, Boullet J, Cardinal T, et al. Photodarkening and photobleaching of an ytterbium-doped silica double-clad LMA fiber [J]. Optics Express, 2007, 15(4): 1606-11.
70. Jauregui C, Limpert J, Tünnermann A. High-power fibre lasers [J]. Nature Photonics, 2013, 7(11): 861-7.
71. Yuanyuan, Fan, Bing, et al. Thermal effects in kilowatt all-fiber MOPA [J]. Optics express, 2011, 19(16): 15162-72.

72. Grukh D A, Kurkov A S, Paramonov V M, et al. Effect of heating on the optical properties of Yb3+-doped fibres and fibre lasers [J]. 2004, Quantum Electron. 34:579.
73. Gao Xuesong, Gao Chunqing, Lin Zhifeng, et al. Numerical analysis of temperature distribution in high-power double-clad fiber lasers [J]. Journal of Beijing University of Technology, 2005, (11): 61-5.
74. Hansen K R, Alkeskjold T T, Broeng J, et al. Thermo-optical effects in high-power Ytterbium-doped fiber amplifiers [J]. Optics Express, 2011, 19(24): 23965-80.
75. Daniel J M O, Simakov N, Hemming A, et al. A double clad ytterbium fibre laser operating at 400°C [C]. SPIE, 2015.
76. Cheng Y, Yang Q, Yu C, et al. Temperature dependence of the spectral properties of Yb^{3+} /P^{5+} /Al^{3+} co-doped silica fiber core glasses [J]. Optical Materials Express,2021,11(8): 2459-2467
77. Lai WC, Ma PF, Xiao H, et al. High power narrow linewidth fiber laser technology [J]. Intense Lasers and Particle Beams, 2020, 32(12): 121001.
78. Ran Y, Wang S, Su R, et al. Advances in the suppression of excited Brillouin scattering in narrow linewidth fiber amplifiers [J]. Advances in Lasers and Optoelectronics, 2015, 52(4): 040003.
79. Wang, J., Liu, M., Li, M., et al. Ten-year review and prospect of fiber laser mode instability research [J]. Intense Lasers and Particle Beams, 2020, 32(12): 121003.
80. Jeong Y, Nilsson J, Sahu J K, et al. Single-frequency, single-mode, plane-polarized ytterbium-doped fiber master oscillator power amplifier source with 264 W of output power [J]. Opt Lett, 2005, 30(5): 459-61.
81. Jeong Y, Nilsson J, Sahu J K, et al. Power Scaling of Single-Frequency Ytterbium-Doped Fiber Master-Oscillator Power-Amplifier Sources up to 500 W [J]. IEEE Journal of Selected Topics in Quantum Electronics, 2007, 13(3): 546-51.
82. Huang Z-M, Shu Q, Tao R-M, et al. >5kW Record High Power Narrow Linewidth Laser From Traditional Step-Index Monolithic Fiber Amplifier [J]. IEEE Photonics Technology Letters, 2021, 33(21): 1181-4.
83. Ma P, Xiao H, Liu W, et al. All-fiberized and narrow-linewidth 5 kW power-level fiber amplifier based on bidirectional pumping configuration [J]. High Power Laser Science and Engineering, 2021, 1-20.
84. Ma P.F., Song J.X., Wang G.J., et al. High power narrow linewidth fiber laser breakthrough in 6 kW class with near single mode output [J]. Chinese Journal of Lasers, 2022, 49(9): 0916002.
85. Ward B G. Maximizing power output from continuous-wave single-frequency fiber amplifiers [J]. Opt Lett, 2015, 40(4): 542-5.
86. Webb A S, Boyland A J, Standish R J, et al. MCVD in-situ solution doping process for the fabrication of complex design large core rare-earth doped fibers [J]. Journal of Non-Crystalline Solids, 2010, 356(18-19): 848-51.
87. Jeong Y E, Sahu J, Payne D A, et al. Ytterbium-doped large-core fiber laser with 1.36 kW continuous-wave output power [J]. Optics express, 2004, 12(25): 6088-92.
88. Jen C-K, Neron C, Shang A, et al. Acoustic Characterization of Silica Glasses [J]. Journal of the American Ceramic Society, 1993, 76(3): 712-6.
89. Ballato J, Hawkins T, Foy P, et al. On the fabrication of all-glass optical fibers from crystals [J]. Journal of Applied Physics, 2009, 105:053110
90. Dragic P, Law P C, Ballato J, et al. Brillouin spectroscopy of YAG-derived optical fibers [J]. Optics Express, 2010, 18(10): 10055-67.
91. Dragic P D, Liu Y-S, Ballato J, et al. YAG-derived fiber for high-power narrow-linewidth fiber lasers [C]. Fiber Lasers IX: Technology, Systems, and Applications. 2012.
92. Dong L. Stimulated thermal Rayleigh scattering in optical fibers [J]. Optics Express, 2013, 21(3): 2642-56.
93. Zervas M N. Transverse mode instability, thermal lensing and power scaling in Yb 3+-doped high-power fiber amplifiers [J]. Optics Express, 2019, 27(13): 19019-41.

94. Ballato J, Cavillon M, Dragic P. A unified materials approach to mitigating optical nonlinearities in optical fibers. i. Thermodynamics of optical scattering [J]. International Journal of Applied Glass Science, 2018, 9(2): 263-77.
95. Yu N, Hawkins T W, Bui T-V, et al. AlPO 4 in silica glass optical fibers: Deduction of additional material properties [J]. IEEE Photonics Journal, 2019, 11(5): 1-13.
96. Hawkins T, Dragic P, Yu N, et al. Kilowatt power scaling of an intrinsically low Brillouin and thermo-optic Yb-doped silica fiber [J]. JOSA B, 2021, 38(12): F38-F49.
97. Hauge J M, Papior S R, Pedersen J E, et al. Narrow-linewidth all-solid large-mode-area photonic crystal fiber amplifier; proceedings of the Fiber Lasers XVI: Technology and Systems, F, 2019 [C]. SPIE.
98. Pulford B, Ehrenreich T, Holten R, et al. 400-W near diffraction-limited single-frequency all-solid photonic bandgap fiber amplifier [J]. Optics Letters, 2015, 40(10): 2297-300.
99. Pulford B, Holten R, Matniyaz T, et al. kW-level monolithic single-mode narrow-linewidth all-solid photonic bandgap fiber amplifier [J]. Opt Lett, 2021, 46(18): 4458-4461
100. Matniyaz T, Bingham S P, Kalichevsky-Dong M T, et al. High-power single-frequency single-mode all-solid photonic bandgap fiber laser with kHz linewidth [J]. Optics Letters, 2022, 47(2): 377-80.
101. Hochheim S, Brockmüller E, Wessels P, et al. Single-frequency 336 W spliceless all-fiber amplifier based on a chirally-coupled-core fiber for the next generation of gravitational wave detectors [J]. Journal of Lightwave Technology, 2022, 40(7): 2136-43.
102. Shiraki K, Ohashi M, Tateda M. SBS threshold of a fiber with a Brillouin frequency shift distribution [J]. Journal of Lightwave Technology, 1996, 14(1): 50-7.
103. Hansryd J, Dross F, Westlund M, et al. Increase of the SBS threshold in a short highly nonlinear fiber by applying a temperature distribution [J]. Journal of lightwave technology, 2001, 19(11): 1691.
104. Wellmann F, Steinke M, Wessels P, et al. Performance study of a high-power single-frequency fiber amplifier architecture for gravitational wave detectors [J]. Applied Optics, 2020, 59(26): 7945-50.
105. Yoshizawa N, Imai T. Stimulated Brillouin scattering suppression by means of applying strain distribution to fiber with cabling [J]. Journal of Lightwave Technology, 1993, 11(10): 1518-22.
106. Zhang L, Cui S, Liu C, et al. 170 W, single-frequency, single-mode, linearly-polarized, Yb-doped all-fiber amplifier [J]. Optics Express, 2013, 21(5): 5456-62.
107. Huang L, Wu H, Li R, et al. 414 W near-diffraction-limited all-fiberized single-frequency polarization-maintained fiber amplifier [J]. Optics Letters, 2017, 42(1): 1-4.
108. Shiraki K, Ohashi M, Tateda M. Suppression of stimulated Brillouin scattering in a fibre by changing the core radius [J]. Electronics letters, 1995, 31(8): 668-9.
109. Huang L, Ma P, Su R, et al. Comprehensive investigation on the power scaling of a tapered Yb-doped fiber-based monolithic linearly polarized high- peak-power near-transform-limited nanosecond fiber laser [J]. Optics Express, 2021, 29(2): 761-82.
110. Ye Y, Lin X, Yang B, et al. Tapered Yb-doped fiber enabled a 4 kW near-single-mode monolithic fiber amplifier [J]. Optics Letters, 2022, 47(9): 2162-5.
111. Lai W, Ma P, Liu W, et al. 550 W single frequency fiber amplifiers emitting at 1030 nm based on a tapered Yb-doped fiber [J]. Optics Express, 2020, 28(14): 20908-19.
112. Li M-J, Chen X, Wang J, et al. Al/Ge co-doped large mode area fiber with high SBS threshold [J]. Optics Express, 2007, 15(13): 8290-9.
113. Gray S, Liu A, Walton D T, et al. 502 Watt, single transverse mode, narrow linewidth, bidirectionally pumped Yb-doped fiber amplifier [J]. Optics express, 2007, 15(25): 17044-50.
114. Robin C, Dajani I, Pulford B. Modal instability-suppressing, single-frequency photonic crystal fiber amplifier with 811 W output power [J]. Optics Letters, 2014, 39(3):666-669
115. Marciante J R. Gain Filtering for Single-Spatial-Mode Operation of Large-Mode-Area Fiber Amplifiers [J]. IEEE Journal of Selected Topics in Quantum Electronics, 2009, 15(1): 30-6.

116. John R. Marciante R G R, Vladimir V. Shkunov, Rockweli D A. Near-diffraction-limited operation of step-index large-mode-area fiber lasers via gain filtering [J]. Optics Letters, 2010, 35(11): 1828-30.
117. S. Ikoma, H. K. Nguyen, M. Kashiwagi, et al. 3 kW single stage all-fiber Yb-doped single-mode fiber laser for highly reflective and highly thermal conductive materials processing [C]. Proceedings Volume 10083, Fiber Lasers XIV: Technology and Systems; 100830Y (2017)
118. Ikoma S, Uchiyama K, Takubo Y, et al. 5-kW single stage all-fiber Yb-doped single-mode fiber laser for materials processing; proceedings of the Fiber Lasers XV: Technology and Systems, F, 2018 [C].
119. Wang Y, Kitahara R, Kiyoyama W, et al. 8-kW single-stage all-fiber Yb-doped fiber laser with a BPP of 0.50 mm-mrad, Proceedings Volume 11260, Fiber Lasers XVII: Technology and Systems; 1126022 (2020) [C].
120. Fang, Fang, Zhang, Yibo, et al. Gain-tailored Yb/Ce codoped aluminosilicate fiber for laser stability improvement at high output power [J]. Optics express, 2019, 27(15): 20824-36.
121. Huang Z, Shu Q, Luo Y, et al. 3.5 kW narrow-linewidth monolithic fiber amplifier at 1064 nm by employing a confined doping fiber [J]. JOSA B, 2021, 38(10): 2945-52.
122. Wu H, Li R, Xiao H, et al. High-power tandem-pumped fiber amplifier with beam quality maintenance enabled by the confined-doped fiber [J]. Optics Express, 2021, 29(20): 31337-47.
123. Wu Henshuo, Song Jiaxin, Ma Pengfei, et al. High beam quality 6 kW class narrow linewidth fiber laser [J]. Infrared and Laser Engineering, 2022, 51(3):20210859-1.
124. Cooper M A, Gausmann S, Antonio-Lopez J E, et al. Confined doping LMA fibers for high power single frequency lasers, Proceedings Volume 11981, Fiber Lasers XIX: Technology and Systems; 1198106 (2022)[C].
125. Tsvetkov S V, Khudyakov M M, Lobanov A S, et al. SBS gain suppression in a passive single-mode optical fiber by the multi-mode acoustic waveguide design [J]. Journal of Lightwave Technology, 2020, 39(2): 592-9.
126. Wang Y, Ke W, Sun Y, et al. Research of high brightness 1018nm ytterbium doped fiber laser; proceedings of the XX International Symposium on High- Power Laser Systems and Applications 2014, F, 2015 [C]. International Society for Optics and Photonics.
127. Gu G, Liu Z, Kong F, et al. Efficient ytterbium-doped phosphosilicate double-clad leakage -channel-fiber laser at 1008-1020 nm; proceedings of the Fiber Lasers XIII: Technology, Systems, and Applications, F, 2016 [C]. SPIE.
128. Liu Ying, Cao Jianqiu, Xiao Hu, et al. Current status and prospects of 980 nm fiber laser development [J]. Advances in Laser and Optoelectronics, 2012, 49(8): 5.

Chapter 5
Ytterbium-Doped Large Mode Area Silica Photonic Crystal Fiber and Its Applications

With the application development of ultrafast lasers in the field of precision processing, how to overcome nonlinear effects and further improve the performance of ultrafast lasers has become a research hotspot. The larger the mode area of the fiber, the higher its nonlinear threshold [1]. Therefore, using large mode area fibers is one of the most direct and effective ways to overcome nonlinear effects [2–5]. Compared with traditional double-clad fibers, photonic crystal fibers can possess many unique optical properties such as endless single-mode operation, adjustable dispersion, and large mode area [6, 7], which can overcome many difficulties faced by traditional fiber lasers. Foreign research institutions and companies such as the University of Bath, the University of Jena, and NKT Photonics have conducted extensive researches on the design, preparation, and application of key technologies for large mode area photonic crystal fibers [8, 9]. In 2000, the Russell group at the University of Bath [10] first achieved laser output in Yb^{3+} doped photonic crystal fiber. Currently, the maximum core diameter of Yb^{3+} doped large mode area photonic crystal fiber has reached 135 μm [11]. Chinese companies and institutions such as Wuhan Fenghuo, Huazhong University of Science and Technology, Yanshan University, Tianjin University, Nankai University, 46th Institute of Chinese Electronics, and Shanghai Institute of Optics and Mechanics have successively carried out related researches on large mode area photonic crystal fiber, achieving certain results.

L. Hu, *Rare Earth Doped Silica Fiber and Its Applications*,
https://doi.org/10.1007/978-981-96-5306-5_5

5.1 Classification and Research Progress of Ytterbium-Doped Large Mode Area Silica Photonic Crystal Fiber

Amplifying pulsed light with rare earth-doped large mode area fiber is an important way to achieve high peak power and large pulse energy. The fiber mode area can be increased by enlarging the fiber core diameter and reducing the numerical aperture (NA) of the fiber core. However, in double-clad step-index fiber, too low numerical aperture of the fiber core will cause a sharp increase in bending loss, increasing the difficulty of use. Therefore, for double-clad step-index fiber, to ensure good mode characteristics, the fiber core diameter is generally less than 35 μm. Large mode area photonic crystal fiber can achieve high beam quality laser output with a large fiber core diameter through special structural design.

5.1.1 Classification of Ytterbium-Doped Large Mode Area Photonic Crystal Fiber

Since the introduction of the first ytterbium-doped silica photonic crystal fiber in 2000, researchers have successively developed various types of ytterbium-doped large mode area silica photonic crystal fiber to meet specific application requirements. Based on their guiding mechanisms, they can be divided into index-guidance photonic crystal fiber (IG-PCF) and photonic bandgap fiber (PBGF).

5.1.1.1 Index-Guidance Photonic Crystal Fiber

In the Ytterbium-doped index-guidance silica photonic crystal fiber, the fiber core is ytterbium-doped silica, and the inner cladding is the silica which refractive index controlled by air holes or different refractive index media. The refractive index of the fiber core must exceed the equivalent refractive index of the cladding. This type of fiber does not require strict periodicity of the cladding structure. Currently, index-guidance photonic crystal fibers mainly include equivalent filling photonic crystal fiber, large pitch photonic crystal fiber, Leakage channel photonic crystal fiber, etc.

Equivalent Filling Photonic Crystal Fiber

Equivalent filling photonic crystal fibers are typically composed of rare earth-doped fiber cores and periodically arranged air holes. Figure 5.1 shows the structure of a typical rare earth-doped equivalent filling silica photonic crystal fiber, where the fiber core is doped with the rare earth ions, the hexagonally packed air holes serve as the inner cladding, and the large air holes form the outer cladding, which can

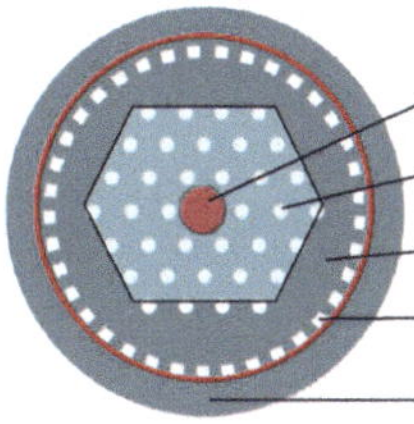

Fig. 5.1 Equivalent filling large mode area photonic crystal fiber structure

achieve a large cladding numerical aperture. The effective refractive index of the fiber cladding changes with wavelength. Single-mode transmission occurs when the effective refractive index of the fundamental mode of fiber core is greater than the cladding refractive index, while the refractive index of the higher-order modes is lower than the cladding refractive index.

The single-mode transmission condition of the equivalent filling photonic crystal fiber can be represented by the normalized frequency (V):

$$V = \frac{2\pi a \times NA}{\lambda} \tag{5.1}$$

$$\mathrm{NA} = \sqrt{n_{core}^2 - n_{eff}^2} \tag{5.2}$$

In the formula, a is the radius of the photonic crystal fiber core; λ is the laser wavelength; NA is the numerical aperture of the rare earth-doped fiber core; n_{core} is the refractive index of the fiber core; n_{eff} is the effective refractive index of the inner cladding.

Similar to traditional fibers, when $V < 2.405$, the core laser can maintain single-mode transmission. However, in traditional double-clad fibers, due to the limitations of the manufacturing process, it is difficult to control the refractive index difference between the fiber core and the inner cladding within the order of 5×10^{-4}, i.e., NA<0.038. Consequently, the single-mode core diameter cannot exceed 21 μm at the 1 μm wavelength band. Even with advanced laser technologies, quasi-single-mode 1.08 μm laser output can only be achieved in double-clad fibers with a 30 μm core [12]. In contrast, the effective refractive index of the cladding in photonic crystal fibers can be precisely controlled by adjusting the diameter d of the air holes and the pitch Λ between adjacent air holes, thus achieving a very small core NA. If the core NA is controlled to about 0.01, single-mode laser output can be achieved in ultra-large mode area fibers with a core diameter of 80 μm. However photonic crystal fiber with such ultra-low core NA is very sensitive to external disturbances such as bending and vibration, so it needs to be made into a rod-shaped fiber with a larger outer diameter to overcome these problems.

In 2000, the Russell group at the University of Bath [10] first achieved 14 mW laser output in a Yb^{3+} doped equivalent filling photonic crystal fiber, as shown in Fig. 5.2. In 2003, Limpert et al. from the University of Jena [8] first achieved

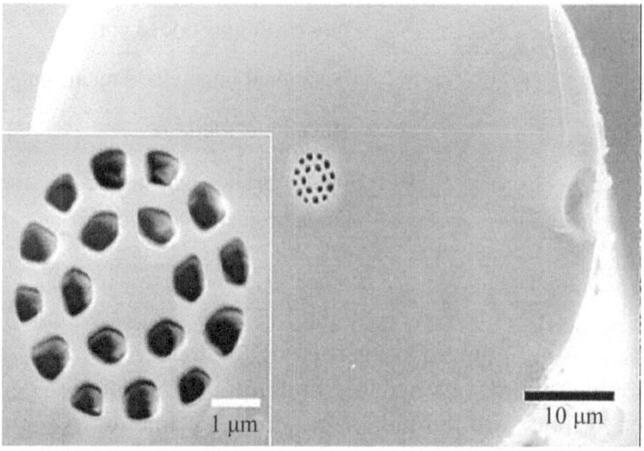

Fig. 5.2 The end face of the first Yb^{3+} doped photonic crystal fiber [10]

high-power laser output in a Yb^{3+} doped equivalent filling photonic crystal fiber with a core diameter of 28 µm. Using a 2.3 m long fiber, they achieved 80 W of laser output, with a slope efficiency of 78%, and a beam quality factor M^2 of about 1.2. In 2006, Limpert et al. [13] used a 0.58 m long Yb^{3+} doped equivalent filling photonic crystal fiber to achieve 320 W of continuous laser output, with a core diameter of 60 µm, corresponding to a mode area of about 2000µm^2. In 2009, the Limpert group [14] used an 80 µm core diameter Yb^{3+} doped equivalent filling photonic crystal fiber and chirped pulse amplification (CPA) technology to achieve 100 W of average power laser output, with a peak power of 1GW.

Large Pitch Photonic Crystal Fiber

To obtain a larger mode field diameter, a research group from the University of Jena [15] first proposed a large pitch photonic crystal fiber (LPF), as shown in Fig. 5.3. The large pitch photonic crystal fiber utilizes the air hole structure to differentiate between the fundamental mode and higher-order modes through the delocalization effect on the higher-order modes. In the study of modes in active fibers, many scholars [16, 17] used the size of the overlap between the mode area and the core area to study the gain of the mode. The percentage of this overlap relative to the total mode area is defined as the overlap factor (OF).

The calculation formula for the overlap factor is as follows:

$$\mathrm{OF} = \frac{\int IdA_{core}}{\int IdA_{clad}} \tag{5.3}$$

In the formula, dA_{core} is the core area element; dA_{clad} is the cladding area element; I is the electromagnetic field energy density of each order mode. Therefore, the numerator in formula (5.3) calculates the energy occupied by each order mode in the fiber core,

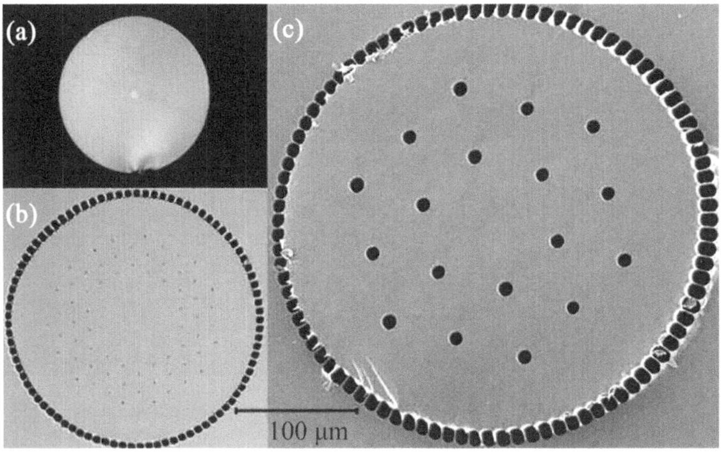

Fig. 5.3 End face diagram of large pitch photonic crystal fiber proposed by the University of Jena [15]

and the denominator calculates the energy occupied by each order mode in the cladding structure. In large pitch photonic crystal fibers, by designing the diameter and pitch of air holes, the overlap area between each mode and the core can be effectively adjusted, making the overlap area of the fundamental mode very high and the overlap area of the high-order mode very low, thereby promoting the fundamental mode in the fiber core. The gain obtained is much greater than that of the high-order mode, realizing the single-mode laser output of the super-large mode photonic crystal fiber.

In 2011, Eidam [15] reported the use of large pitch photonic crystal fiber with a core diameter of 108 μm, and achieved a peak power output of 3.8GW through femtosecond chirped pulse amplification, with a pulse width of 480 fs, an average power of 11 W, and a single pulse energy of 2.2 mJ. In 2012, Stutzki from the University of Jena [11] reported that a pulsed laser output with an average power of 130 W, a pulse width of less than 60 ns, a single pulse energy of 26 mJ, and a beam quality factor M^2 less than 1.3 was obtained using a large pitch photonic crystal fiber with a core diameter of 135 μm. In 2014, Otto et al. from the University of Jena [18] reported that a pulsed laser output of 2 kW was obtained in a main oscillator power amplifier (MOPA) system using a large pitch photonic crystal fiber with a core diameter of 90 μm, with a single pulse energy of 100 μJ, a peak power of 0.8 MW, and a beam quality factor M^2 less than 3.

Leakage Channel Photonic Crystal Fiber

In order to reduce the bending loss of photonic crystal fiber, Dong et al. [19] proposed a new type of leakage channel photonic crystal fiber (LCF) in 2005. Leakage channel fibers are generally formed by 1–2 layers of large-diameter air holes arranged periodically. By designing suitable air hole diameters and hole pitch, it is

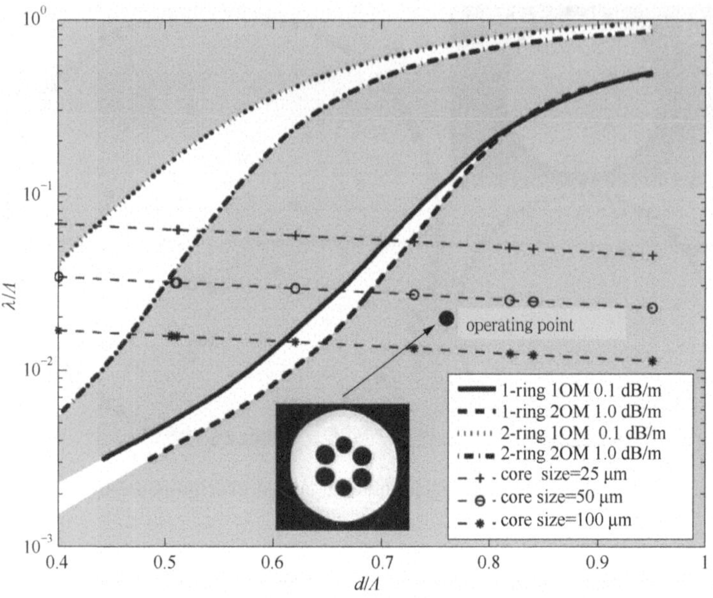

Fig. 5.4 Comparison of fundamental mode and higher-order mode limiting losses of leakage channel fiber

possible to achieve very small leakage loss for the fundamental mode, generally less than 0.1 dB/m, while the leakage loss of the higher-order mode is relatively much larger, generally greater than 1 dB/m. Thus the higher-order modes generated in the fiber core are gradually lost after transmitting a certain distance, and only the fundamental mode is retained, so the leakage channel fiber can have good bending resistance. Figure 5.4 is a comparison of the fundamental mode and higher-order mode confinement losses of the leakage channel fiber designed by Dong et al. [19] In 2007, Dong et al. [20] reported a doped core area 40 μm air hole leakage channel fiber, achieving a laser output of 4.3 W, with a factor M^2 approximately 1.3. In 2009, Dong et al. [21] reported an all-solid-state Yb-doped leakage channel photonic crystal fiber, achieving a laser output of 98 W using a non-polarized Yb-doped leakage channel fiber with a length of 4.8 m and a core diameter of 46 μm.

5.1.1.2 Photonic Bandgap Fiber

Unlike refractive index guiding photonic crystal fibers, the core refractive index of photonic bandgap fibers is lower than the cladding refractive index. Its guiding mechanism is through the bandgap effect of the two-dimensional photonic crystal structure, that is, when the cladding structure pitch meets certain conditions, the light wave in the design frequency range is prohibited from propagating in the

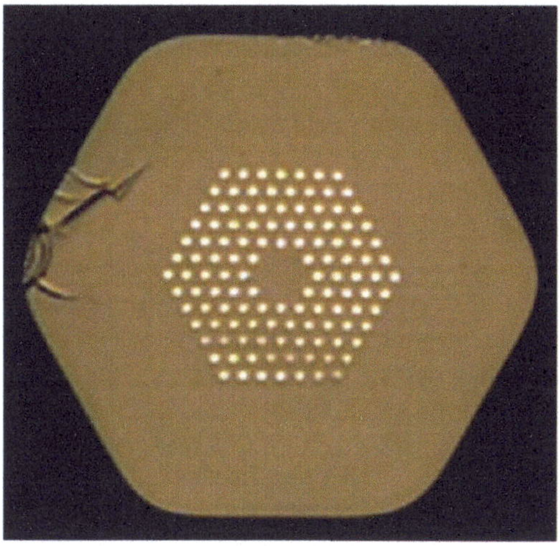

Fig. 5.5 Yb-doped photonic bandgap fiber [22]

cladding direction and only transmitted in the defect core. The core of the photonic bandgap fiber can be air or solid material, and the structural periodicity of the cladding is relatively strict. In the ytterbium-doped photonic bandgap fiber, the core is ytterbium-doped silica, and the cladding is formed by periodically arranging high refractive index rods in the low refractive index base glass, as shown in Fig. 5.5.

In 2006, Isomäki [23] first reported ytterbium-doped photonic bandgap fibers. In 2011, Laurila obtained a near-diffraction limit of 110 W laser output in a single-mode ytterbium-doped photonic bandgap fiber with a mode diameter of 59 μm. In 2015, Pulford [24] achieved a laser output of 400 W using a 50 μm core diameter Yb-doped photonic bandgap fiber, and the M^2 factor is 1.2–1.3. In 2017, Kong [25] reported a 60 μm core diameter photonic bandgap fiber, using an 8 m long fiber to achieve a laser output of 1050 W.

5.1.2 *Research Progress of Ytterbium-Doped Large Mode Area Photonic Crystal Fiber*

Compared with traditional double-clad fibers, Yb^{3+} doped photonic crystal fibers have a larger mode area, which, on the one hand, can significantly increase the nonlinear effect threshold of the fiber, making it conducive to obtaining large pulse energy output; on the other hand, due to the scattering effect of the cladding structure on the pump light, can reduce the proportion of spiral light in the fiber cladding, and improve the absorption efficiency of the pump light, allowing reduction of nonlinear effects in shorter fibers.

In 2000, Wadsworth from the Russell group [10] first achieved a laser output of 14 mW in Yb^{3+} doped photonic crystal fiber, marking the advent of the first photonic crystal fiber laser. The following year they [26] used a 915 nm titanium sapphire laser for pumping to obtain a 315 mW single-mode laser with an output wavelength of 1040 nm. In 2003, the group [27] obtained a laser output of 3.9 W in a photonic crystal fiber with a core diameter of 15 μm, and a slope efficiency was 30%.

In 2003, Limpert [8] first achieved a higher power laser output in a Yb^{3+} doped photonic crystal fiber with a core diameter of 28 μm (mode area of about 350μm^2), using a 2.3 m long fiber to obtain a laser output of 80 W, with a slope efficiency of 78%, and a beam quality factor M^2 of about 1.2.

In 2005, Bonati [28] achieved a laser output of 1.53 kW using a large mode area photonic crystal fiber with a core diameter of 50 μm (mode area is about 1589.6 μm^2). The beam quality factor M^2 was less than 3, and it was expected to achieve a laser output of 4.6 kW after increasing the pump power. This was the first time that the output power of a large mode area photonic crystal fiber had broken through the kilowatt level.

In 2007, Teodoro [29] amplified a laser with a pulse width of 1 ns in a rod-shaped photonic crystal fiber with a core diameter of 100 μm in the Master Oscillator Power Amplifier (MOPA) system, achieving a laser output of 4.3 mJ pulse energy, 4.5 MW peak power, and 42 W average power. The beam quality factor M^2 was about 1.3.

In 2008, Hansen [30] used a rod-shaped polarization-maintaining photonic crystal fiber with a core diameter of 70 μm to achieve efficient coupling pumping through an air cladding combiner, achieving a single-mode laser output of 350 W. The beam quality factor M^2 was less than 1.1.

In 2009, the Limpert research group [14] used a rod-shaped large mode area photonic crystal fiber with a core diameter of 80 μm and a mode area of 4000 μm^2 and used Chirped Pulse Amplification (CPA) technology to achieve an average power output of 100 W, with a peak power reaching 1GW.

In 2011, Eidam [15] used a femtosecond chirped pulse amplification system to achieve a peak power output of 3.8GW in a large-pitch photonic crystal fiber with a core diameter of 108 μm. The pulse width was 480 fs, the average power was 11 W, and the single pulse energy was 2.2 mJ.

In 2012, Stutzki [11] used a Q-switched MOPA system to achieve a single pulse energy of 26 mJ laser output through three-stage amplification in a large-pitch photonic crystal fiber with a core diameter of 135 μm. The average power was 130 W, the pulse width was less than 60 ns, and the beam quality factor M^2 was less than 1.3.

In 2014, Otto [18] used a MOPA system to amplify picosecond pulses in a large-pitch photonic crystal fiber with a length of 1 m and a core diameter of 90 μm, achieving a pulsed laser output of 2 kW. The single pulse energy was 100 μJ, the peak power was 0.8 MW, and the beam quality factor M^2 was less than 3.

In 2015, the University of Limoges in France [31] reported a 40 μm core diameter Yb^{3+} doped silica solid-state non-periodic large-pitch fiber (fully aperiodic large-pitch fiber, FA-LPF), achieved a continuous laser output of 230 W, and the beam quality factor M^2 reached 1.3 at 100 W.

In 2017, Kong from Clemson University [25] reported a solid-state bandgap fiber that outputs nearly 1 kW of laser power. To suppress mode instability, they designed a multi-layer cladding resonant structure, prepared a solid-state bandgap fiber with a core diameter of 60 μm, and obtained a maximum laser output of 910 W under a bending diameter of 60 cm with a 9 m fiber, and no mode instability was observed; additionally, a maximum laser output of 1050 W was obtained under a bending diameter of 80 cm with an 8 m fiber, and the laser efficiency reached 90%.

Chinese research on large mode area photonic crystal fibers started relatively late. After years of effort, Tianjin University, Nankai University, University of Shanghai for Science and Technology, East China Normal University, Shanghai Institute of Optics and Fine Mechanics, etc., have carried out research on high-power lasers, amplifiers, etc., using imported large mode area photonic crystal fibers. Yanshan University, Wuhan Fenghuo Company, Huazhong University of Science and Technology, and Shanghai Institute of Optics and Fine Mechanics have successively carried out the design and preparation of Yb^{3+} doped large mode area photonic crystal fibers since 2010.

In 2009, Wei Chen from Huazhong University of Science and Technology [32] prepared Yb^{3+} doped core rods using the MCVD method, and drew Yb^{3+} doped photonic crystal fibers, with a core NA of 0.06 and an effective mode area of about 1465.7 μm^2.

In 2012, Gui Chen from Huazhong University of Science and Technology [33] used the MCVD method combined with gas-liquid mixed doping technology to prepare large-core Yb^{3+} doped silica fiber preforms, and prepared photonic crystal fibers with a core diameter of about 90 μm based on this, with a mode area of about 1330 μm^2, a core NA of 0.065, and for the first time achieved high-power laser output of domestically produced Yb^{3+} doped silica photonic crystal fibers, obtaining 102 W continuous wave (CW) output in a 1 m long fiber, with a slope efficiency of 76%. However, due to the difficulty in controlling the refractive index of the core rod, this fiber did not achieve single-mode laser output.

In 2013, Jiantao Liu from Yanshan University [34] prepared Yb^{3+} doped large mode area photonic crystal fibers using powder sintering technology, with a mode diameter of 26 μm and a mode area of 550μm^2, but no laser experiment results were reported.

In 2016, Ziwei Wang from Shanghai Institute of Optics and Fine Mechanics [35] used a self-developed 105 μm core diameter ultra-large mode area photonic crystal fiber in the MOPA system to achieve a 255 W average power laser output, with a pulse width of 21 ps, a repetition rate of 10 MHz, and a peak power of 1.2 MW.

In 2016, the research group of Jinyan Li from Huazhong University of Science and Technology [36] developed 50 μm and 127 μm core diameter all-solid large mode area photonic crystal fibers. The 50 μm core diameter fiber maintained quasi-single mode transmission characteristics at the 1064 nm band, the beam quality factor M^2 was 1.37, and an amplified laser output of 8 W was obtained, the seed laser pulse width was 100 ps, and the repetition rate was 500 kHz; the 127 μm core diameter fiber obtained a 16 W amplified laser output, but the beam quality was not ideal.

In 2017, Shikai Wang from Shanghai Institute of Optics and Fine Mechanics [37] reported the preparation of large-size Yb^{3+} doped core rods through the sol-gel method combined with high-temperature sintering method, and drew a 100 μm core diameter large mode field photonic crystal fiber, achieving a laser output with an average power of 310 W in the MOPA amplification system, and the peak power reached 1.5 MW. However, due to the high refractive index of the fiber core, the beam quality of the laser was not ideal.

In 2017, Fan Wang from Shanghai Institute of Optics and Fine Mechanics [38] reported a 50 μm core diameter Yb^{3+} doped photonic crystal fiber, achieving quasi-single mode amplified laser output in the MOPA system, with an average power of 97 W, M^2 factor was 1.4, seed pulse width was 21 ps, repetition rate was 49.8 MHz, peak power was 93 kW.

In 2019, Meng Wang from Shanghai Institute of Optics and Fine Mechanics [39] obtained a laser output with an average power of 272 W in a 50 μm core diameter photonic crystal fiber by improving the uniformity of the core rod, the single pulse energy was 5.6 μJ, the peak power was 266 kW (the seed light wavelength was 1030 nm, the repetition rate was 48.7 MHz, and the pulse width was 21 ps). The M^2 factors for output laser powers of 120 W and 272 W were 1.6 and 2.2, respectively. In addition, an average power of 102 W picosecond pulse amplification output was obtained in a 75 μm core diameter photonic crystal fiber (repetition rate 1 MHz, pulse width 100 ps), the peak power exceeded 1 MW, the slope efficiency was 62.2%, the M^2 factor was 2.1 [40].

In 2021, Fan Wang [41] reported the preparation of a 40 μm core diameter polarization-maintaining photonic crystal fiber through high-concentration F ion doping. In a 2 m long fiber, a laser output with an average power of 103 W was obtained (repetition rate 30 MHz, pulse width 30 ps), the slope efficiency was 52%, the M^2 factor was 1.46.

Table 5.1 summarizes the laser results of Yb^{3+} doped large mode area photonic crystal fibers at home and abroad. From the table, it can be seen that research institutions have made good progress in the performance indicators of large core diameter, high power, large energy, and high beam quality of Yb^{3+} doped large mode area photonic crystal fibers.

Table 5.1 Laser results of Yb^{3+} doped large mode area photonic crystal fibers at home and abroad

Core diameter/μm	Average power	Pulse energy	Peak power	Output pulse width	Repetition rate	M^2	Research unit	References
100	42 W	4.3 mJ	4.5 MW	~1 ns	9.6 KHz	1.3	Arkwright company	[29]
100	100 W	1 mJ	1 GW	800 fs	100 KHz	1.2	University of Jena	[14]
108	11 W	2.2 mJ	3.8 GW	480 fs	78 MHz	1.8	University of Jena	[15]
135	130 W	26 mJ	500 kW	<60 ns	5 kHz	1.3	University of Jena	[11]
90	2000 W	100 μJ	0.8 MW	150 ps	20 MHz	<3	University of Jena	[18]
60	1050 W						Clemson university	[25]
26							Yanshan university	[34]
90	102 W						Huazhong University of Science and Technology	[33]
105	255 W	25.5 μJ	1.2 MW	21 ps	10 MHz	Poor	Shanghai Institute of Optics and Fine Mechanics	[35]
50	8 W	16 μJ	16 kW	100 ps	500 kHz	1.4	Huazhong University of Science and Technology	[36]
100	310 W	31 μJ	1.48 MW	21 ps	10 MHz	Poor	Shanghai Institute of Optics and Fine Mechanics	[37]
50	97 W	1.95 μJ	92.7 kW	21 ps	49.8 MHz	1.4	Shanghai Institute of Optics and Fine Mechanics	[38]
50	272 W	5.6 μJ	266 kW	21 ps	48.7 MHz	2.2	Shanghai Institute of Optics and Fine Mechanics	[39]
75	102 W	102 μJ	1 MW	100 ps	1 MHz	2.1	Shanghai Institute of Optics and Fine Mechanics	[40]
40	103 W	3.43 μJ	144 kW	30 ps	30 MHz	1.46	Shanghai Institute of Optics and Fine Mechanics	[41]

5.2 Preparation and Performance of Ytterbium-Doped Large Mode Area Photonic Crystal Fiber

To meet the needs of ultrafast laser technology, the amplified laser obtained through large mode area photonic crystal fiber must maintain good beam quality. To achieve this application goal, it is first necessary to design the structure of the ytterbium-doped large mode area photonic crystal fiber, and then prepare it. The preparation of ytterbium-doped large mode area photonic crystal fiber is more difficult than ordinary cladding structure ytterbium doped fiber, especially due to the high requirements for refractive index distribution and core uniformity. The conventional method for preparing double-cladding fiber preforms - MCVD combined with solution doping method [42], it is difficult to meet the preparation requirements of high-brightness large mode area photonic crystal fiber in terms of core rod size, uniformity, refractive index value and distribution. The publicly reported methods that can be used to prepare ytterbium-doped large mode area photonic crystal fiber core rods include powder sintering method [43], porous glass method [44], sol-gel method combined with high-temperature sintering method [45]. In addition, due to the special design of the microstructure, the preparation and fiber drawing of the ytterbium-doped large mode area photonic crystal fiber preform are quite different from those of ordinary cladding structure ytterbium doped fiber. The preparation and performance of ytterbium-doped large mode area photonic crystal fiber are introduced in detail below.

5.2.1 *Fiber Preparation*

The preparation process for photonic crystal fiber is: first prepare the preform by stacking method, and then draw it into photonic crystal fiber on a special fiber drawing tower. Since the cladding of the photonic crystal fiber contains a periodic structure, the focus and difficulty of its preparation lie in maintaining the consistency of the periodic structure area and design parameters during the drawing process. This section focuses on the different materials of the cladding structure area, dividing it into ytterbium-doped air hole photonic crystal fiber and ytterbium-doped all-solid photonic crystal fiber, and introduces their preparation process respectively.

5.2.1.1 Ytterbium-Doped Air Hole Photonic Crystal Fiber

Preparation of Ytterbium-Doped Air Hole Photonic Crystal Fiber Preform

According to the structural design requirements, the preform of the ytterbium-doped air hole photonic crystal fiber is composed of five parts: a Yb^{3+} doped core rod, inner cladding capillaries, an inner cladding tube, outer cladding capillaries and an outer cladding tube, as shown in Fig. 5.6. The Yb^{3+} doped core rod is a highly uniform Yb^{3+} doped silica glass prepared based on the sol-gel method combined with high-temperature sintering method (preparation methods and processes are seen in Chap. 3 of this book). Then the silica glass is ground and polished to achieve the required size of the core rod; the inner and outer cladding tubes can use the appropriate specifications of F300 quartz tubes; the inner and outer cladding capillaries are drawn from the corresponding size of F300 quartz tubes.

Through the high-order mode off-domain model [46] and the restricted loss model [47], the optimal aperture and outer diameter ratio (d/Λ) of the inner cladding capillary for a certain core diameter of photonic crystal fiber can be simulated. Assuming that the optimal d/Λ for the 19-core photonic crystal fiber is 0.125, the size of the inner cladding capillary is designed to be an outer diameter of 1.14 mm/inner hole 143 μm. The corresponding Yb^{3+} doped core rod has an outer diameter of 4.9 mm, the inner cladding tube size is an outer diameter of 19.5 mm/inner hole 15 mm, the outer cladding tube size is an outer diameter of 32 mm/inner hole 21 mm, and the outer cladding capillary size is an outer diameter of 750 μm/inner

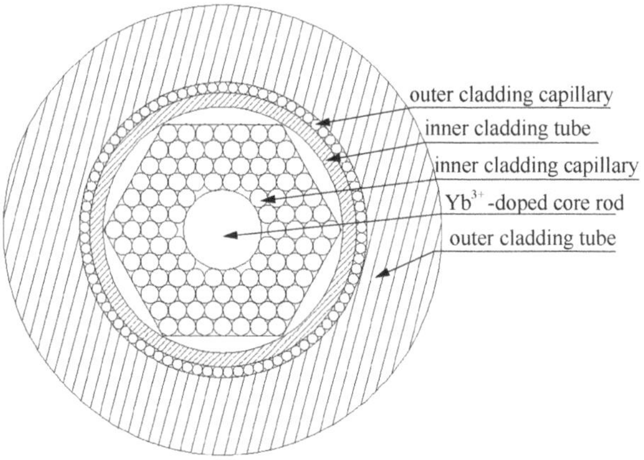

Fig. 5.6 Schematic diagram of 19-core photonic crystal fiber preform

hole 700 μm. The sizes of the five components of the preform are shown in Table 5.2. Therefore, the preparation of the 19-core photonic crystal fiber preform requires four specifications of Heraeus F300 quartz tubes, which are the outer diameter 30 mm/inner hole 3.75 mm thick wall quartz tube for drawing the inner cladding capillary, the outer diameter 30 mm/inner hole 28 mm thin wall quartz tube for drawing the outer cladding capillary, the outer diameter 19.5 mm/inner hole 15 mm quartz tube for the inner cladding tube, and the outer diameter 32 mm/inner hole 21 mm quartz tube for the outer cladding tube. The preform preparation process is shown in Fig. 5.7.

In the production of the preform, it is necessary to draw two specifications of F300 silica tubes into inner and outer cladding capillaries. During the process of drawing capillaries, to avoid contamination caused by manual cutting, an automatic cutting and collection device for capillaries was designed on the drawing tower, as shown in Fig. 5.8. When the capillary length reaches the set value in the drawing process, the blade in the automatic cutting device cuts the capillary under pneumatic drive, breaks it via moving guide wheel, and then it falls into the collection box.

Table 5.2 Sizes of the components of the 19-core photonic crystal fiber preform

Component	Outer Diameter	Inner Hole Diameter	d/Λ	Quartz Tube Outer Diameter	Quartz Tube Inner Diameter
Yb^{3+} doped core rod	4.9 mm				
Inner cladding capillary	1.14 mm	~143 μm	0.125	30 mm	3.75 mm
Inner cladding tube	19.5 mm	15 mm			
Outer cladding capillary	750 μm	700 μm	0.933	30 mm	28 mm
Outer sheath tube	32 mm	21 mm			

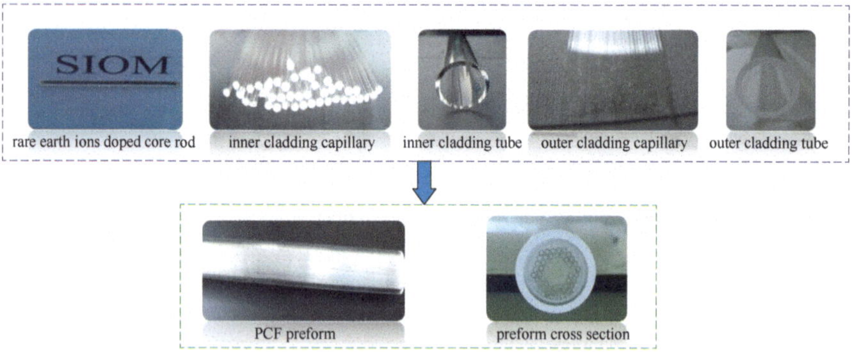

Fig. 5.7 Process flow chart for the preparation of photonic crystal fiber preform

The 19-core photonic crystal fiber preform is prepared by the stacking method. During stacking, 108 pieces of inner cladding capillaries need to be arranged into a hexagonal close-packed structure, and nearly 100 pieces of outer cladding capillaries are built into a circular structure between the inner and outer cladding tubes. To facilitate capillary construcion, a preform stacking device was designed and manufactured, as shown in Fig. 5.9. Polytetrafluoroethylene is used as the main material to avoid impurity contamination when handling the capillaries. The four polytetrafluoroethylene components are assembled into a hexagonal structure, and different

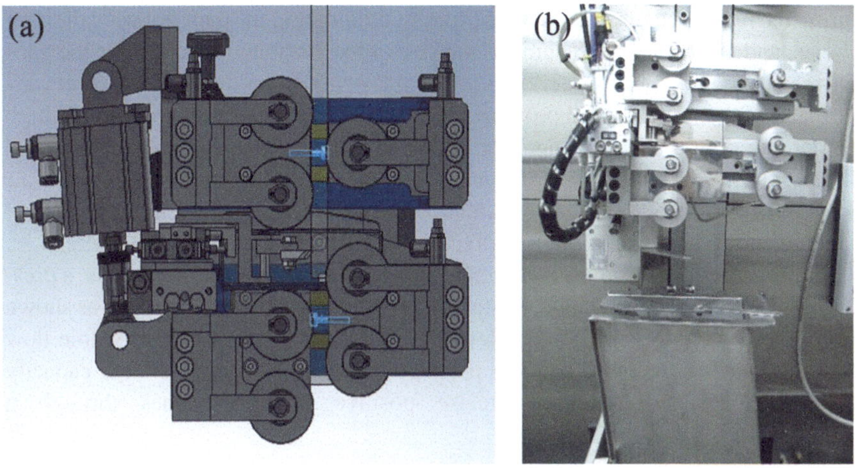

Fig. 5.8 Automatic cutting and collection device for silica capillaries, (**a**) Design diagram (**b**) Physical diagram

Fig. 5.9 Stacking device for photonic crystal fiber preform

sizes of hexagonal structures can be obtained by adjusting the fixed position, so that preforms of different sizes can be prepared. The polytetrafluoroethylene structure is fixed on the slider, and the position of the slider on the guide rail can be adjusted to prepare different lengths of preforms.

Drawing of Ytterbium-Doped Air Hole Photonic Crystal Fiber

After fabrication the photonic crystal fiber preform, the most critical part in drawing the photonic crystal fiber is maintaining the microstructure of the fiber unchanged during the high-temperature drawing process, eliminating the gaps between the capillaries, and maintaining the number and size of the air holes. To eliminate the gaps between the capillaries, it is necessary to apply negative pressure to the gaps in various parts of the preform, and promote the closure of the gaps in the preform at high temperatures through a certain degree of vacuum. To maintain the number and diameter of the air holes, it is necessary to provide a certain air pressure for each air hole, and ensure that each air hole can open to the designed diameter at high temperature and remain stable through precise positive pressure control.

To provide precisely controlled positive and negative pressure conditions, a pressure control device for photonic crystal fibers was designed and fabricated, as shown in Fig. 5.10. The pressure control device provides nitrogen with controllable flow through a mass flow meter, reduces pressure fluctuations through a large-capacity pressure tank, and monitors precise positive pressure values through a

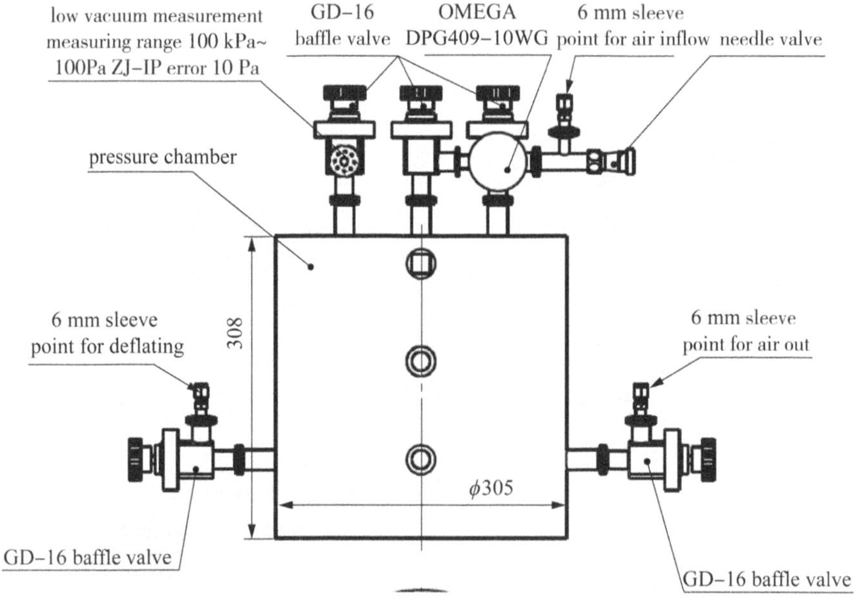

Fig. 5.10 Pressure control device for the preparation of photonic crystal fibers

high-precision pressure gauge. Finally, it is provided to the capillaries of each part through a specially designed preform chuck. At the same time, this device provides a stable vacuum through a diaphragm pump and a needle valve. Using a high-precision vacuum gauge to monitor the negative pressure value of the pressure tank, and then provide it to the preform. The pressure control device can provide precisely controlled positive and negative pressure at the same time, meeting the preparation requirements of photonic crystal fibers.

To apply the positive and negative pressure of the pressure control device to the air hole part and the gap part of the preform at the same time, a special preform chuck was designed, as shown in Fig. 5.11a. Through this preform chuck, positive pressure can be provided for each capillary hole in the preform, prompting the capillary air holes, especially the outer layer capillary air holes, to fully open and maintain a certain size. At the same time, negative pressure can eliminate the gaps between the capillaries and between the capillaries and the silica tube, prompting the various parts of the preform to fully close and form a whole, which is conducive to improving the strength of the fiber. Figure 5.11b is a partial view of the outer layer air holes of the prepared photonic crystal fiber. By precisely controlling the positive pressure conditions, the outer layer air holes of the fiber are expanded from round to elliptical, and the wall thickness between the holes is less than 1 μm, which can achieve a numerical aperture of more than 0.5 in the inner layer, which can greatly improve the coupling efficiency of the pump in the photonic crystal fiber.

Due to the complexity of the large mode area photonic crystal fiber structure and the high consistency requirements, a special design has been made for the temperature field of the drawing furnace. Fig. 5.12 shows the design of the graphite heater and furnace structure of the photonic crystal fiber drawing furnace. By shortening the height of the graphite heating zone, reducing the range of the high temperature zone, and reducing the radiation effect of the high temperature zone on the two low temperature zones, the preform rod is only heated by a very small high temperature zone, avoiding the temperature field outside the high temperature zone destroying

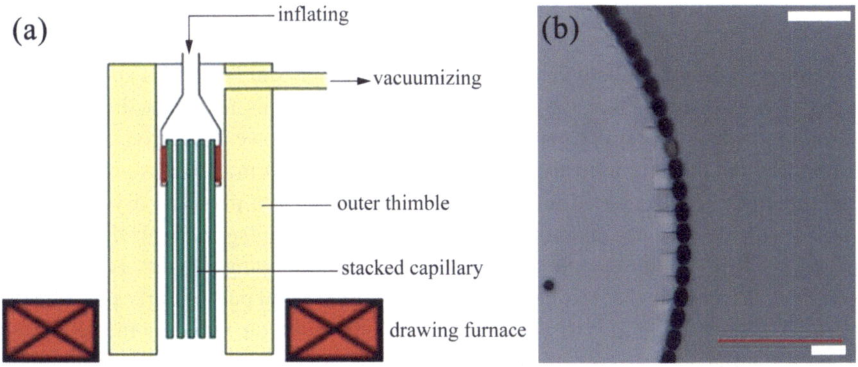

Fig. 5.11 (**a**) Structure of the preform chuck for photonic crystal fiber and (**b**) the outer layer air holes of the fiber

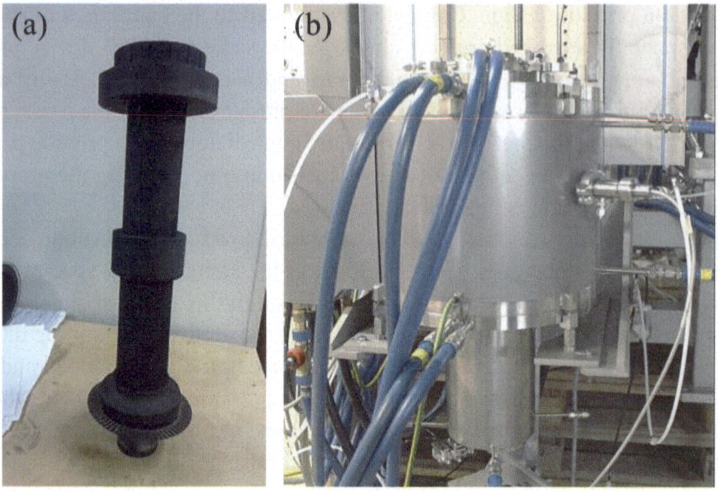

Fig. 5.12 The furnace for drawing photonic crystal fiber (**a**) graphite heater and (**b**) the photograph of furnace

the preform structure. In addition, the cooling tube of the drawing furnace has been extended. After forming the photonic crystal fiber structure and leaving from the high temperature zone, the fiber obtains a relatively long cooling time through the long cooling tube, avoiding the stress caused by rapid cooling, which is beneficial to improve the strength of the fiber.

Through the design and application of the above-mentioned air pressure control device, preform rod chuck, drawing furnace structure, etc., and the improvement and optimization of drawing temperature, drawing speed, drawing tension, and the setting of preform air pressure and vacuum, a complete and design-compliant large mode area photonic crystal fiber is obtained.

5.2.1.2 Ytterbium-Doped Solid-State Photonic Crystal Fiber

The biggest difference between solid-state photonic crystal fibers and air hole photonic crystal fibers is that the former uses solid materials to adjust the photonic crystal structure. Generally, solid materials with much lower refractive index than the inner cladding are used to replace the air holes to form a solid-state photonic crystal structure. Compared with air hole large mode area photonic crystal fibers, solid-state photonic crystal fibers have the following advantages: one is that it is not necessary to maintain the air hole structure, so the drawing process is relatively simple; the second is that it is easy to fuse with commercial pump sources to achieve all-fiberization; the third is to avoid structural contamination, reducing the difficulty of preservation and use cost; the fourth is to avoid thermal accumulation in the fiber core due to the presence of air holes, which is beneficial to fiber heat dissipation at high power operation.

In 2003, the group of University of Southampton [48] used two different refractive index borosilicate glasses to prepare passive solid-state refractive index guiding photonic crystal fibers; in 2010, the group of the Xi'an Institute of Optics and Fine Mechanics [49] reported a passive solid-state refractive index guiding polarization maintaining photonic crystal fiber. In 2012, the group of the Shanghai Institute of Optics and Fine Mechanics [50] further developed this type of fiber and was the first in the world to successfully develop a solid-state refractive index guiding active photonic crystal fiber, which was used in laser experiments: a power of 7.9 W and a slope efficiency of 38.1% were achieved in the Nd^{3+} doped phosphate solid-state photonic crystal fiber, and then Yb, Yb/Er doped more than 10 W phosphate and silicate solid-state photonic crystal fiber near single-mode laser output was achieved [51–53], and it opened up a new active fiber structure for high-brightness fiber lasers. In 2015, the group of University of Limoges [31] reported a 40 μm core diameter Yb^{3+} doped silica all-solid non-periodic large-pitch fiber (FA-LPF), which achieved 230 W continuous laser output, and the beam quality factor M^2 reached 1.3 at 100 W. In 2016, the group of Huazhong University of Science and Technology [36] reported 50 μm and 127 μm core diameter ytterbium doped all-solid photonic crystal fibers, respectively achieving 8 W and 16 W pulsed amplified laser output, verifying that the 50 μm core diameter fiber had no significant impact on the transmission mode of 1064 nm seed light, M^2 remained at 1.37, but the 127 μm core diameter fiber can no longer maintain quasi-single mode transmission.

As mentioned earlier, due to the high refractive index of the Yb/Al doped core rod, the beam quality of the ytterbium doped air hole large mode area photonic crystal fiber is not ideal. Therefore, by increasing the refractive index of the cladding material, reducing the refractive index difference between the core and the cladding, and reducing the numerical aperture of the core, high-brightness laser can be achieved. In addition, by replacing the air holes with low refractive index materials to form an all-solid photonic crystal fiber structure, the difficulty of splicing with commercial pump sources is reduced, making it easier to achieve all-fiberization and more promising for device applications.

Structure of Ytterbium-Doped All-Solid Photonic Crystal Fiber

Based on the Yb^{3+} doped core rod prepared by the sol-gel method, Wang Meng from Shanghai Institute of Optics and Fine Mechanics [54] designed and fabricated the all-solid ytterbium-doped large mode are photonic crystal fiber shown in Fig. 5.13. The fiber structure consists of four parts: the core uses a Yb^{3+} doped core rod; the hexagonal structure of the cladding uses high refractive index materials, such as Ge or Al doped silica glass; the cladding structure holes use materials with a lower refractive index than the cladding material, such as F or B doped silica glass, or pure silica glass; the outside of the cladding uses an F300 tube as a protective layer. The initial research fiber cladding structure uses a large size Ge doped silica glass prepared by the axial vapor deposition (VAD) method, the cladding structure holes use

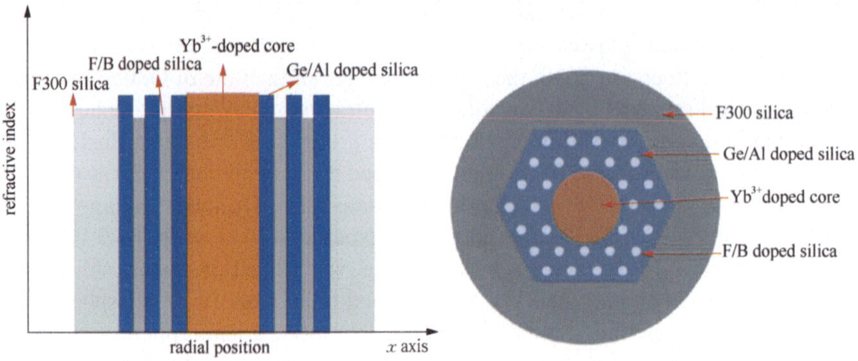

Fig. 5.13 Structure of all-solid ytterbium-doped large mode area photonic crystal fiber

pure silica material, and the core uses a 7-core replacement structure with a core diameter of about 50 μm.

In the process of preparing a 50 μm core diameter fiber, the Ge doped glass rod is first prepared by the VAD method as the cladding material, and then according to the refractive index of the Ge doped glass rod, it matches the Yb^{3+} doped core rod, so that the refractive index difference between the two parts reaches 3×10^{-4}. Then the stacking method is used to prepare fiber preform rod, which finally draws into a solid-state ytterbium-doped photonic crystal fiber.

Preparation of the Cladding Material for the Ytterbium-Doped Solid-State Photonic Crystal Fiber

The VAD method was first proposed by Nippon Telegraph and Telephone Corporation (NTT) [55]. It is currently mainly used for the preparation of Ge-doped core rods for communication fibers, as shown in Figure 5.14a. The refractive index profile of the fiber core prepared by the VAD method is usually not flat, and Figure 5.14b is a typical refractive index distribution diagram of the fiber core cross-section. However, as the inner cladding material for the solid-state ytterbium-doped photonic crystal fiber, the inner cladding with an uneven refractive index distribution will affect the transmission of the laser, so it is necessary to further treat the Ge-doped glass rod with a high-temperature homogenization method.

The Ge-doped glass rod prepared by VAD is homogenized at high temperature using an MCVD lathe, then drawn into a thin rod, and the uniformity of the Ge-doped thin rod is tested. After the Ge-doped thin rod is drawn into a fiber, the refractive index distribution of the Ge-doped fiber core is tested using the IFA-100 fiber refractive index analyzer, as shown in Fig. 5.15. The inset is a microscopic end-face photograph of the thin rod. It can be seen that after homogenized at high temperature, the relative average refractive index of the Ge-doped glass rod is about 1.85×10^{-3}, and the refractive index fluctuation is within 5×10^{-5}. This indicates that for the Ge-doped glass rod with the uneven refractive index distribution, high-temperature

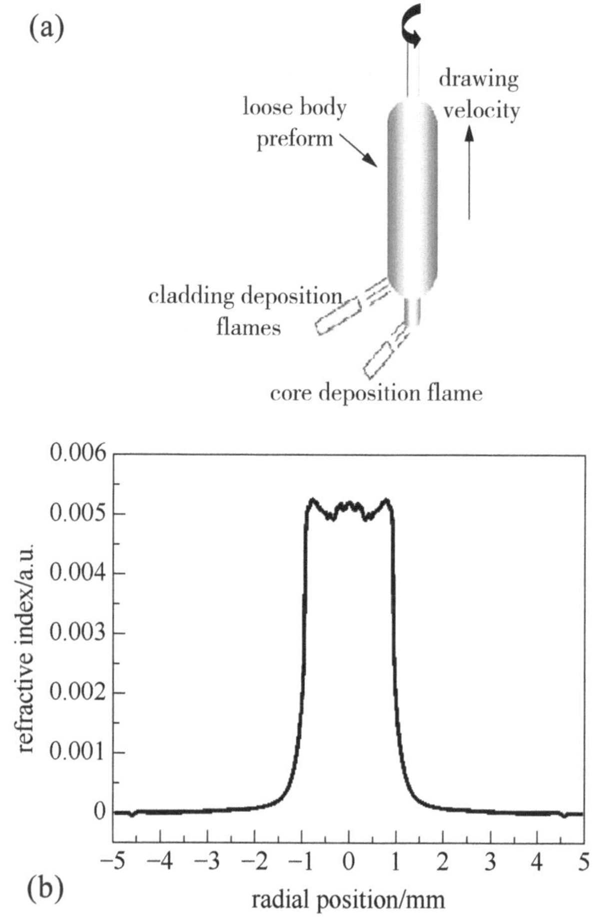

Fig. 5.14 (**a**) VAD preparation of Ge-doped fiber preform rod process and (**b**) typical cross-sectional refractive index distribution of preform

homogenization can flatten the refractive index distribution. The loss of the Ge-doped fiber was tested, as shown in Fig. 5.16. It can be found that the loss of the homogenized Ge-doped fiber at 1200 nm is about 19 dB/km. Compared with the single-mode fiber prepared by VAD, the loss at 1200 nm is about 0.5 dB/km, indicating that the high-temperature homogenization process has increased the loss by tens of dB/km, which may be due to the introduction of impurities at the interface during the homogenization process, resulting in an increase in absorption loss. Using the stacking method, the homogenized Ge-doped thin rod and F300 silica rod are used to make a preform rod at a ratio of d/Λ of 0.41, and then drawn into a Ge-Si thin rod with an outer diameter of 1.14 mm as the inner cladding structural unit of the solid-state ytterbium-doped photonic crystal fiber, and its end-face microscopic photograph is shown in Fig. 5.17.

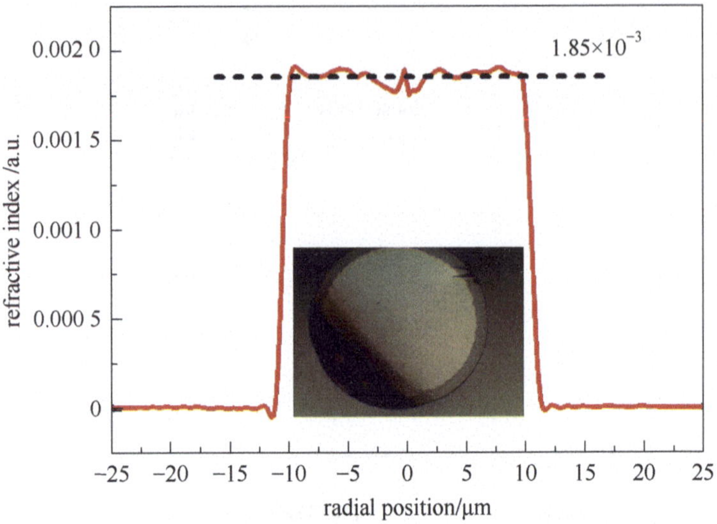

Fig. 5.15 Refractive index distribution of Ge-doped fiber core after high-temperature homogenization

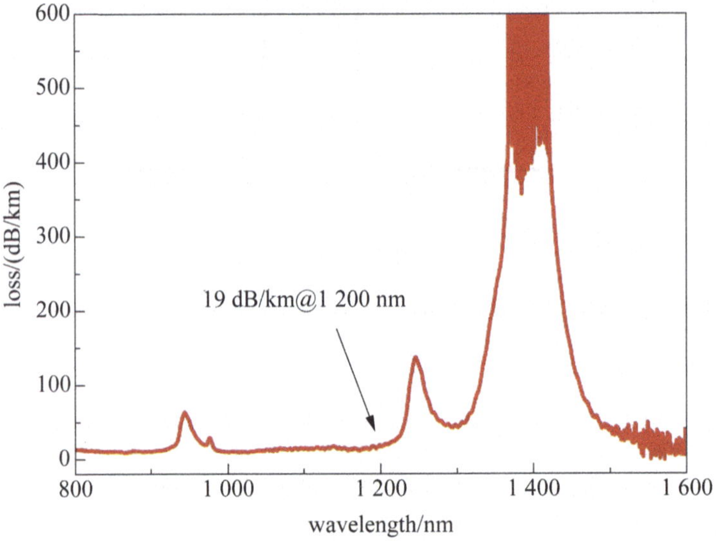

Fig. 5.16 Loss spectrum of Ge-doped fiber

Fig. 5.17 End-face microscopic photograph of Ge-Si thin rod

Preparation of Ytterbium-Doped Solid-State Photonic Crystal Fiber

After high-temperature homogenization, the refractive index difference of the Ge-doped inner cladding thin rod relative to pure silica is 1.85×10^{-3}. According to the structural simulation results of the high-brightness solid-state photonic crystal fiber, it is necessary to match the Yb^{3+} doped core rod with the relative refractive index of 2.15×10^{-3}. Based on Xu et al. [56] study on the influence of Yb^{3+} and Al^{3+} doping amount on the refractive index of silica glass, 0.095 mol%Yb_2O_3–0.95 mol%Al_2O_3 doping composition was selected to prepare the core rod, which was labeled as YA-0.095-0.95. The YA-0.095-0.95 core rod was prepared by sol-gel method combined with high-temperature sintering method, and then processed into a $\phi 2.8 \times 90$ mm thin rod. The refractive index distribution of the core after high-temperature drawing of the YA-0.095-0.95 core rod was tested by the rod-tube method, and the test results are shown in Fig. 5.18. The inset is a photo of the YA-0.095-0.95 core rod. From the figure, the relative average refractive index of the YA-0.095-0.95 core rod is 2.18×10^{-3}, and the refractive index fluctuation is within 5×10^{-5}. The refractive index of the core rod is about 3.3×10^{-4} larger than that of the inner cladding, so the relative NA of the core is 0.031, which basically meets the structural design requirements of the high-brightness solid-state ytterbium-doped photonic crystal fiber.

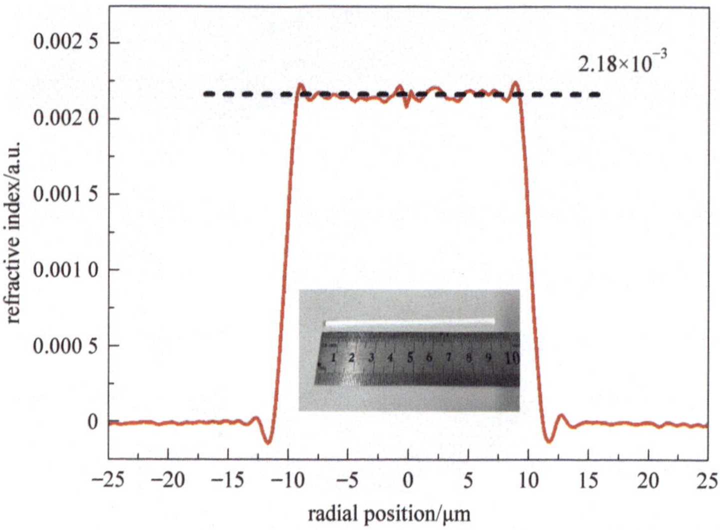

Fig. 5.18 Refractive index distribution of YA-0.095-0.95 core rod

Using 54 pieces of Ge-Si thin rods as the inner cladding, and YA-0.095-0.95 core rod as the fiber core, the solid-state ytterbium-doped photonic crystal fiber preform is stacked. Then the preform is drawn into a solid-state photonic crystal fiber at a high temperature of 2050 °C. Unlike the precise control of air pressure and vacuum required to maintain the structure of the air hole photonic crystal fiber, the solid-state photonic crystal fiber does not have an air hole structure, so it only needs to maintain a vacuum to eliminate the gap between cladding and core rods. Therefore, the process of preparing solid-state ytterbium-doped photonic crystal fiber is greatly simplified. Figure 5.19 is the microscopic photograph of the end face of the prepared solid-state ytterbium-doped photonic crystal fiber. The fiber core diameter is about 50 μm, the outer diameter is about 310 μm, the structure unit diameter d is about 6.6 μm, the unit pitch Λ is about 16 μm, and the number of structure unit layers is 3.

Using the IFA-100 fiber refractive index analyzer, the two-dimensional refractive index distribution and one-dimensional refractive index distribution of the 50 μm core diameter solid-state ytterbium-doped photonic crystal fiber were tested, as shown in Fig. 5.20. Among them, Fig. 5.20 a is the two-dimensional refractive index distribution of the solid-state photonic crystal fiber at a wavelength of 633 nm, from which the refractive index of the fiber core, inner cladding, and structure unit can be distinguished, and the refractive index distribution of the fiber core is relatively uniform; Figure 5.20b is the one-dimensional refractive index distribution of the solid-state photonic crystal fiber, from which the distribution of the three-layer inner cladding structure units can be clearly seen. The difference between the average refractive index of the fiber core and the effective refractive index of the Ge-doped inner cladding is calculated to be about 3.6×10^{-4}, and the relative NA of the fiber core is 0.032.

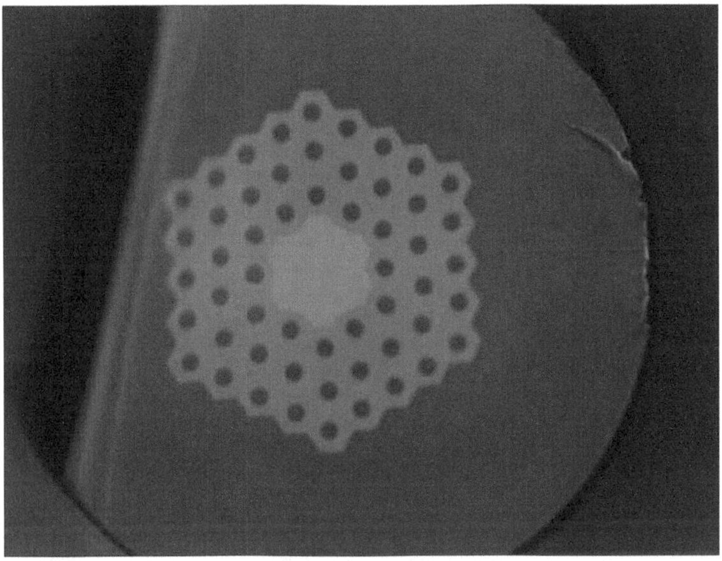

Fig. 5.19 Microscopic photograph of the end face of the 50 μm core diameter solid-state Ytterbium-doped photonic crystal fiber

5.2.2 Fiber Performance

Photonic Crystal Fibers (PCFs) are a unique class of optical fibers distinguished by their periodic arrangement of microstructures in the cladding, which can be either regularly or irregularly structured, endowing them with distinct optical properties compared to conventional double-clad fibers. These properties include endless single-mode characteristics and ultra-large cladding numerical apertures. Below is a detailed introduction to the characteristics of PCFs in terms of absorption, loss, and mode propagation.

5.2.2.1 Absorption

The absorption characteristic is a critical parameter for active fibers, determining the extent to which the fiber effectively utilizes pump light and influencing the laser efficiency. For double-clad fibers, adopting non-circular symmetrical structures such as octagonal shapes can significantly reduce the proportion of skew ray when pump light transmits along fiber, thereby improve the absorption of the pump light. However, some pump light with lower numerical aperture may still not be absorbed. The structured region in photonic crystal fibers can alter the transmission direction of light, allowing it to pass through the core and be absorbed sufficiently, as shown in Fig. 5.21. NKT Corporation compared the absorption of double-clad step-index fibers and photonic crystal fibers at different numerical apertures in a report, as

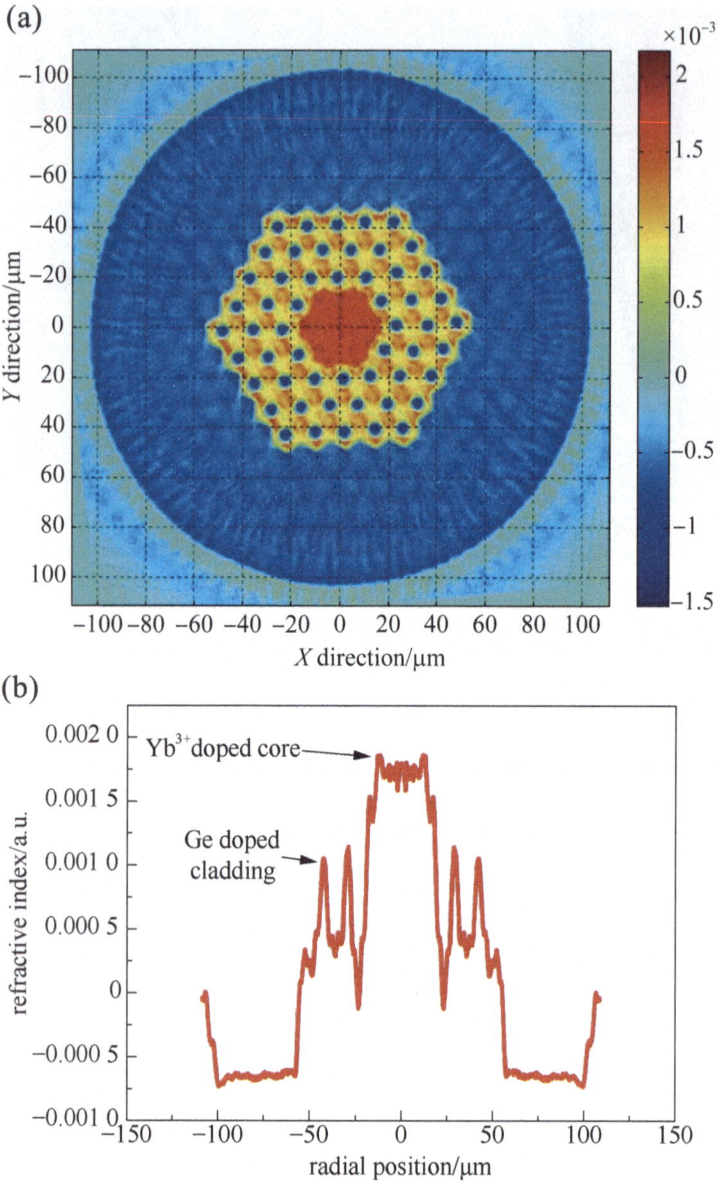

Fig. 5.20 Refractive index distribution of 50 μm core diameter solid-state ytterbium-doped photonic crystal fiber. (**a**) Two-dimensional (**b**) One-dimensional

depicted in Fig. 5.22. It can be observed that the absorption of photonic crystal fibers changes little when the pump light has different numerical apertures; in contrast, for double-clad step-index fibers, the numerical aperture of the pump light significantly affects their absorption. Therefore, when having the same rare-earth

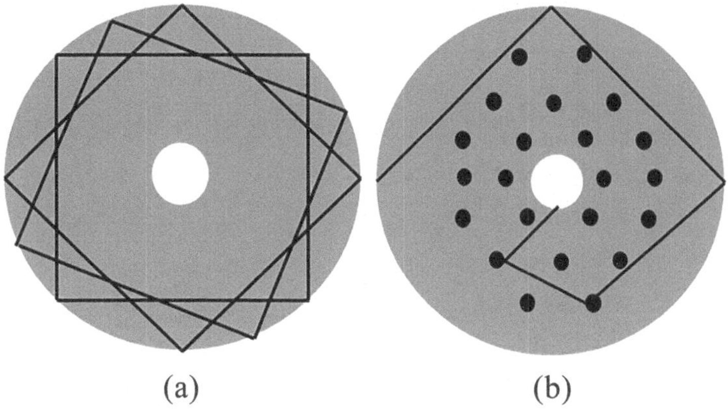

Fig. 5.21 Light transmission in step index fiber (left) and Photonic crystal fiber (right)

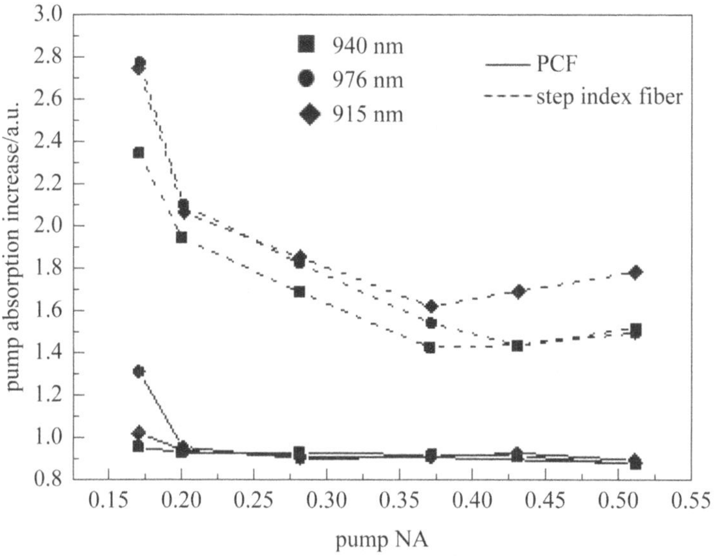

Fig. 5.22 Absorption of pumping with different divergent angle in step index fiber and Photonic crystal fiber

doping concentration, photonic crystal fibers absorb pump light more fully than double-clad step-index fibers.

Rare-earth doped large mode area photonic crystal fibers (PCFs) are characterized by their large mode field area (exceed 30 μm), high gain per unit length, and favorable mode properties, making them directly applicable in pulse amplification systems. This allows for the achievement of large pulse energies (mJ), high peak powers (MW), and high-brightness laser output. To further reduce nonlinearity, these fibers are typically used in short lengths, thus requiring higher absorption

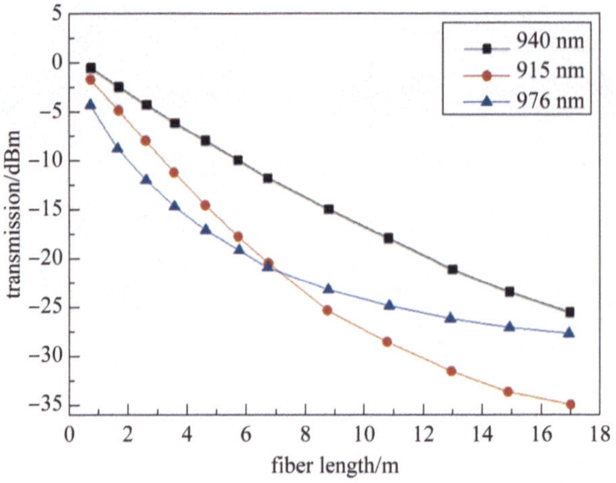

Fig. 5.23 Absorption of ytterbium-doped photonic crystal fiber at different lengths

coefficient. The ytterbium-doped PCF products provided by NKT Photonics, the absorption coefficient of fibers with a 40-micrometer core size is approximately 10 dB/m at 975 nm, while the absorption coefficient with an 85-micrometer core size can reach 15 dB/m at 975 nm. The absorption coefficient of photonic crystal fibers can be measured using the cut-back method, but it is important to ensure light transmits within the inner clad of the fiber. Additionally, due to the large absorption and emission cross-sections of ytterbium ion at 976 nm, the choice of measured fiber length is crucial for fibers with high ytterbium ion doping. The absorption coefficient will be no longer linearly changing with length if the he fiber length used in measurement exceeds a certain value, especially at 976 nm as shown in Fig. 5.23. This means that when measured fibers with high ytterbium doping concentrations, the length of fiber should be chosen properly, and it should also be changed with measured wavelength.

5.2.2.2 Numerical Aperture of Air Hole Cladding

The numerical aperture (NA) of the photonic cladding should be larger to ensure sufficient couple of pump light A large NA can be obtained using air hole cladding, generally greater than 0.5, and can even reach 0.8. This is currently unachievable with conventional fiber coating materials (which are usually 0.46 ~ 0.48). The NA of the photonic crystal fiber is determined by the air holes spacing between the air holes. The relationship between the spacing and the NA is shown in Fig. 5.24 [57]. The spacing is usually less than half of its transmission wavelength. The size of the NA of the inner cladding of the photonic crystal fiber can be tested by the far-field scanning method.

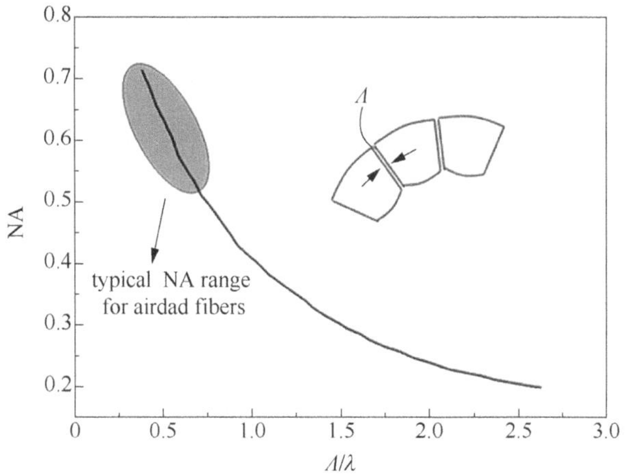

Fig. 5.24 Numerical aperture of the air hole cladding as a function of the hole spacing

5.2.2.3 Mode Properties

The commonly fiber is composite of core made by rare-earth-doped silica glass, and the cladding by pure silica glass. The dispersion curves of these two materials are essentially identical. Therefore, the numerical aperture (NA) of the core in the double-clad fiber does not change significantly at different wavelengths. When the fiber used at a wavelength longer than the single-mode cutoff wavelength, only a single mode can be transmitted, and if used at a wavelength shorter than that, multiple modes can be transmit. The core NA of commercially available ytterbium-doped cladding fibers is about 0.06, and the lowest value reported in the literature can be reduced to about 0.03 [58]. For photonic crystal fibers, the effective refractive index n_{eff} of the cladding is related to the filling rate of the air holes, so the effective refractive index of the core is determined changes with the structural parameters of the cladding. Figure 5.25 shows the change of the core NA with the cladding structural parameters, where d represents the diameter of the air hole, and Λ represents the distance between the centers of two neighboring air holes [59]. As shown, when the d/Λ increases, the refractive index of the core gradually increases.

Due to the NA of core can change significantly with the structure and wavelength, the photonic crystal fiber can maintain single-mode characteristics at a large wavelength range. Figure 5.26 shows the NA of core at different wavelength of NKT's photonic crystal fiber product. It can be seen from the figure that the NA of core of the photonic crystal fiber increases with the increase of wavelength. When the fiber operates in the long-wave band, the effective refractive index of the cladding will decrease, and the NA of core will increase; when the fiber operates in the short-wave band, the effective refractive index of the cladding will increase, and the NA of core will decrease. Therefore, photonic crystal fibers can maintain single-mode characteristics within a large wavelength range, which is impossible for step-index fibers.

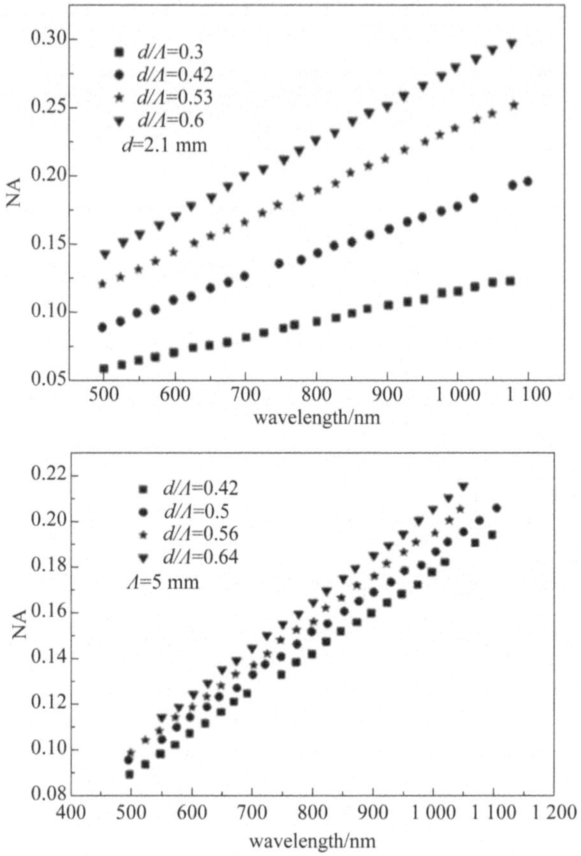

Fig. 5.25 Numerical aperture of fiber core as a function of with air holes parameters

When core size increases the NA of core should be smaller to make single-mode transmittances. For a photonic crystal fiber with 40 μm core, its NA of core is about 0.03. If fibers have larger diameters of core, their NA of core are even smaller.

The normalized frequency V determines the number of modes propagating in the fiber. For step-index fibers, the condition for single-mode transmission is as follows:

$$\mathrm{V} = \frac{2\pi a}{\lambda}\sqrt{n_{core}^2 - n_{clad}^2} < 2.405 \tag{5.4}$$

where, a is the core radius; n_{core} and n_{clad} are the refractive indices of the core and cladding respectively; λ is the output wavelength.

The normalized constant of photonic crystal fiber can be defined as

$$\mathrm{v} = \frac{2\pi a}{\lambda}\sqrt{n_{core}^2 - n_{eff}^2} \tag{5.5}$$

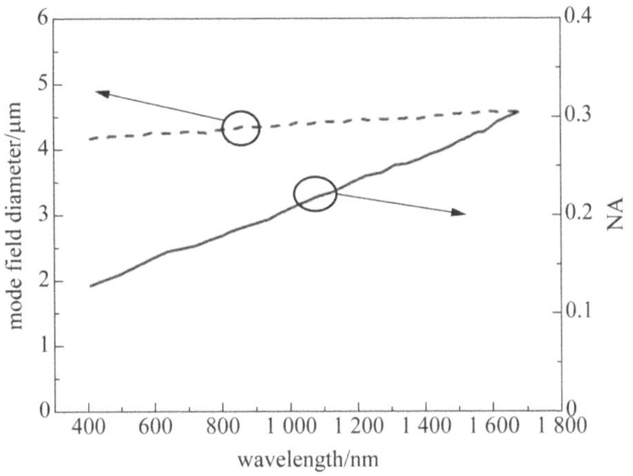

Fig. 5.26 Mode field diameters and Numerical apertures at different wavelength

n_{FSM} can be used to replace n_{eff}. FSM is defined as the basic spatial filling mode transmitted by the cladding of a photonic crystal fiber with an infinitely large structure. The effective refractive index of the cladding can be defined as

$$n_{\text{eff}} = \beta_{\text{FSM}} \cdot k \tag{5.6}$$

The transverse wave vector of the core k is effectively transmitted only between 0 and $k_{\max} = (k^2 n_{core}^2 - \beta_{FSM}^2)^{1/2}$.

The core radius of the photonic crystal fiber a can have different definitions, it can be defined as 0.5Λ, 0.625Λ, Λ and others [60–62], therefore the normalized constant of the single-mode photonic crystal fiber can be defined at different values. As can be seen from the above formula, the normalized constant V of PCF is not only related to the wavelength. As shown in Fig. 5.27 that when Λ/λ tends to infinity, single-mode transmission can be achieved under every d/Λ, the photonic crystal fiber can have single-mode transmission characteristics in a wide wavelength range (known as endless single-mode fiber).

In order to pursue a larger mode field area, the core radius will be expanded, and its single-mode wavelength range will be reduced. Figure 5.28 shows the single-mode area of 7-core fiber and 1-core fiber under the same structural parameters. As shown, the single-mode area of the 7-core fiber is much smaller than that of the 1-core fiber [63].

Figure 5.29 shows the normalized constant of 7-core photonic crystal fiber with different air hole and hole spacing conditions. When *d/Λ is* small (air filling factor), the fiber has sing-mode characteristics in a larger wavelength range. But small air filling factor may lead to large waveguide mode limiting loss. Indeed, it is critical to balance different parameters.

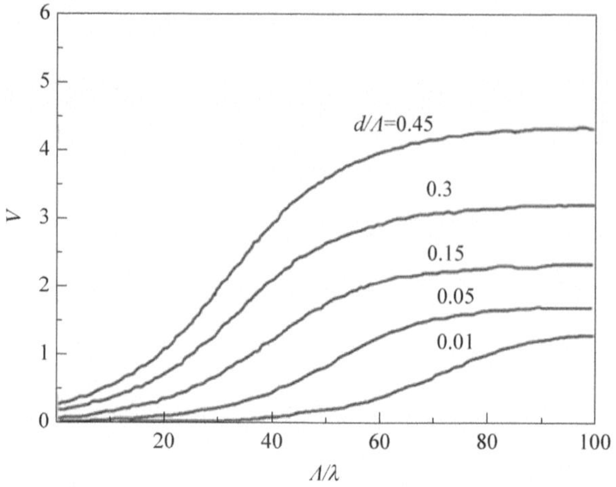

Fig. 5.27 Normalized constant of photonic crystal fiber as a function of hole spacing [62]

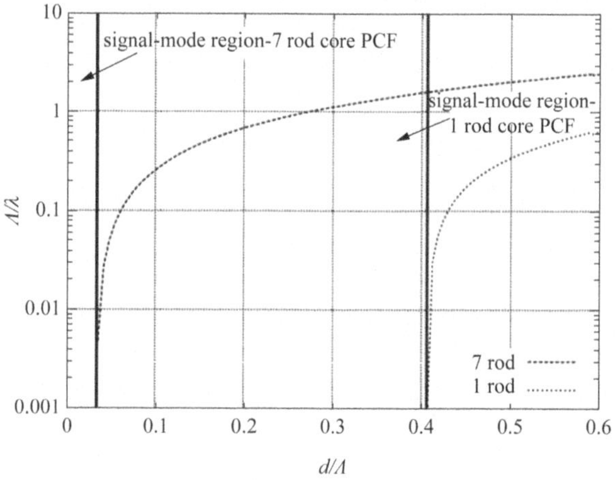

Fig. 5.28 Single-mode area of 7-core PCF fiber and 1-core PCF fiber

5.2.2.4 Mode Instability

Mode instability effect refers to the phenomenon where the laser mode will exhibit time-related random changes when the output power of the fiber exceeds a certain threshold, mainly manifested a deterioration of the beam quality. In 2010, Eidam from the University of Jena in Germany observed the phenomenon of mode instability during the experiment, as shown in Fig. 5.30 [64], which attracted wide attention from researchers around the world. The phenomenon of mode instability is currently believed to originate from the interference between higher-order modes and

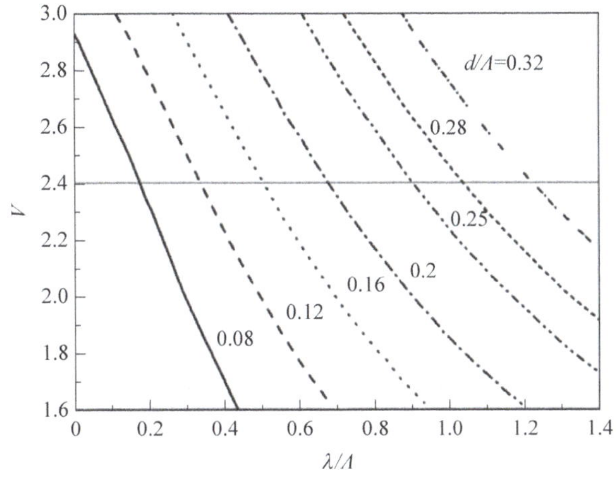

Fig. 5.29 The influence of air holes parameters on normalized constant of 7-core PCF fiber [63]

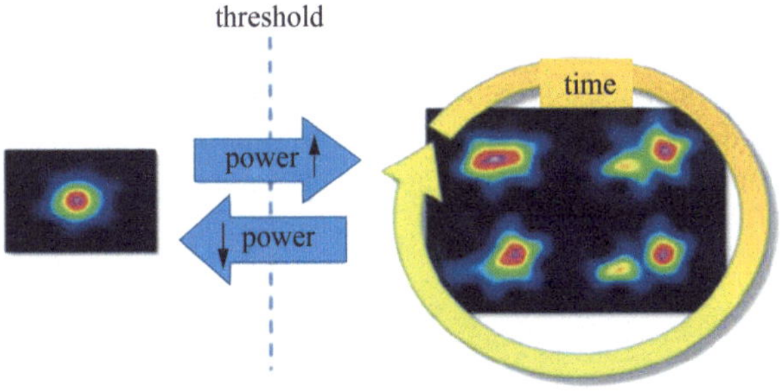

Fig. 5.30 Mode time-related instability

the fundamental mode within the fiber. This interference leads to a periodic distribution of light intensity in the core during the amplification process, and results in periodic gain extraction and the formation of a periodic thermal distribution in the doped region of the core. Due to the thermo-optic effect, a long-period refractive index grating is formed in the core. This grating facilitates the coupling of energy between lower-order and higher-order modes.

Understanding the mechanism behind mode instability, researchers have proposed various methods to increase the threshold for this phenomenon in photonic crystal fibers (PCFs). A study from Friedrich Schiller (Jena University in Germany) found that photonic darkening of fiber can decrease threshold for mode instability significantly. The research indicates that heat generated by quantum defect is not the only source of heat in the fiber, the photonic darkening effect could be another significant source of heat [65]. When fiber is pumped at 976 nm and emits at 1030 nm, the heat load on the fiber caused by photonic darkening is higher than that caused by quantum defect. Even if the photonic darkening effect only results in a small power loss, its impact on the threshold for mode instability is substantial. Therefore, reducing the photonic darkening effect in the fiber can increase the threshold of mode instability. Researchers of Jena University proposed a large pitch fiber (LPF) to increasing the loss of higher-order modes and increase the threshold for mode instability in the fiber. Due to its open waveguide structure, it has a strong non-localizing effect on higher-order modes, significantly reducing the gain of higher-order modes. Thus, the large pitch fiber can increase the power threshold for mode instability, with experimental results showing an increase of about three times [66]. Researchers from the Technical University of Denmark proposed a distributed mode filtering (DMF) fiber [67]. By designing the refractive index distribution of the fiber, higher-order modes are coupled into the high refractive index cladding and do not propagate in the core, thereby suppressing higher-order modes. This structure can increase the threshold power for mode instability in fiber amplifiers based on large mode area (LMA) fibers, with experiments showing an increase of 1.5 times. Researchers at Jena University also proposed that the increasing threshold mode instability can be achieved through partial doping [68]. Partial doping refers to a doping area that is smaller than the core area. Such a doping structure can result in different gains for the fundamental mode and higher-order modes, achieving the effect of suppressing higher-order modes and thus increasing the threshold power for the onset of mode instability. In addition to changing the inherent properties of the fiber to increase the threshold for mode instability, it is also possible to increase the threshold by optimizing the heat management [69], using tandem pumping [70], and increasing the power of seed [71].

5.2.2.5 Loss

Loss is one of the most important fiber parameters. For cladding fiber, the loss of its core mainly depends on the loss of the core material itself and the boundary loss between the core and the cladding. The background loss of rare earth-doped fiber is usually less than 10 dB/km@1200 nm, while the loss of passive fiber is close to the theoretical limit, about 0.2 dB/km at 1.5 μm.

For photonic crystal fibers, in addition to the background loss, there is also structural limitation loss. Its background loss can be described by the following formula:

$$\alpha_{dB} = A/\lambda^4 + B + \alpha_{OH} + \alpha_{IR} + \alpha_{uv} \tag{5.7}$$

In the formula, the first term represents the loss results from Rayleigh scattering, the second term represents the loss caused by the waveguide defects of the fiber, the third and fourth terms represent the loss of hydroxyl and infrared absorption, and the last term represents the loss caused by ultraviolet absorption. Hydroxyl absorption, ultraviolet absorption, and waveguide defects composite the background loss of fibers mainly at near infrared wavelengths.

The loss caused by hydroxyl absorption in photonic crystal fibers basically exceeds 10 dB/km at 1380 nm, and the loss caused by this is about 0.1 dB/km at 1550 nm, which is similia to cladding fiber. Therefore, for photonic crystal fibers working at 1.5 μm, the dehydroxylation process can effectively reduce the loss. The loss caused by hydroxyl absorption has a small effect on the wavelength of 1 μm. However, the tail of ultraviolet absorption at the wavelength of 1 μm has a significant impact on the loss of fibers, so it is necessary to reduce the Fe ions in the fiber material as much as possible, thereby reducing the impact of ultraviolet absorption on the loss at the wavelength of 1 μm.

Waveguide confinement loss (leakage loss) must also be considered in photonic crystal fibers. The confinement loss is caused by the finiteness of the structural layers in photonic crystal fibers. Therefore, all waveguide modes are "leaking" during transmission along the photonic crystal fiber. The structure of photonic crystal fibers should to guarantee the leakage loss of high-order modes are large enough and the leakage loss of fundamental mode is small, to ensure the single-mode characteristics of the fiber. The larger the core diameter, the greater the structural loss. The structural loss can reach 0.5 dB/m in large mode area photonic crystal fibers. Therefore, the absorption of photonic crystal fibers is usually large, and the length of fiber used is short.

If the confinement of the air holes is not enough, fundamental mode can escape from the core to the cladding. Therefore, designing reasonable air hole size and hole spacing to ensure that fundamental mode have low confinement loss is important. Reasonable design can reduce the loss to a negligible level. Theoretical studies have shown that confinement loss strongly depends on the number of layers of air holes which is shown in Fig. 5.31. Hence, the number of layers of air holes can be appropriately increased to reduce confinement loss. In Fig. 5.31a, the confinement loss is significantly reduced with increasement of air holes layers. Figure 5.31b shows that, the confinement loss decreases as the diameter of the air hole increases when the air hole spacing is fixed. However, as the air filling factor increases, it is also to significantly reduce the loss of high-order modes, degrading the mode properties of the fiber.

The loss of photonic crystal fibers can also be tested using the cutback method. However, since it usually has an air hole clad, the light transmitted in its cladding cannot be stripped easily, and hence the test results maybe include the loss of both the core and the cladding. The loss in two polarization directions maybe different for polarization-maintaining photonic crystal fibers.

5.2.2.6 Bending Loss

Loss will be significantly increase if the bending radius of the fiber is small enough. Bending can give rise to mod couple from fundamental mode to higher-order modes, thereby leaking into the cladding. For total internal reflection photonic crystal fibers, large air filling factor (i.e., large d/Λ) can reduce the loss caused by bending. But air

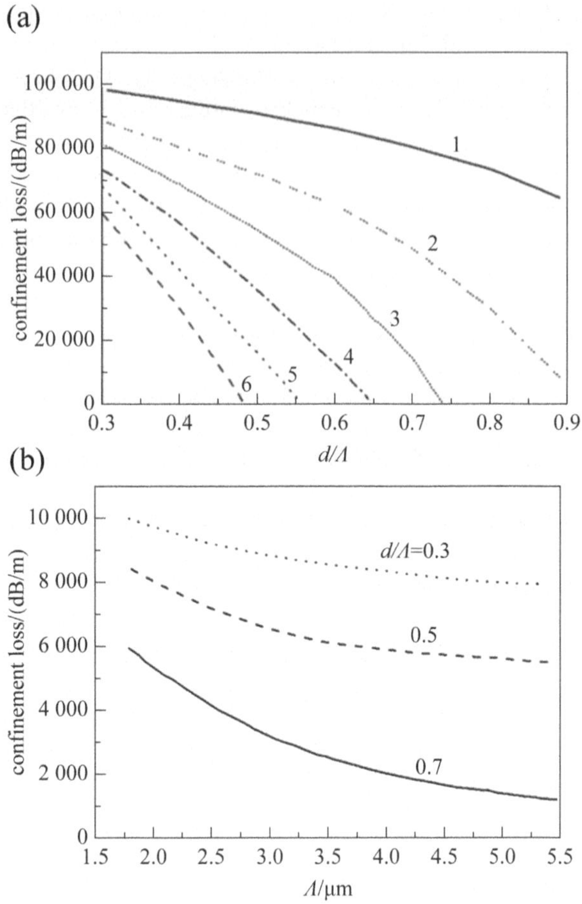

Fig. 5.31 The effect of air holes layers (**a**) and parameters of air holes (**b**) on confinement loss [72]

filling factor is limited in larger mode field photonic crystal fibers to maintain good modal properties.

There have been many studies about the bending loss of cladding fibers. But these methods are based on the circularly symmetric core distribution of the fiber, and can't be applied for photonic crystal fibers. Literature [73] reported a theoretical method to effectively simulates the bending loss of photonic crystal fibers, and proposed two mechanisms that affect bending loss: transition loss and pure bending loss. Transition loss mainly occurs in the fiber where bending starts and ends. Transition loss can be considered as a kind of coupling loss, mainly because the mode field does not completely match when the fiber is bent and when the fiber is straightened. Pure bending loss occurs in the middle bending part of the fiber, which is caused by the transmission difference of the mode between peripheral region and central region. Mode in two regions cannot maintain a stable phase difference. Literature [74] gives the expression of bending loss of hexagonal photonic crystal fiber:

$$\alpha \simeq \frac{1}{8\sqrt{6\pi}} \frac{1}{n_s} \frac{\Lambda}{A_{eff}} \frac{\lambda}{\Lambda} F\left(\frac{R}{6\pi^2 n_s^2 \Lambda} \left(\frac{\lambda}{\Lambda} \right)^2 V_{PCF}^3 \right) \tag{5.8}$$

$$F(x) = x^{-1/2} \exp(-x) \tag{5.9}$$

Power attenuation:

$$P(z) = P(0) \exp(-2\alpha z) \tag{5.10}$$

where, n_s is the refractive index of the fiber core; A_{eff} is the effective mode field area; λ is the wavelength in use; Λ is the air hole span; V_{PCF} is the normalized constant of the photonic crystal fiber; R is the bending radius. It has approved that experiment results agree well with the calculated results.

For photon crystal fibers with smaller core diameters, their numerical aperture of core is larger, so the bending loss is small. For large mode area photon crystal fibers, their numerical apertures of core are around 0.03 even smaller. Figure 5.32 introduces the bending resistance characteristics of NKT's 40 μm core diameter ytterbium-doped photon crystal fiber products. It can be seen that when the bending diameter is greater than or equal to 30 cm, the loss caused by bending is small.

5.2.2.7 Birefringence

An ideal fiber has symmetrical cross-section and refractive index, and the orthogonal HE $_{x11}$ mode and HE $_{y11}$ mode have the same transmission constants and mutually degenerate, hence can keep the polarization direction of linearly polarized light unchanged without external disturbances. But the fiber is not so perfect, or under external disturbances, two modes are coupled to each other, reducing the polarization of the transmitted light. Therefore, the difference between the propagation

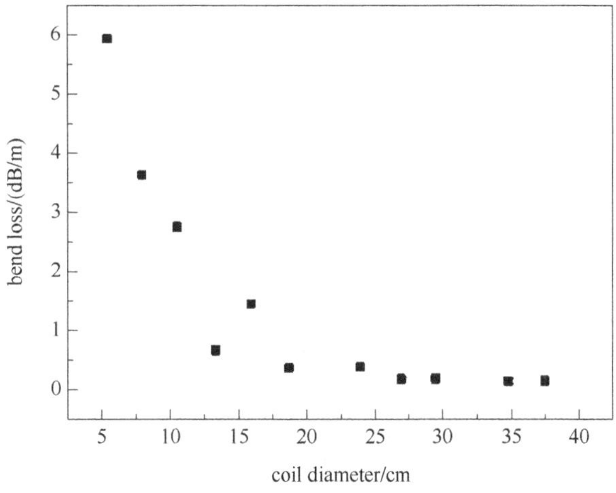

Fig. 5.32 Bend loss as function of coil diameter

constants can be artificially increased to reduce the coupling efficiency and make birefringence-polarization maintaining fiber. The one with a larger propagation constant is called the slow axis, and the one with a smaller propagation constant is called the fast axis. The value of birefringence can be represented by birefringence *B*:

$$B = \frac{\Delta\beta}{k_0} = \frac{|\beta_x - \beta_y|}{k_0} = \frac{|n_{eff}^x k_0 - n_{eff}^y k_0|}{k_0} = |n_{eff}^x - n_{eff}^y| \tag{5.11}$$

B represents the difference of effective refractive indices of the two orthogonal modes. The larger, the better to polarization maintaining. In photonic crystal fibers, high birefringence can be achieved by increasing the asymmetry of the geometric dimensions or changing the stress in a certain direction. The symmetry can be break by altering the shape, diameter, or arrangement of the air holes near the fiber core. Anisotropic materials are filled around the fiber core to also increase the stress in a direction.

Boron-doped silica rods are usually used as the material for the preparation of polarization maintaining fibers. For cladding fibers, the diameter and position of the boron rod need to be considered for their impact on birefringence. For photonic crystal fibers, the number and structure of boron rods need to be considered for their impact on birefringence. Figure 5.33 shows the impact of the number of boron rods on the birefringence. When the diameter of the boron rod is 14 μm, the impacts of the birefringence introducing by 20 boron rods and 32 boron rods on core are shown. It can see that birefringence introduced by 32 boron rods is more flat and higher. Since refractive index of the fiber core has strict fluctuation limitation, appropriate number of boron rods should be selected to obtain the ideal birefringence in photonic crystal fibers.

Due to the high stress of the boron rod, it is difficult to completely remove the outer pure silica deposition tube during processing. The thickness of the residual silica layer on the outside of the boron rod also affects the birefringence of the photonic crystal fiber. Figure 5.34 shows the effect of different silica layer thicknesses on the birefringence (the light-colored circular area in the illustration represents the

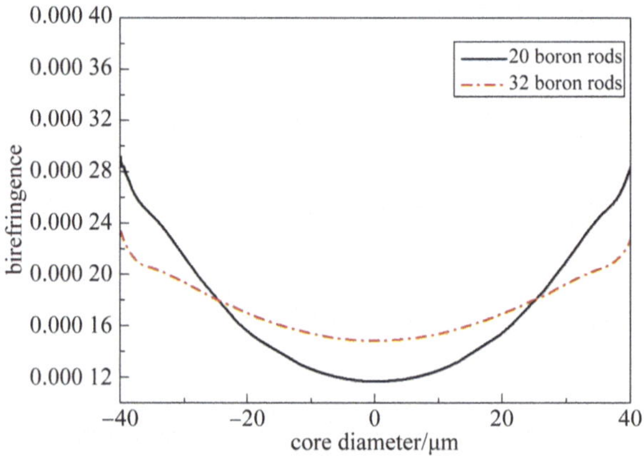

Fig. 5.33 The impact of the boron rods number on the birefringence of photonic crystal fibers

boron rod, and the dark hexagonal area represents the residual silica, with the hexagonal diameter fixed at 15 μm). As shown, the thickness of the silica layer has a significant effect on the birefringence when it changes from 0 to 3 μm. When the thickness of the silica layer is large, the fiber may not be able to obtain sufficient birefringence.

Because the effective refractive index of the core of the large-mode-area photonic crystal fiber is small compare with silica, the birefringence can cause a larger limiting loss on its fast axis, thereby forming a polarization effect. The introduction of NKT's DC-200/40-PZ-Yb type photonic crystal fiber indicates that under certain usage conditions, its loss bands on the fast and slow axes has significant difference during 1000 ~ 1200 nm as shown in Fig. 5.35. The polarizing fiber structure yields a strong attenuation for the fast axis (y-polarization) for wavelengths <1200 nm, while the slow axis (x-polarization) is well guided down to <1000 nm.

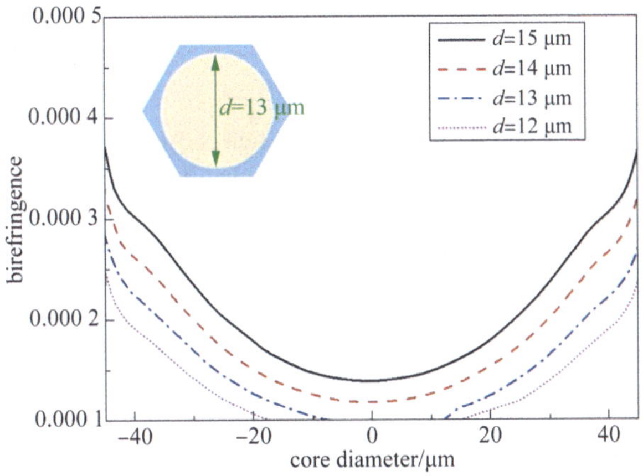

Fig. 5.34 The effect of boron rod size on the birefringence of photonic crystal fiber

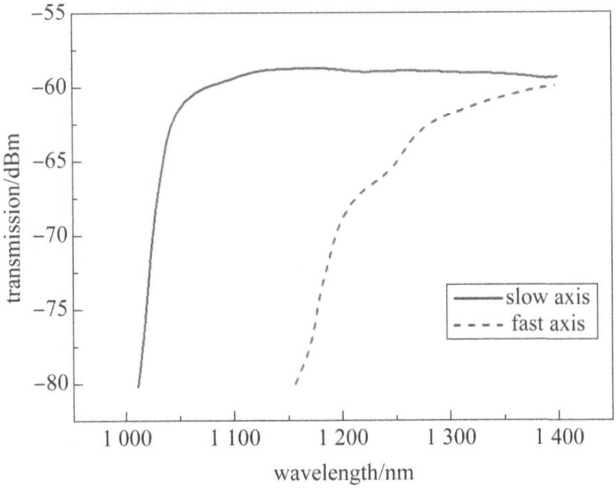

Fig. 5.35 Transmission scan for the two polarizations

5.3 Processing of Air Hole Structure Ytterbium-Doped Large Mode Area Photonic Crystal Fiber

Air hole structure photonic crystal fiber can use a single material, and various properties can be obtained by adjusting the arrangement and diameter of air holes, there are many researches on air hole photonic crystal fiber. A well-processed fiber can reduce fusion loss, improve coupling efficiency, and thus improve system stability. However, the presence of air holes brings new challenges to processing of photonic crystal fiber, which can easily cause tearing, fragmentation, and other phenomena, leading high fusion loss, and even end face damage. Therefore, the processing of photonic crystal fiber is very critical. The next section will introduce cleaving, coupling, fusion, and damage of air hole photonic crystal fiber.

5.3.1 End Face Processing

The condition of fiber end face must be good when is used. The end face processing to photonic crystal fiber is same with that to cladding fiber. Both fibers can be cleaved or polished by machine. Photonic crystal fiber has many air holes inside, its processing technology is relatively complex, hence it is difficult to obtain an ideal end face. High temperature increasing even damage will be induced by imperfect end face.

The slivering and polishing are two methods to get good end face of crystal fiber.

5.3.1.1 Cleaving

The vibration waves propagate along the direction perpendicular to the fiber until the tension breaks the fiber. The air holes in photonic crystal fibers would prevent the transmission of vibration waves, and the vibration wave may transmit along the edge of the air holes or be deviated the direction. This phenomenon can be observed from the cross section of fiber shown in Fig. 5.36. There are several gray "traces" origin from the air hole. If the depth of the "trace" is small, it has basically no effect on the next processing. If the depth of the "trace" is large, it may be the original of bubbles during fusion.

Cleavers such as Furukawa CT series can deal the photonic crystal fiber which the diameters are below 125 micrometers, and get high quality. Other cleavers such as the Vytran LDC series can adjust many parameters like tension, so are suitable to deal with the fibers which the diameters are large. For photonic crystal fibers, the tension should be lower, and the tension needs to be adjusted according to the structure of the photonic crystal. Overall, if the area of air holes occupies a larger proportion of the entire fiber cross section, the tension needs to be lower. The obtained

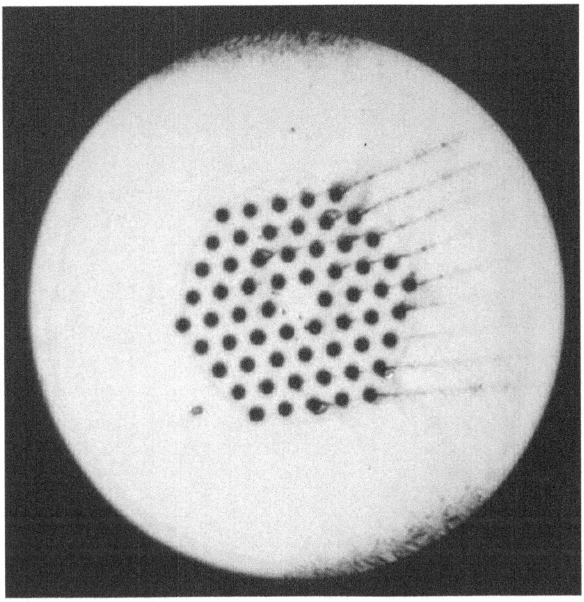

Fig. 5.36 The impact of air hole effect on slivering of photonic crystal fiber

fiber end face should have the angle less than 1°, and there is no damage when the impact of air holes is decreased as possible.

If the outer diameters of photonic crystal fibers exceed 125 μm, many parameters need to be optimized to obtain a good end face. Especially for photonic crystal fibers with "large and thin" air holes, it will heighten challenge. Cleaving may be fall with if the parameters are not optimized carefully. You may get big angle, teared, even crashed end face. The difficulty of cleaving is further increased when the photonic crystal fiber contains stress rods. The stress rods in the fiber have large stress to achieve polarization maintaining characteristic, meanwhile they make the fiber "fragile". The tension should be lowest. The end face of NKT's DC-200/40-PZ-Yb fiber is shown in Fig. 5.37 with optimized cleaving parameters. There are few "traces" along the air holes in the outer clad, and there are superficial "traces" in the inner clad.

Using a small cutting tension may not achieve the cut; using a large cutting tension may cause the cladding to bear a large stress before the cutting oscillation wave has reached the cladding, causing the cladding to be torn under high tension, forming a large cutting angle, and even causing the end face to appear "shattered". When the oscillation wave converges on the other side of the cutting point, a large "convergence point" will be formed due to its different transmission distance and direction. The difficulty of cutting is further increased when the photonic crystal fiber still contains a stress area. The stress area in the fiber usually has a large stress to achieve good polarization maintaining performance. For cutting, large stress means

Fig. 5.37 Microscope image of fiber slivered by optimized parameters

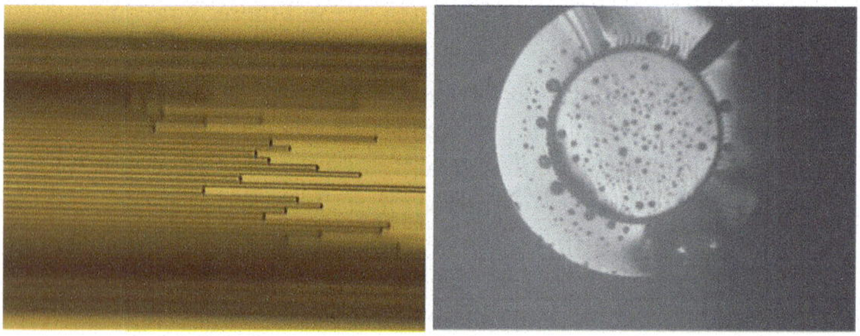

Fig. 5.38 Capillary siphon effect of air holes photonics crystal fibers

"easy to shatter". Therefore, for fibers with stress areas, lower cutting tension should be used to avoid "shattering" of the stress area. The optimized cutting end face of NKT's DC-200/40-PZ-Yb type fiber is shown in Fig. 5.37. As can be seen from the figure, there are "drawings" around the air holes in the outer cladding, and there are shallow "drawings" in the inner cladding.

Alcohol will be sucked into the air holes due to the capillary siphon effect when cleaning the fiber with alcohol. The alcohol may not evaporate completely, hence the residual alcohol would contaminate the fiber, as shown in Fig. 5.38 [75]. Therefore, further processing is needed to ensure that there is no residue after cleaning.

5.3.1.2 Polishing of Photonic Crystal Fiber end Face

Polishing the fiber can obtain a perfect end face which may have arbitrary angle. For air-hole photonic crystal fibers, polishing powder is easy to enter the air holes but is hard to clean. Therefore, before polishing of photonic crystal fibers, the air holes need to be collapsed or blocked first. It is hoped not to destroy the structure of the air holes which determine the mode characteristics of the fiber. It is necessary to explore a suitable process to collapse or block the air holes when polishing is used.

The collapsing of air holes can be achieved by heat source supplied by electrode or CO_2 laser or Hydrogen-oxygen flame. After precisely controlling of the polishing length, the fiber can have collapsed length in the order of tens of micrometers, which is convenient for cleaning the end face after polishing. The resident collapsed fiber can be consider as the end cap of the fiber.

The collapsed length of the fiber should be short, and the change of air holes along the radial direction can be consistent. Figure 5.39 shows the changes of air holes obtained with different collapse processes. As can be seen from Figure 5.39a–c, the change of air holes is becoming consistent with the improvement of the collapse processes. Figure 5.39d–f show the side and end images of the photonic crystal fiber obtained with optimized parameters [75], from which we can see that there are no "traces" of the air holes in the inner and outer cladding. Compared with cleaving, the process of fiber collapse and polish is relatively complex, requiring more steps and time.

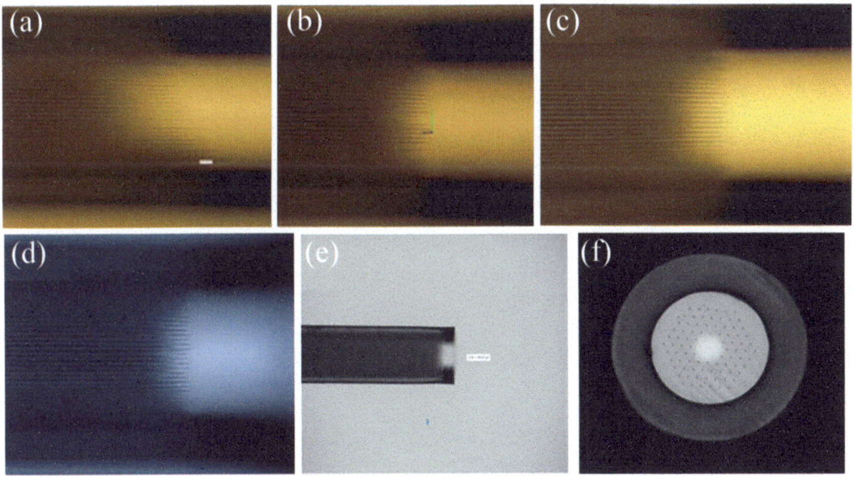

Fig. 5.39 The microscope images of photonic crystal fiber processed by collapsing and polishing: (**a–c**) side images of fiber with different air holes collapsing parameters; (**d–f**) side and end faces of the photonic crystal fiber prepared by the polishing method with optimized processing parameters [75]

Regardless of cleaving or polishing the photonic crystal fiber, the angle of end face is important. When the end face of the fiber is directly used as a mirror of a laser cavity, the angle end face must be small, because the core NA of the photonic crystal fiber is usually small, and a large end face reflection angle will cause the reflected light to enter the cladding, which is equivalent to increase the loss of the laser cavity and reducing the laser efficiency. The larger the mode field diameter of the fiber, the greater impaction of the end face angle on loss. Figure 5.40 shows the trend of reflection loss caused by the end face angle when the end face is used as the mirror of the laser cavity. As can be seen, the back reflection loss of fiber with a 30 μm core is much more sensitive to the angle than that with 15 μm core.

The smaller angle of the end face, the smaller splicing loss with the same process. So, the end face angle used for fusion is generally less than 1 ° is considered acceptable. When the photonic crystal fiber is used in a spatial amplification system, more attention is paid to the damage of the fiber end face, and the impact of angle can be eliminated by adjusting the frame. In order to avoid self-oscillation feedback between the two end faces, it is necessary to coat the two end faces with an anti-reflection film, or make the end face have a certain angle.

5.3.2 *Coupling*

Photonic crystal fibers with smaller core diameters, Gaussian light can be coupled into the core by butt-joint, splicing, and spatial coupling. For photonic crystal fibers with large mode field area, spatial coupling is usually used.

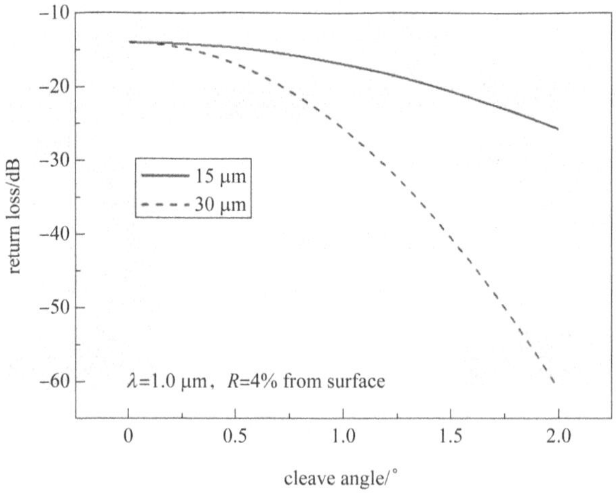

Fig. 5.40 The impact of slivering angle on back reflection loss

When beam coupling is performed by butt-joint, a metal connector (FC/PC), or a sub-Miniature A connector (SMA) can be chosen. For FC/PC connectors, the transmission power is limited at the watt level. For SMA connectors, it can withstand tens of watts of power. SMA connectors with stripped light windows can withstand hundreds of watts of power. Different connectors are shown in Fig. 5.41.

When coupling with a connector (Fig. 5.42), the factors affecting the coupling efficiency are the mode field diameters ω_1, ω_2 of the two fibers, and the lateral offset Δx, the Fresnel reflection of the two coupled fiber and the angles of end faces $\Delta\theta$.

The coupling efficiency of the fibers with mode field diameters ω_1, ω_2 is shown as follow:

$$\eta_1 = \frac{4\omega_1^2\omega_2^2}{\omega_1^2 + \omega_2^2} \tag{5.12}$$

The affection of lateral offset Δx on coupling efficiency is shown as follow:

$$\eta_2 = \exp\left[\frac{2(\Delta x)^2}{\omega_1^2 + \omega_2^2}\right] \tag{5.13}$$

If there is a gap between the two fibers, the Fresnel reflection of the two coupled fiber ends affects the coupling efficiency is shown as follows, In the formula, n is the refractive index of the fiber core.

$$\eta_3 = 1 - \left(\frac{n-1}{n+1}\right)^2 \tag{5.14}$$

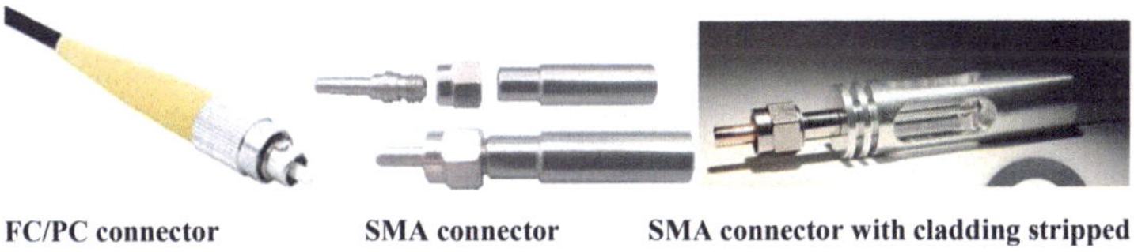

Fig. 5.41 Different types of fiber connectors. (**a**) FC/PC connector. (**b**) SMA connector. (**c**) SMA connector with cladding stripped

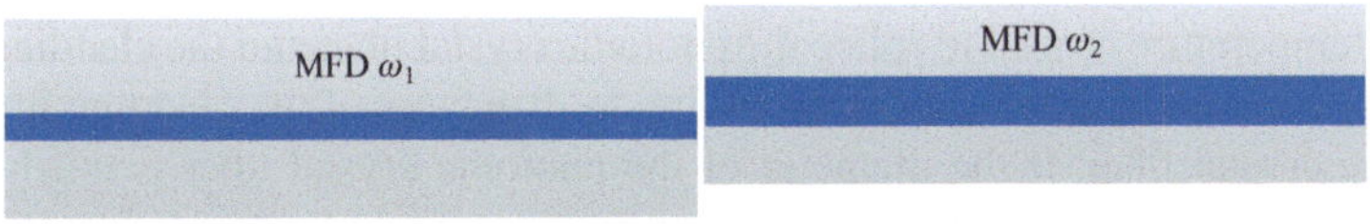

Fig. 5.42 The schematic of coupling between two fibers

If the two fibers are cut at an angle $\Delta\,\theta$ (assuming both fibers are should be flat), the impact on the coupling efficiency is as follow:

$$\eta_4 = \exp\left[-\left(\frac{\pi\Delta\theta\omega}{\lambda / n}\right)^2\right] \tag{5.15}$$

Considering the above factors, when two optical fibers are coupled by butt-joint, the coupling efficiency is determined by $\eta = \eta_1\,\eta_2\,\eta_3\,\eta_4$. Therefore, to achieve high coupling efficiency, it is necessary to minimize the impact of each factor.

The appropriate lens can be selected to shape the coupling light when using spatial coupling. If coupling the light output from HI1060 fiber into the core of NKT's 40/200 photonic crystal fiber. The mode field diameter of the HI1060 fiber is about 6 μm, the numerical aperture is 0.12, the core NA of the NKT40–200 fiber is 0.03, and the mode field diameter is about 30 μm. Choosing the focal length $\frac{f_1}{f_2} = \frac{\omega_{PCF}}{\omega_{HI}} = 5$ lens group for light shaping can efficiently couple the light emitting from HI1060 fiber into the photonic crystal fiber.

The main difficulty is to achieve the low-loss splicing between the photonic crystal fiber and the cladding fiber when photonic crystal fibers are used in MOPA(Master Oscillator-Power Amplifier) system. Alpha company in France cooperated with NKT company to make a full fiber amplifier with NKT's 40–200 Yb^{3+} doped photonic crystal fiber, which can be conveniently spliced with the cladding fiber, and added a self-developed mode matcher, and got the splicing loss is about 0.5 dB, and achieved 75% amplification efficiency and 250 W amplification output. There are many studies on the splicing of photonic crystal fibers and cladding fibers in China. The research group of Hou Jing from the National University of Defense Technology has done a lot of work in the post-processing of photonic crystal fibers, but mainly focused on nonlinear photonic crystal fibers [76]. The research group led by Xiao Limin from Fudan University has achieved a 0.25 dB splicing loss between single-mode fiber and 25 μm photonic crystal fiber [77].

The angle of the splicing end face must be small dues to the lower splicing temperature in order to avoid destroying the air holes. To obtain a splicing end face with small angle, the cleaving process is changed according to outer diameter of the photonic crystal fiber, the proportion of the air holes, and whether there are stress rods. When air holes occupy a small proportion, the cleaving process is like that of cladding fiber with the same size. When the air holes occupy a larger proportion, more adjustment is needed to obtain an ideal splicing end face.

The temperature is need to splice the photonic crystal fiber and the cladding fiber. The proper temperature is chosen according to diameters of the cladding fiber and photonic crystal fiber. If the diameter of the photonic crystal fiber is much larger than that of cladding fiber, the splicing temperature should be decided by cladding fiber. If the diameter of the photonic crystal fiber is comparable to cladding fiber or smaller, the splicing temperature should be decided by photonic crystal fiber. The fusion time should be short. The fusion strength of the photonic crystal fiber is relatively weak, and the fusion point needs to be protected carefully. Figure 5.43 shows the splicing image of the photonic crystal fiber and the cladding fiber. As shown, the air holes of the photonic crystal fiber are not collapsed through optimized splicing processes.

Figure 5.44 shows the temperature increasement of the splicing point with the pump power. The splicing between the homemade Yb^{3+} doped 40–260 photonic crystal fiber and 25/250 μm polarization maintained cladding fiber is optimized carefully. As shown, the temperature of splicing point is about 45 °C at 100 W@976 nm pump power, which is acceptable even without active cooling.

Because of the low temperature when splicing, if the cleaving angle of fused fiber is large or there are "traces" on the fusion surface, the bubbles are likely to form at fusion point, and bubbles cannot be eliminated by subsequent heating. The bubbles will reduce the coupling efficiency, cause temperature rising, and even damage. Without considering factors such as offset and bubbles, the fusion loss is

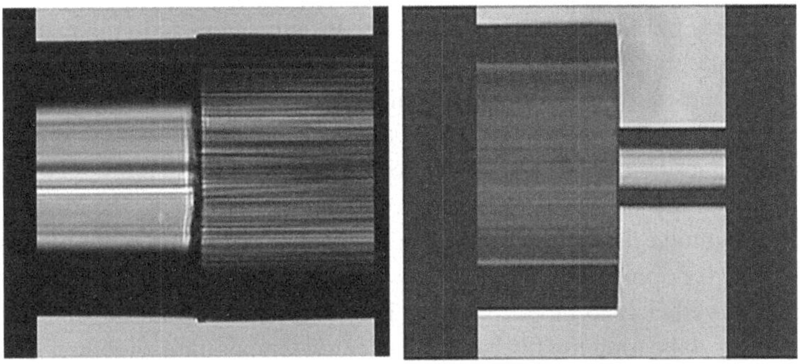

Fig. 5.43 Microscope images of splicing between photonic crystal fiber and cladding fiber

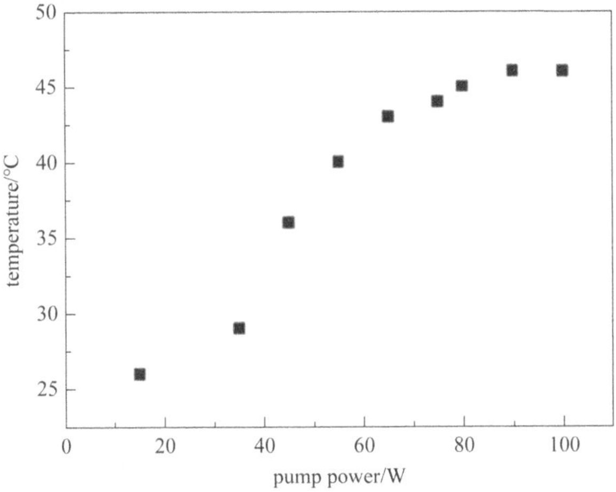

Fig. 5.44 Splicing point temperature under different pumping power

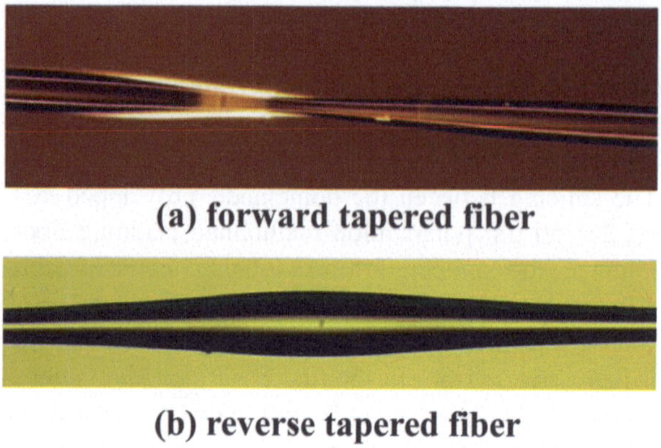

Fig. 5.45 Microscope images of forward and reverse tapered fibers [77]. (**a**) forward tapered fiber. (**b**) reverse tapered fiber

induced by mode field mismatch. In order to reduce the mode field mismatch, methods such as enlarging the mode field diameter of the cladding fiber or reducing the mode field diameter of the photonic crystal fiber can be adopted.

To increase the mode field diameter of the cladding fiber, methods such as core expansion or tapering can be adopted. Core expansion of the cladding fiber refers to the diffusion of the doped ions in the core to the cladding by heating, thereby increasing the mode field diameter of the fiber. This method is more effective for fibers with high doping concentration. Another method to change the mode field diameter of the fiber is to taper the fiber, including tapering and reverse tapering. Tapering refers to heating the fiber to the softening temperature, and then stretching it along the fiber. During the tapering process, the mode field diameter of the fiber's fundamental mode decreases first and then increases, so the tapering parameters can be determined according to the fiber that needs to be matched. Tapering can be used to prepare couplers, mode field adapters, fiber combiners, etc. Reverse tapering refers to heating the fiber to the softening temperature, and then advancing along the fiber. Figure 5.45 shows the appearance of tapered and reverse tapered fibers.

The methods to change mode field diameter of photonic crystal fibers are flexible. It can be achieved by change the air holes, thereby changing the mode field area of the photonic crystal fiber. Photonic crystal fibers can also be tapered. The post-processing of photonic crystal fibers is more flexible, but it requires more precise control.

5.3.3 Damage

Large mode field photonic crystal fibers are mainly used in high peak power, high energy laser systems, so the probability of damage is higher. The material damage caused by picosecond or nanosecond lasers is usually due to the energy transfer between the light-induced electrons and the fiber material, and the released energy

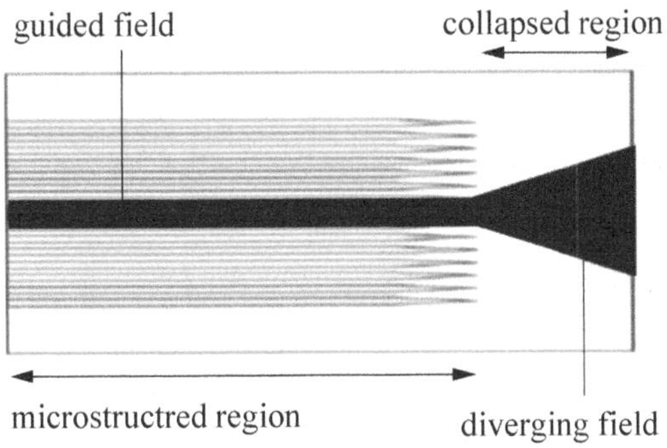

Fig. 5.46 Schematic diagram of collapsed photonic crystal fiber as endcap

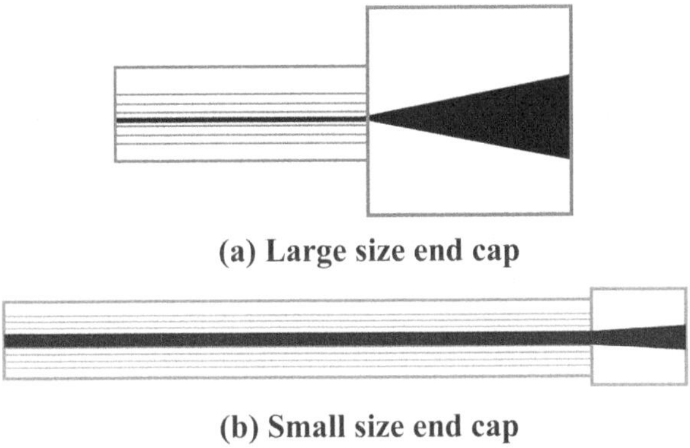

Fig. 5.47 Schematic diagram of different diameters endcaps. (a) Large size end cap. (b) Small size end cap

causes the fiber to melt or break, so increasing the output laser area can improve the end face damage power threshold of the fiber.

To enlarge the output laser area, an endcap needs to be prepared for the fiber. The most convenient method is to directly collapse the fiber as endcap at the output end of the photonic crystal fiber. When the laser transmits in the structure-collapsed area, there is no constraint of the waveguide, the laser spot gradually enlarges with the propagation distance, and the damage power threshold at the output end face increases. Its structure is shown in Fig. 5.46.

Splicing a section of pure silica as an end cap is a good method, and the out diameters of pure silica can be customized as needed. As shown in Fig. 5.47a, for photonic crystal fibers with ultra-large core diameters, the diameter of the output end cap is usually larger in order to obtain a larger output laser spot and increase the damage threshold. For photonic crystal fibers with smaller core, as shown in Fig. 5.47b, an end cap may have almost the same diameters. Larger diameter end cap can achieve higher damage threshold, but it is relatively difficult to splice with the photonic crystal fiber, while a small end cap is relatively easy to splice. Therefore. The fiber end cap can be coated with an anti-reflection film to reduce reflection loss.

(a) End cap damage **(b) Fiber damage**

Fig. 5.48 Microscope images of damaged endcap and fiber. (**a**) End cap damage. (**b**) Fiber damage

Even the fiber end cap can withstand larger energy, damage may occur, especially when the end cap is unclean. Therefore, it should keep the end cap clean.

Some damages may occur inside the fiber, such as core damage. The reasons of the damage are roughly as follows: one is the fiber damage caused by the self-pulse effect due to too low seed power in the amplification system; the second is the damage caused by the nonlinear effect. In addition, defects in the fiber during the preparation process, may cause fiber damage, as shown in Fig. 5.48.

5.4 Laser Performance and Applications of Ytterbium-Doped Large Mode Area Photonic Crystal Fiber

Russell's research group reported the first ytterbium-doped photonic crystal fiber at 2000 year, with 1.2 μm air hole spacing and 0.6 ~ 0.8 μm hole size. The core diameter is 1.6 μm, and the Yb doped area is 0.9 μm. Using a titanium sapphire laser as the pump, a 14 mW laser output power at 1040 nm was obtained [10]. Since then, the researches on ytterbium-doped photonic crystal fibers increased rapidly. Double-clad structure ytterbium-doped photonic crystal fibers, large-span ytterbium-doped photonic crystal fibers, and bandgap ytterbium-doped photonic crystal fibers are reported. The fibers with large effective mode field area can still maintain single-mode characteristics which greatly expanding the application in the field of ultra-short pulse amplification. The next sections introduce the laser performance of the photonic crystal fibers and their application.

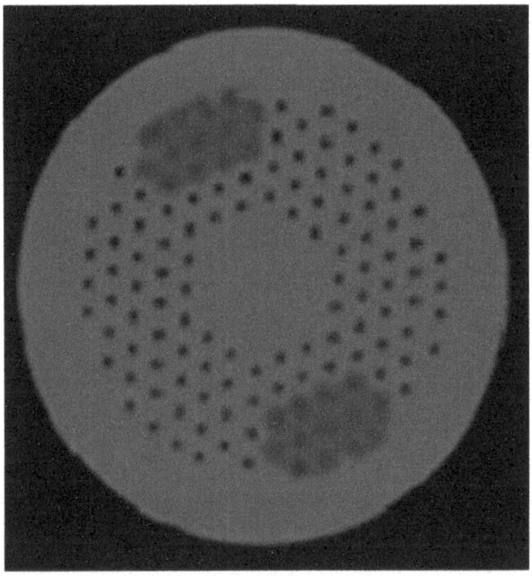

Fig. 5.49 Microscope images of 100 μm core diameter Yb^{3+} doped photonic crystal fiber [78]

5.4.1 Laser Performance of Large Mode Area Photonic Crystal Fiber with Air Holes

Starting from 2003, the University of Jena in Germany carried out many researches on ytterbium-doped photonic crystal fibers. They cooperated with the then Crystal Fiber A/S company, now NKT company, developed an ytterbium-doped photonic crystal fiber with a mode field diameter of 21 μm and a double-clad air hole structure. The air hole is 2 μm, and the hole spacing is 11.5 μm. Subsequently, the core diameter of the ytterbium-doped photonic crystal fiber continued to increase. At 2006 year, a double-clad air hole ytterbium-doped photonic crystal fiber with a core diameter of 100 μm was reported, with a mode field diameter of 75 μm, and the beam quality factor $M^2 = 1.3$. In 2011, Northrop Grumman Corporation achieved 60% amplification efficiency used this fiber with 85 cm long as the main amplification medium in its MOPA system, as shown in Fig. 5.49. 1.1 mJ pulse energy at 780 ps pulse width was achieved and corresponded peak power was about 1.35 MW. The beam quality factor $M^2 = 1.1$, which is shown in Fig. 5.50 [78].

Due to the influence of the stress on the refractive index of the fast and slow axes, this fiber shows single polarization characteristics in a straight state, as shown in Fig. 5.51. The power attenuated significantly when laser at1064nm polarized along fast axis injected in the core, while the power attenuation is small along slow axis.

In order to further increase the core diameter, the University of Jena [11] proposed a large-span air-hole structure photonic crystal fiber, and reported a large-span ytterbium-doped photonic crystal fiber with 135 μm core diameter and corresponded mode field diameter was about 100 μm at 2012 which is shown in Fig. 5.52. Using three-stage amplification, 26 mJ pulse energy and 130 W average power were obtained. Figure 5.52 shows the laser profile under different pulse energies, indicating that when average power was below 50 W, due to the anti-cross

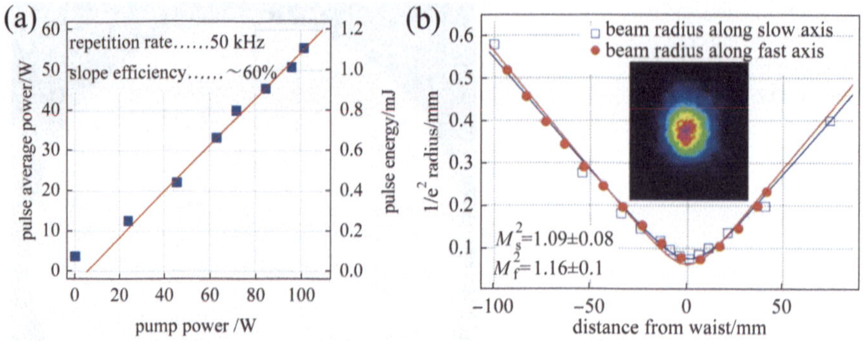

Fig. 5.50 (**a**) Pulse amplification efficiency curve and (**b**) beam quality factor of slow and fast axes

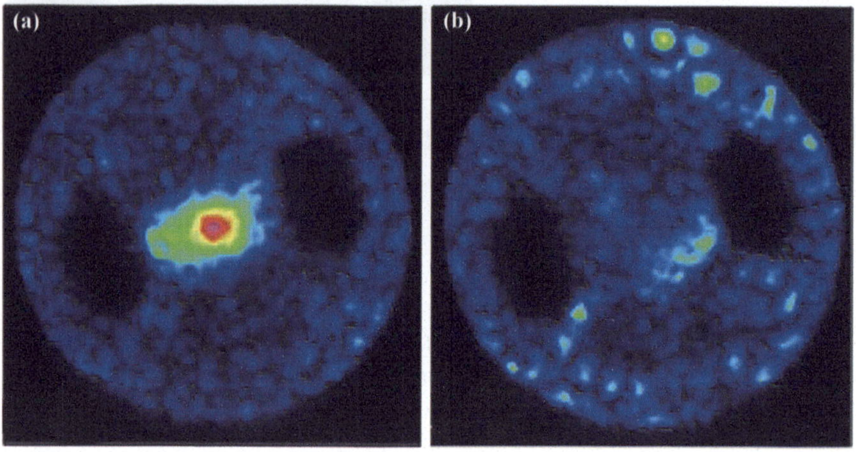

Fig. 5.51 Transmission characteristics of 1064 nm wavelength laser in the slow axis (**a**) and fast axis (**b**) of the fiber [78]

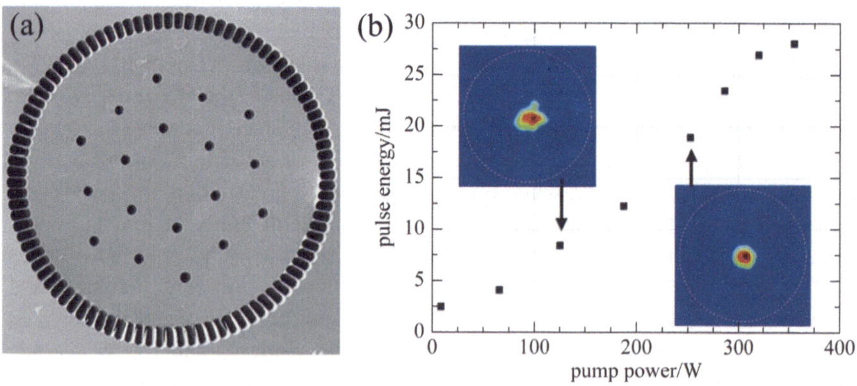

Fig. 5.52 (**a**) Electron scanning image of 135 μm large-span fiber and (**b**) change of beam quality with pulse energy [11]

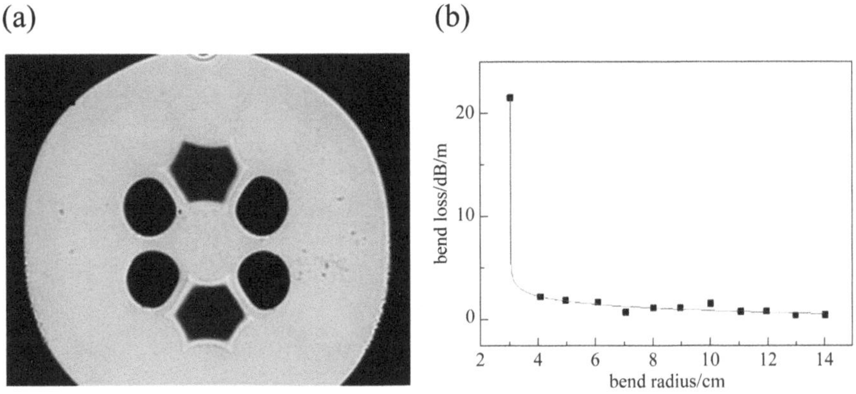

Fig. 5.53 (**a**) Cross section of fiber (**b**) Bend loss as function of coil diameter [77]

effect, part of the fundamental mode in the core leaks to the cladding area, causing the beam quality to deteriorate. When the amplification power was increased to above 50 W, the fundamental mode leaking to the cladding is reduced, and the beam quality is improved.

The results show that the energy transmitted in the core was 26 mJ when the output pulse energy was 28.5 mJ. The beam quality factor is 1.6 and was 1.3 with adding a pinhole to remove the energy in the cladding. This result indicates that the energy transmitted in the cladding would cause a significant degradation of the beam quality. When the amplification energy was less than 20 mJ, the tested mode field diameter of the fiber was about 100 μm. When the output energy reaches 26 mJ, the tested mode field diameter of the fiber was 90 μm. This is mainly due to the change of the core refractive index caused by the thermal optical effect.

The principle of leaky mode fiber is allowing light to leak out from the core, but controlling the loss of different modes through structure. The first ytterbium-doped leaky mode photonic crystal fiber was reported at 2006 [79]. At 2007, a polarization-maintaining ytterbium-doped leaky mode photonic crystal fiber with a mode field diameter of 42 μm was reported. The cross-section of this fiber was shown in Fig. 5.53. The stress birefringence was about 2.1×10^{-4}. This fiber has excellent bending resistance, and the induced loss was small when the bending radius was greater than 4 cm as shown in Fig. 5.53. As shown in Fig. 5.54, a 3 m long fiber was pumped by a 980 nm LD, and the amplification efficiency at 1025 nm is 60%. The laser profile was shown as the inset, and the beam quality factor was about 1.2 [80].

Yanshan University researched on photonic crystal fibers early. In 2009, Yue'e Chen et al. reported the home-made Yb^{3+} doped double-clad photonic crystal fiber [81]. The research groups of Jinyan Li from Huazhong University of Science and Technology and Lili Hu from Shanghai Institute of Optics and Fine Mechanics also carried out the researches on Yb^{3+} doped silica photonic crystal fibers. The research group of Jinyan Li prepared the core rod of photonic crystal fiber by porous glass method [82]. The research group of prepared the core rod of Yb^{3+} doped photonic crystal fiber by sol-gel method, and prepared Yb^{3+} doped silica photonic crystal fiber

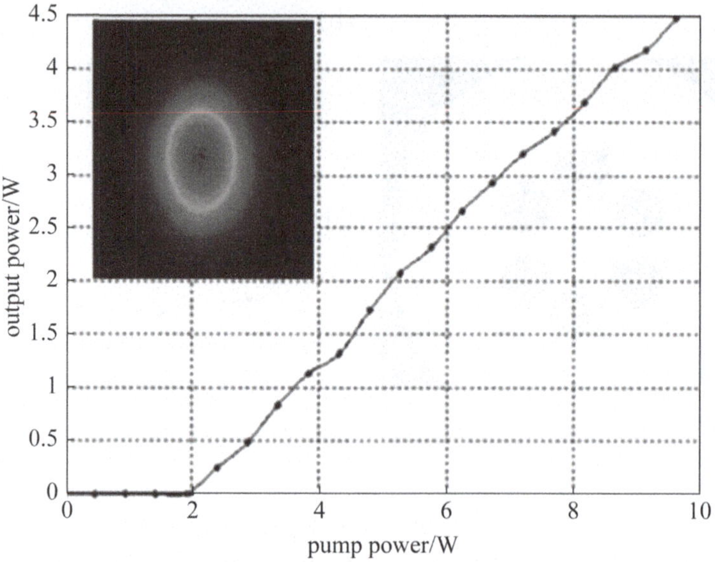

Fig. 5.54 Laser efficiency curve of the fiber [80]

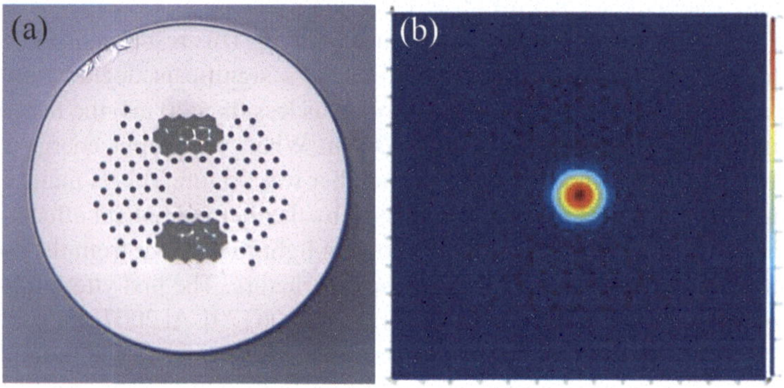

Fig. 5.55 (**a**) Microscope images of 40 μm core diameter Yb^{3+} doped photonic crystal fiber (**b**) Simulated intensity distribution of the fundamental mode [41]

with air holes and all-solid Yb^{3+} doped silica photonic crystal fiber [37–41, 83]. The following section focuses on the works of Huazhong University of Science and Technology and Shanghai Institute of Optics and Fine Mechanics on Yb^{3+} doped large mode field photonic crystal fibers.

When the diameter of core is 40 μm or less, the photonic crystal fiber still retains its "flexible" characteristics which means can be coiled. Hu Lili group has made Yb^{3+} doped polarization-maintaining silica photonic crystal fibers with a diameter of 40/260 μm by the sol-gel method [41]. The cross section of the fiber is shown in Fig. 5.55. The diameter of the core is 40 μm, the inner cladding is 260 μm, and a 6-layer micro-holes are arrayed to keep low confinement loss. A low refractive

Table 5.3 Parameters of 40 μm core diameter Yb^{3+} doped photonic crystal fiber [41]

Core diameter	≈40 μm
Inner cladding diameter	260 ± 10 μm
Pump absorption@915 nm	≈2 dB/m
Pump absorption @976 nm	≈6.5 dB/m
NA of core @1030 nm	≈0.03
NA of clad @976 nm	≈0.46
Mode area diameter@1030 nm	29 ± 2 μm
BirefringenceΔn@1080 nm	$\geq 1*10^{-4}$

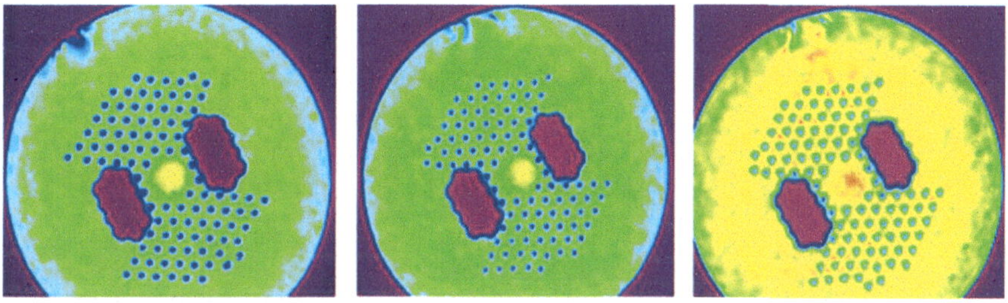

Fig. 5.56 Output Near-field images of the fiber for optimum alignment and gradually misalignment input beam

index acylate coating is used as the outer cladding. The result of theoretical simulation of the fiber shows that the overlap factor of the LP01 mode in the fiber with the core reaches 77%, the leakage loss of fundamental mode is small. Table 5.3 lists some parameters of this fiber.

Figure 5.56 shows the near-field images of the output from a fiber of 1.5 m length which was coiled with 30 cm diameter around 1 μm. The input single beam is about 6 μm and was moved along slow axis direction with the purposed of exciting high order modes. The output near-field images show that the fundamental mode for all alignment positions and no high order modes can be excited.

The amplification and diagnostics setups are shown in Fig. 5.57. The test involves measurements of: optical efficiency, near-field images, beam quality and time stability. The seed power is 2.4 W with 30 MHz repetition rate and 30 ps pulse width at 1030 nm. The pump and seed are coupled into the fiber from different fiber end faces. The length of fiber is 2 m and the bending diameter is 30 cm.

Figure 5.58 gives the amplification power curve. At 250 W pump power, average output power is 143 W, and the light-to-light efficiency is about 57%. The output beam profiles changes with output power are shown in Fig. 5.59. When the output power is higher than 100 W, part of the fundamental mode in the core leaks to the cladding area.

Figure 5.60 gives the beam quality factor M^2 under different output powers. When the amplification power is below 60 W, the beam quality factor does not change much, $M^2 = 1.1 \sim 1.2$. When the amplification power is higher than 60 W, the M^2 factor gradually increases. There are two reasons for the deterioration of the

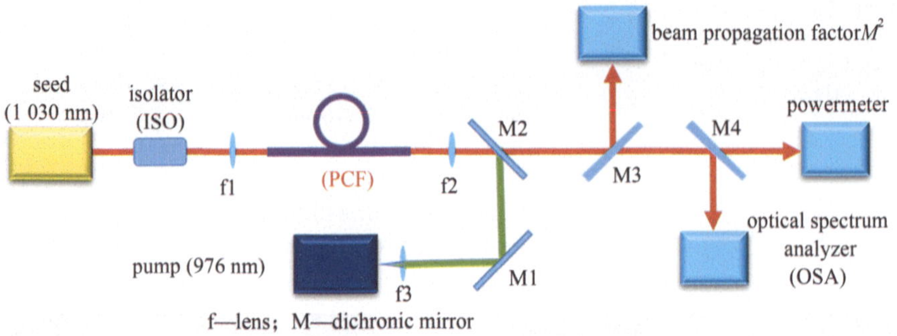

Fig. 5.57 Schematic diagram of the amplification optical path [41]

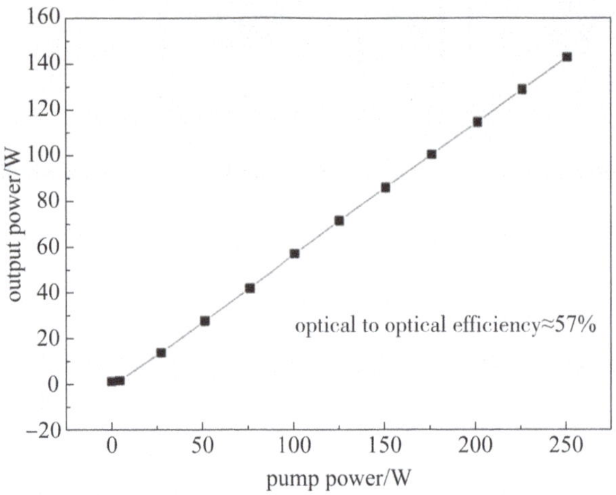

Fig. 5.58 Output power as a function of pumping power [41]

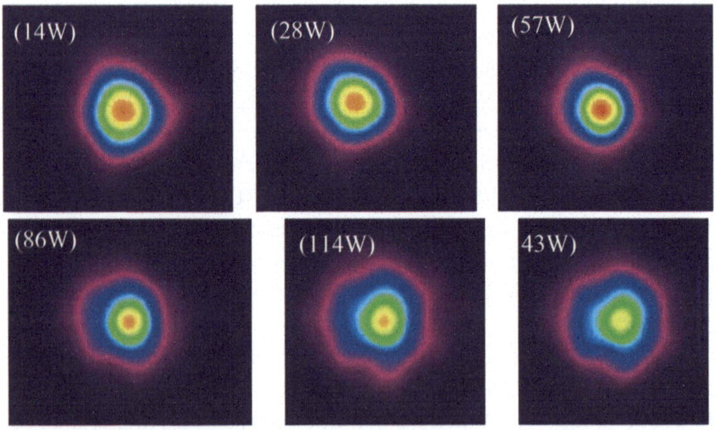

Fig. 5.59 Beam profiles at different powers [41]

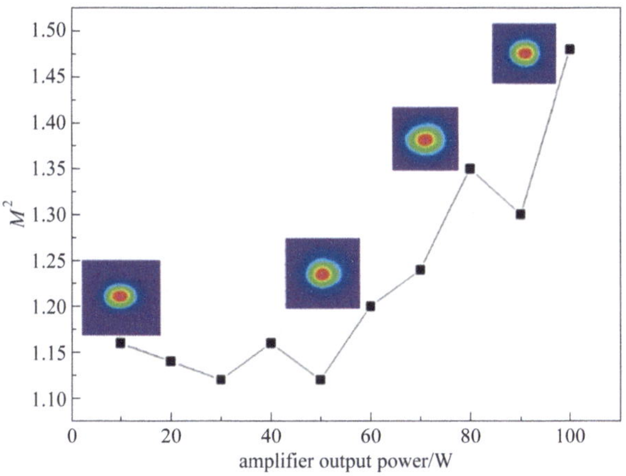

Fig. 5.60 Beam quality factor at different amplification power [41]

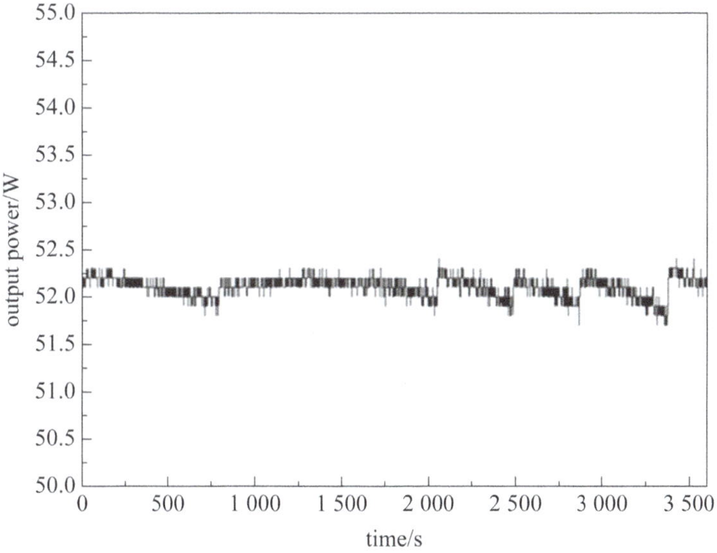

Fig. 5.61 Power fluctuation over time at 52 W

beam quality factor as the amplification power increases: one reason is the proportion of cladding light increases whit the amplification power increases; second reason is the loading heat in the fiber. Due to the thermal-optic effect, the refractive index of the fiber core will change accordingly, leading to an increase of gain of high-order modes, thereby degrading the beam quality.

When pump power is less than 100 W, output power still maintains stability without active cooling. When the pump power is higher than 100 W, active cooling is required. Fig. 5.61 shows the fluctuation of output power within 1 h under the condition of no active cooling at 100 W pump power. The power fluctuates in the range of

51.8 ~ 52.2 W, and the fluctuation in the latter part of the curve is due to the instability of the adjustment frame. When the pump power is higher than 100 W, through the infrared thermometer, it can be observed that the temperature of the fiber continues to rise for long time, which will cause the fiber coating to be damaged.

Polarization maintaining properties can be achieved either by adding stress regions or by structural asymmetry in photonic crystal fibers. The attenuation characteristics of the fast and slow axes under the straightened state of 1 m fiber were tested at 1 μm. As shown in Fig. 5.62, although the transmission intensity is different when laser polarized along fast and slow axis injected in the core. Amplifications along two axes also show that the fiber can maintain polarization but cannot polarize under test condition. The stress birefringence difference is about $\Delta n \approx 1.4 \times 10^{-4}$ measured by the interference method. The polarization extinction ratio PER measured at different amplification powers is all greater than 12 dB, as shown in Fig. 5.63.

In order to reduce heat loading in the fiber, the doping concentration of Yb^{3+} in the core material was reduced, and a 50 μm core silica photonic crystal fiber was

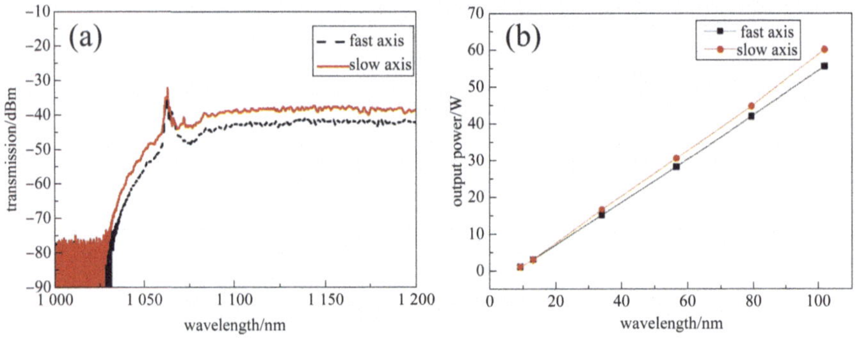

Fig. 5.62 (**a**) Transmission scan for the two polarization directions. (**b**) amplification curves along slow and fast axis of fiber

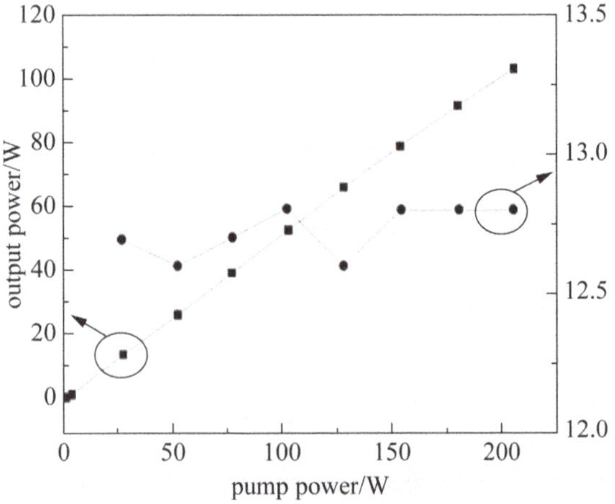

Fig. 5.63 Output laser power and PER as a function of the pumping power

prepared. In addition, in order to improve the tolerance to temperature, the outer air holes cladding is manufactured. Figure 5.64 shows the scanning electron microscope image.

Due to the low doping concentration of Yb^{3+}, the absorption coefficient of pump is small, so big fiber length is needed for amplification. Fig. 5.65 shows the power amplification curve with a 4.7 m fiber under a bending radius of 23.5 cm. The beam profiles

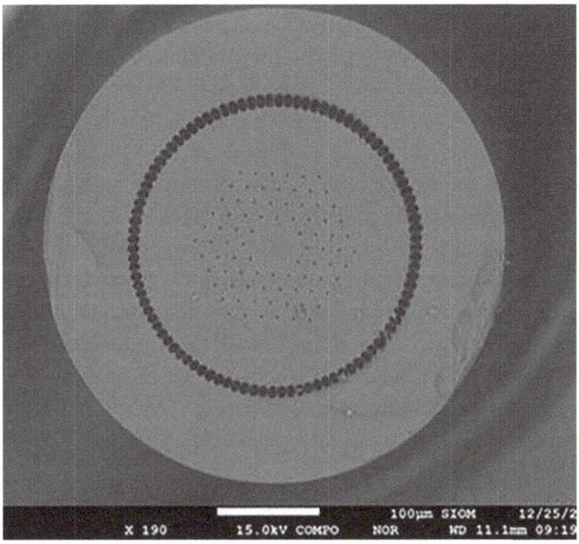

Fig. 5.64 Scanning electron microscope image of the fiber [39]

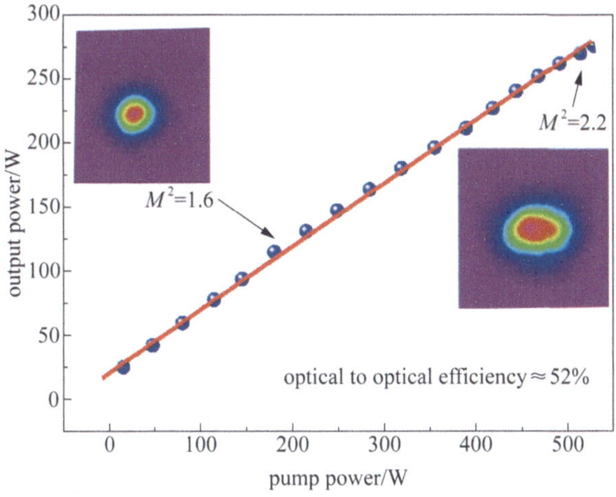

Fig. 5.65 Output laser power as a function of the pumping power and beam quality at different output power [39]

under different powers are also shown. At 520 W pump power of, 272 W amplified output power was obtained, and corresponding optical-optical efficiency is 52%. The beam quality factor M^2 is 1.6 at 120 W and 2.2 at 272 W. As the output power increases, the heat loading in fiber core increases and hence causes increase of refractive index of the fiber ultimately affecting the beam quality. Stability test at 120 W shows that output power fluctuation was 1.3 W during the nearly 2 h which is shown in Fig. 5.66.

Concludes from the above results, low Yb^{3+} doping concentration can resistance higher pupu power and get higher output power. At the same time, small of absorption coefficient means great fiber length is needed and the nonlinear threshold is low.

For larger core photonic crystal fibers, the fundamental mode loss is relatively large, so bending should be avoided. The Shanghai Institute of Optics and Fine Mechanics [40] manufactured 75 μm core Yb- doped photonic crystal fiber. The inner cladding diameter is 260 μm, and outer diameter is 450 μm. The numerical aperture of the cladding is about 0.55 to insure enough collection of pump. The absorption coefficient is about 15 dB/m@976 nm, hence 1 m fiber length can insure enough absorption of pump. The injected seed power is 5 W with 1 MHz repetition rate, 100 ps pulse width at1030 nm. the amplification characteristics are shown in Fig. 5.67. The amplification efficiency is 61%. 102 W amplified output was obtained at 166 W pump power, and corresponding pulse energy reached 102 μJ, the peak power exceeded 1 MW. The beam quality factor M^2 was about 2.1. Figure 5.68 shows the laser spectrum when the amplified output is 102 W, with no obvious nonlinear effects and a signal-to-noise ratio is greater than 30 dB.

Wang Pu of Beijing University of Technology [84] used a 100 μm core diameter Yb-doped photonic crystal fiber developed by Shanghai Institute of Optics and Mechanics in CPA amplification system. They achieved 103 W average power with 1 MHz a repetition rate 1 ps pulse width. Large core diameter photonic crystal fiber requires higher uniformity of the fiber core and consistency of the air hole structure, making it more difficult to manufacture.

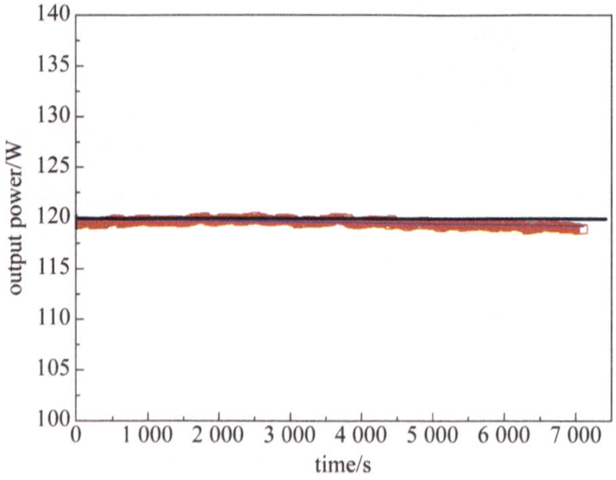

Fig. 5.66 Output power fluctuation over time at 120 W [39]

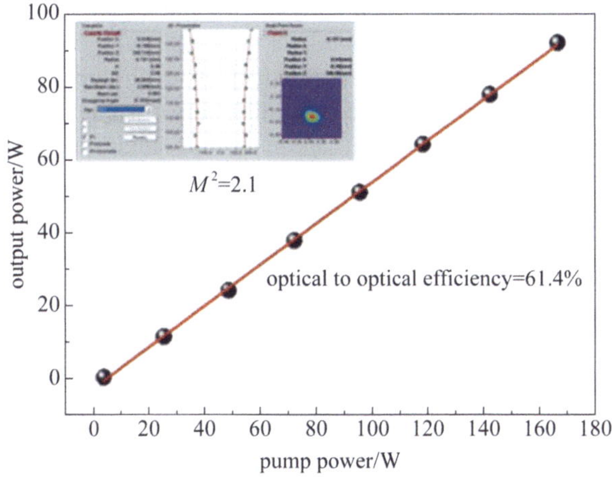

Fig. 5.67 Output laser power as a function of the pumping power. The inset is the beam quality factor [40]

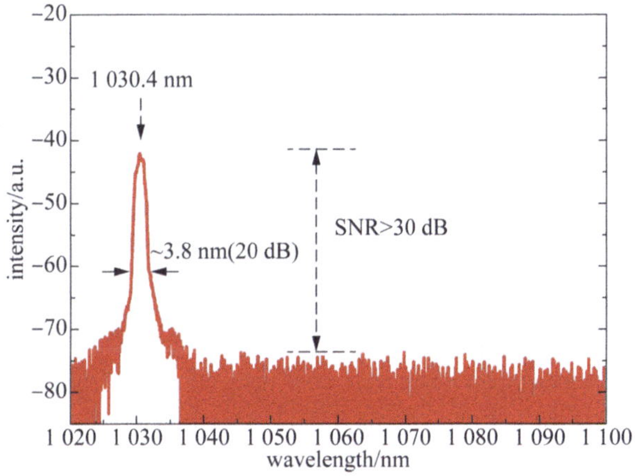

Fig. 5.68 Measured spectrum at 102 W output power [40]

5.4.2 Laser Characteristics of All-Solid-State Ytterbium-Doped Photonic Crystal Fiber

Solid-state materials are arrayed in all-solid-state photonic crystal fiber to optimize the fiber characteristics. Compared with air-holes photonic crystal fiber, all-solid-state photonic crystal fibers have some advantages: first, the all-solid structure is easy to slive and splice; second, it is not fragile, reducing storage difficulty; third, its high heat conduction compare with air holes is beneficial for fiber heat dissipation at high power operation.

The research group of Dong Liang at Clemson University in the United States [85, 86] has done many works on Yb-doped all-solid-state photonic crystal fiber, and reported different structural types. Bandgap type Yb-doped photonic crystal fiber cross section was shown in Fig. 5.69, with a core diameter of 50 μm, the fiber has ultra-high amplification efficiency and excellent single mode characteristics. There was no high order mode was excited during 1025 ~ 1040 nm [22].

In 2009, Dong Liang et al. [21] reported 46 μm core diameter all-solid Ytterbium-doped leakage mode photonic crystal fiber, as shown in Fig. 5.70a. Amplification characteristics were tested with different fiber length as shown in Fig. 5.70b. Single-pass gain can reach 33 dB.

They also reported a polarization-maintaining Ytterbium-doped leakage mode photonic crystal fiber structure with 80 μm core (Fig. 5.71a) [21]. the mode field diameter was 62 μm. 230 kW peak power was achieved with 4 m fiber length. The beam quality factor of the output pulse was about 1.35, and the polarization degree was about 15 dB.

In 2018, NKT company launched 30 μm core Yb-doped all-solid silica photonic crystal fiber. Its cross section and waveguide images was shown in Fig. 5.72a. As shown in the figure, the air holes are replaced by low refractive index rods making the fiber easy to sliver and splice. The splice loss of this fiber with 14/135 fiber can be less than 0.3 dB [87]. The near field image of output from fiber was shown in Fig. 5.73b. Under 25cm coiled diameter, he amplification efficiency was 83% with 3.5 m fiber. Output power stability test has been conducted. The power fluctuation was less than 1.5% over 10,000 h.

Domestic researchers focus on refractive index-guided all-solid ytterbium-doped photonic crystal fibers. Huazhong University of Science and Technology [36] manufacture the ytterbium-doped silica glass through gas phase doping MCVD technology. 50 μm core diameter all-solid photonic crystal fibers was made by replaced air holes with fluorine-doped silica rods. Figure 5.73 shows the fiber design schematic and the fiber microscopic image.

As design, fluorine-doped silica rods were arrayed around the core. The numerical aperture of the fiber core was about 0.03. Figure 5.74 shows the near-field beam profiles of output from fiber when a 100 mW 1064 nm single-mode insert light is moved across the core. There was no higher-order mode is excited under all situation. The beam quality factor is about 1.37.

The 4 m all-solid fiber was spliced to the fiber amplification system, and under the conditions of 100 ps pulse width, 500 kHz repetition rate, and 16.5 cm coiled radius, 8 W output power at 1064 nm was achieved and corresponding slope efficiency was 77% Fig. 5.75 shows the output beam profile and intensity distribution along x direction.

They also manufactured 127 μm core diameter solid-state photonic crystal fiber [36]. The core mode distribution pattern of the fiber was recorded as Fig. 5.76, which showed a fundamental-like distribution.

Different from the design of all solid-state photonic crystal fiber of Huazhong University of Science and Technology, Shanghai Institute of Optics and Fine Mechanics design solid-state photonic crystal fibers by increasing the refractive

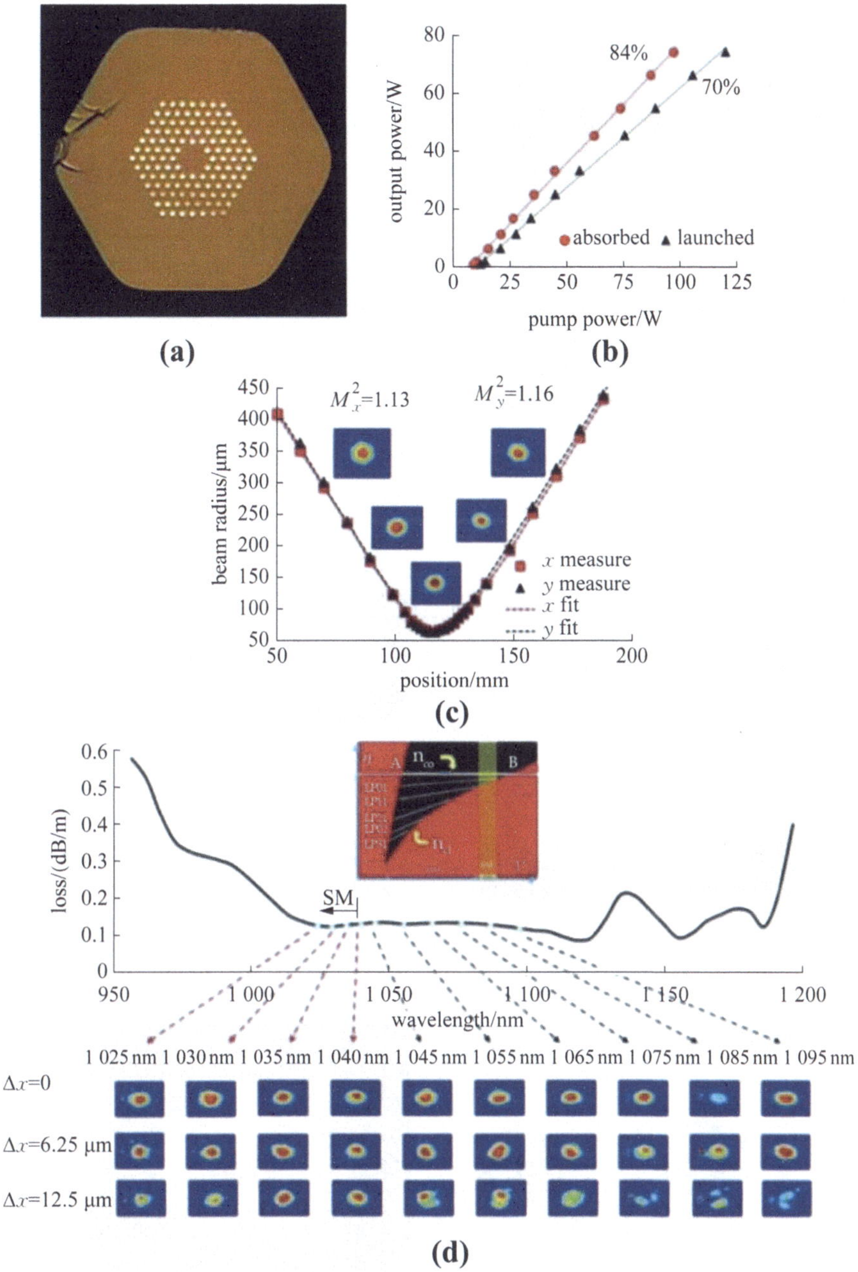

Fig. 5.69 Characteristics of bandgap-type Ytterbium-doped photonic crystal fiber. (**a**) Cross-section of 50 μm core-doped ytterbium bandgap fiber. (**b**) Laser efficiency. (**c**) Beam quality factor. (**d**) Mode characteristics at different wavelengths

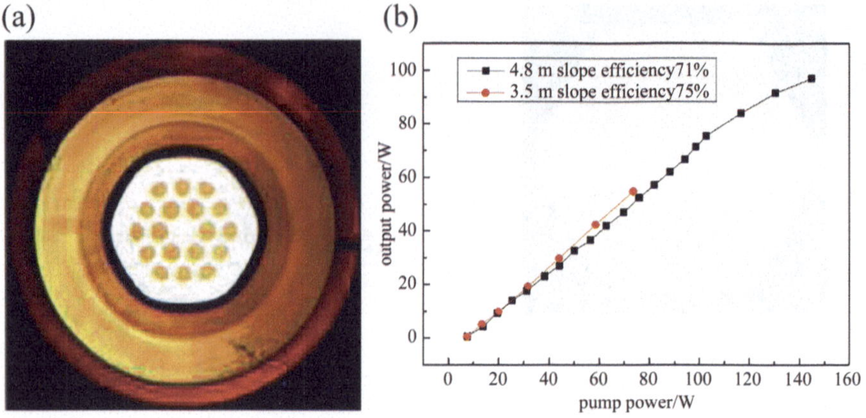

Fig. 5.70 (**a**) Cross-section of a 46 μm core all-solid leakage mode photonic crystal fiber and (**b**) amplification efficiency curve [21]

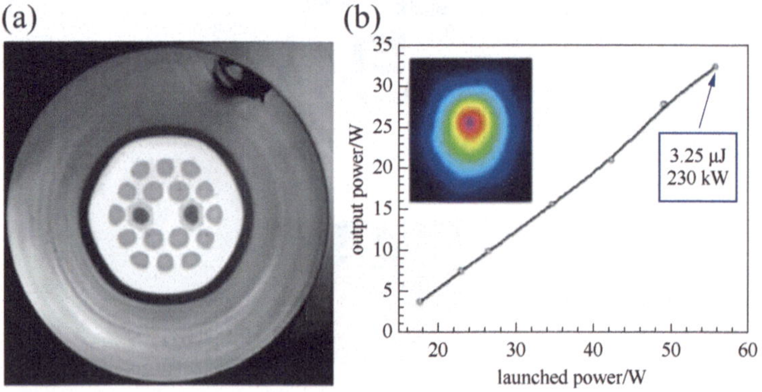

Fig. 5.71 (**a**) Cross-section of an 80 μm core polarization-maintaining all-solid leakage mode photonic crystal fiber and (**b**) amplification efficiency curve [21]

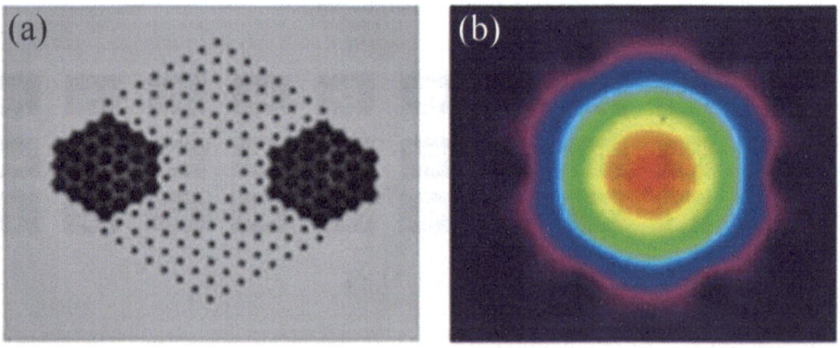

Fig. 5.72 (**a**) cross section of all-solid photonic crystal fiber structure and (**b**) near-field image [84]

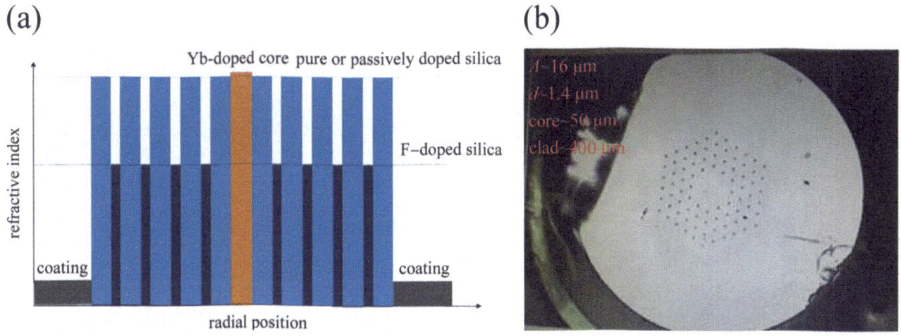

Fig. 5.73 (**a**) Schematic refractive index profile of the designed fiber. (**b**) Microscope images of the fiber [36]

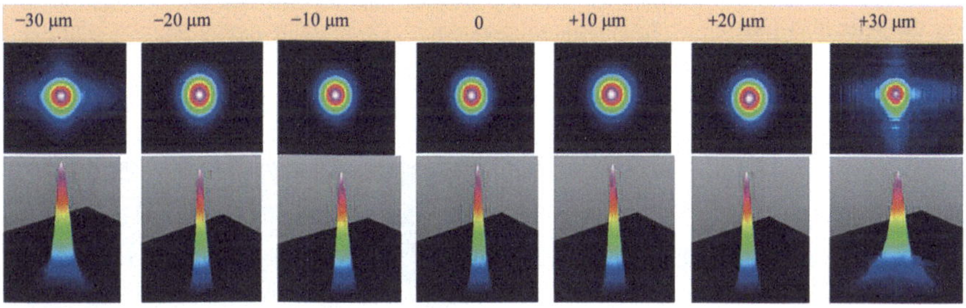

Fig. 5.74 Near field mode profile of 50 μm core fiber at 1064 nm under misaligning the launching condition [36]

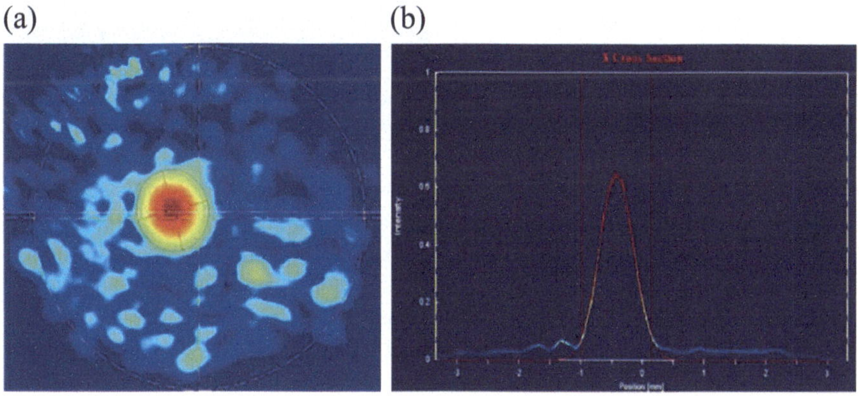

Fig. 5.75 (**a**) Near field mode profile and (**b**) its intensity distribution along x direction of 50 μm core diameter fiber [36]

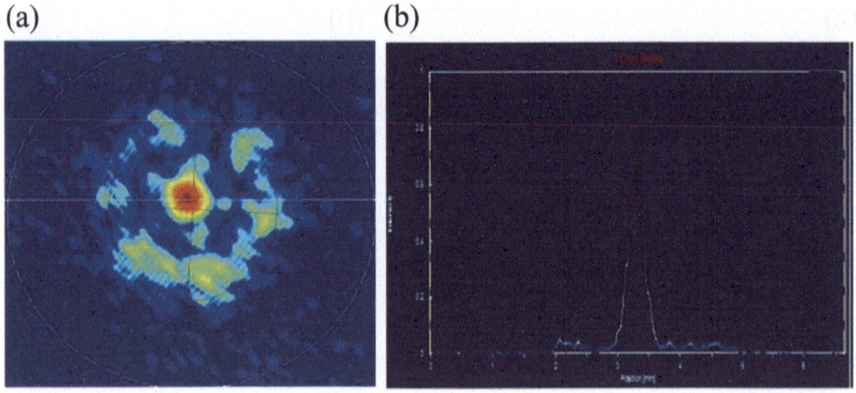

Fig. 5.76 (**a**) Near field mode profile and (**b**) its intensity distribution along x direction of 127 μm core diameter fiber [36]

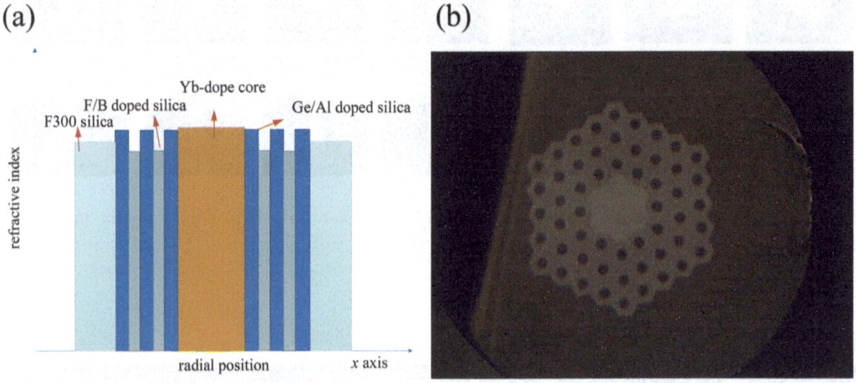

Fig. 5.77 (**a**) Schematic refractive index profile of the designed fiber. (**b**) Microscope images of the fiber [54]

index of the cladding [54]. Figure 5.77 shows the fiber design diagram and cross section of the manufactured 50 μm core diameter solid-state photonic crystal fiber. The refractive index of the fiber core is approximately equal to germanium-doped or aluminum-doped silica glass and fluorine-doped or boron-doped silica glass rods are arrayed around the core. The outer clad glass is Heraeus's F300 type silica glass. The absorption coefficient of the fiber is about 4 dB/m at 976 nm.

The amplification performances of this solid-state ytterbium-doped photonic crystal fiber under different repetition rate with 3 m fiber were tested as shown in Fig. 5.78. The wavelength of seed is 1030 nm, the pulse width is 100 ps, the repetition rates are 1 MHz, 10 MHz and 30 MHz, and the power is 1.6 W. When the repetition rate is 1 MHz, 28 W of amplified output power was achieved at 62 W pump power. if the pump power was further increased, the amplified power no longer increases, and the efficiency dropped significantly. When the repetition frequencies

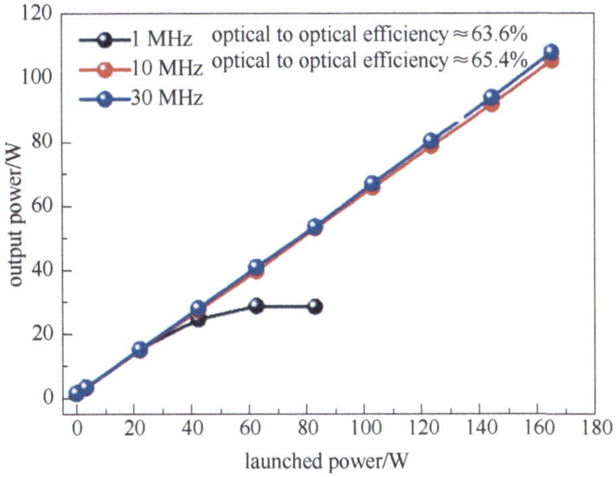

Fig. 5.78 Amplification power curves at different repetitions [54]

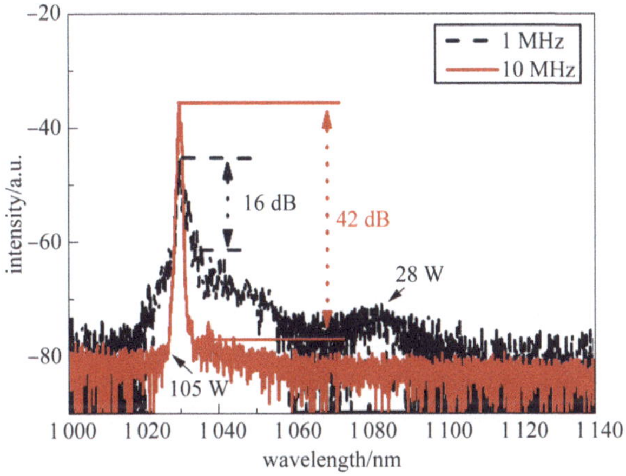

Fig. 5.79 Amplified output spectrum at different repetition rates. Red line is measured at 105 W with 10 MHz repetition, black line is measured at 28 W with 1 MHz repetition [54]

are 10 MHz and 30 MHz, 105 W and 108 W laser output power were obtained respectively when 165 W pump light was coupled in, and the light-to-light efficiency were 63.6% and 65.4% respectively. No drops happened during measurements.

The spectrum of the amplified laser is shown in Fig. 5.79. At 1 MHz repetition rate, the stimulated Raman scattering (SRS) effect appeared at 28 W output power, and the intensity ratio of the amplified signal to the amplified spontaneous emission (ASE) was less than 16 dB. when pump light increased further, the power of ASE significantly increased, indicating that the increased pump light power is no longer converted into signal wavelength, but is consumed by ASE or SRS. Figure 5.79 also

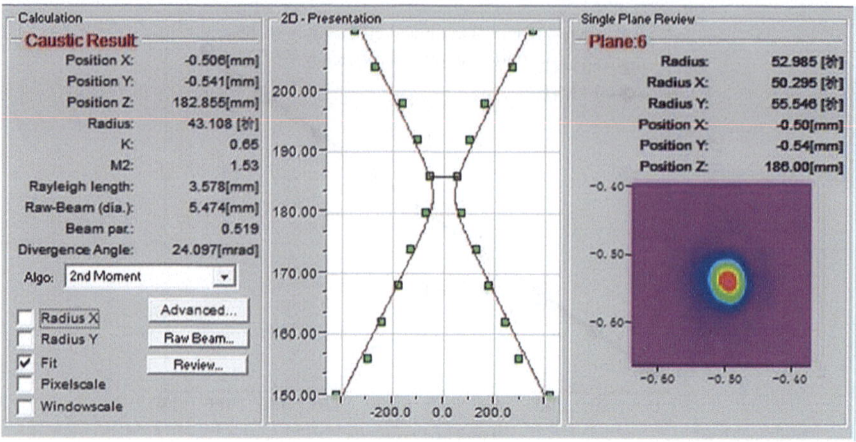

Fig. 5.80 Beam quality factor at 105 W output power [54]

shows the spectrum of 105 W output power at 10 MHz repetition frequency. There is no SRS single, and the signal-to-noise ratio of the laser is greater than 42 dB. The beam quality factor is 1.53 at 105 W as shown in Fig. 5.80.

5.4.3 Applications of Ytterbium-Doped Large Mode Area Photonic Crystal Fiber

When high-power ultra-short pulses propagate in the fiber, high peak power will induce nonlinear effects, causing pulse waveform distortion. Improving the diameter of fiber core is an effective method to eliminate the nonlinear effects. Therefore, Ytterbium-doped large mode field photonic crystal fiber is mainly used in pulse amplification systems to achieve high power, large pulse energy, and high beam quality output. The mode field diameter of the photonic crystal fiber is continuously increasing, and the core diameter of the current gain fiber has reached the level of hundreds of micrometers, greatly reducing the nonlinear accumulation of the fiber. Mourou (winner of Nobel Prize in Physics) proposed Coherent Amplification Network plan [88] to produce large pulse energy, high peak power lasers, which can be used for laser ion accelerators, laser-driven nuclear waste treatment, orbital debris cleaning and laser-driven fusion energy, etc. University of Jena reported amplifications with 72 μm and 81 μm core diameter ytterbium-doped photonic crystal fiber [89]. Then an average power of 700 W, pulse width of 262 fs, and pulse energy of 12 mJ near the diffraction limit pulse output were obtained at the 1032 nm through 4-way beam combining, as shown in Fig. 5.81. The next goal is to achieve hundreds of millijoules or even joule-level output by increasing the number of combination paths.

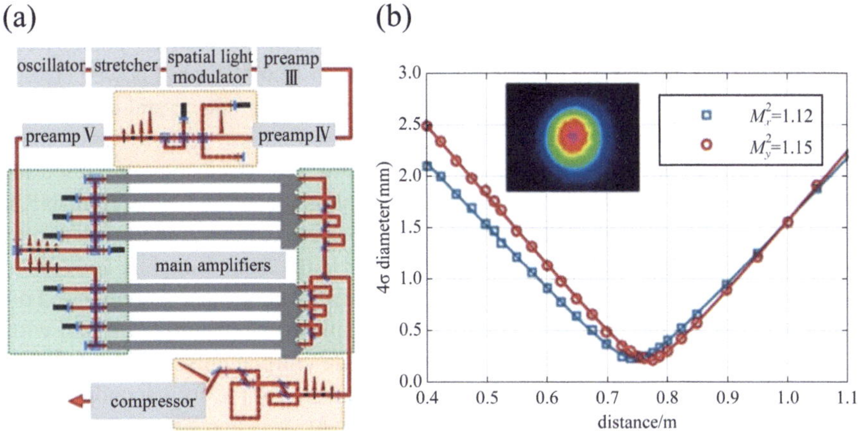

Fig. 5.81 4-way beam coherent combination optical path schematic (**a**) and output mode, beam quality factor (**b**) [89]

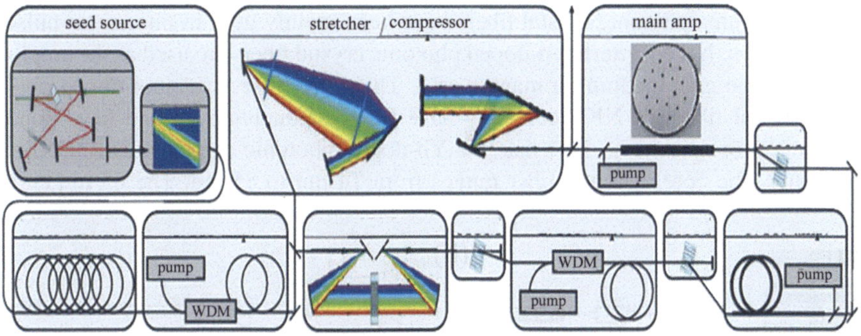

Fig. 5.82 Schematic drawing of the experimental setup. AOM: Acousto-Optical Modulator, WDM: Wavelength Division Multiplexer, SLM: Spatial Light Modulator [15]

Eidam et al. [15] using an Ytterbium-doped large-pitch fiber with108 μm core diameter as the main amplifier, reported on the experimental demonstration of a fiber chirped- pulse amplification system capable of generating nearly transform-limited sub 500 fs pulses with 2.2 mJ pulse energy at 11 W average power. The peak power of 3.8 GW could be achieved by combining active phase shaping with an efficient reduction of the acquired nonlinear phase. The schematic diagram is shown in Fig. 5.82.

Röser et al. [90] reported on an ytterbium-doped fiber chirped-pulse amplification system delivering millijoule level pulse energy at repetition rates above 100 kHz corresponding to an average power of more than 100 W. As the main amplifier, an 80 μm core diameter short length photonic crystal fiber was employed, which allows the generation of pulse energies up to 1.45 mJ at a stretched pulse duration of 2 ns. Liu et al. from Tianjin University in China [91] demonstrated a high pulse energy femtosecond laser amplifier with Yb^{3+} -doped single polarization large-mode-area photonic crystal fibers. Laser pulses with 124 fs duration (the shortest pulse) were generated with 1.56 μJ pulse energy, corresponding to 12.6 MW peak power. Huang et al. from Tianjin University [92] reported a multicore photonic crystal fiber (PCF) amplification system delivering about 110-W laser pulses with a duration of sub-100 fs. Li Feng from Xi'an Institute of Optics and Mechanics [93] used an 85 μm core diameter rod-like ytterbium-doped photonic crystal fiber from NKT company as main amplifier, achieved 124 μJ single pulse energy and corresponding peak power was 139.8 MW after supress. Zhang Haitao from Tsinghua University [94] constructed a single-channel chirped pulse amplification laser system based on Yb-doped photonic crystal fiber, which achieved a pulse energy output of 1 mJ with a beam quality close to the diffraction limit. The de-chirped pulse width of 231 fs was achieved by precise systematic dispersion control, and the corresponding peak power reached 3.85 GW. The beam quality factor M^2 was approximately 1.2, the schematic diagram of the magnified optical path and the mode characteristics are shown in Fig. 5.83.

After decades of development, ytterbium-doped photonic crystal fibers have developed many types with different mechanisms, structures. Ytterbium-doped large mode area photonic crystal fibers have been mainly used in ultra-short pulse amplification. NKT's ytterbium-doped photonic crystal fibers are used as the amplification stage gain medium in many works. This is because ytterbium-doped photonic crystal fibers of NKT company have been commercialized and have good photonic properties. Fig. 5.84 lists the Yb-doped photonic crystal fibers of NKT company. The core diameter fiber ranges from 14 μm to 85 μm. The 85 μm core

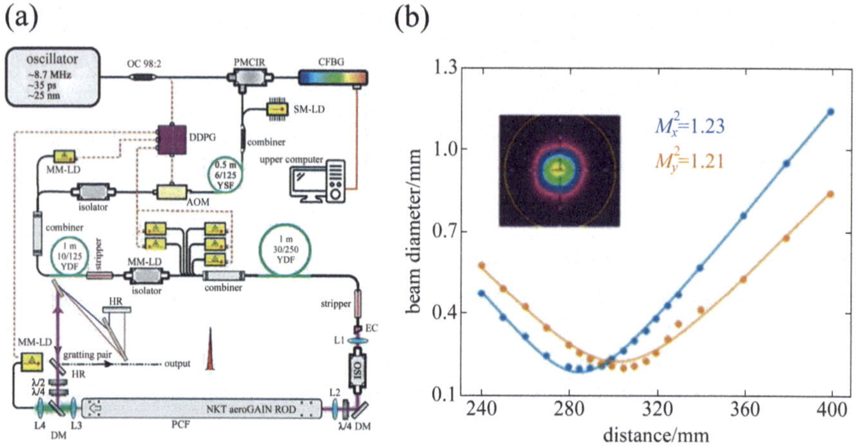

Fig. 5.83 Schematic of the FCPA system. *PMCIR* polarization maintaining optical circulator, *OC* optical coupler, *CFBG* chirped fiber Bragg grating, *WDM* wavelength-division multiplexer, *SM/MM-LD* single/multimode laser diode, *AOM* acoustic optical modulator, *DDPG* digital delay and pulse generator, *ISO* isolator, YDF means Ytterbium doped fiber, *PCF* photonics crystal fiber, *DM* dichroic mirror, *EC* end cap, *HR* high-reflectivity mirror, *λ/4* quarter-wave plate, *λ/2* halfwave plate. (**a**) and beam quality factor (**b**) [91]

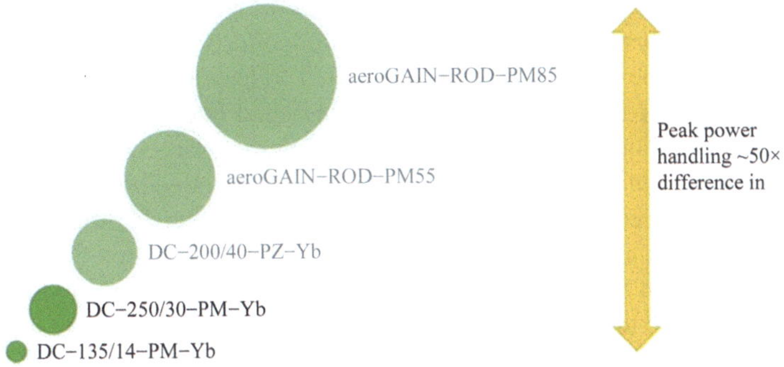

Fig. 5.84 Ytterbium-doped photonic crystal fiber products of NKT company

Table 5.4 Application of ytterbium-doped photonic crystal fibers

Fiber	Application
DC-135/14-PM-Yb	Femtosecond (fs): (200 fs ~ 400 fs,10 ~ 20 W) Picosecond (ps): (10 ~ 20 kW peak power, 10 ~ 20 W) Wavelength: 1030 ~ 1064 nm
DC-200/40-PZ-Yb	Femtosecond (fs): (150 fs ~ 400 fs, 20 ~ 50 W) Picosecond (ps): (100 ~ 200kWpeak power, 40 ~ 100 W) Nanosecond(ns):(100 μJ ~ 1000 μJ pulse energy, 40 ~ 100 W) Wavelength: 1030 ~ 1064 nm
GAIN-ROD-PMx5-Yb	Femtosecond(fs):(~400 fs, 100 ~ 400 μJ, 20 ~ 100 W) Picosecond(ps): (MW level peak power, 50 ~ 100 W) Nanosecond (ns): (mJ level pulse energy, 50 ~ 100 W) Wavelength: 1030 ~ 1045 nm

diameter photonic crystal fiber optimizes the mode characteristics with two mechanisms. Table 5.4 lists the recommended application range some Yb-doped photonic crystal fiber mentioned above.

With the application of chirped amplification technology, ultra-short lasers achieve small pulse width, large pulse energy, and high peak power. They have been widely used in basic research fields such as ultrafast nonlinear optics, terahertz wave generation, and femtosecond chemistry. Especially the high-energy femtosecond laser, due to its small thermal effect when interacting with matter, has been successfully applied in ultra-precision processing, micro-photonics device manufacturing, precision medical surgery, nanomedicine, and defense laser weapons. Foreign companies represented by IPG Photonics are focusing on the field of ultra-fast fiber lasers, and many ultra-fast laser companies in China are also catching up. Ytterbium-doped large-mode-field photonic crystal fibers are gradually being applied to ultra-fast fiber laser systems. At present, only the Danish NKT company can provide Yb-doped photonic crystal fibers commercially, and the prices are expensive. In recent years, China has made significant progress in developing of ytterbium-doped large-mode-field silica fibers. Developing commercialized large-mode-field silica fiber is still the direction of future efforts.

3C fibers and tapered fibers have also attracted widespread attention and are the research hotspots of large-mode-field ytterbium-doped fibers in recent years. The 3C fiber can still maintain single-mode characteristics in a larger core size when the numerical aperture of the core is around 0.06, and the reported biggest core diameter is 85 μm. nLight company can supply 33 μm core diameter 3C fiber product and has been used in many researches. The Canadian Optical Research Center and the Russian Academy of Sciences have both launched polarization-maintaining tapered fiber products. With the development of application requirements, more ytterbium-doped large-mode-field fibers will join the "competition".

This chapter reviews the research progress of different mechanisms and types of ytterbium-doped large-mode-field photonic crystal fibers. Introducing the preparation methods and unique properties of ytterbium-doped photonic crystal fibers, detailly analyzing of the processing and use of air hole structure photonic crystal fibers and the optimized solution. Briefly introduced the application trend of ytterbium-doped large-mode-field photonic crystal fibers. It can be foreseen that due to the unique nature of photonic crystal fibers, it will promote its application in more fields such as ultrafast lasers.

References

1. Mortensen N A. Effective area of photonic crystal fibers [J]. Optics Express, 2002,10(7):341-348.
2. Limpert J, Schreiber T, Liem A, et al. Thermo-optical properties of air-clad photonic crystal fiber lasers in high power operation [J]. Optics Express, 2003,11(22):2982-2990.
3. Knight J, Birks T, Cregan R, et al. Large mode area photonic crystal fibre [J]. Electronics Letters, 1998,34(13):1347-1348.
4. Limpert J, Liem A, Reich M, et al. Low-nonlinearity single-transverse-mode ytterbium-doped photonic crystal fiber amplifier [J]. Optics Express, 2004,12(7):1313-1319.
5. Mortensen N, Nielsen M D, Folkenberg J R, et al. Improved large-mode-area endlessly single-mode photonic crystal fibers [J]. Optics Letters, 2003,28(6):393-395.
6. Russell P. Photonic crystal fibers [J]. Sciences, 2003,299(5605):358-362.
7. Wang Wei. Present situation and future development in photonic crystal fibers [J]. Laser Optoelectronics Progress, 2008,45(2):43-58.
8. Limpert J, Schreiber T, Nolte S, et al. High-power air-clad large-mode-area photonic crystal fiber laser [J]. Optics Express, 2003,11(7):818-823.
9. Eidam T, Hanf S, Seise E, et al. Femtosecond fiber CPA system emitting 830W average output power [J]. Optics Letters, 2010,35(2):94-96.
10. Wadsworth W, Knight J, Reeves W, et al. Yb^{3+} -doped photonic crystal fibre laser [J]. Electronics Letters, 2000,36(17):1452-1454.
11. Stutzki F, Jansen F, Liem A, et al. 26mJ, 130W Q-switched fiber-laser system with near-diffraction-limited beam quality [J]. Optics Letters, 2012,37(6):1073-1075.
12. Beier F, Hupel C, Kuhn S, et al. Single mode 4.3kW output power from a diode-pumped Yb-doped fiber amplifier [J]. Optics Express, 2017,25(13):14892-14899.
13. Limpert J, Schmidt O, Rothhardt J, et al. Extended single-mode photonic crystal fiber lasers [J]. Optics Express, 2006,14(7):2715-2720.

14. Limpert J, Roser F, Schimpf D N, et al. High repetition rate gigawatt peak power fiber laser systems: challenges, design, and experiment [J]. IEEE Journal of Selected Topics in Quantum Electronics, 2009,15(1):159-169.
15. Eidam T, Rothhardt J, Stutzki F, et al. Fiber chirped-pulse amplification system emitting 3.8GW peak power [J]. Optics Express, 2011,19(1):255-260.
16. Dauliat R, Gaponov D, Benoit A, et al. Inner cladding microstructuration based on symmetry reduction for improvement of singlemode robustness in VLMA fiber [J]. Optics Express, 2013,21(16):18927-18936.
17. Stutzki F, Jansen F, Otto H J, et al. Designing advanced very-large-mode-area fibers for power scaling of fiber-laser systems [J]. Optica, 2014,1(4):233-242.
18. Otto H J, Stutzki F, Modsching N, et al. 2kW average power from a pulsed Yb-doped rod-type fiber amplifier [J]. Optics Letters, 2014,39(22):6446-6449.
19. Wong W S, Peng X, McLaughlin J M, et al. Breaking the limit of maximum effective area for robust single-mode propagation in optical fibers [J]. Optics Letters, 2005,30(21):2855-2857.
20. Dong Liang, Peng Xiang, Li Jun. Leakage channel optical fibers with large effective area [J]. Journal of the Optical Society of America B, 2007,24(8):1689-1697.
21. Dong Liang, McKay H A, Fu Libin, et al. Ytterbium-doped all glass leakage channel fibers with highly fluorine-doped silica pump cladding [J]. Optics Express, 2009,17(11):8962-8969.
22. Gu Guancheng, Kong F, Hawkins T, et al. Ytterbium-doped large-mode-area all-solid photonic bandgap fiber lasers [J]. Optics Express, 2014,22(11):13962-13968.
23. Isomäki A, Okhotnikov O G. Femtosecond soliton mode-locked laser based on ytterbium-doped photonic bandgap fiber [J]. Optics Express, 2006,14(20):9238-9243.
24. Pulford B, Ehrenreich T, Holten R, et al. 400-W near diffraction-limited single-frequency all-solid photonic bandgap fiber amplifier [J]. Optics Letters, 2015,40(10):2297-2300.
25. Kong F, Gu Guancheng, Hawkins T W, et al. ~1 kilowatt Ytterbium-doped all-solid photonic bandgap fiber laser [C]//Proceedings of the Fiber Lasers XIV: Technology and Systems, F, 2017.
26. Wadsworth W, Knight J, Russell P S J. Large mode area photonic crystal fibre laser [C]// Proceedings of the Technical Digest Summaries of papers presented at the Conference on Lasers and Electro-Optics Postconference Technical Digest (IEEE Cat No 01CH37170),F, 2001.
27. Wadsworth W, Percival R, Bouwmans G, et al. High power air-clad photonic crystal fibre laser [J]. Optics Express, 2003,11(1):48-53.
28. Bonati G, Voelckel H, Gabler T, et al. 1.53kW from a single Yb-doped photonic crystal fiber laser [C]//Proceedings of the Photonics West, San Jose, Late Breaking Developments, Session, F, 2005.
29. Teodoro D F, Brooks C D. Multi-MW peak power single-transverse mode operation of a 100micron core diameter, Yb-doped photonic crystal rod amplifier [C]//Proceedings of the Fiber Lasers IV: Technology, Systems, and Applications, F, 2007.
30. Hansen K P, Olausson C B, Broeng J, et al. Airclad fiber laser technology [C]//Proceedings of the Fiber Lasers V: Technology, Systems, and Applications, F, 2008.
31. Roy P, Dauliat R, Benoît A, et al. Ultra large mode area fibers with aperiodic cladding structure for high power single mode lasers [C]//Proceedings of the Workshop on Specialty Optical Fibers and Their Applications, F, 2015.
32. Chen Wei, Li Shiyu, Wang Yanliang, et al. Design and fabrication of high numerical aperture and large mode-field Ytterbium-doped photonic crystal fibers [J]. Research on Optical Communication, 2009,35(2):45-46.
33. Chen Gui, Jiang Zuowen, Peng Jinggang, et al. Study of air-clad large-mode-area ytterbium doped photonic crystal fiber[J]. Acta Physica Sinica, 2012,61(14):144206-144206.
34. Liu Jiantao, Zhou Guiyao, Xia Changming. Fabrication of Yb^{3+}/Al^{3+} co doped large mode area photon crystal fiber based on powder sintering technology[J]. Acta Photonica Sinica, 2013,42(5):552-554.

35. Wang Ziwei, Li Qiurui, Wang Zhaokun, et al. 255W picosecond MOPA laser based on self-made Yb-doped very-large-mode-area photonic crystal fiber [J]. Chinese Optics Letters, 2016,14(8):081401-081404.
36. Wei Huifeng, Chen Kangkang, Yang Yucheng, et al. All-solid very large mode area ytterbium-doped silica microstructured fiber based on accurate control on cladding index [J]. Optics Express, 2016,24(8):8978-8987.
37. Wang Shikai, Xu Wenbin, Wang Fan, et al. Yb^{3+} -doped silica glass rod with high optical quality and low optical attenuation prepared by modified sol-gel technology for large mode area fiber [J]. Optical Materials Express, 2017,7(6):2012-2022.
38. Wang Fan, Hu Lili, Xu Wenbin, et al. Manipulating refractive index, homogeneity and spectroscopy of Yb^{3+} -doped silica-core glass towards high-power large mode area photonic crystal fiber lasers [J]. Optics Express, 2017,25(21):25960-25969.
39. Wang Meng, Wang Fan, Feng Suya, et al. 272W quasi-single-mode picosecond pulse laser of ytterbium-doped large-mode-area photonic crystal fiber [J]. Chinese Optics Letters, 2019,17(7):0714011-0714015.
40. Wang Meng, Wang Fan, Yu Chunlei, et al. Megawatt peak power output of ultra-low core NA large mode field photonic crystal fiber [J]. Acta Optica Sinica, 2019,39(05):05360011-05360013.
41. Wang Fan, Wang Meng, Shao Chongyun, et al. Highly fluorine and ytterbium doped polarization maintaining large mode area photonic crystal fiber via the sol-gel process [J]. Optics Express, 2021,29(25):41882-41893.
42. Aljamimi S M, Anuar M S K, Muhd-Yasin S Z, et al. On the fabrication of aluminum doped silica preform using MCVD and solution doping technique [C]//Proceedings of the 2012 IEEE 3rd International Conference on Photonics, F, 2012.
43. Langner A, Schotz G, Such M, et al. A new material for high power laser fibers [C]// Proceedings of the Fiber Lasers V: Technology, Systems, and Applications, F, 2008.
44. Liu Shuang, Wang Meng, Zhou Qinling, et al. Ytterbium-doped silica photonic crystal fiber laser fabricated by the nanoporous glass sintering technique [J]. Laser Physics, 2014,24(6):0658011-0658015.
45. Hu Lili, Wang Shikai, Lou Fengguang, et al. Preparation method of Yb-doped silica fiber preform core rod: China, CN103373811B [P/OL]. 2015-05-13.
46. Limpert J, Stutzki F, Jansen F, et al. Yb-doped large-pitch fibres: effective single-mode operation based on higher-order mode delocalisation [J]. Light: Science and Applications, 2012,1(4):1-5.
47. White T, McPhedran R, de Sterke C M, et al. Confinement losses in microstructured optical fibers [J]. Optics Letters, 2001,26(21):1660-1662.
48. Feng Xian, Monro T M, Petropoulos P, et al. Solid microstructured optical fiber [J]. Optics Express, 2003,11(18):2225-2230.
49. Zou Kuaisheng, Wei Deliang, Lu Min, et al. Development of all-glass solid-core polarization-maintaining photonic crystal fiber [J]. Acta Photonica Sinica, 2010,39(2):251-254.
50. Zhang Guang, Zhou Qinling, Yu Chunlei, et al. Neodymium-doped phosphate fiber lasers with an all-solid microstructured inner cladding [J]. Optics Letters, 2012,37(12):2259-2261.
51. Wang Longfei, Li Wentao, Sheng Qiuchun, et al. All-solid silicate photonic crystal fiber laser with 13.1W output power and 64.5% slope efficiency [J]. Journal of Lightwave Technology, 2014,32(6):1116-1119.
52. Wang Longfei, He Dongbing, Feng Suya, et al. Yb/Er co-doped phosphate all-solid single-mode photonic crystal fiber [J]. Scientific Reports, 2014,4(1):61391-61393.
53. Wang Longfei, He Dongbing, Feng Suya, et al. Heavily Yb-doped phosphate large-mode area all-solid photonic crystal fiber operating at 990nm [J]. Laser Physics Letters, 2015,12(7):0751021-0751025.
54. Wang Meng. High brightness Yb-doped large mode field photonic crystal fiber preparation and performance research [D]. Beijing: University of Chinese Academy of Sciences, 2019.
55. Izawa T, Miyashita T, Hanawa F. Continuous optical fiber preform fabrication method: United States, US4062665A [P/OL]. 1977-12-13.

56. Xu Wenbin, Wang Meng, Feng Suya, et al. Fabrication and laser amplification behavior of Yb $^{3+}$ /Al $^{3+}$ co-doped photonic crystal fiber [J]. IEEE Photonics Technology Letters, 2016,28(4):391-393.
57. Wadsworth W J, Percival R M, Bouwmans G, et al. Very high numerical aperture fibers [J]. IEEE Photonics Technology Letters, 2004,16(3):843-845.
58. Petit V, Tumminelli R P, Minelly J D, et al. Extremely low NA Yb doped preforms (<0.03) fabricated by MCVD [C]//Proceedings of the Fiber Lasers : Technology, Systems, and Applications, F, 2016.
59. Guo Yanyan, Hou Lantian. Measurement and numerical study of the numerical aperture of photonic crystal fibers [J]. Spectroscopy and Spectral Analysis, 2010,30(7):1908-1912.
60. Ferrando A, Silvestre E, Miret J J, et al. Full-vector analysis of a realistic photonic crystal fiber [J]. Optics Letters, 1999,24(5):276-278.
61. Birkst T A, Mogilevtsev D, Knight J C, et al. The analogy between photonic crystal fibres and step index fibres [C]//Proceedings of the Optical Fiber Communication Conference, F, 1999.
62. Koshiba M. Full-vector analysis of photonic crystal fibers using the finite element method [C]// Proceedings of the IEICE Transactions on Electronics, F 04/01,2002.
63. Foroni M, Poli F, Rosa L, et al. Cutoff properties of large-mode-area photonic crystal fibers [C]//Proceedings of the Proceedings of 2005 IEEE/LEOS Workshop on Fibres and Optical Passive Components, 2005, F, 2005.
64. Smith A V, Smith J J. Overview of a Steady-Periodic Model of Modal Instability in Fiber Amplifiers [J]. IEEE Journal of Selected Topics in Quantum Electronics, 2014,20(5):472-483.
65. Otto H J, Modsching N, Jauregui C, et al. Impact of photodarkening on the mode instability threshold [J]. Optics Express, 2015,23(12):15265-15277.
66. Jauregui C, Eidam T, Otto H J, et al. Physical origin of mode instabilities in high-power fiber laser systems [J]. Optics Express, 2012,20(12):12912-12925.
67. Laurila M, Saby J, Alkeskjold T T, et al. Q-switching and efficient harmonic generation from a single-mode LMA photonic bandgap rod fiber laser [J]. Optics Express, 2011,19(11):10824-10811.
68. Eidam T, Wirth C, Jauregui C, et al. Experimental observations of the threshold-like onset of mode instabilities in high power fiber amplifiers [J]. Optics Express, 2011,19(14):13218-13224.
69. Robin C, Dajani I A, Zeringue C M, et al. Gain-tailored SBS suppressing photonic crystal fibers for high power applications [C]//Proceedings of the Fiber Lasers IX: Technology, Systems, and Applications, F, 2012.
70. Richardson D J, Nilsson J, Clarkson W A. High power fiber lasers: current status and future perspectives [J]. Journal of the Optical Society of America B, 2010, 27(11): B63-B92.
71. Laurila M, Jørgensen M M, Hansen K R, et al. Distributed mode filtering rod fiber amplifier delivering 292W with improved mode stability [J]. Optics Express, 2012, 20(5): 5742-5711.
72. Ferrarini D, Vincetti L, Zoboli M, et al. Leakage properties of photonic crystal fibers [J]. Optics Express, 2002, 10(23): 1314-1319.
73. Mortensen N A, Folkenberg J R. Low-loss criterion and effective area considerations for photonic crystal fibers [J]. Journal of Optics A: Pure and Applied Optics, 2003, 5(3): 163–164.
74. Nielsen M, Mortensen N, Albertsen M, et al. Predicting macrobending loss for large-mode area photonic crystal fibers [J]. Optics Express, 2004, 12(8): 1775-1779.
75. Zhang Guodong. Research on the end collapse of large mode area photonic crystal fiber [D]. Beijing: University of Chinese Academy of Sciences, 2018.
76. Chen Zilun, Xi Xiaoming, Sun Guilin, et al. Studies on the Hole Inflation and Tapering Techniques of Photonic Crystal Fibers [J]. Journal of National University of Defense Technology, 2011, 33(2): 1-4.
77. Yu R, Wang C, Benabid F, et al. Robust mode matching between structurally dissimilar optical fiber waveguides [J]. ACS Photonics, 2021, 8(3): 857-863.
78. Teodoro D F, Hemmat M K, Morais J, et al. High peak power operation of a 100μm-core Yb-doped rod-type photonic crystal fiber amplifier [C]. proceedings of the Fiber Lasers VII: Technology, Systems, and Applications, F, 2010.

79. Dong L, Li J, Peng X. Bend-resistant fundamental mode operation in ytterbium-doped leakage channel fibers with effective areas up to 3160μm2 [J]. Optics Express, 2006, 14(24): 11512-11519.
80. Peng X, Dong L. Fundamental-mode operation in polarization-maintaining ytterbium-doped fiber with an effective area of 1400 micro m2 [J]. Optics Letters, 2007, 32(4): 358-360.
81. Chen Yuee, Hou Lantian. Preparation of Yb^{3+} Doped Double-clad Photonic Crystal Fiber [J]. Opto-Electronic Engineering, 2009, 36(2): 62-64.
82. Chu Yingbo, Liu Yongguang, Liu Changbo, et al. Extra-Large-Core Yb^{3+} Doped Fiber and Its Laser Research Based on Glass Phase-Separation Technique [J]. Chinese Journal of Lasers, 2018, 45(12): 29-34.
83. Liu S, Li H, Tang Y, et al. Fabrication and spectroscopic properties of Yb3+-doped silica glasses using the sol-gel method [J]. Chinese Optics Letters, 2012, 10(8): 081601.
84. Xu Yan, Peng Zhigang, Cheng shaochen, et al. Research Progress of Ytterbium-Doped Fiber-Solid High-Power Ultrashort Pulse Amplification [J]. Chinese Journal of Lasers, 2021, 48(5): 38-60.
85. Jauregui C, Limpert J, Tünnermann A. High-power fibre lasers [J]. Nature Photonics, 2013, 7(11): 861-867.
86. Dong Liang, Peng Xiang, Li Jun, Leakage channel optical fibers with large effective area [J]. J. Opt. Soc. Am. B, 2007,24(8):1689-1697.
87. Papior S R, Weirich J, Johansen M M, et al. Photonic crystal fiber technology for high-performance all-fiber monolithic ultrafast fiber amplifiers [C]. proceedings of the LASE, F, 2018.
88. Li Hongxun, Zhang Rui. Progress of Fiber Amplification Network and Its Application [J]. Laser & Optoelectronics Progress, 2017, 54(1): 11-22.
89. Kienel M, Müller M, Klenke A, et al. 12mJ kW-class ultrafast fiber laser system using multidimensional coherent pulse addition [J]. Optics Letters, 2016, 41(14): 3343-3346.
90. Röser F, Eidam T, Rothhardt J, et al. Millijoule pulse energy high repetition rate femtosecond fiber chirped-pulse amplification system [J]. Optics Letters, 2007, 32(24): 3495-3497.
91. Liu Bowen, Hu Minglie, Song Youhui, et al. Photonic Crystal Fiber Femtosecond Laser Amplifier with millijoules and 100 fs Level Output [J]. Chinese Journal of Lasers, 2010, 37(9): 2415-2418.
92. Huang Lili, Hu Minglie, Fang Xiaohui, et al. Generation of 110-W sub-100-fs Pulses at 100 MHz by nonlinear amplification based on multicore photonic crystal fiber [J]. IEEE Photonics Journal, 2016, 8(3): 1-7.
93. Li Feng, Yang Zhi, Zhao Wei, et al. Hundred microjoule level femtosecond fiber laser amplification system [J]. Chinese Journal of Lasers, 2015,42(12):1202005.
94. Zhang Haitao, Zu Jiaqi, Deng Decai, et al. Yb-doped fiber chirped pulse amplification system delivering 1mJ, 231fs at 1kHz repetition rate [J]. Photonics, 2022,9(67):9020067.

Chapter 6
Erbium-Doped Silica Fiber and Its Applications

In 1985, the University of Southampton in the UK reported the production of low-loss rare earth-doped silica fiber by using the MCVD method [1]. Within the fellowing two years, Payne et al. [2, 3] reported the first fiber laser and amplifier (EDFA) near 1.54 μm based on erbium-doped silica fiber. Due to its good mechanical properties, high tensile strength, and ease of fusion with communication fibers, erbium-doped silica fiber can meet the high reliability requirements of optical amplifiers, thus becoming the preferred gain medium for commercial optical amplifier. Since then, erbium-doped fiber amplifiers have undergone rapid development, with research work continuously achieving significant progress, perfectly solving the limitation of communication fiber loss on optical transmission distance, and truly realizing long-distance optical communication. Today, EDFA has become the most widely used optical module in optical communication systems, mainly used for device insertion loss compensation, long-distance transmission relay, and as a power amplifier or preamplifier, amplifying the signal output by optical transmitter or recevied by optical receiver. Due to the strong demand, erbium-doped fiber, as the core component of EDFA, is currently the most deeply researched and widely used rare earth-doped fiber. With the development of the research, erbium-doped silica fiber in the field of high-power lasers has also made significant progress, represented by ytterbium-erbium co-doped double-clad fiber, which has important applications in 1.5 μm LIDAR and laser communication.

This chapter will focus on the basic properties of erbium ion (Er^{3+}) energy level, spectral properties and its structure and gain amplification performance in silica glass, and introduce the basic characteristics of different types of erbium-doped fibers for optical communication and high-power fiber laser applications.

L. Hu, *Rare Earth Doped Silica Fiber and Its Applications*,
https://doi.org/10.1007/978-981-96-5306-5_6

6.1 Energy Level Structure of Erbium Ions

Due to the shielding of outer electrons, the energy level structure of rare earth ions mainly depends on the rare earth ions themselves but it also be affected by the matrix material. Analyzing the energy level structure characteristics of Er^{3+}, and the effects of different co-doping elements and coordination environments in the silica glass matrix on the spectral properties, are the basis for studying the laser performance of erbium-doped silica fiber.

6.1.1 Energy Level Structure of Erbium Ions and Absorption Spectrum of Erbium-Doped Silica Glass

At present, researchers have conducted extensive researches on the energy level structure and spectral properties of Er^{3+} in various matrices. Figure 6.1 is the energy level structure diagram of Er^{3+} [4], from which it can be seen that the energy level structure of Er^{3+} is very complex, containing more than 10 energy levels from low to high, with the highest energy level exceeding 25000 cm^{-1}. The main application of erbium-doped silica fiber is to use the metastable state $^4I_{13/2}$ energy level of Er^{3+} to transit to the ground state $^4I_{15/2}$. The radiative transition of the two energy levels

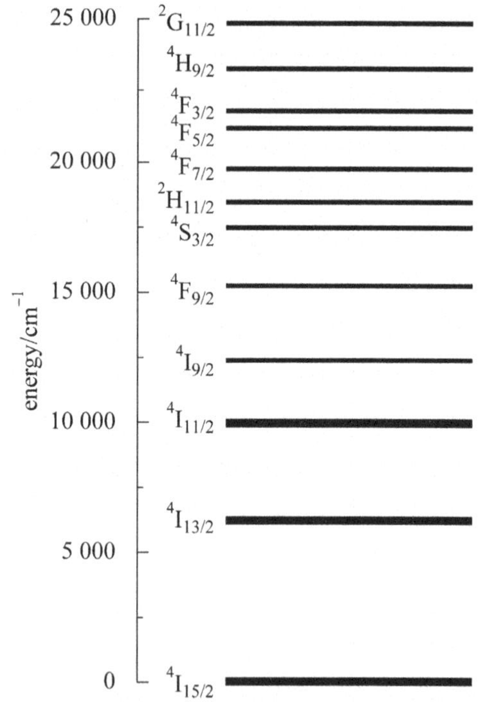

Fig. 6.1 Schematic diagram of Er^{3+} energy level structure

is used as a 1.5 μm band optical fiber amplifier or laser. Therefore, the main discussion in this section will concentrate on this radiative transition process and the factors that may affect it.

Figure 6.2 is the absorption spectrum of erbium-doped silica glass, and some of the characteristic absorption peaks of Er^{3+} corresponding to the energy levels are presentedin the figure. The characteristic absorption peaks of Er^{3+} are located at 1530 nm, 976 nm, 808 nm, 650 nm, 540 nm, 521 nm, 488 nm, 451 nm, 406 nm and 379 nm, respectively corresponding to the absorption transition process of Er^{3+} from the ground state $^4I_{15/2}$ energy level to the excited state $^4I_{13/2}$, $^4I_{11/2}$, $^4I_{9/2}$, $^4F_{9/2}$, $^4S_{3/2}$, $^2H_{11/2}$, $^4F_{7/2}$, $^4F_{5/2}$, $^2H_{9/2}$ and $^4G_{11/2}$ energy levels. Most of the absorption peaks can be used to pump erbium-doped silica fibers to introduce 1.5 μm band emission. In the early research of erbium-doped silica fibers, due to the limitation of available pump sources, researchers mainly used 800 nm laser diodes (LD) to study erbium-doped fibers. However, as can be seen from Fig. 6.2, the absorption capacity of Er^{3+} at 800 nm is weak. Compared with 980 nm or 1480 nm, to achieve the same absorption coefficient, the required 800 nm pump density is about an order of magnitude higher. In addition, when pumped at 800 nm, the excited state pump absorption is severe, especially at high population inversion. Therefore, nowadays, the main pump wavelength of EDFA is 980 nm or 1480 nm. When Er^{3+} is pumped by a 980 nm LD, Er^{3+} will first be excited from the ground state $^4I_{15/2}$ energy level to the $^4I_{11/2}$ energy level. Since the lifetime of the $^4I_{11/2}$ energy level is very short, Er^{3+} quickly relaxes to the $^4I_{13/2}$ energy level. After accumulation of enough population at the energy level, Er^{3+} radiative transition to the $^4I_{15/2}$ energy level produces light in the 1.5 μm band.

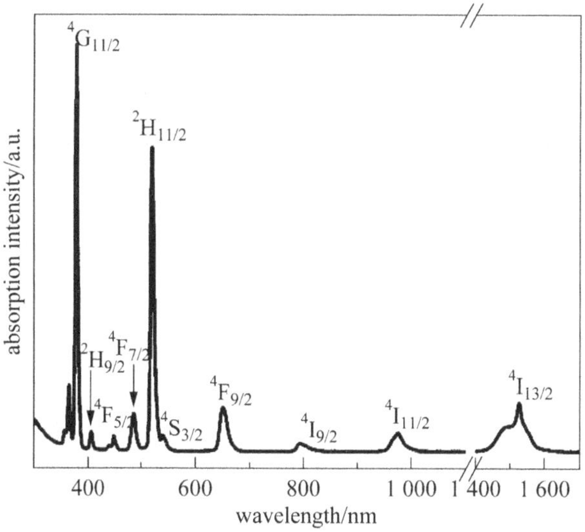

Fig. 6.2 Absorption spectrum of erbium-doped silica glass

6.1.2 Stark Splitting of the 1.5 μm Laser Energy Level of Erbium Ions

In the optical glass, rare earth ions are affected by the matrix local field, resulting in Stark splitting. For a specific *LSJ* energy level structure, when *J* is a half-integer, the maximum number of Stark energy levels are $J + 1/2$. For Er^{3+}, as shown in Fig. 6.3, its excited state $^4I_{13/2}$ can be split into up to 7 sub-energy levels, and the ground state $^4I_{15/2}$ can be split into up to 8 sub-energy levels. In different matrix glasses, with the change of the symmetry of the matrix material, the possible Stark energy levels may also change accordingly. The lower the symmetry, the more the Stark splitting levels. Therefore, studying the number of Stark splits of Er^{3+} energy levels can infer the symmetry of the local field of the matrix material. In addition, in different matrix materials, the positions of the sub-energy levels within the 4f configuration of Er^{3+} differ by tens to hundreds of cm^{-1}, which is due to the different electric field distributions around Er^{3+}. In addition, the electric field distribution around Er^{3+} in silica glass has a significant impact on the distance of the Er-O bond and the coordination number of Er^{3+}.

Figure 6.4 shows the transition between the Stark energy levels of Er^{3+} in erbium-doped alumino-silicate glass [5, 6]. In this matrix, the upper state of $^4I_{13/2}$ is split into 4 sub-levels, and the ground state $^4I_{15/2}$ is split into 5 sub-levels, which is less than the maximum possible number of Stark splits that each level can produce. This indicates that there is a certain symmetry in the glass matrix, and the degeneracy of the sub-levels of Er^{3+} has not been completely lifted. Figure 6.5 shows the absorption and emission cross-section spectra of the erbium-doped aluminum silica glass at room temperature [5, 6], where the positions of the transitions between the above-mentioned Stark levels are marked. It can be seen that the positions of absorption and emission transitions between sub-levels are very close. The maximum width of

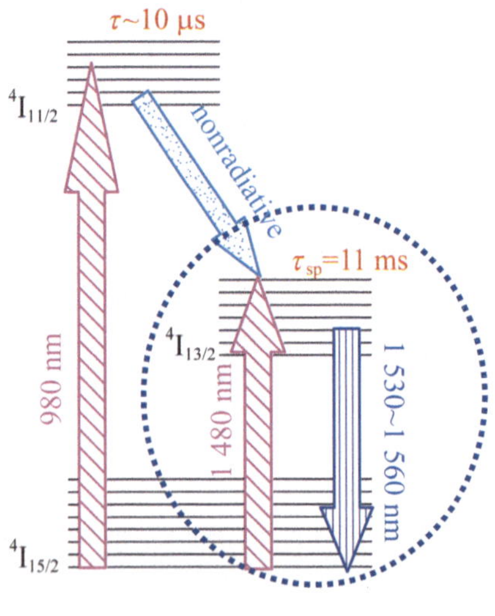

Fig. 6.3 Stark splitting of Er^{3+}: $^4I_{13/2}$ and $^4I_{15/2}$ energy states

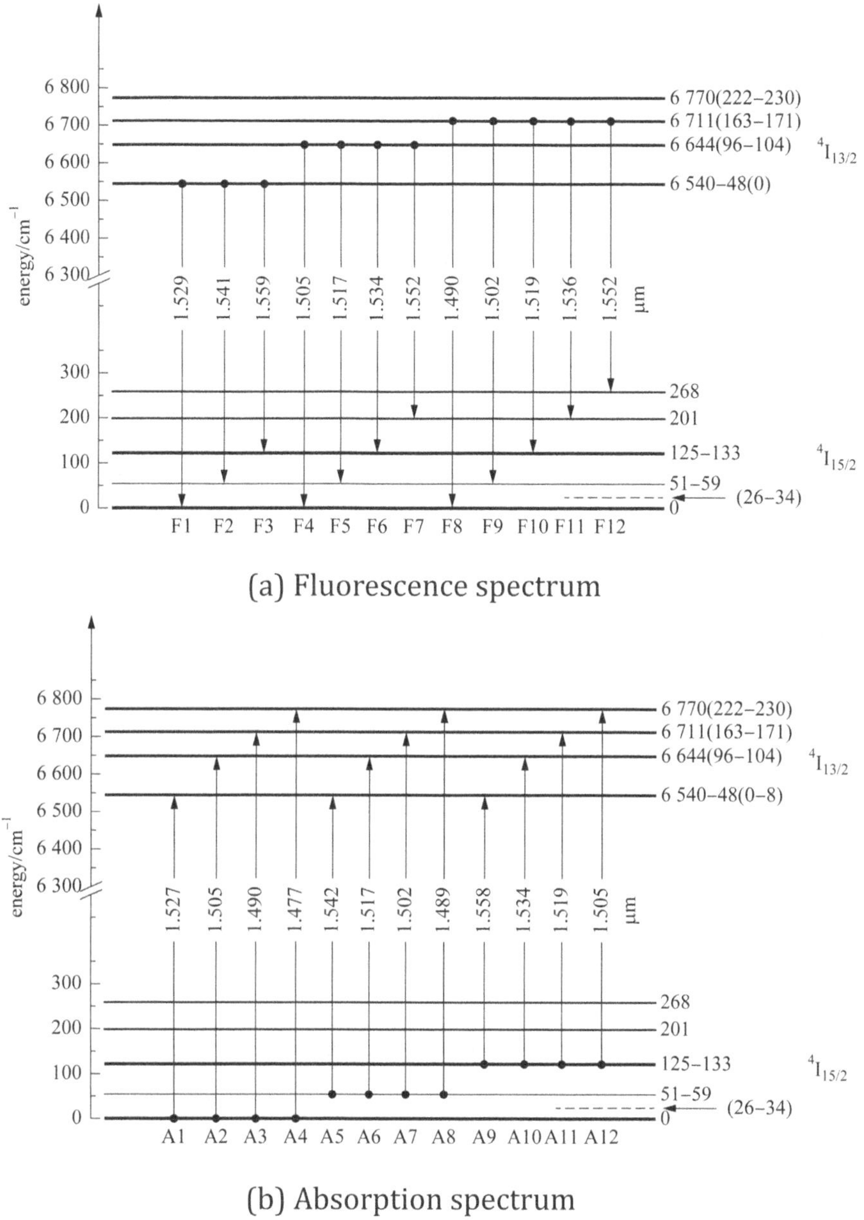

Fig. 6.4 Schematic diagram of Stark splitting of Er^{3+} in aluminum silica glass. (**a**) Fluorescence spectrum. (**b**) Absorption spectrum

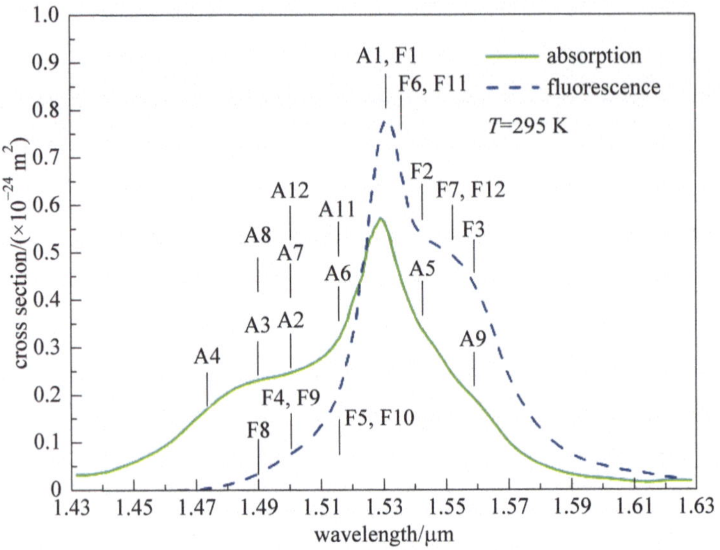

Fig. 6.5 Absorption and emission cross-section of Er^{3+} in aluminum silica glass,corresponding to the transitions between Stark levels in Fig. 6.4

the Stark split levels of the ground state and the upper state is about 250 cm^{-1}. The average interval between each sub-level is about 50 cm^{-1}, such a small energy level interval means that there is a rapid thermal equilibrium process of population redistribution between the Stark sub-levels under room temperature. The interval between the upper state and the ground state is larger, and the probability of non-radiative transition decreases exponentially with the increase of the interval between the two levels. Therefore, the thermal relaxation rate between the Stark sub-levels within the Er^{3+} energy state is much greater than the non-radiative transition process between the upper and lower levels.

6.1.3 *Spectral Characteristics of Erbium Ions in the 1.5 μm Band*

The lifetime of Er^{3+} mainly depends on its energy state. Among all the excited states of Er^{3+}, $^4I_{13/2}$ has the largest energy gap with the lower energy level and the longest lifetime, generally in the order of 10 ms, so Er^{3+} can be used for introducing 1.5 μm band luminescence. In contrast, for the commonly used 980 nm pump structure, Er^{3+} is pumped to the $^4I_{11/2}$ energy state, which has a smaller energy gap with the $^4I_{13/2}$ energy state. The excited ions will quickly relax to the $^4I_{13/2}$ state in the form of non-radiative transitions. Its lifetime at the $^4I_{11/2}$ state is only about 10 μs, so the 1.5 μm band EDFA can use 980 nm LD as the pump source.

Table 6.1 Lifetime of Er^{3+} $^4I_{13/2}$ energy level in different glass matrices

Matrix glass	Lifetime/ms	References	Matrix glass	Lifetime/ms	References
Sodium potassium barium silicate glass	14	[7]	LGS-E phosphate glass	7.7	[7]
Aluminum phosphorus doped silica	10.8	[8]	Phosphate glass	10.7	[10]
Aluminum germanium doped silica	9.5 ~ 10	[9]	Tellurite glass	4	[11]

The energy level lifetime, radiative transition, and non-radiative transition probability of the excited state of Er^{3+} are significantly affected by the composition of the matrix glass and the Er^{3+} doping concentration. Table 6.1 shows the lifetime of the $^4I_{13/2}$ of Er^{3+} in different glass matrices reported in the literatures. It can be seen that the lifetime of the $^4I_{13/2}$ of Er^{3+} in silicate glass can reach up to 14 ms, in phosphate it is about 8 ~ 10 ms, and in tellurite glass it is only 4 ms. This is because the refractive index of the tellurite glass matrix is higher than that of phosphate, and the refractive index of phosphate is higher than that of silicate. The high refractive index glass matrix makes the local field around Er^{3+} larger, thus the emission cross-section is larger, and the fluorescence lifetime of the $^4I_{13/2}$ energy level decreases. In addition, when Er^{3+} is doped in high concentrations, upconversion or concentration quenching effects can occur, leading to a decrease in the lifetime of the excited state, for more information as seen in Sects. 6.3.4 and 6.3.5.

The influence of the matrix material on the spectral characteristics of Er^{3+} mainly manifests in two aspects: first, it causes Stark splitting. Due to the influence of the non-uniform distribution of the electric field, the original degeneracy of the energy levels is lifted. Therefore, for a given electronic energy level transition, the spectrum presents a fine structure, making the substructure of the spectrum obvious. Second, it causes energy level broadening. The mechanism of activation ion energy level broaden in glasses is complex, and multiple factors can affect it. It is generally believed that the disordered structure of the glass is an important reason for the broadening of the rare earth ion energy states. Firstly, it comes from the perturbation of the rare earth ion energy states by the matrix electric field. This perturbation is different for different ions, depending on the surrounding environment. The broadening caused by the matrix electric field is a non-uniform broadening, which is temperature-independent. Another broadening mechanism is the uniform spectral broadening caused by phonons. When electrons transition between energy states, in addition to absorbing or emitting photons, they can also exchange energy with the matrix environment in some form, including the creation and annihilation of phonons. The phonon energy mainly depends on the matrix glass material and temperature. For a given temperature, this will cause uniform broadening of the absorption and emission spectrum.

In different glass matrices, affected by the surrounding local environment coordination field, the Er^{3+} absorption and emission spectra will present different forms. Figure 6.6 shows the absorption and emission spectra of Er^{3+} in three common silica

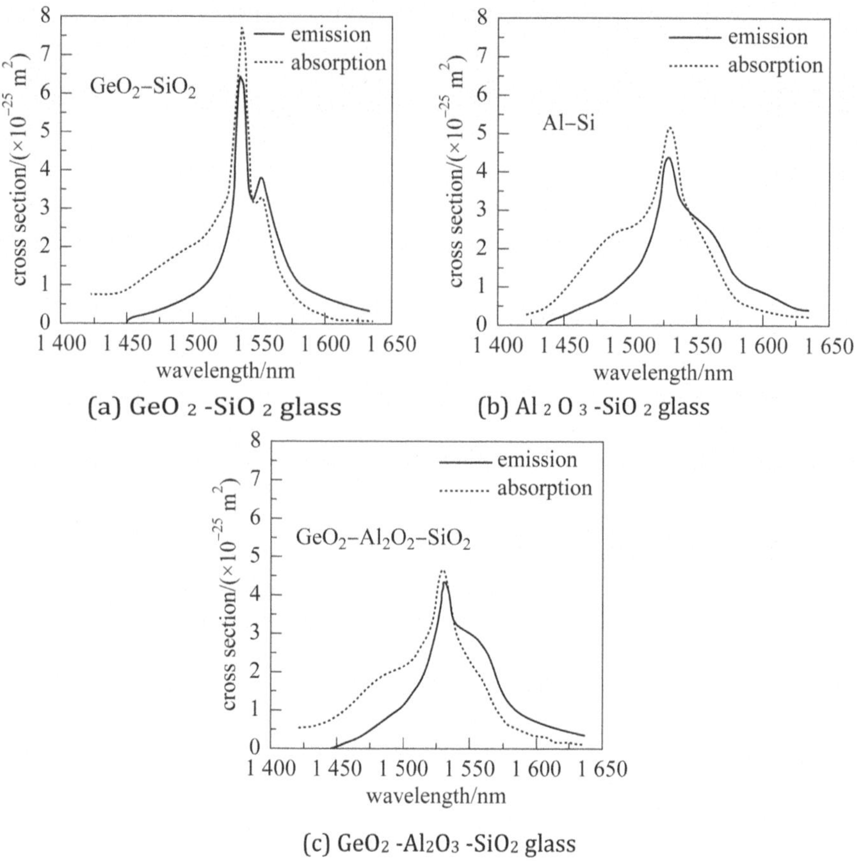

Fig. 6.6 Absorption and emission spectra of three different erbium-doped silica glass. (**a**) GeO_2-SiO_2 glass. (**b**) Al_2O_3-SiO_2 glass, (**c**) GeO_2-Al_2O_3-SiO_2 glass

glass matrices [12]. It can be seen that in GeO_2-SiO_2 glass, the peak values of the absorption and emission cross-sections of Er^{3+} are larger than those in Al_2O_3-SiO_2 glass, but the width of the absorption and emission spectra is smaller than that of Al_2O_3-SiO_2 glass. For practical consideration, when using EDFA, it is hoped that there will be a relatively flat gain coefficient within a certain wavelength range. So Al_2O_3-SiO_2 glass is more suitable for practical applications than GeO_2-SiO_2 glass. In addition, doping Al ions in silica glass can help to improve the dispersion of rare earth ions and increase the numerical aperture of the fiber. This can reduce the size of the fiber core and help to obtain a higher pump density, so as to obtain as high gain as possible under limited pump power conditions. The addition of Ge elements can also effectively increase the numerical aperture of the fiber core, so in practical applications, the matrix composition of erbium-doped silica fiber is generally GeO_2-Al_2O_3-SiO_2 glass.

Table 6.2 Absorption and emission cross sections and corresponding peak wavelengths of Er^{3+} $^4I_{15/2} \leftrightarrow ^4I_{13/2}$ level transition in different glass

Matrix Glass	Peak Wavelength /nm	Absorption Cross Section $\sigma_{abs}/(\times10^{-21}\ cm^2)$	Emission Cross Section $\sigma_{em}/(\times10^{-21}\ cm^2)$	References
Al-P doped silica	1531	6.60	5.70	[8]
L-22 silicate glass	1536	5.80	7.27	[8]
L11 fluorophosphate glass	1533	6.99	7.16	[8]
Ge-doped silica	1530	7.9 ± 0.3	6.7 ± 0.3	[12]
Al-doped silica	1530	5.1 ± 0.6	4.4 ± 0.6	[12]
Ge-Al co-doped silica	1530	4.7 ± 1.0	4.4 ± 1.0	[12]

The absorption cross-section and stimulated emission cross-section are the most important parameters of activated ions. The absorption cross-section reflects the absorption efficiency of the material to the pump, and the stimulated emission cross-section reflects the laser output capability. LikeYb^{3+}, the absorption cross section σ_{abs} (λ) and emission cross section σ_{emi} (λ) of Er^{3+} can also be calculated by measuring its absorption spectrum, as described in Sect. 2.1.3. In different glass matrices, the spectral shape of Er^{3+} is different, and accordingly, its absorption and emission cross sections also change. Table 6.2 shows the absorption and emission cross sections and corresponding peak wavelengths of the Er^{3+} $^4I_{15/2} \leftrightarrow ^4I_{13/2}$ level transition reported in literatures for different glass.

6.2 Properties and Structure of Erbium -Doped Silica Glass

In the matrix material, the coordination number of rare earth ions is large and generally requiring 7 ~ 8 coordinating ions. The tetrahedral spatial network structure of silica glass results in little non-bridging oxygen ions, which easily leads to the aggregation of multiple rare earth ions to share non-bridging oxygen ions. Therefore, in pure silica glass, the solubility of Er^{3+} is very low, generally not exceeding 100 ppm [13]. The further increase in the concentration of Er^{3+} easily leads to clustering and even phase separation.

In order to increase the doping concentration of Er^{3+} in silica glass without clustering, it is necessary to add co-dopants to increase the rare earth ions solubility. In silica glass, the most common and effective co-dopant is Al_2O_3 and P_2O_5. Aluminum ions can exist in the silica glass network structure in the form of four coordination ($[AlO_{4/2}]^-$, network former) or eight coordination ($[AlO_{6/2}]^-$, network modifier). The four-coordinated alumina can share non-bridging oxygen with rare earth ions, so rare earth ions can exist in the aluminum-rich environment with a larger solubility. In silica glass, the role of phosphorus is similar to that of aluminum elements. Phosphorus elements exist in the form of regular tetrahedral structure $[PO_{5/2}]^0$ (including P = O double bond) or $[PO_{4/2}]^+$ (requiring cations to achieve charge balance), so that rare earth ions can also largely dissolve in phosphorus-rich environment.

The above discussion means that different co-dopants (including the type and concentration of co-dopants) will cause differences in the glass network structure environment, which will lead to differences of the spectral properties of rare earth ions, including fluorescence lifetime, absorption and emission cross-section, and excited state absorption. Therefore, studying the network structure of erbium-doped silica glass, analyzing the local environment of Er^{3+} and its impact on the energy states of Er^{3+}, is the basis for studying the luminescence performance of erbium-doped silica glass.

6.2.1 The Influence of Aluminum Doping on the Structure and Properties of Erbium-Doped Silica Glass

The introduction of aluminum in erbium-doped silica glass and fiber has the following advantages [14, 15]:① Increasing the refractive index of glass and fiber to meet the required numerical aperture (NA); ② It is also conducive to increasing the solubility of Er^{3+}, reducing the clustering phenomenon of Er^{3+} in doped fiber; ③ Increasing the fluorescence bandwidth of erbium-doped silica glass to achieve wide and flat gain in the C + L band. In order to more clearly characterize the impact of Al elements on the network structure of silica glass and the luminescence performance of Er^{3+}, a series of Er^{3+} doped silica glass with different Al contents(AE0, AE10, AE50, AE100, AE150, AE200) were prepared by sol–gel method combined with high-temperature sintering method, as seen in Table 6.3.

The effect of Al concentration on the local environment of Er^{3+} was detected by Electron Paramagnetic Resonance (EPR). As the pulse EPR test of Er^{3+} must be carried out under extremely low temperature conditions (<3 K), in order to facilitate the characterization of the effect of Al concentration on the local environment of

Table 6.3 Composition of Er-doped and Yb-doped silica glass with different Al/RE ratios, unit: mol%

Sample	Al_2O_3	Er_2O_3	Yb_2O_3	SiO_2	Al/RE ratio
AE0	— [a]	0.05	—	99.95	0
AE10	0.5	0.05	—	99.45	10
AE50	2.5	0.05	—	97.45	50
AE100	5	0.05	—	94.95	100
AE150	7.5	0.05	—	92.45	150
AE200	10	0.05	—	89.95	200
AY10	0.5	—	0.05	99.45	10
AY50	2.5	—	0.05	97.45	50
AY100	5	—	0.05	94.45	100

a. "—" Indicates that this element is not doped in the sample rare earth ions, Yb^{3+}, which is also a rare earth ion and easier to dectect, is used as a probe ion instead of Er^{3+}. Silica glass doped with Yb^{3+} (AY10、 AY50、 AY100; as seen Table 6.3) was prepared using the same method. Figure 6.7a is the echo detected field-swept (EDFS) electron paramagnetic resonance spectrum of silica glass doped with Yb^{3+}. The EDFS spectrum shows a very wide asymmetric peak in the magnetic field range of 150 ~ 1200mT, which is related to the disordered structure of the glass. As the aluminum concentration increases, the EDFS peak moves towards higher magnetic field. This indicates that the local environment of Yb^{3+} in the glass has changed. The interaction between paramagnetic ions and surrounding magnetic nuclei can be detected through two-dimensional hyperfine sublevel correlation spectroscopy (2D-HYSCORE). Figures 6.7b–d are the 2D-HYSCORE spectra of Yb-doped silica glass codoped with different Al concentration under 600mT magnetic field. In the HYSCORE spectrum, the Larmor frequency υ_n of the magnetic nucleus is directly proportional to the nuclear gyromagnetic ratio γ_n and the applied magnetic field B_0, that is, $\upsilon_n = B_0 \gamma_n/2\pi$. This means that the two diagonal

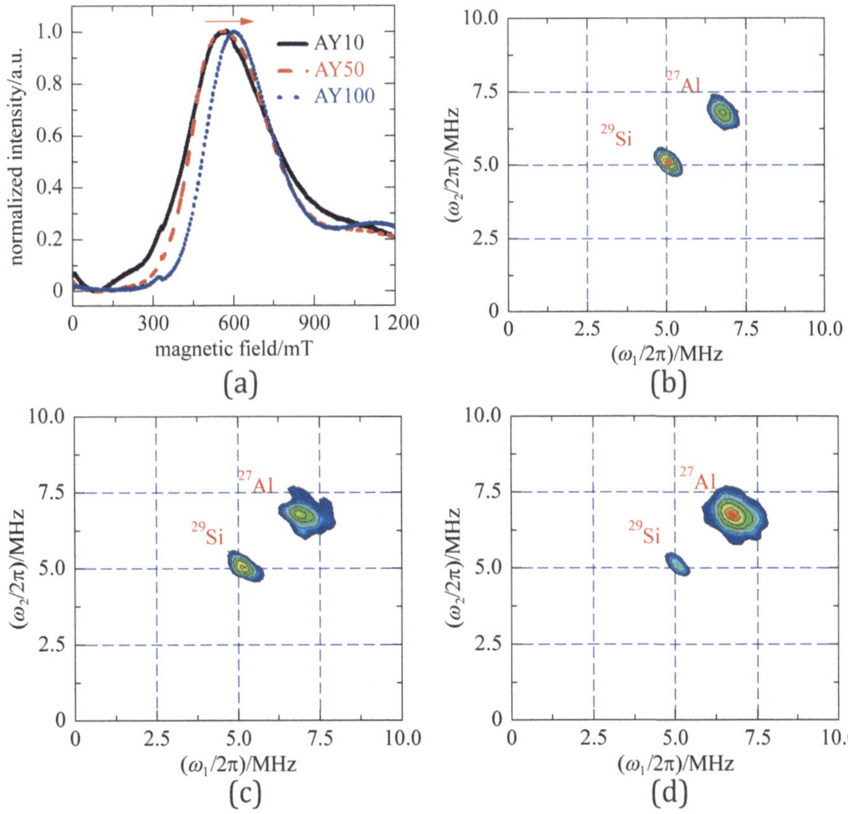

Fig. 6.7 Yb^{3+} Echo detection field scan (EDFS) electron paramagnetic resonance spectrum (**a**) and Sample AY10 (**b**), AY50 (**c**), AY100 (**d**) 2D-HYSCORE spectrum [16]

peaks at 5 MHz and 6.7 MHz are correspond to the isotopes ^{29}Si (natural abundance 4.68%) and ^{27}Al (natural abundance 100%), respectively. It can be seen that the signal of ^{27}Al gradually increases with the concentration of Al_2O_3, while the signal of ^{29}Si decreases. As in AY100 glass, Yb^{3+} is mainly located in the Al-rich environment. This indicates that the proportion of Al elements around Yb^{3+} increases with the concentration of Al_2O_3. The above results are consistent with the results obtained by Saitoh et al. [14, 15] through electron spin echo envelope modulation (ESEEM) in Er-Al co-doped silica glass.

Figure 6.8a, b are the Raman spectra and Fourier Transform Infrared spectroscopy (FTIR) of the AE series glasses, respectively. When the Al concentration is relatively low, the Raman and infrared spectra of AE10 and AE50 glasses are nearly similar to the sample without Al (AE0). As Al is higher than 5 mol%, the intensity of the vibration peaks attributed to the Si—O—Si plane 4-ring (D_1) and plane 3-ring (D_2) in the Raman spectrum decreases, and a new vibration peak corresponding to Si—O—Al appears at 1150 cm^{-1}. However, the coordination number of Al is

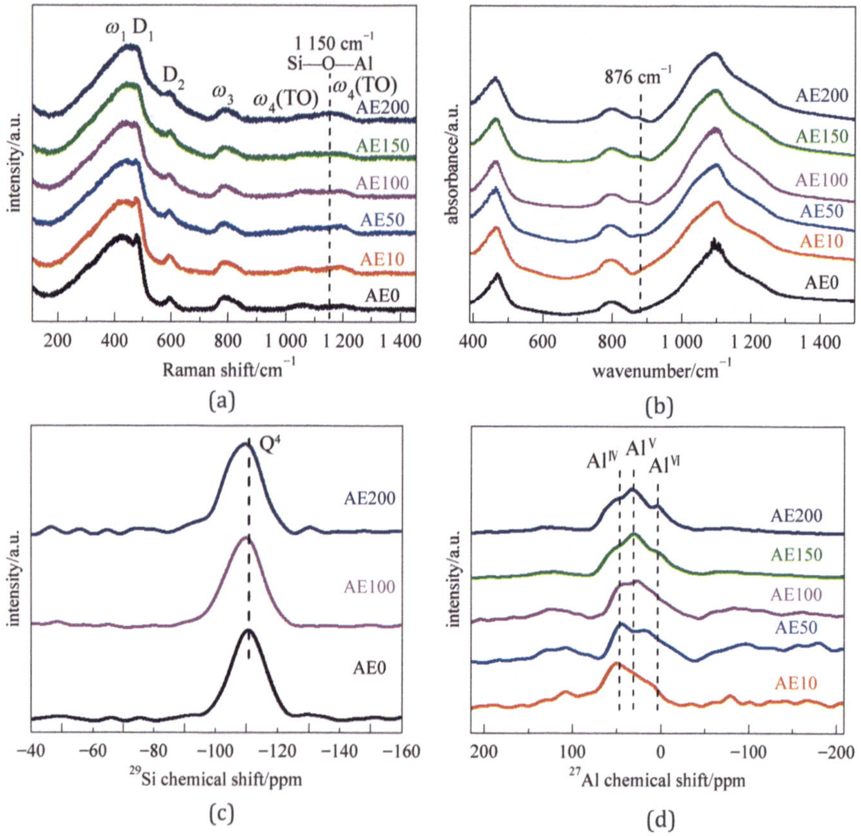

Fig. 6.8 Raman spectra (**a**), FT-IR spectra (**b**), ^{29}Si MAS NMR spectra (**c**) and ^{27}Al MAS NMR spectra (**d**) of Erbium-doped silica glass with different Al/Er ratios [16]

currently controversial [17]. The peak of the glass D_1 and D_2 decreases with the concentration of Al_2O_3; when the Al content increases to 10 mol%, the D_1 peak almost disappears, and a new vibration peak attributed to Si—O—Al vibration appears in the infrared spectrum at 876 cm^{-1}. However, as the Al content is low, there are no infrared vibration peaks corresponding to AlO_6 and AlO_5 at positions 556 cm^{-1} and 670 cm^{-1} [18], respectively.

Figure 6.8c, d are the magic angle spinning nuclear magnetic resonance (MAS NMR) spectra of ^{29}Si and ^{27}Al, respectively, in the AE series silica glass. Si exists in the glass network structure in the form of four coordination; as the Al content gradually increases, the center of gravity of the spectrum gradually shifts to the high frequency direction. This means that the Q^4_{mAl} (*m* represents the number of Si—O—Al around each Si atom) structure appears, leading to an increase in Si—O—Al structure [19]. Through the deconvolution of the overlapping components of the three different aluminum coordination environments (Al^{IV}, Al^{V} and Al^{VI}) in the MAS NMR spectrum of ^{27}Al, the content of aluminum with different coordination numbers can be obtained. The results show that as the concentration of Al_2O_3 increases, the relative content of Al^{IV} gradually decreases, while the proportions of Al^{V} and Al^{VI} are gradually increasing. This observation is consistent with the research results reported by Sen et al [20].

Rocca et al. [21] demonstrated, based on the results of extended X-ray absorption fine structure (EXAFS) measurements, that there is a clear preferential bonding between Er and Al ions, and the higher the Al concentration, the higher the order of Er-Al ions. Haruna et al. [22] had studied the local structure around Er^{3+} through X-ray absorption fine structure measurements, in order to clarify the relationship between the glass network and chemical composition in Er^{3+} doped fibers. The results show that Er^{3+} exists in the form of Er—O, and the number of oxygen coordination around Er^{3+} increases with the increase of Al concentration. The distance of Er—O bond changes significantly with the increase of Al concentration.

Figure 6.9a shows the absorption spectra of the AE series glasses. Compared with the AE0 sample (without aluminum), as the concentration of Al_2O_3 increases, the peak absorption intensity of Er^{3+} gradually decreases, and the absorption bands at 980 nm and 1530 nm blue shift. This is related to the local structural changes of Er^{3+}. By calculating the J-O parameters, it can be known [16] that as the Al concentration increases, the asymmetry around Er^{3+} increases and the covalency of the Er—O bond gradually weakens.

Figure 6.9b shows the normalized fluorescence spectra of the AE series glass. Al^{3+} co-doping significantly affects the fluorescence spectra of Er^{3+}. Compared with the AE0 sample without Al^{3+}, as the concentration of Al_2O_3 increases, the following main changes are observed: the full width at half maxima (FWHM) increases from 27.2 nm (AE0) to 54.3 nm (AE200), the main peak position moves towards a shorter wavelength (from 1535 nm of AE0 to 1528 nm of AE200), and the shoulder peak at 1550 nm only exists in AE0 and AE10 samples, and disappears in other samples with high concentration of Al_2O_3. This is because as Al^{3+} gradually increases, the amount of Al^{3+} around Er^{3+} gradually increases (as shown in Figure 6.7a), and the three types of coordination of Al^{3+} coexist (as shown in Fig. 6.8d). At the same time,

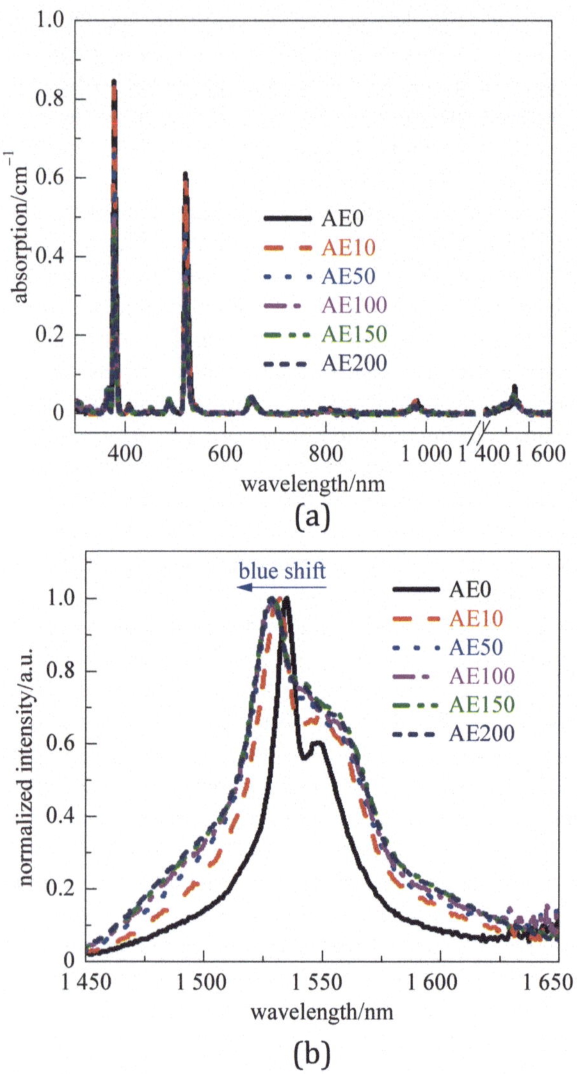

Fig. 6.9 Absorption spectrum (**a**) and fluorescence spectrum (**b**) of Er^{3+} doped silica glass with different Al/Er ratios [16]

the Er—O bond length and the number of Er—O coordination both increase, leading to the diversity of sites around Er^{3+} and the broadening of Er^{3+} spectrum.

Table 6.4 lists the peak absorption and stimulated emission cross-sections, fluorescence full width at half maxima (FWHM), and fluorescence lifetime of Er^{3+} in the AE series glasses for the transition between $^4I_{13/2}$ and $^4I_{15/2}$. As the Al concentration increases, the absorption cross-section of the glass sample near 980 nm decreases from 0.34×10^{-20} cm^2 to 0.20×10^{-20} cm^2. Due to spectral broadening, the

Table 6.4 Fluorescence FWHM, lifetime near 1530 nm, absorption and emission cross-sections, and gain bandwidth (FWHM× σ_{emi}) of AE series glasses [16]

Sample	Fluorescence Full Width at Half Maxima /nm	Lifetime @1530 nm /ms	Absorption Cross Section			Emission Cross Section		Gain Bandwidth /(pm ^{2} × nm)
			Peak Wavelength /nm	Cross Section @976 nm /pm^2	Cross Section @1480 nm /pm^2	Peak Wavelength /nm	Cross Section /pm^2	
AE0	27.2	13.25	983	0.34	0.17	1535	0.83	22.6
AE10	45.0	11.77	978	0.34	0.18	1532	0.63	28.3
AE50	50.0	11.17	978	0.27	0.20	1529	0.58	29.0
AE100	51.9	10.93	977	0.26	0.24	1528	0.58	30.1
AE150	53.2	10.67	976	0.22	0.26	1528	0.56	29.8
AE200	54.3	10.07	974	0.20	0.28	1528	0.55	29.8

absorption cross-section at 1480 nm increases from 0.17 × 10^{-20} cm^2 to 0.28 × 10^{-20} cm^2. The peak stimulated emission cross-section near 1530 nm decreases from 0.83 × 10^{-20} cm^2 to 0.55 × 10^{-20} cm^2, and then remains uncharged as the Al/Er ratio increases from 50 to 200.

The gain bandwidth (the product of FWHM and the stimulated emission cross-section σ_{emi}) is an important parameter for characterizing the broadband gain amplification of an optical amplifier. The larger the product, the better the performance of the broadband amplifier. As can be seen from Table 6.4, the gain bandwidth increases with Al_2O_3 concentration in AE glasses. The maximum FWHM× σ_{emi} of AE100 glass is 30.1 pm^2 × nm, which is higher than that of phosphate glass (~23.7 pm^2 × nm) [23], and comparable to that of germanate glass (~30.1 pm^2 × nm) [24].

6.2.2 The Influence of Phosphorus Doping on the Structure and Properties of Erbium-Doped Silica Glass

To achieve high-power low-noise amplification at the 1.5 μm band of erbium-doped fiber, 980 nm LDs are generally used as the pump source for commercial erbium fiber lasers. As the absorption cross-section of Er^{3+} near 980 nm is relatively small, it limits the absorption; while a longer fiber can absorb enough pump light, it lowers the nonlinear threshold [25, 26], severely preventing the output laser power scaling.

Since Er^{3+} and Yb^{3+} have a large absorption overlap at 980 nm, as shown in Fig. 6.10, and, there is strong absorption for Yb^{3+}at 980 nm, so the absorption of the erbium-doped fiber under 980 nm pump can be significantly improved by co-doping with Yb^{3+}. This can effectively shorten the length of the erbium-doped fiber, which is an important route to achieve high-power laser output at 1.5 μm. In order to improve the energy transfer efficiency of Yb^{3+} $^2F_{5/2}$ →Er^{3+} $^4I_{11/2}$, a large amount of phosphorus is often introduced into the Yb/Er co-doped silica fiber.

Saitoh et al. [14] have proved that the co-doping of P enhances the solubility of Er^{3+} in silica glass by using Electron Spin Echo Envelope Modulation Spectroscopy (ESEEM). The results showed that, similar to phosphate glass, the doped P element preferentially coordinates with Er^{3+} to form a "solvent shell structure", thereby "wrapping" Er^{3+}. In order to explore the local environment around the rare earth ions in P-doped silica glass, Shao et al. [27] reported the characteristics of the 2D-HYSCORE spectrum of Er/Yb co-doped silica glass with the increase of P/Al ratio by Electron Paramagnetic Resonance, as shown in Fig. 6.11. It was found that when the P/Al ratio is less than 1, the sub-nearest neighbor ions around the rare earth ions are Si and Al; as the P content gradually increases, when the P/Al ratio is greater than 1, the rare earth ions tend to be in a P-rich environment, being "wrapped" by P ions, thereby dispersing the rare earth ions in the glass and reducing their clusters. This indicates that the increase in the P/Al ratio directly changes the local

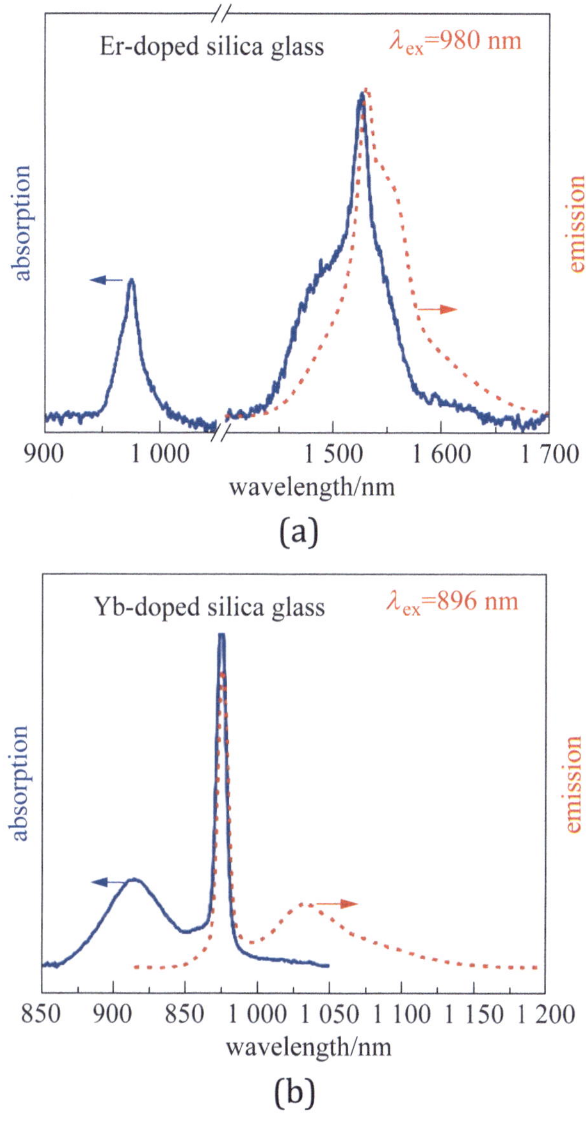

Fig. 6.10 Er^{3+} (**a**) and Yb^{3+} (**b**) absorption and emission spectra in the near-infrared region of silica glass [27]

environment around the rare earth ions; as the P/Al ratio increases, the environment around the rare earth gradually changes from Si and Al to a P-rich environment.

De Oliveira et al. [28] have conducted a structural study on binary SiO_2-P_2O_5 glass through NMR spectra, the results showed that these glasses are composed of randomly interconnected $P^{(3)}$ and Si^{IV} units (Fig. 6.12). When the P_2O_5 content is between 0 ~ 25 mol%, as the P_2O_5 content increases, the glass transition

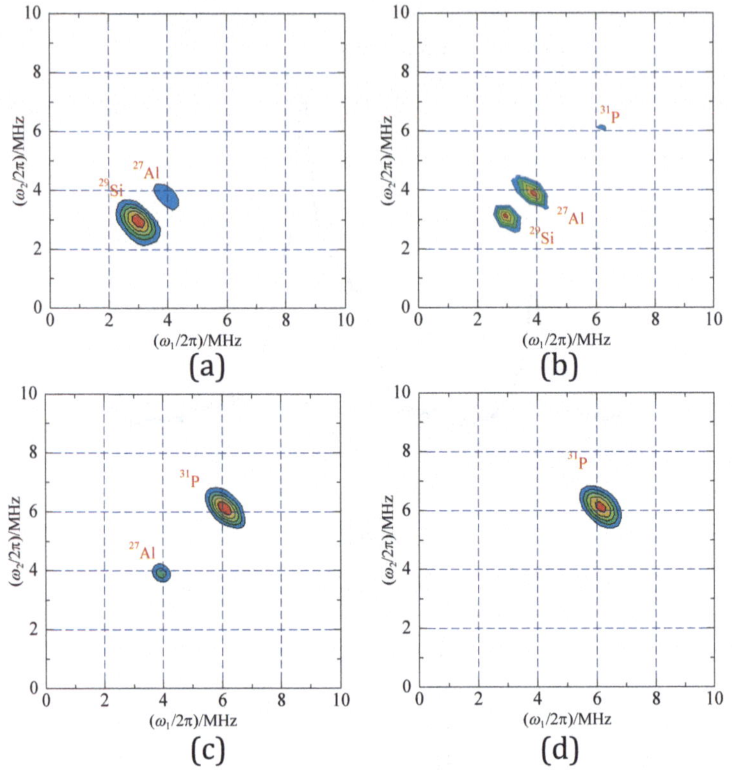

Fig. 6.11 2D-HYSCORE spectra of Al single-doped (**a**), P/Al < 1 (**b**), P/Al = 1 (**c**) and P/Al > 1 (**d**) silica glass [27]

temperature decreases due to the decrease in the average network coordination number (i.e., the degree of polymerization and packing density decrease). In phosphosilicate glass with high P content (P_2O_5 > 30 mol%), high coordination number silicon (Si^V and Si^{VI}) appears. From ^{31}P static NMR, when P_2O_5 > 25 mol%, an additional isotropic component $P^{(0)}$ appears; this may be related to the appearance of high coordination number silicon, which also leads to an increase in the glass transition temperature. Li et al. [29], by using EXAFS, have proved that the Si—O—Si connection is gradually replaced by Si—O—P (as shown in the Raman spectrum given in Fig. 6.13) as the P_2O_5 content increases. When $P_2O_5 \leq 5$ mol%, P mainly exists as orthophosphate and pyrophosphate groups; when P_2O_5 > 10 mol%, phosphorus mainly exists as metaphosphate units of different chain lengths.

In order to better characterize the energy transfer efficiency between Yb^{3+} and Er^{3+}, Er-Yb-Al-P(EYAPS) series glass samples were prepared using the sol–gel method combined with high-temperature sintering. Their theoretical compositions are in Table 6.5.

Figure 6.13 shows the Raman spectra of the EYAPS series samples. For samples with P/Al < 1, their Raman spectra are similar to pure silica, and their maximum

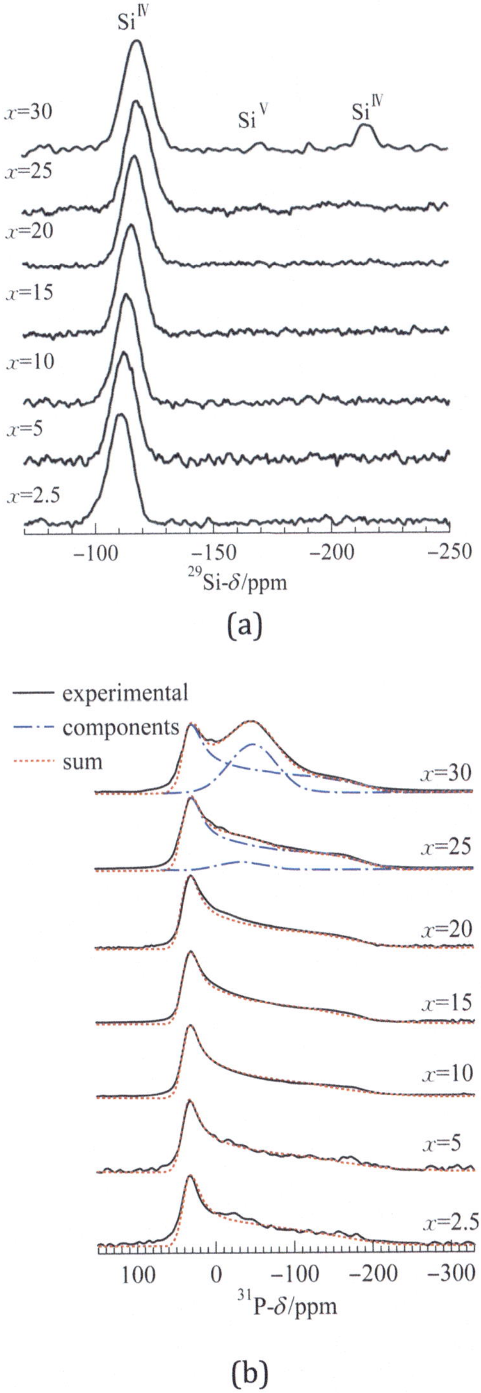

Fig. 6.12 ^{29}Si MAS (**a**) and ^{31}P static (**b**) NMR spectra of silica glass with different P contents (100-x) SiO_2 - x P_2O_5 [28]

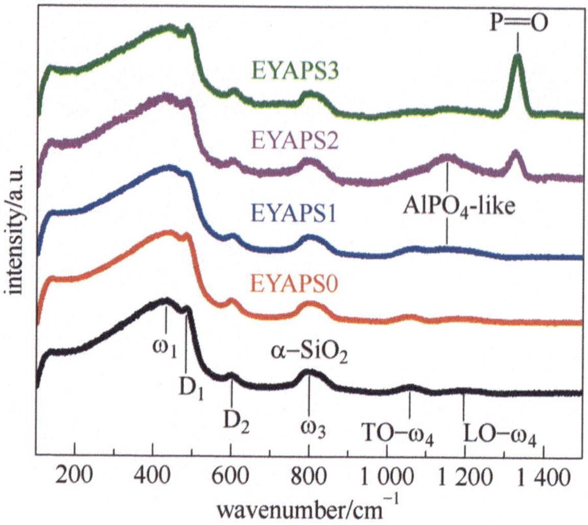

Fig. 6.13 Raman spectra of silica glass doped with different Al/Er ratios [27]

Table 6.5 Theoretical composition of EYAPS series samples, unit: mol%

Sample	Er_2O_3		Al_2O_3	P_2O_5	SiO_2	P/Al ratio
EYAPS0	0.1	0.5	4	0	95.4	0
EYAPS1	0.1	0.5	4	1	94.4	0.25
EYAPS2	0.1	0.5	4	8	87.4	2
EYAPS3	0.1	0.5	0	12	87.4	∞

phonon energy is about 1200 cm^{-1}; and for EYAP2 and EYAP3 samples, a broad peak is observed at 1145 cm^{-1}, indicating the existence of AlO_4 structural units [30]; While for samples with P/Al > 1, a narrow and sharp vibration peak is observed near 1326 cm^{-1}, this peak corresponds to the stretching vibration of the P=O bond in the P(3) unit [31]. The increase of phonon energy will accelerate the non-radiative transition rate from the Er^{3+} $^4I_{11/2}$ to the $^4I_{13/2}$ level [32], shortening the lifetime of the $^4I_{11/2}$ energy level and reducing the number of ions at the $^4I_{11/2}$ state. Therefore by suppressing the Er^{3+} $^4I_{11/2}$ state to Yb^{3+} $^2F_{5/2}$ state's reverse energy transfer, this can improving the energy transfer efficiency for Yb^{3+} to Er^{3+}.

Figure 6.14 shows the absorption spectra of the EYAPS series samples in the ultraviolet–visible band. Early research results showed that the absorption peak at 330 nm was caused by Yb^{2+} [33]. It can be seen that the absorption peak of Yb^{2+} is the strongest in the sample without P (EYAPS0). With the P/Al ratio increasing, the absorption of Yb^{2+} gradually decreasing; While for P > Al, there are almost no absorption peak of Yb^{2+} in the EYAPS3 and EYAPS4 samples, indicating that the addition of P can effectively reduce the content of Yb^{2+}.

Figure 6.15a shows the absorption spectrum of the EYAPS samples near 1 μm, with a typical Yb^{3+} absorption characteristics. Due to the low doping concentration

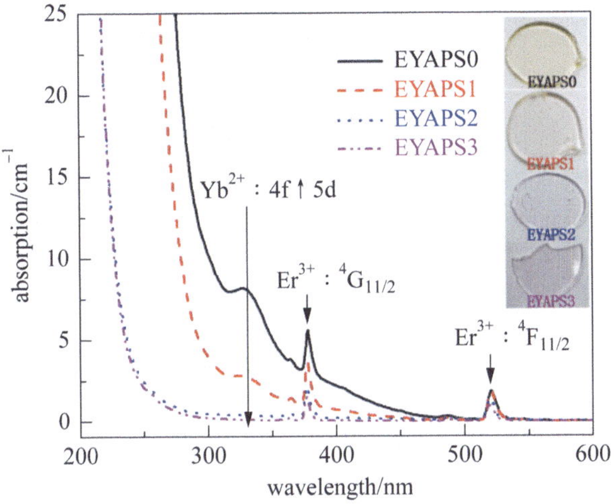

Fig. 6.14 Absorption spectra of EYAPS samples in the UV–visible band [27]

of Er^{3+}, the absorption of Er in this band is very weak and is completely covered by the Yb absorption. It can be seen that the absorption spectrum of Yb^{3+} shows different features for P < Al and P > Al: for P < Al, the absorption peak at 915 nm in the EYAPS0 and EYAPS1 samples is wide and relatively full; while for P > Al, the absorption peak at 915 nm in the EYAPS2 and EYAPS3 samples is significantly decreased becoming narrower but relatively flat. Similarly, the absorption peak intensity at 975 nm is also significantly decreased. This is due to the change of the Yb^{3+} local environment This is consistent with that in only Yb-doped silica glass [34].

Figure 6.15b is the absorption spectrum of EYAPS around 1.5 μm, it can be seen that the absorption also show different shapes for P < Al and P > Al. Compared with EYAPS0 and EYAPS1 samples, the absorption of EYAPS2 and EYAPS3 samples at 1480 nm significantly decreases, and the peak absorption near 1530 nm shows a red shift. For P < Al, the peak absorption is at 1528 nm, and the full width at half maxima, (FWHM) is about 40 nm; while for P > Al, the peak absorption is at 1532 nm, and the FWHM is about 18 nm. This indicates that the change in the P/Al ratio causes a change in the local environment around Er^{3+}, which has a significant impact on its absorption spectrum.

Figure 6.16a is the normalized fluorescence spectrum of EYAPS samples under 896 nm Xe lamp excitation. The fluorescence spectra near 1 μm and 1.5 μm are typical Yb^{3+} and Er^{3+} emission. With the increase of the phosphorus-aluminum ratio, the photoluminescence intensity at 1 μm significantly decreases. Compared with P/Al < 1 samples (EYAPS0 and EYAPS1), the 1.5 μm fluorescence bandwidth of P/Al > 1 samples (EYAPS2 and EYAPS3) significantly narrows, and the photoluminescence peak shows a red shift. For P/Al < 1 samples (EYAPS0 and EYAPS1), the fluorescence peak is at 1530 nm, and the FWHM is about 46 nm. While for P/Al > 1 samples (EYAPS2 and EYAPS3), the photoluminescence peak shifts to 1535 nm

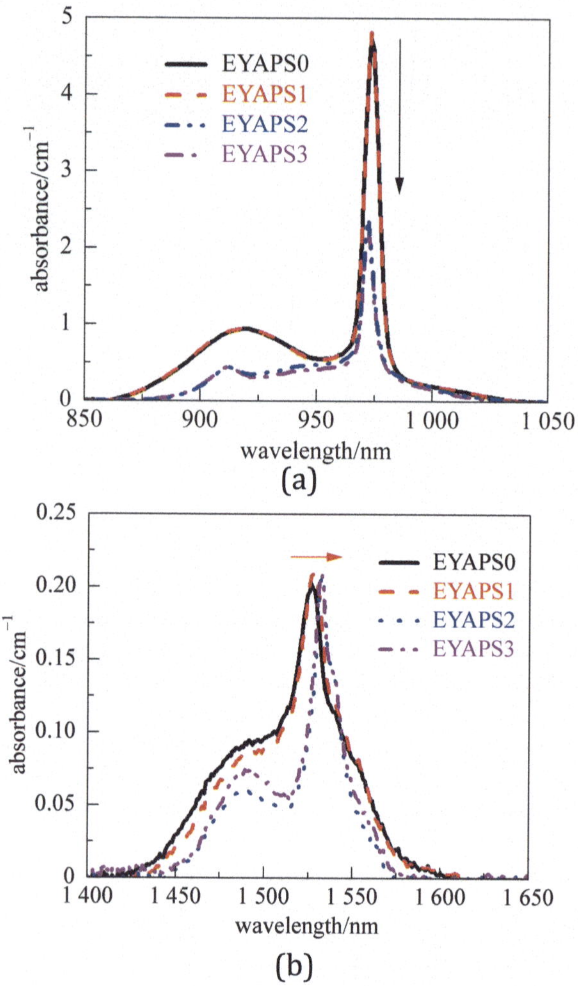

Fig. 6.15 Absorption spectra of EYAPS samples near 1 μm (**a**) and 1.5 μm (**b**) [27]

and the FWHM of the photoluminescence band narrows to 21 nm (Table 6.6), which means that the erbium-doped silica fiber with P/Al > 1 is not suitable for broadband amplification in the C band (1530 ~ 1565 nm). However, in the samples with P/Al > 1, the L band above 1580 nm is relatively flat, which is conducive to improving the gain flatness of the L band (1565 ~ 1625 nm) erbium-doped fiber [35, 36].

Figure 6.16b shows the ratio of fluorescence intensity of Er^{3+}: $^4I_{13/2} \rightarrow {}^4I_{15/2}$ at 1530 nm to that of Yb^{3+}: $^2F_{5/2} \rightarrow {}^2F_{7/2}$ at 975 nm (expressed as I_{1530}/I_{975}). With the increasing the P/Al ratio, the I_{1530}/I_{975} ratio increases from 4 to 14. This indicates that the addition of P greatly suppresses the luminescence of Yb^{3+} and improves the energy transfer from Yb to Er.

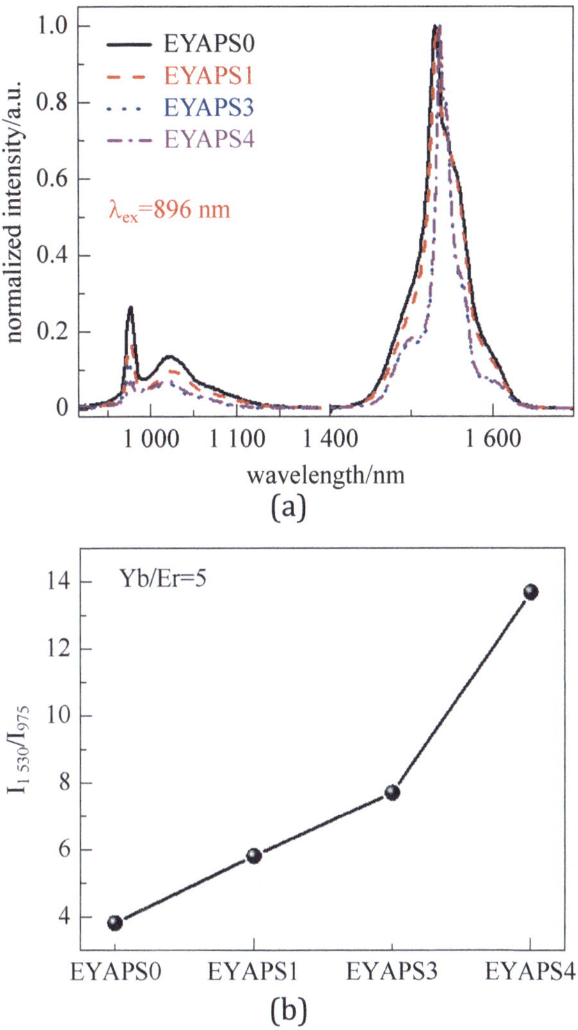

Fig. 6.16 (**a**) Normalized fluorescence spectra of EYAPS samples and (**b**) ratio of fluorescence intensities at 1530 nm of Er^{3+} to 975 nm of Yb^{3+} [27]

Table 6.7 shows the fluorescence lifetime τ_{Yb} of the Yb^{3+}: $^2F_{5/2}$ energy level and the fluorescence lifetime τ_{Er} of the Er^{3+}: $^4I_{13/2}$ energy level in the EYAPS samples. After removing Er^{3+} from the sample (i.e., only Yb^{3+} is doped in the above EYAPS samples, Er^{3+} is not doped, and the Er_2O_3 in the composition is replaced by SiO_2), the Yb^{3+} $F_{5/2}$ fluorescence lifetime is set as τ_{Yb}^0, and the energy transfer efficiency is calculated by the following formula [37]:

Table 6.6 Er^{3+} fluorescence full width at half maxima, FWHM and fluorescence peaks, absorption and emission cross-sections near 1530 nm of EYAPS series glasses [27]

	Absorption			Emission		
Sample	Peak wavelength /nm	Absorption FWHM /nm	Cross-section /pm^2	Peak wavelength /nm	Fluorescence FWHM /nm	Cross-section /pm^2
EYAPS0	1526	43.26	0.52	1530	46.10	0.58
EYAPS1	1528	37.31	0.53	1532	45.62	0.59
EYAPS2	1532	18.41	0.59	1535	21.49	0.64
EYAPS3	1532	18.40	0.60	1535	21.39	0.66

Table 6.7 EYAPS glass Er^{3+}: $^4I_{13/2}$ and Yb^{3+}: $^2F_{5/2}$ energy level lifetimes τ_{Er} and τ_{Yb}

Sample	Er^{3+} Lifetime τ_{Er}/ms	Yb^{3+} Lifetime τ_{Yb}/μs	Yb^{3+} Lifetime τ_{Yb}^0 /μs	Transfer efficiency η/%
EYAPS0	11.98	172	1253	86.3
EYAPS1	11.9	128	1267	89.9
EYAPS2	11.84	183	1842	90.1
EYAPS3	10.6	158	2011	92.1

Note: τ_{Yb}^0 is the lifetime of the Yb^{3+}: $^2F_{5/2}$ level in the above samples after Er_2O_3 is replaced by SiO_2 [27]

$$\eta = 1 - \frac{\tau_{Yb}}{\tau_{Yb}^0} \tag{6.1}$$

With the increase of P/Al ratio, there is little change of the fluorescence lifetime of the Er^{3+}: $^4I_{13/2}$ energy level, always around 11 ms. The energy transfer efficiency of the sample without P (EYAPS0) is the lowest, only 86.3%; while the energy transfer efficiency of the only P doped sample (EYAPS3) is the highest 92.1%. With increasing P/Al ratio, the energy transfer efficiency from Yb to Er gradually increases from 86.3% (EYAPS0) to 92.1% (EYAPS3). This indicates that the introduction of P helps to improve the energy transfer from Yb to Er in silica glass. On the one hand, as described in Chap. 2, the introduction of P increases the solubility of rare earth ions, making Yb ions in a P-rich environment and forming P-O-Yb leakage. The covalency of P-O bonds is strong and the influence of O atoms on Yb ions is weak, which means that the symmetry of coordination field around Yb ions is less broken, decreasing the f → f transition (forbidden transition in a symmetric environment) between the upper and lower states of Yb^{3+}. This can reduce the probability of radiative transition from the excited state of Yb^{3+} to the ground state, suppressing the 1 μm emission from Yb^{3+}; on the other hand, the introduction of P increases the phonon energy of the glass matrix. It can accelerate the non-radiative transition rate of Er^{3+} from $^4I_{11/2}$ to $^4I_{13/2}$ [38], reducing the population of Er^{3+} $^4I_{11/2}$ and suppressing the reverse energy transfer from Er^{3+} $^4I_{11/2}$ to Yb^{3+} $^2F_{5/2}$. Ultimately, the energy transfer probability from Yb^{3+} $^2F_{5/2}$ to Er^{3+} $^4I_{11/2}$ is improved.

6.3 Gain Characteristics of Erbium-Doped Silica Fiber Amplifiers

This section will discuss and establish the three-level rate equations used for the simulation of 1.5 μm band erbium-doped silica fiber amplifiers. And the characteristics of small signal gain and saturation region gain will be simply discussed in combination with the light transmission equation. On this basis, the impact of non-uniform broadening, excited state absorption, up-conversion and cluster effects, temperature variation, and overlap factors on the performance of erbium-doped silica fiber amplifiers will be further discussed from the perspective of fiber materials.

6.3.1 Rate Equations Under Three-Level Approximation

The energy structure of Er^{3+} is very complex (Fig. 6.1), making it difficult to obtain the accurate theoretical model, which is not conducive to the analysis of the characteristics of erbium-doped fibers. Therefore, generally, its energy states is usually simplified in approximation according to the actual application scenario, which is convenient for theoretical analysis and numerical simulation. The erbium-doped fiber amplifier usually uses a 980 nm laser diode as the pump source, and then radiates light in the 1.5 μm band. In this application, the Er^{3+} ion can be approximated as a three-level system, as shown in Fig. 6.17.

Below is a brief discussion of the rate equation under the three-level approximation. In Fig. 6.17, the ground state $^4I_{15/2}$, is donated as level 1, and the laser upper state $^4I_{13/2}$, is donated as level 2, and the pump state $^4I_{11/2}$is donated as level 3; the particle number densities of levels 1, 2, and 3 are represented as N_1, N_2, N_3 respectively. The pump photon flux is represented by φ_{p}, that is, the number of photons

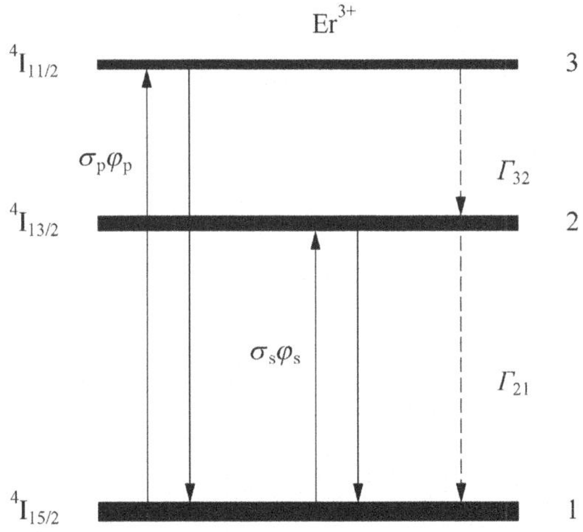

Fig. 6.17 Schematic of simplified three-level system for Er^{3+} ion [39]

passing through unit area per unit time; the signal photon flux is represented by φ_s. For particles at the level 3, there are two ways to return to the lower levels, including radiative transitions from level 3 to level 2, donated as A_{32}^{R}, and radiative transitions from level 3 to level 1, donated as A_{31}^{R} and the non-radiative transition from energy level 3 to energy level 2, donated as A_{32}^{NR}. In erbium-doped silica glass, the non-radiative transition rate of particles at energy level 3 is much greater than the radiative transition rate; this is due to the small energy difference ~ 3670 cm^{-1} between level 3 and level 2. The maximum phonon energy is about 1100 cm^{-1} in silica glass. Thus, only 3 phonons are needed to assist the non-radiative transition. According to the literature [6], the non-radiative transition rate from level 3 to level 2 is A_{32}^{NR} =5.53 × 10^4 s^{-1}, while the total radiative transition rate of the $^4I_{11/2}$ state is A_{31}^{R} + A_{32}^{R} ≈125 s^{-1}. Therefore, it can be roughly considered that particles at energy level 3 mainly relax to level 2 through non-radiative transition, and the non-radiative transition rate is much greater than the radiative transition rate from level 2 to level 1. That is, the particle density at level 3 is close to 0. Due to the large energy difference between levels 1 and 2, the transition from level 2 to level 1 is mainly radiative transition. The transition rates from level 3 to level 2 and from level 2 to level 1 are represented as Γ_{32} and Γ_{21}, respectively, of which Γ_{21} is $1/\tau_2$ (τ_2 (represented as the lifetime of level 2). The pump light absorption cross-section from level 1 to level 3 is σ_p, and the signal light emission cross-section from level 2 to level 1 is σ_s, then the variation of particle density at each level can be represented by the following rate equations:

$$\frac{dN_3}{dt} = -\Gamma_{32}\ N_3 + (N_1 - N_3)\ \sigma_p\ \varphi_p \tag{6.2}$$

$$\frac{dN_2}{dt} = -\Gamma_{21}\ N_2 + \Gamma_{32}\ N_3 - (N_2 - N_1)\ \sigma_s\ \varphi_s \tag{6.3}$$

$$\frac{dN_1}{dt} = \Gamma_{21}\ N_2 - (N_1 - N_3)\ \sigma_p\ \varphi_p + (N_2 - N_1)\ \sigma_s\ \varphi_s \tag{6.4}$$

For a steady state, the variation of particles at each level with time is 0, that is,

$$\frac{dN_1}{dt} = \frac{dN_2}{dt} = \frac{dN_3}{dt} = 0 \tag{6.5}$$

The total particle density is N (i.e., the doping concentration of rare earth ions), then we have

$$N = N_1 + N_2 + N_3 \tag{6.6}$$

The calculation results are as follows:

$$N_2 - N_1 = \frac{\varphi_p \sigma_p - \Gamma_{21}}{\Gamma_{21} + 2\varphi_s \sigma_s + \varphi_p \sigma_p} N \tag{6.7}$$

In an ideal situation where no other losses are considered, when the number of particles in level 2 and level 1 are equal, i.e. N1 = N2, The particle inversion degree between the upper and lower levels is 0, which means the optical amplification threshold is reached. Thus, the required pump light flux threshold φ_{th} can be expressed as:

$$\varphi_{th} = \frac{\Gamma_{21}}{\sigma_p} = \frac{1}{\tau_2 \sigma_p} \tag{6.8}$$

When the pump light flux is greater than φ_{th}, the particle inversion degree of the upper and lower levels is greater than 0, that is, N1 > N2. The signal light increases after passing through the gain medium, manifested as amplification effect. In contrast, when the pump light flux is less than φ_{th}, the particle inversion degree of the upper and lower levels is less than 0, i.e., N2 < N1, the light decreases after passing through the gain medium, which is manifested as an absorption effect. The intensity of the pump light (i.e., the energy of the pump light per unit time and unit area) is $I_p = h\upsilon_p \varphi_p$, and its threshold I_{th} can be expressed as

$$I_{th} = \frac{h\nu_p}{\tau_2 \sigma_p} \tag{6.9}$$

According to the above formula, in order to reduce the threshold of pump light intensity, the pump absorption cross-section of Er^{3+} should be increased or the lifetime of level 2 should be improved. Larger absorption cross-section means a higher probability of pump photons absorption, and longer upper level lifetime means more energy stored in the upper level per unit time.

6.3.2 Small Signal Gain and Saturation Region Gain

For analysis of the longitudinal amplification characteristics of the fiber, the erbium-doped fiber can be regarded as a one-dimensional waveguide, only considering the distribution of Er^{3+} and the light field along the longitudinal direction of the fiber. The light field intensity at the fiber position z can be expressed as.

$$I(z) = \frac{P(z)\Gamma}{A_{eff}} \tag{6.10}$$

$P(z)$ is the optical field power; Γ is the overlap factor of the optical field and Er^{3+} in the fiber core; A_{eff} is the effective area of the Er^{3+} distribution. After the pump light and signal pass through the Er^{3+} doped fiber, the variation of the photon flux can be represented as.

$$\frac{d\varphi_s}{dz} = (N_2 - N_1)\ \sigma_s\ \varphi_s \tag{6.11}$$

$$\frac{d\varphi_p}{dz} = (N_2 - N_1)\ \sigma_p\ \varphi_p \tag{6.12}$$

Therefore, the variation of the intensity of the signal and pump along the fiber can be represented as.

$$\frac{dI_s(z)}{dz} = (N_2 - N_1)\ \sigma_s\ I_s(z) = \frac{\sigma_p \frac{I_p}{h\nu_p} - \Gamma_{21}}{\Gamma_{21} + 2\sigma_s \frac{I_s}{h\nu_s} + \sigma_p \frac{I_p}{h\nu_p}} N\sigma_s\ I_s(z) \tag{6.13}$$

$$\frac{dI_p(z)}{dz} = (N_3 - N_1)\ \sigma_p\ I_p(z) = -\frac{\sigma_s \frac{I_s}{h\nu_s} + \Gamma_{21}}{\Gamma_{21} + 2\sigma_s \frac{I_s}{h\nu_s} + \sigma_p \frac{I_p}{h\nu_p}} N\sigma_p\ I_p(z) \tag{6.14}$$

The condition for the signal to be amplified is

$$I_p \geq I_{th} = \frac{h\upsilon_p}{\tau_2 \sigma_p} \tag{6.15}$$

Define the population inversion degree $D = (N_2 - N_1)/N = (2N_2 - N)/N$. When $D = -1$, it indicates that all the populations are in the ground state, when $D = 1$, all the populations are in the excited state. When the signal is very small, far less than the amplified threshold, it can be approximated that the population inversion degree in the fiber is close to 1. In this situation, the pump is almost unchanged along the fiber, and the population are all in the upper level $N_2 \approx N$. Thus $N_1 \approx 0$, and we have.

$$\frac{dI_s(z)}{dz} = N\sigma_s\ I_s(z) \tag{6.16}$$

where

$$I_s(z) = I_s(0)\exp(N\sigma_s z) \tag{6.17}$$

The gain coefficient $\alpha = N\sigma_s$, is a constant. Therefore, when the pump power is much greater than the signal, and the population inversion degree of the fiber is close to 1, the gain coefficient remains constant, only related to the Er^{3+} doping concentration and its emission cross-section. However, when the power of the signal increases to a certain extent, the population inversion decreases and is no longer close to 1, the above situation no longer applies.

The saturation light intensity is defined as.

$$I_{sat}(z) = \frac{1 + I_p / I_{th}}{2\eta} = \frac{1}{2}\frac{h\nu_s}{h\nu_p}\frac{\sigma_p}{\sigma_s}\left(1 + I_p \frac{\sigma_p \tau_p}{h\nu_p}\right) \tag{6.18}$$

The saturation light intensity I_{sat} increases with the pump intensity, while decreases with the emission cross-section. When the pump light remains constant, the intensity of the signal increases approximately linearly with the length of the fiber, and the gain coefficient per unit length gradually decreases. The 3 dB saturated output power is defined as the signal output power when the small signal gain drops by 3 dB, which is an important parameter of EDFA. When the input signal light is relatively small, the gain coefficient of EDFA is large. As the power of the signal light increases, the gain decreases after EDFA enters the saturation region. Figure 6.18 shows the variation of signal light gain with output power when the pump power is constant. When the signal light power is small, the output power is small, the erbium-doped fiber works in the linear region, and the output power increases rapidly with the increase of signal light. As the power of the signal increases to a certain

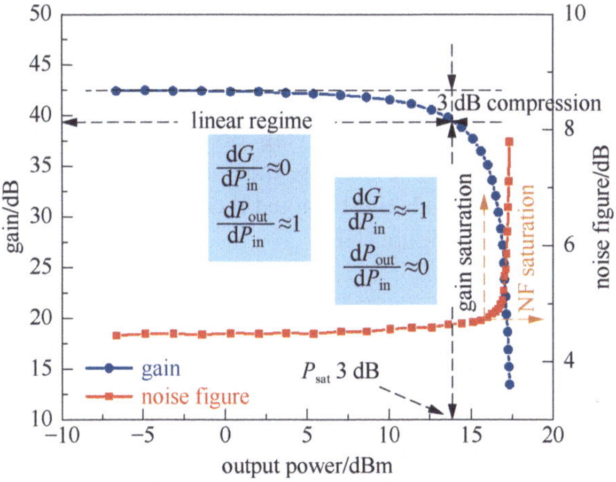

Fig. 6.18 Gain and noise coefficient variation with output power [40]

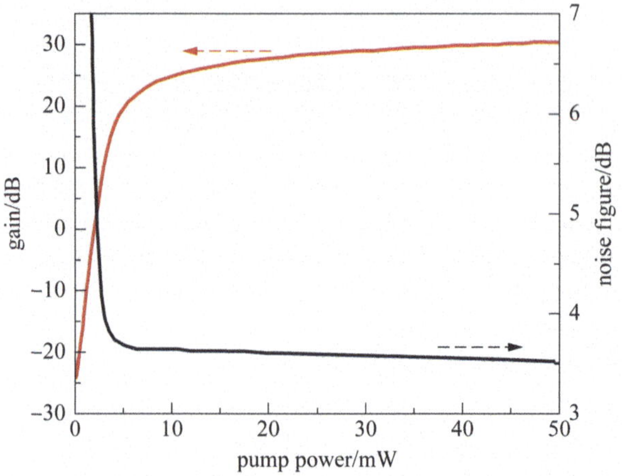

Fig. 6.19 Gain, NF variation with pump power [39]

extent, the output power is large, the erbium-doped fiber enters the saturation region, the gain drops rapidly, the increase of output power becomes slow, and gradually approaches the saturated output power. For this situation, the output power no longer increases with signal power, and for much higher output signal further increase of pump power is needed.

Figure 6.19 shows the variation of gain with pump power where the input signal power is constant. When the pump power is small, a relative large signal light significantly consumes the populations in the excited state, and the erbium-doped fiber works in the deep saturation region, with small gain. As the pump power increases, the population inversion degree rapidly increases, and the gain linearly increases with the pump. When the pump power increases to a very large amount, the population inversion degree is close to 100%, and the erbium-doped fiber works in small signal region, with constant gain. Further increase the output power can only be achieved by increasing the power of the signal. Accurately understanding the small signal region and saturation region characteristics of erbium-doped fiber and selecting the appropriate working range in practical applications can fully utilize the characteristics of erbium-doped fiber, achieving optimal performance. In practical optical communication, in order to increase the output power, EDFA mostly works in the saturation region, but it should be noted that oversaturation will bring about the deterioration of the noise figure (NF).

6.3.3 Spectral Broadening and Spectral Hole Burning

The spectral broadening of transitions between rare earth ion states are generally divided into homogeneous and inhomogeneous. Inhomogeneous broadening mainly stems from variation of the local environment of rare earth ions. This spectral inhomogeneous broadening can be seen as the superposition of several sub-homogeneous broadenings, each representing a possible local coordination environment of the rare earth ions. In addition, another possible spectral broadening mechanism for rare earth ions comes from the Stark splitting of the energy state. Due to the influence of the local environment, the energy state of rare earth ions introduce Stark splitting, and the radiative transition center energies between different Stark levels are close and overlap with each other. Generally, the more Stark splitting of energy states, the wider the energy state broadening, and the wider the spectral broadening of radiative transitions between the two states. If the ion rebalancing process between the Stark states of rare earth ions is fast enough, that is, the process of rebalancing is faster than the interaction process between the signal and the rare earth ions, then the spectral broadening caused by Stark splitting will show homogeneous broadening characteristics. Specifically, for Er^{3+} doped silica glass fiber, giving the large amount of Stark splittings of the upper and lower states and the large phonon energy at room temperature, the 1.5 μm radiation spectrum line of Er^{3+} will show strong homogeneous broadening. In the above discussion of the three-level system, it is assumed that the energy states are homogeneously broadened, and the properties of each Er^{3+} in the matrix are the same. They all have the same center frequency, line width, cross-section, fluorescence lifetime. These assumptions are valid in the small signal gain region of the erbium-doped fiber amplifier. However, in practical, the network structure of the glass is disordered, the lattice positions occupied by Er^{3+} are chaotic, and the coordination environments of Er^{3+} at different positions are different. Er^{3+} in the glass may occupy several different coordination environments. While the properties of the same type Er^{3+} are similar, the properties of different Er^{3+} are quite different. Therefore, it is also necessary to consider the inhomogeneous broadening of the Er^{3+} radiative transition spectral, especially in the saturated gain region. For the inhomogeneous broadening radiative transition spectral, some Er^{3+} ions in the matrix only contribute to a specific frequency and do not contribute to other frequencies. That means a specific frequency in the spectrum is attributed to a specific Er^{3+} in the matrix. Therefore, when the erbium-doped fiber amplifier operates in the saturated gain region, the gain curve is likely to show the inhomogeneous broadening characteristics resulting in spectral hole burning (SHB) effect occurs, as shown in Fig. 6.20 [40]. The dashed line in the figure has a saturation wavelength of 1531 nm, the solid line has a saturation wavelength of 1553 nm, and the inset shows the difference between the two curves. During amplification, when a saturated signal of a specific frequency is injected, due to the strong non-uniform broadening characteristics of the erbium-doped silica fiber, the signal only strongly interacts with a small amount of the Er^{3+} ions in the fiber, causing these ions to undergo stimulated transitions. The population inversion degree decreases sharply, resulting in a local gain saturation and a decrease of the gain coefficient at the signal frequency, which is presented as a "burn hole" on the signal gain curve. This is known as the spectral hole burning effect. As uniform broadening is simultaneously in the fiber, the burn hole has a certain width. The stronger the signal light intensity, the deeper the spectral hole burning. In dense wavelength division multiplexing systems, the actual process of adding and

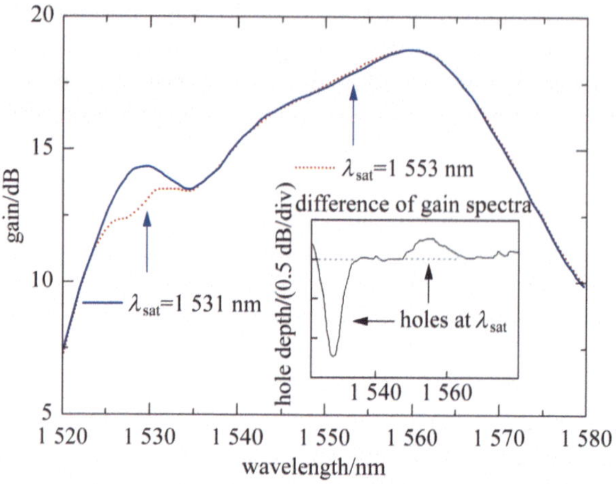

Fig. 6.20 The hole burning effect of the erbium-doped fiber gain spectrum

dropping channels often results in few-wave amplification, i.e., the signal amplification is concentrated in a few channels. This will cause a significant spectral hole burning effect, resulting in a local decrease in gain. This will make the accurate simulation of the gain spectrum very difficult. Especially after multiple stages of cascading, this effect will gradually accumulate, causing a decline in system performance. So it is necessary to reduce the impact of spectral hole burning. Spectral hole burning is derived from the complexity of the Er^{3+} coordination field in erbium-doped silica glass. Reports have pointed out [41–43] that changing the doping concentration of Er, Al, and Ge has a certain impact on the spectral hole burning effect. The elements and concentrations of dopants in commercial erbium-doped silica fibers are relatively fixed, so it is very difficult to suppress the spectral hole burning effect by optimizing the composition of the fiber. Studies have shown that the depth and width of the spectral hole burning are affected by the wavelengths and power of the signal and pump, temperature, and other factors [44–47]. Spectral hole burning is most obvious near 1530 nm. And 980 nm pumping is more severe than 1480 nm pumping and increasing temperature helps to reduce the spectral hole burning effect. For this reason, researchers have proposed a series of improvements for suppressing spectral hole burning [48–51], such as reducing the saturation depth, making the wavelength interval of each channel uniformly distributed under the condition of fewer waves, compensating for the gain reduction caused by the hole-burning effect through rapid gain control, using pseudo-wave filling to avoid the concentration on a few channels and the sudden drop of signal power, wavelength-related path allocation and so on. However, there are still problems such as high complexity, difficulty in implementation, mismatch with practial scenarios, and high cost, which require the exploration of simpler and more economical solutions.

6.3.4 Excited State Absorption

The above discussion simplifies the structure of Er^{3+} into a three-level system, ignoring the possibility that the upper state of the 1.5 μm laser may transition to other higher energy state due to the pump or signal. Therefore, a simple three-level system cannot describe all the transition processes that occur in erbium-doped fiber. In some special cases, the complex higher states of Er^{3+} needs to be considered. For example, when using a 980 nm LD to pump the erbium-doped fiber, green fluorescence can usually be observed. This is because when the pump power density is large, the population inversion of the erbium fiber is close to 100%. Although the lifetime of the $^4I_{11/2}$ energy level is very short, a small amount of the Er^{3+} on $^4I_{11/2}$ can further absorb the 980 nm pump and transition to a higher state $^4F_{7/2}$, which is called excited state absorption (ESA). The populations on the $^4F_{7/2}$ state transition to the $^2H_{11/2}$ and $^4S_{3/2}$ states through non-radiative transition, and then emit green light of ~530 nm and ~ 550 nm through radiative transition to the ground state, as shown in Fig. 6.21 [52]. The green luminescence will gradually increase with the pump power.

Unlike up-conversion process and cluster effect, even when the doping concentration of Er^{3+} is very low and uniformly distributed, significant green luminescence can still be observed in erbium-doped fiber as long as the pump power is large enough. This phenomenon indicates that in addition to the normal 1.5 μm band luminescence, Er^{3+} also accompanies other energy transfer processes pumped by 980 nm LD, as shown in Fig. 6.22. The process of Er^{3+} ions on the $^4I_{11/2}$ state absorbing 980 nm pump light and transitioning to a higher state is known as pump excited state absorption (pump ESA). For higher pump power, this process will accelerate the consumption of pump power and reduce the pump efficiency. When using 800 nm pump, Er^{3+} ions on the $^4I_{13/2}$ state can absorb 800 nm pump light and transition to the $^2H_{11/2}$ state. Since the lifetime of the $^4I_{13/2}$ state is much longer than that of the $^4I_{11/2}$ [52_55], this excited state absorption process will be more severe. Similarly,

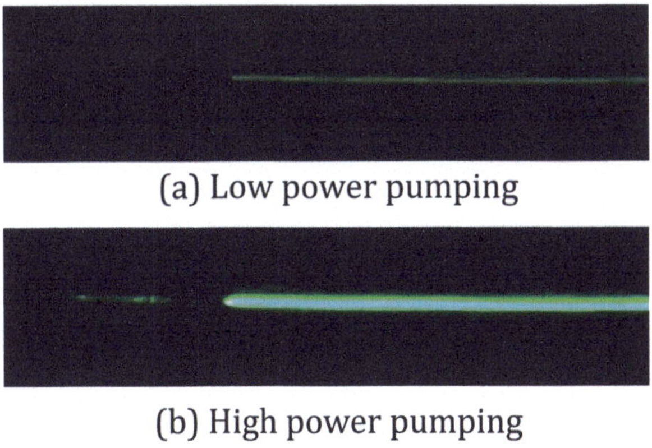

Fig. 6.21 Green light in erbium-doped silica fiber under 980 nm pumping. (**a**) Low power pumping. (**b**) High power pumping

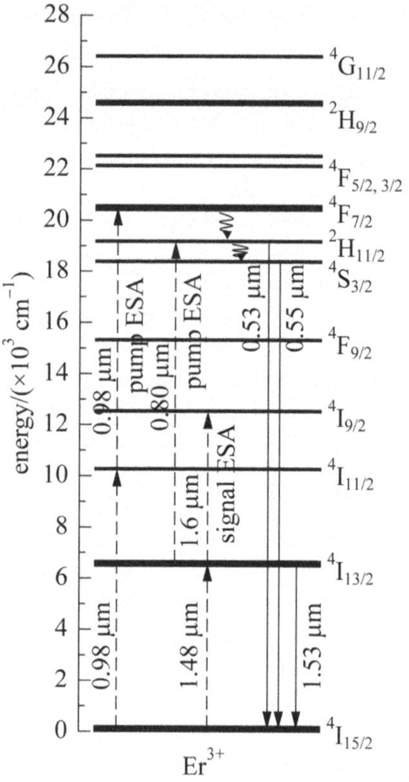

Fig. 6.22 Possible energy transfer processes under different pump conditions for Er^{3+} ion

Er^{3+} ions on the $^4I_{13/2}$ energy level can also absorb signal light and transition to the $^4I_{9/2}$ state, which is known as signal excited state absorption (signal ESA). During this process, the excited state absorption consumes additional population of $^4I_{13/2}$, reducing the population inversion, which is equivalent to increasing the signal light loss. When the signal wavelength is greater than 1600 nm, the emission cross-section of Er^{3+} is very small. While the excited state absorption can significantly reduce the fiber gain coefficient, the noise coefficient increases. This will seriously decline the performance of the extended L-band EDFA.

According to the Giles model [56], the gain coefficient g of the fibercan be expressed as

$$g(\lambda) = g^*(\lambda)\, N_2 - \alpha(\lambda)\, N_1 \tag{6.19}$$

N_1 and N_2 are the population of the $^4I_{15/2}$ and $^4I_{13/2}$ state, respectively, $N_1 + N_2 = 1$; $\alpha(\lambda)$ is the absorption coefficient; $g^*(\lambda)$ is the gain coefficient when Er^{3+} is fully inverted, which can be obtained by testing the small signal gain of the Er^{3+} doped fiber with a population inversion degree close to 100%, and then dividing by the fiber length. Bolshtyansky et al. [57] have proposed a method for simulating and testing the excited state absorption spectrum of L-band erbium-doped fiber based on the Giles model. By introducing excited state absorption coefficient g_{ESA} into the rate equation, the gain coefficient g^* can be expressed as

$$g^*(\lambda) = g_{21}(\lambda) - g_{\mathrm{ESA}}(\lambda) \tag{6.20}$$

$g_{21}(\lambda)$ represents the gain coefficient without excited state absorption, which is only related to the emission process from $^4I_{13/2}$ to $^4I_{15/2}$, and can be obtained by describing the spontaneous emission in the fiber:

$$\frac{\partial P_{ASE}}{\partial z} = \left(g^*(\lambda)\ N_2 - \alpha(\lambda)\ N_1\right)\ P_{\mathrm{ASE}} + \frac{2hc^2}{\lambda^3}\ \Delta\lambda g_{21}(\lambda)\ N_2 \tag{6.21}$$

where P_{ASE} is the power of amplified spontaneous emission (ASE); $\Delta\lambda$ is the linewidth.

$g_{21}(\lambda)$ can be obtained by the following two methods:

The first method is to directly measure the spontaneous emission spectrum of a short optical fiber with high population inversion pumped by a 980 nm LD. Since it is necessary to collect somewhat power of output spectrum, the fiber cannot be very short, so it is almost impossible to obtain an accurate spontaneous emission spectrum. To solve this problem, one can test the ASE spectrum of a certain length of fiber (~20 cm), and the spontaneous emission is partially amplified in the fiber, then the ASE output can be obtained by solving the equation:

$$S_{\mathrm{out}}(\lambda) = \frac{hc^2}{\lambda^3}\ N_2\ \Delta\lambda \frac{g_{21}(\lambda)}{\gamma(\lambda)}\left(e^{\gamma(\lambda)L} - 1\right) \tag{6.22}$$

$$\gamma(\lambda) \equiv g^*(\lambda)\ N_2 - \alpha(\lambda)\ N_1 \tag{6.23}$$

Then.

$$P_{\mathrm{ASE}} = AS_{\mathrm{out}}(\lambda)\ \frac{\lambda^3\gamma(\lambda)}{e^{\gamma(\lambda)L} - 1} \tag{6.24}$$

where A is a constant that is independent of wavelength; L is the length of the fiber. When the wavelength is less than 1580 nm, there is no excited state absorption of the signal light, so we can let $g_{21}(\lambda) = g^*(\lambda)$ to calculate the constant A first, and then calculate the spectrum of $g_{21}(\lambda)$ for wavelengths greater than 1580 nm, that is

$$g_{\mathrm{ESA}}(\lambda) = g_{21}(\lambda) - g^*(\lambda) \tag{6.25}$$

This method can effectively reduce the errors caused by the length of the fiber and the degree of population inversion.

The second method is based on McCumber theory [58, 59] to calculate $g_{21}(\lambda)$, but its accuracy is not as well as the aforementioned "direct method", so it will not be discussed in detail here.

6.3.5 Up-Conversion and Cluster Effects

As is well known, the length of active fibers used in erbium-doped fiber amplifiers is about 1 ~ 100 m, and the overall performance of low-concentration doped erbium fiber amplifiers is better. This is because as the doping concentration of rare earth ions increases, the average distance between adjacent ions decreases and the probability of interaction between adjacent rare earth ions increases, resulting in an enhanceed up-conversion effect of excited state ions. Figure 6.23 shows the energy transfer and up-conversion process between two Er^{3+} ions, where one Er^{3+} loses energy transition from the excited state $^4I_{13/2}$ to the ground state $^4I_{15/2}$, transferring energy to another Er^{3+} in the excited state $^4I_{13/2}$, causing it to transition to a higher energy state $^4I_{9/2}$. The up-conversion effect is related to the degree of population inversion, the higher the degree of particle inversion, the more obvious the up-conversion effect. When the distance between adjacent rare earth ions is further reduced (i.e., when 2 or more rare earth ions occupy the same lattice site), it will result in a sharp drop in fiber optical conversion efficiency, which is called concentration quenching effect.

The solubility of Er^{3+} in pure silica glass is very small, and clusters will form when its concentration exceeds 100 ppm. However, even if there is no clustering effects in a Er^{3+} uniform distributed fiber, up-conversion processes can still occur. Researchers have reported a theoretical model of the effect of Er^{3+} concentration on the luminescent performance in erbium-doped silica fibers [60, 61], as shown in Fig. 6.24. The main network structure of silica glass is composed of silicon-oxygen tetrahedra. The open structure of the matrix leads to relatively large space between atoms, where Er^{3+} ions can be doped. Homogeneous up-conversion (HUC)

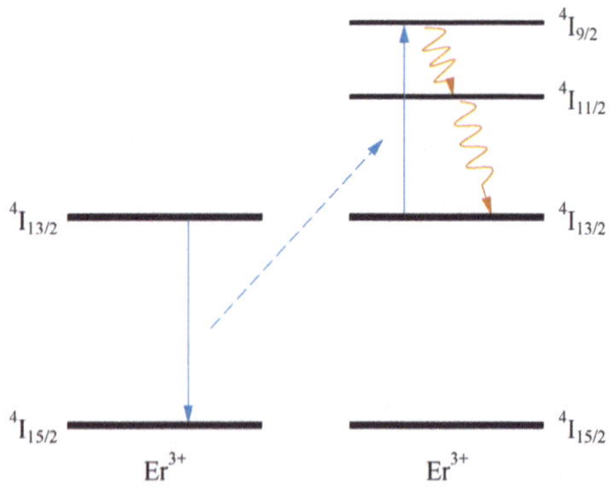

Fig. 6.23 Schematic diagram of the up-conversion process of adjacent Er^{3+}

Fig. 6.24 Two-dimensional network model of Er/Al co-doped quartz glass

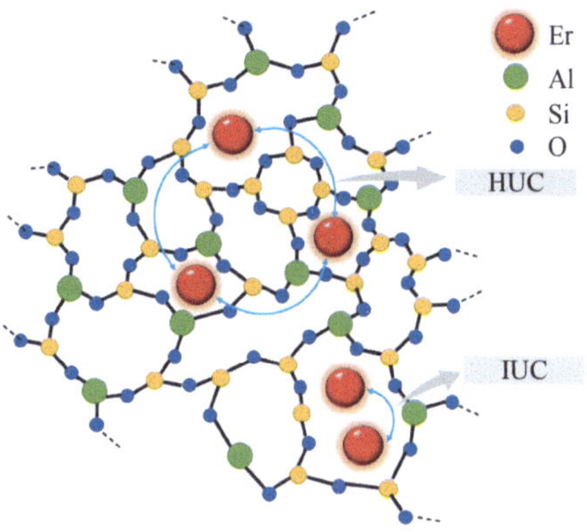

originates from the cooperative energy transfer process of two or three Er^{3+} ions, assuming that Er^{3+} is randomly distributed in the glass network. Homogeneous up-conversion is a reflection of the average distance between all Er^{3+}. In fact, the distribution of Er^{3+} in the silica glass network is not completely random, but due to residual interactions such as local charge compensation, it is easy to form ion clusters. In the cluster structure, at least two Er^{3+} occupy the same lattice site. The inhomogeneous up-conversion process (IUC) induced by the cluster enhances non-saturated absorption, thereby reducing the efficiency of the lasers and amplifiers. The interaction strongly depends on the distance between the Er^{3+} ions, so the inhomogeneous up-conversion process (in sub-microsecond region) of Er^{3+} is faster than the homogeneous up-conversion (in millisecond region) [62]. For these two situations simultaneously occur when Er^{3+} is doped in high concentrations, the impact on the performance of erbium-doped fiber amplifiers should be analyzed by combining rate equation simulations. In the three-level rate equation in Sect. 6.3.1, it has been mentioned that the lifetime of the $^4I_{11/2}$ level of Er^{3+} is very short, mainly non-radiatively transitioning to the $^4I_{13/2}$ level, so it can be approximated that the population of the $^4I_{11/2}$ level is 0. In this consideration, the three-level rate equation can be reduced to a two-level rate equation.

In the following paragraph, we will discuss the impact of the up-conversion process on the performance of erbium-doped fiber amplifiers. Considering the simplest and possible situation, as shown in Fig. 6.23, during the up-conversion process, two adjacent Er^{3+} in the excited state $^4I_{13/2}$, one loses energy and transitions to the ground state, and the other obtains energy and transitions to the $^4I_{9/2}$ state. The ions that transition to the $^4I_{9/2}$ state quickly return to the $^4I_{13/2}$ state through a non-radiative transition process due to it short lifetime. Therefore, even if considering this up-conversion process, the rate equation of two level system can still be used. In addition, as mentioned above, the up-conversion process mainly occurs in uniformly doped Er^{3+} fibers. The adjacent Er^{3+} ions in the excited state are independent of each other. That is, the state of one Er^{3+} is unrelated to its adjacent Er^{3+}. The up-conversion process can only occur when these two ions are in the excited state at the same time. In this case, the change in the number of excited state ions caused by the up-conversion effect is - $C\ N_2^2$, where N_2 is the population at the $^4I_{13/2}$ state, and C is a parameter related to the up-conversion process. Now, the rate equations are rewritten as.

$$\frac{dN_2}{dt} = -\frac{N_2}{\tau} + \left(\sigma_p^{(a)}\ N_1 - \sigma\,\mathrm{p}(\mathrm{e})\ \sigma_p^{(e)}\ N_2\right)\varphi_\mathrm{p} + \left(\sigma_s^{(a)}\ N_1 - \sigma_s^{(e)}\ N_2\right)\varphi_s - C\ N_2^2 \tag{6.26}$$

$$\frac{dN_2}{dt} = -\frac{dN_1}{dt} \tag{6.27}$$

where, N_1 represents the population in the groud state; φ_p and φ_s are the intensities of the pump and the signal respectively; τ is the upper state lifetime; $\sigma_p^{(a)}$, ${}_p^{(e)}$, $\sigma_s^{(a)}$ and $\sigma_s^{(e)}$ are the absorption and emission cross sections of the pump and the signal respectively. Compared with the rate equation in Sect. 6.3.1, only one term - $C\ N_2^2$ is added to describe the decrease of population at the upper state caused by the up-conversion process. In 1989, Blixt et al. [63] used a 1480 nm laser to pump Er^{3+} doped fiber, and observed that the up-conversion luminescence at 980 nm was proportional to the square of the pump power at low pump power. This indicates that in the rate equation describing the variation of the upper state population, it is reasonable to introduce the term - $C\ N_2^2$ to describe the up-conversion process. Of course, if the pump is strong enough or the population at the $^4I_{13/2}$ state is saturated, the upconversion process will no longer maintain a quadratic relationship with the pump power. For a weak pumppower, the up-conversion coefficient C is calculated to be approximately 10^{-22} m^3/s. And for every 10,000 Er^{3+} ions excited to the $^4I_{11/2}$ state, only one emits a 980 nm photon through radiative transition, while the rest transition to the $^4I_{13/2}$ energy level through non-radiative transition processes [63], further confirming the reasonableness of the above rate equation assumption. In addition, Blixt et al. [63] have observed that under appropriate fiber length conditions (different rare earth ion concentrations in different fibers), as the up-conversion luminescence intensity increases, the optimal amplification performance of the fiber significantly decreases.

Although the up-conversion model can be used to partially analyze and explain the impact of the increase in Er^{3+} concentration on the amplification performance of the fiber. As the Er^{3+} concentration further increases, it cannot analyze and explain the non-saturated absorption even under very high pump power. To explain this phenomenon, based on the rare earth ion clustering effect, researchers [15] proposed the pair-induced quenching model: two Er^{3+} occupy the same lattice site, closely combined to form an ion pair. The energy transfer time between them is about 5 ~ 50 μs, much faster than the practical achievable pump rate. Therefore, no matter how strong the pump power, the two Er^{3+} in this ion pair cannot be pumped to the excited state at the same time. That is one ion is always remaining in the ground state. The difference between this model and the up-conversion effect is that: in the latter, the two adjacent ions are independent of each other. That is, one ion being in the ground state or excited state does not affect the state of the adjacent ion. While in the pair-induced quenching model, if one ion is in the excited state, the other ion must be in the ground state. Therefore, in this situation, before applying the rate equation to analyze the amplification performance of Er^{3+} doped fiber, the Er^{3+} ions must be divided into two categories: independent single Er^{3+} ion and paired Er^{3+} ions. By using 980 nm pump, $N_1^{(s)}$ and $N_2^{(s)}$ are the independent Er^{3+} population in level 1 and level 2, respectively. $N_1^{(p)}$ and $N_2^{(p)}$ are the paired Er^{3+} population in level 1 and level 2, respectively. If only considering pump absorption, the population variation of the upper and lower states satisfy

$$\frac{\mathrm{d}N_2^{(S)}}{\mathrm{d}t} = -\frac{\mathrm{d}N_2^{(S)}}{\tau} + \sigma_P^{(a)}\varphi_P N_1^{(S)} - C\left(N_2^{(S)}\right)^2 \tag{6.28}$$

$$\frac{dN_2^{(p)}}{dt} = -\frac{dN_2^{(p)}}{\tau} + 2\ \sigma_p^{(a)}\ \varphi_\mathrm{p}\ N_1^{(p)} \tag{6.29}$$

$$\frac{dN_2^{(s)}}{dt} = -\frac{dN_1^{(s)}}{dt} \tag{6.30}$$

$$\frac{dN_2^{(p)}}{dt} = -\frac{dN_1^{(p)}}{dt} \tag{6.31}$$

In Eq. (6.29), the 2 on the right side of the second term indicates that the absorption of the Er^{3+} ion pair in the ground state is twice that of a single Er^{3+} ion. Assuming that the doped Er^{3+} concentration is N, and the proportion of Er^{3+} in ion pair state is 2 k, then we have.

$$N_1^{(s)} + N_2^{(s)} = (1 - 2\ k)\ N \tag{6.32}$$

$$N_1^{(p)} + N_2^{(p)} = kN \tag{6.33}$$

The signal, pump absorption, and spontaneous emission amplification observed in the experiment are all the sum of the two types of Er^{3+} (independent Er^{3+} ion and Er^{3+} ion pair). Therefore, it is known that the pump absorption of the Er^{3+} doped fiber amplifier containing Er^{3+} ion pair is.

$$\frac{dP(z)}{dz} = -\sigma_p^{(a)}\ \left(N_1^{(s)} + 2\ N_1^{(p)} + N_2^{(p)}\right)\ P(z) \tag{6.34}$$

where P (z) is the pump power along the fiber at position z. If the pump is strong enough, all individual Er^{3+} ions are pumped to the excited state, that is, $N_1^{(S)}$ is approximately equal to 0. Therefore, the absorption of Er^{3+} in Eq. (6.29) can be used to represent the non-saturated absorption coefficient of optical fiber. Researchers

[15] have applied the above model to analyze the silica fiber amplifier with a Er^{3+} concentration of 350 ~ 2000 ppm. The analysis showed that as the concentration of Er^{3+} increases, the ratio of Er^{3+} coupled in pairs in the optical fiber increases from 0 to 20%. The green luminescence phenomenon of Er^{3+}doped silica fiber pump at 980 nm also confirms the above Er^{3+} pair model [64]. The green fluorescence in the optical fiber mainly comes from the $^4S_{3/2}$ state transition to the ground state. Under 980 nm pumping, in order to excite Er^{3+} to the $^4S_{3/2}$ state, Er^{3+} needs to first absorb a pump photon to reach the $^4I_{11/2}$ state, and then absorb another 980 nm photon to reach the $^4F_{7/2}$ state, and finally reach the $^4S_{3/2}$ state through non-radiative transition. Assuming that all Er^{3+} in the doped fiber exist independently, calculations show that as the pump power increases, if the all the Er^{3+} reaches the $^4I_{13/2}$state, the green fluorescence should be linearly related to the pump power; however, the experiment observed that the green fluorescence is proportional to the square of the pump power. This is because, since the above-mentioned Er^{3+} pair or cluster phenomenon exists, some Er^{3+} ions are always in the ground state even under strong pumping condition. Using the ion pair quenching model, researchers successfully explained the above phenomenon [64].

6.3.6 Temperature Dependence of Gain Spectrum

The absorption and emission cross-sections of erbium-doped silcia glass vary with temperature, resulting in the variation of the gain spectrum of the optical fiber with temperature. By detecting the absorption spectrum of Er/Al doped silica glass at different temperatures, the variation of the absorption cross-section of Er^{3+} with temperature can be calculated [65]. Figure 6.25 shows the variation of absorption cross-section at 980 nm and 1480 nm wavelengths with temperature.

As the temperature rises from 20 °C to 140 °C, the absorption cross-section at the 980 nm wavelength decreases from 0.226×10^{-20} cm ^{2} to 0.196×10^{-20} cm^2. Meanwhile, the 980 nm absorption peak is redshift. As for the 1480 nm band, the absorption cross-section also decreases from 0.21×10^{-20} cm^2 to 0.18×10^{-20} cm^2. For high-power erbium-doped fiber lasers, regardless of the pump wavelength, the absorption efficiency of the fiber laser will significantly decrease under high temperature. From Fig. 6.25, it can also be seen that as the temperature rises from 20 °C to 140 °C, the absorption cross-section at 1528 nm decreases from 0.52×10^{-20} cm^2 to 0.47×10^{-20} cm^2, while the absorption cross-section at 1558 nm increases from 0.17×10^{-20} cm^2 to 0.22×10^{-20} cm^2, indicating that the absorption spectrum of Er^{3+} at the 1.5 μm wavelength is also redshift with temperature increasing.

Figure 6.26 shows the variation of the stimulated emission cross-section with temperature. As the temperature rises, the emission cross-section at the short-wave region of 1480 nm gradually increases from 0.065×10^{-20} cm ^{2} to 0.082×10^{-20} cm^2. As for the zero-phonon line, the emission cross-section at 1530 nm decreases from 0.46×10^{-20} cm^2 to 0.44×10^{-20} cm^2. And for 1558 nm, it increases from 0.45×10^{-20} cm^2 to 0.52×10^{-20} cm^2. Figure 6.27 shows the fluorescence spectrum of

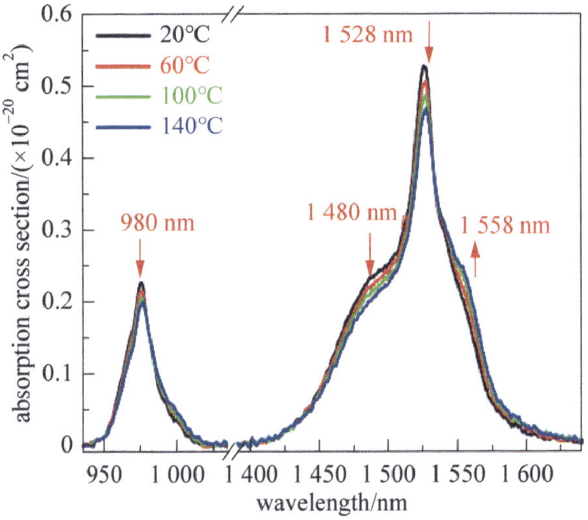

Fig. 6.25 Er/Al doped silica glass absorption cross-sectionunder different temperatures [65]

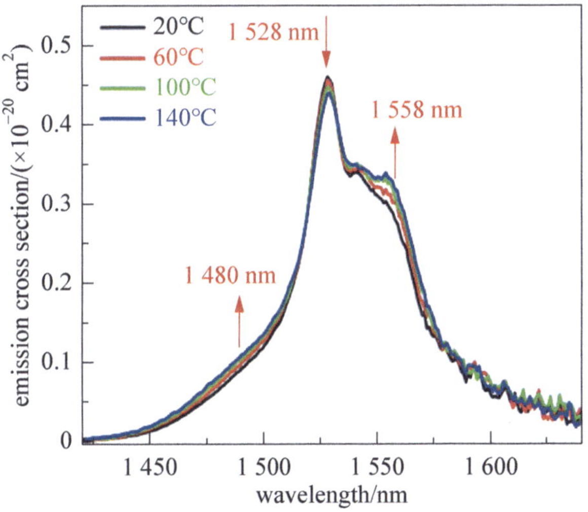

Fig. 6.26 Emission cross-section of Er^{3+} in silcia matrix at different temperatures [65]

erbium-doped silica glass at different temperatures in the 1400 ~ 1700 nm range. There are four main fluorescence emission bands, namely the narrow-band fluorescence peak at 1530 nm and the wide-band fluorescence peaks near 1480 nm, 1550 nm, and 1600 nm. Among them, the 1480 nm wide-band fluorescence peak slightly increases with increasing temperature, while the other three fluorescence peaks all decrease with temperature. Figure 6.28 shows the fluorescence lifetime of Er^{3+} at 1.56 μm in the silica glass. At a normal temperature of 20 °C, the

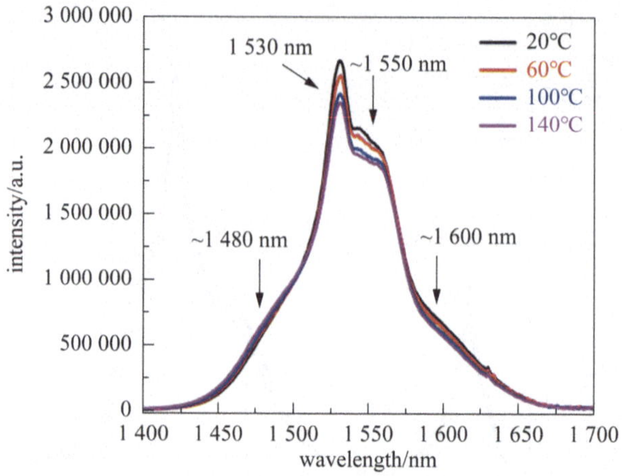

Fig. 6.27 Fluorescence spectrum of Er^{3+} in silica matrix at different temperatures [65]

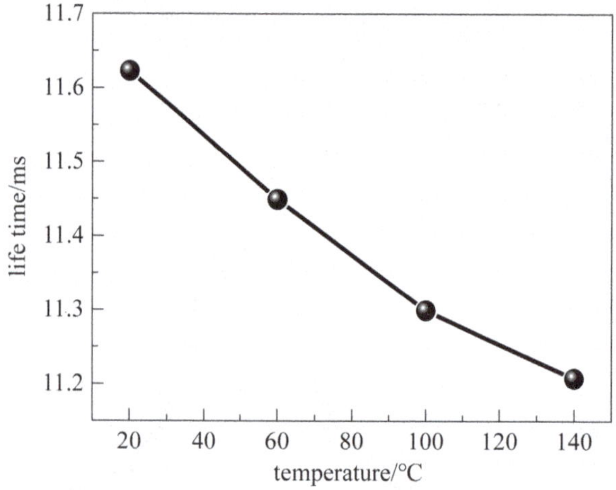

Fig. 6.28 Temperature-dependent fluorescence lifetime of Er^{3+} at 1.56 μm [65]

fluorescence lifetime is 11.63 ms, and when the temperature reaches 140 °C, the lifetime decreases to 11.2 ms. Within this temperature variation range, no temperature quenching occurs. By using linear fitting, it is known that the rate of decreasing of fluorescence lifetime with temperature is approximately −0.003 ms/°C.

Reference [57] simulated the variation of the gain spectrum of an EDFA with an average gain of 25dB and flat gain at 20°C as the temperature increased from −40°C to 80°C, as shown in Fig. 6.29. The gain spectrum is no longer flat with the temperature varying. In the 1528~1540nm band, the gain increases with temperature. While in the 1540~1563nm band the gain decreases with temperature. Researchers have also showed [66] that Sb-Er co-doped fiber has different temperature-dependence

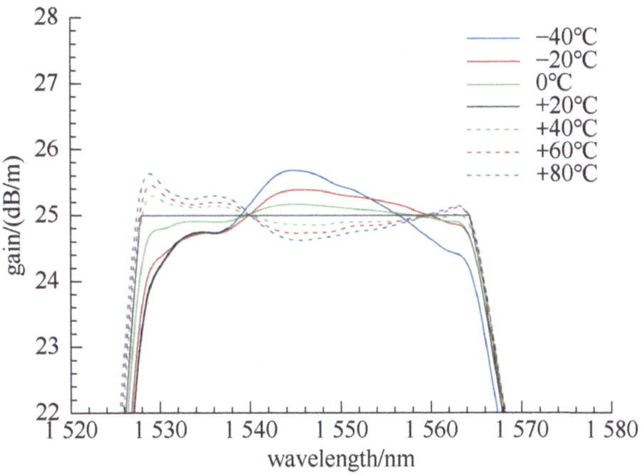

Fig. 6.29 Variation of the gain spectrum with temperature for a fiber with an average gain of 25 dB and flat gain at 20 °C [57]

properties compared to that of Al-Er co-doped fiber. Using Sb-Er co-doped fiber in combination with Al-Er co-doped fiber can help reduce the gain spectrum fluctuation caused by the temperature variation. In the practial application, variation of environmental temperature can lead to degradation of the EDFA gain flatness. Thus, built-in temperature control devices are usually used to ensure that the optical fibers are within a relatively constant temperature range, or a temperature-compensated gain flat filter is used to reduce the fluctuation of the output signals.

6.3.7 Overlap Factor and Partial Doping Effect

In Sect. 6.3.1, the Er^{3+} doped fiber is considered as a one-dimensional waveguide, only considering its interaction with the pump or signal longitudinally. However, on the cross-section of the fiber, the distribution of the optical field can also affect the performance of the Er^{3+} doped fiber. The following paragraph will briefly discuss the impact of the transverse optical field distribution on the Er^{3+} doped fiber performance.

In a step-index Er^{3+} doped fiber, the core is doped with Er^{3+}, Al^{3+} and others and the cladding is usually pure silica glass. Due to the diffusion of core elements during fiber drawing, actual Er^{3+} doped fibers are no long a strict step-index fibers. However, this does not affect the use of the step-index distribution to simply analyze the fiber characteristics. In order to reduce the pump power threshold and increase the pump intensity, the core diameter is generally small. Thus, for commonly used communication Er^{3+} doped fibers, its cut off wavelength is mostly less than 980 nm. Therefor, the optical field of 980 nm pump and 1550 nm signal light can be approximated as Gaussian. That is, in practical, the light waves in the fiber are not all bound in the core, and part of the light waves are transmitted in the cladding, which is called evanescent wave.

On the cross-section of the optical fiber, the intensity distribution of the pump and signal is proportional to the mode field, and can be expressed as

$$I(r,\ \phi) = P\varphi(r,\ \phi) \tag{6.35}$$

where, P is the total power of the mode field; r is the distance from the optical field to the center of the fiber core; ϕ is the angle; φ (r, ϕ) is the normalized intensity distribution. Therefore, we have.

$$\int_0^{2\pi}\int_0^{\infty}\varphi(r,\ \phi)\ r\ \mathrm{d}\ r\ \mathrm{d}\ \phi = 1 \tag{6.36}$$

Supposing the distribution of Er^{3+} concentration in the fiber core is n (r, ϕ), it is generally believed that the Er^{3+} concentration is equal at the same radius. So the Er^{3+} concentration distribution can be converted to a uniform distribution with an average concentration N within the fiber core,

$$N = \frac{2\int_0^{\infty} n(r,\ \phi) rdr}{R^2} \tag{6.37}$$

where R is the fiber core radius. The overlap factor represents the overlap between the transverse light field and the Er^{3+} distribution in the fiber core, which can be expressed as.

$$\Gamma = \int_0^{2\pi}\int_0^{\infty}\varphi(r,\ \phi)\ \frac{n(r,\ \phi)}{N}\ r\ \mathrm{d}\ r\ \mathrm{d}\ \phi \tag{6.38}$$

In step-index single-mode fibers, the distribution of the fundamental mode follows the Bessel function. And near the cutoff wavelength, the distribution of the fundamental mode approximates a Gaussian distribution. In practical applications, the Gaussian function is more convenient than the Bessel function, expressed as follows:

$$\varphi(r,\ \phi) = \frac{2}{\pi\omega^2} e^{-\frac{2r^2}{\omega^2}} \tag{6.39}$$

where ω is the mode field radius, which is approximately expressed according to the empirical formula:

$$\omega = R\left(0.65 + \frac{1.619}{V^{1.5}} + \frac{2.879}{V^6}\right) \tag{6.40}$$

where, $V = \frac{2\pi R \cdot NA}{\lambda}$ is the normalized frequency of the fiber. The pump and the signal can only interact with the overlapping part of the Er^{3+} ions, and the degree of overlap can be represented by the overlap factor.

$$\Gamma = 1 - e^{-\frac{2R^2}{\omega^2}} \tag{6.41}$$

Due to different doping processes, Er^{3+} can either fill the entire core, as shown in Fig. 6.30a, which is called uniform doping; or it can also partially occupy the core, as shown in Fig. 6.30b, which is called partial doping. As can be seen from Fig. 6.30b, if partially doped, Er^{3+} is more concentrated in the center of the light field. and its "perceived" pump light intensity is much higher. Thus it can achieve a higher population inversion degree, which is beneficial to reduce the pump power threshold. That can help to improve the gain per unit pump power and reduce the noise figure. Figure 6.31 compares the noise figure of a 2 μm fiber core which is uniformly doped with that of 1 μm partially doped [4]. It can be seen that the noise

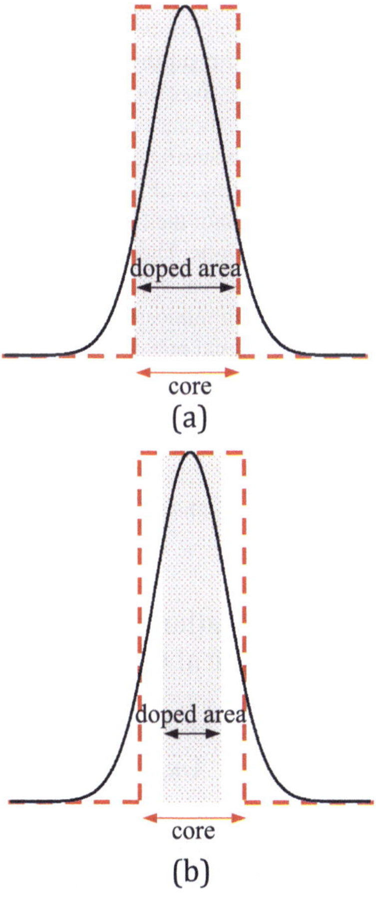

Fig. 6.30 Schematic diagram of Er^{3+} ions fullly doping (**a**) and partially doping (**b**) in the fiber core [4]

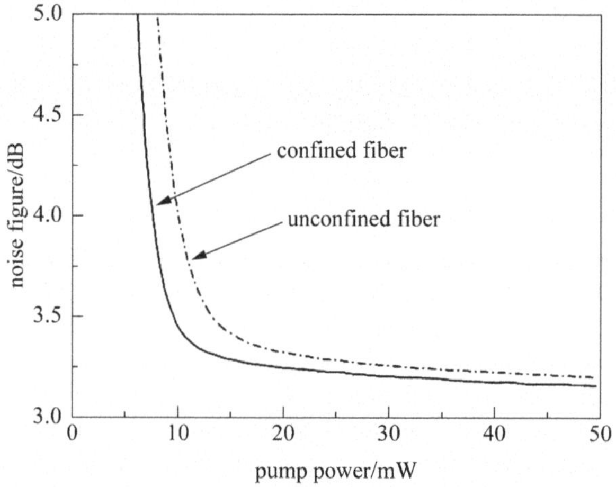

Fig. 6.31 Simulation curve of the noise figure in partially doped and uniformly doped erbium fibers with pump power [4]

figure of partially doped fiber is smaller. However, partial doping will reduce the overlap factor, resulting in decreasing of the absorption coefficient, leading to an increase of the fiber length. In this case, if we want to keep the absorption coefficient constant, the rare earth doping concentration have to be increased, which may bring about clustering effects and reduce the fiber conversion efficiency. However, it is worth noting that if the pump power is very large, the population inversion of Er^{3+} at the edge of the light field is also very high, and the effect of partial doping is no longer obvious. Therefore, it is necessary to design the fiber parameters according to the practical application. Generally, the single-mode laser diode (LD) used for optical communication is power-limited due to its high reliability requirements. If the input signal power is large, the erbium-doped fiber will be in the deep saturation zone, which means partial doping can effectively increase the population inversion and reduce the noise figure. In special application such as undersea, the pump power of optical amplifiers using remote pumping technology is very valuable. Partially doped erbium fiber is beneficial for reducing the pump threshold and improving the signal gain efficiency.

6.4 Applications of Erbium-Doped Fiber in Optical Communication

6.4.1 Gain and Noise Characteristics of Erbium-Doped Fiber Amplifiers

Gain and noise figure are the two most important parameters of optical fiber amplifiers. Ignoring the other factors such as gain efficiency, fiber loss, polarization mode dispersion, and gain dynamics, if a signal is injected into the fiber, the variation of of photons in the transition mode can be calculated. The interaction between light and rare earth ions can be expressed by the following equation [67]:

$$\mathrm{d}N_{\mathrm{m}} = \left[N_2 B_{21} g_{21}(v)\rho(v) - N_1 B_{12} g_{12}(v)\rho(v) + hvN_2 B_{21} g_{21}(v)\right] \mathrm{d}z \tag{6.42}$$

where ν is the photon frequency; N_{m} is the photon flux in the mode; N_2 and N_1 are the number of population in the upper and lower states, respectively; B_{21} and B_{12} are the Einstein coefficients for stimulated emission and stimulated absorption, respectively; $g_{ij}(\nu)$ is the normalized line shape of the transition between the two states; $\rho(\nu)$ is the radiation density (energy per unit frequency and unit volume). The first term on the right side of eq. (6.42) is stimulated emission, the second term is stimulated absorption, and the third term is spontaneous emission. The radiation density equals the photon density times the photon energy, that is, $\rho(\nu) = N_{\mathrm{m}} h\nu$. If we do not considering energy state degeneracy, i.e. $B_{21} = B_{12}$. Equation (6.42) can be rewritten as.

$$\frac{dN_m}{N_m + n_{sp}} = B_{21}\ hv(N_2 - N_1)\ g(v)\ \mathrm{d}z \tag{6.43}$$

where, n_{sp} is the inversion parameter or spontaneous factor:

$$n_{\mathrm{sp}} = \frac{N_2}{N_2 - N_1} \tag{6.44}$$

However, the actual absorption and emission cross-sections vary with wavelength, and due to energy state broadening, the energy state presents a thermal distribution. Therefore, in Eq. (6.42), stimulated emission should be rewritten as $\sigma_{\mathrm{e}}(\lambda) N_2$, and stimulated absorption should be rewritten as $\sigma_{\mathrm{a}}(\lambda) N_1$, where $\sigma_{\mathrm{e}}(\lambda)$ and $\sigma_{\mathrm{a}}(\lambda)$ are the stimulated emission and absorption cross-sections, respectively. λ is the transition wavelength. In this consideration, the inversion parameter is.

$$n_{\mathrm{sp}} = \frac{\sigma_e(\lambda)N_2}{\sigma_e(\lambda)N_2 - \sigma_a(\lambda)N_1} \tag{6.45}$$

Integrate both sides of Eq. (6.43). The integration range on the right is the fiber length from 0 to L, and the integration range on the left is from $N_{\mathrm{m}}(0)$ the number of injected photons to $N_{\mathrm{m}}(L)$ the number of output photons. The relationship between the input and output photon density can be solved as

$$N_{\mathrm{m}}(L) = GN_{\mathrm{m}}(0) + n_{\mathrm{sp}}(G-1) \tag{6.46}$$

where

$$G = \mathrm{e}^{gL} = e^{B_{21}hv(N_2 - N_1)g(v)L} \tag{6.47}$$

where g is the amplifier gain coefficient; G is the total gain. Equation (6.46) contains the most important parameters of the amplifier: gain and noise. The first term on the right side is the amplified signal, and the second term is the amplified spontaneous radiation output. As the amplifier gain is G, the output noise power within the bandwidth $\Delta \nu$ is

$$P_{\text{ASE}} = n_{\text{sp}} \left(G - 1 \right) h\nu \ \Delta\nu \tag{6.48}$$

Equation (6.48) is the expression of the noise of the erbium-doped fiber amplifier, which represents the noise power of the amplifier per unit space for single polarization. A single-mode fiber has 2 polarizations, so the total noise power output of single-mode fiber needs to be multiplied by 2. A multimode fiber amplifier has many modes, and its total noise power is much larger.

Equation (6.46) shows that when a signal passes through the fiber amplifier, it will get an amplified signal and amplified spontaneous emission output. The amplified spontaneous emission will reduce the signal-to-noise ratio of the output signal. The factor representing the reduction in signal-to-noise ratio is the noise figure, which is defined as the ratio of the signal-to-noise ratio of the input signal to that of the output signal:

$$NF = \frac{SNR_{in}}{SNR_{out}} \tag{6.49}$$

The main sources of noise in erbium-doped fiber amplifiers include shot noise, thermal noise, signal-spontaneous emission beat noise, and spontaneous emission-spontaneous emission beat noise. It can be proven that the main source of noise in fiber amplification is signal-spontaneous emission beat noise. According to Eq. (6.46), the noise figure expression for single-mode fiber can be derived as.

$$NF \approx 2 \ n_{\text{sp}} \frac{G-1}{G} + \frac{1}{G} \tag{6.50}$$

where$G$ is the linear gain coefficient. In practical, the noise figure is represented by $NF = \log_{10} \left(2n_{sp} \dfrac{G-1}{G} + \dfrac{1}{G} \right)$ in dB. Under the conditions of small signal high gain and sufficient population inversion, $NF \approx 2\, n_{\text{sp}}$, $n_{\text{sp}} = 1$,the noise figure is 3 dB. So the lowest noise figure in erbium-doped fiber amplifiers is 3 dB.

In practical use of erbium-doped fiber amplifiers, the measurement of gain and noise figure is a basic job. In the following paragraph, by using the input signal spectrum and the output spectrum as seen in (Fig. 6.32), the measurement methods and steps of the gain and noise figure are presented, [4].

According to the definition of gain coefficient:

$$G\left(\text{dB}\right) = 10\lg\left(\frac{P_{signal-out}}{P_{signal-in}} \right) \tag{6.51}$$

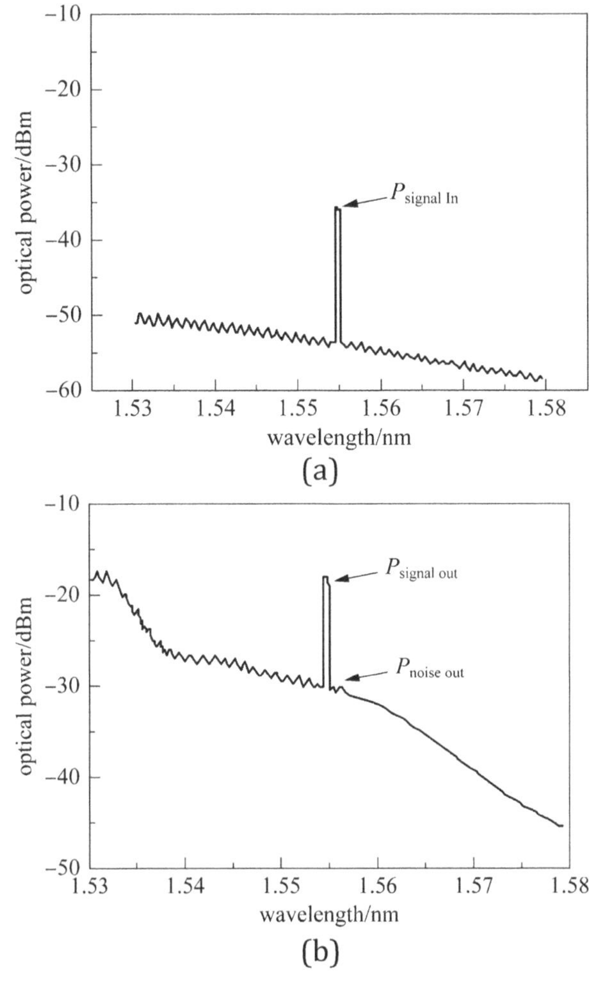

Fig. 6.32 Input (**a**) and output (**b**) signal of erbium-doped fiber amplifier measured by spectrometer

For using a spectrometer for measurement, $P_{\text{signal-out}}$ also includes a portion of the ASE, so it needs to be deducted when calculating the gain coefficient. ASE is included in the signal and cannot be directly measured. Since the width of the signal is generally narrow, a common method to determine the ASE power is by using interpolation. That is, in the absence of a signal, it is assumed that the ASE noise spectrum is a smooth curve, so the signal's ASE noise power can be determined by simply linear fitting the adjacent $P_{\text{noise-out}}$. Now, the expression of the gain coefficient is rewritten as

$$G(\text{dB}) = 10\lg\left[\frac{\left(P_{signal-out} + P_{noise-out}\right) - P_{noise-out}}{P_{signal-in}}\right] \tag{6.52}$$

Combining Eqs. (6.48) and (6.50), the noise figure is expressed as follows:

$$NF(\text{dB}) \approx 10\lg\left(\frac{P_{ASE}}{h\nu\Delta\nu G}+\frac{1}{G}\right) \tag{6.53}$$

where P_{ASE} is the ASE power within the frequency interval $\Delta\nu$; h is the Planck constant; ν *is the frequency of the signal photon; G is the linear gain coefficient (not in dB units, i.e.,* $G = \frac{P_{signal-out}}{P_{signal-in}}$ *). During the spectrometer test process, the wavelength interval* $\Delta\lambda$ *is often used, and its relationship with the frequency interval* $\Delta\nu$ *is*

$$\Delta\nu = \frac{c}{\lambda}\left[\frac{\Delta\lambda}{\lambda}\right] \tag{6.54}$$

Therefore, by accurately testing the gain coefficient and the amplified spontaneous emission P_{ASE}, the noise figure can be obtained. Taking Fig. 6.32 as an example, according to the above measurement method and operation steps, the gain coefficient is 14.34 dB, the linear gain coefficient is 26.96, $P_{ASE} = 0.7 \times 10^{-6}$ W, and its noise figure is 3.56 dB according to formula (6.53).

6.4.2 Erbium-Doped Silica Fiber Amplifier

The most popular application of erbium-doped silica fiber is in the production of EDFA optical amplifiers. A typical EDFA includes erbium-doped silica fiber, wavelength division multiplexer, semiconductor pump source, optical isolator, and optical filter, etc. The wavelength division multiplexer is used to combine the signal light and pump source into the erbium-doped fiber. The optical isolator is used to suppress reflections at the device fusion point or other points that may bring stray light, ensuring the stable operation of the device. The optical filter is used to filter out the amplified spontaneous emission (ASE) noise generated in the amplifier and ensure a flat gain within the working wavelength range. Due to their function and position in the communication systems, there are three types of EDFAs: (1) Power amplifier, located after the transmitter, is used to boost the power of multiple wavelength signals before transmission. Since the power of the combined signal is generally quite large, the noise figure and gain requirements for the power amplifier are not very high. It is only required to have a large output power after amplification. (2) In-line (or relay) amplifier located in the transmission system after the power amplifier, is used to periodically compensate for transmission losses, generally requiring a smaller noise figure and larger output optical power. (3) Pre-amplifier, located before the optical receiver and after the line amplifier, is used to amplify signals and improve the sensitivity of the receiver. Though, under the condition that optical signal to noise ratio (OSNR) meets the requirement, a larger input power can suppress the noise of the receiver and improve the sensitivity of reception, it requires the noise figure to be as small as possible, but no requirements for the output power. A larger input power can suppress the noise of the receiver itself and improve the sensitivity of reception. It requires the noise figure to be as small as possible, and there are no high requirements for output power. According to the pumping directions, there are three EDFAs: co-directional pumping, counter-directional pumping, and bidirectional pumping (Fig. 6.33). Under the co-directional pumping arrangement, the signal and pump are injected into the input end of the erbium-doped silica fiber in the same direction through the wavelength division multiplexer, and the amplified signal is output through the optical isolator and filter in turn. In the counter-directional pumping mode, the signal and pump are injected into the two ends of the erbium-doped silica fiber separately. Bidirectional pumping combines the co-directional and counter-directional pumping methods. The signal is injected into the input end of the erbium-doped silica fiber, and the pumps are injected into both ends of the erbium-doped silica fiber at the same time.

In the small signal gain condition, i.e., when the injected signal power is below -30 dBm, there is almost no difference between using co-directional or counter-directional pumping arrangements, and the same gain coefficient can be obtained. This is because a very small injected signal power is not enough to affect the Er^{3+} population inversion. Since the number of particles at the upper state of the forward and backward pumped optical fiber is symmetrically mirrored, the corresponding ASE is also symmetrically mirrored. That is, the forward ASE generated by co-directional pumping is consistent with the backward ASE generated by

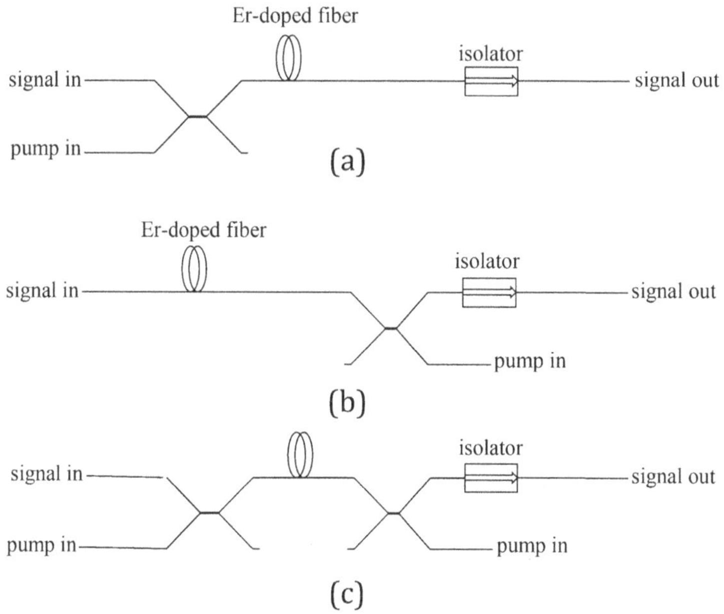

Fig. 6.33 Three pumping arrangements of EDFA [4] (**a**) Co-directional pumping (**b**) Counter-directional pumping (**c**) Bidirectional pumping

counter-directional pumping. However, bidirectional pumping is completely different., The distribution of the population at the upper state is completely different from that of single-end pumping, so the ASE spectrum generated is also different. Especially when the optical fiber is long enough, under the same pumping power condition, bidirectional pumping can produce higher gain than that of single-end pumping. In addition, if there is Er^{3+} pump excited state absorption or up-conversion luminescence in the fiber, bidirectional pumping can produce much higher gain than that of single-end pumping. This is mainly because bidirectional pumping can make the distribution of the population at upper state more uniform, reducing the local pump power density, and thus reducing the occurrence probability of the above-mentioned excited state absorption and up-conversion luminescence.

As the signal power increases, when the input signal power is comparable to the ASE power, different pumping arrangements will have a significant impact on the gain coefficient of the amplifier. Taking the online amplifier with a signal input power between $-20 \sim -10$ dBm as an example, choosing counter or bidirectional pumping can obviously obtain a larger gain coefficient than that of the co-directional pumping, especially when the optical fiber is relatively long. As shown in Fig. 6.34a, when the injected signal power is −20 dBm and the 980 nm pump power is about 50 mW, it can be seen that: with the increase of fiber length, compared with the co-directional pumping, counter-directional pumping can obviously obtain a much larger signal output power. This is mainly because that the distribution of the population at the upper state under counter-directional pumping is more conducive to signal power amplification. Large signal and pump power are conducive to suppressing the generation of ASE, thus enabling more energy to be converted from the pump wavelength to the signal wavelength. Figure 6.34a also shows that with the increase of the active fiber length, compared with single-end pumping, the signal output obtained by bidirectional pumping is the largest. But when the fiber length is relatively short, no matter what kind of pumping arrangement is used, there is little different of the distribution of the population at the upper state in the fiber, and the ASE is also relatively weak, so the impact of the pumping arrangement on the signal gain coefficient can be almost ignored.

When the injected signal power is relatively large, generally greater than 0 dBm, the signal in the entire fiber is much stronger than ASE, so no matter what kind of pumping arrangement is used, it will not have a significant impact on the output power, as shown in Fig. 6.34b. The variation of noise figure of small signal gain amplifier with pump power under co- and counter-directional pumping arrangements [4].

In addition to the gain coefficient, the noise figure is also a problem that must be considered in the design and use of EDFA. When considering small signal gain amplifier, the variation of the noise figure with the pump power under the conditions of 980 nm pump with fiber lengths of 8 m and 12 m in the co- and counter-directional pumping arrangements are shown in Fig. 6.35. It can be seen that: under the same conditions, the noise figure of the co-directional pumping is much lower than that of counter-directional pumping. Qualitatively speaking, the reason for the above situation is still related to the distribution of the population in the upper state in the fiber

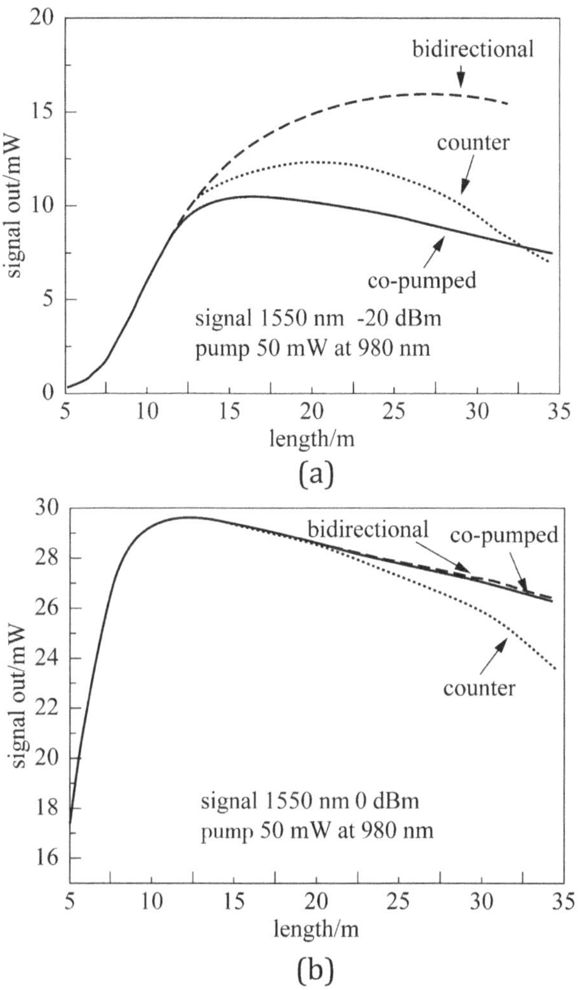

Fig. 6.34 The variation of signal output power under different pumping methods when the pump power is 50 mW and the signal injection power is −20 dBm (**a**) and 0 dBm (**b**), with the increase of the fiber length [4]

under different pumping arrangements. Under the condition of co-directional pumping, there are many Er^3+ in the upper state at the signal injection end and fewer at the output end, so the gain coefficient obtained by the signal at the injection end is much greater than that at the output end. Under the condition of counter-directional pumping, the situation is just the opposite, the gain coefficient obtained by the signal at the injection end is less than that at the output end. The gain coefficient at the input end is small. That is equivalent to some degree of loss before the signal enters the gain fiber, and any form of loss will degrade the noise figure of the signal. Therefore, in the small signal gain region, without considering other factors, the

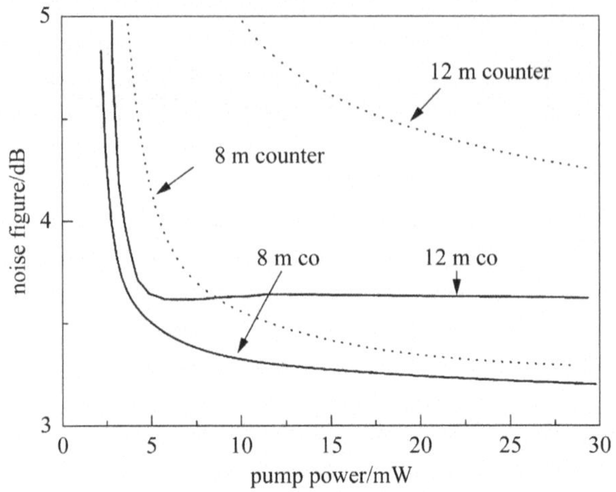

Fig. 6.35 The change variation of amplifier noise coefficient figure of small signal gain amplifier with pump power under co- and counter-directional pumping arrangements

noise figure of the co-directional pumping is better than that of counter-directional pumping. However, the above situation does not apply to the case where the gain fiber is relatively short, because: a short fiber is equivalent to a relative high pump source. And under the condition of sufficient pump power, the Er^{3+} population inversion in the whole fiber is strong enough. Thus, the signal amplified by both the co- and counter-directional pumping can obtain a lower noise figure.

6.4.3 L-Band Erbium-Doped Fiber

With the development of computer networks and data services, the demand for communication system capacity is rapidly growing. Therefore, improving the capacity of existing fiber optic communication systems has become a research focus in the field of optical communication. At present, there are three main ways to achieve the goal: ① increasing the transmission rate of single channel; ② reducing the channel interval and increasing the number of channels; ③ fully utilizing the broadband transmission potential of optical fibers, extending the transmission band from the C-band (1530 ~ 1565 nm) to the L-band (1565 ~ 1610 nm), and achieving simultaneous transmission of signals in the C + L band. Increasing the transmission rate of a single channel will increase the impact of dispersion on the system, thereby increasing the requirements for system dispersion management and compensation, and increasing the system cost. Increasing the number of channels will enhance the nonlinear effect. Thus, the requirements for the wavelength stability of system devices will be much stricter, which will also increase the system cost. Expanding the transmission bandwidth does not encounter the technical problems that the first two ways must face, and can fully exploit the potential bandwidth resources of optical fibers, so it is a more direct and fundamental method to solve the problem.

As early as 2002, the NTT Corporation in Japan [68] achieved L-band optical amplification in erbium-doped tellurite multi-component glass fiber from 1581 to 1616 nm, and exploited related products. However, the tellurite fiber also has excited state absorption [69] at wavelengths greater than 1620 nm, which limits the further expansion of the gain bandwidth. In addition, there are many disadvantages of the multi-component glass fiber: it has extremely poor mechanical performance, thus, the fiber is easy to break and its long-term reliability is not high. Finally, it is difficult to match and splice with the existing silica fiber, which hinders its commercial application. Therefore, the most likely scheme to expand the gain bandwidth is to adjust the composition and the ratio of co-doped elements based on the erbium-doped silica fiber.

For a long time, the research on erbium-doped silica fiber materials is mainly focused on the C-band, and there is relatively less research on further expanding the L-band amplification. It is generally believed that only when the excited state stimulated absorption cross-section of Er^{3+} ($^4I_{13/2} \rightarrow {}^4I_{9/2}$) is not greater than the stimulated emission cross-section ($^4I_{13/2} \rightarrow {}^4I_{15/2}$), it is possible to achieve net gain in the L band. Literature [70] had reported the effect of the proportion of co-doped elements on the signal excited state absorption cross-section in erbium-doped silica fiber: the three erbium-doped fibers a, b, and c were a- single doped with aluminum, b- co-doped with phosphorus and aluminum P/Al = 2.18, c- co-doped with phosphorus and aluminum P/Al = 23.2 respectively. As shown in Fig. 6.36, the intersection points of the

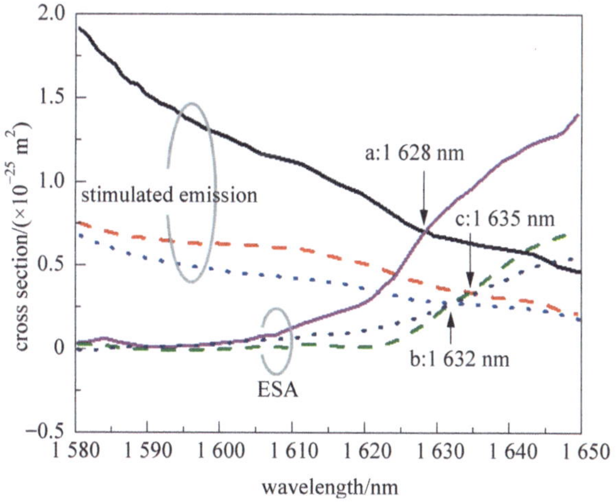

a-Single doped Al; b-Co-doped P/Al=2.18;

c- Co-doped with phosphorus and aluminum P/Al=23.2

Fig. 6.36 The effect of co-doping elements on the signal excited state absorption cross-section of erbium-doped quartz fiber [70]. (**a**) Single doped Al; (**b**) Co-doped P/Al = 2.18; (**c**) Co-doped with phosphorus and aluminum P/Al = 23.2

stimulated emission cross-section and the excited state absorption cross-section of the three fibers are different. For fiber a, the intersection point is at 1628 nm, and for fiber b it moves to 1632 nm, and for fiber c it further moves to 1635 nm. The above results show that the introduce of phosphorus helps to move the net gain cut-off point of erbium-doped fiber towards the longer wavelength direction, thereby expanding the gain bandwidth of erbium-doped silica fiber in the L-band.

In recent years, driven by the demand of market applications, research and development of L-band erbium-doped silica fiber materials have been carried out by some top research institutions, including the University of Southampton in the UK, OFS Company in the US, Laval University in Canada, Huazhong University of Science and Technology in China and etc. [71, 72]. Research by Sumitomo Electric in Japan [73] shows that by increasing the doping concentration of P elements in silica fiber, the L-band gain bandwidth can be expanded; however, due to the limitations of the improved chemical vapor deposition preparation process, the doping concentration of P in silcia fiber is difficult to exceed 12. wt%, and the background loss of the fiber also gradually increases with the increase of P doping concentration. Research from Huazhong University of Science and Technology [72] shows that by co-doping Ce ions to increase the energy transfer process, the absorption edge of the excited state of Er^{3+} in P-doped silcia fiber can be red-shifted, but the fiber still shows a significant gain decline in the band greater than 1620 nm; in addition, too high co-doping Ce ions will lead to the deterioration of fiber gain performance. Laval University in Canada improves the L-band gain bandwidth by optimizing the pump wavelength of the optical amplification system [71]. Based on the two-wavelength pumped two-stage optical amplification system at 1480 nm and 1545 nm, a 20 dB gain was obtained in the 1570 ~ 1620 nm range. This scheme significantly increases the system complexity, which is not conducive to miniaturization and cost reduction. Meanwhile, due to the bandwidth limitation of Er-doped silica fiber, the gain also significantly decreases in the band greater than 1620 nm.

The above works show that the gain bandwidth limit of Er-doped silcia fiber in L-band is around 1620 nm [35, 36, 74, 75]. This is mainly due to two reasons: one is that at wavelength greater than 1620 nm, it is already at the boundary of the fluorescence spectrum of Er-doped silcia glass. Thus, the emission cross-section of Er^{3+} is extremely low, resulting in a very low gain coefficient. The second is that Er-doped silica fiber has strong excited state absorption at wavelength greater than 1620 nm, causing a sharp drop in net gain of the fiber. Therefore, to improve the gain performance of Er-doped fiber near 1620 nm, efforts need to be contributed to increase the emission cross-section of Er^{3+} in the long-wavelength band and reduce the ESA effect at the same time.

6.5 High-Power Erbium-Doped Fiber Laser

High-power 1.5 μm lasers can be widely used in laser radar, remote sensing, free-space optical communication, material processing, etc. The atmosphere has a transmission window between 1.5 ~ 1.75 μm, which coincide with the emission wavelength of Er^{3+} doped fiber lasers. The 1.5 μm Er^{3+} laser is eye-safe. The human cornea has a strong absorption for lasers with wavelengths greater than 1.4 μm, only a small amount of light can reach the retina, which can avoid permanent damage to the eyes from stray light. In material processing fields, due to the special absorption of some material, the1.5 μm Er^{3+} Lasers can be much higher efficiency.

With the development of EDFA and the proposal of double-clad fiber structure [76], as well as the development of related fiber devices, semiconductor pump sources, and fiber laser technology, the output power of rare-earth-doped fiber lasers has rapidly increased. Currently, commercial available ytterbium-doped silica fiber lasers have achieved laser output as large as 100 kW [77], while the maximum laser output of Erbium-doped fibers, as the earliest studied rare-earth-doped fibers and widely used in optical communication, can only be in hundred-watt level [78, 79], far lower than tthat of Ytterbium-doped fiber lasers. This is mainly because that, Er^{3+} has a much smaller absorption cross-section at 9xx nm than Yb^{3+}, resulting in single-doped Er^{3+} fiber of the same length to be unable to achieve sufficient absorption for high-power laser output. Using a longer fiber to increase absorption will lead to an increase in the background loss, affecting the laser conversion efficiency. In addition, increasing of fiber length can easily lead to nonlinear effects, which will also affect the fiber laser output performance.

One solution to the above problems is to use erbium-ytterbium co-doped fibers, where Yb^{3+} sensitizes Er^{3+}, utilizing the much larger absorption cross-section and wide absorption band of Yb^{3+} at 9xx nm to indirectly increase the absorption of Er ions. In addition, co-doping with Yb^{3+} can also improve the dispersion of Er^{3+}, reducing the concentration quenching problem caused by an increase of Er^{3+} concentration. Although Er-Yb co-doped fibers can increase the pump absorption rate, they still face problems such as limited fiber laser output efficiency (limited by the quantum defect between the pump and signal wavelengths), Yb^{3+} laser self-excitation [80], and fiber fuse [79] that occurs during high-power laser output.

This section focuses on different schemes for high power 1.5 μm Er-doped fiber lasers, including the use of Er-Yb co-doped fibers, single Er doped fibers, 9xx nm pump and in-band pump. Introduce the research progress of erbium-doped fiber and erbium-doped fiber lasers, and focus on explaining the bottleneck and future development direction that further improve the laser output of 1.5 μm erbium-doped fiber lasers from the perspective of fiber materials.

6.5.1 High-Power Erbium-Ytterbium Co-Doped Fiber

Using erbium-ytterbium co-doped fiber as the gain medium and a 9xx nm laser diode as the pump source to achieve high power 1.5 μm laser output is currently the most popular scheme, and there are many commercialized fibers and fiber laser products. The reason for choosing erbium-ytterbium co-doped fiber to develope 1.5 μm fiber laser is mainly determined by the absorption and emission characteristics of Er^{3+} itself. A large amount of research shows that Er^{3+} is prone to clustering in silica glass, leading to concentration quenching. In Er^{3+} doped silcia fiber, even if Al or P can be co-doped to increase the doping concentration of Er^{3+}, the maximum doping amount is only about 1000 ppm [39]. According to the absorption spectrum of Er^{3+} shown in Fig. 6.2, all pump absorption bands can be used to pump Er^{3+} to achieve 1.5 μm laser output, but so far the most mature development and the largest output power are mainly 9xx nm laser diodes. Therefore, 1.5 μm Er^{3+} laser output mainly uses 9xx nm laser diodes as pump sources. In addition, although the absorption and emission cross-sections of rare earth ions will vary with different co-dopants (Al, P, Ge), the absorption cross-section of Er^{3+} at the 9xx nm band is much smaller than that of Yb^{3+}, which leads to low absorption. If erbium-doped double-clad fiber wants to fully absorb the pump power, usually a fiber length of more than a hundred meters is required. Although the background loss of germanium-doped communication fiber has been reduced to be as low as 0.15 dB/km@1550 nm, the background loss of erbium-doped silica fiber is still generally about 10 dB/km. The background loss of ~100 m erbium-doped gain fiber will significantly reduce the output efficiency of the fiber laser and easily lead to nonlinear effects. There is a significant spectral overlap between Yb^{3+} luminescence and and Er^{3+} absorption. In the erbium-ytterbium co-doping system, energy can be transferred from Yb^{3+} to Er^{3+}. The use of Er-Yb co-doping can effectively solve the problem of low absorption coefficient of Er^{3+} under cladding pumping conditions. Moreover, due to the two-level system of Yb^{3+}, there is basically no clustering problem in Yb-doped silica fiber. The large introduction of Yb^{3+} can to some extent separate the Er^{3+} ions, solving the concentration quenching problem that is prone to occur when high concentration Er^{3+} ions are doped.

Therefore, using a 9xx nm laser diode to pump Er-Yb co-doped fiber has become the best choice for achieving high power 1.5 μm fiber laser.

The energy diagram of the 1.5 μm laser output achieved by pumping erbium-ytterbium co-doped fiber with 9xx nm pump are shown in Fig. 6.37. The main energy transfer processes are as follows: The ground state Yb^{3+} ($^2F_{7/2}$) is excited to the upper state ($^2 F_{5/2}$) by absorbing the pump photons. Then the excited state $^2 F_{5/2}$ Yb^{3+} can transfer energy to Er^{3+} through resonant transfer, exciting it from the ground state to the $^4I_{11/2}$ state, while the Yb^{3+} returns to the ground state. The excited state Er^{3+} $^4I_{11/2}$ has a short lifetime, approximately 10 μs, and quickly relaxes to the $^4I_{13/2}$ state through non-radiative transition; The Er^{3+} $^4I_{13/2}$ state has a longer lifetime ~10 ms. Continuous pumping forms population inversion between Er^{3+} $^4I_{13/2}$ and $^4I_{15/2}$ state. Adding an appropriate resonator forms positive feedback, or when a 1.5 μm signal is injected, stimulated radiation occurs to produce a 1.5 μm laser output.

In addition to the above mentioned processes of generating 1.5 μm laser, the following energy transfer or transition processes also occur in erbium-ytterbium co-doped fibers, affecting the laser efficiency of the fiber. Firstly, the reversed transfer process from Er^{3+} $^4I_{11/2}$ to Yb^{3+} $^2F_{5/2}$, which can weaken or limit the pump absorption of the fiber. Secondly, there are possible excited state absorption process and up-conversion process that may occur in the erbium-ytterbium co-doped fiber.

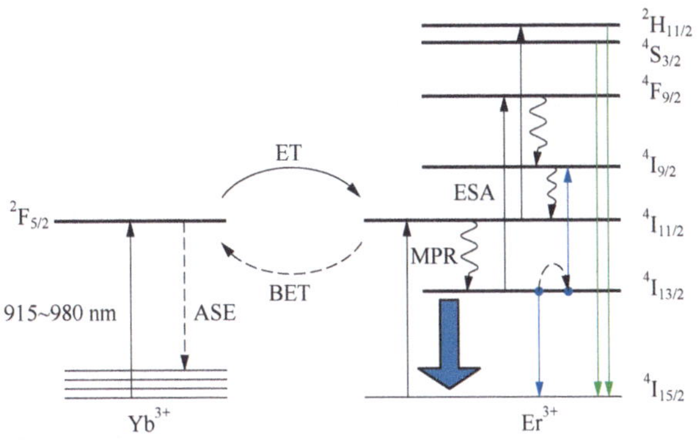

Fig. 6.37 Schematic diagram of energy levels in erbium-ytterbium co-doped fiber pumped at 9xx nm

Excited state absorption includes the Yb^{3+} in the $^2F_{5/2}$ excited state transferring energy to the Er^{3+} at the $^4I_{11/2}$ and $^4I_{13/2}$ states. The former excites Er^{3+} to the $^2H_{11/2}$ and $^4S_{3/2}$ states, which may be related to the green emission in erbium-ytterbium co-doped fibers [81, 82] (radiative transition from $^2H_{11/2}$ and $^4S_{3/2}$ states to the ground state $^4I_{15/2}$, producing green light with a wavelength of 540 nm or 520 nm). The latter excites Er^{3+} to the $^2F_{9/2}$ state, then returns to the $^4I_{13/2}$ state through non-radiative transition via $^4I_{9/2}$ and $^4I_{11/2}$ states. The possible cooperative up-conversion process mainly refers to the Er^{3+} at the $^4I_{13/2}$ state transferring energy to another Er^{3+} at the $^4I_{13/2}$ state. The Er^{3+} giving the energy returns to the ground state $^4I_{15/2}$, and the Er^{3+} receiving the energy is excited to the $^4I_{9/2}$ energy level, and then returns to the $^4I_{13/2}$ state.

Promoting the laser generation process and inhibiting unwanted energy transfer processes are the key to improving the efficiency of the 1.5 μm laser in erbium-ytterbium co-doped fiber. This is also the basis for the selection of the glass matrix in erbium-ytterbium co-doped fiber. The energy transfer processes that may occur in erbium-ytterbium co-doped fiber include energy transfer between erbium and ytterbium ions, and the cooperative up-conversion process within Er^{3+} ions.

The energy transfer between erbium and ytterbium ions mainly aims to improve the energy transfer efficiency from Yb^{3+} to $Er^{3+}$and suppress the reverse energy transfer process from Er^{3+} to Yb^{3+}. As described in Section 6.2.2, on one hand, adding P to the glass matrix suppresses the direct transition of excited Yb^{3+} to its lower state. And on the other hand, the addition of P element introduces P = O bond in the glass matrix, which increases the maximum phonon energy of the glass to 1300 cm^{-1} (the maximum phonon energy in silica glass is about 1100 cm^{-1}) [83], which can promote the Er^{3+} at the $^4I_{11/2}$ state to transfer to the $^4I_{13/2}$ state through non-radiative transition, thereby suppressing the reverse energy transfer process from Er^{3+} to Yb^{3+} [38], which can greatly improve the energy transfer efficiency from Yb^{3+} to Er^{3+}. However, the excited Yb^{3+} at the $^2F_{5/2}$ state transferring energy to the Er^{3+} at the $^4I_{11/2}$ and $^4I_{13/2}$ states is inevitable in erbium-ytterbium co-doped fiber.

Increasing the dispersion of Er^{3+} is beneficial to suppress the up-conversion process of Er^{3+} at the $^4I_{13/2}$ state.

Figure 6.38 shows the variation of the Er^{3+} absorption coefficient at 1535 nm with pump power for both Er-Yb co-doped fibers and Er-doped fibers [81]. Apart from Yb^{3+} doping, the Er^{3+} doping concentration, the preparation process, and the fiber length used for absorption testing are all the same for both fibers. It can be seen that the Er-Yb co-doped fiber reaches saturation at about 100 mW pump power and no longer absorb pump power, while the Er-doped fiber can still furtherabsorb pump light. This indicates that the presence of Yb^{3+} in the fiber can reduce the clustering of Er^{3+}, possibly because Yb^{3+} is distributed around Er^{3+}, shielding the interaction between Er^{3+}. Therefore, Er-Yb co-doped silica fibers should be highly doped with P and a large amount of Yb^{3+} ions, with a high Yb/Er ratio

With the development of glass matrix of Erbium-Ytterbium co-doped fibers and the maturation of high-power laser diodes, high-power 1.5 μm band of Er^{3+}laser output is getting increasing. In 2001, researchers [32] first demonstrated the laser output of an Er-Yb co-doped fiber with a cladding pump structure. Two types of Er-Yb co-doped fiber were used: Fiber A and Fiber B. Both were single-mode fibers, with a core size of 6 ~ 7 μm, numerical aperture ≤0.17, inner cladding diameter 300 μm, and the doping ratio is Yb:Er = 5.2:0.43. The difference is that the co-doping amount of P in Fiber A is only 1/5 of that in Fiber B. Using a 976 nm LD pump, a 1.6 μm high-reflectance mirror as the input, and a vertically cut fiber end face as the output, a spatial optical path was set up for the fiber laser experiment. The results showed that Fiber A began to output Er^{3+} ~ 1.6 μm and Yb^{3+} ~ 1.08 μm dual-wavelength lasers under only about 2 W pump. As the pump power increased

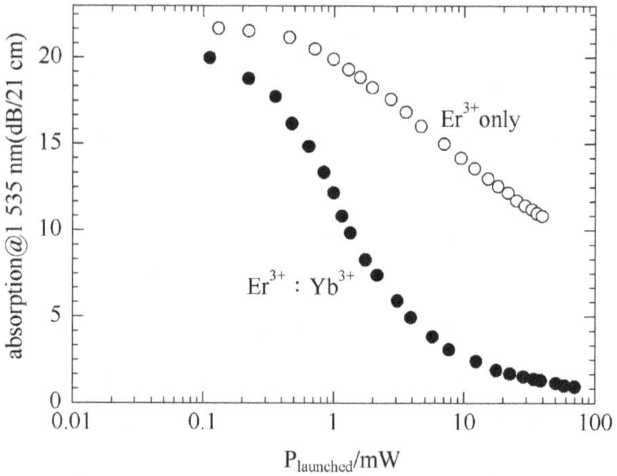

Fig. 6.38 The variation of absorption coefficient at 1535 nm with pump power for single erbium doped and erbium ytterbium co doped (with the same Er doping concentration) phosphorus silicon fibers

further, the 1.6 μm laser output became saturated, while only the Yb^{3+} ~ 1.08 μm laser output power increased, as shown in Fig. 6.39.

On the contrary, in fiber B, the Yb^{3+} ~ 1.08 μm laser signal was never observed. Limited by the potential damage to the fiber end face due to further increase in pump power, the maximum power output of the 1.6 μm laser in fiber B is about 2 W. Considering only the absorbed pump power, the slope efficiency of the output laser is about 56%, close to the theoretical limit of the laser efficiency when pumped at 976 nm ($\lambda_{pump}/\lambda_{laser}$ = 976/1600 = 61%). The above results confirm that in the erbium-ytterbium co-doped fiber, high doping of P element is beneficial to enhance the energy transfer process from Yb^{3+} to Er^{3+}, suppress the self-excitation of Yb^{3+}, and promote the output of Er^{3+} fiber laser. The Optoelectronics Research Centre (ORC) at the University of Southampton, UK, has conducted extensive research on high-power erbium-ytterbium co-doped fibers and fiber laser output performance [80, 84–88]. From 2001 to 2003, ORC reported its first erbium-ytterbium co-doped double-clad fiber [84–85]. The fiber has a size of 12/125μmwith a circular cladding. The core numerical aperture of the fiber is 0.175, and the core absorption coefficient is about 60 dB/m@1530 nm. By using cladding mode disturbance, the maximum cladding pump absorption coefficient of the fiber is about 13 dB/m@980 nm. The free oscillation laser output of this fiber under 915 nm and 980 nm pumping are: when pumped at 915 nm, the optimal length of the fiber is 2 m, and the slope efficiency of the fiber laser is about 50% (considering only the absorbed pump power, the same below); when pumped at 980 nm, the optimal length of the fiber is 1.4 m, and the slope efficiency of the fiber laser is about 38%; when using a 915 nm double-end pumping, a fiber length of 4 m was used to achieve a maximum laser output of 16.8 W.

In 2003, by increasing the core and cladding size of the rare-earth-doped fiber, ORC first reported a 1.5 μm laser output of over 100 W from an erbium-ytterbium

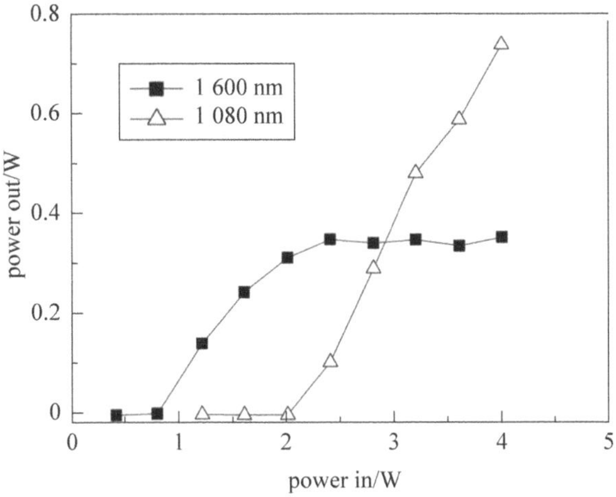

Fig. 6.39 Laser output of Fiber A with increasing pump power [32]

co-doped fiber laser [80]. The fiber core size was 24 μm, the numerical aperture was 0.2, the pump cladding was D-shaped, the long and short side dimensions were 400/360 μm, the cladding numerical aperture was 0.48, and the fiber cladding absorption coefficient was about 2.5 dB/m@975 nm. Using this 5 m fiber for high-power laser experiments, a slope efficiency of 40% for low-power laser output and 30% for high-power laser output was achieved under 975 nm single-end pumping. The reason for the decrease in slope efficiency during high-power output is that as the pump power increases, the 1.06 μm laser of Yb^{3+} begins to self-oscillate, as shown in Fig. 6.40. When the maximum 103 W 1.56 μm laser output was obtained, the 1.06 μm Yb^{3+} laser also reached about a hundred-watt.

In order to suppress the self-oscillation of Yb^{3+} during high-power laser output, in 2005, based on an improved rare-earth-doped fiber design, ORC achieved a 120 W 1.5 μm laser output under the same pumping conditions, while the self-oscillation laser of Yb^{3+} was only 1.5 W [86]. The fiber core size used was 30 μm, the numerical aperture was 0.2, the pump cladding was D-shaped, the long and short side dimensions were 400/360 μm, and the cladding numerical aperture was 0.48. Compared with the fiber mentioned above, the increased core size increased the cladding absorption of the fiber to be 4.5 dB/m@975 nm, shortening the required fiber length. Using a 4-meter-long erbium-ytterbium co-doped fiber, an Er^{3+} laser output of about 120 W was achieved under a maximum 310 W 976 nm pumping, with a slope efficiency of about 40%. Researchers believe that increasing the core size, shortening the length of the fiber, and using a more efficient erbium-ytterbium co-doped fiber matrix can help suppress Yb^{3+} self-oscillation. In the same year, by further increasing the absorption coefficient of the fiber, the ORC research group achieved a much higher erbium-ytterbium co-doped fiber laser output [87]. The fiber core diameter was still 30 μm, the numerical aperture was 0.22, the pump

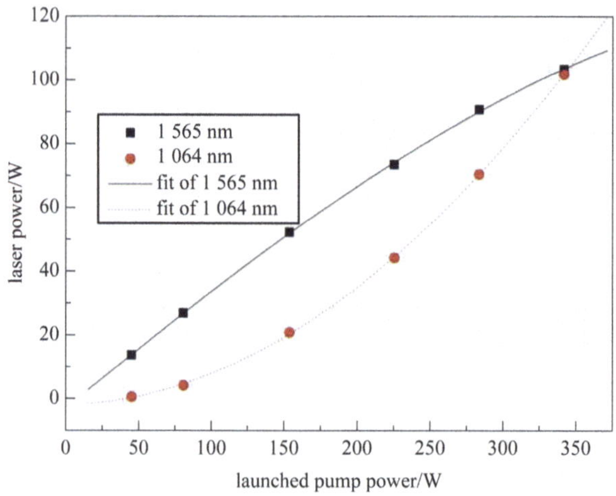

Fig. 6.40 The variation of laser output of erbium-ytterbium co-doped fiber laser with pump power [80]

cladding was D-shaped, the long and short side dimensions were 400/360 μm, and the cladding numerical aperture was 0.44, but the cladding absorption coefficient of the fiber was increased to be 6.9 dB/m@975 nm, further shortening the required fiber length to be 3 m. An maximum output power was about 188 W by using 976 nm double-end pumped. The slope efficiency of the fiber laser was about 43.1% at low output, and it dropped to 36.8% at high laser power due to the self-oscillation of Yb^{3+}. When the maximum 188 W 1.5 μm band laser output was achieved, the 1 μm Yb laser was less than 20 W.

In 2007, ORC reported the highest output power of the erbium-ytterbium co-doped fiber laser [88]. The core diameter of the gain fiber was 30 μm, the numerical aperture was 0.21, and a D-type inner cladding with a diameter of about 600 μm was used, with a cladding numerical aperture of 0.48. By increasing the cladding diameter to inject more pump power, the cladding absorption coefficient of the fiber was decreased to 3.8 dB/m@976 nm. The laser experiment was conducted with 6 m gain fiber by using 976 nm LD double-end pumped. A maximum output power of 297 W 1.5 μm fiber laser was achieved under active cooling condition. The slope efficiency of the output laser is about 40% at low power output, and it decreases to 19% at high power output due to the self-excitation of Yb^{3+}. At the maximum pump power, the Yb^{3+} self-excited laser output exceeded that of the 1.5 μm laser, reaching 338 W. Researchers tried to increase the cavity loss to suppress the parasitic oscillation of Yb^{3+}. Although this can reduce the 1 μm laser, it can affect the stability of the 1.5 μm laser output, and may even cause the fiber burnout.

The above results show that by increasing the doping concentration of rare earth ions or the core-to-cladding ratio of erbium-ytterbium fibers, increasing the absorption coefficient of the fiber, and shortening the use length of the fiber, the parasitic oscillation of Yb^{3+} can be suppressed to a certain extent, and the 1.5 μm wavelength laser output capability of the fiber can be improved. However, there are two problems: first of all, as the pump power further increases, the erbium-ytterbium co-doped fiber will inevitably experience Yb^{3+} parasitic oscillation (this is because the energy transfer rate from Yb^{3+} to Er^{3+} is limited due to the lifetime of the $E\,r^{3+4}\,I_{11/2}$ level. For a specific fiber, as the pump power increases, when the population inversion of Yb^{3+} accumulates to a certain level, it will trigger the ASE and even self-emission of Yb^{3+}); secondly, in order to increase the energy transfer rate from Yb^{3+} to Er^{3+}, the erbium-ytterbium co-doped fiber is doped with a large amount of P, which makes the numerical aperture (NA) of the fiber core very large, about 0.2 (for a common 25 μm core diameter fiber, its normalized frequency V value is about 10.1, and it can accommodate up to 14 modes), becoming a typical multimode fiber, which is not suitable for improving the beam quality of the fiber laser output.

In order to improve the beam quality of the output laser from erbium-ytterbium co-doped optical fibers, the American company Nufern [89] proposed the structure of erbium-ytterbium co-doped silica optical fiber with a germanium-doped pedestal as shown in Fig. 6.41. By adding a layer of germanium-doped silica between the rare earth-doped fiber core and the pure silica cladding. Adjusting the thickness of

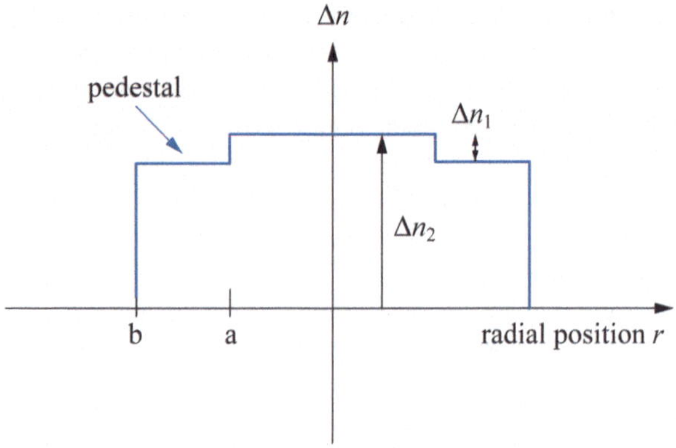

Fig. 6.41 Refractive Index Distribution of erbium-ytterbium co-doped with germanium-based pedestal [89]

the germanium-doped silica layer, it will not affect the modes (especially the fundamental mode) in the fiber core. In this situation, the germanium-doped silica layer becomes the real cladding of the rare earth-doped fiber core. And by adjusting the refractive index of the germanium-doped silica layer, the numerical aperture of the fiber core is brought to around 0.1. Thus, the V-value of the optical fiber is greatly reduced, turning out to be a few-mode fiber, which is fundamentally beneficial to improving the beam quality of the fiber laser output. On the other hand, the numerical aperture of the fiber core can also be reduced by doping other elements in the fiber core. In 2019, by doping a large amount of F in traditional erbium-ytterbium co-doped optical fiber, Russian researchers [90] successfully decreased the numerical aperture of the fiber core to 0.07.

Regarding the problem of Yb^{3+} self-excitation under high-power laser conditions, apart from suppression through the design and manufacture of optical fiber materials, researchers have also tried to improve the design of optical fiber lasers or amplifiers, including introducing special structures to increase the transmission loss of the Yb^{3+} laser band to suppress the emission of Yb^{3+}; allowing Yb^{3+} to actively lase forming a secondary pump for Er^{3+}, thereby improving the stability and slope efficiency of the system. Among all these schemes, the most effective and simplest is to adopt a non-peak pumping scheme, including the use of 915 nm or 940 nm pumping. Before 2016, in order to reduce the self-excitation of Yb^{3+}, researchers tried to minimize the active fiber length of and adopted 976 nm pumping scheme to maximize the Yb^{3+} pump absorption. When using 915 nm or 940 nm pumping, since the absorption coefficient of Yb^{3+} at these two wavelengths is 3 ~ 5 times lower than that of 976 nm, it can reduce the population inversion of Yb^{3+} and thus increase the self-excitation threshold. In 2016, researchers [91] experimentally compared the output of 1.5 μm lasers from erbium-ytterbium co-doped fibers pumped by 940 nm and 976 nm. The experiment adopted a scheme of "seed laser + two-stage amplification", with the main amplification medium being a 25/300 μm erbium-ytterbium co-doped fiber with a germanium-doped pedestal, which was designed and manufactured by Nufern in the United States. The fiber core has a numerical aperture of about 0.1 and the cladding is octagonal. When pumped at 940 nm, the length of the gain fiber is 5 m, with a pump absorption of about 12 dB. At the maximum pump power of 390 W, the output laser power is 207 W, with a slope efficiency of 50.5% and an optical-optical efficiency of 49.3%. When pumped at 976 nm, the length of the fiber is reduced, but the same pump absorption of 12 dBas that of 940 nm pump is maintained. At the maximum pump power of 250 W, the output laser power is 97.6 W, with a slope efficiency of 40.2% and an optical-optical efficiency of 35.6%. The results show that, under the same conditions, the laser output efficiency of 940 nm pumping is higher than that of 976 nm pumping. In addition, the spectral show that when pumped at 940 nm, the signal-to-noise ratio of the Er^{3+}1.5 μm to that of the Yb^{3+}1 μm is as high as 50 dB at the maximum laser output of 207 W, almost achieving no Yb^{3+} lasing. The reason is that when pumped at 940 nm, the population inversion degree of Yb^{3+} is relatively low, which suppresses the self-lasing of Yb^{3+}. However, when pumped at 976 nm, due to the larger absorption coefficient, quantum defects cause a much higher thermal load per unit length, thereby reducing the efficiency of the fiber. Figure 6.42 shows the beam quality of the fiber at the maximum laser output of 207 W, with beam quality M^2 of 1.05. This indicates that the erbium-ytterbium co-doped fiber structure design by Nufern can effectively improve the beam quality of the laser output.

In 2020, Matniyaz et al. in the United States [79] reported the highest power output of erbium-ytterbium co-doped fiber lasers. The experiment adopted a "seed + amplification" scheme. The aforementioned 25/300 μm erbium-ytterbium

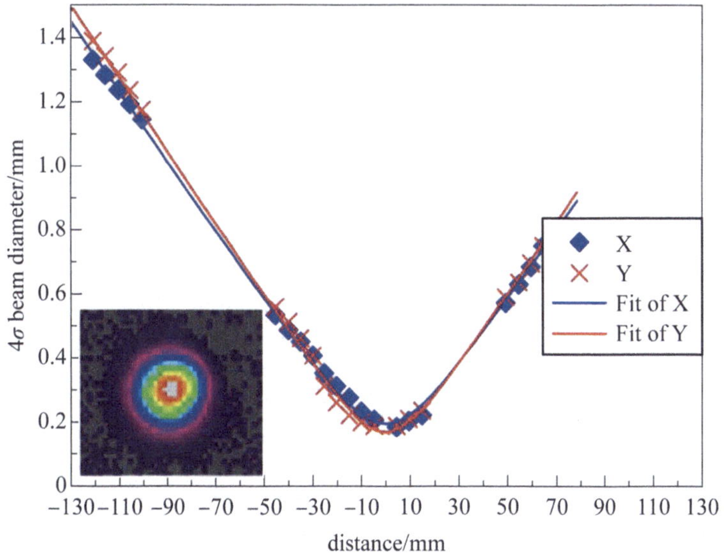

Fig. 6.42 Beam quality M^2 of the erbium-ytterbium co-doped fiber with a germanium-doped pedestal with 207 W laser output [91]

co-doped fiber from Nufern with a fiber length of 6.4 m was used. The fiber laser is 915 nm LD pumped. As the pump power increased, the laser output of the fiber linearly increased, with a maximum power of 302 W and an optical-optical efficiency of 56% (close to the theoretical limit of 58.6%). The spectral test results at the maximum output power show that the ASE of both 1 μm and 1.5 μm bands can be ignored. Further analysis suggests that the laser efficiency of the fiber approaching the theoretical limit is not only related to the use of the 915 nm off-peak absorption pumping scheme, but may also benefit from the higher rare earth ion doping concentration [92], which can better suppress Yb^{3+} self-lasing. Further increase in the fiber laser output is limited by the damage of the coating, which is induced by the thermal effect of quantum defects and the resulting fiber fuse phenomenon.

6.5.2 *In-Band Pumped Erbium-Doped Fiber Lasers or Amplifiers*

Since 2000, due to the limitations of available commercial pump sources, and the small absorption cross-section of Er^{3+} and its tendency to cluster in silica glass, researches on Er^{3+} 1.5 μm high-power laser fibers have mainly focused on erbium-ytterbium co-doped fibers. However, the maximum output power is currently only 300 W. And with the increase in pump power, they will inevitably face the problem of Yb^{3+} self-excitation. More importantly, the theoretical quantum defect of erbium-ytterbium co-doped fiber lasers using 9xx nm pump is as high as about 40%. The large thermal load also limits the possibility of further increasing the fiber laser output. Therefore, researchers have begun to shift their focus to only erbium doped fibers, hoping to reduce the fiber thermal load by using a in-band pumping scheme, in order to achieve higher light-to-light conversion efficiency and fiber laser output.

From 2009 to 2011, researchers [93, 94] successively achieved 1.5 μm fiber laser output by using 1532.5 nm InGaAsP/InP laser diode in-band pumping erbium-doped double-clad fiber. The experiment employed the simplest oscillator structure, using a high-reflective fiber grating and the vertically cut fiber end face to form the laser resonant cavity. The active fiber was the Er60–20/125 μm double-clad fiber from Likkie. By optimizing the length of the active fiber, a maximum of about 88 W 1590 nm laser output was achieved, with a light-to-light efficiency of 69%, far higher than that of erbium-ytterbium co-doped fibers pump by 9xx nm, but still far lower than the theoretical quantum efficiency of about 90%. Analysis suggests that the main reason affecting the output efficiency is the lack of optimized fiber design and fabrication. There may be some Er^{3+} clusters, up-conversion luminescence, and residual hydroxyl groups in the fiber.

Subsequently, Jebali et al. from Canada [95, 96] achieved higher output power and efficiency of in-band pumped erbium-doped fiber laser output by optimizing the fiber laser design. The experimental scheme is shown in Fig. 6.43. First, a 1535 nm fiber laser output is achieved, and then multiple 1535 nm erbium-ytterbium

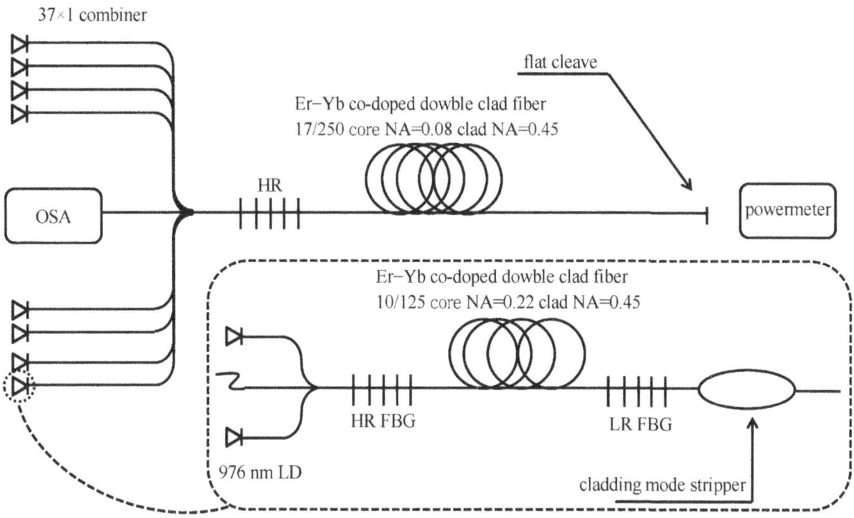

Fig. 6.43 High power laser output of erbium-doped fiber with in-band band pumping [96]

co-doped fiber lasers are combined to pump the erbium-ytterbium fiber to achieve 1585 nm fiber laser output. When for achieving 1535 nm fiber laser, Yb^{3+} is a sensitizing ion responsible for transferring energy to Er^3 + ions. However, when using 1535 nm as pump source to achieve 1585 nm fiber laser output, although the active fiber is still an erbium-ytterbium doped fiber, the Yb^{3+} ions are no longer participating in the energy transfer process and only play a role in dispersing Er^{3+} ions. Therefore, the erbium-ytterbium co-doped fiber with high P doping used for 1535 nm laser output has a core numerical aperture of ~0.22; while the erbium-ytterbium doped fiber used for 1585 nm fiber laser output, in order to reduce the core numerical aperture, is Al/P co-doped, with a core numerical aperture of only about 0.08. At the maximum pump power, this fiber laser finally obtains about 264 W of fiber laser output, with a slope efficiency of about 74%.

When using the in-band band pump source to improve the efficiency of erbium-doped fiber lasers, in addition to the possible clustering and loss factors of the erbium-doped fiber itself, another noteworthy factor is the transmission loss of the 1530 nm pump source induced by the low refractive index coating of the double-clad fiber. Figure 6.44 shows the variation of cladding loss coefficient with wavelength when low refractive index coating and fluorine quartz glass are used as the outer cladding, respectively. It can be seen that at 1535 nm, the former has a loss coefficient of about 20 dB/km, and the latter has a loss coefficient of about 5 dB/km. Considering the absorption cross-section and doping concentration of Er^{3+}, the length of erbium-doped fiber used for in-band pumping is generally 15 ~ 20 m, so the cladding loss coefficient is also an important factor affecting the fiber laser output efficiency (That is currently not considered for the erbium doped double cladding fiber for in-band pumping).

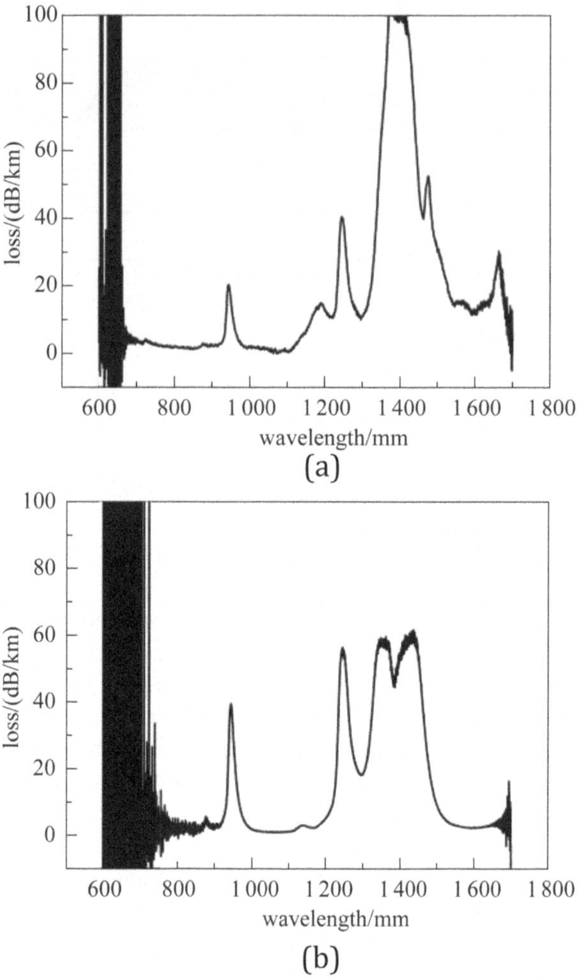

Fig. 6.44 Variation of the cladding loss coefficient with wavelength when low refractive index coating (**a**) and fluorine silica (**b**) are used as the outer cladding

In the future, in order to further improve the laser output power and efficiency of the in-band pumped erbium-doped fiber, fluorine doped silica glass should be used as the outer cladding to reduce the transmission loss coefficient of the pump source.

6.5.3 Erbium-Doped Fiber Lasers Pumped by 980 nm LD

Erbium-ytterbium co-doped fiber lasers pumped at 9xx nm always face the problem of Yb^{3+} self-excitation. Although the in-band pumping scheme can greatly improve the efficiency of erbium-doped fiber laser, high power 1530 nm semiconductor pump sources are not only low electro-optical conversion efficiency, but also very expensive and far less mature than 9xx nm LD. However, when using a 1535 nm fiber laser as the in-band pump source, the systems are complex and expensive. Moreover, the electro-optical efficiency of the laser system has not yet been improved. Therefore, as the simplest and most economical solution, using a 980 nm laser diode to directly pump erbium-doped fiber to achieve 1.5 μm laser wavelength output is still an important research direction. Researchers have been studying the l 976 nm pumped erbium-doped fiber lasers for a long time, but it was not until 2014 that the Fiber Optics Research Center of the Russian Academy of Sciences (RAS) achieved a laser output of over 100 W for the first time [97]. The experiment used a "seed + amplification" scheme. The active fiber was a home-made Er-Al-F co-doped fiber. The core size was 34 μm, the numerical aperture (NA) was about 0.06, and the cut-off wavelength was 1700 nm. As the absorption cross-section of Er is small and the doping concentration is limited, in order to increase the cladding absorption coefficient, the cladding adopts a square structure of 110 μm × 110 μm. Its pump absorption is about 0.6 dB/m@980 nm. An amplifier experiment was conducted using a fiber length of 40 meters. Under 980 nm pumping, a maximum laser output of 103 W was achieved, and the slope efficiency was 37%. Further analyze and compare the difference in the electro-optical efficiency between this scheme and the in-band pumping scheme. It is found that: generally, the electro-optical efficiency of LD is about 50%, and the electro-optical efficiency of Er laser output realized by the 980 nm semiconductor direct pumping scheme is about 19%; while using the in-band pumping scheme, taking the 1535 nm fiber laser as the pump source to achieve 1585 nm output as an example, the electro-optical efficiency of 976 nm LD is 50%, the efficiency of the 1535 nm fiber laser is about 40%, and the efficiency of achieving 1585 nm laser output is about 74%, so the total electro-optical efficiency of this scheme is about 14.8%, which is obviously lower than that of the 980 nm direct pumping scheme. In 2018, the ORC in the UK achieved the highest laser output of Er 1.5 μm by using the above 980 nm direct pumping scheme. Figure 6.45 shows the refractive index distribution of the fiber and the fiber end face used in the experiment The fiber core is Er-Al co-doped silica glass.

The numerical aperture is about 0.08. In order to increase the cladding absorption, the core is 146 μm, and the inner cladding adopts a D-type structure. The long and the short side diameter is about 700 μm and 660 μm, respectively. The fiber cladding absorption is about 0.45 dB/m@980 nm. The experiment was conducted using a 35 m long fiber and a spatially coupled dual end pumping arrange. With a maximum pump power of approximately 1.91 kW, a 1.601 μ m laser with a maximum output of 656 W was achieved, with a slope efficiency of 35.6% Due to using a super-large core diameter, its normalized frequency V value at 1.6 μm is 23, which is a multimode fiber. As the core numerical aperture of the fiber is low, about 25% of the generated laser leaks into the fiber inner cladding. According to the above results, first, although the Er-doped fiber directly pumped by 980 nm LD is currently the most economical solution for achieving high power 1.5 μm laser output, the lower pump absorption coefficient severely restricts the design and manufacture of active fibers. It is necessary to use a larger core diameter to increase the pump

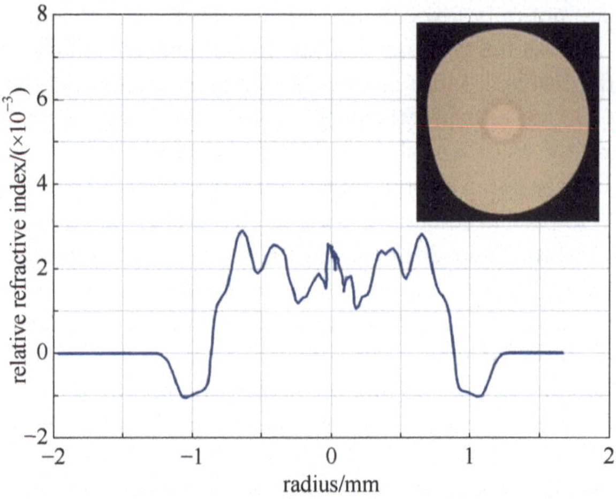

Fig. 6.45 Fiber preform refractive index distribution and the fiber cross-section of the ORC for achieving 656 W 1.5 μm laser output

absorption coefficient, but this also limits the increase in output laser brightness. On the other hand, as mentioned before, the absorption coefficient of Er^{3+} at 980 nm is redshift and decreasing with temperature. Therefore, the thermal effect caused by the large quantum defect can lead to a decrease in pump absorption during high power laser output, which will prevent further power scaling.

6.5.4 Future Challenges in Erbium-Doped 1.5 μm Band Fiber Lasers

In the past 20 years, Er^{3+} 1.5 μm high-power fiber lasers have made a series of important progress, with output power increasing from a few watts to hundreds of watts. There are still many important challenges.

For erbium-ytterbium co-doped fiber lasers pumped by 9xx nm LD, the main challenge is the self-excitation of Yb^{3+} inevitably introduced with the increase of pump power. Moveover, the thermal effect caused by the huge quantum defect fundamentally limits the further power scaling. In addition, Er-doped fibers also suffer the problem of low pump absorption cross sention. These fibers have to increase the fiber absorption by reducing the cladding diameter or enlarging the core diameter. The former limits the pump injection, and the latter will reduce the beam quality.

For in-band pumper Er^{3+} fiber lasers, due to the low quantum defect which can almost be ignored, they should be the best choice for high-power laser output.

However, this scheme still faces the following two problems. Firstly, the acquisition of high-power 1530 nm pump sources. Although significant progress has been made in the research of semiconductor pump sources at this wavelength, they are currently expensive. Using a 1530 nm erbium-ytterbium fiber laser as pump source, although it is higher brightness, makes the system complex and expensive. More importantly, since 1530 nm is at the short-wave edge of the Er^{3+} laser emission band, it is difficult to achieve high-power 1530 nm fiber laser output. On the other hand, the in-band pumping also faces the problem of low pump absorption. It can be alleviated by enlarging the core diameter or reducing the cladding diameter. However, to further increase the output and brightness of the fiber laser, it is necessary to increase the doping concentration of Er^{3+} as much as possible while avoiding clustering, and at the same time find a way to achieve lower numerical aperture of the fiber core.

This chapter introduces the basic knowledge of Er^{3+} doped silica fiber and fiber lasers, including the energy transfer characteristics of Er^{3+} 1.5 μm emission, the network structure and spectral property variation of erbium-doped silica glass, and the basic principles and main applications of erbium-doped fiber in optical communication and high-power fiber laser fields. Since obtaining the first low-loss erbium-doped silica fiber in 1985, in less than 40 years, the optical communication network based on erbium-doped fiber amplifiers has become the main global communication network, greatly improving the speed and efficiency of information transmission. On the other hand, with the orders of magnitude growth of information volume, information transmission, and information processing, optical communication has put forward more demands for transmission bandwidth, which prompts researchers to try to further widen the gain bandwidth of erbium-doped fiber amplifiers. Moreover, the existing ground communication network can no longer meet the increasing needs. And the space optical communication network has also been put on the agenda and has begun to be implemented, such as the Starlink program of SpaceX. The complex and harsh usage environment also puts forward higher requirements on the performance of erbium-doped fibers, such as radiation resistance, longer service life (high replacement cost in complex environments), and higher power. In the field of high-power lasers, compared with ytterbium doped fibers that can achieve a single fiber power output of 10 kW, due to the complex energy level structure and small absorption cross-section of erbium ions, the maximum output power that erbium-doped fibers can currently achieve is still in the hundreds of watts range. It is believed that with the continuous deepening of technical research, the output capacity and stability of erbium-doped fibers will be greatly improved in the future. And their application scope in laser radar, free-space optical communication, material processing, and biomedical and other fields will be further expanded.

References

1. Poole S B, Payne D N, Fermann M E. Fabrication of low-loss optical fibers containing rare-earth ions [J]. Electronics Letters, 1985,21(17):737-738.
2. Poole S, Payne D, Mears R, et al. Fabrication and characterization of low-loss optical fibers containing rare-earth ions [J]. Journal of Lightwave Technology, 1986,4(7):870-876.
3. Mears R J, Reekie L, Jauncey I M, et al. Low-noise erbium-doped fibre amplifier operating at 1.54μm [J]. Electronics Letters, 1987,23(19):1026-1028.
4. Becker P C, Olsson N A, Simpson J R. Erbium-doped fiber amplifiers: fundamentals and technology [M]. San Diego: Academic Press, 1999.
5. Desurvire E, Simpson J R. Evaluation of $^4I_{15/2}$ and $^4I_{13/2}$ Stark-level energies in erbium-doped aluminosilicate glass fibers [J]. Optics Letters, 1990,15(10):547-549.
6. Jacquier B. Rare earth-doped fiber lasers and amplifiers [M]. Boston: Springer, 1997.
7. Gapontsev V P, Matitsin S M, Isineev A A, et al. Erbium glass lasers and their applications [J]. Optics and Laser Technology, 1982,14(4):189-196.
8. Miniscalco W J. Erbium-doped glasses for fiber amplifiers at 1500nm [J]. Journal of Lightwave Technology, 1991,9(2):234-250.
9. Nykolak G, Becker P C, Shmulovich J, et al. Concentration-dependent $^4I_{13/2}$ lifetimes in Er^{3+}-doped fibers and Er^{3+}-doped planar waveguides [J]. IEEE Photonics Technology Letters, 1993,5(9):1014-1016.
10. Nishi T, Nakagawa K, Ohishi Y, et al. The amplification properties of a highly Er^{3+}-doped phosphate fiber [J]. Japanese Journal of Applied Physics, 1992,31(Part 2, No. 2B):L177-L179.
11. Mori A, Ohishi Y, Yamada M, et al. 1.5μm broadband amplification by tellurite-based EDFAs [C]//Proceedings of the Conference on Optical Fiber Communications, Optical Society of America, Dallas, Texas, 1997.
12. Barnes W L, Laming R I, Tarbox E J, et al. Absorption and emission cross section of Er^{3+} doped silica fibers [J]. IEEE Journal of Quantum Electronics, 1991,27(4):1004-1010.
13. Shimizu M, Yamada M, Horiguchi M, et al. Concentration effect on optical amplification characteristics of Er-doped silica single-mode fibers [J]. IEEE Photonics Technology Letters, 1990,2(1):43-45.
14. Saitoh A, Matsuishi S, Se-weon C, et al. Elucidation of codoping effects on the solubility enhancement of Er^{3+} in SiO_2 glass: striking difference between Al and P codoping [J]. The Journal of Physical Chemistry B, 2006,110(15):7617-7620.
15. Delevaque E, Georges T, Monerie M, et al. Modeling of pair-induced quenching in erbium-doped silicate fibers [J]. IEEE Photonics Technology Letters, 1993,5(1):73-75.
16. Jiao Yan, Guo Mengting, Wang Renle, et al. Influence of Al/Er ratio on the optical properties and structures of Er^{3+}/Al^{3+} co-doped silica glasses [J]. Journal of Applied Physics, 2021,129(5):053104.
17. Matson D W, Sharma S K, Philpotts J A. The structure of high-silica alkali-silicate glasses. A Raman spectroscopic investigation [J]. Journal of Non-Crystalline Solids, 1983,58(2):323-352.
18. Poe B T, Mcmillan P F, Angell C A, et al. Al and Si coordination in SiO_2-Al_2O_3 glasses and liquids: A study by NMR and IR spectroscopy and MD simulations [J]. Chemical Geology, 1992,96(3):333-349.
19. Ren Jinjun, Zhang Long, Eckert H. Medium-range order in sol-gel prepared Al_2O_3-SiO_2 glasses: new results from solid-state NMR [J]. The Journal of Physical Chemistry C, 2014,118(9):4906-4917.
20. Sen S, Youngman R E. High-resolution multinuclear NMR structural study of binary aluminosilicate and other related glasses [J]. The Journal of Physical Chemistry B, 2004,108(23):7557-7564.
21. Rocca F, Ferrari M, Kuzmin A, et al. EXAFS studies of the local structure of Er^{3+} ions in silica xerogels co-doped with aluminium [J]. Journal of Non-Crystalline Solids, 2001(293-295):112-117.

22. Haruna T, Iihara J, Yamaguchi K, et al. Local structure analyses around Er^{3+} in Er-doped fiber with Al co-doping [J]. Optics Express, 2006,14(23):11036-11042.
23. Jiang Shibin, Luo Tao, Hwang B-C, et al. Er^{3+} -doped O_5: $70SiO_2$ [J]. Journal of Non-Crystalline Solids, 2000, 263-264:364-368.
24. Lin H, Pun E Y B, Man S Q, et al. Optical transitions and frequency upconversion of Er^{3+} ions in $Na_2O\cdot Ca_3Al_2Ge_3O_{12}$ glasses [J]. Journal of the Optical Society of America B, 2001,18(5):602-609.
25. Liu Dongfeng, Chen Guofu, Wang Xianhua. Experimental study on self-starting passive mode-locked Er^{3+} fiber ring cavity soliton laser [J]. Science in China Series A: Mathematics, 1999,29(7):656-661.
26. Wang Fan. Study on the structure, spectrum and fiber laser performance of Yb^{3+} and Er^{3+} doped quartz glass [D]. Shanghai: Shanghai Institute of Optics and Fine Mechanics, Chinese Academy of Sciences, 2019.
27. Shao Chongyun, Wang Fan, Jiao Yan, et al. Relationship between glass structure and spectroscopic properties in $Er^{3+}/Yb^{3+}/Al^{3+}/P^{5+}$ doped silica glasses [J]. Optical Materials Express, 2020,10(5):1169-1181.
28. De Oliveira Jr M, Aitken B, Eckert H. Structure of P_2O_5 -SiO_2 pure network former glasses studied by solid state NMR spectroscopy [J]. The Journal of Physical Chemistry C, 2018,122(34):19807-19815.
29. Li D, Fleet M E, Bancroft G M, et al. Local structure of Si and P in SiO_2-P_2O_5 and Na_2O-SiO_2-P_2O_5 glasses: a XANES study [J]. Journal of Non-Crystalline Solids, 1995,188(1-2):181-189.
30. Wang Fan, Shao Chongyun, Yu Chunlei, et al. Effect of $AlPO_4$ join concentration on optical properties and radiation hardening performance of Yb-doped Al_2O_3-P_2O_5-SiO_2 glass [J]. Journal of Applied Physics 2019,125(17):173104.
31. Aitken B G, Youngman R E, Deshpande R R, et al. Structure-property relations in mixed-network glasses: multinuclear solid state NMR investigations of the system $x\,Al_2O_3$: $(30-x)$ P_2O_5: $70SiO_2$ [J]. The Journal of Physical Chemistry C, 2009,113(8):3322-3331.
32. Cheo P K, King G G. Clad-pumped Yb: Er codoped fiber lasers [J]. IEEE Photonics Technology Letters, 2001,13(3):188-190.
33. Shao Chongyun, Xie Fenghou, Wang Fan, et al. UV absorption bands and its relevance to local structures of ytterbium ions in $Yb^{3+}/Al^{3+}/P^{5+}$-doped silica glasses [J]. Journal of Non-Crystalline Solids, 2019(512):53-59.
34. Xu Wenbin, Ren Jinjun, Shao Chongyun, et al. Effect of P^{5+} on spectroscopy and structure of $Yb^{3+}/Al^{3+}/P^{5+}$ co-doped silica glass [J]. Journal of Luminescence, 2015(167): 8-15.
35. Qian L, Gupta S, Bolen R. Low-noise, efficient erbium-doped phosphosilicate fiber amplifiers extending to 1620nm [C]//Proceedings of the Conference on Lasers and Electro-Optics/ Quantum Electronics and Laser Science and Photonic Applications Systems Technologies, Optical Society of America, Baltimore, Maryland, 2005.
36. Byriel I P, Palsdottir B, Andrejco M, et al. Silica based erbium doped fiber extending the L-band to 1620+ nm [C]//Proceedings of the Proceedings 27th European Conference on Optical Communication, Amsterdam, Netherlands, 2001.
37. Hwang B C, Jiang S, Luo T, et al. Characterization of cooperative upconversion and energy transfer of Er^{3+} and Yb^{3+}/Er^{3+} doped phosphate glasses [C]//Proceeding of SPIE, 1999(3622): 10-18.
38. Artemyev E, Murzin A, Fedorov Y, et al. Some characteristics of population inversion of the $^4I_{13/2}$ level of erbium ions in ytterbium-erbium glasses [J]. Soviet Journal of Quantum Electronics, 1981,11(9):1266-1268.
39. Dong Liang, Samson B. Fiber lasers: basics, technology, and applications [M]. New York: CRC Press, 2016.
40. Zyskind J. Erbium-doped fiber: amplifiers: what everyone needs to know [C]//Proceedings of the 2016 Optical Fiber Communications Conference and Exhibition (OFC), Anaheim, CA, USA, 2016.

41. Ono S, Tanabe S, Nishihara M, et al. Effect of erbium ion concentration on gain spectral hole burning in silica-based erbium-doped fiber [C]//Proceedings of the Optical Fiber Communication Conference, Optical Society of America, Anaheim, California, USA, 2005.
42. Aizawa T, Sakai T, Wada A, et al. Effect of spectral-hole burning on multi-channel EDFA gain profile [C]//Proceedings of the Optical Fiber Communication Conference, Optical Society of America, San Diego, California, USA, 1999.
43. Zyskind J L, Desurvire E, Sulhoff J W, et al. Determination of homogeneous linewidth by spectral gain hole-burning in an erbium-doped fiber amplifier with GeO_2: SiO_2 core [J]. IEEE Photonics Technology Letters, 2002,2(12):869-871.
44. Sarmani A R, Sheih S J, Adikan F, et al. Spectral hole burning effects initiated by uniform signal intensities in a gain-flattened EDFA [J]. Chinese Optics Letters, 2011,9(2):020603.
45. Ferreira J O M, Rapp L. Dynamics of spectral hole burning in EDFAs: dependence on temperature [J]. IEEE Photonics Technology Letters, 2012,24(1):67-69.
46. Rapp L, Ferreira J. Dynamics of spectral hole burning in EDFAs: dependency on pump wavelength and pump power [J]. IEEE Photonics Technology Letters, 2010,22(16):1256-1258.
47. Yadlowsky M J. Pump wavelength-dependent spectral-hole burning in EDFAs [J]. Lightwave Technology Journal of, 1999,17(9):1643-1648.
48. Pilipetskii A, Abbott S, Kovsh D, et al. Spectral hole burning simulation and experimental verification in long-haul WDM systems [C]//Proceedings of the Optical Fiber Communication Conference, Optical Society of America, Atlanta, Georgia, USA, 2003.
49. Oliveira J, Herbster A F, Oliveira J R, et al. Minimization of gain error due to spectral hole burning using HGC-EDFA with generalized dynamic gain range [C]//Proceedings of Frontiers in Optics 2009/Laser Science XXV/Fall 2009 OSA Optics & Photonics Technical Digest, Optical Society of America, San Jose, California, USA, 2009.
50. Bolshtyansky M, Cowle G. Spectral hole burning in EDFA under various channel load conditions [C]//Proceeding of Optical Fiber Communication Conference, Optica Publishing Group, San Diego, California USA, 2009, OTuH5.
51. Elbers J, Fürst C. Spectral power fluctuations in DWDM networks caused by spectral-hole burning and stimulated Raman scattering, optically amplified WDM networks [M]. Oxford: Academic Press, 2011: 201-219.
52. Barmenkov Y O, Kiryanov A V, Guzman-chavez A D, et al. Excited-state absorption in erbium-doped silica fiber with simultaneous excitation at 977 and 1531nm [J]. Journal of Applied Physics, 2009,106(8):083108.
53. Delevaque E, Georges T, Monerie M, et al. Modeling of pair-induced quenching in erbium-doped silicate fibers [J]. Photonics Technology Letters IEEE, 1993,5(1):73-75.
54. Townsend J E, Barnes W L, Jedrzejewski K P, et al. Yb^{3+} sensitised Er^{3+} doped silica optical fibre with ultrahigh transfer efficiency and gain [J]. MRS Online Proceedings Library, 1991,244(1):143-147.
55. Quimby R S, Minis Ca Lco W J, Thompson B. Clustering in erbium-doped silica glass fibers analyzed using 980nm excited-state absorption [J]. Journal of Applied Physics, 1994,76(8):4472-4478.
56. Giles C R, Desurvire E. Modeling erbium-doped fiber amplifiers [J]. Journal of Lightwave Technology, 1991,9(2):271-283.
57. Bolshtyansky M, Wysocki P, Conti N. Model of temperature dependence for gain shape of erbium-doped fiber amplifier [J]. Journal of Lightwave Technology, 2000,18(11):1533-1540.
58. Payne S A, Chase L L, Smith L K, et al. Infrared cross-section measurements for crystals doped with Er^{3+}, Tm^{3+}, and Ho^{3+} [J]. IEEE Journal of Quantum Electronics, 1992,28(11):2619-2630.
59. Mccumber D E. Einstein Relations Connecting Broadband Emission and Absorption Spectra [J]. Physical Review, 1964,136(4A):A954-A957.
60. Kiryanov A V, Barmenkov Y O, Sandoval-Romero G E, et al. Er^{3+} concentration effects in commercial erbium-doped silica fibers fabricated through the MCVD and DND technologies [J]. IEEE Journal of Quantum Electronics, 2013,49(6):511-521.

61. Myslinski P, Nguyen D, Chrostowski J. Effects of concentration on the performance of erbium-doped fiber amplifiers [J]. Journal of Lightwave Technology, 1997,15(1):112-120.
62. Myslinski P, Fraser J, Chrostowski J. Nanosecond kinetics of upconversion process in EDF and its effect on EDFA performance [C]//Proceedings of the Optical Amplifiers and Their Applications, Optical Society of America, Davos, USA 1995,18: ThE3.
63. Blixt P, Nilsson J, Carlnas T, et al. Concentration-dependent upconversion in Er^{3+} -doped fiber amplifiers: Experiments and modeling [J]. IEEE Photonics Technology Letters, 1991,3(11):996-998.
64. Richard S Q, William J M, Barbara A T. Upconversion and 980nm excited-state absorption in erbium-doped glass [C]//Proceeding of SPIE, 1993(1789): 50-57.
65. He Qiang, Wang Fan, Lin Zhiquan, et al. Temperature dependence of spectral and laser properties of Er^{3+} /Al^{3+} co-doped aluminosilicate fiber [J]. Chinese Optics Letters, 2019,17(10):101401.
66. Ryu U, Im Y, Han W, et al. Suppression of temperature dependence in EDFA gain by hybrid concatenation of antimony-doped silica EDF [C]//Proceedings of the Optical Fiber Communications Conference, Atlanta, GA, USA, 2003,2: 600-601.
67. Gan Fuxi, et al. Photonics Glass and Its Applications [M]. Shanghai: Shanghai Science and Technology Publishing House, 2011.
68. Ono H, Sakamoto T, Mori A, et al. An erbium-doped tellurite fiber amplifier for WDM systems with dispersion-shifted fibers [J]. IEEE Photonics Technology Letters, 2002,14(8):1070-1072.
69. Sakamoto T, Mori A, Masuda H, et al. Wide band rare-earth-doped fiber amplification technologies — gain bandwidth expansion in the C and L bands special feature [J]. NTT Technical Review, 2004,2(12):38-43.
70. Kakui M, Kashiwada T, Onishi M, et al. Optical amplification characteristics around 1.58μm of silica-based erbium-doped fibers containing phosphorous/alumina as codopants [C]// Proceedings of the Optical Amplifiers and Their Applications, Optica Publishing Group, Vail, Colorado, 1998,25.
71. Lei Chengmin, Feng Hanlin, Wang Lixian, et al. An extended L-band EDFA using C-band pump wavelength [C]//Proceedings of the Optical Fiber Communication Conference, Optical Society of America, San Diego, California, USA, 2020.
72. Lou Yang, Chen Yang, Gu Zhimu, et al. Er^{3+} /Ce^{3+} Co-doped phosphosilicate fiber for extend the L-band amplification [J]. Journal of Lightwave Technology, 2021,39(18):5933-5938.
73. Haruna T. Optical fiber for amplification and optical fiber amplifier: United States, US008023181B2 [P]. 2011.
74. Qian Li, Fortusini D, Benjamin S D, et al. Gain-flattened, extended L-band (1570-1620nm), high power, low noise erbium-doped fiber amplifiers [C]//Proceedings of the Optical Fiber Communications Conference, Optical Society of America, Anaheim, California, 2002.
75. Sano A, Masuda H, Yoshida E, et al. 55×86Gb/s CSRZ-DQPSK transmission over 375 km using extended L-band erbium-doped fiber amplifiers [C]//Proceedings of the 2006 European Conference on Optical Communications, Cannes, France, 2006.
76. Snitzer E, Po H, Hakimi F, et al. Double clad, offset core Nd fiber laser [C]//Proceedings of the Optical Fiber Sensors, Optical Society of America, New Orleans, Louisiana, 1988.
77. https://www.ipgphotonics.com/en/products/lasers/high-power-cw-fiber-lasers#[1-micron].
78. Lin Huaiqin, Feng Yujun, Feng Yutong, et al. 656W Er-doped, Yb-free large-core fiber laser [J]. Optics Letter, 2018,43(13):3080-3083.
79. Matniyaz T, Kong F, Kalichevsky-dong M T, et al. 302W single-mode power from an Er/Yb fiber MOPA [J]. Optics Letter, 2020,45(10):2910-2913.
80. Sahu J K, Jeong Y, Richardson D J, et al. A 103W erbium-ytterbium co-doped large-core fiber laser [J]. Optics Communications, 2003,227(1):159-163.
81. Vienne G G, Caplen J E, Liang D, et al. Fabrication and characterization of Yb^{3+}: Er^{3+} phosphosilicate fibers for lasers [J]. Journal of Lightwave Technology, 1998,16(11):1990-2001.
82. Maurice E, Monnom G, Dussardier B, et al. Clustering effects on double energy transfer in heavily ytterbium-erbium-codoped silica fibers [J]. Journal of the Optical Society of America B, 1996,13(4):693-701.

83. Vienne G G, Brocklesby W S, Brown R S, et al. Role of aluminum in ytterbium-erbium codoped phosphoaluminosilicate optical fibers [J]. Optical Fiber Technology, 1996,2(4):387-393.
84. Alam S U, Turner P W, Grudinin A B, et al. High-power cladding pumped erbium-ytterbium co-doped fiber laser [C]//Proceedings of the Optical Fiber Communication Conference and International Conference on Quantum Information, Optical Society of America, Anaheim, California, 2001.
85. Nilsson J, Alam S, Alvarez-chavez J A, et al. High-power and tunable operation of erbium-ytterbium Co-doped cladding-pumped fiber lasers [J]. IEEE Journal of Quantum Electronics, 2003,39(8):987-994.
86. Sahu J K, Jeong Y, Richardson D J, et al. Highly efficient high-power erbium-ytterbium co-doped large core fiber laser [C]//Proceedings of the Advanced Solid-State Photonics, Optical Society of America, Vienna, 2005.
87. Shen Deyuan, Sahu J K, Clarkson W A. Highly efficient Er, Yb-doped fiber laser with 188W free-running and >100W tunable output power [J]. Optics Express, 2005,13(13):4916-4921.
88. Jeong Y, Yoo S, Codemard C A, et al. Erbium: ytterbium codoped large-core fiber laser with 297-W continuous-wave output power [J]. IEEE Journal of Selected Topics in Quantum Electronics, 2007,13(3):573-579.
89. Tankala K, Samson B, Carter A, et al. New developments in high power eye-safe LMA fibers [C]//Proceeding of SPIE, 2006(6102): 610206.
90. Khudyakov M M, Lobanov A S, Lipatov D S, et al. Single-mode large-mode-area Er-Yb fibers with core based on phosphorosilicate glass highly doped with fluorine [J]. Laser Physics Letters, 2019,16(2):025105.
91. Creeden D, Pretorius H, Limongelli J, et al. Single frequency 1560nm Er:Yb fiber amplifier with 207W output power and 50.5% slope efficiency [C]//Proceeding of SPIE, 2016(9728):97282L.
92. Matniyaz T, Kong F, Kalichevsky-dong M, et al. Single-mode Er/Yb fiber MOPA at 1562nm with 302W power and 56% optical efficiency [C]//Proceeding of SPIE, 2021(11665): 116650M.
93. Dubinskii M, Zhang J, Ter-mikirtychev V. Highly scalable, resonantly cladding-pumped, Er-doped fiber laser with record efficiency [J]. Optics Letter, 2009,34(10):1507-1509.
94. Zhang J, Fromzel V, Dubinskii M. Resonantly cladding-pumped Yb-free Er-doped LMA fiber laser with record high power and efficiency [J]. Optics Express, 2011,19(6):5574-5578.
95. Jebali M A, Maran J N, Larochelle S, et al. A 103W high efficiency in-band cladding-pumped 1593nm all-fiber erbium-doped fiber laser [C]//Proceedings of the Conference on Lasers and Electro-Optics, Optical Society of America, San Jose, California, 2012.
96. Jebali M A, Maran J N, Larochelle S. 264W output power at 1585nm in Er-Yb codoped fiber laser using in-band pumping [J]. Optics Letter, 2014,39(13):3974-3977.
97. Kotov L, Likhachev M, Bubnov M, et al. Yb-free Er-doped all-fiber amplifier cladding-pumped at 976nm with output power in excess of 100W [C]//Proceeding of SPIE, 2014(8961):89610X.

Chapter 7
Rare Earth Doped Silica Fiber and Its Applications at 2 μm Band

2 μm band laser has broad application prospects in many fields due to its unique wavelength absorption characteristics. Water molecules have a strong absorption peak near 1.94 μm, making 2 μm band laser safe for human eyes and suitable for applications such as remote sensing detection, laser radar, and other fields, which can detect water vapor, wind speed, etc. in the atmospheric environment. At the same time, the penetration depth of 2 μm band laser into human tissue is shallow, making it a precise "scalpel" and having good hemostatic effect. It has been applied in medical fields such as laser lithotripsy, soft tissue cutting, and urinary system treatment. 2 μm band laser is in the high atmospheric transmittance window, with low scattering loss and strong penetration ability for smoke and other substances, suitable for space communication. 2 μm band laser is also an important pump source for 3 ~ 5 μm band mid-infrared optical parametric oscillators, which has important application prospects in mid-infrared active detection, security and optoelectronic countermeasures. Most transparent organic materials have strong absorption in the mid-infrared 2 μm band, and 2 μm laser has high processing efficiency for cutting, welding, marking, etc. of plastics.

Commonly used 2 μm band luminescent rare earth ions mainly include Tm^{3+} (~1.95 μm) and Ho^{3+} (~2.1 μm) ions. The cross-relaxation process between energy levels of Tm^{3+} ions enables a theoretical quantum efficiency of up to 200% under 790 nm semiconductor laser pumping, and the theoretical optical-to-optical efficiency can exceed 80%. Compared to Tm^{3+} ions, Ho^{3+} ions can achieve a laser wavelength of up to 2.1 μm, which is slightly longer, and offers higher atmospheric transmittance. Compared to thulium and holmiun solid-state lasers, thulium and holmiun fiber lasers possess advantages such as excellent beam quality, high efficiency, compact structure, and easier thermal management, making them a long-standing resecrah hotspot in the field of fiber lasers. This chapter initially elaborates on the structure and spectral properties of thulium and holmium-doped silica glass, introduces the latest research progress in 2 μm band rare earth doped silica fibers, summarizes some challenges faced by the preparation of 2 μm laser fibers, reviews

L. Hu, *Rare Earth Doped Silica Fiber and Its Applications*,
https://doi.org/10.1007/978-981-96-5306-5_7

the development history of 2 μm fiber lasers both domestically and internationally, and provides an outlook on their application prospects and development trends.

7.1 Energy Level Structure of Thulium and Holmium Ions in Silica Glass

Ho and Tm belong to the lanthanide elements, with atomic numbers of 67 and 69 respectively, and relative atomic masses of 164.9 and 168.9. The outer electron configuration of Ho atom is $4f^{11}6s^2$,while that of Tm atom is $4f^{13}6s^2$. The common valence state of the ions is +3. The following analyzes the energy level structure of Tm^{3+} and Ho^{3+} to reveal their basic luminescent properties.

7.1.1 Energy Level Structure of Thulium Ions

The electron configuration of Tm^{3+} is $4f^{12}$, exhibiting a rich set of transitional electron energy levels. Compared with the simple two-level structure of Yb^{3+}, the energy level structure of Tm^{3+} is very complex. Its main energy level structure is shown in Fig. 7.1, where $^3H_6$ is the ground state energy level, and $^3F_4$ is the first excited state energy level. The luminescence in the 1700 ~ 2100 nm band is generated by the transition from the first excited state energy level to the ground state energy level. On the one hand, the complex energy level structure of Tm^{3+} provides multiple pathways for achieving population inversion, enabling laser output at different

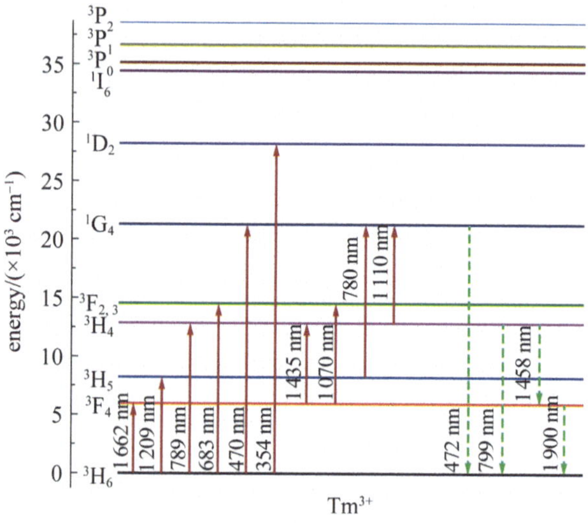

Fig. 7.1 Schematic diagram of the main energy level structure of Tm^{3+}

wavelengths. On the other hand, the complex energy level structure can also bring some "side effects", such as excited state absorption, upconversion processes, etc., which deteriorate the laser output performance.

The Tm^{3+} ion has many absorption bands in different wavelength ranges. Fig. 7.2 shows the absorption spectrum of thulium-doped silica glass, with absorption peaks at 1662 nm, 1209 nm, 789 nm, 683 nm, 470 nm, 354 nm, etc., providing various options for pump wavelength selection. There are three main pumping methods to obtain 2 μm band lasers: $^3H_6 \rightarrow {}^3F_4$ (centred at ~1.6 μm); $^3H_6 \rightarrow {}^3H_5$ (centred at ~1.2 μm); $^3H_6 \rightarrow {}^3H_4$ (centred at ~0.8 μm).

The 1.6 μm pumping method is to pump Tm^{3+} on the $^3H_6$ energy level to the $^3F_4$ energy level, and then the transition of ions from the $^3F_4$ energy level to the ground state $^3H_6$ level, thereby emitting photons in the 2 μm band. This pumping method belongs to the in-band pumping, with small quantum loss and high utilization rate of pumping light, and is an efficient way to obtain lasers in the 2 μm band. However, there is currently a lack of commercial 1.6 μm wavelength high-power semiconductor pumping sources. Typically, long-wavelength erbium-doped lasers or Raman lasers are used as in-band pumping sources, which increases the complexity of the system and reduces the overall electro-optical conversion efficiency.

The 1.2 μm pumping method is to pump Tm^{3+} on the $^3H_6$ energy level to the $^3H_5$ level, then undergo non-radiative relaxation to the $^3F_4$ level, and finally transition from the $^3F_4$ level to the ground state $^3H_6$ level to emit photons in the 2 μm band. In this process, due to the long fluorescence lifetime of the $^3F_4$ excited state energy level, the ions on the $^3F_4$ level can reabsorb pump photons and transition to the $^3F_{2,3}$ levels. Therefore, using 1.2 μm wavelength pumping leads to significant excited-state absorption, making it prone to upconversion luminescence, reducing the utilization efficiency of pump light. However, the pumping sources in this band are very abundant and relatively mature, such as ytterbium-doped fiber lasers and Nd:YAG solid-state lasers.

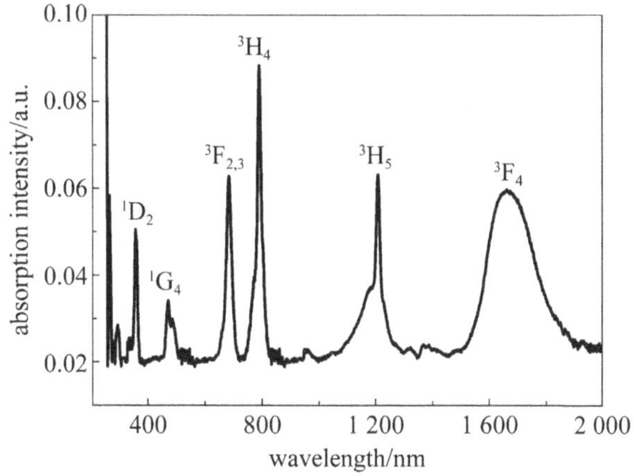

Fig. 7.2 Absorption spectrum of thulium-doped silica glass

The 790 nm pumping method is introduced below. As shown in Fig. 7.3, the 790 nm pumping method is to pump Tm^{3+} from the $^3H_6$ energy level to the $^3H_4$ level. Tm^{3+} ions on the $^3H_4$ level transition to the $^3F_4$ level. The energy difference between the $^3H_4$ and $^3F_4$ levels is similar to that between the $^3H_6$ and $^3F_4$ levels. Thus, when Tm^{3+} on the $^3H_4$ level transitions to the $^3F_4$ level, the released energy is transferred to another Tm^{3+} in the ground state nearby, causing it to transition from the ground state $^3H_6$ level to the $^3F_4$ level. This energy transfer process is called cross-relaxation process. The cross-relaxation process between Tm^{3+} produces two 2 μm photons from one 790 nm photon, with a theoretical quantum efficiency of up to 200%. Under 790 nm pumping, there is no corresponding level to excite particles from the $^3H_4$ and the $^3F_4$ levels to higher levels. Although Tm^{3+} on the $^3H_5$ level can be excited to the $^1G_4$ level, emitting upconversion blue light, the fluorescence lifetime of the $^3H_5$ level is very short, and Tm^{3+} ions on this level quickly transitions to the $^3F_4$ level, resulting in weak upconversion blue light generated by 790 nm pumping. Because commercial ~790 nm laser diodes are mature, this pumping method has become the preferred pumping method for high-power thulium-doped lasers, with a theoretical light-to-light conversion efficiency of over 80%. The similar energy difference between the $^3H_4$ and $^3F_4$ levels and between the $^3H_6$ and $^3F_4$ levels result in an energy transfer upconversion process between Tm^{3+} ions. Tm^{3+} on the $^3F_4$ level transitions to the $^3H_6$ level, transferring energy to neighboring Tm^{3+} ions on the $^3F_4$ level, causing them to transition to the higher $^3H_4$ level. The energy transfer upconversion process between Tm^{3+} ions consumes the population of the upper energy level $^3F_4$, reducing the energy conversion efficiency of the 2 μm band. For 2 μm lasers, upconversion luminescence inhibits the cross-relaxation process, leading to a decrease in energy conversion efficiency, which is unfavorable for high-power laser output.

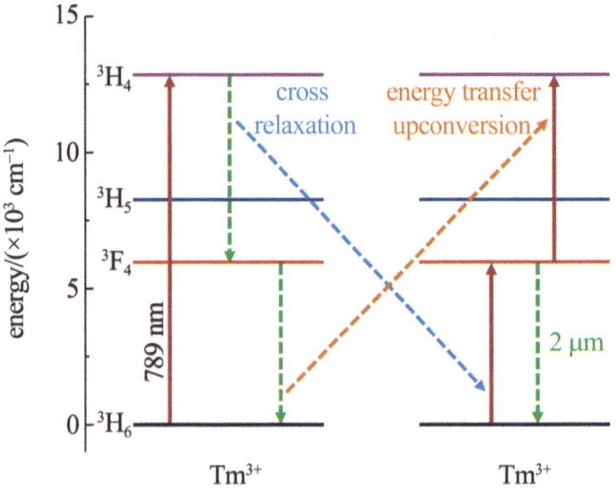

Fig. 7.3 Cross-relaxation process andenergy transfer up-conversion process of Tm^{3+}

7.1.2 Energy Level Structure of Holmium Ions

The electronic configuration of Ho^{3+} is $4f^{10}$, which has a more complex energy level structure than that of Tm^{3+}. Figure 7.4 shows a schematic diagram of the main energy level structure of Ho^{3+}, where $^5I_8$ is the ground state energy level, and $^5I_7$ is the first excited state energy level. The transition from the first excited state energy level to the ground state energy level produces 2 μm band photons. Figure 7.5 shows the absorption spectrum of holmium-doped silica glass, from which the main absorption peaks are located at 1945 nm, 1145 nm, 640 nm, 537 nm, 450 nm, and 360 nm. The absorption peaks of $^5I_5$ energy level and $^5I_4$ energy level to the ground state energy level $^5I_8$ are located at 700 ~ 900 nm. Although these wavelengths fall within the wavelength range of the laser diode, their absorption intensity is very weak and making them unsuitable for use as a pump wavelength. Other short-wavelength pumps cause a large quantum defect, resulting in low energy conversion efficiency and output power. Common pump wavelengths includes 1945 nm and 1150 nm, which can be obtained through thulium-doped lasers and ytterbium-doped lasers, respectively. Compared to the pump wavelengths around 1.1 μm using ytterbium-doped laser, utilizing pump wavelengths around 1.9 μm with thulium-doped lasers in-band pumping Ho^{3+} can significantly reduce the quantum defect, reduce the thermal load of the fiber, and improve the laser slope efficiency and output power.

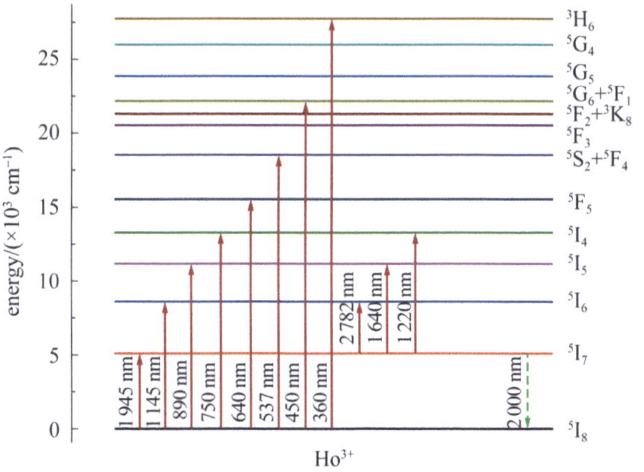

Fig. 7.4 Schematic diagram of the main energy level structure of Ho^{3+}

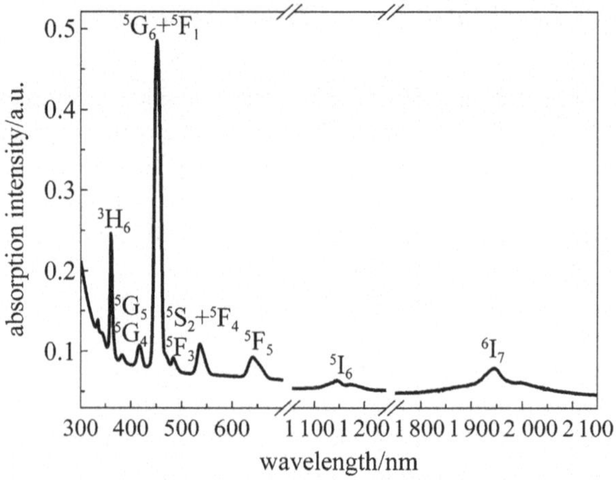

Fig. 7.5 Absorption spectrum of holmium-doped silica glass

7.2 Spectral Properties and Structure of Thulium-Doped Silica Glass

Currently, the laser glass materials used to achieve 2 μm wavelength laser output include silica glass, silicate glass, tellurite glass, germanate glass, sulfur-based glass, fluoride glass, and fluorophosphate glass, etc. Compared with other multi-component glass host, silica glass has the characteristics and advantages of dense network structure, good mechanical strength, low background loss, and high damage threshold. Short-wave infrared silica fiber lasers has an output power greater than 1000 W [1] at 2 μm band, which is an order of magnitude higher than the maximum output power of short-wave infrared lasers based on other glass host. Therefore, rare earth-doped silica glass is an ideal gain medium for 2 μm wavelength fiber laser.

7.2.1 Spectral Properties of Tm^{3+}/Al^{3+} Doped Silica Glass in the 2 μm Band

Compared with other rare earth ions, the cross-relaxation process of Tm^{3+} ions at 790 nm allows the quantum efficiency of the 2 μm laser to reach 200% [2], which far exceeds its Stokes efficiency. High concentration doping is beneficial to reduce the average distance between Tm^{3+} ions, promoting the occurrence of the cross-relaxation process. At the same time, High doping concentration of Tm^{3+} can increase the pump absorption coefficient, shorten the length of the fiber, and benefit the device integration. Therefore, high concentration doping of Tm^{3+} ions is essential for efficient output of 2 μm laser. How to prepare silica glass with high concentration and uniform distribution of Tm^{3+} ions is the key to fabricate silica fiber in

2 μm band. Aluminum is the most common doping element in silica fiber, which can increase the solubility and doping concentration of rare earth ions in the silica glass network. This section introduces a series of Tm^{3+}/Al^{3+} co-doped silica glasses with different Tm^{3+} concentrations prepared by the sol-gel method, and systematically analyzes their physical and spectral properties.

According to reference [3], when the Al^{3+}/Tm^{3+} ratio is 15:1, Tm^{3+}/Al^{3+} doped silica glass can obtain better spectral properties. Based on this, Wang et al. [4] prepared silica glasses with 5 different Tm^{3+} concentrations using the sol-gel process combined with high-temperature powder sintering method, denoted as TAS series glasses, with specific compositions of xTm$_2$O$_3$ - 15xAl$_2$O$_3$ -(100-16x)SiO$_2$ (mol%, $x = 0.1, 0.3, 0.5, 0.8, 1.0$; denoted as TAS1, TAS2, TAS3, TAS4, TAS5). Table 7.1 lists the theoretical doping concentration of Tm^{3+} ions and the actual measured doping concentration. In this series of glasses, the highest doping concentration of Tm^{3+} ions reached 4.26×10^{20} cm^{-3}, which is comparable to the doping concentration of Tm^{3+} ions in tellurite glass [5] and germanate glass [6]. Table 7.1 also lists the density and refractive index of the TAS series glass samples, showing that both the density and refractive index of the glass increase with the increase in the concentration of Al_2O_3 and Tm_2O_3.

According to the Judd-Ofelt theory, utilizing the absorption spectrum of Tm^{3+} and the data of Tm^{3+} concentration, density, and refractive index listed in Table 7.1, the J-O parameters Ω_t (t = 2, 4, 6) of the TAS series glasses can be calculated. Among the three intensity parameters, Ω_2 reflects the intensity of hypersensitive transitions, while Ω_4 and Ω_6 reflect the overall intensity of spectral transitions [7]. Table 7.2 lists the J-O parameters of Tm^{3+} in silica glass and other host glasses [5, 6, 8–11]. The data in the table shows that for most host glasses, the J-O parameters follow the rule of $\Omega_2 > \Omega_4 > \Omega_6$. The J-O parameters of the TAS series glasses also follow this rule. The data in Table 7.2 indicates that the J-O parameters of Tm^{3+} doped silica glass prepared by the sol-gel method are close to those prepared by the MCVD method [8]. Ω_2 in the J-O parameters is related to the structural changes around the rare earth ions in the glass, while Ω_6 is related to the rigidity of the glass [12, 13]. By comparing the parameters of various host glasses in Table 7.2, it is found that TAS4 glass has a higher Ω_2, which indicates that Tm^{3+} ions are in a lower symmetry environment in silica glass. Ω_4/Ω_6 represents the spectral properties of

Table 7.1 Doping concentration, density, and refractive index of Tm^{3+} in TAS series glass samples [4]

Sample	Theoretical Tm^{3+} Concentration /($\times 10^{20}$ cm^{-3})	Actual Tm^{3+} Concentration /($\times 10^{20}$ cm^{-3})	Density/ (g·cm^{-3})	Refractive Index @1800 nm
TAS1	0.44	0.439	2.22	1.44754
TAS2	1.29	1.286	2.25	1.45473
TAS3	2.16	2.160	2.33	1.46239
TAS4	3.40	3.382	2.39	1.47542
TAS5	4.28	4.264	2.42	1.47982

Table 7.2 J-O parameters of Tm^{3+} in different host glasses

Type of Glass	Ω_2(×10^{-20} cm^2)	Ω_4(×10^{-20} cm^2)	Ω_6(×10^{-20} cm^2)	Ω_4/Ω_6
Tellurite glass [5]	3.79	1.10	0.98	1.12
Germanate glass [6]	5.55	2.03	1.26	1.61
ZBLAN [8]	1.96	1.36	1.16	1.17
Silica glass [8]	6.23	1.81	1.36	1.40
Fluorophosphate glass [9]	3.01	2.56	1.54	1.66
Silicate glass [10]	3.36	1.25	0.75	1.67
PbO-Bi_2O_3 -Ga_2O [11]	3.59	0.76	1.27	0.60
TAS4 [4]	6.94	1.80	1.33	1.35

the material [14], and it can be seen from Table 7.2 that the Ω_4/Ω_6 value of TAS4 glass is 1.35, which is higher than that of tellurite glass [5], ZBLAN [8] and PbO-Bi_2O_3 -Ga_2O [11] glass.

To study the effect of Tm^{3+} ion concentration on the 2 μm band luminescent properties of Tm^{3+}/Al^{3+} doped silica glass, a semiconductor laser with a central wavelength of 808 nm was used for excitation. The test range is from 1300 nm to 2200 nm, and the fluorescence spectra of the TAS series glasses is shown in Fig. 7.6. Two fluorescence peaks can be seen from the figure: one peak is generated by the Tm^{3+}: $^3H_4 \rightarrow {}^3F_4$ transition, centered at 1460 nm, with a weak intensity, and the amplified fluorescence spectrum is shown in Figure 7.6b. The other peak is generated by the Tm^{3+}: $^3F_4 \rightarrow {}^3H_6$ transition, centered at 1800 nm. The intensity of the fluorescence peak at 1800 nm is much greater than the peak at 1460 nm, and the relative intensity ratio of the two peaks for the TAS4 sample is 90.3. For the TAS series, the fluorescence intensity at 1800 nm first increases and then decreases with the increase of Tm_2O_3 doping concentration, reaching the highest fluorescence intensity when the Tm_2O_3 doping concentration is 0.8 mol% (TAS4). As the Tm_2O_3 concentration increases, the possibility of cross-relaxation of Tm^{3+} ions ($^3H_4 + {}^3H_6 \rightarrow {}^3F_4 + {}^3F_4$) increases, thereby increasing the number of particles on the $^3F_4$ energy level and leading to an increase in the fluorescence intensity at 1800 nm. However, excessive rare earth ions are prone to concentration quenching, so when the Tm_2O_3 doping concentration exceeds 0.8 mol%, the fluorescence intensity begins to decrease.

Figure 7.7a shows the change in normalized fluorescence intensity at 1800 nm and 1460 nm with the doping concentration of Tm_2O_3 in TAS series silica glass. It can be seen that the normalized intensities at 1800 nm and 1460 nm both reach the maximum when the doping concentration of Tm_2O_3 is 0.8 mol%. Figure 7.7b shows the change in the ratio of fluorescence intensities at 1800 nm and 1460 nm with the doping concentration of Tm_2O_3. It can be seen that as the doping concentration of Tm_2O_3 increases, the ratio of the two fluorescence intensities initially increases, reaching a maximum at 0.8 mol%, and then decreases. The decrease in the ratio of fluorescence intensities at 1800 nm and 1460 nm at a Tm_2O_3 concentration of 1.0 mol% is caused by concentration quenching.

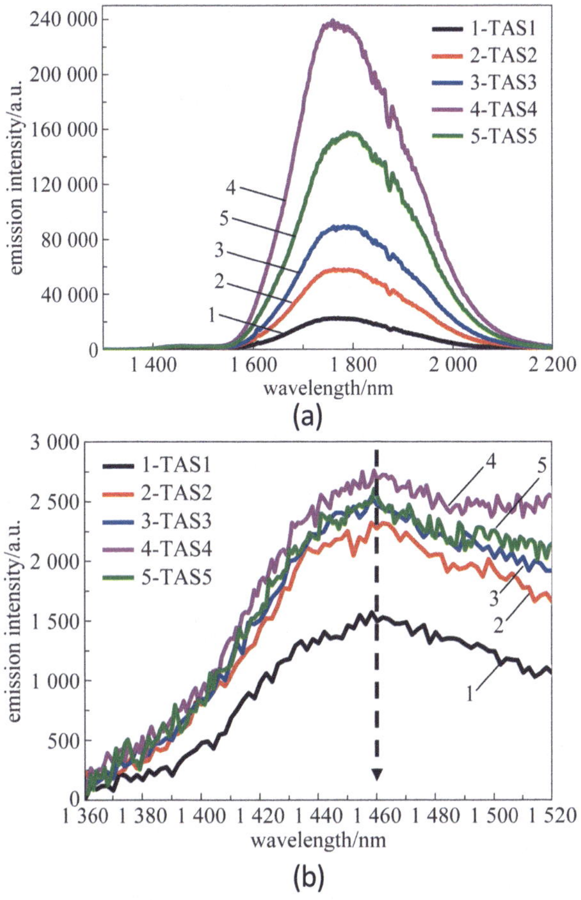

Fig. 7.6 Fluorescence spectra of silica glass doped with different Tm_2O_3 concentrations (**a**) and the amplified spectra at 1360 ~ 1520 nm (**b**)

Figure 7.8 shows the calculated absorption and stimulated emission cross-section of 0.8 mol% Tm_2O_3 doped silica glass (TAS4). The maximum stimulated emission cross-section at 1812 nm is 6.00×10^{-21} cm^2, and the maximum absorption cross-section at 1668 nm is 4.65×10^{-21} cm^2 (a large absorption cross-section indicates high absorption efficiency of the pump light). Table 7.3 lists the stimulated emission cross-section σ_{ems}, radiative lifetime τ_{rad} and $\sigma_{ems} \times \tau_{rad}$ values for Tm^{3+} doped silica glass and multicomponent glass corresponding to the 1.8 μm luminescent transition $^3F_4 \rightarrow {}^3H_6$. $\sigma_{ems} \times \tau_{rad}$ is commonly used to measure the gain performance of laser materials [16]. As can be seen from the table, the stimulated emission cross-section of the TAS4 sample is larger than that of fluorophosphate glass [9], tellurite glass [15], and silicate glass [10], ZBLAN glass [8] and Tm^{3+} doped silica glass prepared by MCVD method [8], but less than germanate glass [6]. TAS glasses represented

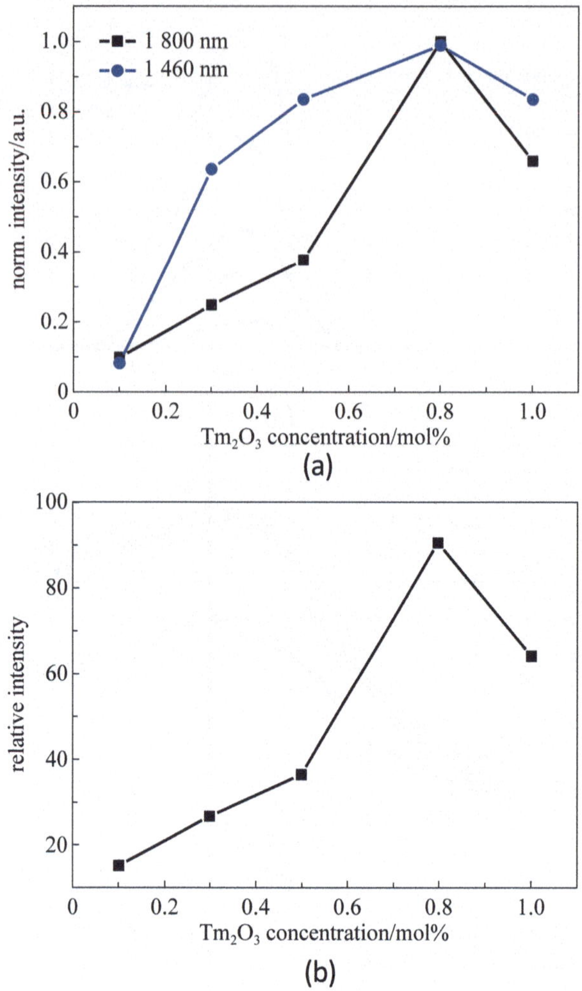

Fig. 7.7 The change in normalized fluorescence intensity (**a**) and its ratio (**b**) at 1800 nm and 1460 nm in TAS series silica glass with doping concentration of Tm_2O_3

by TAS4 have a larger $\sigma_{ems} \times \tau_{rad}$ value, indicating that TAS glass is a promising 2 μm laser material.

Based on the absorption cross-section and stimulated emission cross-section, the gain coefficient G(λ) of rare earth ions at 2 μm and the evaluation of material gain performance can be calculated using eq. 7.1.

$$G(\lambda) = N\left[P\sigma_{ems}(\lambda) - (1-P)\sigma_{abs}(\lambda)\right] \tag{7.1}$$

P is the ratio of the number of inverted particles on the $^3F_4$ level to the total number of particles, where N is the total concentration of rare earth ions.

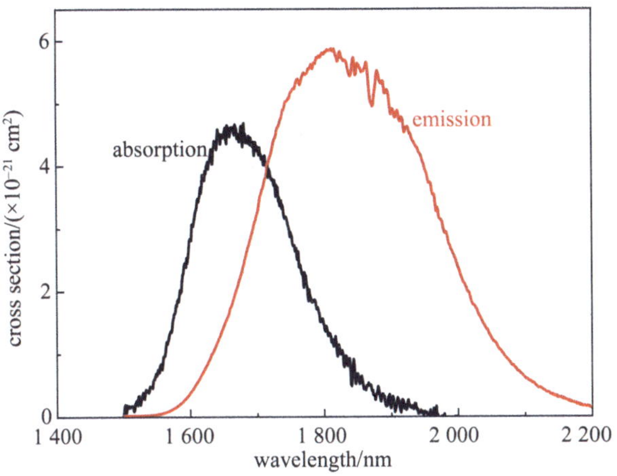

Fig. 7.8 Absorption and stimulated emission cross-section of TAS4 glass in the ~2 μm band

Table 7.3 $^3F_4 \rightarrow {}^3H_6$ level transition stimulated emission cross-section σ_{ems}, radiative lifetime τ_{rad} and $\sigma_{ems} \times \tau_{rad}$ values of different thulium-doped glass host

Type of Glass	σ_{ems}/(×10^{-21} cm^2)	τ_{rad}/ms	$\sigma_{ems} \times \tau_{rad}$ (×10^{-21} cm^2·ms)
ZBLAN [8]	2.40	11.1	26.64
Silica Glass [8]	4.60	4.56	20.98
Germanate Glass [6]	7.70	1.77	13.63
Fluorophosphate Glass [9]	4.11	4.44	18.25
Silicate glass [10]	3.89	5.21	20.27
Tellurite glass [15]	4.00	1.78	7.12
TAS4 [4]	6.00	5.01	30.06

Figure 7.9 shows the change in the gain coefficient of TAS4 glass with wavelength when the value of P ranges from 0 to 0.5. It can be seen that as the proportion of particles in the upper energy level increases, the wavelength corresponding to the maximum gain coefficient blueshifts, showing the typical characteristics of a three-level laser system. When the proportion of inverted particles in the $^3F_4$ level in the glass reaches 20%, gain can be achieved, indicating that TAS4 may have a low laser threshold.

7.2.2 *The Influence of La^{3+}/Al^{3+} Ratio on the Structure and Spectral Properties of High Concentration Thulium-Doped Silica Glass*

Different lanthanide ions have similar ionic radii, and their oxides also have similar crystal structures and lattice constants, so they can replace each other [17]. When different types of rare earth ions are co-doped, the added rare earth ions will

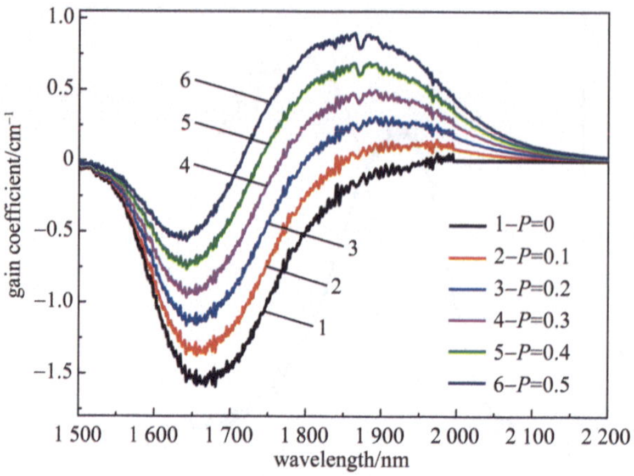

Fig. 7.9 Gain curve of TAS4 glass sample

partially replace the rare earth ions in the clusters in the silica glass, thereby increasing the distance between the rare earth ions in the clusters, that is, improving the dispersion. However, when different luminescent rare earth ions are co-doped, although the clustering of rare earth ions can be suppressed, energy transfer and detrimental up-conversion will occur between different ions due to the rich energy level structure of luminescent rare earth ions [18, 19], thus affecting the spectral properties. Although the method of co-doping inert rare earth ions and luminescent rare earth ions into quartz glass suppresses the clusters of luminescent rare earth ions, the concentration of luminescent rare earth ions is not easy to meet the requirements of high power laser output. Co-doping with aluminum ions can increase the solubility of rare earth ions in silica glass. Adding inert rare earth ions to silica glass to partially replace aluminum ions can not only increase the solubility of rare earth ions in silica glass, but also ensure thermal and chemical stability. Therefore, theoretically, co-doping with inert rare earth and aluminum ions is an effective method to obtain high concentration rare earth doped high quality optical glass.

When the concentration of Tm^{3+} is high, the distance between them is very close, and cross-relaxation processes are likely to occur, reducing the population of particles in $^3H_4$ energy level, and increasing the population of particles in $^3F_4$ energy level, thereby enhancing the fluorescence intensity of Tm^{3+} ions at the 2 μm band [20]. However, high concentration of Tm^{3+} can lead to the deterioration of the anti-crystallization performance of silica glass. Wang et al. [21] explored the effect of La^{3+}/Al^{3+} co-doping on the structure and spectral properties of silica glass doped with high concentration Tm^{3+}ions. Using the sol-gel process combined with high temperature powder sintering method, series glass samples with different La^{3+}/Al^{3+} ratios were prepared, with composition of $0.8Tm_2O_3$-(12-x) Al_2O_3-$x La_2O_3$–$87.2SiO_2$ ($x = 0$, 1.2 mol%, 4.5 mol%), denoted as TLAS1, TLAS2 and TLAS3 respectively.

To explore the role of La^{3+} in silica glass, the Raman spectra and FTIR spectra of the TLAS series samples were tested, and the results are shown in Fig. 7.10a and Fig. 7.10b, respectively. In the Raman spectra, the positions of the silicon structure group Q_n^{Si} ($n = 4, 3, 2, 1, 0$) are approximately at 1200 cm^{-1}, 1100 cm^{-1}, 950 cm^{-1}, 900 cm^{-1} and 850 cm^{-1} [22]. Since the field strength of Al—O is weak, as Si is replaced by Al, the positions of the Q_n^{Si} groups will shift to lower wavenumbers. For the FTIR spectra, the peak at 950 cm^{-1} is due to the stretching vibration of the Si—O— groups [23]. The formation of the Si—O— groups is due to the introduction of network modifiers in glass, causing the Si—O—Si group to break [24]. As

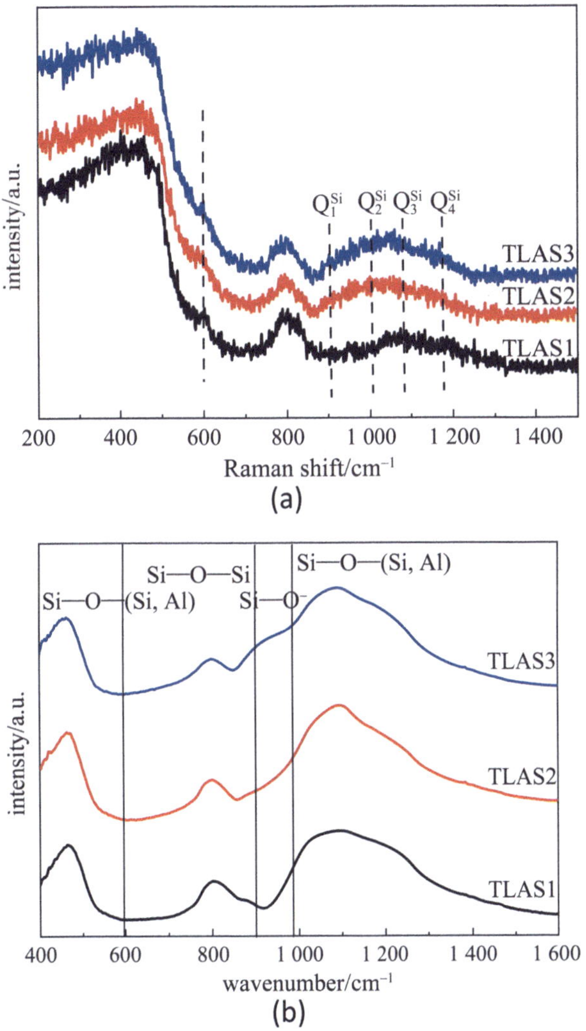

Fig. 7.10 Raman spectra (**a**) and FTIR spectra (**b**) of thulium-doped silica glass with different La^{3+}/Al^{3+} ratios

can be seen from the figure, $La^{3+}/Al^{3+}/Tm^{3+}$ doped silica glass has a wide distribution of Q_n^{Si} groups, and as the La^{3+}/Al^{3+} ratio increases, the main structural group of silicon changes from Q_3^{Si} group to Q_2^{Si} group. The addition of La^{3+} ions leads to the break of Si—O—Si groups and increase of Si—O— groups, hence, Tm^{3+} doped silica glass forms a more loose network structure as the La^{3+}/Al^{3+} ratio increases.

In order to study the local environment of Tm^{3+} in silica glass, the ^{27}Al MAS NMR and ^{29}Si MAS NMR spectra of TLAS series samples with different La^{3+}/Al^{3+} ratios were tested, as shown in Fig. 7.11. It can be seen from the figure that Al in thulium-doped silica glass mainly exists in the form of AlO_4, and the amounts of AlO_5 and AlO_6 decrease with the increase of the La^{3+}/Al^{3+} ratio. The single-pulse ^{27}Al and ^{29}Si MAS NMR spectra of samples containing paramagnetic substances show wide spinning sidebands, while NMR spectra without paramagnetic substances do not exhibit spinning sidebands [25–27]. The TLAS series samples contain the paramagnetic element Tm, so wide spinning sidebands can be seen in Fig. 7.11, which are caused by the influence of the anisotropic magnetic susceptibility and the dipole field, leading to transitions from the 1/2 to −1/2 [28].

To further confirm that these broadened spinning sidebands are caused by the paramagnetic element Tm, YLAS series quartz glasses were prepared using the same method by replacing equivalent amount of Tm with non-paramagnetic Y. The specific molar ratio compositions are $0.8Y_2O_3$ - $(12\text{-}x)Al_2O_3$ - xLa_2O_3–$87.2SiO_2$ ($x = 0$, 1.2, 4.5 mol%), denoted as YLAS1, YLAS2, and YLAS3, respectively. As shown in Fig. 7.12a, the ^{27}Al MAS NMR spectra of YLAS series samples did not exhibit broadened spinning sidebands. In fact, in the MAS NMR spectra of the glass, various spinning sidebands can always be observed due to hyperfine transitions, although the spinning sidebands may not be very obvious when the experimental conditions are chosen appropriately. Due to the low content of Al_2O_3 in TLAS samples (7.5 ~ 12 mol%), the multiple quantum magic angle spinning (MQMAS) signal of ^{27}Al is very weak. Especially, the addition of paramagnetic Tm^{3+} ions in the samples would result in signal broadening or even loss, leading to poor signal-to-noise ratio of MQMAS. The two-dimensional ^{27}Al MQMAS spectrum of the YLAS1 sample was tested as an example due to the largest signal-to-noise ratio. Figure 7.12b shows the ^{27}Al MQMAS spectrum of the YLAS1 sample. AlO_4, AlO_5, and AlO_6 were clearly identified in the F1 dimension with chemical shifts at 65.5, 35.8, and 3.2 ppm, respectively, and second-order quadrupole coupling constants of 5.8, 4.4, and 3.1 MHz, accordingly. These chemical shifts and second-order quadrupole effect parameters are consistent with the those in SiO_2-Al_2O_3 glass [29]. However, the second-order quadrupole effect parameters are very different from those in high-aluminum-doped RE_2O_3-Al_2O_3-SiO_2 (RE = Y, Lu) glass, indicating a distinct local structure of Al^{3+} ions in TLAS glasses [30, 31]. Studies have shown that both the intensity and width of the spinning sidebands in the ^{27}Al MAS NMR spectrum increase with the increase in the number of paramagnetic Tm^{3+} around Al^{3+} [28, 32]. The opposite trends in the spinning sidebands of ^{27}Al MAS NMR and ^{29}Si MAS NMR spectra indicate that as the La^{3+}/Al^{3+} ratio increases, some paramagnetic Tm^{3+} around Al^{3+} has moved to the vicinity of $[SiO_4]^{-4}$.

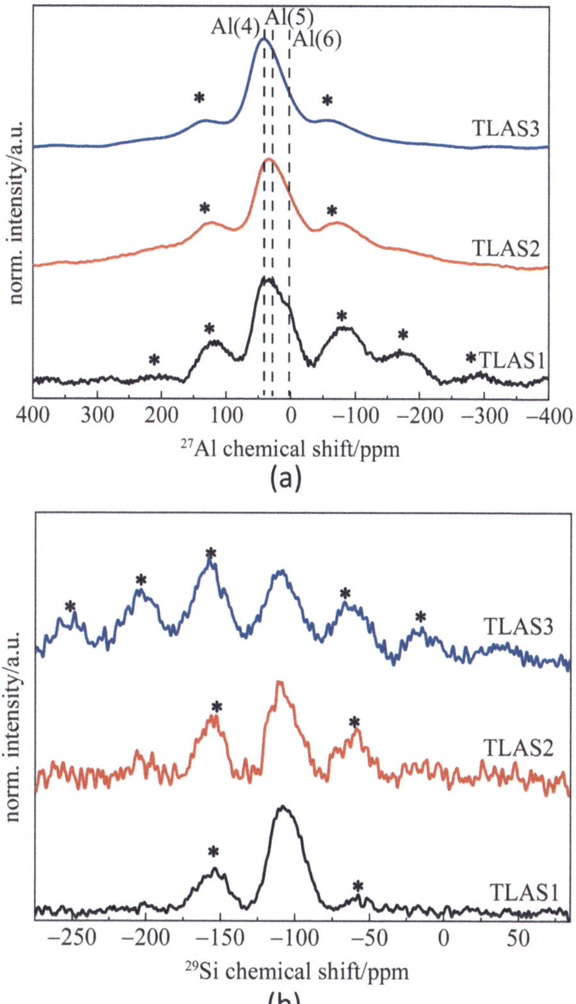

Fig. 7.11 ^{27}Al MAS NMR spectra (**a**) and ^{29}Si MAS NMR spectra (**b**) of thulium-doped silica glass with different La^{3+}/Al^{3+} ratios (* represents spinning sidebands)

The effects of different La^{3+}/Al^{3+} ratios on the structure of high concentration thulium-doped silica glass were studied by NMR, FTIR, and Raman spectroscopy. As the La^{3+}/Al^{3+} ratio increases in the TLAS series samples, the SiO_4 tetrahedron (Q_2 units) groups increases, AlO_4 groups increase, AlO_5 and AlO_6 groups decrease, and more La^{3+} and Tm^{3+} are connected to the Si—O— groups, forming LaO_6 or TmO_6 polymers. In TLAS1 glass, the second coordination of Tm^{3+} is Al^{3+}, while in TLAS3 glass, the second coordination of Tm^{3+} is Al^{3+} or Si^{4+}. Therefore, compared to TLAS1, the spatial distribution of Tm^{3+} in TLAS3 is more uniform.

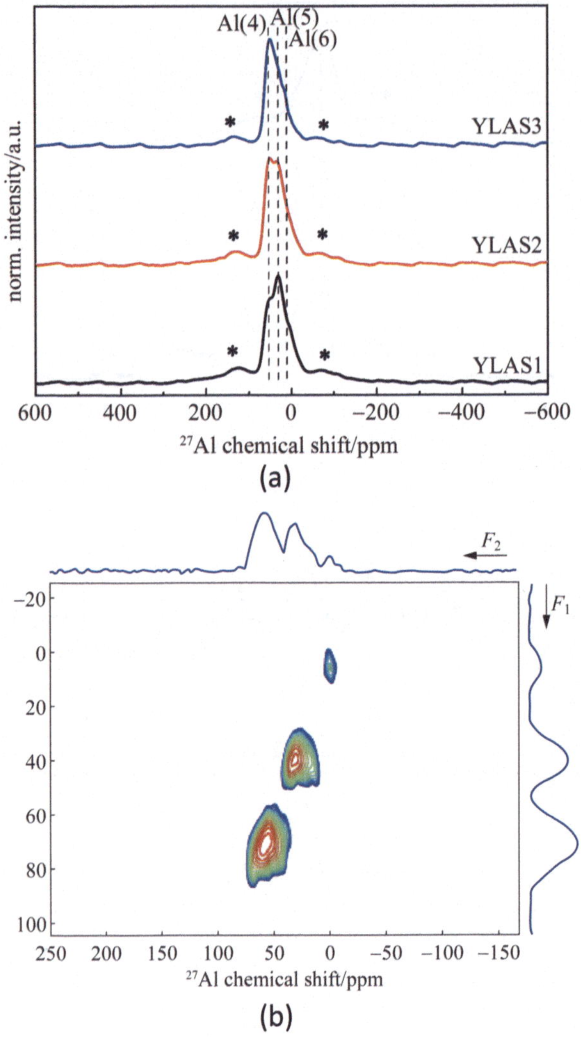

Fig. 7.12 ^{27}Al MAS NMR spectra (**a**) and two-dimensional ^{27}Al MQMAS spectra (**b**) of YLAS series samples and YLAS1 sample

In order to study the effect of the La^{3+}/Al^{3+} ratio on the 2 μm band emission spectrum of Tm^{3+}, the fluorescence spectra of 1300 ~ 2200 nm were tested under the pump of an 808 nm laser diode, and the results are shown in Figs. 7.13. Two emission peaks can be seen from the figure: a weaker peak centered at 1460 nm, originated from the $^3H_4 \rightarrow {}^3F_4$ transition as shown in Fig. 7.13b; and a stronger peak centered at 1800 nm, originated from the $^3F_4 \rightarrow {}^3H_6$ transition. The emission peak intensity of TLAS series samples at 1.8 μm decreases with the increase of the La^{3+}/Al^{3+} ratio. Tm^{3+} ions in the $^3H_4$ level have two potential transitions: one is that Tm^{3+}

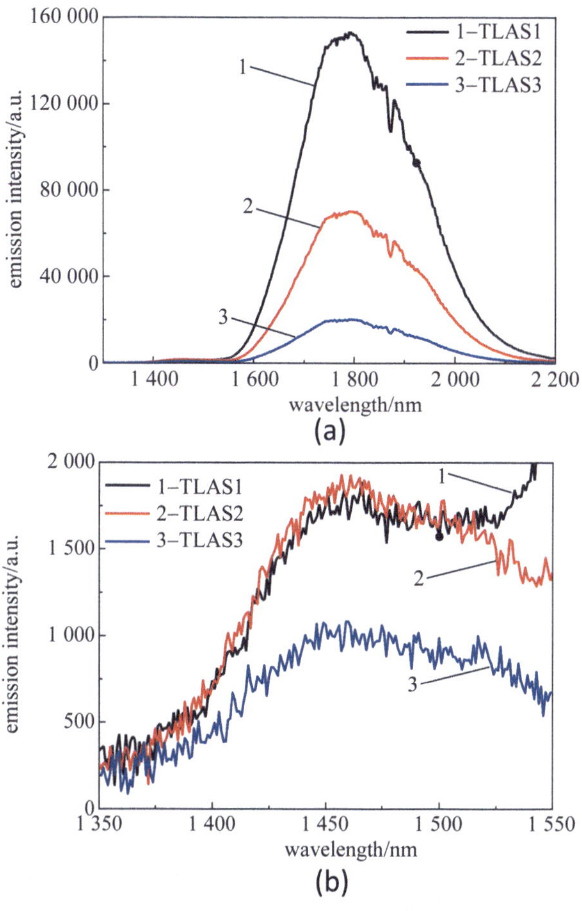

Fig. 7.13 Fluorescence spectra of TLAS glass in 1300 ~ 2200 nm band (**a**) and the magnified fluorescence spectrum in 1350 ~ 1550 nm (**b**)

on the $^3H_4$ energy level transitions to the $^3F_4$ energy level through spontaneous radiation, producing 1.46 μm fluorescence; the other is that Tm^{3+} on the $^3H_4$ energy level transitions to the $^3F_4$ energy level through cross-relaxation process, transferring the released energy to the neighboring Tm^{3+} on the ground state, producing 1.8 μm fluorescence. Therefore, the ratio of 1.8 μm to 1.46 μm fluorescence intensity can reflect the efficiency of cross-relaxation [33]. Figure 7.13b shows the magnified fluorescence spectrum from1350 ~ 1550 nm. Figure 7.14 shows a decrease in fluorescence intensity ratio of 1.8 μm to 1.46 μm with the increase in La^{3+}/Al^{3+} ratio. This indicates a decrease in the probability of cross relaxation with the increase of the La^{3+}/Al^{3+} ratio in TLAS series samples. The emission spectrum of Tm^{3+} in Fig. 7.13 further confirms that with the addition of La_2O_3 into silica glass, the dispersion of Tm^{3+} in the glass improves, and the probability of cross-relaxation between Tm^{3+}

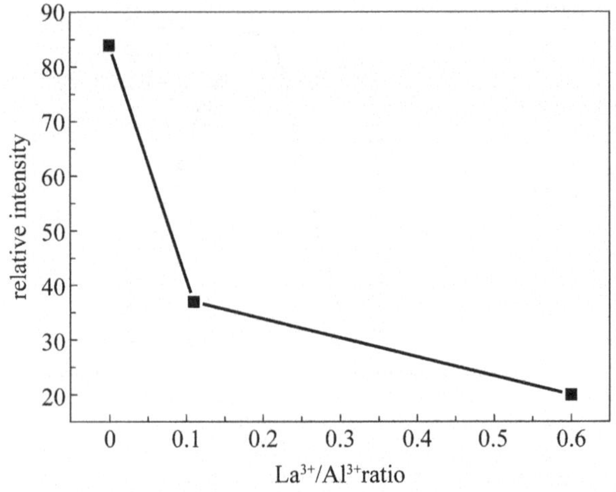

Fig. 7.14 Fluorescence intensity ratio of 1.8 μm to 1.46 μm versus the La^{3+}/Al^{3+} ratio

decreases. Therefore, the 1.8 μm luminescence intensity decreases as the La^{3+}/Al^{3+} ratio increases.

The stimulated emission cross-section of TLAS series samples can be calculated using the F-L formula, and the effective bandwidth ($\Delta\lambda_{eff}$) can be calculated using the following formula:

$$\Delta\lambda_{eff} = \int \frac{I(\lambda)d\lambda}{I_{max}} \tag{7.2}$$

where λ is the wavelength, $I(\lambda)$ is the fluorescence spectrum intensity and I_{max} is the maximum intensity of the fluorescence spectrum.

The stimulated emission cross-section (σ_{ems}) and effective bandwidth ($\Delta\lambda_{eff}$) of TLAS series samples and other matrix glasses doped with Tm^{3+} are listed in Table 7.4 [15, 34]. The data in the table shows that the stimulated emission cross section and effective bandwidth decrease with increase in the La^{3+}/Al^{3+} ratio. From the F-L formula and eq. 7.2, it can be seen that the stimulated emission cross-section is inversely proportional to the effective bandwidth. Therefore, the fluorescence intensity of the TLAS series samples is consistent with the trend of the stimulated emission cross section. The data in Table 7.4 also shows that the stimulated emission cross section of the TLAS series samples at 1.8 μm is greater than that of tellurite glass [15] and fluoride glass [34], but smaller than that of silicate, gallate, and aluminate glasses [34], and its fluorescence effective bandwidth is greater than the other matrix glasses listed in the table.

Table 7.4 Effective bandwidth and stimulated emission cross section of Tm^{3+} at 1.8 μm in different matrix glasses

Type of Glass	TLAS1	TLAS2	TLAS3	Fluoride glas	Silicateglass	Aluminate glass	Gallateglass	Tellurite glass
$\Delta\lambda_{eff}$/nm	285.29	287.34	293.27	240.00	159.00	280.00	278.00	–
$\sigma_{ems}/(\times 10^{-21}\ cm^2)$	6.00	5.78	5.14	3.20	6.10	5.90	6.30	4.00

7.2.3 The Influence of La^{3+} and Y^{3+} on the Structure and Spectral Properties of High Concentration Thulium-Doped Silica Glass

Both Y^{3+} and La^{3+} are optically inert rare earth ions, but their ionic radii are different, leading to different influences on the structure of silica glass. Figure 7.15 shows the FTIR spectra of silica glass co-doped with $Y^{3+}/La^{3+}/Tm^{3+}$. TAS0, TLAS2, and TYAS2 represent $0.8Tm_2O_3$ - $(12\text{-}x)Al_2O_3$ - xRe_2O_3–$87.2SiO_2$ (Re = La, Y; $x = 0$, 4.5 mol%). It can be seen that the relative intensity of the Si—O— group at 950 cm $^{-1}$ increases with the increase in La_2O_3 or Y_2O_3 concentration, indicating an increase in the amount of non-bridging oxygen Si—O—LE (LE = Tm, La or Y) [23]. Since the crystallization of TAS0 glass is minimal, the FTIR spectrum of the TAS0 sample mainly shows information about the glass phase, displaying the most bridging oxygen in the glass network: as La_2O_3 or Y_2O_3 is added to the silica glass, the amount of non-bridging oxygen increases. In the TLAS and TYAS series samples, La^{3+} or Y^{3+} acts as a network modifier occupying the octahedral position, therefore, the addition of La_2O_3 or Y_2O_3 to the silica glass balances the negative charge of AlO_4 groups and improving the anti-crystallization performance of Tm^{3+} doped silica glass. As can be seen from Fig. 7.15, the intensity of the Si—O— group in the TLAS2 sample is greater than that in the TYAS2 sample. Therefore, both La^{3+} or Y^{3+} can depolymerize the glass Si—O—Si bond, with La^{3+} ions showing a stronger ability to break these bonds compared to Y^{3+} ions when the content of La_2O_3 and Y_2O_3 is the same.

Figure 7.16 shows the ^{27}Al MAS NMR spectra of silica glass co-doped with Y^{3+}, La^{3+} and Tm^{3+}. It can be seen that in the silica glass doped with Y^{3+}, La^{3+} and Tm^{3+}, the Al element mainly exists in the form of AlO_4, and the amount of AlO_5 and AlO_6 decreases with the increase of La_2O_3 or Y_2O_3 content. Compared with La_2O_3 doped glass, the relative intensity of AlO_5 and AlO_6 in Y_2O_3 doped glass is larger,

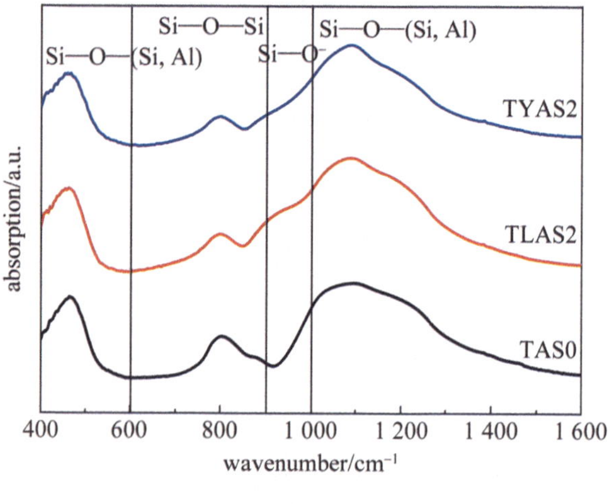

Fig. 7.15 FTIR spectra of silica glass co-doped with Y^{3+}, La^{3+} and Tm^{3+}

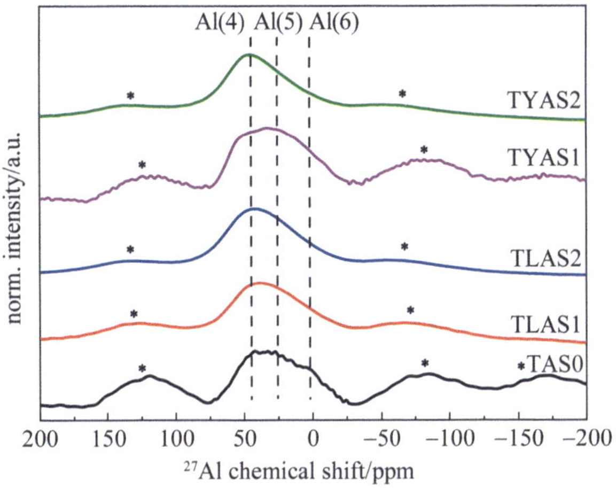

Fig. 7.16 ^{27}Al MAS NMR spectra of silica glass co-doped with Y^{3+}, La^{3+} and Tm^{3+} (* represents rotating sidebands)

indicating that the Al—O groups in La_2O_3 doped glass tends to form low-coordinated AlO_4, that is, La_2O_3 has a stronger charge compensation ability for AlO_4 polyhedron than Y_2O_3. These results are consistent with previous research results [35]. This is because La, Y and Al are both network modifiers, and the ability to form a glass network is in the order of Al > Y > La, while the ability to become glass modifiers is exactly the opposite. In summary, ^{27}Al NMR spectra show that in silica glass doped with La_2O_3, Al atoms tend to act as network formers. The decrease in the intensity and width of the spinning sidebands in the ^{27}Al MAS NMR spectra indicates that the number of Tm^{3+} ions around Al^{3+} ions decreases with increasing La_2O_3 or Y_2O_3 content. The spinning sideband intensity of ^{27}Al NMR spectrum in Y_2O_3 doped glass is greater than that in La_2O_3 doped glass, indicating that when the content of La_2O_3 and Y_2O_3 is the same, the number of Tm^{3+} ions around Al^{3+} ions in Y_2O_3 doped glass is greater than that in La_2O_3 doped glass.

Through the Raman, FTIR and NMR spectroscopic analysis of TLAS and TYAS series samples, La^{3+} and Y^{3+} ions play an important role in adjusting the structure of silica glass. When La_2O_3 or Y_2O_3 is added to silica glass, La^{3+} and Y^{3+} ions can compensate for the negative charge of the network former AlO_4. The second coordination of Tm^{3+} in TAS0 glass is mostly Al^{3+}, so the clusters of Tm^{3+} in silica glass are mainly Tm-O-Tm type connections. When La_2O_3 or Y_2O_3 is added to silica glass, some Tm^{3+} is connected with La^{3+} or Y^{3+}, forming La/Y-O-Tm type connections. Therefore, when La_2O_3 or Y_2O_3 is added to silica glass, the distribution of Tm^{3+} ions in the glass is more uniform, and the dispersing ability of La_2O_3 on Tm^{3+} ions is stronger than that of Y_2O_3.

To study the influence of La^{3+}/Al^{3+} ratio and Y^{3+}/Al^{3+} ratio on the emission spectrum of Tm^{3+} at 2 μm, the fluorescence spectra from 1500 to 2200 nm under the pump of 808 nm LD were tested, as shown in Fig. 7.17. It can be seen that the fluorescence intensity decreases with the increase in La_2O_3 or Y_2O_3 content.

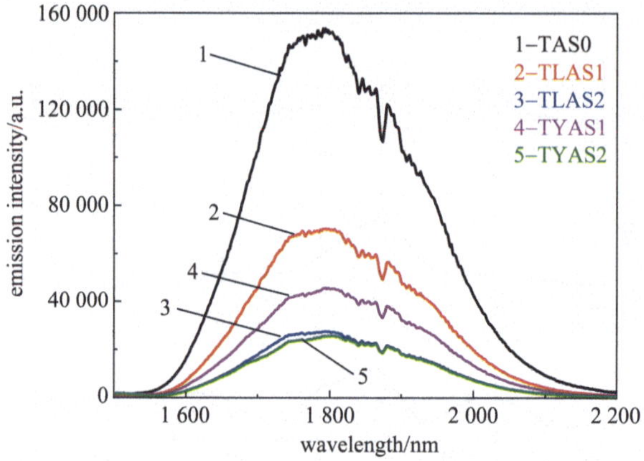

Fig. 7.17 Fluorescence spectra of TLAS and TYAS series glasses from 1500 to 2200 nm

By studying the structure of TLAS and TYAS series samples, it is known that when La_2O_3 or Y_2O_3 is added to silica glass, the dispersion of Tm^{3+} in the glass is improved and the probability of cross relaxation effect between Tm^{3+} decreases. Therefore, the fluorescence intensity at 2 μm decreases with increase of La_2O_3 or Y_2O_3 content. Because La_2O_3 has a stronger ability to disperse Tm^{3+} than Y_2O_3, when the content of La_2O_3 and Y_2O_3 are the same, the probability of cross relaxation process of Tm^{3+} in Y_2O_3 doped glass is greater than that in La_2O_3 doped glass. However, Fig. 7.17 shows that the fluorescence intensity of La_2O_3-doped glass is higher than that of Y_2O_3-doped glass when the La_2O_3 and Y_2O_3 contents are the same. The aforementioned results show that when the content of La_2O_3 and Y_2O_3 are the same, the symmetry of Tm^{3+} in Y_2O_3 doped glass is higher than that in La_2O_3 doped glass. Therefore, the lower symmetry around Tm^{3+} ions in La_2O_3 doped glass is likely the main reason for achieving higher light intensity.

7.3 Spectral Properties and Structure of Holmium-Doped Silica Glass

7.3.1 Structure and Spectral Properties of Silica Glass Doped with Different Al^{3+}/Ho^{3+} Ratios

Silica glass forms a dense three-dimensional network structure from SiO_4 tetrahedra, which limits the doping concentration of rare earth ions. Especially, doping rare earth ions singly into quartz glass can easily cause clustering of rare earth ions, leading to the degradation of spectral properties. Therefore, it is necessary to improve the clustering problem of rare earth ions in silica glass by co-doping with other ions. Studies have shown that doping an appropriate amount of Al^{3+} into the silica matrix can effectively dissolve rare earth ions in silica glass. Although the proper addition of Al^{3+} ions in silica glass can effectively increase the solubility of Ho^{3+} ions, thus suppressing clustering effects and enhancing emission intensity, excessive Al^{3+} ions may lead to a decrease in the emission intensity of Ho^{3+} ions [36, 37]. In order to compare the effects of different Al^{3+} concentration on the structure and spectral properties of Ho^{3+} doped silica glass, Wang et al. [38]. prepared a series of quartz glass samples with different Al^{3+}/Ho^{3+} ratios using the sol-gel method combined with high-temperature powder sintering method, named as HA series. The content of Ho_2O_3 was fixed at 0.2 mol%, and the content of Al_2O_3 continuously increases. The composition of the glass samples is shown in Table 7.5. As can be seen from the table, the HA0 sample without Al_2O_3 doping exhibits severe crystallization, while the rest of the samples are all in a transparent glass state. This indicates that Al_2O_3 in the silica glass plays a role in dispersing Ho^{3+} ions and preventing crystallization, so the crystallized HA0 sample is no longer studied.

To study the form of Al^{3+} in silica glass, the ^{27}Al solid-state NMR spectra of samples HA1, HA3, and HA5 were tested, and the results are shown in Fig. 7.18. According to the isotropic chemical shift values of ^{27}Al with different coordination,

Table 7.5 Composition of Ho^{3+}/Al^{3+} doped silica glass

Sample	Ho_2O_3/mol%	Al_2O_3/mol%	SiO_2/mol%	Al/Ho	Phase and Morphology
HA0	0.20	0	99.8	0	Cristobalite opaque
HA1	0.20	1.0	98.8	5	Glass, transparent
HA2	0.20	2.0	97.8	10	Glass, transparent
HA3	0.20	3.0	96.8	15	Glass, transparent
HA4	0.20	4.0	95.8	20	Glass, transparent
HA5	0.20	5.0	94.8	25	Glass, transparent

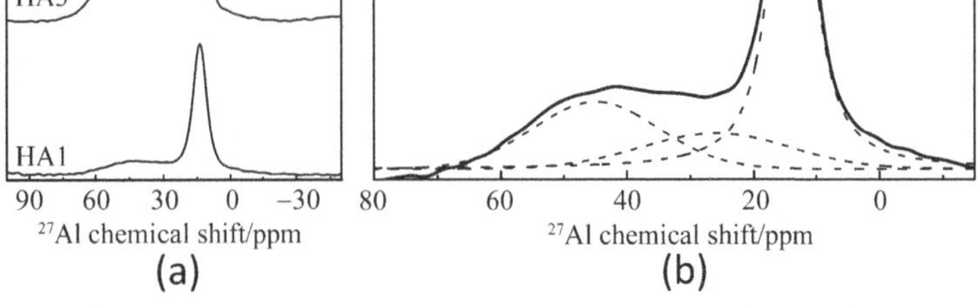

Fig. 7.18 ^{27}Al MAS NMR spectrum of HA series glass (**a**) and ^{27}Al MAS NMR spectrum deconvoluted with the Czjzek model of sample HA5 (**b**)

Table 7.6 Different coordinated Al contents and ^{27}Al isotropic chemical shift values in silica glass with different Al^{3+}/Ho^{3+} ratios

Sample	Al_4/%	Al_5/%	Al_6/%	Al_4/ppm	Al_5/ppm	Al_6/ppm
HA1	11.8	12.7	75.5	45.9	27.1	13.9
HA3	40.4	25.8	33.8	45.7	26.9	13.8
HA5	14.9	23.7	61.4	45.1	26.8	13.9

the approximate content of different coordinating Al in the glass was obtained by deconvolution of the ^{27}Al MAS NMR spectrum in Fig. 7.18a, and the data are listed in Table 7.6. As an example of the HA series samples, the deconvolution of the ^{27}Al MAS NMR spectrum of HA5 is shown in Fig. 7.18b. From Fig. 7.18 and Table 7.6, it can be seen that Al^{3+} mainly exists in the HA series glass as AlO_6 and AlO_5, indicating that aluminum ions in the HA series glass mainly act as network modifiers. From the data in the table, it can be seen that as the Al^{3+}/Ho^{3+} ratio increases, the amount of AlO_4 first increases and then decreases, while the amount of AlO_6 first decreases and then increases. When the Al^{3+}/Ho^{3+} ratio is 15, the content of AlO_4 in the HA3 sample is the highest. The trend of different coordinating Al contents in lanthanum and yttrium doped aluminosilicate glass with the change of Al^{3+}/Ho^{3+} ratio is consistent with that in the HA series glass [39].

According to the Judd-Ofelt theory, using the tested absorption spectroscopy data and the reduced matrix element data of Ho^{3+} ions reported in the literature [40], the J-O intensity parameters Ω_t ($t = 2, 4, 6$) are fitted by the least squares method and listed in Table 7.7. Ω_2 is very sensitive to the composition and structure of glass, reflecting the covalency of the bond between rare earth ions and oxygen, as well as the symmetry around the rare earth ions, with a larger Ω_2 indicating lower symmetry around the rare earth ions. As can be seen from the table, the Ω_2 of the HA series samples increases first and then decreases with the increase of the Al^{3+}/Ho^{3+} ratio. This indicates that the asymmetry of Ho^{3+} in silica glass increases first and then decreases with the increase of the Al^{3+}/Ho^{3+} ratio. The sample with the highest asymmetry of Ho^{3+} in silica glass is HA3 with an Al^{3+}/Ho^{3+} ratio is 15.

To study the effect of different Al^{3+}/Ho^{3+} ratios on the 2 μm spectra, the fluorescence spectra of HA series glass samples under 640 nm LD pumping in the 1800 ~ 2200 nm band were tested, and the results are shown in Fig. 7.19. It can be seen that the wavelength of peak fluorescence of all samples does not change significantly, but the fluorescence intensity first increases and then decreases with the increase of the Al^{3+}/Ho^{3+} ratio, and when the Al^{3+}/Ho^{3+} ratio is 15, the fluorescence intensity reaches its maximum. Florian et al. [39] have shown that rare earth ions tend to be around AlO_4, indicating that the higher the AlO_4 content, the better the dispersibility of rare earth ions. Therefore, AlO_4 can increase the solubility of Ho^{3+} and prevent clustering, thus reducing fluorescence quenching and increasing fluorescence efficiency. Analysis of the ^{27}Al MAS NMR spectra of the HA series samples shows that the amount of AlO_4 first increases and then decreases with the increase of the Al^{3+}/Ho^{3+} ratio, and the content of AlO_4 is the highest in the HA3 sample with an Al^{3+}/Ho^{3+} ratio of 15, indicating good dispersibility of Ho^{3+} ions.

Table 7.7 Ω_λ ($\lambda = 2, 4, 6$) value of Ho^{3+} doped silica glass

Sample	$\Omega_2/(\times 10^{-20}\ cm^2)$	$\Omega_4/(\times 10^{-20}\ cm^2)$	$\Omega_6/(\times 10^{-20}\ cm^2)$
HA1	11.70	2.59	1.19
HA2	12.13	2.64	1.17
HA3	13.20	2.69	1.18
HA4	9.92	2.64	1.09
HA5	7.44	1.49	0.89

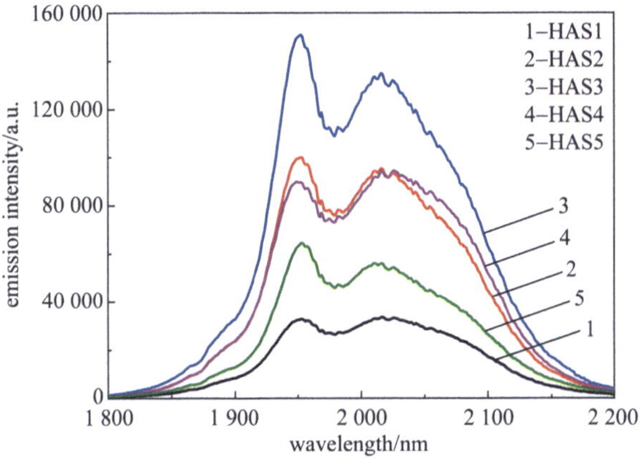

Fig. 7.19 Fluorescence spectra of HA series samples under 640 nm excitation in the 1800 ~ 2200 nm band

This is consistent with the overall change in fluorescence intensity of the HA series samples.

The quantum efficiency [41] of the Ho^{3+} ion level $^5I_7$ can be calculated by the following formula:

$$\eta = \frac{\tau_m}{\tau_c} \times 100\% \tag{7.3}$$

In the formula, τ_c is the radiative lifetime calculated by J-O theory; τ_m is the lifetime of the $^5I_7$ level obtained from the test. Table 7.8 lists the radiative lifetime (τ_c), test fluorescence lifetime (τ_m) and calculated quantum efficiency (η) of Ho^{3+} ions in the $^5I_7$ energy level. As can be seen from the table, the quantum efficiency of Ho^{3+} ions in the $^5I_7$ energy level increases first and then decreases with the increase of the Al^{3+}/Ho^{3+} ratio, and the quantum efficiency is the highest when the Al^{3+}/Ho^{3+} ratio is 15. The calculated stimulated emission cross-section and quality factor of the HA series samples are also listed in Table 7.8. It can be seen that the quality factor of HA3 glass is the largest, indicating that the sample can obtain an increased gain bandwidth when the Al^{3+}/Ho^{3+} ratio is 15. In summary, the results of the analysis of the structure and spectrum of the HA series samples show that when the doping

Table 7.8 radiation lifetime, fluorescence lifetime, quantum efficiency, emission cross-section, and quality factor of Ho^{3+}/Al^{3+} doped silica glass at 2 μm

Sample	τ_c/ms	τ_m/ms	η/%	$\sigma_{em}/(\times 10^{-21}\ cm^2)$	$\sigma_{em} \times \tau_m/(\times 10^{-21}\ cm^2 \cdot ms)$
HA1	13.60	0.91	6.69	3.57	3.25
HA2	13.87	1.04	7.50	3.63	3.77
HA3	13.63	1.19	8.73	3.64	4.33
HA4	14.62	1.23	8.41	3.44	4.23
HA5	17.05	1.33	7.80	2.91	3.87

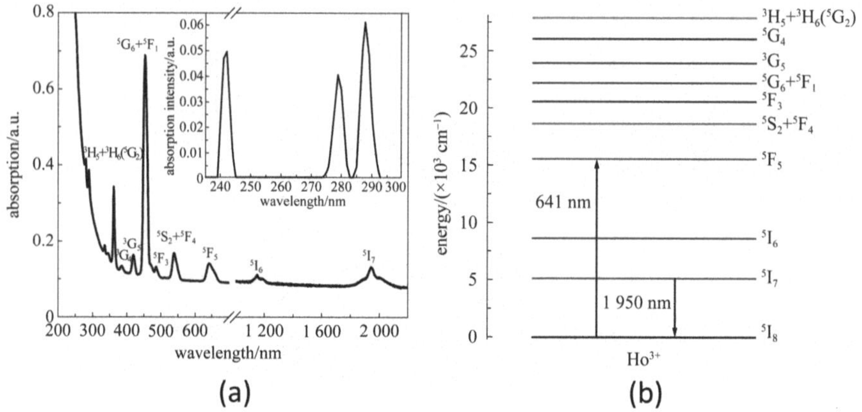

Fig. 7.20 Absorption spectrum (**a**) and Ho^{3+} energy level structure obtained through the absorption spectrum (**b**) of Ho^{3+}/Al^{3+} co-doped silica glass

concentration of Ho_2O_3 is 0.2 mol%, the optimal spectral performance of Ho^{3+}/Al^{3+} co-doped silica glass at 2 μm is when the Al^{3+}/Ho^{3+} ratio is 15.

Al^{3+}/Ho^{3+} ratio was fixed at 15 to investigate the luminescent properties of 2 μm emission in quartz glass with different Ho^{3+} ion concentrations. The compositions of the 5 prepared samples are x Ho_2O_3–15 x Al_2O_3 -(100-16x)SiO_2 (mol%, x = 0.14, 0.2, 0.3, 0.4, 0.5), denoted as HAS1, HAS2, HAS3, HAS4, and HAS5, respectively.

Figure 7.20a shows the absorption spectrum of Ho^{3+}/Al^{3+} co-doped silica glass in the wavelength range of 200 ~ 2200 nm, and the inset shows the absorption spectrum of the HAS3 sample in the ultraviolet region. The absorption energy levels corresponding to each transition of Ho^{3+} are labeled in the figure. Compared with other matrix glasses, silica glass has high ultraviolet transmittance, so the absorption peak of Ho^{3+} ions transitioning from the ground state to excited states higher than $^5G_2$ can be observed in Ho^{3+} doped quartz glass.

Figure 7.20b shows the Ho^{3+} energy level structure derived from the absorption spectrum. It should be noted that the transitions from $^5I_5$ and $^5I_4$ levels to the ground state produce very weak luminescence, hence these levels are not shown in the figure.

Using J-O theory and parameters such as absorption spectra, the values of Ω_2, Ω_4, and Ω_6 for HAS3 glass are 8.83×10^{-20} cm², 2.45×10^{-20} cm², 1.06×10^{-20} cm², respectively. Table 7.9 lists the J-O parameters of Ho^{3+}/Al^{3+} co-doped silica glass

Table 7.9 J-O intensity parameters of different holmium-doped glasses

Sample	$\Omega_2/(\times10^{-20}\ cm^2)$	$\Omega_4/(\times10^{-20}\ cm^2)$	$\Omega_6/(\times10^{-20}\ cm^2)$
Silicate Glass [34]	3.60	2.30	0.65
Germanate Glass [34]	3.30	1.14	0.17
Zn-Al-Bi-B [42]	16.4	4.06	0.81
Fluorophosphate Glass [43]	2.08	3.11	1.50
Tellurite Glass [44]	4.20	2.80	1.10
Phosphate Glass [45]	8.58	4.31	2.88
HAS3 [38]	8.83	2.45	1.06

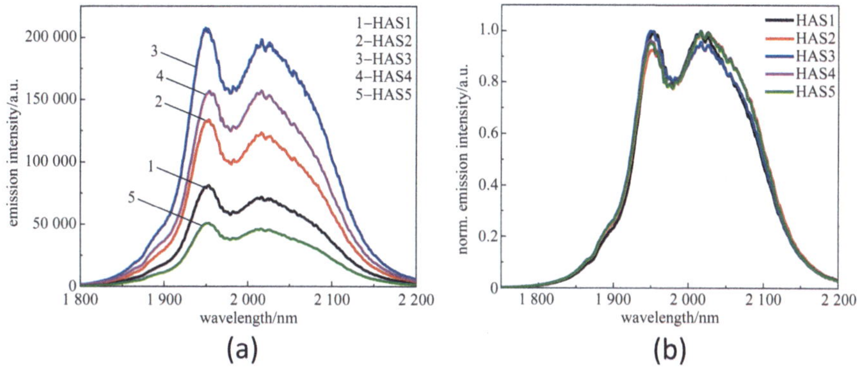

Fig. 7.21 Fluorescence spectrum (**a**) and normalized fluorescence spectrum (b) at 2 μm of Ho^{3+}/Al^{3+} co-doped silica glass

HAS3 and other matrix glass data [32, 42–45]. Generally, Ω_2 value is related to the orderliness of the material structure and the symmetry of the coordination field, and it is more sensitive to changes in material composition. Ω_2 value depends on the coordination field around the rare earth ions. A larger Ω_2 value indicates lower symmetry of the coordination field of rare earth ions and stronger covalent bonding in the material. As shown in Table 7.9, the Ω_2 value of Ho^{3+}/Al^{3+} co-doped silica glass is greater than that of silicate glass [34], germanate glass [34], fluorophosphate glass [43], tellurate glass [44] and phosphate glass [45]. This indicates that the environment of Ho^{3+} in Ho^{3+}/Al^{3+} co-doped silica glass has stronger covalency bonding and lower symmetry.

The fluorescence spectra of the HAS series glasses in the range of 1800 ~ 2200 nm were tested using a 640 nm LD, as shown in Fig. 7.21a, while Fig. 7.21b is the normalized fluorescence spectrum. As can be seen from Fig. 7.21a, the fluorescence intensity first increases and then decreases with the increase in Ho^{3+} concentration. When the Ho_2O_3 concentration is 0.3 mol%, the fluorescence intensity reaches its maximum. In other words, when the concentration of Ho_2O_3 exceeds 0.3 mol%, the concentration quenching becomes very obvious. This is mainly because the cross-relaxation effect between Ho^{3+} ions is enhanced when Ho^{3+} ions are heavily doped. As can be seen from Figure 7.21b, the position and half-width of the fluorescence peak do not change with the concentration of Ho^{3+}. The half-width of the fluorescence peak of the $^5I_7 \rightarrow {}^5I_8$ transition is about 175 nm, which is significantly larger than the reported fluorescence half widths of the $^5I_7 \rightarrow {}^5I_8$ transition of Ho^{3+} ions in germanate, silicate, and phosphate glasses [34].

Figure 7.22 shows the absorption and stimulated emission cross-section of the HAS3 silica glass at the 2 μm band. The calculated maximum absorption and stimulated emission cross-sections are 4.65×10^{-20} cm^2 at 1668 nm and 4.12×10^{-20} cm^2 at 2026 nm, respectively.

The gain bandwidth of the amplifier is determined by the bandwidth of the fluorescence spectrum and the stimulated emission cross-section. The gain bandwidth data of the sample HAS3 and other matrices are listed in Table 7.10. The gain bandwidth of Ho^{3+} doped silica glass is greater than that of germanate [34], silicate [34], and phosphate glass [34], indicating that Ho^{3+} doped silica glass can provide a larger gain bandwidth.

The absorption and stimulated emission cross sections calculated above can be used to evaluate the gain coefficient based on formula 7.1, thereby assessing the gain performance of the material. Figure 7.23a shows the relationship curve between

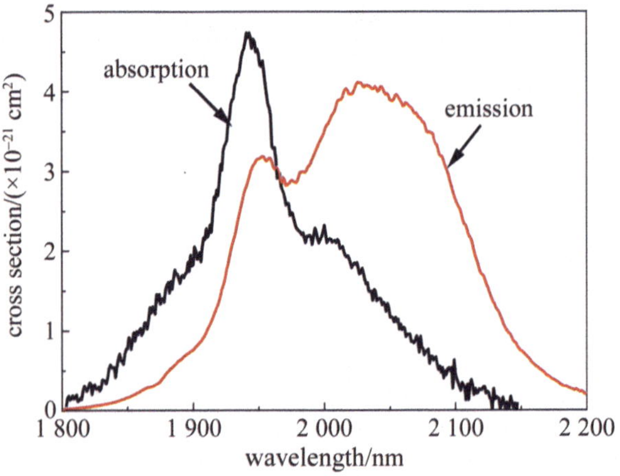

Fig. 7.22 Absorption and stimulated emission cross-section of Ho^{3+}:$^5I_7 \rightarrow {}^5I_8$ Ho^{3+}/Al^{3+} co-doped HAS3 silica glass

Table 7.10 Emission cross-section, fluorescence half-width, and gain bandwidth of different holmium-doped glasses

Sample	$\sigma_{ems}/(\times 10^{-21}$ cm$^2)$	$\Delta\lambda/(\times 10^{-21}$ cm$^2)$	$\sigma_{ems} \times \Delta\lambda/(\times 10^{-21}$ cm^2.nm)
Germanate glass [34]	4.0	84	336
Silicate glass [34]	7.0	82	574
Phosphate glass [34]	6.2	78	484
Fluorophosphate glass [46]	5.5	150	825
HAS3 [38]	4.12	175	721

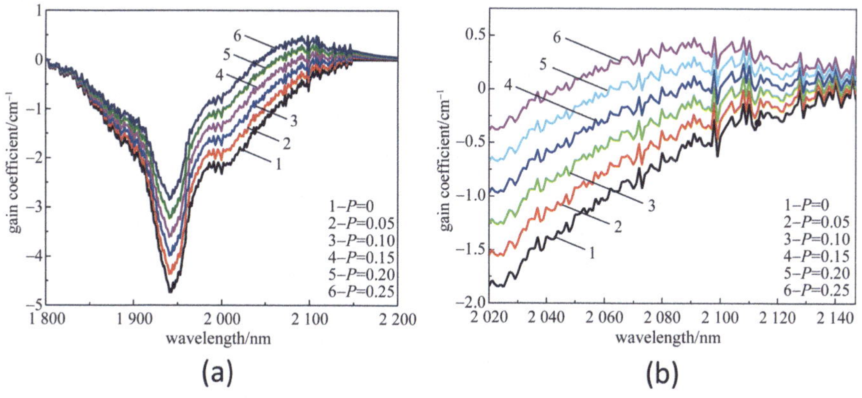

Fig. 7.23 Gain coefficient of holmium-doped silica glass (**a**) and enlarged gain coefficient diagram at 2.1 μm (**b**)

Table 7.11 $^5I_7$ level radiation lifetime, fluorescence lifetime, and quantum efficiency of Ho^{3+}/Al^{3+} doped silica glass

Sample	HAS1	HAS2	HAS3	HAS4	HAS5
τ_c/ms	13.12	13.62	13.89	14.44	15.67
τ_m/ms	1.231	1.189	1.124	1.086	1.022
η/%	9.38	8.73	8.09	7.52	6.52

the gain coefficient and wavelength of HAS3 glass, with the pump inversion ratio ranging from 0 to 0.25. Figure 7.23b is an enlarged view of the gain coefficient in the 2020 ~ 2140 nm band. It can be seen that when the number of particles in the upper energy level of the laser exceeds 15%, a gain of 2.1 μm can be achieved.

The quantum efficiency of the $^5I_7$ energy level can be calculated using the radiative lifetime obtained from the J-O theory calculation and the measured fluorescence lifetime of Ho^{3+} at the $^5I_7$ energy level. Table 7.11 lists the radiative lifetimes, measured fluorescence lifetimes, and quantum efficiency of the $^5I_7$ energy level of Ho^{3+} for all HAS series samples. The measured fluorescence lifetime and quantum efficiency of the HAS series samples are 1.022 ~ 1.231 ms and 6.52% ~ 9.38%, respectively. Compared with other silica glasses reported in the literature [47, 48], the HAS series glass has a longer fluorescence lifetime at the $^5I_7$ energy level. The maximum phonon energy of silica glass is 1050 ~ 1100 cm^{-1}, and the higher phonon energy leads to large non-radiative transitions between the excited energy levels of rare earth ions. Large non-radiative transitions and impurities (such as hydroxyl and other transition metal impurities present in the raw materials) result in a low quantum efficiency of the $^5I_7 \rightarrow {}^5I_8$ transition of Ho^{3+}.

7.3.2 Structure and Spectral Properties of High Concentration Ho^{3+} and Lanthanum/Yttrium Aluminum Co-Doped Silica Glass

To study the influence of La_2O_3 and Y_2O_3 on the structure and spectral properties of high concentration Ho^{3+} doped silica glass, Wang et al. [49] prepared two series of Ho^{3+} doped silica glass samples using the sol-gel method combined with high-temperature powder sintering method. The first series was named HLAS, with a molar composition of $0.8Ho_2O_3$ -(12- x)Al_2O_3 - x La_2O_3–$87.2SiO_2$ (x = 0, 1.2 mol%, 4.5 mol%), referred to as HAS0, HLAS1, and HLAS2, respectively. The second series was named HYAS, with a molar composition of $0.8Ho_2O_3$ -(12- x)Al_2O_3 - x Y_2O_3–$87.2SiO_2$ (x = 0, 1.2 mol%, 4.5 mol%), referred to as HAS0, HYAS1, and HYAS2, respectively.

To study the influence of La^{3+} or Y^{3+} ions on the spectral properties of Ho^{3+} doped silica glass from the structural source, the FTIR spectra and NMR spectra of two glass series were tested. The test results of the FTIR spectra are shown in Fig. 7.24. The shoulder peak at 950 cm^{-1} is attributed to the asymmetric vibration of the terminal Si—O— group [23]. The relative intensity of the Si—O— group increases with the increase of La_2O_3 or Y_2O_3 concentration, indicating that La $^{3+}$ and Y $^{3+}$ can depolymerize the Si—O—Si bond in the glass. When the content of La_2O_3 and Y_2O_3 is the same, the depolymerization ability of La^{3+} on the glass network is greater than that of Y^{3+} ions. This is consistent with the conclusion drawn from the aforementioned TLAS and TYAS series samples.

Figure 7.25 shows the ^{27}Al MAS NMR spectra of the HLAS and HYAS series samples. It can be seen that Al in the HLAS and HYAS series samples mainly exists in the form of AlO_4, indicating that Al mainly acts as a network former in this glass, and AlO_5 and AlO_6 content decreases with the increase of La_2O_3 or Y_2O_3 content.

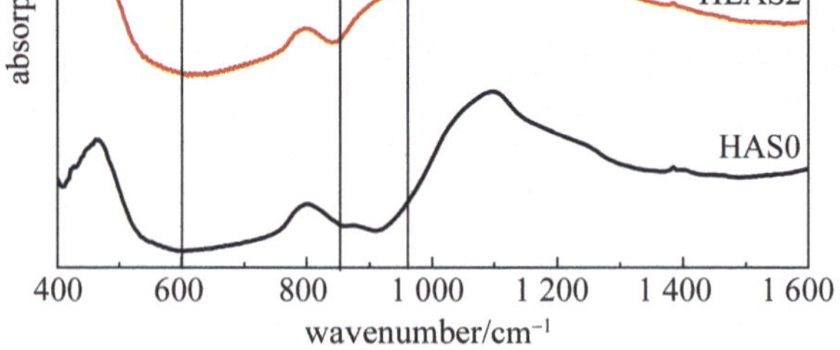

Fig. 7.24 FTIR spectra of silica glass co-doped with of Y^{3+}, La^{3+} and Ho^{3+}

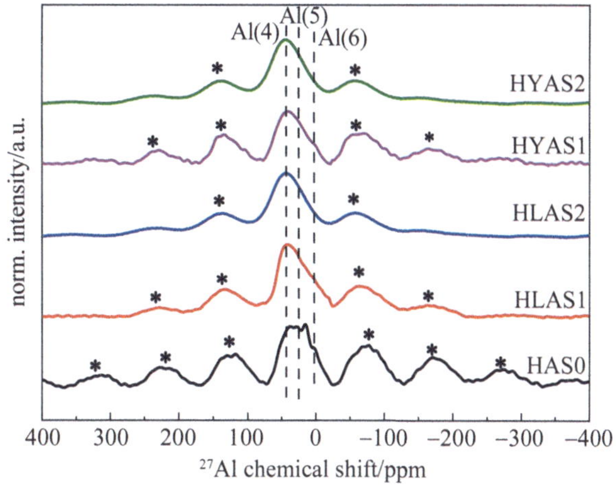

Fig. 7.25 ^{27}Al MAS NMR spectra of Y^{3+} and La^{3+} co-doped with Ho^{3+} in silica glass (* represents spinning sidebands)

Compared to La_2O_3 doped glass, Y_2O_3 doped glass tends to place Al in a higher coordination environment. These results are consistent with the research results in TLAS and TYAS series samples. In Fig. 7.25, the intensity and width of the spinning sidebands in ^{27}Al MAS NMR spectra of the HLAS and HYAS series samples decrease with the increase in La_2O_3 or Y_2O_3 content, indicating a decrease in the number of Ho^{3+} around Al^{3+}. The intensity of the spinning sidebands in the ^{27}Al NMR spectra of Y_2O_3-doped glass is higher than that of La_2O_3 doped glass, indicating that when the content of La_2O_3 and Y_2O_3 is the same, the number of Ho^{3+} ions around Al^{3+} ions in Y_2O_3-doped glass is greater than in La_2O_3-doped glass.

To study the effect of La^{3+}/Al^{3+} ratio and Y^{3+}/Al^{3+} ratio on the 2 μm band emission spectrum of Ho^{3+} ions, the fluorescence spectra in the range of 1800–2200 nm were tested under 640 nm LD pumping, as shown in Fig. 7.26. As can be seen from the figure, the fluorescence intensity increases with the increase in La_2O_3 or Y_2O_3 content, and when the content of La_2O_3 and Y_2O_3 is the same, the fluorescence intensity of La_2O_3 doped glass is greater than that of Y_2O_3 doped glass. Structural studies on the HLAS and HYAS series samples revealed that after the addition of La_2O_3 or Y_2O_3 to quartz glass, the dispersibility of Ho^{3+} ions in the glass increases. Furthermore, when the La_2O_3 and Y_2O_3 content are the same, the dispersibility of Ho^{3+} ions in La_2O_3 doped glass is superior to that in Y_2O_3 doped glass. In the previous section, it was known through the discussion of the spectral properties of Ho^{3+}/Al^{3+} co-doped silica glass that when the doping concentration of Ho_2O_3 is 0.3 mol%, concentration quenching will occur. Therefore, after adding La_2O_3 or Y_2O_3, the dispersion of Ho^{3+} is improved, its spectral properties are also improved, and when the dispersion is further improved, fluorescence quenching will be suppressed, the spectral properties will be further improved. So, when the content of La_2O_3 and Y_2O_3 is the same, the fluorescence intensity of La_2O_3 doped glass is higher than that of Y_2O_3 doped glass.

Using the F-L formula, the stimulated emission cross-section of the $^5I_7 \rightarrow {}^5I_8$ transition of HLAS and HYAS series samples are listed in Table 7.12. The product

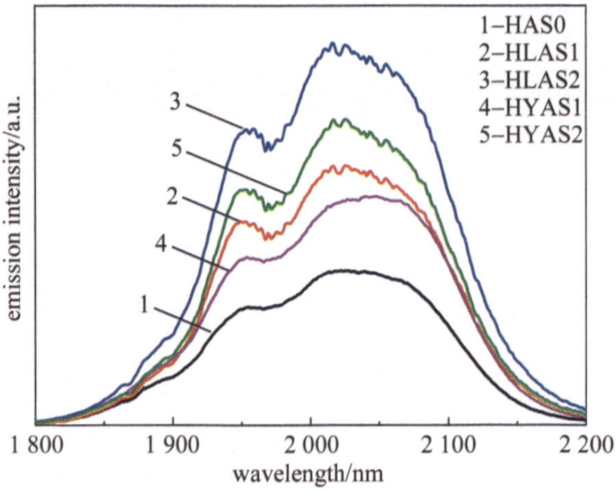

Fig. 7.26 Fluorescence spectra of HLAS and HYAS series samples

Table 7.12 Stimulated emission cross-section and quality factor of HLAS and HYAS samples

Sample	HAS0	HLAS1	HLAS2	HYAS1	HYAS2
τ_{mea}/ms	0.96	1.00	1.47	0.97	1.39
σ_{ems}/($\times 10^{-21}$ cm^2)	3.96	4.01	4.29	3.98	4.13
$\sigma_{ems} \times \tau_{mea}$/($\times 10^{-21}$ cm^2 ·ms)	3.80	4.01	6.31	3.86	5.74

of the stimulated emission cross-section and the fluorescence lifetime is called the quality factor of laser gain, which is an important parameter for measuring laser materials. Table 7.12 lists the quality factors of HLAS and HYAS series samples. As can be seen from the table, the quality factor of the sample increases with the increase in La^{3+}/Al^{3+} or Y^{3+}/Al^{3+} ratio, and when the content of La_2O_3 and Y_2O_3 is the same, the quality factor of La_2O_3 doped glass is higher than that of Y_2O_3 doped glass. This indicates that introducing La^{3+} or Y^{3+} into high concentration Ho^{3+} doped silica glass can improve the spectral properties of holmium ions, help to achieve high laser gain, and La^{3+} is more suitable than Y^{3+} for improving the dispersion of Ho^{3+}, thereby improving its 2 μm band luminescence performance.

7.4 Thulium-Doped Silica Fiber and Fiber Lasers

There are two effective methods to obtain 2 μm lasers: one is to directly generate based on thulium and holmium laser materials; the other is to use 1 μm lasers to generate 2 μm laser through Optical Parametric Oscillation (OPO) technology, but this method produces laser beam with poor quality and broad spectrum, so it is less used. The method of directly generating 2 μm lasers based on thulium and holmium laser materials is divided into two types: one is solid-state laser using Ho:YAG,

Tm:Ho:YLF, and Tm:Ho:GdVO$_4$ as laser media, the other is fiber laser using thulium-doped fiber, holmium-doped fiber, or thulium-holmium co-doped fiber as laser media. Thulium-doped fiber lasers are easy to obtain high slope efficiency laser output due to the large fiber cooling area, good cooling effect, and cross-relaxation characteristics, and their output laser wavelength can be wideband tunable. With the development of double-clad thulium-doped fiber fabrication technology and the maturity of 790 nm semiconductor pump sources, thulium-doped fiber lasers have become the main solution for 2 μm wavelength lasers, especially high-power lasers. The following will introduce the progress of thulium-doped silica fiber and lasers in details.

7.4.1 Research Progress of Thulium-Doped Silica Fiber

As a gain medium, thulium-doped silica fiber plays a decisive role in achieving high-performance laser output, which is the research focus in the field of 2 μm high-power fiber lasers. For thulium-doped fibers, increasing the doping concentration can reduce the average distance between thulium ions, which helps to enhance the cross-relaxation process, improve the laser slope efficiency, reduce quantum defects and fiber heating, and is also conducive to improving the long-wave laser output efficiency.

Figure 7.27 summarizes the laser slope efficiency of silica fibers with different thulium concentrations reported in the literature [3, 8, 50–66]. The Stokes theory limit efficiency without considering the cross-relaxation process is about 40%, and the theoretical limit efficiency after considering the cross-relaxation process can reach 80%. As can be seen from the figure, with the increase of thulium doping concentration, the slope efficiency of the fiber under 790 nm wavelength pumping shows an overall upward trend. The highest reported slope efficiency is 74.5%, with a thulium mass fraction greater than 7.4%. It is worth noting that due to the complex factors such as doping method, background loss, clusters, fiber structures, etc., there is a large difference in the slope efficiency at the same doping concentration. Therefore, in order to ensure high concentration doping, it is necessary to continuously optimize the fiber preparation process.

In 2018, Tumminelli et al. from Coherent (USA) and Hemming et al. from Defense Science and Technology Group (Australia) [65] jointly reported that a double-cladding fiber with a Tm$_2$O$_3$ doping concentration greater than 8.5 wt% was prepared using an all-halide vapor deposition process, with a core numerical aperture (NA) of 0.26 and fiber size of 6/125 μm. When using a 793 nm LD to pump this fiber, the slope efficiency with respect to the absorbed pump power is as high as 74.5%. This fiber is not only the silica fiber with the highest thulium doping concentration reported so far, but also achieves the highest slope efficiency under 793 nm LD pumping. They pointed out that the hydroxyl (OH) content in this fiber is extremely low, and the hydroxyl absorption of other rare earth-doped fibers prepared by the same process is less than 5 dB/km at a wavelength of 1380 nm. In

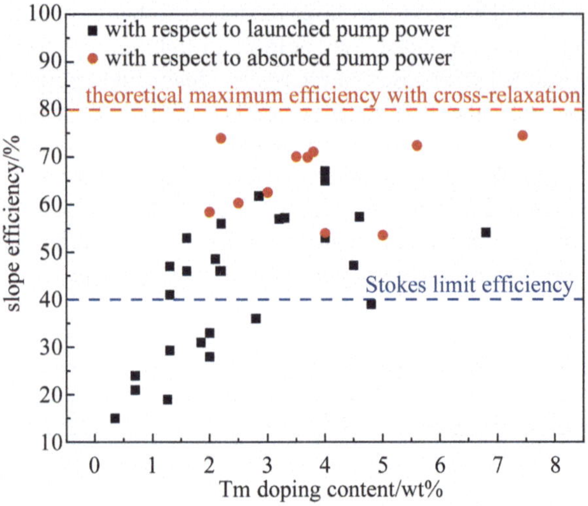

Fig. 7.27 Slope efficiency of silica fibers with different thulium concentrations

2019, the University of Southampton (UK) reported on the preparation of high concentration thulium-doped silica fiber using a hybrid process by combining both vapor-phase and solution-doping techniques [67, 68]. In high concentration thulium-doped silica fiber, it is necessary to co-dope Al elements to increase the solubility of Tm^{3+} in silica glass. It is difficult to achieve an Al_2O_3 doping concentration greater than 4 mol% by traditional solution doping process [69], therefore, they proposed to mix $AlCl_3$ vapor with $SiCl_4$ gas for soot deposition, and then dope thulium elements through solution soaking. This not only significantly increases the doping concentration of Tm and Al, but also makes the element distribution in the fiber core more uniform. They found that as the thulium doping concentration increases from 2 wt% to 5.6 wt%, the slope efficiency with respect to the absorbed pump light gradually increases from 58.5% to 72.4%, as shown in Table 7.13. They pointed out that a thulium doping concentration of 3.5 wt% can already achieve a slope efficiency greater than 70%, and a lower doping concentration is beneficial for short-wavelength laser output.

In 2020, Barber et al. from the University of Southampton reported a nested ring structure of thulium-doped silica fiber, the structural design and refractive index profile are shown in Fig. 7.28 [70]. The central region of the fiber core is free of thulium doping, while high concentration thulium (>3.5 wt%) is distributed in a ring shape. This fiber design can reduce the overlap factor of the LP_{01} mode with the thulium-doped area, reducing the pump absorption in the fiber, thereby reducing the gain coefficient for long wavelengths, which is beneficial to the laser output of short wavelengths. Under 793 nm LD pumping with a power of 101 W, this fiber obtained a single-mode laser output of 62 W at 1907 nm, and the slope efficiency with respect to the absorbed pump power was 70%. In the same year, they increased the laser output power at 1907 nm to 131 W [71].

Table 7.13 Thulium-doped fiber parameters prepared by the University of Southampton using a hybrid doping technology [67]

Fiber Number	Tm Concentration /wt%	Al Concentration /wt%	Fiber Length /m	Laser Wavelength /nm	Slope Efficiency /%	Maximum Output Power /W
HGS-01	3.8	5.6	11	2025	71.1	22.6
HGS-02	5.6	6.1	8	2036	72.4	24.4
HGS-03	3.5	4.9	12	2022	70.1	20.9
HGS-04	3.0	7.8	9	2013	62.6	16.2
HGS-05	2.5	5.1	10	2009	60.4	15.8
HGS-06	2.0	7.2	12	2004	58.5	15.2

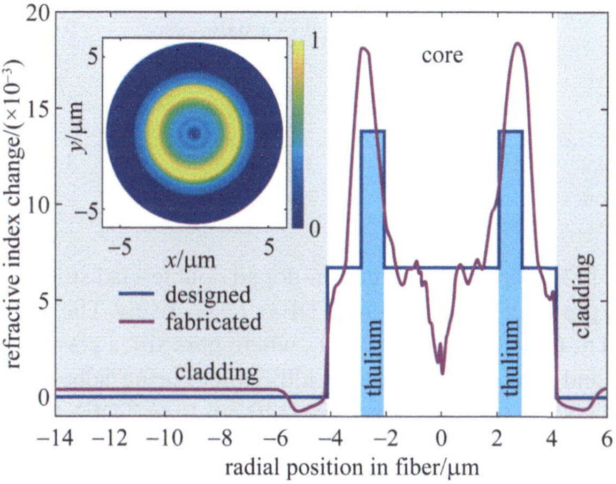

Fig. 7.28 Nested ring structure and refractive index profile of thulium-doped fiber at the University of Southampton [70]

In 2020, Li et al. from Huazhong University of Science and Technology [72] prepared a double-clad thulium-doped silica fiber with a size of 25/400 μm using the MCVD process. The numerical aperture of the fiber core is 0.1, Tm_2O_3 and Al_2O_3 concentrations are 2.6 wt% and 1.01 wt% respectively, and the cladding absorption coefficient at 793 nm is 3 dB/m, and the cladding background loss at 885 nm is approximately 43 dB/km. A first-stage amplification system was built using 8-m-long thulium-doped fiber, achieving a laser output of 530 W with a slope efficiency of 50%.

Hu et al. from Shanghai Institute of Optics and Fine Mechanics [33, 73, 74] started the study on thulium-doped silica fibers using sol-gel process combined with high-temperature powder sintering method since 2013. The parameters of the prepared fiber are shown in Table 7.14. In 2013, Li et al. [33] prepared a 35/280 μm double-clad fiber (TDF1) with a fiber core composition of $0.3Tm_2O_3$–$4.5Al_2O_3$–95.2 SiO_2, obtaining 1.23 W laser output at 1592 nm with a slope efficiency of 11.7%. In

Table 7.14 Parameters and laser results of thulium-doped silica fiber prepared by sol-gel method

Fiber	Core Diameter/ μm	Inner Cladding Diameter/μm	Core Numerical Aperture	Laser Wavelength/ nm	Slope Efficiency	Maximum Output Power/W
TDF1	35	280	0.192	1952	11.7%	1.23
TDF2	38	350	0.102	1969	4.5%	1.11
TDF3	19	170	0.102	1969	4.3%	0.24
TDF4	45	450	0.17	1952	12.2%	2.4

Table 7.15 Parameters of thulium-doped double-clad silica fiber prepared by MCVD method by Hu's research group

Fiber Parameter	10/130 μm	25/400 μm
Core diameter	10 ± 1 μm	25 ± 2 μm
Core NA	0.14 ± 0.01	0.09 ± 0.01
Cladding NA	≥0.46	≥0.46
Cladding diameter	130 ± 3 μm	400 ± 10 μm
Core/clad offset	≤1 μm	≤1.5 μm
Cladding absorption @793 nm	2.2 ± 0.2 dB/m	3.0 ± 0.2 dB/m
Laser slope efficiency	≥40%	≥50%

2014, Lou et al. [73] prepared two thulium-doped double-clad silica fibers with core diameters of 38 μm (TDF2) and 19 μm (TDF3), respectively. The core glass compositions was $0.1Tm_2O_3$–$1.0Al_2O_3$–$98.9SiO_2$, where pure silica glass tube was used as inner cladding and low refractive index ultraviolet curing adhesive as coating. A laser output of 1.11 W was obtained in a 140-cm-long TDF2 fiber, with a slope efficiency of 4.5%, and a beam quality factor $M^2 = 1.99$. A laser output of 0.24 W was obtained in a 125-cm-long TDF3 fiber, with a slope efficiency of 4.3%, and a beam quality factor $M^2 = 1.33$. In 2016, Wang et al. [74] prepared a 45/450 μm double-clad fiber (TDF4), with a core NA of 0.17. A laser output of 2.4 W was obtained in a 50-cm-long TDF4 fiber, with a slope efficiency of 12.2%. Limited by the sol-gel preparation process, the early prepared thulium-doped silica fiber core glass had high hydroxyl content and large background loss, so the laser slope efficiency was very low (less than 2 wt%), and the cross-relaxation of thulium ions was weak, which is also a reason for the low slope efficiency.

The background loss of optical fibers prepared by the sol-gel method is high. In contrast, the use of MCVD method can effectively reduce impurities and hydroxyl content in the fiber core, significantly reducing the background loss of the optical fiber, and improving the laser slope efficiency. In 2019, Hu et al. used the MCVD method combined with solution doping technology to prepare two types of double-clad thulium-doped optical fibers and the fiber parameters are shown in Table 7.15. The cladding absorption coefficient of the 10/130 μm thulium-doped fiber at 793 nm is ~2.2 dB/m, and the laser slope efficiency is greater than 40%; the cladding absorption coefficient of the 25/400 μm thulium-doped fiber at 793 nm is ~3.0 dB/m, and the laser slope efficiency can reach 50%. Meng et al. [75] used a home-made

10/130 μm double-clad thulium-doped fiber (with a cladding absorption coefficient of ~2.5 dB/m at 793 nm) and the 10/130 μm thulium-doped double-clad fiber from Nufern (with a cladding absorption coefficient of ~4.5 dB/m at 793 nm) to build thulium-doped fiber laser oscillators at 1918 nm, 1941 nm, and 2013 nm. They found that under the same conditions, the laser output efficiency of the commercial fiber was significantly improved compared to the self-developed fiber, with a slope efficiency difference of 6% ~ 11%. However, the energy distribution of the laser output beam of the commercial fiber was not ideal, while that of the home-made fiber was more ideal. This is mainly because the thulium ion doping concentration of the self-developed fiber is lower than that of the commercial fiber, resulting in lower cross-relaxation efficiency of Tm^{3+} ions and lower energy conversion efficiency. Additionally, the NA of the home-made fiber is smaller than that of the commercial fiber, resulting in a better beam quality.

7.4.2 Challenges for Thulium-Doped Silica Fiber

During the preparation of silica fiber, a small amount of water vapor or hydroxyl inevitably enters the silica glass network, forming Si—O—H bonds. The fundamental vibration peak of Si—O—H in silica is at 2.72 μm, the second harmonic vibration peak is at 1.38 μm, and the third harmonic vibration peak is at 0.95 μm. The vibration peaks of Si—O—H hydroxyl and [SiO_4] tetrahedra combine to form a series of broadband absorption peaks, known as hydroxyl mixed frequency peaks. Figure 7.29 shows the loss spectra of dry silica glass (with extremely low hydroxyl content) and wet silica glass (with higher hydroxyl content) [76], where ν_3 is the fundamental vibration peak of hydroxyl, and ν_1 is the vibration peak of [SiO_4]

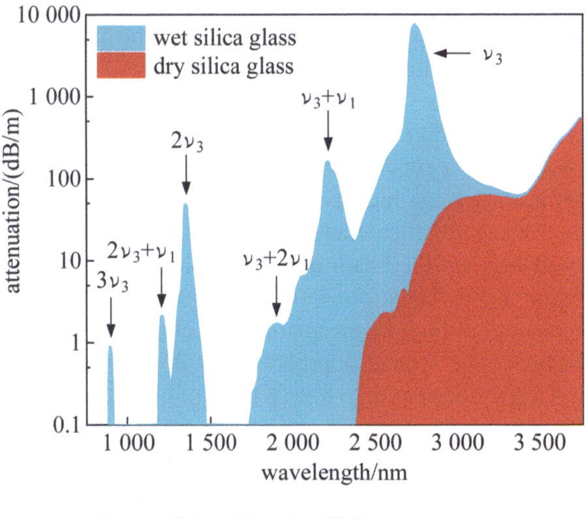

Fig. 7.29 Loss spectrum of wet and dry silica glass [76]

tetrahedra. As shown in the figure, the hydroxyl mixed frequency peaks at ~1894 nm ($\nu_3 + 2\,\nu_1$) and at ~2212 nm ($\nu_3 + \nu_1$) both introduce additional background loss in the 2 μm band. As the wavelength increases, its impact becomes more severe. Heraeus in Germany measured the absorption coefficients of hydroxyl in silica glass and fibers with different hydroxyl contents at different wavelengths, and compared them with the results reported in the literature, as shown in Table 7.16 [76–81]. The absorption coefficient of hydroxyl near 2212 nm is approximately 200 dB/km/ppm (meaning that 1 ppm hydroxyl will introducing an additional background loss of up to 200 dB/km), while near 1894 nm, it is about 0.84 dB/km/ppm. It is evident that the hydroxyl content has a significant impact on the background loss of the fiber in the 2 μm band, and the absorption coefficient at the longer wavelength is much larger than that at shorter wavelength. High background loss can cause a significant decrease in laser efficiency, an increase in fiber thermal load, and other adverse effects, seriously degrading the laser performance of the fiber. Therefore, reducing the hydroxyl content in the fiber core is crucial for preparing high-performance thulium-doped silica fiber.

Shardlo et al. at the University of Southampton [64] shows that the high-temperature process in fiber preparation leads to the diffusion of dopant elements in the fiber core, resulting in different concentrations of Tm^{3+} ions on the cross-section of the core. The Tm and Al concentrations near the interface between the fiber core and the cladding are lower than those in the central area of the fiber core. This leads to a weakening of the cross-relaxation process and a decrease in slope efficiency. Calculations of the overlap factor between the laser mode field and the distribution of Tm^{3+} ions show that the above concentration distribution is the main reason for the laser efficiency being lower than 70%. Therefore, they pointed out that using a high Tm doping level around 3.6 wt% in combination with a more uniform doping profile and the use of a passive alumina-silicate ring around the Tm doped region to increase the core diameter is expected to make the slope efficiency close to the theoretical limit.

For thulium-doped fibers, in addition to increasing the Tm^{3+} doping concentration to promote cross-relaxation, it is also necessary to co-dope with Al elements to increase the solubility of Tm^{3+} in silica glass, reduce clustering and scattering loss, usually with an Al/Tm ratio greater than 10. However, high concentrations of Tm and Al doping will cause a significant increase in the refractive index of the fiber core silica glass, leading to an increase in the refractive index difference between the fiber core and the pure silica cladding and thus an increase in the numerical aperture. This will degrade the beam quality and limit its applications with high beam quality requirements. Therefore, how to balance the dual needs of high energy conversion efficiency and high beam quality is a challenge in the preparation of thulium-doped fibers. The typical approach is to dope germanium on the outer periphery of the Tm/Al-doped core to form a layer of high refractive index pedestal, reducing the effective numerical aperture of the core to achieve good beam quality, as shown in Fig. 7.30. However, related research shows [82, 83] that the germanium-doped pedestal also brings some problems, as shown in Fig. 7.31. On the one hand, a part of the light transmitted through the fiber core may couple into the

Table 7.16 Hydroxyl absorption peak positions and absorption coefficients measured in silica glass and fibers with different hydroxyl contents [76]

OH band ident.	λ (nm)	Wet silica bulk direct melt[a] α_{OH} (dB/km/ppm)	Wet silica bulk soot process[b] α_{OH} (dB/km/ppm)	Wet silica core fiber[c] α_{OH} (dB/km/ppm)	Dry silica core fiber[d] α_{OH} (dB/km/ppm)	Results α_{OH} (dB/km/ppm)	Exp. error
ν_3	2722 ± 1	10,000	10,000	–	–	10,000	–
$\nu_3 + \nu_1$	2212 ± 1	199	202	–	–	201	±5
$\nu_3 + 2\nu_1$	1894 ± 1	0.91	0.77	–	–	0.84	±0.1
$2\nu_3$	1383 ± 1	62.1	63.3	62.7	62.7	62.7	±1
$2\nu_3 + \nu_1$	1246 ± 1	2.62	2.69	2.7	2.9	2.7	±0.5
$2\nu_3 + 2\nu_1$	1139 ± 1	0.067	–	0.062	0.08	0.07	±0.01
$3\nu_3$	943 ± 1	1.55	1.62	1.5	1.66	1.6	±0.1
$3\nu_3 + \nu_1$	878 ± 1	–	–	0.085	0.07	0.08	±0.02
$3\nu_3 + 2\nu_1$	825 ± 1	–	–	0.0038	–	0.0038	±0.0005
$4\nu_3$	724 ± 1	–	–	0.078	–	0.078	±0.005
$4\nu_3 + \nu_1$	685 ± 1	–	–	0.0044	–	0.0044	±0.0005
$4\nu_3 + 2\nu_1$	651 ± 1	–	–	0.00028	–	0.00028	±0.00005
$5\nu_3$	593 ± 1	–	–	0.0064	–	0.0064	±0.0005
$5\nu_3 + \nu_1$	566 ± 1	–	–	0.0003		0.0003	±0.0001
$6\nu_3$	506 ± 1	–	–	0.0005		0.0005	±0.0002
$7\nu_3$	444 ± 1	–	–	0.00012		0.00012	±0.00005

Note: [a] The bulk glass sample (code name: F100) was prepared by flame hydrolysis of $SiCl_4$ n an oxyhydrogen torch and then direct melt, with a hydroxyl content of about 700 ppm

[b] The bulk glass sample (code name: F310) was prepared by flame hydrolysis of $SiCl_4$ in an oxyhydrogen torch to form a porous soot, then sintered in vacuum, with a hydroxyl content of about 250 ppm

[c] The core of this fiber sample was the bulk glass sample[a]

[d] The core of this fiber sample was the bulk glass sample (F300) prepared by flame hydrolysis of $SiCl_4$ in an oxyhydrogen torch to form a porous soot and then a dehydration step was introduced before vacuum sintering, with a hydroxyl content of about 0.2 ppm

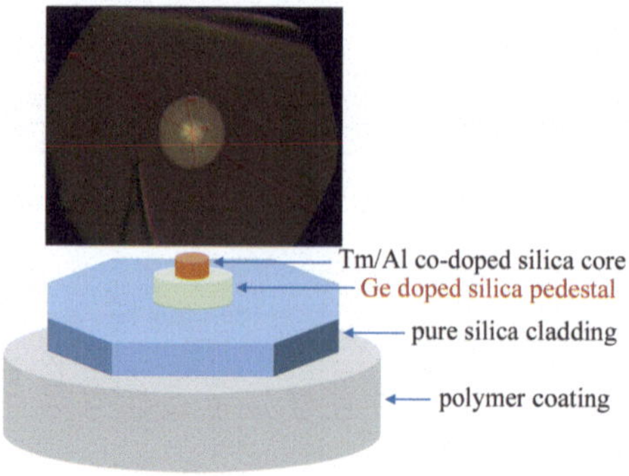

Fig. 7.30 Cross-section and structure of thulium-doped fiber with a germanium-doped pedestal

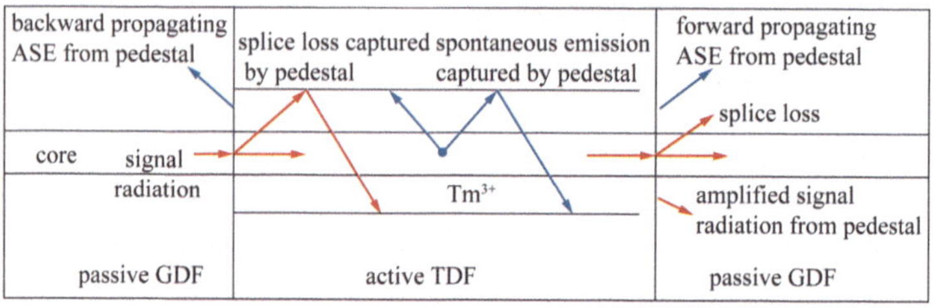

Fig. 7.31 Schematic diagram of fusion splicing of passive fiber without pedestal and thulium-doped fiber with pedestal [83]

pedestal layer due to factors such as fusion splicing loss, coupling loss, mode field mismatch, and amplified spontaneous emission (ASE). Due to the high overlap factor between the pedestal layer and the fiber core, it may lead to a higher gain, making the laser system very sensitive to fusion splicing, especially in pulsed laser systems, and it will also cause strong reverse ASE transmission in the pedestal layer. On the other hand, the light coupled into the pedestal layer will pass through the fusion splice point into the cladding of the passive fiber, not only causing degradation of laser performance, but also possibly causing damage to passive devices such as gratings, beam combiners, isolators, laser diodes, cladding light strippers, etc. Moreover, due to the large NA of the pedestal, there is a lack of effective means to filter out the parasitic light coupled into the pedestal layer. The method of suppressing the high-order modes of the fiber core by bending the thulium-doped fiber will generate additional parasitic light in the pedestal layer, thereby increasing the possibility of coupling between the fundamental mode of the fiber core and the transmission mode of the pedestal layer, thereby affecting the beam quality. Therefore, the design of the pedestal layer has a great impact on the laser performance of the thulium-doped fiber, and it is necessary to reduce the fusion splicing loss as much as possible. Simakov et al. [83] shows that increasing the ratio of the diameter of the pedestal to the diameter of the fiber core can effectively reduce the overlap factor of the light transmitted by the pedestal layer and the fiber core, thereby reducing the impact of fusion quality on laser performance, making the laser output more stable.

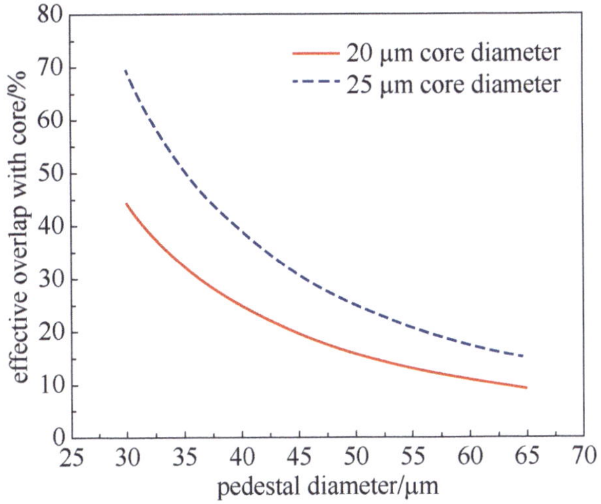

Fig. 7.32 Variation of effective overlap factor with pedestal diameter for different core diameters [83]

As shown in Fig. 7.32, under the same fiber core diameter, as the pedestal diameter increases, the effective overlap factor gradually decreases. Under the same pedestal diameter, the smaller the fiber core diameter, the smaller the overlap factor. Further research by Jollivet et al. [82] found that in the 25/400 μm large mode field thulium-doped fiber, increasing the ratio of the pedestal diameter to the fiber core diameter from 1.7 to 3, after fusion with the matched passive fiber, the fundamental mode coupling efficiency of the fiber core is significantly improved, and the output beam quality is also improved.

7.4.3 Development of Thulium-Doped Fiber Lasers

The thulium-doped fiber laser was first reported in 1988, when Hanna et al. from the University of Southampton [84] used a 797 nm dye laser to pump a 27-cm-long thulium-doped fiber and achieved a 1.94 μm laser with a power of 2.7 mW and a slope efficiency of 13%. Thanks to the development of 790 nm semiconductor lasers, the continuous development of 2 μm high-power fiber laser technology began gradually since 1998. Jackson and King [51] used a 790 nm LD array to pump a thulium-doped fiber with a Tm concentration of 1.85 wt%, achieving a maximum output power of 5.4 W and a slope efficiency of 31%. By adjusting the fiber length, the output laser wavelength can be tuned within the range of 1880 ~ 2033 nm, with an output power greater than 4 W in the range of 1.94 ~ 2.01 μm. In 2000, Hayward et al. [52] used a 787 nm laser diode to bidirectionally pump a double-clad thulium-doped fiber with a Tm concentration of 2.2 wt%, achieving a single-mode laser output power of 14 W and a slope efficiency of 46% at 2 μm. In 2002, Clarkson et al. [85] reported the use of a 787 nm LD to pump a thulium-doped silica fiber, achieving tunable continuous laser output in the range of 1860 ~ 2090 nm through a diffraction grating, with a maximum output power of 7 W at 1940 nm and an output power greater than 5 W in the range of 1860 ~ 2090 nm.

The breakthrough in thulium-doped fiber lasers occurred in 2005 when Jackson et al. from the University of Sydney [53] used a 793 nm laser diode to bidirectionally pump a 1.8-m-long 27/400 μm thulium-doped silica fiber with a thulium doping concentration of 2.2 wt% and an Al to Tm doping ratio of 10, achieving a laser output of 85 W at 2.04 μm and a slope efficiency of 56%. In the same year, they collaborated with the University of Southampton [86], using a 790 nm laser diode to pump a 2-m-long 18/300 μm thulium-doped fiber with a Tm concentration of 2.2 wt% and an Al to Tm doping ratio of 8, achieving an output of 30.8 W at 2025 nm and a slope efficiency of 61%. In 2007, they used a 793 nm laser diode to pump a 4-m-long 20/270 μm thulium-doped fiber with a Tm concentration of 3.2 wt%, achieving a laser output of 120 W at a central wavelength of 2033 nm with a slope efficiency of 57%, and a beam quality factor M^2 of 1.1 at 10 W output [87]. In 2007, Evgueni Slobodtchikov and Peter F. Moulton from Q-Peak company, along with Frith from Nufern [88], used two 350 W 795 nm laser diodes to bidirectionally pump a 5-m-long 25/400 μm thulium-doped fiber, achieving a laser output of 263 W in the 2050 ~ 2070 nm wavelength range with a slope efficiency of 59%. In 2007, Meleshkevich et al. from IPG [89] used a 1567 nm Erbium-doped fiber laser as a pump source, fusing splicing a 5-m-long 20 μm core diameter thulium-doped fiber with a fiber Bragg grating to form an all-fiber system, achieving an output of over 400 W at 1908 nm with M^2 less than 1.1, linewidth less than 1 nm, and a slope efficiency of 60%. In 2009, Moulton et al. from Q-Peak and Frith et al. from Nufern et al. [59] again reported achieving a laser output of 300 W at 2040 nm using a 5-m-long 25/400 μm thulium-doped fiber, M^2 of 1.2, and a slope efficiency of 61.8%. At the same time, they achieved a multimode laser output of 885 W in a 7-m-long 35/625 μm thulium-doped fiber (M^2 = 6–10), with a slope efficiency of 49.2%. In 2009, Goodno et al. [60] used a 2040 nm DFB seed source injected into a 3.1-m-long 25/400 μm thulium-doped fiber after three-stage pre-amplification, achieving a 608 W single-frequency single-mode laser output with a linewidth less than 5 MHz, M^2 of ~1.05, and the slope efficiency of 54%. In 2010, Moulton et al. from Q-Peak reported their result of kilowatt-level thulium-doped fiber laser in collaboration with Nufern [1]. They used six 150 W 790 nm semiconductor pumps and a 10-m-long 20/400 μm thulium-doped fiber, and obtained over 500 W laser output in a single-stage master oscillator power amplifier (MOPA) with a slope efficiency of 61.6%. In a two-stage MOPA amplifier, they achieved over 1000 W laser output at 2045 nm with a slope efficiency of 53.2%.

In the following 10 years, the power growth of thulium-doped fiber lasers stagnated, and most research focused on optimizing the laser system and utilizing the output spectral bandwidth. Thulium-doped fiber lasers with output power greater than 100 W can cover the 1900 nm to 2100 nm wavelength range [90–93]. In 2014, Creeden et al. [94] used a thulium-doped fiber laser with a laser wavelength of

1908 nm to in -band pump thulium-doped fiber, and obtained a 1.43 W laser output at a wavelength of 2005 nm with a slope efficiency of 90.2% with respect to the launched pump power. In the same year, they used the same in-band pumping method to obtain a 41 W laser output at 1993 nm through core pumping, with a slope efficiency of 92.1%, and a 123.1 W laser output at 1993 nm through cladding pumping with a slope efficiency of 91.6%. This is the highest slope efficiency reported for thulium-doped fiber lasers [95]. Using short-wavelength thulium lasers to in-band pump thulium-doped fibers results in quantum defect of less than 10%, significantly reducing fiber heating and having the potential to achieve higher power output. In 2016, Walbaum et al. from the University of Jena in Germany [96] obtained a 567 W laser output at 1970 nm using an oscillating thulium-doped fiber laser, with a slope efficiency of 49.4% with respect to the absorbed pump power, and a beam quality factor M^2 of 2.6. In 2018, Limpert et al. from the University of Jena [97] used chirped pulse amplification technology to obtain a pulsed laser output with an average power of up to 1150 W (before compression) in a 6-m-long 50/250 μm photonic crystal fiber, setting a new record for the average power of thulium-doped fibers. After pulse compression, the output wavelength was 1960 nm, the pulse width was 256 fs, the repetition rate was 80 MHz, the single pulse energy was 13.2 μJ, the peak power was 50 MW and M^2 was less than 1.1. In 2019, Burns et al. from the University of Southampton [98] used a erbium-doped fiber laser at 1580 nm to core pump a thulium-doped fiber, achieving a laser power output of 47 W at 1726 nm, with a slope efficiency of 80% with repect to the absorbed pump power. In 2021, Anderson et al. from Air Force Research Laboratory [99] in USA used a 9-m-long 20/400 μm fiber to achieve a narrow linewidth amplified output with an average power of 1.1 kW at 1.95 nm, a linewidth of 5GHz, a slope efficiency of 51% and M^2 less than 1.1.

Due to the limitations of active fibers, key components, and pump sources, the research on thulium-doped double-clad fiber lasers in China started relatively late. In 1998, Liu et al. from Xi'an Institute of Optics and Precision Mechanics [100] first reported the experimental results of near 2 μm laser output in China. They used a Nd^{3+}:YLF laser at 1053 nm to pump a home-made thulium-doped single-mode silica fiber. When the gain fiber length was 1 m and the pump power was 187 mW, the maximum laser output power of 153 μW at the central wavelength of 1871 nm was obtained, with a slope efficiency of 0.32%. In 2006, Zhang et al. from Harbin Institute of Technology [56] achieved a laser output of 2.2 W in a 3.7-m-long double-clad thulium-doped fiber. In 2007, the Shanghai Institute of Optics and Fine Mechanics achieved a continuous laser output of 70 W using a thulium-doped silica double-clad fiber [101]. In 2008, Xu et al. from the Shanghai Institute of Optics and Fine Mechanics [102] reported a tunable thulium-doped fiber laser, achieving a tunable laser output in the range of 1866 ~ 2107 nm, with a maximum output power of 32 W and a slope efficiency of up to 70%. Xia et al. from Harbin Institute of Technology [103] reported a 2 μm laser output of 6.1 W using a 5-m-long 20/300 μm D-shaped fiber, with a slope efficiency of 55.6%. Zhang et al. [104] reported an amplified output power of 3 W using a 2-m-long 27.5/400 μm thulium-doped fiber. In 2009, they [105] reported a thulium-doped fiber laser with an all-fiber structure, with a maximum continuous laser power output of 39.4 W and a slope efficiency of 34%. In the same year, Zhang et al. from Harbin Institute of Technology [106] used a 793 nm laser diode to pump a 3.2-m-long 25/250 μm home-made fiber, producing a seed light with a power of 2.4 W and a spectral width of 0.1 nm, which was then amplified by a 5.7-m-long 25/400 μm thulium-doped fiber from Nufern to achieve a narrow linewidth laser output of 30.6 W at 1947.6 nm, with a slope efficiency of 39.1% and a spectral width of 0.2 nm. In 2010, Tang et al. from the Shanghai Institute of Optics and Fine Mechanics [107] used a 793 nm laser diode as a pump source to bidirectionally pump a 4-m-long double-clad thulium-doped fiber with a core diameter of 30 μm, achieving a maximum laser output power of 150 W at 2.04 μm, with a slope efficiency of 56.3%. In the same year, they [108] obtained an average power output of 100 W at the center wavelength of 2020 nm using a gain-switch and a two-stage thulium-doped fiber amplification system, with a slope efficiency of 52%, an M^2 of 1.01. When the repetition rate is 500 Hz, the pulse energy reaches 10.4 mJ, and the peak power reaches 138 kW. In 2011, Yu et al. from the Shanghai Institute of Optics and Mechanics [109] reported a system based on Tm:Ho:LuLiF master oscillating seed source and thulium-doped fiber power amplification. The seed with a repetition rate of 11.3 kHz, a linewidth of 0.4 nm, and a power of 0.66 W was amplified by a two-stage 25/400 μm polarization-maintaining thulium-doped fiber to obtain a laser output power of 32.4 W, with a linewidth of 0.4 nm and a central wavelength located at 2058.5 nm.

In 2013, Wang et al. from the National University of Defense Technology [110] reported a high-power single-frequency thulium-doped fiber amplifier with an all-fiber structure. They used a MOPA structure to amplify the single-frequency seed laser in two stages. The first stage of amplification used a 2.5-m-long 9/125 μm single-mode fiber, and the second stage used an 8-m-long 10/125 μm double-clad fiber. The maximum laser output power at 1971 nm was 102 W, with a slope efficiency of 50% relative to the absorbed pump power. In 2014, Hu et al. from Tsinghua University [111] utilized an all-fiber oscillation structure with a pair of fiber Bragg gratings centered at ~1908 nm as the laser resonator cavity, 7 fiber-coupled 790 nm laser diodes as the pump source, and a 3-m-long 25/400 μm double-clad thulium-doped fiber as the gain medium. Under a pump power of 443 W, an output laser power of 227 W was obtained, with a slope efficiency of 54.3%, a light-light conversion efficiency of 51.2%, and a beam quality factor M^2 of 1.56. In the same year, Liu et al. from Beijing University of Technology [112] reported a 200 W-level single-frequency, single-polarization thulium-doped fiber laser. They used polarization-maintaining fibers and devices to build a four-stage all-fiber MOPA system. A stable single-frequency distributed feedback diode laser with a linewidth less than 2 MHz and a maximum power of 3.5 mW was used as the seed source. After two stages of pre-amplification in a 10/130 μm polarization-maintaining thulium-doped fiber, the power reached 5 W. It was then further amplified to 75 W in a 4.5-m-long 25/400 μm polarization-maintaining thulium-doped double-cladding fiber, and finally amplified to 210 W in a 25/400 μm polarization-maintaining thulium-doped double-clad fiber, with a slope efficiency of ~53%, M^2 of 1.6, a central wavelength of 2000.92 nm, a laser linewidth less than 0.8 pm, and a polarization extinction ratio greater than

17 dB. In 2015, they reported the first domestically produced 300 W-level high-power, narrow-linewidth output thulium-doped fiber laser [113]. The experimental system used a one-stage oscillation plus two-stage amplification all-fiber MOPA structure. The oscillator provided a seed light with a power of 560 mW and a central wavelength of 2000.3 nm. After two stages of power amplification in a polarization-maintaining thulium-doped amplifier, a maximum output power of 342 W was finally obtained, with a slope efficiency of 56%, a spectral 3 dB bandwidth of 90 pm, and M^2 less than 1.15.

In 2016, Yin et al. from the National University of Defense Technology [114] reported a wavelength-tunable high-power thulium-doped fiber laser at 2 μm band. The system uses an all-fiber MOPA structure, achieving a laser output of 300 W-level. The laser oscillator uses a tunable band-pass filter for wavelength selection. The laser amplifier amplifies the power of a 5 W single-mode seed laser at 1910 nm to 327.5 W, with a slope efficiency of 57.4%. The amplifier can achieve over 270 W laser output in the 1910 ~ 2050 nm range, and the system shows good output power stability during a 20-minute operation test. When operating in the 1910 ~ 1930 nm band, the laser output spectrum exhibited a noticeable ASE generation.

In 2017, Yao et al. [115] from Fudan University used a 25/400 μm thulium-doped cladding fiber to construct an all-fiber MOPA system, achieving a narrow linewidth laser output of 400 W at 1941 nm, with a laser linewidth of 67 pm and an amplification stage slope efficiency of 53%. In 2018, they used a 3 × 1 signal combiner to non-coherently combine three thulium-doped fiber MOPA systems, achieving a maximum laser output power of 790 W, with a slope efficiency of 52.2% with respect to the launched pump power, and a beam quality factor M^2 of 2.7 [116].

In 2019, Li et al. [117] from Huazhong University of Science and Technology reported a home-made 25/400 μm thulium-doped double-clad fiber with a cladding absorption coefficient of 3 dB/m at 793 nm. Using a MOPA structure, they achieved a 406 W narrow linewidth laser output with a central wavelength of 1980.89 nm, a 3 dB spectral width of 84 pm, and an amplification stage slope efficiency of 54%. In 2020, they used this fiber to amplify a 57 W seed laser to 530 W through one stage amplification, with a corresponding amplification stage slope efficiency of 50% [72].

In summary, in the past decade, with the acquisition of high-power LDs, advancement of fiber preparation processes, and improvements of laser thermal management technologies, the development of high-power thulium-doped fiber lasers has progressed rapidly and the output power has been continuously improved. The wavelength of high-power thulium-doped fiber lasers is mainly concentrated in the range of 1940 ~ 2050 nm, but in the wavelength range closer to 1.9 μm, the reabsorption of Tm^{3+} is serious, making it relatively difficult to achieve high power, high efficiency, and narrow linewidth output.

Compared with ytterbium-doped fiber lasers, thulium-doped fiber lasers have higher SBS and SRS thresholds and also have certain advantages in the threshold of mode instability. It is reported that when the mode instability threshold is reached, the average thermal load of ytterbium-doped fiber is 30 ~ 34 W/m, while thulium-doped fiber can reach 98 W/m [97]. Theoretically, thulium-doped fiber lasers can achieve higher power single-mode laser output, so there is still a lot of room for improvement in terms of power.

For thulium-doped fiber lasers, there are two main issues that need to be addressed to achieve high-power output. The first is the thermal management of thulium-doped fibers. For a 2 μm thulium-doped fiber laser pumped by a ~ 790 nm semiconductor laser, the quantum defect is large, and the fiber heats up severely during high power output, leading to degradation in beam quality. The thermal stability of the polymer coating is also a limiting factor. The second is how to increase the pump power coupled into the fiber. For high-power fiber lasers, pump coupling has always been a challenge, and increasing the brightness of the semiconductor laser is crucial. Using a 793 nm laser diode to pump thulium-doped fiber relies on the cross-relaxation process of Tm^{3+}, which is beneficial for miniaturizing the laser system. While using short-wavelength thulium-doped lasers to in-band pump thulium-doped fibers reduces the difficulty of fiber thermal management but increases the complexity of the laser system. Currently, using low-power 2 μm solid/semiconductor or fiber lasers with good spectral characteristics and high beam quality as seed sources, followed by amplification in a multi-stage all-fiber MOPA structure with a ~ 790 nm LD pump or a short-wavelength thulium-doped laser with in-band pumping double-clad large-mode-area thulium-doped fiber, is the preferred solution for achieving high-power, high-beam quality 2 μm laser output.

7.5 Holmium-Doped Silica Fiber and Fiber Lasers

Compared with Tm^{3+} doped lasers, Ho^{3+} doped lasers have longer wavelengths and higher atmospheric transmittance. Fig. 7.33 shows the atmospheric transmittance curves of Tm^{3+} doped lasers and Ho^{3+} doped lasers [118]. Due to the overlap of the

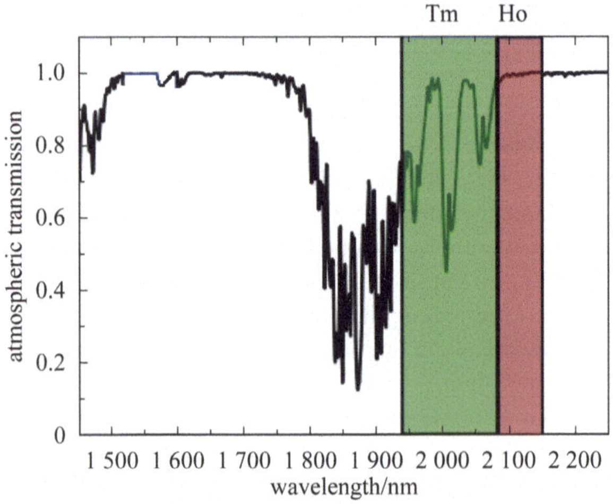

Fig. 7.33 Atmospheric transmittance curves [118]

emission wavelength of Tm^{3+} ions with a water absorption peak, it is limited when used in environments with higher humidity, resulting in poor atmospheric transmittance. In addition, when the power of Tm^{3+} doped fiber lasers reaches the kilowatt level, the thermal light oscillations in the fiber core are significant, which may lead to degraded beam quality. The atmospheric transmission window has higher transmittance in the 2.1 ~ 2.3 μm, thanks to the smaller scattering loss of long waves in the atmosphere. In the 2.1 ~ 2.15 μm band, the atmospheric transmittance is about 90% at a distance of 5 km, and about 55% at a distance of 23 km [119]. The quantum efficiency of Ho-doped fibers in-band pumped with thulium lasers is very high, with a theoretical slope efficiency of up to 90%. This makes the thermal load of Ho-doped fibers very low. Additionally, the high beam quality of thulium-doped fiber lasers can solve the coupling problem of high-power pump light, which is beneficial for achieving high-power laser output from holmium-doped silica fibers.

7.5.1 Research Progress of Holmium-Doped Silica Fiber

In 2013, Hemming et al. [120, 121] from the Defense Science and Technology Group (Australia) used fluorine-doped silica glass as the outer cladding of the fiber, replacing the traditional low-refractive-index polymer coating, effectively solving the absorption problem of coating for 1.9 μm pump light. The prepared fiber consists of 5 layers, from inside to outside, including holmium-doped core, pure silica inner cladding, fluorine-doped silica outer cladding, pure silica outer cladding, and polymer coating. The core diameter is 40 μm, the numerical aperture is 0.08. An average power of 140 W was obtained under the 1.95 μm thulium-doped fiber laser in-band pumping, with a slope efficiency of 55% with respect to the launched pump power. In 2014, Friebele et al. from the Naval Research Laboratory (USA) [118] prepared a holmium-doped fiber with hydroxyl content as low as 0.14 ppm using an atmospheric exclusion technique for solution doping, with a slope efficiency of 62% under the pumping of a 1940 nm thulium-doped fiber laser. In 2016, Hemming et al. [122] reported a 6/125 μm fiber with a Ho^{3+} doping concentration of 0.5 wt%, a core NA of 0.22, and a slope efficiency of up to 87% under the pumping of a 1.95 μm thulium-doped fiber laser. This is also the highest slope efficiency achieved for holmium-doped fiber by in-band pumping. In 2018, the Naval Research Laboratory, the Defense Science and Technology Group, and Clemson University [123] jointly reported that holmium-doped fibers with a hydroxyl content of less than 0.5 ppm can be prepared by combining atmospheric exclusion technique for solution doping with chlorine gas dehydration process, which is very important for achieving high energy conversion efficiency. As shown in Fig. 7.34, when the hydroxyl content is extremely low, the slope efficiency of the holmium-doped fiber can exceed 76%. At the same time, Baker et al. from the Naval Research Laboratory [123–125] introduced Lu_2O_3 and LaF_3 into the local environment of Ho^{3+} ions using nanoparticle doping technology, reducing the phonon energy of Ho^{3+} ions surrounding local environment and lowering the non-radiative transition probability. The prepared

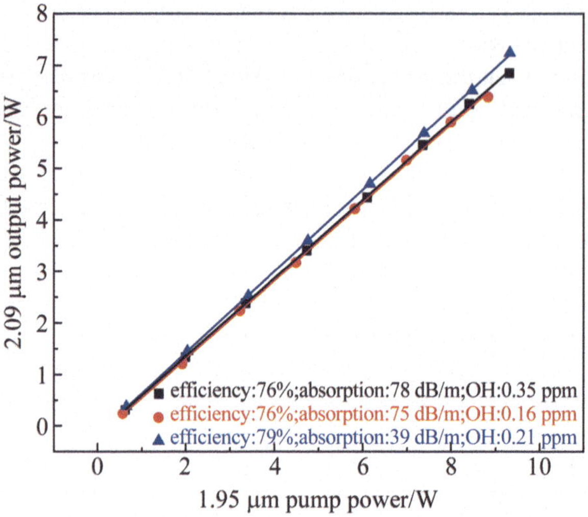

Fig. 7.34 Ultra-low hydroxyl content at different doping concentrations slope efficiency of holmium-doped fiber [123]

high-quality holmium-doped fiber has a hydroxyl content of less than 0.5 ppm, and the background loss is as low as 0.02 dB/m. The slope efficiencies of holmium-doped fibers cooping with Lu_2O_3 and LaF_3 nanoparticles under in-band pumping are 85.2% and 82.3%, respectively, as shown in Fig. 7.35.

7.5.2 Challenges for Holmium-Doped Silica Fiber

Similar to Er^{3+} ions, the inherent complex upper energy level of Ho^{3+} ions can easily lead to clustering effects, and due to small energy levels spacing, they can easily interact with phonons, degrading laser performance. Therefore, even in optical fibers with very low doping concentrations, clustering effects can be observed, and their influence cannot be completely avoided. Doping with elements such as Al and La can effectively increase the solubility of rare earth ions in silica glass, improve the dispersion of Ho^{3+} ions, and weaken the clustering effect. Wang et al. from the Korea Institute of Science and Technology studied the performance of holmium-doped fibers in the absence and presence of clusters through theoretical simulations [126, 127], the results of which are consistent with experiments. The clustering effect in holmium-doped fibers is very similar to that in erbium-doped fibers introduced in Chap. 6 of this book, which will be briefly discussed below.

Figure 7.36 shows the brief energy level structure of Ho^{3+} ions, with the energy levels $^5I_8$, $^5I_7$, $^5I_6$, $^5I_5$, $^5I_4$ denoted as energy levels 1, 2, 3, 4, 5, respectively. The population density of Ho^{3+} ions in each energy level is represented as N_1, N_2, N_3, N_4, N_5. Among them, GSA1 represents the ground state absorption process from energy

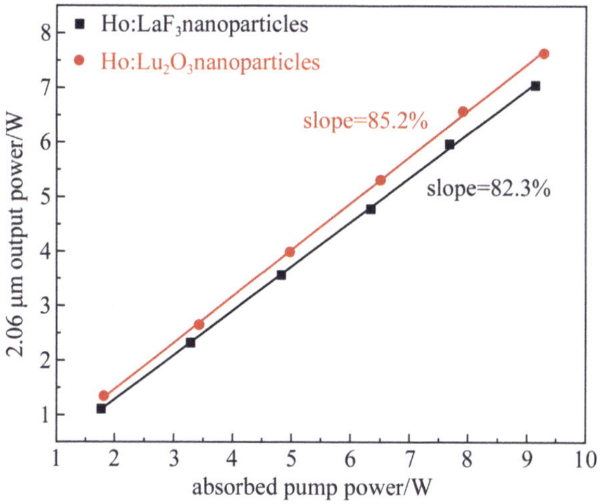

Fig. 7.35 Slope efficiency of Holmium-doped fibers prepared by LaF_3 and Lu_2O_3 nanoparticle doping process [124]

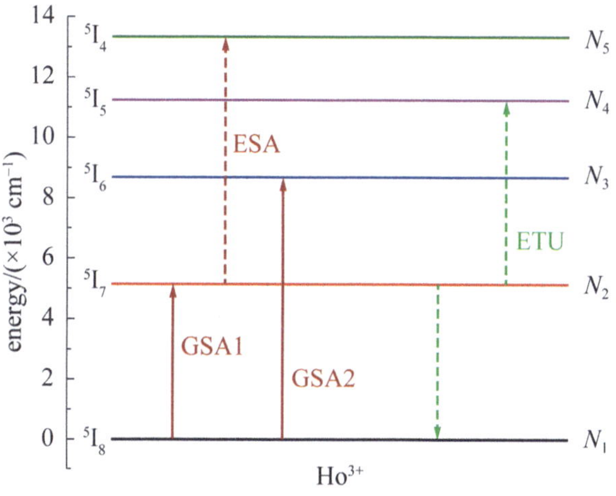

Fig. 7.36 Brief energy level diagram of Ho^{3+} ion

level $^5I_8$ to energy level $^5I_7$, GSA2 represents the ground state absorption process from energy level $^5I_8$ to energy level $^5I_6$. ESA represents the excited state absorption process from energy level $^5I_7$ to energy level $^5I_4$. Since the energy level difference is close to the GSA2 process, the ESA process will compete with the GSA2 process for pump light. The energy level difference of the GSA1 process does not match, and the ESA process will be suppressed during the in-band pumping. The ETU in the figure represents the cooperative upconversion process of $^5I_7$, $^5I_7 \rightarrow$ $^5I_5$, $^5I_8$. Energy transfer occurs between neighboring Ho^{3+} ions at $^5I_7$ level, causing one Ho^{3+}

ion to transition to a higher energy level, and the other to transition to the ground state level, completing the cooperative upconversion. The upconversion process is greatly affected by the energy transfer rate between ions, so when clustering occurs, the proximity of neighboring ions results in a high energy transfer rate, significantly increasing the probability of upconversion.

When not considering ion clusters (that is, the Ho^{3+} in the fiber is uniformly distributed), the change in population density of each energy level can be represented by the following rate equation:

$$\frac{dN_1}{dt}=\Gamma_{41}\ N_4+\Gamma_{31}\ N_3+\Gamma_{21}\ N_2-\left(\sigma_p^{(a)}\ N_1-\sigma_p^{(e)}\ N_2\right)\varphi_{\mathrm{p}} -\left(\sigma_s^{(a)}\ N_1-\sigma_s^{(e)}\ N_2\right)\varphi_{\mathrm{s}}+C\ N_2^2 \tag{7.4}$$

$$\frac{dN_2}{dt}=\Gamma_{42}\ N_4+\Gamma_{32}\ N_3-\Gamma_{21}\ N_2+\left(\sigma_p^{(a)}\ N_1-\sigma_p^{(e)}\ N_2\right)\varphi_{\mathrm{p}} +\left(\sigma_s^{(a)}\ N_1-\sigma_s^{(e)}\ N_2\right)\varphi_{\mathrm{s}}-2\ C\ N_2^2 \tag{7.5}$$

$$\frac{dN_3}{dt}=\Gamma_{43}\ N_4-\Gamma_{32}\ N_3-\Gamma_{31}\ N_3 \tag{7.6}$$

$$\frac{dN_4}{dt}=-\Gamma_{43}\ N_4-\Gamma_{42}\ N_4-\Gamma_{40}\ N_4+C\ N_2^2 \tag{7.7}$$

$$N=N_1+N_2+N_3+N_4 \tag{7.8}$$

In the formula, N represents the total population density of Ho^{3+} particles (under the in-band pumping conditions, the population density of level 5 is almost 0 and can be ignored). C is the cooperative upconversion rate. φ_{s} and φ_{p} represent the photon flux of the signal light and the pump light respectively. Γ_{ij} represents the radiative transition rate from level i to level j.

When considering the existence of ion clusters, the Ho^{3+} in the fiber can be divided into two parts: one part is the uniformly distributed individual Ho^{3+} ions, denoted as N^i. The other part is the Ho^{3+} ions forming clusters, denoted as N^c. Here, we only discuss the simple case where each cluster contains two Ho^{3+} ions. According to the cluster model, because the distance between Ho^{3+} in the cluster is extremely small, the energy transfer rate is extremely large, and the situation where both Ho^{3+} are in the $^5I_7$ upper level can be ignored. Therefore, there are two states of Ho^{3+} clusters: state a is that both Ho^{3+} are at the ground state $^5I_8$ level, and the cluster density is represented as N_a^c; state b is that the two Ho^{3+} are at the $^5I_8$ and $^5I_7$ levels respectively, and the cluster density is represented as N_b^c. k is the number of Ho^{3+} ion pairs (N_a^c + N_b^c) as a proportion of the total number of Ho^{3+} ions (N).

$$N^c=2\left(N_a^c+N_b^c\right)=2\ kN \tag{7.9}$$

$$N^i = (1-2k)\ N \tag{7.10}$$

The rate equation for Ho^{3+} clusters can be written as

$$\frac{dN_a^c}{dt} = \Gamma_{21}\ N_b^c + \left(\sigma_s^{(e)} N_b^c - 2\ \sigma_s^{(a)} N_a^c\right)\ \varphi_s - 2\ \sigma_p^{(a)}\ N_a^c \varphi_p \tag{7.11}$$

The transmission equations of pump light and signal light can be written as

$$\frac{dP_p^{\pm}}{dz} = \mp\left\{\Gamma_p \sigma_p^{(a)}\left[N_1^i + 2N_a^c + N_b^c\right] + \xi_p\right\} P_p^{\pm} \tag{7.12}$$

$$\frac{dP_s^{\pm}}{dz} = \pm\left\{\Gamma_s \sigma_s^{(e)}\left[N_2^i + N_b^c\right] - \Gamma_s \sigma_s^{(a)}\left[N_0^i + 2N_a^c + N_b^c\right] - \xi_s\right\} P_s^{\pm} \tag{7.13}$$

In the equation $P_p^{\pm}$ and $P_s^{\pm}$ represent the power of the pump light and the signal light, respectively. The positive sign indicates forward transmission, and the negative sign indicates reverse transmission; $\mathit{\Gamma}_p$ and $\mathit{\Gamma}_s$ represent the overlap factors of the pump light and the signal light, respectively; N_1^i and N_2^i represent the population density of uniformly distributed Ho^{3+} particles located at energy levels 1 and 2. ξ_p and ξ_s represent the loss of pump light and signal light, respectively.

Wang et al. from the Korean Institue of Science and Technology [127] simulated the performance of the holmium-doped fiber (SM-HDF-10/130) produced by Nufern under two non-clustering and clustering conditions and compared it with experimental results. They found that when $k = 0.05$, the simulation results match the experimental results, indicating that about 10% of the holmium ions exist in cluster form in the fiber. The rapid energy transfer up-conversion processes occur between the holmium ions in the clusters, and they quickly relax to the ground state energy level, which leads to their absorption of pump light without contributing to laser output, but instead consuming additional pump light energy. Therefore, even in high-quality holmium-doped fibers with low doping concentrations, the clustering effect can still lead to intensified unsaturated absorption, decreased slope efficiency, and other adverse effects. It is well known that co-doping with aluminum in silica glass can improve the dispersion of rare earth ions and reduce the clustering effect. Kamrádek et al. [128] found that when the Al/Ho ratio is greater than 60 in holmium-doped fiber, a slope efficiency greater than 80% can be achieved.

Using thulium-doped fiber lasers to pump holmium-doped silica fibers is the preferred solution for obtaining high-power lasers in the 2.1 μm band, but traditional double-clad active fibers often use polymer as the outer cladding. The C—H stretching vibration absorption of the low-refractive-index polymer outer cladding is located at 1.9 μm, which can strongly absorb 1.9 μm pump light and generate a lot of heat. Moreover, the poor heat load capacity of the organic polymer cladding makes it susceptible to burnout under high-power pumping, limiting the long-term reliability of holmium-doped silica fibers at high power operation. Therefore, when using thulium doped lasers to pump holmium-doped fibers, organic polymer cladding should be avoided. A common solution is to use fluorine-doped silica glass as the outer cladding on the outside of the pure silica glass octagonal inner cladding, as shown in Fig. 7.37, which was reported by Hemming et al. [121, 129] By replacing the low refractive index organic cladding with fluorine-doped silica glass cladding, with the outermost layer being pure silica glass, can effectively prevent the aging and burning of the outer cladding under high-power pumping, thus improving the long-term reliability of the fiber.

Compared to thulium-doped fibers, holmium-doped fibers generate lasers with longer wavelengths and the laser performance is more severely affected by the hydroxyl content. Hydroxyl absorption can lead to a sharp increase in background loss and an increase in the thermal load of the fiber core. Therefore, strictly controlling the hydroxyl content in the holmium-doped fiber core is key to improving the laser performance of holmium-doped fibers. As shown in Fig. 7.38 [118], calculations shows that at an output power of 2.7 kW, the axial heat distribution of a holmium-doped fiber with a core hydroxyl content of 1 ppm is much higher than that of a holmium-doped fiber without hydroxyl. Hydroxyl-induced fiber core loss will significantly reduce the laser slope efficiency, exacerbate core heating, and increase the difficulty of fiber thermal management. The Naval Research Laboratory [118] applied an in-situ solution doping technique to the preparation process of holmium-doped core rods. During solution doping, the solution is introduced into a sealed deposition tube through a polytetrafluoroethylene tube. After soaking for a period of time, the liquid is pumped out externally without removing the deposition tube, reducing contact with moisture in the atmosphere. The MCVD setup is shown in Fig. 7.39. At the same time, they strictly control the water content in the raw materials and carrier gas, and combine the atmospheric exclusion solution doping

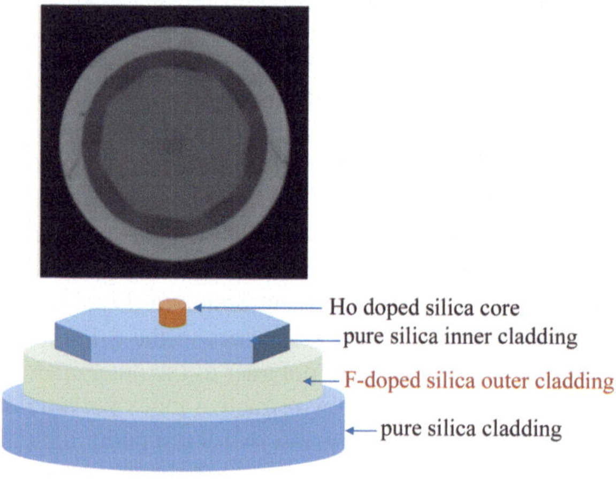

Fig. 7.37 Cross-section and structure of Holmium-doped fiber with fluorine-doped glass cladding structure [129]

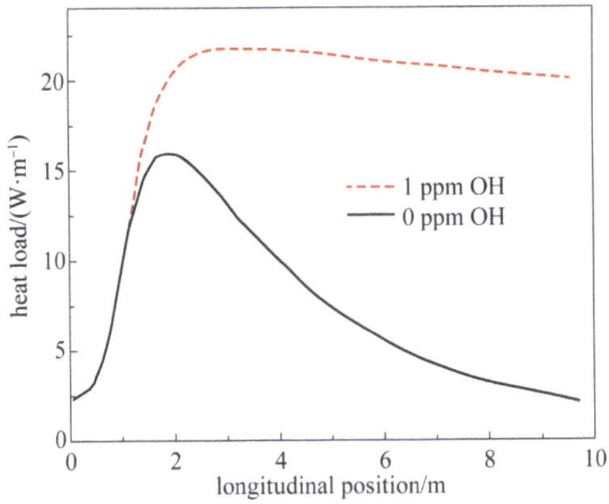

Fig. 7.38 Axial distribution of core thermal load of holmium-doped fiber with 1 ppm hydroxyl content and holmium-doped fiber without hydroxyl at 2.7 kW output [118]

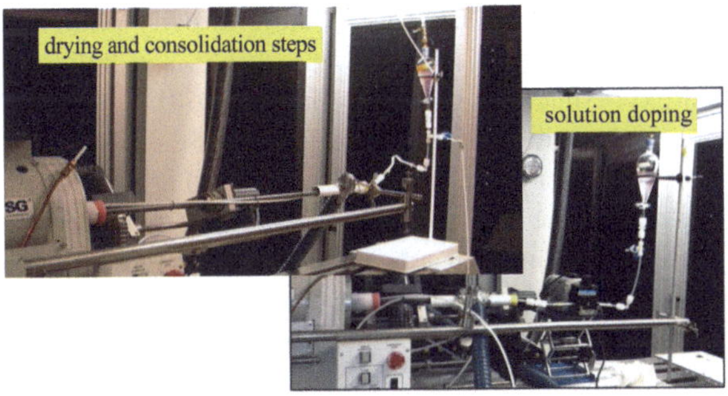

Fig. 7.39 Picture of atmospheric exclusion solution doping setup proposed by the Naval Research Laboratory [118]

technology with the drying process to stably obtain a core with a hydroxyl content of less than 0.2 ppm.

When in-band pumping holmium-doped fiber, the cladding needs to transmit 1.9 μm pump light. Therefore, not only the hydroxyl in the core will cause additional loss at the laser wavelength, but the hydroxyl in the cladding will also cause additional loss at the pump wavelength, absorb pump light to generate heat, and greatly increase the thermal load of the fiber. To address this issue, P.C. Shardlow et al. from the University of Southampton [130–132] used a pulsed CO_2 laser to process the silica inner cladding in the preparation of triple-clad holmium-doped fiber preforms instead of traditional mechanical grinding and polishing methods to obtain better surface processing quality, as shown in Fig. 7.40. Therefore, before the jacketing of fluorine-doped silica outer cladding, the step of flame polishing the preform surface with a hydrogen-oxygen flame was eliminated, avoiding the introduction of additional hydroxyl groups into the silica cladding, effectively reducing

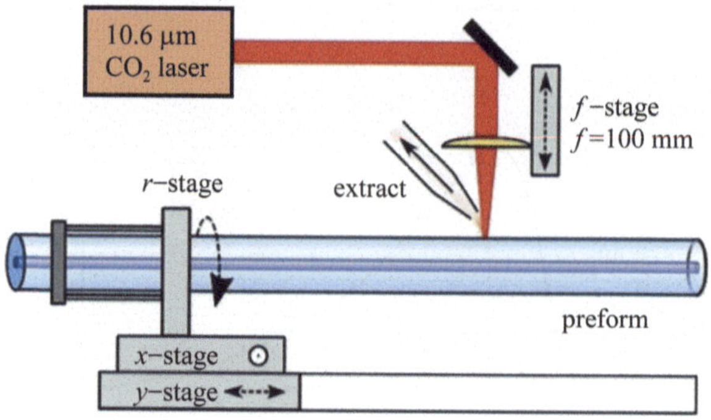

Fig. 7.40 Schematic diagram of CO_2 laser processing device [132]

the hydroxyl content in the cladding. They processed the preform rod prepared by the MCVD lathe with hydrofluoric acid to an outer diameter of 6 mm, removing a part of the silica cladding containing hydroxyl groups. Then, they processed the octagonal inner cladding with a CO_2 laser before directly jacketing the fluorine-doped silica tube. They draw a 14/65/180 μm triple-clad holmium-doped fiber, reducing the hydroxyl content of the inner cladding to 0.3 ppm [132]. It is worth noting that Jasim et al. from the Czech Academy of Sciences [133] strictly compared the effects of CO_2 laser processing and mechanical processing on the hydroxyl content of the cladding in the preform under the same process. They found that CO_2 laser heat treatment greatly increases the surface temperature of the preform, causing hydroxyl in the surface layer to diffuse into the center of the preform, while mechanical cold processing can better remove hydroxyl from the preform surface.

They pointed out that compared with mechanical cold processing, CO_2 laser thermal processing can achieve better surface quality, reduce scattering loss, facilitate full automation and processing of irregular claddings, but this method is not as effective as mechanical cold processing in removing the original hydroxyl groups on the surface of the preform rod. Therefore, to obtain an inner cladding with ultra-low hydroxyl content, it is necessary to choose the appropriate process flow and strictly control each step in the preparation of the preform.

The 2.1 μm band is close to the multi-phonon absorption edge of silica glass, causing the inherent background loss of the silica glass matrix to increase sharply. At the same time, due to the small energy difference between the excited state level $^5I_7$ and the ground state level $^5I_8$, which is about 4760 cm^{-1}, and the phonon energy of the silica glass matrix is about 1100 cm^{-1}. Therefore, Ho^{3+} ions in the excited state level $^5I_7$ only need the participation of 4 phonons to undergo multi-phonon relaxation, without radiative transition to the ground state level $^5I_8$, causing quenching of the upper level fluorescence lifetime. The measured fluorescence lifetime of Ho^{3+} in silica glass is about 1 ms, while in crystals its fluorescence lifetime can be as high as 16 ms [134]. The multi-phonon relaxation process consumes additional population at the upper energy level, generates heat, reduces the slope efficiency of the laser and is detrimental to high-power thulium-doped lasers. Therefore, by doping to reduce the phonon energy of the silica matrix, the slope efficiency of holmium-doped silica fiber can be effectively improved. Baker et al. from the Naval Research Laboratory [123, 124] introduced low phonon energy Lu_2O_3 and LaF_3 into the local environment of Ho^{3+} through nanoparticle doping process, reducing the probability of non-radiative transition, which can improve the slope efficiency of holmium-doped fiber.

7.5.3 Development of Holmium-Doped Fiber Lasers

The experimental result of continuous laser with holmium-doped silica fiber were first reported in 1989 by Hanna et al. from the University of Southampton [47], using a 457.9 nm argon ion laser as the pump source to achieve a laser output at 2.04 μm. One of the main challenges faced by holmium-doped fiber lasers is the lack of mature semiconductor laser pump sources, and early research often used ~1.1 μm as the pump wavelength. In 2000, Kurkov et al. from the Russian Academy of Sciences [135] used a 1.15 μm ytterbium-doped laser to pump holmium-doped fiber, achieving a laser output of 280 mW with a slope efficiency of 20%. In 2007, Jackson et al. [136] used a 1.15 μm laser diode to directly clad pump low-concentration holmium-doped fiber, achieving a laser output of 2.2 W with a slope efficiency of up to 29%. In 2009, Kurkov et al. [137] reported using a ytterbium-doped GTWave fiber to generate a 1.125 μm laser and pump holmium-doped fiber, achieving a laser output of 4.2 W at 2.05 μm, with a slope efficiency of 34% with repect to the absorbed pump power. In 2010, they used the same laser setup to increase the 2.05 μm laser output power to 10 W, with a slope efficiency of 30% with respect to the absorbed pump power [138]. In 2011, they optimized low-concentration holmium-doped fiber, reduced the concentration of Ho^{3+} clusters, and further increased the slope efficiency of the holmium laser pumped at 1.15 μm to 45.5%, with an output power of 3.5 W [139]. Using ~1.15 μm laser to pump thulium-doped fiber results in a large quantum defect, and the theoretical energy conversion efficiency is only 56%, which poses a great challenge to fiber thermal management under high power output.

In 2011, A. Hemming et al. [140] first use a 1.95 μm thulium-doped fiber laser to in-band pump a holmium-doped fiber, achieving a 99 W output of 2.12 μm laser, with a slope efficiency of 65% with respect to the absorbed pump power. Subsequently, due to the rapid development of thulium-doped fiber lasers, the power rapidly increased. The use of ~1.95 μm thulium-doped laser to in-band pump holmium-doped fiber has the advantages of small quantum defect, high energy conversion efficiency, and low fiber thermal load, making it the first choice for achieving high-power laser output. In 2013, Hemming et al. [120, 121] coupled six 1.95 μm thulium-doped fiber lasers through a combiner into a holmium-doped silica fiber with a fluorine-doped glass cladding, and obtained an average power of 407 W of ~2.12 μm laser output under 1000 W pump power [129], with a slope efficiency of about 40% with respect to the pump power. This is also the highest output power achieved by holmium-doped silica fiber so far. In 2019, Holmen et al. from the University of Southampton collaborated with Hemming et al. [141], achieving watt-level tunable laser output in the wavelength range of 2025 ~ 2200 nm, with the slope efficiency of 58% at 2050 nm and a slope efficiency of 27% at 2200 nm.

In summary, Ho^{3+} doped fiber has a low absorption coefficient at 1 μm and a large quantum defect, resulting in a low slope efficiency of 2 μm laser in thulium-doped fiber lasers when pumped at 1 μm wavelength. The development of thulium-doped fiber lasers has brought high-brightness and high-power pump sources for holmium-doped fiber lasers and the use of in-band pumping scheme has effectively reduced the quantum defect and thermal management difficulty of holmium-doped fiber lasers, leading to a rapid increase in the output power of holmium-doped fiber lasers. However, compared with direct pumping schemes of laser diodes, the cascaded in-band pumping scheme also brings about the inconvenience of complexity and large size of laser system. In addition, the clustering effect of holmium ions, the multi-phonon relaxation process, and the loss caused by hydroxyl also limit the further increase of the power of holmium-doped fiber lasers to the kilowatt level.

7.6 Thulium-Holmium Co-Doped Silica Fiber and Fiber Lasers

In addition to directly pumping, indirectly pumping Ho^{3+} ions can also obtain ~2 μm laser by adding sensitizing ions such as Yb^{3+}, Er^{3+}, or Tm^{3+} into the medium through energy transfer [142–144]. Among them, thanks to the mature and inexpensive ~790 nm laser diode pump source of Tm^{3+} ions, co-doping thulium with holmium has been widely studied. Compared with the scheme of cascading pumping holmium-doped fiber with thulium-doped fiber laser, the scheme of pumping thulium-holmium co-doped fiber with ~790 nm laser diode pump source is conducive to reducing the complexity of the laser system and facilitating miniaturization. At the same time, due to the polymer coating does not have obvious absorption for ~790 nm pump light, compared with the holmium-doped fiber that needs to use fluorine-doped glass cladding pumped by thulium-doped fiber laser, thulium-holmium co-doped fiber can use polymer coating, which to some extent reduces the difficulty of fiber preparation process.

7.6.1 Development of Thulium-Holmium Co-Doped Silica fiber and Fiber Lasers

In 1994, Oh et al. from Brown University in the USA [145] first reported the Tm-Ho co-doped silica fiber prepared by MCVD method, using a Ti:sapphire laser at 820 nm as the pump source, and obtained a tunable output in the range of 2037 ~ 2096 nm by changing the length of the laser resonator cavity, with a maximum output power of 12.5 mW. In the same year, Ghisler et al. from the University of Bern in Switzerland [146] first achieved laser output using Tm-Ho co-doped silica fibers pumped by an 809 nm semiconductor laser. In 2007, Jackson et al. [147] used a 793 nm laser diode to pump thulium-holmium co-doped silica fiber, achieving a laser output of 83 W at 2105 nm with a beam quality factor M^2 ~ 1.5, and a slope efficiency of 42%. In 2008, Eichhorn et al. from the French-German Research Institute of Saint-Louis ISL [148] used a 20/300 μm double-clad Tm-Ho co-doped fiber to achieve a single pulse energy of 264 μJ, a pulse width of 45 ns, and an average output power of 12.3 W. In 2010, Hemming et al. [149] used Tm-Ho co-doped fiber to obtain a 2 μm laser output with a linewidth less than 0.5 nm and a wavelength tuning width greater than 280 nm, and achieved a maximum slope efficiency of 25% and an output power of 6.8 W at 2040 nm. In 2013, Yang et al. from the National University of Defense Technology in China [150] reported a gain-switched mode-locked Tm-Ho co-doped fiber laser with a central wavelength of 1958 nm, and achieved a maximum power of 2.17 W supercontinuum output. In 2020, Ramírez-Martínez et al. from the University of Southampton [151] studied the impact of different Tm/Ho doping ratios on the performance of fiber lasers. When Tm/Ho doping ratio is 10 ~ 20, the energy transfer efficiency from Tm to Ho can reach 75%, and an average power of 38 W with a slope efficiency of 56% was obtained for the 2.1 μm laser output. Due to cross-relaxation processes between Tm^{3+} and energy transfer processes from Tm to Ho in Tm-Ho co-doped fibers, as well as the occurrence of energy transfer up-conversion effects at high power operation, the slope efficiency and output power of Tm-Ho co-doped fibers are lower compared to thulium-doped or holmium-doped fibers, and related research is relatively insufficient. However, because it can use mature ~790 nm laser diodes as pump sources and reduce the size of the laser system, thulium-holmium co-doped fiber is of great significance for extending the laser wavelength to above 2.1 μm. In addition, the concept of using Tm^{3+} laser to in-band pump Ho^{3+} ions has also been widely used in laser system design and fiber structure design. For example, Yao et al. from Fudan University in China [152] used an all-fiber system of thulium-doped fiber and holmium-doped fiber to achieve 2.1 μm laser amplification output. Boccuzzi et al. from the University of Rochester in the USA [153] designed a coaxial double-layer thulium-holmium doped fiber, where the fiber core consists of two different doping regions, the center doped with Ho^{3+} and the outer ring doped with Tm^{3+}, as shown in Fig. 7.41. Through simulation calculations, the energy transfer efficiency of this fiber can reach 54% at a pump wavelength of 805 nm.

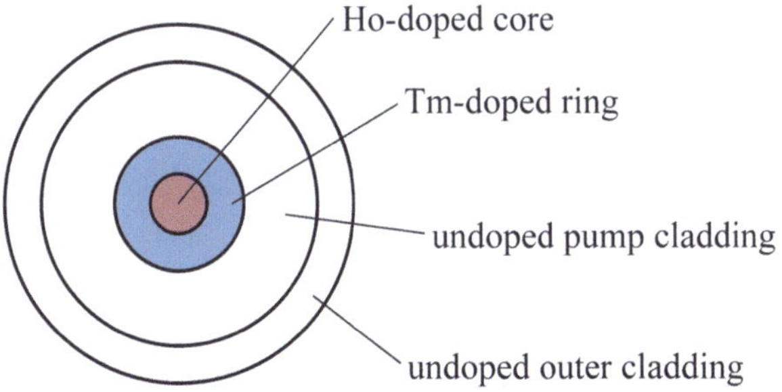

Fig. 7.41 Cross-section design of coaxial double-layer of thulium-holmium doped fiber [153]

7.6.2 Challenges for Thulium-Holmium Co-Doped Silica Fiber

Similar to ytterbium-erbium co-doped fibers, thulium-holmium co-doped fibers are based on energy transfer between Tm^{3+} and Ho^{3+} ions. This luminescent process based on donor (sensitizing ion)-acceptor (luminescent ion) energy transfer has its own drawbacks: when operating at high power, the stimulated radiation probability of the donor ions will be greater than their energy transfer probability, causing almost all of these co-doped systems to have self-excitation effects of sensitizing ions and serious energy up-conversion processes, leading to low energy conversion efficiency and large thermal effects. At the same time, due to the co-doping of Tm^{3+} and Ho^{3+} ions, not only are there issues as mentioned earlier with the individual doping of Tm^{3+} or Ho^{3+} ions, but the complex energy level interactions between Tm^{3+} and Ho^{3+} also bring more serious adverse effects such as energy transfer up-conversion, which degrade the laser performance and pose a more severe challenge for the preparation of high-performance fibers. Starting from the energy level structure, the energy transfer process in thulium-holmium co-doped fibers is briefly analyzed below.

The main energy transfer process for producing 2 μm lasers in Tm-Ho co-doping is shown in Fig. 7.42. Tm^{3+} ions absorb ~793 nm pump photons, excited from the ground state $^3H_6$ to $^3H_4$ level, and then through multiple transition processes (mainly non-radiative transitions) reach $^3F_4$ level. Due to the small energy gap between the $^3F_4$ level of Tm^{3+} ions and the $^5I_7$ level of Ho^{3+} ions level, energy transfer between them is easily achieved. Ho^{3+} at the ground state $^5I_8$ level absorbs ~2 μm photons from Tm^{3+} ions and transit to $^5I_7$ level, leading to population inversion between the upper level $^5I_7$ and the ground state $^5I_8$, and generating emission at ~2.1 μm. In addition, in Tm-Ho co-doped silica glass, when the concentration of Tm^{3+} ions reaches

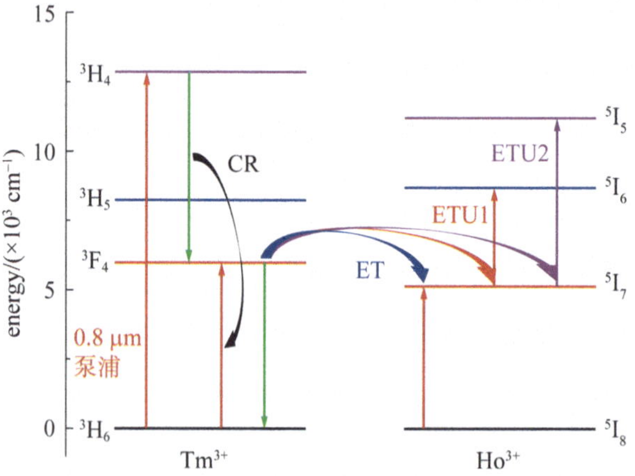

Fig. 7.42 Main energy transfer process between Tm^{3+} and Ho^{3+}

a certain level, significant cross relaxation processes will occur between Tm^{3+} ions, greatly increasing the population of $^3F_4$ level. Therefore, under 790 nm pumping, the laser efficiency of Tm-Ho co-doped fiber is limited by both the cross-relaxation process between Tm^{3+} and the energy transfer process from Tm^{3+} to Ho^{3+}.

In the Tm-Ho co-doped system, there are two kinds of energy transfer upconversion (ETU) processes, Tm^{3+} ($^3F_4$), Ho^{3+} ($^5I_7$) → Tm^{3+} ($^3H_6$), Ho^{3+} ($^5I_6$) and Tm^{3+} ($^3F_4$), Ho^{3+} ($^5I_7$) → Tm^{3+} ($^3H_6$), Ho^{3+} ($^5I_5$), denoted as ETU1 and ETU2, respectively. The energy transfer upconversion process additionally consumes the population of the $^5I_7$ energy level, and the ions that transition to the $^5I_5$ and $^5I_6$ energy levels will return to the $^5I_7$ energy level through non-radiative transition, causing significant thermal effects, reducing energy conversion efficiency, increaseing the pump threshold, and affecting the normal output of 2 μm laser [154]. Due to close energy levels, the approximate resonant energy transfer, coupled with the larger population of the $^3F_4$ and $^5I_7$ energy levels, results in a larger energy transfer upconversion rate [155]. In materials with high phonon energy such as silica glass and silicate glass, the sites occupied by rare earth ions have a large degree of heterogeneity, and the rates of ETU1 and ETU2 may be very close. The non-radiative transition of ions at the upper energy levels $^5I_5$ and $^5I_6$ makes it difficult to measure the rates of ETU1 and ETU2 separately. Therefore, when using the rate equation analysis, ETU1 and ETU2 processes can be considered as the same energy transfer upconversion process.

According to the above analysis, it can be seen that in the Tm-Ho co-doped silica fiber, the cross-relaxation process of Tm^{3+} ions will greatly increase the population in $^3F_4$ level, and this process gradually strengthens with the increase of Tm^{3+} doping concentration. A large number of Tm^{3+} ions are in the $^3F_4$ energy level, which on one hand accelerates the energy transfer rate from Tm^{3+} to Ho^{3+} ions in the $^5I_7$ level, increasing the population in $^5I_7$ energy level, facilitating the transition from the $^5I_7$ energy level to the $^5I_8$ ground state energy level, beneficial for the 2 μm laser output of Ho^{3+}. On the other hand, it will also enhance the up-conversion effect from the $^5I_7$ energy level to the upper energy levels $^5I_5$ and $^5I_6$, increase fiber heating due to non-radiative transition, and reduce energy conversion efficiency, which is not conducive to the 2 μm laser output. Therefore, choosing appropriate Tm^{3+} doping concentration and Tm/Ho doping ratio to promote the effective energy transfer from Tm^{3+} to Ho^{3+} and inhibit energy transfer up-conversion, is the key to improving the laser performance of Tm-Ho co-doped silica fiber. Ramírez-Martínez et al. from the University of Southampton [151] have studied in detail the influence of thulium and holmium doping concentrations on energy transfer efficiency. As shown in Fig. 7.43, when the Tm/Ho doping ratio is 5:1 and 10:1, the fiber with ~5 wt% thulium doping has a faster decrease in lifetime compared to the fiber with ~2 wt% thulium doping, and also has higher energy transfer efficiency. They pointed out that a Tm/Ho doping ratio between 10 and 20 can achieve higher energy transfer efficiency from thulium to holmium. Ultimately, they prepared a thulium-holmium co-doped silica fiber with a Tm/Ho doping ratio of 10 and a thulium doping concentration of 5.5 wt%, achieving a laser output of 37.7 W at 2105 nm. The energy transfer efficiency from Tm to Ho reached 78%, and the slope efficiency with respect to the absorbed pump light was 56%.

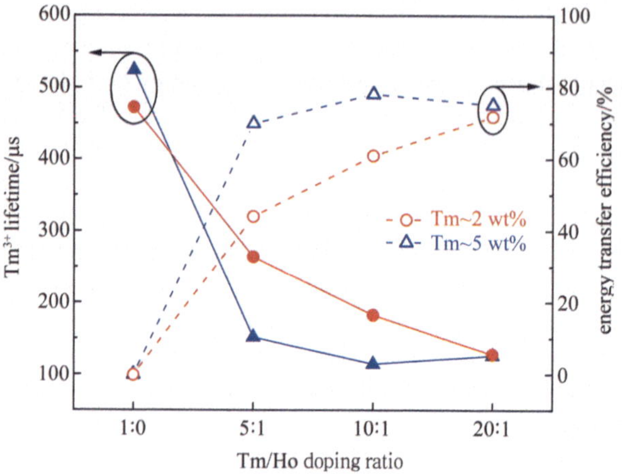

Fig. 7.43 The change of the $^3F_4$ energy level lifetime and energy transfer efficiency with the Tm/Ho doping ratio when pump at 790 nm pumping [151]

7.7 Applications of 2 μm Fiber Lasers

The 2 μm fiber laser, represented by the thulium-doped fiber laser, not only has the advantages of simple structure, high efficiency, good beam quality, and easy operation, but also its laser wavelength is in the absorption band of some important molecules, which can be applied to important fields such as military, medical, detection, and scientific research. The 2 μm fiber laser, due to its important application value, has become a hot research topic worldwide. Compared with the 1 μm ytterbium-doped fiber laser, the 2 μm thulium-doped fiber laser has several advantages, such as: a wider tunable spectral range (over 200 nm, from below 1850 nm to over 2100 nm), including water absorption peak near 1940 nm, which has a shallow tissue penetration depth, and thus becomes a powerful tool for non-invasive surgery and other medical procedures. The holmium-doped fiber laser has a longer wavelength, which has certain advantages in free-space communication and ranging.

Figure 7.44 shows absorption and penetration depth in water and other biological tissues constituents for different wavelengths given by LISA Laser Company [134]. Water accounts for about 70% of the human body, and the absorption characteristics of lasers interacting with water-containing tissues vary greatly at different wavelengths. The absorption coefficient of water molecules to light is an important factor in the interaction between lasers and biological tissues. Water molecules are nearly "transparent" to visible light and near-infrared light, with a very low absorption coefficient. At a wavelength of 500 nm, the absorption coefficient is at its lowest, only 10^{-4} cm^{-1}. Above 500 nm, the absorption coefficient of water molecules increases with wavelength, reaching a strong absorption at 2 μm with a coefficient of 200–600 cm^{-1}. The penetration depth of a 2 μm laser in tissue can reach 0.1 ~ 0.3 mm, providing high surgical precision, good coagulation effect, and eye

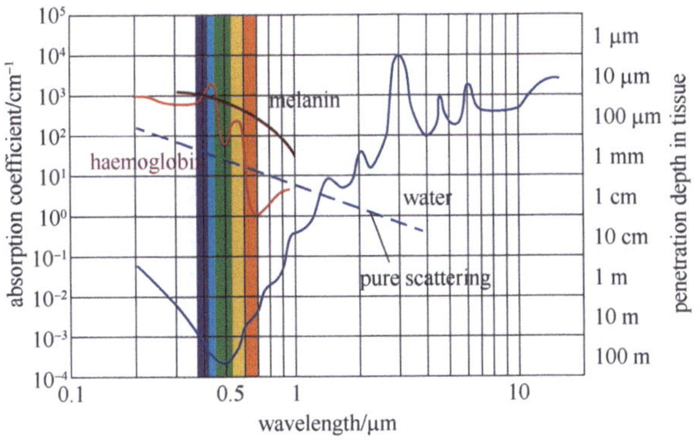

Fig. 7.44 Absorption and penetration depth in water and various "target tissues" for different wavelengths [134]

safety. The 2 μm laser also shows good transmission characteristics in silica fibers, with melanin and hemoglobin having very low absorption rates in this wavelength range. When laser interacts with biological tissue, there are four basic light ablation mechanisms depending on the wavelength, namely photochemical effects, photothermal effects, photomechanical effects, and photoelectric effects. Different wavelengths and radiation energies have different effects on biological tissues: high-intensity thermal effects can cause tissue coagulation, carbonization, vaporization or cutting; low-power lasers can be used for tissue coagulation or hemostasis; focused laser beams can be used as surgical knives for tissue cutting; unfocused lasers can be used for physiotherapy and pathological research. The photothermal effect produced by the 2 μm laser causes the tissue and water to absorb the laser energy, leading to a sharp increase in temperature at the irradiated tissue site, resulting in ablation, vaporization, and cutting effects. In clinical medicine, 2 μm laser lithotripsy has become the most advanced lithotripsy treatment method (holmium laser lithotripsy is fast and safe, with an efficiency of over 90%), while early methods such as medication stone expulsion and extracorporeal shock wave lithotripsy have limitations and are prone to infection or side effects. The 2 μm laser has received high praise in prostatectomy, soft tissue fine cutting, etc., and is considered an excellent safe medical laser source. The 2 μm laser can achieve flexible transmission through silica fibers, and can be combined with various endoscopes or puncture needles to perform minimally invasive surgery, and has been widely recognized and gradually accepted for use in various clinical departments.

In natural conditions, the atmosphere is composed of mixed gases, water vapor, CO_2 and impurities, and only electromagnetic waves of specific frequencies can propagate with low loss in it, and these low-loss transmission bands are called "atmospheric windows". Compared with the 1 μm band, the 2 μm band has lower Rayleigh scattering loss and a lower atmospheric extinction ratio, as shown in Fig. 7.45. In addition, the 2 μm laser is within the safe range for the human eye retina, and is used as an emission source in laser radar remote sensing systems

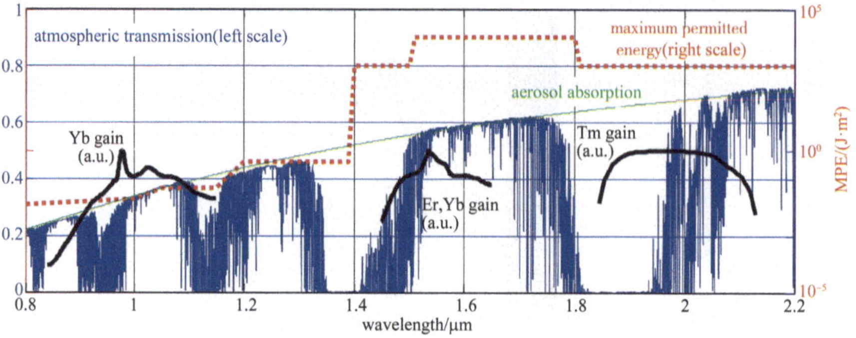

Fig. 7.45 Atmospheric transmission spectrum [156]

[156], which can improve its working distance and safety and accurately and quickly obtain ground or atmospheric three-dimensional spatial information, playing an important role in meteorology, environment and military communication fields. On the other hand, the 2 μm band covers the characteristic absorption lines of molecules such as water, CO_2, and N_2O. Using differential absorption LIDAR, the concentration of atmospheric carbon dioxide and aerosols, cloud layer and water vapor distribution can be monitored, and the accuracy of weather and climate change forecasts can be improved. The National Aeronautics and Space Administration (NASA) proposed as early as 1995 to use a miniaturized 2 μm coherent LIDAR system for global climate monitoring in space [157]. In remote LIDAR detection systems, the analysis is conducted based on the collected laser echo scattering signals. Therefore, the low-loss property of the 2 μm band located in the atmospheric window can better reduce the measurement error of the system.

With the development of modern industry, laser processing has gradually become the preferred processing method in manufacturing, with a huge market demand. The 2 μm band laser is particularly suitable for the processing of polymer materials. Many plastic materials are transparent in the visible and near-infrared bands. If the widely used 1 μm laser in industry is used for processing, it is necessary to add additives to enhance the absorption of laser energy. However, these additives undoubtedly complicate the processing process, and materials containing additives are explicitly prohibited in some special medical devices. On the contrary, most transparent organic materials have strong absorption in the 2 μm band, so the cutting, welding, and marking of plastics can be directly completed with a 2 μm laser, which is particularly important for the manufacture of devices such as bio microfluidic chips. It is evident that the 2 μm laser has great market potential in various plastic processing.

In terms of confidentiality and military, 2 μm band fiber lasers also have many important applications. Battlefield short-distance network communication through the human-eye-safe 2 μm band atmospheric window can reduce signal scattering and interference during transmission, providing a concealed, reliable, and efficient optical communication network. Many dangerous explosive compounds (TNT,

TATP) and biochemical poisons (Sarin, Soman) can be detected in the 7 ~ 12 μm band range, and their characteristic components released by gas volatilization or decomposition (such as ammonia, acetone, etc.) can also be detected in the 2 μm band. In modern high-tech local wars, infrared optoelectronic countermeasures technology mainly use laser beams in the 3 ~ 5 μm mid-infrared band to saturate, blind, and disable infrared detection/tracking/guidance devices. In addition, using a 2 μm band laser pump source to obtain mid-infrared light source is a common technical method: through the 2 μm band high-power fiber laser pumping method, effective Cr ion lasers, super-continuous spectrum lasers, and optical parametric oscillation lasers can be obtained. Broadband mid-infrared lasers are used in fields such as laser guidance, laser-directed infrared interference blinding, active infrared remote sensing, infrared laser ranging and aiming. In summary, the 2 μm band fiber laser is increasingly widely used in many fields, attracting a large number of researchers to continuously explore and have achieved a series of results.

In summary, this chapter starts with the energy level structure of Tm^{3+} and Ho^{3+} ions, discusses the spectral and structural properties of thulium and holmium-doped silica glass, introduces the challenges faced in the preparation of thulium, holmium single-doped and thulium-holmium co-doped optical fibers, and reviews the development of 2 μm band fiber lasers. Unlike the more mature Yb-doped and Er-doped silica fibers, research on thulium and holmium-doped silica fibers at 2 μm band started relatively later with limited reports. From the perspective of laser output power alone, Yb-doped silica fiber has unparalleled advantages, with a current realization of 20 kW single-fiber single-mode laser output,while 2 μm laser fiber still faces many challenges and difficulties, such as the laser wavelength is close to the multi-phonon absorption edge of the silica glass matrix, the absorption coefficient of the hydroxyl group at this band is larger, and the quantum defect is larger, etc. These factors not only greatly increase the difficulty of preparing thulium and holmium-doped silica fibers, but also increase the difficulty of thermal management of such lasers. The highest output power of thulium-doped silica fiber is about 1 kW, and the slope efficiency under 793 nm laser diode pumping can reach 74.5%; the highest output power of holmium-doped silica fiber is about 400 W, and the slope efficiency under 1950 nm thulium-doped laser co-pumping can reach 87%; the highest output power of thulium-holmium co-doped silica fiber is about 80 W, and the slope efficiency under 793 nm laser diode pumping can reach 56%. Although there are still many challenges at present, with the continuous deepening of the research and development of 2 μm silica fiber materials and the continuous progress of laser technology, it is believed that the laser output power and beam quality of the 2 μm band will continue to improve. At present, there is still a gap between domestic 2 μm fiber materials and laser technology and foreign countries. The core gain fiber of the 2 μm band fiber laser mainly relies on imports. Therefore, it is urgent to accelerate the research on high-power thulium-doped, holmium-doped fiber materials and fiber lasers, and realize the localization of core materials and key components. The development of practical, highly reliable 2 μm band high-power fiber lasers is of great significance for promoting the sustainable development of industries, medical treatment, scientific research, national defense and other fields.

References

1. Ehrenreich T, Leveille R, Imtiaz M, et al. 1-kW, all-glass Tm: fiber laser [C]//Proc. SPIE. 7580, Fiber Lasers VII: Technology, Systems, and Applications, 2010.
2. Stutzki F, Gaida C, Gebhardt M, et al. Tm-based fiber-laser system with more than 200 MW peak power [J]. Optics Letters, 2015,40(1):9-12.
3. Jackson S D, Mossman S. Efficiency dependence on the Tm^{3+} and Al^{3+} concentrations for Tm^{3+} -doped silica double-clad fiber lasers [J]. Applied Optics, 2003,42(15):2702-2707.
4. Wang Xue, Lou Fengguang, Wang Shikai, et al. Spectroscopic properties of Tm^{3+}/Al^{3+} co-doped sol-gel silica glass [J]. Optical Materials, 2015,42(1):287-292.
5. Li Kefeng, Zhang Guang, Hu Lili. Watt-level ~2 μm laser output in Tm^{3+} -doped tungsten tellurite glass double-cladding fiber [J]. Optics Letters, 2010,35(24):4136-4138.
6. Balda R, Lacha L M, Fernández J, et al. Optical spectroscopy of Tm^{3+} ions in GeO_2 -PbO -Nb_2O_5 glasses [J]. Optical Materials, 2005,27(11):1771-1775.
7. C.K. Jorgensen, R. Reisfeld. Judd-Ofelt parameters and chemical bonding [J]. Journal of the Less-Common Metals, 1983(93):107-112.
8. Walsh B M, Barnes N P. Comparison of Tm: ZBLAN and Tm: silica fiber lasers; spectroscopy and tunable pulsed laser operation around 1.9 μm [J]. Applied Physics B, 2004,78(3-4):325-333.
9. Tian Ying, Zhang Liyan, Xu Rongrong, et al. 2 μm emission properties in Tm^{3+}/Ho^{3+} -codoped fluorophosphate glasses [J]. Applied Physics B, 2010,101(4):861-867.
10. Wang Xin, Li Kefeng, Yu Chunlei, et al. Effect of Tm_2O_3 concentration and hydroxyl content on the emission properties of Tm doped silicate glasses [J]. Journal of Luminescence, 2014(147):341-345.
11. Heo J, Shin Y B, Jang J N. Spectroscopic analysis of Tm^{3+} in PbO-Bi_2O_3 -Ga_2O_3 glass [J]. Applied Optics, 1995,34(21):4284-4289.
12. Özen G, Aydinli A, Cenk S, et al. Effect of composition on the spontaneous emission probabilities, stimulated emission cross-sections and local environment of Tm^{3+} in TeO_2 -WO_3 glass [J]. Journal of Luminescence, 2003,101(4):293-306.
13. Babu P, Seo H J, Jang K H, et al. Optical spectroscopy and energy transfer in Tm^{3+} -doped metaphosphate laser glasses [J]. Journal of Physics: Condensed Matter, 2005,17(32):4859-4876.
14. Pramod R W, Seongmin J, Seongjae B, et al. Linear and non-linear optical properties of Yb^{3+}/Tm^{3+} co-doped alumino-silicate glass prepared by sol-gel method [J]. Journal of Non-Crystalline Solids, 2005,351(30-32):2446-2452.
15. Balda R, Fernández J, García-revilla S, et al. Spectroscopy and concentration quenching of the infrared emissions in Tm^{3+} -doped TeO_2 -TiO_2 -Nb_2O_5 glass [J]. Optics Express, 2007,15(11):6750-6761.
16. Zhou Bo, Pun E Y-B, Lin Hai, et al. Judd-Ofelt analysis, frequency upconversion, and infrared photoluminescence of Ho^{3+} -doped and Ho^{3+}/Yb^{3+} -codoped lead bismuth gallate oxide glasses [J]. Journal of Applied Physics, 2009,106(10):103-105.
17. Wu Xiaoli, Li Jiguang, Zhu Qi, et al. One-step freezing temperature crystallization of layered rare-earth hydroxide ($Ln_2(OH)_5NO_3 \cdot nH_2O$) nanosheets for a wide spectrum of Ln (Ln=Pr-Er, and Y), anion exchange with fluorine and sulfate, and microscopic coordination probed via photoluminescence [J]. Journal of Materials Chemistry C, 2015,3(14):3428-3437.
18. Cao Ruijie, Lu Yu, Tian Ying, et al. 2 μm emission properties and nonresonant energy transfer of Er^{3+} and Ho^{3+} codoped silicate glasses [J]. Science Report, 2016,6(37873):1-11.
19. Lin J H, Liou H Y, Wang C-D, et al. Giant enhancement of upconversion fluorescence of NaYF4:Yb^{3+}, Tm^{3+} nanocrystals with resonant waveguide grating substrate [J]. ACS Photonics, 2015,2(4):530-536.
20. Liu Xueqiang, Li Ming, Wang Xin, et al. ~2 μm luminescence properties and nonradiative processes of Tm^{3+} in silicate glass [J]. Journal of Luminescence, 2014(150):40-45.
21. Wang Xue, Kang Shuai, Fan Shaohua, et al. Influence of La/Al ratio on the structure and spectroscopy of Tm^{3+} doped Al_2O_3 -La_2O_3 -SiO_2 glasses [J]. Journal of Alloys and Compounds, 2017,690: 583-588.

22. Tsunawaki Y, Iwamoto N, Hattori T, et al. Analysis of CaO-SiO_2 and CaO-SiO_2 -CaF_2 glasses by Raman spectroscopy [J]. Journal of Non-Crystalline Solids, 1981,44(2-3):369-378.
23. Deschamps T, Ollier N, Vezin H, et al. Clusters dissolution of Yb^{3+} in codoped SiO_2 -Al_2O_3 -P_2O_5 glass fiber and its relevance to photodarkening [J]. Journal of Chemical Physics, 2012,136(1):372.
24. Handke M, Mozgawa W, NocuńM. Specific features of the IR spectra of silicate glasses [J]. Journal of Molecular Structure, 1994(325):129-136.
25. Mokaya R, Zhou W, Jones W. Restructuring of mesoporous silica: high quality large crystal MCM-41 via a seeded recrystallisation route [J]. Journal of Materials Chemistry, 2000,10(5):1139-1145.
26. Munoz-aguado M J, Gregorkiewitz M, Bermejo F J. Structural characterization of silica xerogels [J]. Journal of Non-Crystalline Solids, 1995,189(1-2):90-100.
27. Simonutti R, Comotti A, Bracco S, et al. Surfactant organization in MCM-41 mesoporous materials as studied by ^{13}C and ^{29}Si solid-state NMR [J]. Chemistry of Materials, 2008,13(3):771-779.
28. Bakhmutov V I, Shpeizer B G, Clearfield A. Solid-state NMR spectra of paramagnetic silica-based materials: observation of ^{29}Si and ^{27}Al nuclei in the first coordination spheres of manganese ions [J]. Magnetic Resonance in Chemistry, 2006,44(9):861-867.
29. Ren Jinjun, Zhang Long, Eckert H. Medium-range order in sol-gel prepared Al_2O_3 -SiO_2 glasses: new results from solid-state NMR [J]. The Journal of Physical Chemistry C, 2014,118(9):4906-4917.
30. Iftekhar S, Pahari B, Okhotnikov K, et al. Properties and local structures of RE_2O_3 -Al_2O_3 -SiO_2 (RE=Y, Lu) glasses probed by molecular dynamics simulations and solid-state NMR: the role of aluminium and rare-earth ions [J]. Journal of Physical Chemistry, 2012(116):18394-18406.
31. Jaworski A, Stevensson B, Pahari B, et al. Local structures and Al/Si ordering in lanthanum aluminosilicate glasses explored by advanced ^{27}Al NMR experiments and molecular dynamics simulations [J]. Physical Chemistry Chemical Physics, 2012,14(45):15865-15878.
32. Deters H, De Camargo A S S, Santos C N, et al. Structural characterization of rare-earth doped yttrium aluminoborate laser glasses using solid state NMR [J]. The Journal of Physical Chemistry C, 2009,113(36):16216-16225.
33. Li Zhilan, Wang Shikai, Wang Xin, et al. Spectral properties of Tm^{3+} -doped silica glasses and laser behaviors of fibers by sol-gel technology [J]. Chinese Journal of Lasers, 2013,40(8):136-142.
34. Peng B, Izumitani T. Optical properties, fluorescence mechanisms and energy transfer in Tm^{3+}, Ho^{3+} and Tm^{3+} -Ho^{3+} doped near-infrared laser glasses, sensitized by Yb^{3+} [J]. Optical Materials, 1995,4(6):797-810.
35. Marchi J, Morais D S, Schneider J, et al. Characterization of rare earth aluminosilicate glasses [J]. Journal of Non-Crystalline Solids, 2005,351(10):863-868.
36. Nogami M, Abe Y. Fluorescence spectroscopy of silicate glasses codoped with Sm^{2+} and Al^{3+} ions [J]. Journal of Applied Physics, 1997,81(9):6351-6356.
37. Qiao Yanbo, Wen Lei, Wu Botao, et al. Preparation and spectroscopic properties of Yb-doped and Yb-Al-codoped high silica glasses [J]. Materials Chemistry and Physics, 2008,107(2):488-491.
38. Wang Xue, Xu Wenbin, Wang Shikai, et al. Effects of aluminum on the structure and spectroscopic properties of holmium-doped sol-gel silica glasses [J]. Journal of Alloys and Compounds, 2016(657):478-482.
39. Florian P, Sadiki N, Massiot D, et al. ^{27}Al NMR study of the structure of lanthanum- and yttrium-based aluminosilicate glasses and melts [J]. The Journal of Physical Chemistry B, 2007,111(33):9747-9757.
40. Carnall W T, Fields P R, Rajnak K. Electronic energy levels in the trivalent lanthanide aquo ions. I. Pr^{3+}, Nd^{3+}, Pm^{3+}, Sm^{3+}, Dy^{3+}, Ho^{3+}, Er^{3+}, and Tm^{3+} [J]. The Journal of Chemical Physics, 1968,49(10):4424-4442.

41. Gao Song. Laser properties of 2~3 μm spectra and fibers doped with rare earth in germanium tellurite glasses [D]. Beijing: University of Chinese Academy of Sciences, 2016.
42. Mahamuda S, Swapna K, Packiyaraj P, et al. Visible red, NIR and Mid-IR emission studies of Ho^{3+} doped Zinc Alumino Bismuth Borate glasses [J]. Optical Materials, 2013,36(2):362-371.
43. Wang Meng, Yi Lixia, Wang Guonian, et al. 2 μm emission performance in Ho^{3+} doped fluorophosphate glasses sensitized with Er^{3+} and Tm^{3+} under 800nm excitation [J]. Solid State Communications, 2009,149(29):1216-1220.
44. Gao Guojun, Wang Guonian, Yu Chunlei, et al. Investigation of 2.0 μm emission in Tm^{3+} and Ho^{3+} co-doped oxyfluoride tellurite glass [J]. Journal of Luminescence, 2009,129(9):1042-1047.
45. Seshadri M, Ratnakaram Y C, Naidu D T, et al. Investigation of spectroscopic properties (absorption and emission) of Ho^{3+} doped alkali, mixed alkali and calcium phosphate glasses [J]. Optical Materials, 2010,32(4):535-542.
46. Yi Lixia, Wang Meng, Feng Suya, et al. Emission properties of Ho^{3+} $^5I_7 \rightarrow {}^5I_8$ transition sensitized by Er^{3+} and Yb^{3+} in fluorophosphate glasses [J]. Optical Materials, 2009,31(11):1586-1590.
47. Hanna D C, Percival R M. Continuous-wave oscillation of holmium-doped silica fibre laser [J]. Electronics Letters, 1989,25(9):593-594.
48. Jackson S D, King T A. CW operation of a 1.064 - μm pumped Tm-Ho-doped silica fiber laser [J]. IEEE Journal of Quantum Electronics, 1998,34(9):1578-1587.
49. Wang Xue, Li Haiming, Hu Lili. Influence of Y_2O_3 /Al_2O_3 ratio on structure and enhanced emission properties of Ho^{3+} -doped sol-gel high silica glasses [J]. Journal of Luminescence, 2020,220:117020.
50. Jackson S D. Cross relaxation and energy transfer upconversion processes relevant to the functioning of 2 μm Tm^{3+} -doped silica fibre lasers [J]. Optics Communications, 2004,230(1):197-203.
51. Jackson S D, King T A. High-power diode-cladding-pumped Tm-doped silica fiber laser [J]. Optics Letters, 1998,23(18):1462-1464.
52. Hayward R A, Clarkson W A, Turner P W, et al. Efficient cladding-pumped Tm-doped silica fibre laser with high power singlemode output at 2 μm [J]. Electronics Letters, 2000(36):711-712.
53. Frith G, Lancaster D G, Jackson S D. 85 W Tm^{3+} -doped silica fibre laser [C]//IEEE LEOS Annual Meeting Conference Proceedings, 2005:762-763.
54. Frith G, Lancaster D. Power scalable and efficient 790-nm pumped Tm^{3+} -doped fiber lasers [C]//Proc. SPIE 6102, Fiber Lasers III: Technology, Systems, and Applications, 2006,610208.
55. Jackson S D. High power and highly efficient mid-infrared fiber lasers [C]//Proc. SPIE 6453, Fiber Lasers IV: Technology, Systems, and Applications, 2007,64530B.
56. Zhang Yunjun, Wang Yuezhu, Ju Youlun, et al. Experimental investigation of Tm^{3+} -doped silica double-cladding fibre laser [J]. Chinese Physics Letters, 2006,23(9):2452-2454.
57. Eichhorn M, Jackson S D. Comparative study of continuous wave Tm^{3+} -doped silica and fluoride fiber lasers [J]. Applied Physics B, 2008,90(1):35-41.
58. Frith G, Carter A, Samson B, et al. Design considerations for short-wavelength operation of 790-nm-pumped Tm-doped fibers [J]. Applied Optics, 2009,48(27):5072-5075.
59. Moulton P F, Rines G A, Slobodtchikov E V, et al. Tm-doped fiber lasers: fundamentals and power scaling [J]. IEEE Journal of Selected Topics in Quantum Electronics, 2009,15(1):85-92.
60. Goodno G D, Book L D, Rothenberg J E. Low-phase-noise, single-frequency, single-mode 608W thulium fiber amplifier [J]. Optics Letters, 2009,34(8):1204-1206.
61. Frith G, Carter A, Samson B, et al. Mitigation of photodegradation in 790 nm-pumped Tm-doped fibers [C]//Proc. SPIE 7580, Fiber Lasers VII: Technology, Systems, and Applications, 2010,75800A.
62. Mccomb T S, Sims R A, Willis C C C, et al. High-power widely tunable thulium fiber lasers [J]. Applied Optics, 2010,49(32):6236-6242.
63. Ndebeka W I, Heidt A, Schwoerer H, et al. Efficiency of Tm^{3+} -doped silica triple clad fiber laser [C]//Proceedings of the International Summer Session: Lasers and Their Applications, (Optica Publishing Group), 2011, paper Th4.

64. Shardlow P C, Jain D, Parker R, et al. Optimising Tm-doped silica fibres for high lasing efficiency [C]. 2015 European Conference on Lasers and Electro-Optics-European Quantum Electronics Conference, (Optica Publishing Group), 2015, paper CJ_14_3.
65. Tumminelli R, Petit V, Carter A, et al. Highly doped and highly efficient Tm doped fiber laser [C]//Proc. SPIE 10512, Fiber Lasers XV: Technology and Systems, 2018,105120M.
66. Sincore A, Bradford J D, Cook J, et al. High average power thulium-doped silica fiber lasers: review of systems and concepts [J]. IEEE Journal of Selected Topics in Quantum Electronics, 2018,24(3):1-8.
67. Ramírez-martínez N J, Núñez-velázquez M, Umnikov A A, et al. Highly efficient thulium-doped high-power laser fibers fabricated by MCVD [J]. Optics Express, 2019,27(1):196-201.
68. Ramírez-martínez N J, Núñez-velzquez M, Umnikov A A, et al. Novel fabrication technique for highly efficient Tm-doped fibers [C]. Conference on Lasers and Electro-Optics, OSA Technical Digest (online) (Optica Publishing Group), 2018, paper SF3I.2.
69. Strelov K K, Kashcheev I D. Phase diagram of the system Al_2O_3 -SiO_2 [J]. Refractories, 1995,36(8):244-246.
70. Barber M J, Shardlow P C, Barua P, et al. Nested-ring doping for highly efficient 1907 nm short-wavelength cladding-pumped thulium fiber lasers [J]. Optics Letters, 2020,45(19):5542-5545.
71. Barber M, Shardlow P, Barua P, et al. Cladding-pumped nested-ring Tm fiber laser with 131 W single-mode output at 1907 nm [C]//Proc. SPIE 11260, Fiber Lasers XVII: Technology and Systems, 2020,112600F.
72. Liu Yinz, Xing Yingbin, Liao Lei, et al. 530 W all-fiber structure continuous thulium-doped fiber laser [J]. Acta Physica Sinica, 2020(18):105-111.
73. Lou Fengguang, Kuan Peiwen, Zhang Lei, et al. 2 μm laser properties of Tm^{3+} -doped large core sol-gel silica fiber [J]. Optical Materials Express, 2014,4(6):1267-1275.
74. Wang Xue. Study on the structure and 2 μm spectrum of modified silica glass and its fiber properties [D]. Beijing: University of Chinese Academy of Sciences, 2017.
75. Meng Jia, Zhang Wei, Zhao Kaiqi, et al. Performance study of domestic thulium-doped fiber laser oscillator [J]. Chinese Optics, 2019(5):9.
76. Humbach O, Fabian H, Grzesik U, et al. Analysis of OH absorption bands in synthetic silica [J]. Journal of Non-Crystalline Solids, 1996(203):19-26.
77. Keck D B, Maurer R D, Schultz P C. On the ultimate lower limit of attenuation in glass optical waveguides [J]. Applied Physics Letters, 1973,22(7):307-309.
78. Keck D B, Tynes A R. Spectral response of low-loss optical waveguides [J]. Applied Optics, 1972,11(7):1502-1506.
79. Kaiser P, Tynes A R, Astle H W, et al. Spectral losses of unclad vitreous silica and soda-lime-silicate fibers [J]. Journal of the Optical Society of America, 1973,63(9):1141-1148.
80. Elliott C R, Newns G R. Near infrared absorption spectra of silica: OH overtones [J]. Appl. Spectrosc., 1971,25(3):378-379.
81. Clasen R. Optical impurity measurements on silica glass prepared via the colloidal gel route [J]. Glastechnische Berichte, 1990,63(10):291-299.
82. Jollivet C, Farley K, Conroy M, et al. Design optimization of Tm-doped large-mode area fibers for power scaling of 2 μm lasers and amplifiers [C]//Proc. SPIE 10083, Fiber Lasers XIV: Technology and Systems, 2007,100830I.
83. Simakov N, Hemming A V, Carter A, et al. Design and experimental demonstration of a large pedestal thulium-doped fiber [J]. Optics Express, 2015,23(3):3126-3133.
84. Hanna, D. C, Jauncey, et al. Continuous-wave oscillation of a monomode thulium-doped fiber laser [J]. Electronics Letters, 1988,24(19):1222-1223.
85. Clarkson W A, Barnes N P, Turner P W, et al. High-power cladding-pumped Tm-doped silica fiber laser with wavelength tuning from 1860 to 2090nm [J]. Optics Letters, 2002,27(22):1989-1991.
86. Shen D Y, Mackenzie J I, Sahu J K, et al. High-power and ultra-efficient operation of a Tm^{3+} -doped silica fiber laser [C]. Advanced Solid-State Photonics (TOPS), Vol. 98 of OSA Trends in Optics and Photonics (Optica Publishing Group, 2005), paper 516.

87. Jackson S D, Sabella A, Lancaster D G. Application and development of high-power and highly efficient silica-based fiber lasers operating at 2μm [J]. IEEE Journal of Selected Topics in Quantum Electronics, 2007,13(3):567-572.
88. Slobodtchikov E, Moulton P F, Frith G. Efficient, high-power, Tm-doped silica fiber laser [C]. Advanced Solid-State Photonics, OSA Technical Digest Series (CD) (Optica Publishing Group, 2007), paper MF2.
89. Meleshkevich M, Platonov N, Gapontsev D, et al. 415 W single-mode CW thulium fiber laser in all-fiber format [C]. 2007 European Conference on Lasers and Electro-Optics and the International Quantum Electronics Conference, 2007:1.
90. Mccomb T S, Sims R A, Willis C, et al. Tunable high power thulium fiber lasers [J]. 2010 23rd Annual Meeting of the IEEE Photonics Society, 2010:618-619.
91. Yin Ke, Zhang Bin, Xue Guanghui, et al. High-power all-fiber wavelength-tunable thulium doped fiber laser at 2 μm [J]. Opt. Express, 2014,22(17), 19947-19952.
92. Wang Xiong, Jin Xiaoxi, Zhou Pu, et al. High power, widely tunable, narrowband superfluorescent source at 2μm based on a monolithic Tm-doped fiber amplifier [J]. Optics Express, 2015,23(3):3382-3389.
93. Yin Ke, Zhu Rongzhen, Zhang Bin, et al. 300 W-level, wavelength-widely-tunable, all-fiber integrated thulium-doped fiber laser [J]. Opt. Express, 2006,24(10), 11085-11090.
94. Creeden D, Johnson B R, Setzler S D, et al. Resonantly pumped Tm-doped fiber laser with >90% slope efficiency [J]. Optics Letters, 2014,39(3):470-473.
95. Creeden D, Johnson B R, Rines G A, et al. High power resonant pumping of Tm-doped fiber amplifiers in core- and cladding-pumped configurations [J]. Optics Express, 2014,22(23):29067-29080.
96. Walbaum T, Heinzig M, Schreiber T, et al. Monolithic thulium fiber laser with 567 W output power at 1970nm [J]. Optics Letters, 2016,41(11):2632-2635.
97. Gaida, Gebhardt, Heuermann, et al. Ultrafast thulium fiber laser system emitting more than 1 kW of average power [J]. Optics letters, 2018,43(23):5853-5856.
98. Burns M D, Shardlow P C, Barua P, et al. 47W continuous-wave 1726 nm thulium fiber laser core-pumped by an erbium fiber laser [J]. Optics Letters, 2019,44(21):5230-5233.
99. Anderson B M, Soloman J, Flores A. 1.1 kW, beam combinable thulium doped all-fiber amplifier [C]//Proc. SPIE 11665, Fiber Lasers XVII: Technology and Systems, 2021,116650B.
100. Liu Dongfeng. Preliminary study on the generation of 1.871 μm laser from Tm^{3+} doped single-mode silica fiber [J]. Chinese Journal of Lasers, 1998,25(11):973-975.
101. Shanghai Institute of Optics and Fine Mechanics achieved 70 W output power from 2 μm band fiber laser [J]. Opto-Electro-Mechanical Information, 2007(12):1.
102. Tang Yulong, Yang Yong, Xu Jianqiu. High power Tm^{3+} -doped fiber lasers tuned by a variable reflective output coupler [J]. Research Letters in Optics, 2008,919403.
103. Xia Linzhong, Du Geguo, Ruan Shuangchen, et al. High power thulium-doped fiber laser with cladding pumping [J]. Journal of Photonics, 2008,37(6):1089-1092.
104. Zhou Hui, Jing Tao, Zhang Yunjun. The LD-cladding-pumped high power Tm^{3+} -doped silica fibre amplifying at approximately 1.99 μm [J]. Chinese Physics Letters, 2008,25(4):1291-1292.
105. Zhang Yunjun, Song Shifei, Tian Yi, et al. Ld-clad-pumped all-fiber Tm^{3+} -doped silica fiber laser [J]. Chinese Physics Letters, 2009,26(8):164-166.
106. Zhang Y, Jing T. All-fiber clad-pumped Tm^{3+} -doped fiber amplifier using high power fiber combiner [J]. Laser Physics, 2009,19(12):2197-2199.
107. Tang Yulong, Xu Jianqiu, Chen Wei, et al. 150-W Tm^{3+} -doped fiber lasers with different cooling techniques and output couplings [J]. Chinese Physics Letters, 2010,27(10):114-116.
108. Tang Yulong, Xu Lin, Xu Jianqiu. High-power gain-switched Tm^{3+} -doped fiber laser [J]. Optics Express, 2010,18(22):22964-22972.
109. Yu Ting, Shu Shijiang, Chen Weibiao. High repetition rate Tm:Ho:LuLiF master-oscillator and Tm-doped fiber power-amplifier system [J]. Chinese Optics Letters, 2011,9(4):64-66.

110. Wang Xiong, Zhou Pu, Wang Xiaolin, et al. 102 W monolithic single frequency Tm-doped fiber MOPA [J]. Optics Express, 2013,21(26):32386-32392.
111. Hu Zhenyue, Yan Ping, Xiao Qirong, et al. 227-W output all-fiberized Tm-doped fiber laser at 1908-nm [J]. Chinese Physics B, 2014,23(10):104206.
112. Liu Jiang, Shi Hongxing, Liu Kun, et al. 210 W single-frequency, single-polarization, thulium-doped all-fiber MOPA [J]. Optics Express, 2014,22(11):13572-13578.
113. Liu Jiang, Shi Hongxing, Liu Chen, et al. High-power narrow-linewidth thulium-doped all-fiber MOPA. 2015 11th Conference on Lasers and Electro-Optics Pacific Rim (CLEO-PR), 2015:1-2.
114. Yin Ke, Zhu Rongzhen, Zhang Bin, et al. 300 W-level, wavelength-widely-tunable, all-fiber integrated thulium-doped fiber laser[J]. Optics Express, 2016, 24(10): 11085-11090.
115. Yao Weichao, Shan Zhenhua, Shen Chongfeng, et al. 400 W all-fiberized Tm-doped MOPA at 1941nm with narrow spectral linewidth [C]. Laser Congress 2017 (ASSL, LAC), (Optica Publishing Group), 2017, paper JTu2A.33.
116. Yao Weichao, Shen Chongfeng, Shao Zhenhua, et al. 790 W Incoherent beam combination of a Tm-doped fiber laser at 1941nm using a 3×1 signal combiner [J]. Applied Optics, 2018,57(20):5574-5577.
117. Liu Yinzi, Cao Chi, Xing Yingbin, et al. 406W narrow-linewidth all-fiber amplifier with Tm-doped fiber fabricated by MCVD [J]. IEEE Photonics Technology Letters, 2019,31(22):1779-1782.
118. Friebele E J, Askins C G, Peele J R, et al. Ho-doped fiber for high Energy Laser applications [C]//Proc. SPIE 8961, Fiber Lasers XI: Technology, Systems, and Applications, 2014,896120.
119. Hemming A, Simakov N, Haub J, et al. A review of recent progress in holmium-doped silica fibre sources [J]. Optical Fiber Technology, 2014,20(6):621-630.
120. Hemming A, Bennetts S, Simakov N, et al. Development of resonantly cladding-pumped holmium-doped fibre lasers [C]//Proc. SPIE 8237, Fiber Lasers IX: Technology, Systems, and Applications, 2012,82371J.
121. Hemming A, Bennetts S, Simakov N, et al. High power operation of cladding pumped holmium-doped silica fibre lasers [J]. Optics Express, 2013,21(4):4560-4566.
122. Hemming A, Simakov N, Oermann M, et al. Record efficiency of a holmium-doped silica fibre laser [C]. 2016 Conference on Lasers and Electro-Optics (CLEO), 2016:1-2.
123. Baker C, Friebele E J, Burdett A, et al. Recent advances in holmium doped fibers for high-energy lasers [C]//Proc. SPIE 10637, Laser Technology for Defense and Security XIV, 2018,1063704.
124. Baker C, Friebele E J, Burdett A A, et al. Nanoparticle doping for high power fiber lasers at eye-safer wavelengths [J]. Optics Express, 2017,25(12):13903-13915.
125. Friebele E J, Baker C, Burdett A, et al. Ho-nanoparticle-doping for improved high-energy laser fibers [C]//Proc. SPIE 10100, Optical Components and Materials XIV, 2017,1010003.
126. Wang Jiachen, Bae N, Lee S B, et al. Effects of ion clustering and excited state absorption on the performance of Ho-doped fiber lasers [J]. Optics Express, 2019,27(10):14283-14297.
127. Wang Jiachen, Yeom D I, Simakov N, et al. Numerical modeling of in-band pumped Ho-doped silica fiber lasers [J]. Journal of Lightwave Technology, 2018,36(24):5863-5880.
128. Kamrádek M, Aubrecht J, Jelínek M, et al. Holmium-doped fibers for efficient fiber lasers at 2100 nm [C]. OSA High-brightness Sources and Light-driven Interactions Congress 2020 (EUVXRAY, HILAS, MICS), OSA Technical Digest (Optica Publishing Group, 2020), paper MTh3C.5.
129. Hemming A, Simakov N, Davidson A, et al. A monolithic cladding pumped holmium-doped fibre laser [C]//Proceedings of the CLEO OSA Technical Digest (online) (Optica Publishing Group), 2013, paper CW1M.1.
130. Shardlow P C, Simakov N, Billaud A, et al. Holmium doped fibre optimised for resonant cladding pumping [C]. 2017 Conference on Lasers and Electro-Optics Europe and European Quantum Electronics Conference (CLEO/Europe-EQEC), 2017:1-1.

131. Shardlow P C, Standish R, Sahu J, et al. Cladding shaping of optical fibre preforms via CO_2 laser machining [C]. 2015 European Conference on Lasers and Electro-Optics-European Quantum Electronics Conference, (Optica Publishing Group, 2015), paper CJ_P_29.
132. Shardlow P, Standish R, Velazquez M, et al. Cladding shaping of optical fiber preforms via CO_2 laser machining [C]//Proc. SPIE 11355, Micro-Structured and Specialty Optical Fibres VI, 2020,113550A.
133. Jasim A A, Podrazk O, Peterka P, et al. Impact of shaping optical fiber preforms based on grinding and a CO_2 laser on the inner-cladding losses of shaped double-clad fibers [J]. Optics Express, 2020,28(9):13601-13615.
134. Scholle K, Lamrini S, Koopmann P, et al. 2 μm laser sources and their possible applications [M]//Frontiers in Guided Wave Optics and Optoelectronics. London: IntechOpen, 2010.
135. Kurkov A S, Dianov E M, Medvedkov O I, et al. Efficient silica-based Ho^{3+} fibre laser for 2 μm spectral region pumped at 1.15 μm [J]. Electronics Letters, 2000,36(12):1015-1016.
136. Jackson S D, Bugge F, Erbert G. High-power and highly efficient diode-cladding-pumped Ho^{3+} -doped silica fiber lasers [J]. Optics Letters, 2007,32(22):3349-3351.
137. Kurkov A S, Sholokhov E M, Medvedkov O I, et al. Holmium fiber laser based on the heavily doped active fiber [J]. Laser Physics Letters, 2009,6(9):661-664.
138. Kurkov A S, Dvoyrin V V, Marakulin A V. All-fiber 10W holmium lasers pumped at λ=1.15 μm [J]. Optics Letters, 2010,35(4):490-492.
139. Kurkov A S, Sholokhov E M, Tsvetkov V B, et al. Holmium fibre laser with record quantum efficiency [J]. Quantum Electronics, 2011,41(6):492-494.
140. Hemming A, Bennetts S, Simakov N, et al. Resonantly pumped 2 μm Holmium fibre lasers [C]//Proceedings of the Advanced Photonics, OSA Technical Digest (CD) (Optica Publishing Group, 2011), paper SOMB1.
141. Holmen L G, Shardlow P C, Barua P, et al. Tunable holmium-doped fiber laser with multiwatt operation from 2025 nm to 2200 nm [J]. Optics Letters, 2019,44(17):4131-4134.
142. Johnson L F, Geusic J E, Van Uitert L G. Efficient, high-power coherent emission from Ho^{3+} ions in yttrium aluminum garnet, assisted by energy transfer [J]. Applied Physics Letters, 2004,8(8):200-202.
143. Li Kefeng, Zhang Qiang, Fan Sijun, et al. Mid-infrared luminescence and energy transfer characteristics of Ho^{3+}/Yb^{3+} codoped lanthanum-tungsten-tellurite glasses [J]. Optical Materials, 2010,33(1):31-35.
144. Rao P R, Venkatramaiah N, Gandhi Y, et al. Role of modifier oxide in emission spectra and kinetics of Er-Ho codoped Na_2SO_4 -MO -P_2O_5 glasses [J]. Spectrochimica Acta Part A Molecular and Biomolecular Spectroscopy, 2012,86(3):472-480.
145. Oh K, Morse T F, Kilian A, et al. Continuous-wave oscillation of thulium-sensitized holmium-doped silica fiber laser [J]. Optics Letters, 1994,19(4):278-280.
146. Ghisler C, Lüthy W, Weber H P, et al. A Tm^{3+} sensitized Ho^{3+} silica fibre laser at 2.04 μm pumped at 809nm [J]. Optics Communications, 1994,109(3):279-281.
147. Jackson S D, Sabella A, Hemming A, et al. High-power 83 W holmium-doped silica fiber laser operating with high beam quality [J]. Optics Letters, 2007,32(3):241-243.
148. Eichhorn M, Jackson S D. High-pulse-energy, actively Q-switched Tm^{3+}, Ho^{3+} -codoped silica 2 μm fiber laser [J]. Optics Letters, 2008,33(10):1044-1046.
149. Hemming A, Jackson S D, Sabella A, et al. High power, narrow bandwidth and broadly tunable Tm^{3+}, Ho^{3+} -co-doped aluminosilicate glass fibre laser [J]. Electronics Letters, 2010,46(24):1617-1618.
150. Yang Weiqiang, Zhang Bin, Hou Jing, et al. Gain-switched and mode-locked Tm/Ho -codoped 2 μm fiber laser for mid-IR supercontinuum generation in a Tm-doped fiber amplifier [J]. Laser Physics Letters, 2013,10(4):045106.
151. Ramírez-martínez N J, Núñez-velzquez M, Sahu J K. Study on the dopant concentration ratio in thulium-holmium doped silica fibers for lasing at 2.1 μm [J]. Optics Express, 2020,28(17):24961-24967.

152. Yao Weichao, Shen Chongfeng, Shao Zhenhua, et al. High-power nanosecond pulse generation from an integrated Tm-Ho fiber MOPA over 2.1 μm [J]. Optics Express, 2018,26(7):8841-8848.
153. Boccuzzi K A, Newburgh G A, Marciante J R. Tm/Ho -doped fiber laser systems using coaxial fiber [J]. Optics Express, 2019,27(20):27396-27408.
154. Jackson S D. The effects of energy transfer upconversion on the performance of Tm^{3+}, Ho^{3+} -doped silica fiber lasers [J]. IEEE Photonics Technology Letters, 2006,18(17):1885-1887.
155. Jackson S D. The spectroscopic and energy transfer characteristics of the rare earth ions used for silicate glass fibre lasers operating in the shortwave infrared [J]. Laser and Photonics Reviews, 2009,3(5):466-482.
156. Cariou J-P, Augere B, Valla M. Laser source requirements for coherent lidars based on fiber technology [J]. Comptes Rendus Physique, 2006,7(2):213-223.
157. Baker W E, Emmitt G D, Robertson F, et al. Lidar-measured winds from space: A key component for weather and climate prediction [J]. Bulletin of the American Meteorological Society, 1995,76(6):869-888.

Chapter 8
New Neodymium-Doped and Neodymium-Ytterbium Co-Doped Silica Fiber and Applications

Neodymium ions (Nd^{3+}) have multiple absorption bands in the 400~900 nm range, and laser output can be achieved through simple xenon or krypton lamp pumping; at the same time, its lower energy level has four radiative transition bands, which theoretically enable 0.9 μm, 1.06 μm, 1.33 μm and 1.8 μm laser output. Among them, the 1.06 μm transition is a four-level structure, with a low laser threshold and a large emission cross-section, making the Nd^{3+} earliest activated ion in achieving laser output at room temperature. Therefore, 1.06 μm laser has been widely studied, and extensive research results have been achieved in silicate, borate, phosphate, germanate, tellurate and other matrices. In 1961, Snitzer et al. [1] achieved laser output by Nd^{3+}-doped silicate glass. After that, the research on Nd^{3+}-doped laser glasses in China was systematically carried out under the leadership of Mr. Gan Fuxi, and in 1962, a 1.06 μm laser output was realized by Nd^{3+}-doped silicate glass. To this day, Chinese Nd^{3+}-doped laser glass has undergone a transition from silicate glass to phosphate glass, and has been applied in high-energy lasers and high-power lasers. The most representative example is the mass application of Nd^{3+}-doped phosphate laser glass for laser fusion devices.

With the development of laser technology, fields such as laser shock strengthening and laser fusion energy have put forward significant demands for high-energy, high-repetition-rate lasers [2]. High-energy, high-repetition-rate lasers require laser materials to have good thermal shock resistance and a large emission cross-section. Compared to multi-component glass systems, Nd^{3+}-doped silica glass has excellent thermal shock resistance and chemical stability, and is a kind of promising candidate [3]. On the other hand, the 0.9 μm laser of Nd^{3+} can be directly applied to Yb^{3+} laser pumping, atmospheric detection, and differential absorption radar, among other fields [4, 5,] and its second harmonic can be used in underwater communication, biological imaging, quantum optics, atomic cooling and capture, among other fields [6–9]. Therefore, in recent years, research on Nd^{3+}-doped silica fibers aimed at 900 nm laser output has received a lot of attention. In addition, Nd^{3+} -doped fiber lasers and amplifiers at 1.3 μm, as well as Nd^{3+}/Yb^{3+} co-doped dual-wavelength,

L. Hu, *Rare Earth Doped Silica Fiber and Its Applications*,
https://doi.org/10.1007/978-981-96-5306-5_8

tunable silica fiber lasers, have also attracted widespread attention in recent years due to the potential application in fiber optic communication, distributed fiber sensing, gas detection, terahertz radiation. This chapter first introduces the spectral characteristics of Nd^{3+}, and then elaborates on the application of the 900 nm laser. The spectral and laser characteristics of Nd^{3+}-doped silica glasses and optical fibers are discussed, followed by a review of recent advances in 1.3 μm Nd^{3+}-doped fiber lasers and amplifiers. Finally, the spectral properties and laser performance as well as the applications of Nd^{3+}/Yb^{3+} co-doped silica fibers are introduced.

8.1 Spectral Properties of Nd^{3+}-Doped Silica Glass

8.1.1 Energy Level Structure and Spectrum of Neodymium Ions

The electronic configuration of Nd^{3+} is $1s^2\ 2s^2\ 2p^6\ 3s^2\ 3p^6\ 3d^{10}\ 4s^2\ 4p^6\ 4d^{10}\ 5s^2\ 5p^6\ 4f^3$. Its 4f orbit is not fully occupied, which allows electrons transition within it (i.e., f-f transition), forming the characteristic absorption and emission of Nd^{3+}. The 4f electrons are shielded by the outer 5s and 5p electrons, resulting in narrow-band absorption and emission spectra. The energy level structure of Nd^{3+} $4f^3$ electrons is determined by the Hamiltonian of the multi-electron system. The basic interactions of trivalent rare earth ions in the free state mainly are the Coulomb interaction and spin-orbit interaction between electrons in the $4f^n$ orbit. In addition, in solids such as glass, the influence of the external structural environment on the outer electrons of rare earth ions must be considered. The Hamiltonian operator H_{fi} for the i-th electron on the 4f orbit of a rare earth ion can be written as

$$\begin{aligned}\hat{H}_{fi} &= \hat{H}_{H\cdot like} + \hat{H}_{coul} + \hat{H}_{SO} + \hat{H}_{CF} \\ &= \left(\frac{-h^2}{2m} \sum_{i}^{n} \nabla_i^2 - Z_{ef} \sum_{i}^{n} \frac{e^2}{r_i} \right) + \sum_{i<j}^{n} \frac{e^2}{r_{ij}} + \lambda L \cdot S + \sum_i V_{C_i}(r,\theta,\phi)\end{aligned} \tag{8.1}$$

In the formula, $\hat{H}_{H\cdot like}$ represents the action of the central field; e is the electron charge; m is the electron mass; r_i is the distance between the electron and the shielded nucleus; Z_{ef} is the nuclear charge number of the atomic nucleus; $\hat{H}_{coul}$ represents the Coulomb interaction between electrons; r_{ij} represents the distance between electrons; $\hat{H}_{SO}$ represents the spin-orbit interaction; λ is the spin-orbit coupling constant; $\hat{H}_{coul}$ represents the perturbation of the crystal field on the free ion; L is the total orbital angular momentum quantum number; S is the total spin angular momentum quantum number.

When no influences are considered, the electron energy of the same electron configuration is degenerate; when considering the Coulomb interaction between electrons ($\hat{H}_{coul}$), the degenerate energy levels split, forming ^2P, ^2D, ^2F, ^2G, ^2H, ^2I, ^2K,

^2L, ^4S, ^4D, ^4F, ^4G, ^4I multiple spectral energy levels; When further considering the spin-orbit interaction ($\hat{H}_{SO}$), each spectral term will split into $2S + 1$ or $2L + 1$ spectral sub-terms; when $L > S$, it can be split into $2S + 1$ energy levels; when $L < S$, it can be split into $2L + 1$ energy levels. Each spectral sub-term is represented as $^{2S+1}L_J$, where S is the total spin angular momentum quantum number, L is the total orbital angular momentum quantum number, J is the total angular quantum number. The ground state spectral term of Nd^{3+} is ^4I, and the ground state spectral sub-term is $^4I_{9/2}$. Under the action of the crystal field, each spectral sub-term of Nd^{3+} will split into $J + 1/2$ energy levels, known as Stark level splitting. Figure 8.1 shows the energy level splitting displacement and splitting degree of Nd^{3+} under the above effects.

When used as the active ion in laser materials, the upper energy level of Nd^{3+} is $^4F_{3/2}$, the lower energy level is $^4I_{J'}$. Under the excitation of pump light, the Nd^{3+} in the ground state is excited to $^4F_{3/2}$ and above, then quickly relax to the metastable $^4F_{3/2}$ level before radiative transition. During the radiative transition, particles at the $^4F_{3/2}$ level transit down to the $^4I_{J'}$ level, producing a photon. Figure 8.2 shows the fluorescence spectrum of Nd^{3+} under 808 nm laser diode (LD) pumping, $^4F_{3/2} \rightarrow$ $^4I_{9/2}$, $^4I_{11/2}$, $^4I_{13/2}$ transitions correspond to fluorescence peaks at 900 nm, 1060 nm, and 1330 nm, respectively. The fluorescence of $^4F_{3/2}$ $\rightarrow$ $^4I_{15/2}$ transition is usually difficult to

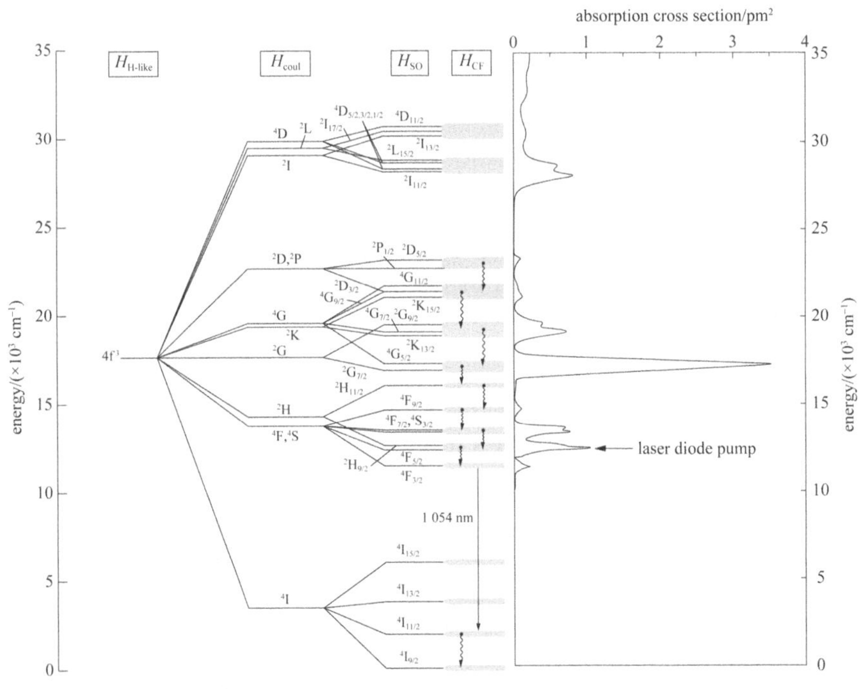

Fig. 8.1 4f³ electron energy level structure of Nd^{3+} (left) and absorption spectrum of Nd^{3+}-doped silica glass (right) [10]

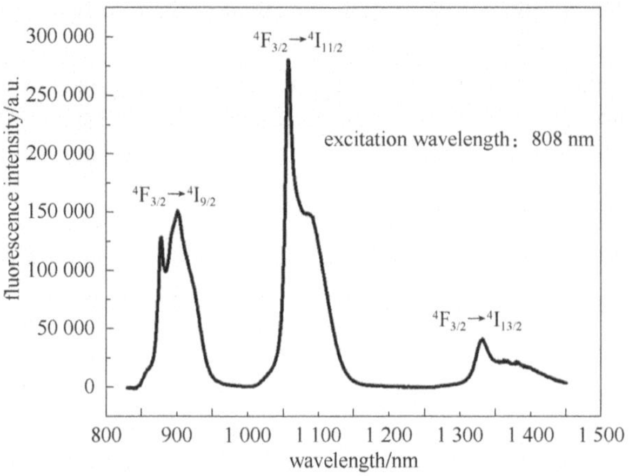

Fig. 8.2 Fluorescence spectrum of Nd^{3+}-doped silica glass under the excitation of 808 nm LD

observe, due to its energy level difference being close to that of $^4I_{9/2} \rightarrow {}^4I_{15/2}$, which easily causes cross-relaxation.

8.1.2 Calculation of Spectral Parameters of Neodymium Ions

According to the selection rule of spectra, the f-f electric dipole transition of lanthanide ions with $\Delta L = 0$ is parity forbidden. However, under the influence of the crystal field of the matrix, the anti-parity $4f^n$ -5d state is mixed into the $4f^n$ configuration of rare earth ions, the f-f transition prohibition is lifted, resulting in a relatively weak electric dipole transition, known as forced or induced electric dipole transition. The Judd-Ofelt theory, starting from configuration mixing, based on the single configuration approximation of the electrostatic field, using perturbation treatment, derives the relationship between transition strength and crystal field, and proposes a method to obtain the optical transition strength parameters by fitting the absorption spectrum [11], is an important theory for studying the f-f transition of rare earth ions [12, 13]. According to the Judd-Ofelt theory, the 4f electron of Nd^{3+} from $(S, L)\,J$ state to $(S', L')\,J'$ state's spontaneous radiation transition probability $(A_{JJ'})$ can be expressed as [14]

$$A_{JJ'} = \frac{64\pi^4 e^2}{3h\lambda^3(2J+1)} \times \frac{n(n^2+2)^2}{9} \sum_{t=2,4,6} \Omega_t |<(S,L)J|U^{(t)}|(S',L')J'>|^2 \quad (8.2)$$

In the formula, e is the electronic charge; n is the refractive index of the material; h is the Planck constant; λ is the central wavelength of the electronic transition; J is the total angular momentum quantum number of the initial state;$| < (S, L)\ J\ |\ U^{(t)}$

$|(S', L') J' > |^2$ is the reduced matrix element, which is only related to the rare earth ions and their energy level structure, and does not change with the matrix material, Nd^{3+} The reduced matrix elements between each energy level have been previous reported [15]; Ω_t ($t = 2, 4, 6$) is called the Judd-Ofelt intensity parameter, which is only related to the material and does not change with the specific transition.

The fluorescence branching ratio ($\beta_{JJ'}$) is

$$\beta_{JJ'} = \beta\left[(S,L)J;(S',L')J'\right] = \frac{A\left[(S, L)J;(S', L')J'\right]}{\sum_{bJ'} A\left[(S, L)J;(S', L')J'\right]} \tag{8.3}$$

The theoretical radiative transition lifetime (τ_{rad}) is

$$\tau_{\mathrm{rad}} = \left\{\sum_{S', L', J'} A\left[(S, L)J;(S', L')J'\right]\right\}^{-1} \tag{8.4}$$

According to the experimental measurement of the corresponding transition fluorescence lifetime (τ_{exp}), the quantum efficiency can also be obtained

$$\eta = \tau_{\mathrm{exp}} / \tau_{\mathrm{rad}} \tag{8.5}$$

8.1.2.1 Judd-Ofelt Intensity Parameters Ω_t ($t = 2, 4, 6$)

As shown in formula (8.2), the intensity of the rare earth ion f-f transition is mainly affected by the reduced matrix element $|<(S,L)J| U^{(t)}| (S',L')J'>|^2$ and the Judd-Ofelt intensity parameters Ω_t ($t = 2, 4, 6$). The reduced matrix element is only related to the energy level structure of the rare earth ions, and has nothing to do with the matrix material, which means the properties of the f-f transition of rare earth ions can be obtained by calculating and analyzing the three Judd-Ofelt parameters. Therefore, it is particularly important to explore the trend and physical significance of the changes in the Judd-Ofelt intensity parameters. The Judd-Ofelt intensity parameters can be expressed as [12]

$$\Omega_t = (2t+1)\sum_{p,\, s}\left|A_{sp}\right|^2 \Xi^2(s,t)(2s+1)^{-1} \tag{8.6}$$

In the formula, A_{sp} is the matrix structure parameter, which is related to the odd terms of the coordination field, and can be expressed as

$$A_{sp} = (-1)^{p+1}\sum_i eq_i R_i^{-s-1} c_{-p}^{-s-1}(\Theta_i, \phi_i) \tag{8.7}$$

where, e is the electron charge; q_i is the ion charge;(R_i, Θ_i, ϕ_i) are the polar coordinates of the ion.

$$\Xi(s,t)=2\sum_{n,l}(2f+1)(2l+1)(-1)^{(f+1)}\times\begin{Bmatrix}1ts\\flf\end{Bmatrix}\begin{Bmatrix}f1l\\000\end{Bmatrix}\begin{Bmatrix}lsf\\000\end{Bmatrix}\times\frac{\langle 4f|r|nl\rangle\langle nl|r^s|4f\rangle}{\Delta E(\psi)} \tag{8.8}$$

$\Xi(s,t)$is related to orbital hybridization, which can be expressed as.

In the formula, ΔE (ψ) represents the energy difference between the $4f^{N-1}$ electron orbit and the intermediate state $4f^{N-1}$ nl^1 mixed electron orbit. For Nd^{3+}, $4f^{N-1}$ nl^1 The mixed electronic orbit is $4f^{N-1}$ $5d^1$ orbit;< nl $|r^s|4f$> is $\int_0^{\infty} R(nl)\, r^k R(n'l')$ dr abbreviation, is the hybrid function of the 4f orbit and the 5d orbit. The above t, s values follow: $t = 2$, $s = 1, 3$; $t = 4$, $s = 3, 5$; $t = 6$, $s = 5, 7$.

When Nd^{3+} is used as the activating ion of laser material, the most commonly used upper energy level is $^4F_{3/2}$ level, the lower energy level is $^4I_{J'}$ level, the reduced matrix elements of the transition from $^4F_{3/2}$ level to each lower level are shown in Table 8.1.

Clearly, since the reduced matrix elements $|<U^{(2)}>|^2$ for these transitions are all 0, the corresponding Ω_2 does not contribute to the intensity of these transitions. However, as can be seen from Eq. (8.6), when the value of t is small, Ω_t is greatly influenced by the matrix structure parameter A_{sp}, which means that the structural information of the matrix can be understood through the analysis of Ω_2. Researchers call the transitions mainly affected by Ω_2 as structure hypersensitive transitions, and found that Ω_2 increases with the increase of asymmetry [16].

On the other hand, the $|<U^{(6)}>|^2$ values corresponding to the four transitions in Table 8.1 are large, which means that the corresponding Ω_6 contributes greatly to the transition strength, so the manipulation of Ω_6 receives a lot of research attention. As can be seen from Eq. (8.6), when the value of t becomes larger, Ω_t is more affected by Ξ (s, t), so Ω_6 is related to the overlap of the outer electron cloud of the nucleus. Studies have shown that as the number of 4f electrons increases, the repulsion to 5d electrons increases, the overlap integral decreases, and thus Ω_6 decreases. At the same time, as the covalency of Nd^{3+} with adjacent atoms increases, the density of the 6 s electron cloud increases, the shielding effect on 4f and 5d electrons increases, the overlap integral decreases, which also leads to a decrease in Ω_6 [17, 18]. In addition, researchers infer that Ω_4 is related to the medium and long-range structure around Nd^{3+}, but no relevant experimental studies have been reported [19, 20].

Table 8.1 Reduced matrix elements of $^4F_{3/2} \rightarrow {}^4I_{J'}$ transitions of Nd^{3+} [15]

Energy Level	Peak Position/cm^{-1}	$\vert<U^{(2)}>\vert^2$	$\vert<U^{(4)}>\vert^2$	$\vert<U^{(6)}>\vert^2$
$^4I_{15/2}$	6148	0	0	0.0280
$^4I_{13/2}$	4098	0	0	0.2093
$^4I_{11/2}$	2114	0	0.1423	0.4083
$^4I_{9/2}$	235	0	0.2283	0.0554

8.1.2.2 Fluorescence Branching Ratio

As previously mentioned, the Judd-Ofelt intensity parameters are related to the asymmetry and covalency of the coordination environment around Nd^{3+}, and the reduced matrix element of each transition is different and does not change with the matrix. This means that the intensity of each transition responds differently to changes in Judd-Ofelt parameters, that is, it responds differently to changes in asymmetry and covalency. This makes it possible to change the trend of Nd^{3+} radiative transitions by adjusting the coordination environment around Nd^{3+}.

The definition of the theoretical fluorescence branching ratio is given by Eq. (8.3). For Nd^{3+}, the initial state energy level $(S, L)\,J$ is the $^4F_{3/2}$ level, and the final state energy level $(S', L')\,J'$ is all Stark splitting sub-energy levels of the $^4I_{J'}$ level. $A[(S,L)J;(S',L')J']$ represents all radiative transition probabilities from all components of the $^4F_{3/2}$ level to all components of the ^4I_J level. Based on Eqs. (8.2) and (8.3), the fluorescence branching ratio of each energy level transition can be written as the variable $X = \Omega_4/\Omega_6$ [21, 22]:

$$\beta_{0.9\mu m} = \beta\left[{}^4F_{3/2};{}^4I_{9/2}\right] = \frac{0.3194X + 0.0767}{0.4247X + 0.4567} \tag{8.9}$$

$$\beta_{1.06\mu m} = \beta\left[{}^4F_{3/2};{}^4I_{11/2}\right] = \frac{0.1053X + 0.2874}{0.4247X + 0.4567} \tag{8.10}$$

$$\beta_{1.35\mu m} = \beta\left[{}^4F_{3/2};{}^4I_{13/2}\right] = \frac{0.0902}{0.4247X + 0.4567} \tag{8.11}$$

$$\beta_{1.8\mu m} = \beta\left[{}^4F_{3/2};{}^4I_{15/2}\right] = \frac{0.0023}{0.4247X + 0.4567} \tag{8.12}$$

The above results ignore the effect of dispersion, and the central wavelength of the emission is taken from the Nd:YAG experimental results.

Figure 8.3 shows the relationship between the branching ratio of each energy level transition and the variable $X = \Omega_4/\Omega_6$. When X approaches zero, $\beta_{0.9\mu m} = 0.17$, $\beta_{1.06\mu m} = 0.63$, $\beta_{1.35\mu m} = 0.2$, $\beta_{1.8\mu m} = 0.005$; as X increases, only the fluorescence branching ratio of the $^4F_{3/2} \rightarrow {}^4I_{9/2}$ transition increases, while the branching ratios of all other transitions decrease; when X approaches positive infinity, $\beta_{0.9\mu m} = 0.75$, $\beta_{1.06\mu m} = 0.25$, $\beta_{1.35\mu m}$ and $\beta_{1.8\mu m}$ all tend to 0. $\beta_{0.9\mu m}$ intersects with $\beta_{1.06\mu m}$ at $X = 0.984$, which means that when X is greater than 0.984, the spontaneous emission probability of 0.9 μm will exceed 1.06 μm, becoming the transition with the highest radiation probability among several transitions. Similarly, $\beta_{0.9\mu m}$ and $\beta_{1.35\mu m}$ intersects at $X = 0.042$, meaning that when X is less than 0.042, the probability of spontaneous radiation at 1.35 μm is greater than at 0.9 μm.

Clearly, the value of X is closely related to the radiative transition trend of Nd^{3+}, which is a very important parameter, but its calculation process is cumbersome. The ratio of $\beta_{1.06\mu m}$ to $\beta_{1.35\mu m}$ (R_{th}) has the following linear relationship with X [22]:

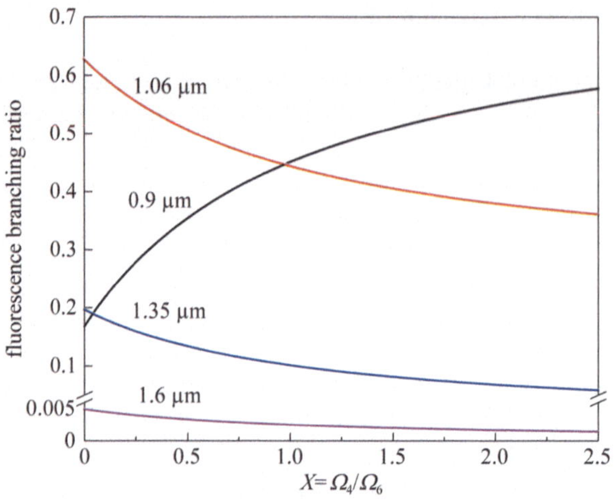

Fig. 8.3 The relationship between fluorescence branching ratio and X in Nd:YAG [22]

$$R_{th} = \beta_{1.06\mu m} / \beta_{1.35\mu m} = 1.167X + 3.185 \tag{8.13}$$

Assuming that R_{th} is equal to the ratio of the integrated fluorescence intensities at 1.06 μm and 1.35 μm (R_{exp}), then based on the measured the fluorescence spectrum of Nd^{3+} and calculated the R_{exp} value, it can conveniently derive the X value. The error in calculating the X value by this method is within 15%.

8.1.3 The Influence of Co-Dopants on the Structure and Spectral Properties of Nd^{3+}-Doped Silica Glass

Nd^{3+}-doped silicate and phosphate laser glasses have been widely studied due to their excellent laser characteristics at 1.06 μm. However, these multi-component glasses have poor thermal shock resistance and chemical stability, which severely restricts their application in high-energy, high-repetition-rate lasers and high-power fiber lasers. In comparison, silica glass has good chemical stability and thermal shock resistance, excellent environmental adaptability, processing performance, and heat dissipation ability, and is widely used in fiber optic communication, biomedical, space lasers, and other fields. Therefore, the relationship between the composition-preparation-structure-performance of Nd^{3+}-doped silica glass is of great significance for new Nd^{3+}-doped laser glasses and fibers.

Nd^{3+} has a high positive electron field strength, thus, more anions are needed for coordination in the matrix glass to reduce the overall energy of the system. In silica glass, the Si-O network structure is rigid, making it difficult to deform the network to provide more O^{2-} for Nd^{3+} coordination. Therefore, when the concentration of Nd^{3+} in silica glass exceeds 1000 ppm, Nd^{3+} tends to form clusters or even crystallize to consume less O^{2-}, achieving a reduction in the overall energy of the system. Clustering will cause concentration quenching between Nd^{3+}, shortening the fluorescence lifetime, reducing the laser conversion efficiency, and hindering laser operation. Typically, co-doping network-forming elements such as Al and P could reduce clustering.

8.1.3.1 The Effect of Co-Doping Al on the Structure and Properties of Nd^{3+}-Doped Silica Glass

Co-doping Al is an important way to improve the optical performance of rare earth doped silica glass. Arai et al. [23] prepared silica glass samples with different Al/Nd ratios, studied the effect of Al on the spectral performance of Nd^{3+}-doped silica glass, as shown in Fig. 8.4. Co-doping Al has a good dispersion effect, which can effectively suppress the concentration quenching of Nd^{3+}, enhancing the luminescence intensity of Nd^{3+}-doped silica glass; as the Al/Nd ratio increases, the fluorescence spectrum blue-shifts overall (Figure 8.4a), and the 1.06 μm fluorescence intensity of Nd^{3+} gradually increases relative to the 0.9 μm fluorescence intensity (Figure 8.4b). In addition, the introduction of a small amount of Al will also lead to an increase in the density of silica glass.

Sen et al. [24] performed X-ray absorption fine structure spectrum analysis on Nd^{3+}-doped silica glass and Al/Nd co-doped silica glass, and found that after co-doping with Al, the Nd-O bond length becomes longer and the distance between Nd^{3+} becomes larger, which is the direct evidence of eliminating Nd^{3+} clustering.

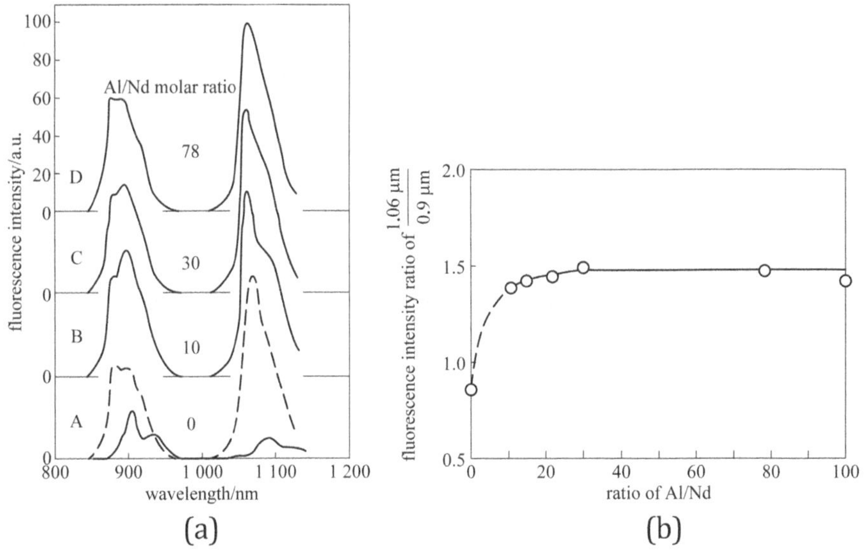

Fig. 8.4 The effect of Al/Nd ratio on the spectrum (**a**) and fluorescence intensity ratio (**b**) of Nd^{3+}-doped silica glass [23]

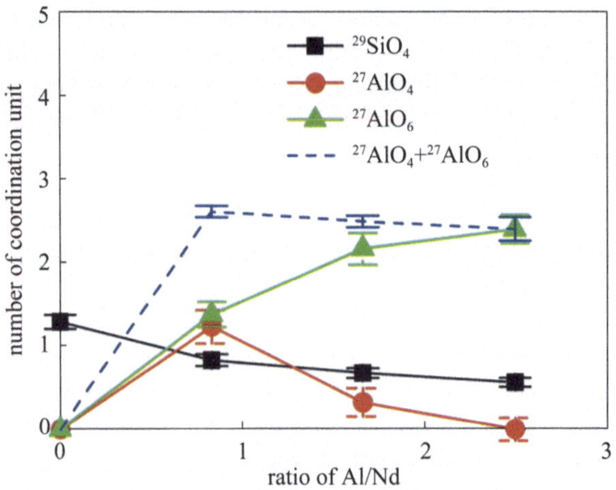

Fig. 8.5 The effect of co-doping Al on the number of Nd^{3+} coordination units in silica glass

They also pointed out that in Al/Nd co-doped silica glass, Nd^{3+} preferentially coordinates with $[AlO_4]^-$ to balance its valence. Qiao et al. [25] prepared Nd^{3+}-doped silica glasses with different Al/Nd ratios using porous glass and studied their spectral properties. The results showed that as the Al/Nd ratio increased, the effective linewidth decreased, and the Judd-Ofelt parameter Ω_2 increased while $\Omega_{4,6}$ decreased. The emission cross-section of Nd^{3+} at 1.06 μm increases. Funabiki et al. [26] prepared Nd^{3+}-doped silica glasses with different Al/Nd ratios using the sol-gel method, and studied its microstructure using various methods such as Electron Spin Echo Envelope Modulation (ESEEM) spectroscopy and ESE. They found that Al has two coordination structures, AlO_4 and AlO_6; they estimated the changes in the coordination units and number of Nd^{3+}, and the results showed (Fig. 8.5) that as the relative content of Al increases, the coordination unit of Nd^{3+} gradually changes from AlO_4 coordination to AlO_6 coordination. They also pointed out that this transition increases the asymmetry of the local environment around a single Nd^{3+}, but the differences between the local environments of different Nd^{3+} will become smaller.

8.1.3.2 The Effect of Co-Doping P on the Structure and Properties of Nd^{3+}-Doped Silica Glass

Co-doping P into Nd^{3+}-doped silica glass will also disperse Nd^{3+}, cause a blue shift in the fluorescence peak, and increase the 1.06 μm branching ratio [23]. To achieve a similar dispersion effect, the content of co-doped P is much higher than that of co-doped Al, and the characteristics of P being easily volatile at high temperatures and easy to phase separate in silica glass make the preparation of high P doped silica glass difficult. Therefore, P/Al co-doping is more often used to regulate the spectral properties of rare earth doped silica glass. Xu et al. [27] prepared P/Al/Nd co-doped silica glass with high optical quality, using the sol-gel method combined with high-temperature sintering. They fixed the Al and Nd contents at 4 mol% and 0.2 mol% respectively, introduced different amounts of P into the Al/Nd co-doped silica glass, prepared PA*x* (*x* is the P/Al ratio, $x = 0, 0.5, 1, 1.25, 2, 2.5$) series samples and studied the change in structure and spectral properties.

Figure 8.6 shows the Raman spectra and refractive index of Nd^{3+}-doped silica glass with different P/Al ratio. Apparently, the increase in P content leads to a continuous increase in the intensity of the vibration peak near 1145 cm^{-1}, which saturates when P/Al > 1, related to the formation of $AlPO_4$; In addition, the maximum phonon energy of silica glass continues to increase, especially when P/Al > 1, a Raman vibration peak at 1320 cm^{-1} appears, corresponding to the formation of P=O bond. The increase in phonon energy will lead to an increase in the probability of multi-phonon relaxation and related non-radiative transitions in laser glass. Correspondingly, as the P/Al ratio increases, the refractive index of silica glass first decreases and then increases, with a turning point appearing near P/Al = 1, because the $AlPO_4$ group has similar structure to the silica glass network. When P/Al > 1, the increased P=O bond will lead to an increase in the refractive index.

Figure 8.7 shows the changes in the absorption spectra and emission spectra of Nd^{3+}-doped silica glass with different P/Al ratios. When P/Al > 1, the hypersensitive transition structure near 580 nm $^4I_{9/2} \rightarrow {}^4G_{5/2}$, $^2G_{7/2}$ absorption peak splits into two peaks, and blue-shifts simultaneously; the 1.06 μm fluorescence peak appears blue shift and spectral narrowing. These phenomena indicate that as the P/Al ratio increases, the energy level structure of Nd^{3+} changes. To study the changes in energy level structure and symmetry, the Judd-Ofelt intensity parameters calculation and Nd-O covalency analysis are carried out. Figure 8.8 shows the changes in the Judd-Ofelt intensity parameters and Nd-O bond covalency of Nd^{3+}-doped silica glass with different P/Al ratios, with parameter α evaluating Nd-O covalency:

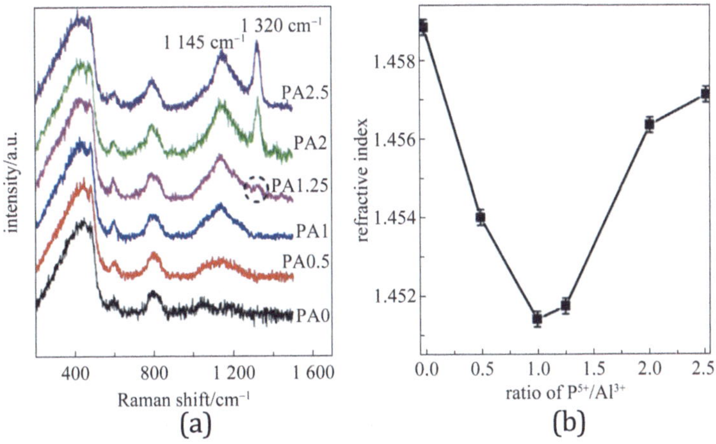

Fig. 8.6 Raman spectra (**a**) and refractive index (**b**) of Nd^{3+}-doped silica glass with different P/Al ratios [27]

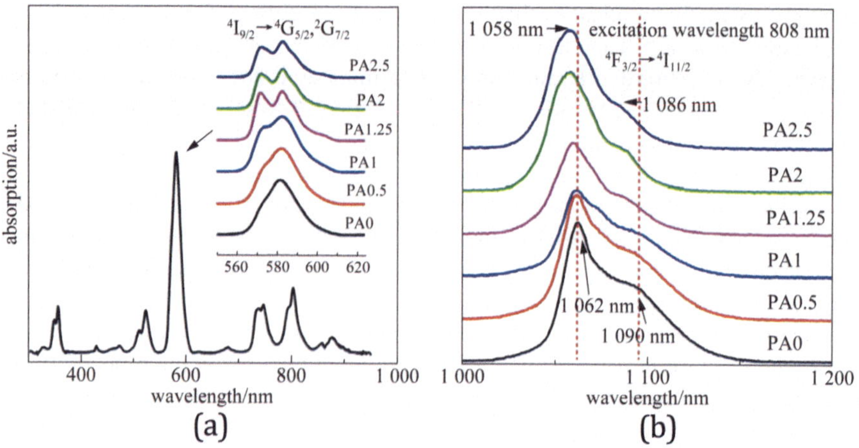

Fig. 8.7 Absorption spectrum (**a**) and emission spectrum (**b**) of Nd^{3+}-doped silica glass with different P/Al ratios [27]

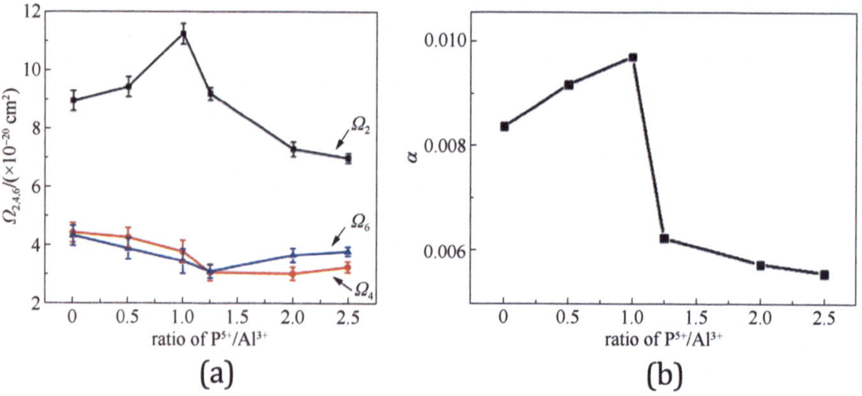

Fig. 8.8 Judd-Ofelt intensity parameters (**a**) and α parameters (**b**) of Nd^{3+}-doped silica glass with different P/Al ratios [27]

$$\alpha = \frac{\sigma_f - \sigma_g}{\sigma_f} \tag{8.14}$$

In the formula, σ_f represents the energy of free ion transition; σ_g represents the energy of ion transition in silica glass.

As can be seen from the Judd-Ofelt intensity parameters calculation results in Fig. 8.8, with the increase of the P/Al ratio, Ω_2 first increases and then decreases, with the turning point appearing near P/Al = 1. This indicates that the asymmetry of the Nd^{3+} coordination environment first increases and then decreases, which is related to the formation and transformation of $AlPO_4$ and P=O; while Ω_4 and Ω_6 first decrease and then increase, with the turning point appearing near 1.25. This indicates that the covalency of Nd–O first increases and then decreases; α parameter changes are consistent with this. When P/Al > 1, the covalency of Nd–O drops sharply, which is attributed to the formation of the P=O bond with strong electron-donating ability and its preferential coordination to the surroundings of Nd^{3+}. When P/Al ≤ 1, $\Omega_4 > \Omega_6$ (i.e., $X > 1$), it means that the probability of the three-level radiative transition from $^4F_{3/2}$ to $^4I_{9/2}$ is greater than the probability of the four-level radiative transition from $^4F_{3/2}$ to $^4I_{11/2}$.

Table 8.2 gives the peak emission cross-section, fluorescence effective linewidth, quantum efficiency, and fluorescence branching ratio, etc. of the $^4F_{3/2} \rightarrow {}^4I_{11/2}$ transition of Nd^{3+}-doped silica glasses with different P/Al ratios. As the content of P continues to increase, the effective linewidth continues to decrease. At the same time, Ω_6 decreases first and then increases as the relative content of P increases, the spontaneous radiation probability of Nd^{3+} at 1.06 μm is affected. Therefore, the emission cross-section of Nd^{3+} at 1.06 μm also shows a trend of first decreasing and then increasing.

8.1.3.3 The Impact of Co-Doping of Non-oxygen Anions on the Structure and Properties of Nd^{3+}-Doped Silica Glass

The rigid network structure of silica glass makes it extremely difficult to control the spectral behavior of Nd^{3+}-doped silica glass. Co-doping with conventional elements such as Al and P into the near-neighbor coordination of Nd^{3+} can only indirectly affect the spectral behavior of Nd^{3+}, and the spectral regulation effect is weak. In order to better control the spectral performance of Nd^{3+}-doped silica glass, researchers tried to directly change the anions of the nearest neighbor coordination of Nd^{3+}, such as introducing halide ions, S^{2-} and other non-oxygen anions into the Nd^{3+}-silica glass network to replace the position of O^{2-}.

Wang et al. [28] used the sol-gel method combined with high-temperature sintering technology to prepare a series of non-oxygen anion groups co-doped Nd^{3+}-doped silica glass such as Nd_2S_3, $NdBr_3$, NdI_3, using Nd^{3+} halides and sulfides (or sulfates) to replace Nd_2O_3, changing the coordination environment of Nd^{3+} . To ensure the uniformity of Nd^{3+}, a certain amount of Al element is also doped. Taking the introduction of I^- as an example, Figure 8.9a compares the calculation results of

Table 8.2 Spectral parameters of Nd^{3+}-doped silica glass with different P/Al ratios [27]

Sample	Emission cross-section/ ($\times 10^{-20}$ cm^2)	Effective linewidth /nm	Quantum efficiency /%	Fluorescence branching ratio/%			
				$^4I_{15/2}$	$^4I_{13/2}$	$^4I_{11/2}$	$^4I_{9/2}$
PA0	1.78	44.35	74	2.6	19.6	51.0	26.8
PA0.5	1.62	44.5	59	2.5	19.1	50.6	27.8
PA1	1.39	45.77	50	2.5	19.1	50.6	27.8
PA1.25	1.48	37.81	45	2.6	19.8	51.1	26.5
PA2	1.7	37.19	51	2.8	20.9	51.8	24.5
PA2.5	1.79	37.03	57	2.7	20.7	51.8	24.8

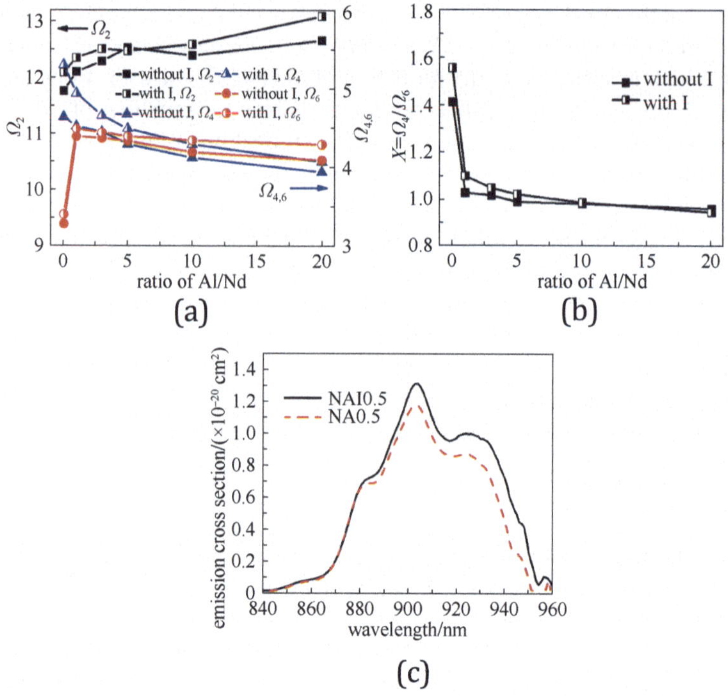

Fig. 8.9 The effect of non-oxygen anions on the Judd-Ofelt parameters and fluorescence spectra of Nd^{3+}-doped silica glass

the Judd-Ofelt intensity parameters of Nd^{3+}-doped silica glass before and after the introduction of I^-. Obviously, the Judd-Ofelt intensity parameters are larger after the introduction of I^-, which means that the introduction of I^- leads to an increase in the asymmetry of the coordination environment of Nd^{3+} and the covalency of the chemical bond between Nd^{3+} and the nearest neighbor atom. At the same time, Figure 8.9b shows that the value of *X* increases as Al/Nd decreases, and under the same Al/Nd ratio, the glass with introducing I^- shows higher *X* value, which is beneficial to improve the fluorescence branching ratio of the three-level transition from $^4F_{3/2}$ to $^4I_{9/2}$. Figure 8.9c is emission cross-sections of $^4F_{3/2} \rightarrow {}^4I_{9/2}$ transition before and after the introduction of I^-, compared with NA0.5(without I^-), the emission cross-section is elevated in NAI0.5 (with I^-), indicating that the introduction of I^- increases the emission cross-section and bandwidth of the three-level transition of Nd^{3+}, which is beneficial for obtaining 0.9 μm laser.

8.2 Nd^{3+}-Doped Silica Fiber Laser and Applications at 0.9 μm and 1.3 μm

In recent years, Nd^{3+}-doped lasers operating at 0.9 μm have attracted a lot of research attention. The laser source at this band can be used as a pump source for Yb^{3+}-doped laser materials; at the same time, this band corresponds to the atmospheric transmission window, which can be directly applied to atmospheric detection and differential absorption radar and other fields [4, 5]. In addition, in the field of biological imaging, because commonly used green fluorescent proteins and their derivatives have stronger two-photon absorption at this band, the two-photon imaging system using 0.9 μm ultrafast laser as the light source shows advantages in deep tissue imaging and in vivo imaging [29]. More importantly, the 0.9 μm laser can produce ~450 nm blue light laser by frequency doubling, which is located at the underwater communication window, and has significant application in deep-sea communication and marine military [6]; single-frequency blue light laser has high coherence, which can be applied to quantum optics, precision optics, and atomic cooling and capture [7, 30]; in addition, blue light laser also has important applications in medical diagnosis [8], laser storage [31], metal processing [32], laser display [33] and deep ultraviolet laser generation [34] .

Currently, laser diodes [35], solid (crystal and ceramic) lasers [36, 37] and Nd^{3+}-doped fiber laser are the commonly used methods for achieving 0.9 μm laser output [38, 39], among which 0.9-μm Nd^{3+}-doped fiber laser has become a research focus for the development of 0.9 μm lasers due to its high pump power density, excellent thermal management ability, long gain distance, waveguide designability and flexibility. At the same time, Nd^{3+}-doped silica fiber has good chemical stability, environmental adaptability and thermal shock resistance compared to other fibers.

8.2.1 The Technology for Enhancing the Luminescence of Nd^{3+}-Doped Silica Glass and Fiber

The aforementioned Fig. 8.2 shows several common transitions of Nd^{3+}. Among them, the four-level transition ${}^4F_{3/2} \rightarrow {}^4I_{11/2}$ is the strongest, its emission cross-section is nearly an order of magnitude higher than the other two transitions (0.9 μm, 1.35 μm), which is related to its larger reduced matrix element (Table 8.1). Therefore, Nd^{3+}-doped laser materials are usually used in 1.06 μm lasers. ${}^4F_{3/2} \rightarrow {}^4I_{9/2}$ is a three-level transition. To achieve 0.9 μm laser operation in Nd^{3+}-doped laser materials, there are problems of reabsorption from the ground state (${}^4I_{9/2}$) and competition with the four-level transition of ${}^4F_{3/2} \rightarrow {}^4I_{11/2}$. This makes achieving 0.9 μm laser output a great challenge. Low-temperature operation is a method for the realizing the 0.9 μm laser, which can reduce the thermal population of the ground state energy level to weaken the reabsorption effect. Additionally, waveguide design and filtering methods are also generally used to suppress 1.06 μm laser [40–42]. The following is several technical means to improve the 0.9 μm luminescence of Nd^{3+}-doped glass and fiber.

8.2.1.1 Overcoming Reabsorption

The $^4I_{9/2}$ energy level of Nd^{3+} is the ground state energy level. The thermal population of this ground state energy level causes serious reabsorption during laser oscillation, limiting the increase in laser output power and efficiency; at the same time, it also brings additional thermal effects. Figure 8.10a shows the absorption band from the ground state $^4I_{9/2}$ energy level to the $^4F_{3/2}$ energy level of Nd^{3+}-doped silica glass and the emission band from the $^4F_{3/2}$ energy level to the $^4I_{9/2}$ energy level, which has significant overlap, especially in the part with a wavelength less than 920 nm. Taking the PM-NDF-5/125 type Nd^{3+}-doped silica fiber from Nufern as an example, Figure 8.10b shows the forward amplified spontaneous emission (ASE) spectrum at different fiber lengths when the 808 nm LD pump power is fixed. As the fiber length increases, the ASE intensity at the 0.9 μm band continuously decreases and a significant redshift occurs. When the fiber length is 7 cm, the peak is at 907 nm, and when the silica fiber length increases to 200 cm, the peak moves to 940 nm, a redshift of nearly 33 nm. This indicates that the longer the fiber, the more severe the reabsorption at the 0.9 μm band. Therefore, to improve the 0.9 μm laser performance of Nd^{3+}-doped silica fiber, it is necessary to weaken the reabsorption effect as much as possible.

A direct way to overcome the reabsorption effect is to decrease the fiber operation temperatures, which reduces the thermal population of the ground state energy level and turns the three-level structure into a quasi-four-level transition. In 2004, Dawson et al. from the Lawrence Livermore National Laboratory [43] compared the gains of 938 nm and 1088 nm of Nd^{3+}-doped silica fiber at temperatures of 77 K, 195 K, and 293 K, pointing out that low temperatures can effectively reduce the reabsorption effect, the gain difference between the 938 nm and 1080 nm bands, and suppress competition between the bands. By cooling the gain fiber with liquid nitrogen and using a 25 m long Nd^{3+}-doped silica fiber, a 938 nm laser output with a

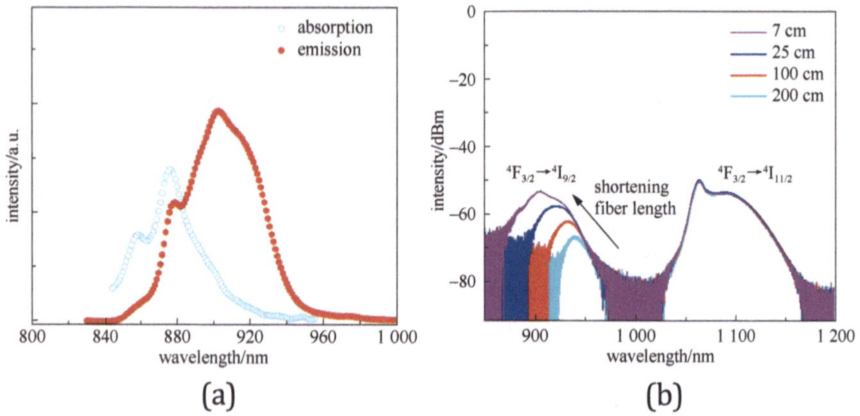

Fig. 8.10 Absorption and emission of Nd^{3+}-doped silica glass (**a**) and ASE spectra of Nd^{3+}-doped silica fiber with different lengths (**b**)

conversion efficiency of 32% and an output power of 11 W was achieved under the pumping of a 35 W multimode 808 nm LD. However, from a practical point of view, the use of liquid nitrogen cooling will seriously reduce the compactness of the laser, make it difficult to put into practical use.

In 2013, Laroche et al. from the University of Caen [44] pointed out that achieving a high particles invertion in the Nd^{3+}-doped fiber, i.e., increasing the pump power density in per unit gain fiber can improve the laser performance of 0.9 μm fiber laser, and proposed to increase the core/cladding area ratio of Nd^{3+}-doped fiber, that is, the use of large mode area (LMA) gain fiber. Combined the theoretical simulation of the effect of fiber size on laser performance, they designed a Nd^{3+}-doped silica double-clad fiber with a core diameter of 20 μm and a cladding of 80 μm using the MCVD method. Under multimode 808 nm LD pumping, a 910 nm fiber laser of 20.1 W was successfully obtained, with a slope efficiency of 44%. The NA of fiber core is 0.08, and the NA of inner cladding is 0.46; this result broke the records made by Dawson et al. in terms of power and efficiency. In 2015, Leconte et al. [45] continued to optimize the fiber structure, keeping the core diameter unchanged at 20 μm and reducing the cladding size to 60 μm. The NA of the core and an inner cladding are 0.07 and 0.45 respectively. A 910 nm fiber laser with a power of 22 W was achieved under an absorbed pump power of 47 W, with a slope efficiency of 47%. In addition, a 915 nm laser output with a power of up to 24 W can also be achieved. By using a volume Bragg grating as a wavelength selection element, a wide-band tunable laser from 872 nm to 936 nm was achieved. By adding an acousto-optic modulator in the cavity, pulsed laser output was also achieved, with a pulse width between 50 ~ 100 ns and a peak power of 2.5 kW.

8.2.1.2 Overcoming Competition Between Different Transitions

There is also intense gain competition between different bands, especially between 0.9 μm and 1.06 μm because the multi-band emission of Nd^{3+} all comes from the same upper energy level, causing the 0.9 μm laser to face transition competition with 1.06 μm. To address this issue, the commonly used method is to increase the fiber loss at 1.06 μm through the design of the waveguide structure, thereby suppressing competition and parasitic laser oscillation at 1.06 μm. Among them, one typical representative is the Nd^{3+}-doped silica fiber with a W-type refractive index distribution. The so-called W-type means introducing an inner cladding outside the core with a refractive index lower than both the core and the outer cladding, as shown in Fig. 8.11. The cutoff wavelength of the W-type fiber depends on the thickness of the inner cladding, and the number of guided modes in the core depends on the refractive index difference between the core and the outer cladding [40]. Therefore, by adjusting the refractive index and diameter of the inner cladding of the Nd^{3+}-doped silica fiber, the loss of the fiber at 1.06 μm can be increased, effectively suppressing parasitic laser oscillation at this wavelength. In 2005, Fu et al. [46] based on an 8 m long W-type Nd^{3+}-doped silica double-clad fiber, achieved a 926.7 nm fiber laser with a slope efficiency of 46.3% and a maximum output power

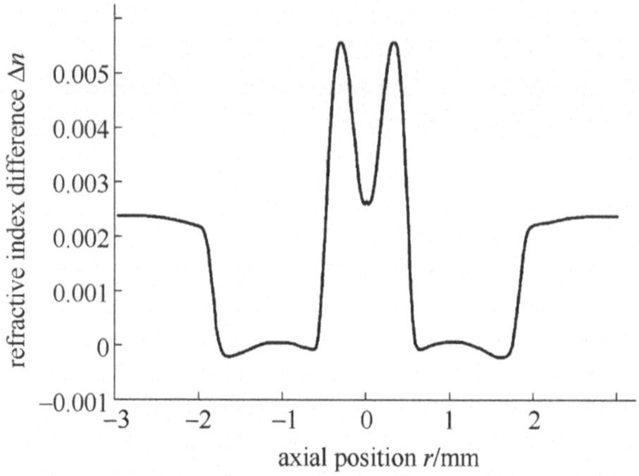

Fig. 8.11 Refractive index distribution of a W-type Nd^{3+}-doped silica fiber preform [40]

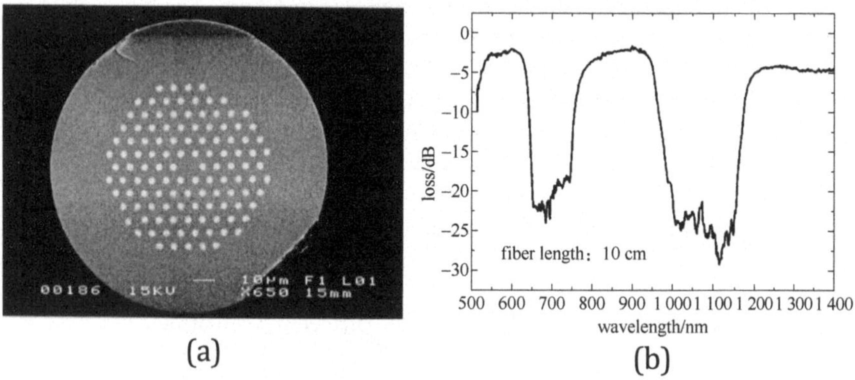

Fig. 8.12 Backscattered electron image of end facet of the PBF (**a**) and fiber transmission loss (**b**)

of 810 mW. In the experiment, the gain fiber was also bent with a radius of 2.5 cm to better suppress 1.06 μm ASE.

However, the W-type fibers face the problem of high fusion point loss with other fiber components (such as fiber Bragg gratings); in the report by Fu et al., the single-point fusion loss is as high as 1.4 dB. Currently, the iXblue has launched a series of commercial Nd^{3+}-doped silica fibers for 0.9 μm lasers; among them, the IXF-2CF-Nd-O-5-125-D type fiber with a "W" structure can effectively suppress 1.06 μm ASE and weaken gain competition [47, 48].

Another typical waveguide structure design is photonic crystal fibers. In 2006, Knight et al. from the University of Bath [49] reported a kind of passive photonic bandgap fiber (PBF), as shown in Figure 8.12a, the periodic structure in the cladding is made of Ge-doped silica glass, with a refractive index difference of 3.3×10^{-2} compared to pure silica in the cladding. As shown in Figure 8.12b, based on this design, 1060 ~ 1100 nm ($^4F_{3/2} \rightarrow {}^4I_{11/2}$) band is located at the high transmission loss

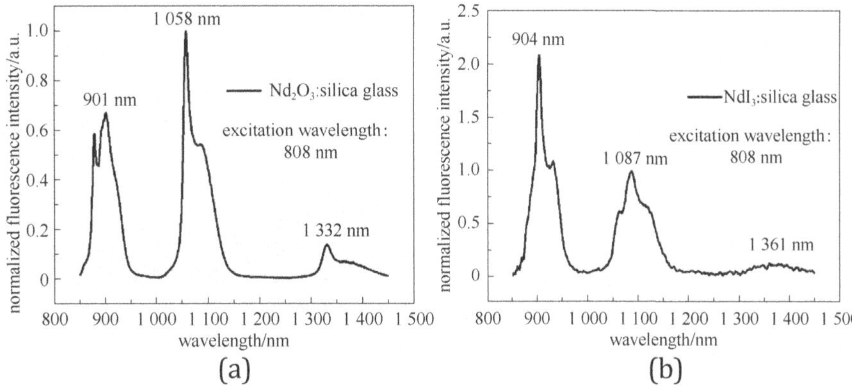

Fig. 8.13 Fluorescence spectra of Nd_2O_3-doped silica glass (**a**) and NdI_3-doped silica glass (**b**)

window of the fiber. By splicing the PBF with a 2.5 m long Nd^{3+}-doped silica fiber, they achieved a 907 nm laser with over 200 mW output power and a slope efficiency of 32%. Using only 10 cm of PBF can significantly suppress 1.06 μm ASE, improving the signal-to-noise ratio of the laser [49] .

8.2.1.3 Improving the Fluorescence Branching Ratio of 0.9 μm

To overcome the transition competition problem, apart from passively suppressing the four-level transition, actively improving the 0.9 μm emission cross-section and fluorescence branching ratio is also a promising strategy. In 2019, Wang et al. [50] studied the luminescent properties and microstructure of Nd^{3+}-doped germanophosphate, and pointed out that the increase of Ge content is beneficial to improve 0.9 μm fluorescence branching ratio.

As previously mentioned, non-oxygen anions doping has a significant effect on improving the 0.9 μm fluorescence branching ratio of Nd^{3+}-doped silica glass. In 2021, Wang et al. [28] prepared NdI_3 doped silica glass; the introduction of I^- changed the coordination environment of Nd^{3+}, and the spectral intensity at 900 nm was greatly increased, or even "reversed" and surpassed the spectral intensity at 1060 nm, as shown in Fig. 8.13. Figure 8.13a is the fluorescence spectrum of Nd_2O_3-doped silica glass, and Figure 8.13b is the fluorescence spectrum of NdI_3-doped silica glass. The significant difference is that the fluorescence intensity of Nd_2O_3-doped silica glass near 0.9 μm is less than that at 1.06 μm, and its fluorescence branching ratio is below 0.4, while the fluorescence intensity of NdI_3-doped silica glass near 0.9 μm is greatly increased, and its fluorescence branching ratio is above 0.5. Specifically, the fluorescence intensity of NdI_3-doped silica glass near 0.9 μm has surpassed the fluorescence intensity near 1.06 μm, which is beneficial for achieving 0.9 μm laser output.

Chen et al. [51, 52] prepared an octagonal silica fiber by NdI_3-doped silica glass with a size of 20/125. Using a 915 nm LD as the seed source, and 808 nm LD as the

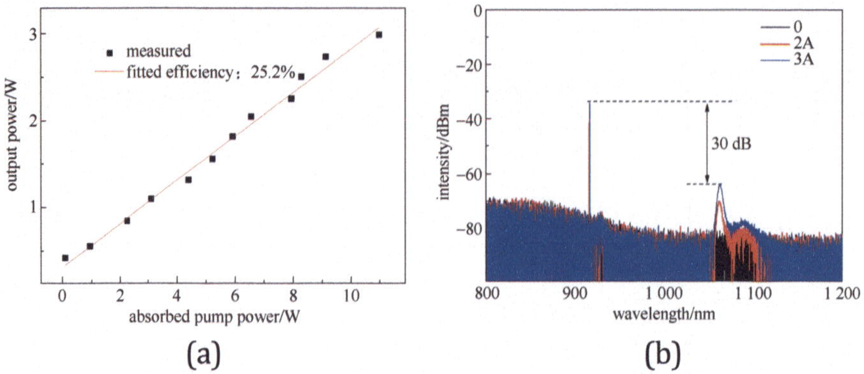

Fig. 8.14 Laser output power (**a**) and spectrum (**b**) by a 20/125 NdI_3-doped silica fiber

pump source and a 1 m long fiber as the gain medium, a simple amplification test was conducted to explore the fiber performance. The seed injection power at 915 nm was 0.72 W, and a 3 W laser output was obtained under pump with a 3-A current; the corresponding slope efficiency is 9.5%. As shown in Figure 8.14a, after deducting the residual pump, the laser efficiency is 25.2%. Figure 8.14b shows the laser spectrum under different pump currents. The spectral intensity at 915 nm is about 30 dB higher than the ASE peak at 1064 nm, indicating that the 1064 nm ASE is effectively suppressed in this fiber.

8.2.2 0.9 μm Continuous Wave and Pulsed Laser by Nd^{3+}-Doped Silica Fiber

8.2.2.1 Continuous Wave High Power Laser and its Frequency Doubling

Blue light generation by frequency doubling based on 0.9-μm laser is a major driving force for the development of 0.9 μm Nd^{3+}-doped high power fiber laser technology. Table 8.3 summarizes the relevant research progress. According to the output power, it can be divided into three stages: the first stage is before 2000, Nd^{3+}-doped single-mode silica fiber was mainly used as the gain medium, the output power is below 100 mW, and the laser conversion efficiency is low [4, 5, 53]. The second stage is from 2000 to 2010, researches mainly concentrate on reducing gain competition and reabsorption. Through liquid nitrogen cooling, waveguide structure design and other solutions, the output power has been successfully increased to about 10 W, and the laser conversion efficiency has also been significantly improved [46, 49, 54]. The third stage is after 2010, based on the large mode field Nd^{3+}-doped silica fiber, the output power exceeds 20 W [44, 45]. So far, people have obtained a power of 83 W at room temperature [55]. In terms of frequency doubling, the power of blue light laser has reached the level of 10 W, and the power of quadrupled ultraviolet light has reached the watt level [34].

Table 8.3 Research progress of 0.9 μm Nd^{3+}-doped fiber laser

Year	Researcher	Research Institution	Fiber Type	Technical Solution	Research Level
1986	Alcock	University of Southampton	Single-mode fiber	590 nm dye laser pumping	900 ~ 945 nm tunable
2004	Nilsson	University of Southampton	9/100 μm, W-type double cladding	808 nm laser diode double-end pumping	2.4 W, 922 ~ 942 nm tunable, efficiency 41%
2006	Nilsson	University of Southampton	Ring-shaped Nd^{3+} doped hollow-core fiber	Laser diode pumping, water-cooled	4.6 W@927 nm, M^2 ~ 1.08
2006	Knight	University of Bath	Photonic bandgap fiber	808 nm titanium sapphire laser pumping	250 mW, efficiency ~32%
2015	Pax	Livermore laboratory	Photonic crystal fiber	808/880 nm laser diode pumping	11.5 W, efficiency ~55%, M^2 ~ 1.35
2018	Laroche	University of Normandy	20/60 μm, double-clad fiber	Butterfly cavity frequency doubling	7.5 W@452 nm, M_x^2 ~1.0, M_y^2 ~1.5
2021	Laroche	University of Normandy	20/60 μm and 30/130 μm	Amplified Q-switched laser oscillator	24 W@905 nm, 510 mW@226 nm
2022	Laroche	University of Normandy	30/130 μm polarization-maintaining fiber	808 nm laser diode pumping	83 W@905 nm

With the increase in the fluorescence branching ratio at 0.9 μm of the Nd^{3+}-doped silica glass, it is believed that the continuous wave high-power fiber laser will reach the kilowatt level, the corresponding frequency-doubled laser power will also be improved, and more applications will be expanded.

8.2.2.2 Narrow-Linewidth Single-Frequency Laser

Among different fiber lasers, the laser output from the resonator with a single longitudinal mode and a single transverse mode is called a single-frequency fiber laser. Its significant feature is ultra-narrow linewidth (10^{-9} nm) and low noise. Narrow linewidth lasers have important application in many fields such as high power coherent combining, coherent optical communication, laser radar, sensing, nonlinear frequency conversion, etc. Since Ball et al. realized a 1.5 μm single-frequency laser by Er^{3+}-doped silica fiber in 1991, research on single-frequency fiber lasers has attracted the extensive research attention, and gradually developed into an important branch in the field of fiber laser technology [56–58]. Among them, the research on single-frequency lasers at various wavelengths has always been a research focus in this field. Currently, Yb^{3+}, Er^{3+} and Tm^{3+}-doped silica single-frequency fiber lasers have been widely studied, but, Nd^{3+}-doped silica single-frequency fiber laser at

0.9 μm has relatively less research due to the limitations of the aforementioned issues of reabsorption and gain competition. In 2016, the research group in Tianjin University [59] achieved a single-frequency 930 nm Nd^{3+}-doped silica fiber laser with a linewidth of 44 kHz based on a distributed Bragg reflector (DBR) cavity. The length of the gain fiber was 2.5 cm, and the cavity length was only 4 cm. However, due to the low fiber gain, the laser efficiency and output power are relatively low, the slope efficiency is only 1.5%, and the output power is 1.9 mW. In 2020, Wang et al. [39] conducted research on the laser performance of Nd^{3+}-doped silica fiber at 915 nm. Figure 8.15a is the test setup for fiber gain coefficient. A commercial Nd^{3+}-doped silica fiber (Nufern, PM-NDF-5/125) was used and the length of the fiber was 4.5 cm.The gain fiber was fusion spliced with adjacent devices; as shown in Figure 8.15b, at different signal powers of 2.6 mW, 5.6 mW and 7.8 mW, the fiber provided a saturated gain of about 4.5 dB, that is, the net gain per unit length of the fiber is 1 dB/cm.

Wang et al. [39] achieved an optimal conversion efficiency of 5.3% by 5.1 cm long Nd $^{3+}$-doped silica fiber and a FBG with 70% reflectivity. Figure 8.16a shows the output spectrum of the DBR short-cavity laser built based on 2.5 cm long Nd^{3+}-doped silica fiber, with a laser center wavelength of 915.34 nm; the inset shows the output power, with the maximum power of the 915 nm laser being only 1 mW, and a conversion efficiency of 0.6%. Figure 8.16b shows the analysis results of the

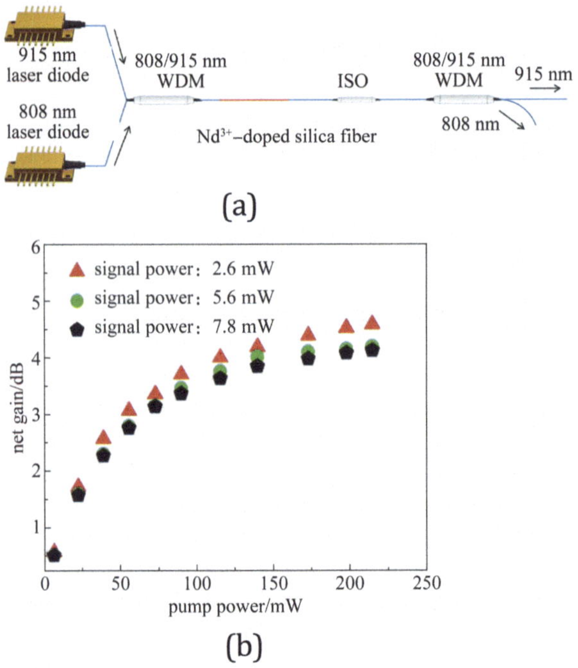

Fig. 8.15 (**a**) Experimental setup for 915 nm gain coefficient measurement of Nd^{3+}-doped silica fiber; (**b**) Net gain of fiber with different incident signal power [39]

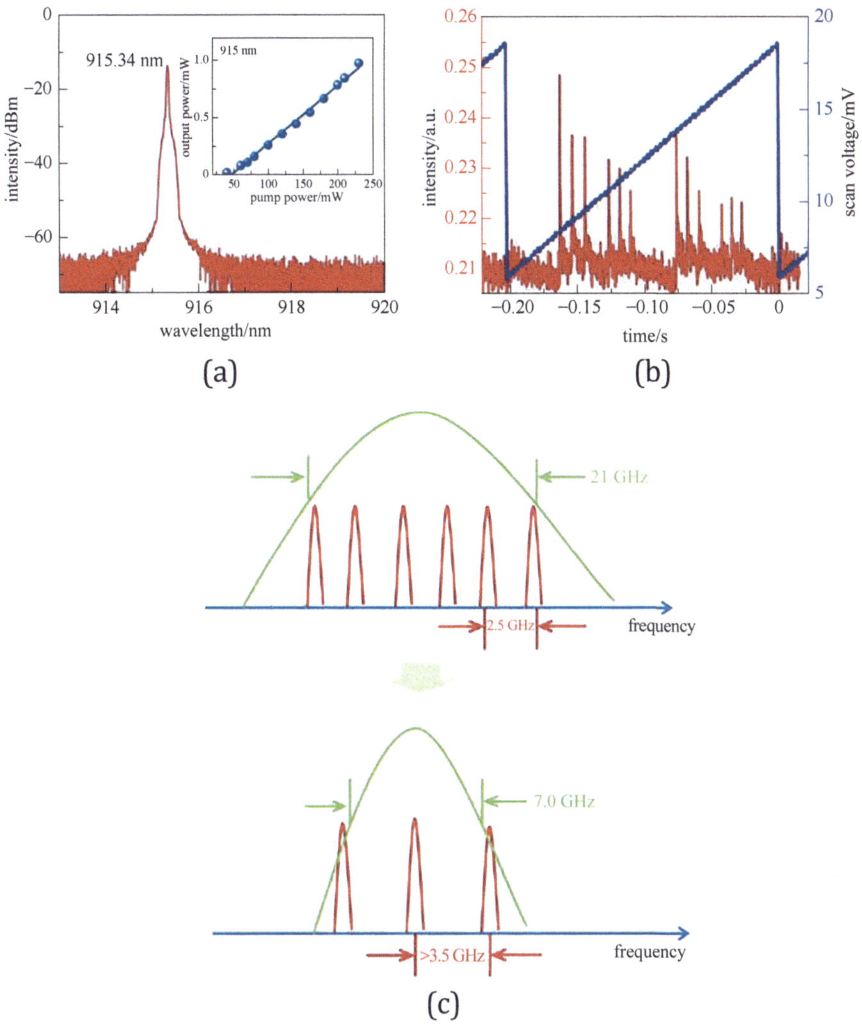

Fig. 8.16 915 nm laser spectrum (inset is the output laser power curve) (**a**); Longitudinal mode situation in the cavity(**b**); Schematic diagram of reducing the longitudinal mode in the DBR cavity(**c**) [39]

cavity longitudinal mode number through the Fabry-Perot interferometer, indicating that the resonant cavity is still multi-longitudinal mode oscillation. Combined with the theoretical analysis of single-frequency mode selection (Fig. 8.16c), a fiber with higher gain coefficient and shorter length is needed to achieve single-frequency 915 nm laser output.

8.2.2.3 Mode-Locked Fiber Laser

The emergence of mode-locking technology is to meet the application needs of high peak power and short pulse lasers. Ultra-short pulse output with pulse widths in the picosecond to femtosecond range can be obtained through mode-locking technology. In the late 1980s, with the emergence of low-loss fibers and laser diodes, fiber lasers developed rapidly; mode-locked fiber lasers, as a typical one, developed and applied rapidly in high-speed optical communication systems with the emergence of optical soliton communication technology in the 1990s [60–63]. To date, mode-locked fiber lasers have been widely used in basic scientific research, biological imaging, biomedicine, and industrial processing.

In recent years, due to two obvious advantages, mode-locked ultrafast lasers at 0.9 μm have attracted widespread attention from researchers in the field of biological imaging: First, the commonly used green fluorescent proteins and their derivatives have stronger two-photon absorption at this wavelength [64] . Generally, a two-photon imaging system using a 0.9 μm ultrafast laser as a excitation source can be used for deep tissue imaging and in vivo imaging, and has lower phototoxicity; Second, compared with traditional solid-state lasers (such as titanium-doped sapphire lasers), fiber lasers have obvious advantages in beam quality, compactness, and stability, which can improve the reliability, portability, and cost-effectiveness of the imaging system, and are conducive to clinical applications.

At present, there are mainly three ways to obtain 0.9 μm mode-locked lasers, including all-solid-state 0.9 μm mode-locked lasers with titanium sapphire as the gain medium, mode-locked lasers at the 0.9 μm band directly produced by Nd^{3+} -doped fiber, and fiber lasers based on nonlinear frequency conversion. Among them, the 0.9-μm Nd^{3+} -doped mode-locked fiber laser has the advantages of simple structure and high stability, and is a commonly used method. At present, researchers can obtain 0.9 μm mode-locked lasers through passive mode-locking methods such as semiconductor saturable absorber mirror (SESAM) mode-locking, nonlinear polarization rotation mode-locking, and nonlinear amplifying loop mirror mode-locking.

In 2005, the University of Tampere [65] achieved a 920 nm femtosecond laser output. As shown in Fig. 8.17, the Nd^{3+}-doped fiber oscillator uses SESAM as mode-locker and a 0.8 m Nd^{3+}-doped silica fiber was used as the gain fiber. The pulse width directly output by the oscillator is 6 ps with a repetition rate of 50 MHz. In the fiber amplifier, a 9 m long W-type Nd^{3+}-doped fiber is used as the gain medium and simultaneously suppresses the 1.06 μm ASE. After amplification, the output power reachs 1.4 W, and the pulse width is compressed to 170 fs by a pair of grating.

The research group from Peking University [66] has systematically studied the 0.9 μm Nd^{3+}-doped mode-locked fiber lasers and their applications in two-photon bioimaging; in 2014, a femtosecond laser with a central wavelength of 910 nm, a repetition rate of 46 MHz, a single pulse energy of 1.3 nJ, and a pulse width of 198 fs was achieved using nonlinear polarization rotation mode-locking. The experimental setup is shown in Fig. 8.18, with the gain medium being a 3.5 m long Nd^{3+}-doped silica fiber. The laser operates in the all-normal dispersion region.

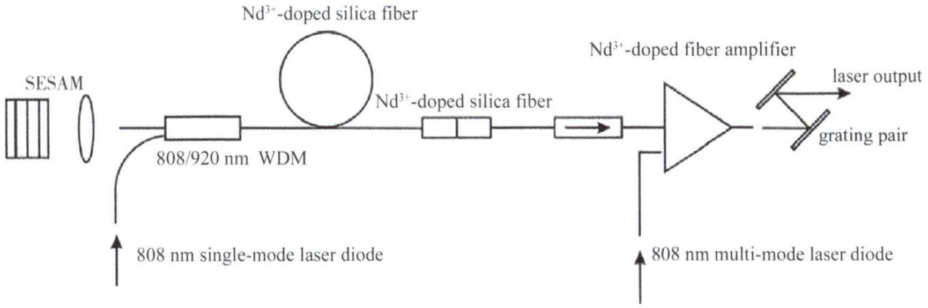

Fig. 8.17 Experimental setup of passively mode-locked 920 nm Nd^{3+}-doped fiber oscillator and a stage of amplification and compression

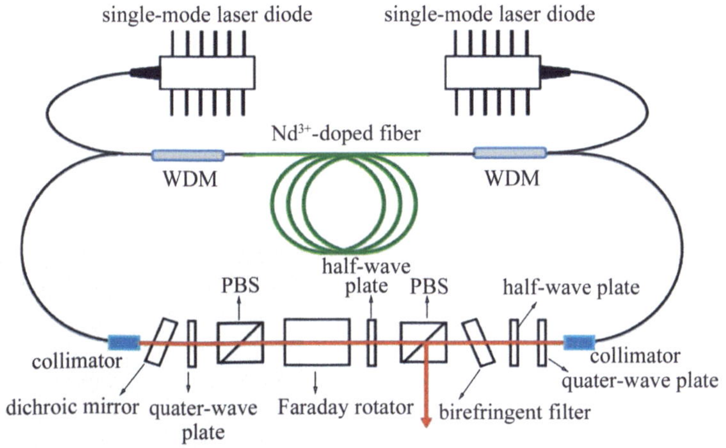

Fig. 8.18 Passively mode-locked Nd^{3+}-doped fiber laser at 910 nm using nonlinear polarization rotation

In 2016, the research group used pre-chirp technology to amplify the oscillator. The experimental setup is shown in Fig. 8.19. The average power of the seed source is 46 mW and the repetition rate is 27.5 MHz; the seed enters a pre-chirped grating pair after the isolator, which broadens the pulse, and then after amplification by a 7.2 m long Nd^{3+}-doped silica fiber, the pulse is compressed by the grating pair, finally achieving a pulse width of 218 fs, an average power of 121 mW, and a single pulse energy of 4.4nJ at 910 nm. Using this as a light source, as shown in Fig. 8.20, clear imaging of Flk:GFP zebrafish blood vessels was successfully demonstrated [67] .

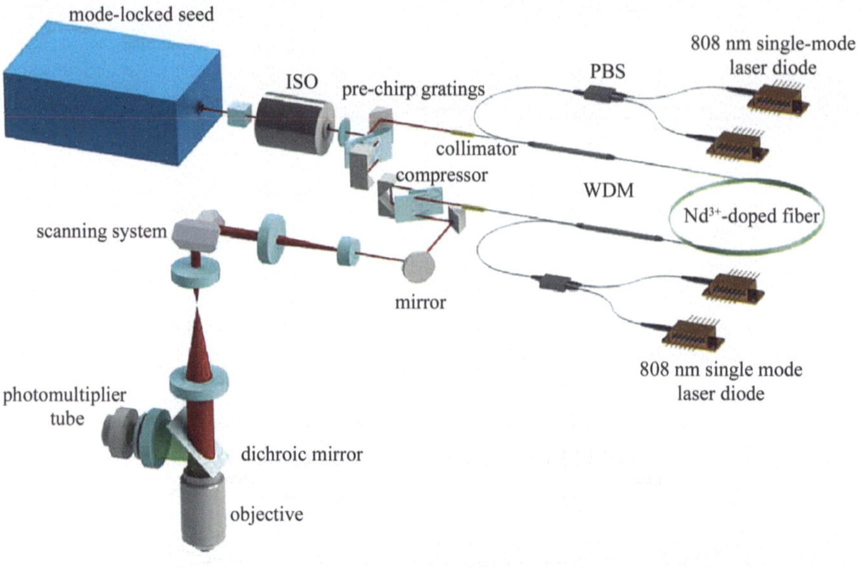

Fig. 8.19 910 nm Nd^{3+}-doped femtosecond fiber laser and two-photon microscope setup [67]

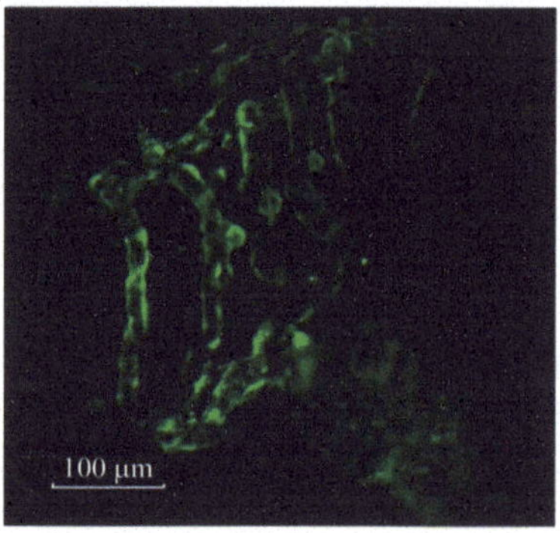

Fig. 8.20 Imaging of Flk:GFP zebrafish blood vessels excited by a 910 nm Nd^{3+}-doped femtosecond fiber laser

8.2.3 *Nd^{3+}-Doped Silica Fiber Lasers and Amplifiers at 1.3 μm*

The 1.3 μm band is located at the second window of fiber communication and the fluorescence band of $^4F_{3/2} \rightarrow {}^4I_{13/2}$ of Nd^{3+} can cover this band (Fig. 8.2), therefore, 1.3 μm Nd^{3+}-doped fiber lasers and amplifiers attracted a lot of early research. In 1986, Alcock et al. at [68] the University of Southampton take the first attempt to realize 1.3 μm laser by Nd^{3+}-doped fiber, however, the 1.3 μm laser was not achieved due to the excited state absorption (ESA) of the $^4F_{3/2} \rightarrow {}^4G_{7/2}$ transition, as shown in Fig. 8.21. In the same year, Po et al. achieved a 1.4 μm Nd^{3+}-doped silica fiber laser. The laser center wavelength red-shifted from the 1.34 μm fluorescence center of the glass, confirming the excited state absorption observed by Alcock et al. Since the communication window is at 1.3 μm, and the excited state absorption causes the

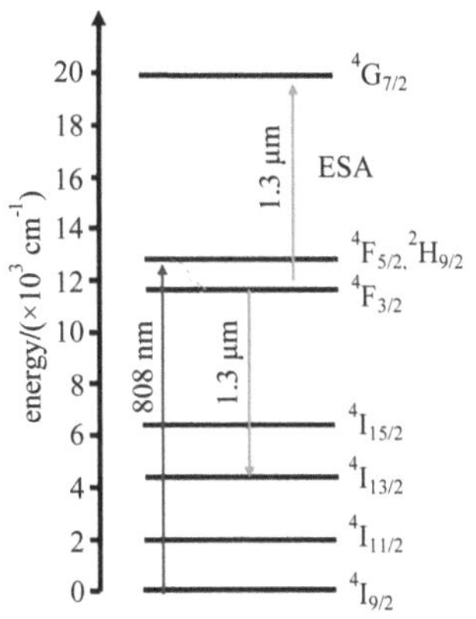

Fig. 8.21 1.3 μm excited state absorption

prominent red-shift of laser wavelength. To solve this problem, many subsequent studies focused on manipulating the components or matrix of Nd^{3+} -doped glass fiber to reduce excited state absorption and achieve laser output closer to the communication window. For example, in 1989, Hakimi et al. [69] added 14 mol% P_2O_5 to the core of Nd^{3+} -doped silica fiber to narrow the fluorescence band near 1.3 μm and reduce long-wave absorption; the laser output at 1.362 μm was achieved under a fiber length of 20 cm, but the laser output power was only 1.9 mW, with an efficiency of 0.95%. In 1990, Grubb et al. [70] used Nd^{3+}-doped phosphate glass fiber as the matrix and achieved a 1.363 μm laser output. The fiber length was 10 cm, with a threshold as low as 5 mW, and a slope efficiency of 10.8%. This is the highest slope efficiency reported for this band. In 1992, Ishikawa et al. [71] achieved a 1.323 μm laser output of 10 mW in a 15 mm long Nd^{3+}-doped fluorophosphate fiber.

In the research of 1.3 μm Nd^{3+}-doped fiber amplifiers, in 2013, Htein et al. [72] prepared a Nd^{3+}-doped silica fiber with a core of 73 μm and a cladding of 700 μm. Using a high-power LED as the pump source, light amplification in the range of 1400 ~ 1450 nm was achieved. Among them, the gain at 1400 nm can reach 6 dB. Dawson et al. from the Lawrence Livermore National Laboratory [73] designed a Nd^{3+}-doped silica fiber with a special waveguide structure and achieved a small signal gain of 19.3 dB at 1427 nm, amplifying the seed light power of 43 mW to 1.2 W. In 2017, Dawson et al. [74] designed a Nd^{3+}-doped microstructured fiber that can suppress 1.06 μm ASE and successfully demonstrated light amplification in the range of 1376 ~ 1466 nm, with a small signal gain of 13.3 dB at 1402 nm and a noise figure of 7.6 dB. Figure 8.22 shows the transmission loss curve of this fiber (the inset is the fiber end facet), and the transmission loss of the fiber at 1 μm is much higher than that at 1.3 μm.

Overall, although the performance of 1.3 μm Nd^{3+}-doped fiber lasers and amplifiers has witnessed progress in the past few years, the laser performance and amplification performance are limited due to the low fluorescence branching ratio and ESA. Since the 1.3 μm Pr^{3+}-doped fluoride fiber was successfully developed [75, 76], there has been relatively less research on 1.3 μm Nd^{3+}-doped fiber lasers and fiber amplifiers.

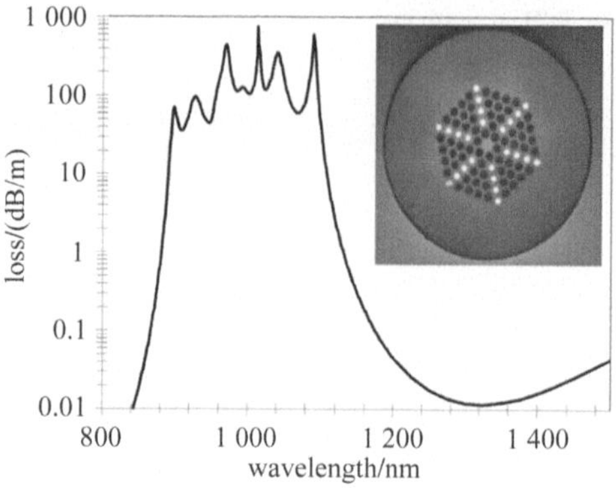

Fig. 8.22 Transmission loss curve of Nd^{3+} -doped microstructured fiber

8.3 Characteristics and Applications of Nd^{3+}/Yb^{3+} Co-Doped Fiber Laser

Nd^{3+} and Yb^{3+} have been widely studied as independent activators in high-energy solid-state lasers and high-brightness high-power fiber lasers. The earliest research on Nd^{3+}/Yb^{3+} co-doping can be traced back to the 1960s. Due to no laser diodes at 915 nm and 980 nm at that time, Nd^{3+}/Yb^{3+} co-doping was the main technical route to obtain lasing of Yb^{3+} through $Nd^{3+} \rightarrow Yb^{3+}$ energy transfer. Although this technology is no longer adopted currently, the research interest in Nd^{3+}/Yb^{3+} co-doping has not decreased. The main reason is that Nd^{3+}/Yb^{3+} co-doping provides a possibility for the development of new materials, due to the great complementarity of the two rare earth ions in terms of fluorescence lifetime, emission cross-section, absorption bandwidth, energy level structure, and other spectral characteristics.

In recent years, Lin et al. [77] conducted an in-depth study on the laser characteristics of Nd^{3+}/Yb^{3+} co-doped silica fibers by 808 nm and 975 nm dual-wavelength pumping. The following section first focuses on the energy transfer efficiency of $Nd^{3+} \rightarrow Yb^{3+}$ in Nd^{3+}/Yb^{3+} co-doped silica fibers, establishes rate equations and conducts simulations, then discusses the laser characteristics under different experimental conditions, and finally introduces the dual-wavelength laser output characteristics and potential applications of Nd^{3+}/Yb^{3+} co-doped silica fibers.

8.3.1 Rate Equations of Nd^{3+}/Yb^{3+} Co-Doping

Table 8.4 shows the parameters of the two types of Nd^{3+}/Yb^{3+} co-doped silica fibers. The fibers were prepared by the sol-gel method and are labeled as NYF-1 and NYF-2, respectively. The Nd^{3+}/Yb^{3+} co-doped silica fibers contain different doping concentrations of Nd^{3+} and Yb^{3+}, and the molar components of the fiber core glass are $95.6SiO_2$-$4Al_2O_3$–$0.3Nd_2O_3$–$0.1Yb_2O_3$ and $95.7SiO_2$-$4Al_2O_3$–$0.1Nd_2O_3$–$0.2Yb_2O_3$. In addition, the two types of fibers have

Table 8.4 Nd^{3+}/Yb^{3+} co-doped silica fibers and parameters [77]

Fiber Type	Size	Cladding	Core Molar Composition	Loss @1200 nm
NYF-1	6/125	Octagonal	$95.6SiO_2$-$4Al_2O_3$–$0.3Nd_2O_3$–$0.1Yb_2O_3$	0.7 dB/m
NYF-2	8/125	Circular	$95.7SiO_2$-$4Al_2O_3$–$0.1Nd_2O_3$–$0.2Yb_2O_3$	1.2 dB/m

slight differences in size and cladding structure, but their impact on the study of laser characteristics can be ignored.

The refractive index distribution of the NYF-1 fiber is shown in Figure 8.23a. The fiber cladding is pure silica and its refractive index at 650 nm is 1.4564, from which the numerical aperture of the fiber core is calculated to be approximately 0.14. The inset shows the cross-section of the NYF-1. The inner cladding was originally designed to be octagonal, but due to surface tension during the preform drawing process, a rounding phenomenon occurred. The concentricity of the fiber core is not high, slightly flat.

Figure 8.23b presents the transmission loss of the NYF-1 fiber. Taking the average of two measurements, the transmission loss of this fiber at 1200 nm is 0.7 dB/m. The transmission loss of the NYF-2 fiber is 1.2 dB/m, slightly larger than that of NYF-1. The absorption coefficients of the cladding pump at 808 nm and the core pump at 975 nm of NYF-1 were measured using the cut-off method, which were 1.3 dB/m and 40 dB/m, respectively.

For Nd^{3+}/Yb^{3+} co-doped fibers, firstly, Nd^{3+} transitions from the ground state to the excited state under pump excitation, and the excited Nd^{3+} returns to the ground state through spontaneous radiation, stimulated radiation, and $Nd^{3+} \rightarrow Yb^{3+}$ energy transfer processes. In actual situations, there are cases where Nd^{3+} and Yb^{3+} are pumped simultaneously. Therefore, a theoretical model that fits the actual situation is considered below, that is, the situation where lasers are pumped by 808 nm and 975 nm simultaneously.

N_{Nd1} and N_{Nd2} represent the particle densities of Nd^{3+} in the ground state and excited state respectively, the total particle density is N_{Nd}, satisfying $N_{Nd} = N_{Nd1} + N_{Nd2}$; N_{Yb1} and N_{Yb2} represent the particle densities of Yb^{3+} in the ground state and excited state respectively, the total particle density is N_{Yb}, satisfying $N_{Yb} = N_{Yb1} + N_{Yb2}$. Under the conditions of dual-wavelength pumping at 808 nm and 975 nm, the following rate equations are established [78]:

$$\frac{dN_{Nd2}}{dt} = W_{808}N_{Nd1} - \frac{N_{Nd2}}{\tau_{Nd}} - W_{Nd}N_{Nd2} - W_{ET}N_{Yb1}N_{Nd2} \tag{8.15}$$

$$\frac{dN_{Yb2}}{dt} = W_{975}N_{Yb1} + W_{ET}N_{Yb1}N_{Nd2} - \frac{N_{Yb2}}{\tau_{Yb}} - W_{Yb}N_{Yb2} \tag{8.16}$$

W_{808} and W_{975} are the pump rates of 808 nm and 975 nm respectively; W_{Nd} and W_{Yb} are the stimulated radiation rate of Nd^{3+} and Yb^{3+}; τ_{Nd} and τ_{Yb} are the fluorescence lifetimes of Nd^{3+} and Yb^{3+} respectively; W_{ET} is the energy transfer rate, which is related to the distance between the two rare earth ions, the absorption and emission

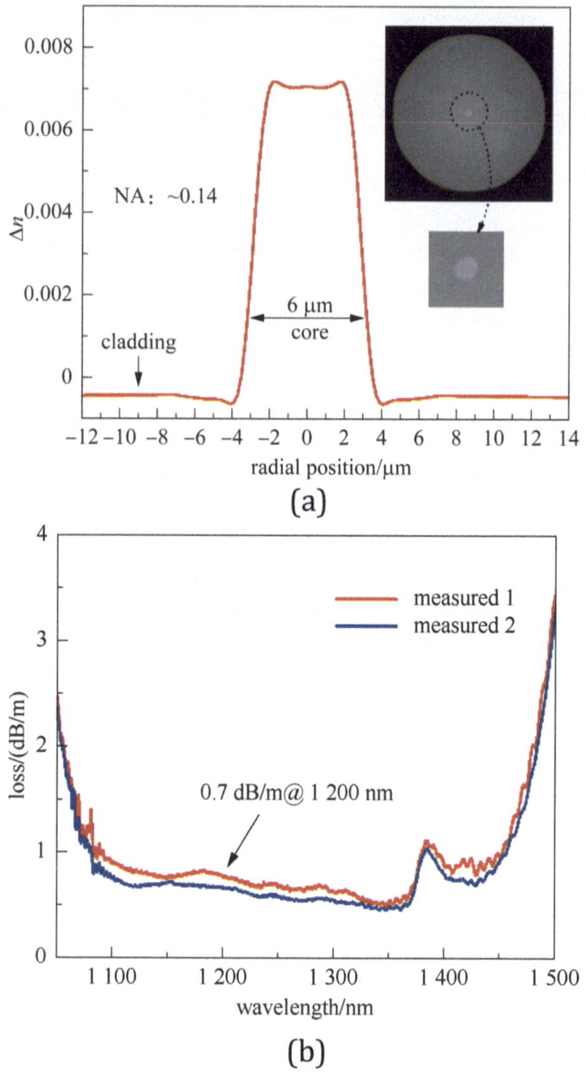

Fig. 8.23 Measurement of NYF-1 fiber parameters [77]. (**a**) Refractive index distribution (inset shows the fiber end facet). (**b**) Transmission loss

spectrum functions, etc., and are introduced in [79, 80]. Here, the fourth term on the right side of eq. (8.15) is explained: according to the quantum theory, the greater the probability that Nd^{3+} is in the excited state and the greater the probability that Yb^{3+} is in the ground state, the greater the probability that Nd^{3+} will transfer energy to Yb^{3+}, hence the inclusion of the product relationship of $N_{Yb1}N_{Nd2}$. Furthermore, it is worth emphasizing that eqs. (8.15) and (8.16) ignore the energy transfer process from Yb^{3+} to Nd^{3+}, because it is a very minor rate process. The mathematical expression of the energy transfer efficiency from Nd^{3+} to Yb^{3+} obtained from eq. (8.15) is represented as $\eta_{Nd\to Yb}$.

$$\eta_{Nd\to Yb}=\frac{W_{ET}N_{Yb1}}{1/\tau_{Nd}+W_{Nd}+W_{ET}N_{Yb1}}\times 100\% \tag{8.17}$$

It can be seen that $\eta_{Nd\to Yb}$ is not a fixed value, but a variable related to W_{Nd} and N_{Yb1}.

Under steady-state conditions, from formulas (8.15) and (8.16), we can respectively get

$$N_{Nd2}=\frac{W_{808}}{W_{808}+\left(1/\tau_{Nd}+W_{Nd}\right)+W_{ET}N_{Yb1}}N_{Nd} \tag{8.18}$$

$$N_{Yb2}=\frac{W_{975}+W_{ET}N_{Nd2}}{W_{975}+\left(1/\tau_{Yb}+W_{Yb}\right)+W_{ET}N_{Nd2}}N_{Yb} \tag{8.19}$$

When substitute eq. (8.18) into eq. (8.19) and use $N_{Yb2}/N_{Yb}=1-N_{Yb1}/N_{Yb}$, the following equation can be obtained

$$\frac{N_{Yb1}}{N_{Yb}}=\frac{-b+\sqrt{b^2-4ac}}{2a} \tag{8.20}$$

Where

$$a=W_{ET}N_{Yb}\left(W_{975}+k\right),k=1/\tau_{Yb}+W_{Yb} \tag{8.21}$$

$$c=k\left(W_{808}+f\right),f=1/\tau_{Nd}+W_{Nd} \tag{8.22}$$

$$b=\left(W_{975}+k\right)\left(W_{808}+f\right)+W_{ET}N_{Yb}\left(W_{808}N_{Nd}/N_{Yb}-k\right) \tag{8.23}$$

For Nd^{3+}, W_{808} and the absorption cross-section $\sigma_{ab/808}$, W_{Nd} and the emission cross-section σ_{em} have the following relationships:

$$W_{808}=\frac{\Gamma_p\sigma_{ab/808}P_{808}}{hv_{p/808}A_c},\quad W_{Nd}=\frac{\Gamma_s\sigma_{em}P_s}{hv_sA_c} \tag{8.24}$$

In the formula, Γ_p is the pump light filling factor; P_{808} is the 808 nm pump power; P_s is the signal light power; h is the Planck constant; ν_p and ν_s are the pump light frequency and signal light frequency respectively; A_c is the cross-sectional area of the fiber core. Similarly, for Yb^{3+}

$$W_{975}=\frac{\Gamma_p\sigma_{ab/975}P_{975}}{hv_{p/975}A_c},\quad W_{Yb}=\frac{\Gamma_s}{hA_c}\left(\frac{\sigma_{em}P_s}{v_s}+\frac{\sigma_{em/975}P_{975}}{v_{s/975}}\right) \tag{8.25}$$

In the equation, W_{Yb} includes the amplification effect of the energy transfer from Nd^{3+} to Yb^{3+} on the 975 nm pump light; P_{975} is the 975 nm pump power. Under the condition of weak excitation at 808 nm, that is, $W_{Nd} \rightarrow 0$ and $N_{Yb1} \rightarrow N_{Yb}$, from eq. (8.17) we can derive the conversion relationship of $W_{ET}\ N_{Yb}$:

$$W_{ET} N_{Yb} \approx \frac{\eta_0}{1-\eta_0} \frac{1}{\tau_{Nd}}, \eta_0 = \left(1 - \frac{\tau_{Nd/Yb}}{\tau_{Nd}}\right) \times 100\% \tag{8.26}$$

Where, η_0 is the initial energy transfer efficiency; $\tau_{Nd/Yb}$ is the fluorescence lifetime of Nd^{3+} in Nd^{3+}/Yb^{3+} co-doped fiber, which can be obtained by testing the fluorescence lifetime of the glass sample. By substituting eqs. (8.20) and (8.26) into eq. (8.17), the influencing factors of $\eta_{Nd \rightarrow Yb}$ under laser operation can be discussed, and the steady-state changes in the energy transfer of $Nd^{3+} \rightarrow Yb^{3+}$ can be analyzed. Table 8.5 provides the parameters related to the calculation of the NYF-1 fiber and the rate equation.

Using the rate equations of 808 nm and 975 nm dual-wavelength pumping, the following simulation elaborates on the effects of 808 nm and 975 nm pumping power, Nd^{3+} stimulated radiation, signal light wavelength and power, and other factors on the energy transfer efficiency of $Nd^{3+} \rightarrow Yb^{3+}$. The calculation parameters are shown in Table 8.5. The 808 nm and 975 nm are cladding and core pumping, respectively.

Table 8.5 NYF-1 Fiber and its Spectral Parameters

Parameter	Value	Unit	Note
Γ_s	0.85	/ [a]	Signal light filling factor [b]
Γ_p	0.0023	/	Pump light filling factor
α_s	0.0016	cm^{-1}	Core loss
d	6	μm	Core diameter
A_c	2.83×10^{-7}	cm^2	Core cross-sectional area
h	6.63×10^{-34}	J·s	Planck's constant
τ_{Yb}	870	μs	Yb^{3+} fluorescence lifetime
τ_{Nd}	370	μs	Nd^{3+} fluorescence lifetime
η_0	60%	/	Initial $Nd^{3+} \rightarrow Yb^{3+}$ transfer efficiency
N_{Yb}	3.78×10^{19}	cm^{-3}	Yb^{3+} doping concentration
N_{Nd}	11.3×10^{19}	cm^{-3}	Nd^{3+} doping concentration
$\sigma_{ab/808}$	1.12×10^{-20}	cm^2	Nd^{3+} absorption cross-section at 808 nm
$\sigma_{em/1061\_Nd}$	1.66×10^{-20}	cm^2	Nd^{3+} emission cross-section at 1061 nm
$\sigma_{ab/975}$	2.44×10^{-20}	cm^2	Yb^{3+} absorption cross-section at 975 nm
$\sigma_{ab/915}$	0.51×10^{-20}	cm^2	Yb^{3+} absorption cross-section at 915 nm
$\sigma_{em/975}$	2.44×10^{-20}	cm^2	Yb^{3+} emission cross-section at 975 nm
$\sigma_{em/1036}$	0.52×10^{-20}	cm^2	Yb^{3+} emission cross-section at 1036 nm
$\sigma_{em/1061\_Yb}$	0.29×10^{-20}	cm^2	Yb^{3+} emission cross-section at 1061 nm

Note: [a] "/" represents dimensionless
[b] For core pumping, $\Gamma_{p/975} = \Gamma_{s/975}$

8.3.1.1 The Effect of 808 nm Pump Power

Figure 8.24a gives the relationship between $\eta_{Nd \to Yb}$, N_{Yb1}/N_{Yb} and the pump power P_{808}. As shown in Fig. 8.24a, when $P_{808} = 0$, $\eta_{Nd \to Yb}$ is numerically equal to η_0, which is 60%; at this time $N_{Yb1}/N_{Yb} = 100\%$, that is, all Yb^{3+} are in the ground state. With the increase of P_{808}, under the action of $Nd^{3+} \to Yb^{3+}$ energy transfer, some ground state Yb^{3+} are converted to the excited state, N_{Yb1} decreases, thus leading to a decrease in transfer efficiency. When P_{808} increases to 3.2 W, $\eta_{Nd \to Yb}$ decreases to

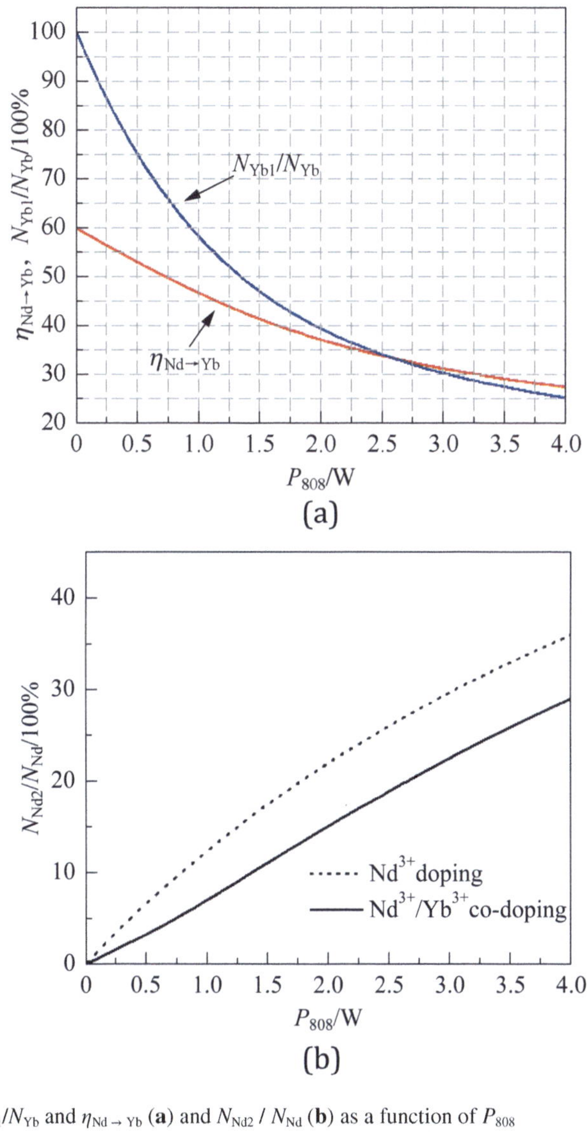

Fig. 8.24 N_{Yb1}/N_{Yb} and $\eta_{Nd \to Yb}$ (**a**) and N_{Nd2} / N_{Nd} (**b**) as a function of P_{808}

30%, at this time, N_{Yb1}/N_{Yb} drops to 28%. It can be seen that under 808 nm pumping, the saturation of Yb^{3+} itself will inhibit the $Nd^{3+} \rightarrow Yb^{3+}$ energy transfer process. As shown in Fig. 8.24b, N_{Nd2}/N_{Nd} gradually increases with the increase of P_{808}, but under the same pumping conditions, unlike the case of Nd^{3+} doping, N_{Nd2} is lower when Nd^{3+}/Yb^{3+} co-doping, which is consistent with the actual situation, this is the manifestation of $Nd^{3+} \rightarrow Yb^{3+}$ energy transfer.

8.3.1.2 The Effect of 975 nm Pump Power

Figure 8.25a shows the relationship between $\eta_{Nd\rightarrow Yb}$ and P_{808} under 808 nm and 975 nm dual-wavelength pumping, with P_{975} fixed at 0, 5 mW, 10 mW, and 80 mW. When $P_{808} = 0$, the change in $\eta_{Nd\rightarrow Yb}$ is due to the effect of P_{975}, which also causes $\eta_{Nd\rightarrow Yb}$ to decrease. The marked $P_{808} = 1.3$ W is the intersection of the four curves. Within the range of $P_{808} < 1.3$ W, P_{975} has the effect of reducing $\eta_{Nd\rightarrow Yb}$; but within the range of $P_{808} > 1.3$ W, P_{975} has the effect of increasing $\eta_{Nd\rightarrow Yb}$. Overall,

The effect of P_{975} makes $\eta_{Nd\rightarrow Yb}$ towards the result of $P_{975} = 80$ mW. In the case of $P_{975} = 80$ mW, $\eta_{Nd\rightarrow Yb}$ stabilizes around 43%, which basically unaffected by P_{808}. The relationship between N_{Yb1}/N_{Yb} and P_{808} is shown in Figure 8.25b, the intersection point of the four cases corresponds to N_{Yb1}/N_{Yb} being 50%, which is exactly the saturation state of Yb^{3+} (simulation calculation has assumed $\sigma_{ab/975} = \sigma_{em/975}$). Therefore, $P_{808} = 1.3$ W is a critical condition. When $P_{808} > 1.3$ W, the fiber will exhibit the phenomenon of amplifying 975 nm pump light (because 975 nm is both pump light and signal light), the number of excited Yb^{3+} particles decreases, N_{Yb1}/N_{Yb} increases, thus leading to an increase in $\eta_{Nd\rightarrow Yb}$.

In actual applications, 915 nm can also be used to pump Yb^{3+}, at this time the stimulated emission of Yb^{3+} at this wavelength can be ignored. Figure 8.26a shows the change in $\eta_{Nd\rightarrow Yb}$ and N_{Yb1}/N_{Yb} with P_{808} under dual-wavelength pumping at 808 nm and 915 nm, respectively, with the 915 nm pump power (P_{915}) fixed at 0, 10 mW, 20 mW, 30 mW, 40 mW and 60 mW. Obviously, it is different from 975 nm pumping. Here $\eta_{Nd\rightarrow Yb}$ changes with P_{808} and P_{915}. It increases and then gradually decreases. Figure 8.26b provides a comparison of the two pumping methods at 915 nm and 975 nm. As the pump power increases, in both cases, $\eta_{Nd\rightarrow Yb}$ gradually decreases and tends to a certain value. For 975 nm pumping, this value is 43%, while for 915 nm pumping, this value is 0 (at this time Yb^{3+} is in the excited state entirely), which is determined by the pump saturation characteristics of Yb^{3+}.

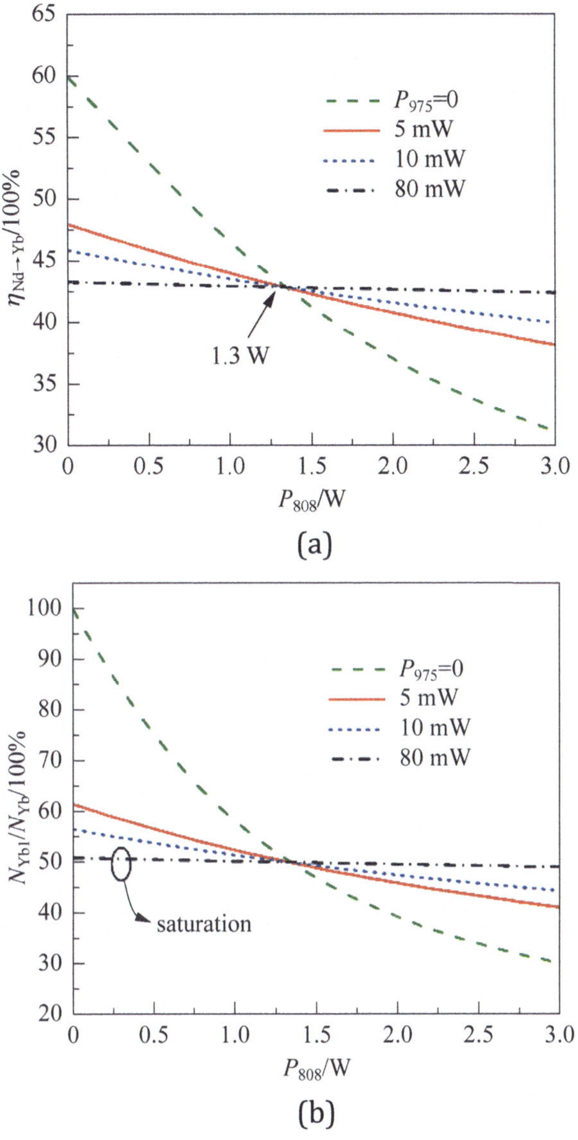

Fig. 8.25 The relationship between $\eta_{Nd\rightarrow Yb}$ (**a**) and N_{Yb1}/N_{Yb} (**b**) as a function of P_{808} under 808 nm and 975 nm dual-wavelength pumping

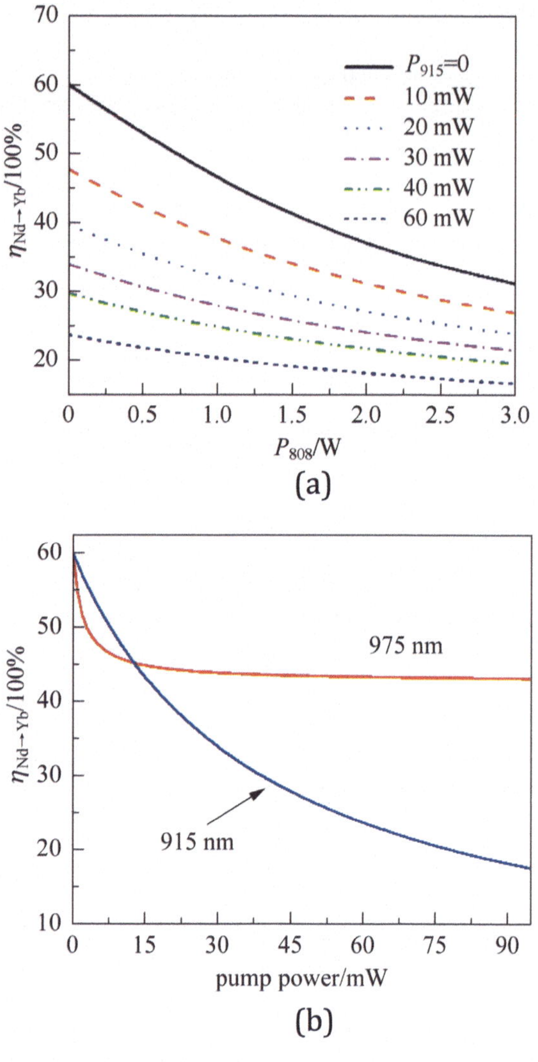

Fig. 8.26 Changes in $\eta_{Nd\to Yb}$ with P_{808} under 808 nm and 915 nm dual-wavelength pumping (**a**) and comparison of 915 nm and 975 nm pumping (**b**)

8.3.1.3 The Effect of Stimulated Radiation of Nd^{3+}

Next, the effect of stimulated radiation of Nd^{3+} on the energy transfer of $Nd^{3+} \to Yb^{3+}$ was further discussed. At this time, it is assumed that Yb^{3+} is basically in the ground state, that is, $N_{Yb1} \to N_{Yb}$, in the calculation, P_{808} and P_{975} are treated as zero. In addition, it is assumed that the signal laser wavelength is 1061 nm, which is the emission peak of Nd^{3+}. Figure 8.27 shows the relationship between $\eta_{Nd\to Yb}$ and the power of the 1061 nm signal laser. When the signal light power increases from 0 to 50 mW, $\eta_{Nd\to Yb}$ decreases from the initial 60% to 20%. When the signal power is further increased to 300 mW, $\eta_{Nd\to Yb}$ will drop to 5%. It can be seen that the stimulated radiation of Nd^{3+} has a strong inhibitory effect on the energy transfer of $Nd^{3+} \to Yb^{3+}$. Because Nd^{3+} is a four-level structure, and dominant in the emission cross-section,

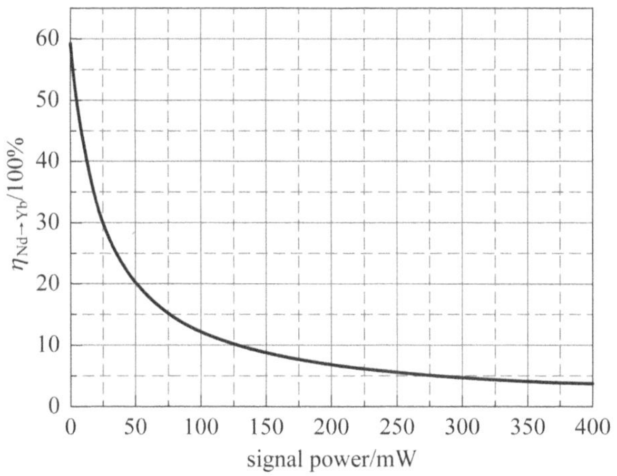

Fig. 8.27 The effect of 1061 nm signal power on $\eta_{Nd \rightarrow Yb}$

it is difficult to take effective measures to control the spontaneous emission of Nd^{3+} in the actual process; however, on the other hand, this is also the source of the unique laser characteristics of Nd^{3+}/Yb^{3+} co-doped fiber.

8.3.1.4 The Influence of Signal Light Wavelength and Power

Next, the situation where the pump light and stimulated radiation act were further explored simultaneously. For stimulated radiation, two representative wavelengths - 1036 nm and 1061 nm were selected. The former is in the spectral region of Yb^{3+}, and the latter is in the spectral overlap region of Yb^{3+} and Nd^{3+}. Figure 8.28a shows the relationship between $\eta_{Nd \rightarrow Yb}$ and P_{808} under 808 nm pumping, including the situations of $P_s = 5$ mW and 10 mW. The dashed line in the figure shows the case of $P_s = 0$ (which has been discussed in Fig. 8.24), which is used here again for comparative analysis. In the case of $\lambda_s = 1036$ nm, compared with the case of $P_s = 0$, P_{s_1036} has the effect of increasing $\eta_{Nd \rightarrow Yb}$, that is, promoting the energy transfer from Nd^{3+} to Yb^{3+} . This is easy to understand, because the stimulated radiation of Yb^{3+} needs to consume excited state particles, causing N_{Yb1}/N_{Yb} to increase. Even so, overall, $\eta_{Nd \rightarrow Yb}$ still decreases with the increase of P_{808}. For the case of $\lambda_s = 1061$ nm, at this time both Nd^{3+} and Yb^{3+} produce stimulated radiation. As known from the 1036 nm case, the 1061 nm stimulated radiation produced by Yb^{3+} has the effect of increasing $\eta_{Nd \rightarrow Yb}$, but the calculation results show that $\eta_{Nd \rightarrow Yb}$ in this case is reduced. This indicates that the stimulated radiation of Nd^{3+} has an effect on $Nd^{3+} \rightarrow Yb^{3+}$. The inhibitory effect of energy transfer is primary. In addition, under this circumstance, $\eta_{Nd \rightarrow Yb}$ also decreases as P_{808} increases.

Figure 8.28b shows the relationship between $\eta_{Nd \rightarrow Yb}$ and P_{808} under 808 nm and 975 nm dual-wavelength pumping, where $P_{975} = 20$ mW is a fixed value. The dashed line in the figure is for comparison when $P_s = 0$. Affected by P_{975}, when $P_{808} = 0$, $\eta_{Nd \rightarrow Yb}$ has decreased to less than 45%. Compared with the results of

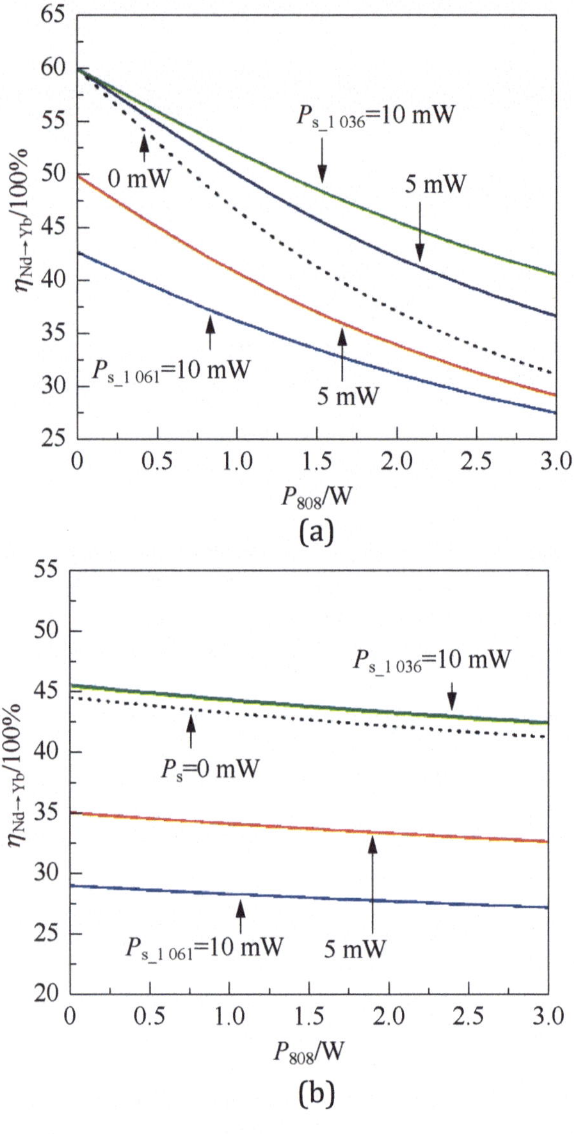

Fig. 8.28 The effect of signal light wavelength and power on $\eta_{Nd\to Yb}$. (**a**) 808 nm single-wavelength pumping ($P_{975} = 0$). (**b**) 808 nm and 975 nm dual-wavelength pumping ($P_{975} = 20$ mW)

Figure 8.28a, the change of $\eta_{Nd\to Yb}$ with P_{808} also tends to be flat, which is the influence of P_{975}. In addition, the increasing effect of P_{s_1036} on $\eta_{Nd\to Yb}$ becomes very limited, on the contrary, the decreasing effect of P_{s_1061} on $\eta_{Nd\to Yb}$ becomes significant.

In summary, the pump light power (P_{808} and P_{975}), signal light power (P_s) and signal light wavelength (λ_s) all affect the energy transfer of $Nd^{3+} \to Yb^{3+}$. Among them, the signal light power associated with the stimulated radiation of Nd^{3+} has a strong inhibitory effect on energy transfer, mainly manifested as affecting the $Nd^{3+} \rightarrow Yb^{3+}$ energy transfer in a manner of rate competition. In addition, the pump power P_{808} and P_{975} also have an inhibitory effect on energy transfer. Although the stimulated radiation of Yb^{3+} has a promoting effect on the energy transfer process, its effect is very limited under the dual-wavelength pumping.

8.3.2 *Laser Characteristics of Nd^{3+}/Yb^{3+} Co-Doped Silica Fiber*

Due to the overlap of the luminescent bands of Nd^{3+} and Yb^{3+}, and the advantages of four-level operations and large emission cross-sections of Nd^{3+}, it is difficult to control the spontaneous emission of Nd^{3+}. As known from the simulation results of the above section, the energy transfer of $Nd^{3+} \rightarrow Yb^{3+}$ in the fiber is inhibited. Based on this understanding, the following focuses on the laser characteristics of Nd^{3+}/Yb^{3+} co-doped silica fiber under 808 nm and 975 nm dual-wavelength pumping, including gain characteristics, laser efficiency, thermal resistance, and photodarkening, etc.

8.3.2.1 Gain Characteristics

Figure 8.29 shows the experimental setup for characterizing the gain characteristics of Nd^{3+}/Yb^{3+} co-doped silica fiber, which is composed of a seed laser and a fiber amplifier. The seed laser is composed of NYF-1 fiber, fiber Bragg grating, 980 nm laser diode, and optical isolator, with a gain fiber length of 30 cm. By using fiber Bragg gratings at different working wavelengths and controlling the driving current of the 980 nm laser diode, seed lasers with wavelengths of 1036 nm, 1053 nm, 1061 nm, and 1080 nm can be output, all with a power of 2.2 mW. The fiber amplifier is composed of a 30 cm long NYF-1 fiber, a forward-pumped 808 nm laser diode, and a backward-pumped 975 nm laser diode. The signal input fiber and output fiber of the fiber combiner are both 10/125 μm fibers, with a numerical aperture of 0.08. For this reason, two mode field adapters are used to connect the 10/125 μm

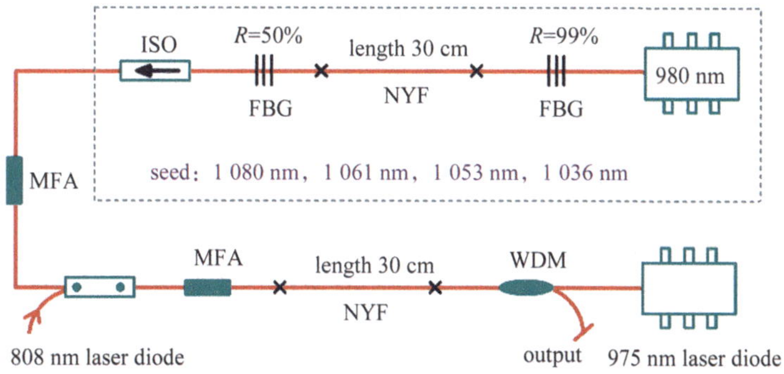

Fig. 8.29 Experimental setup for Nd^{3+}/Yb^{3+} co-doped silica fiber gain measurement

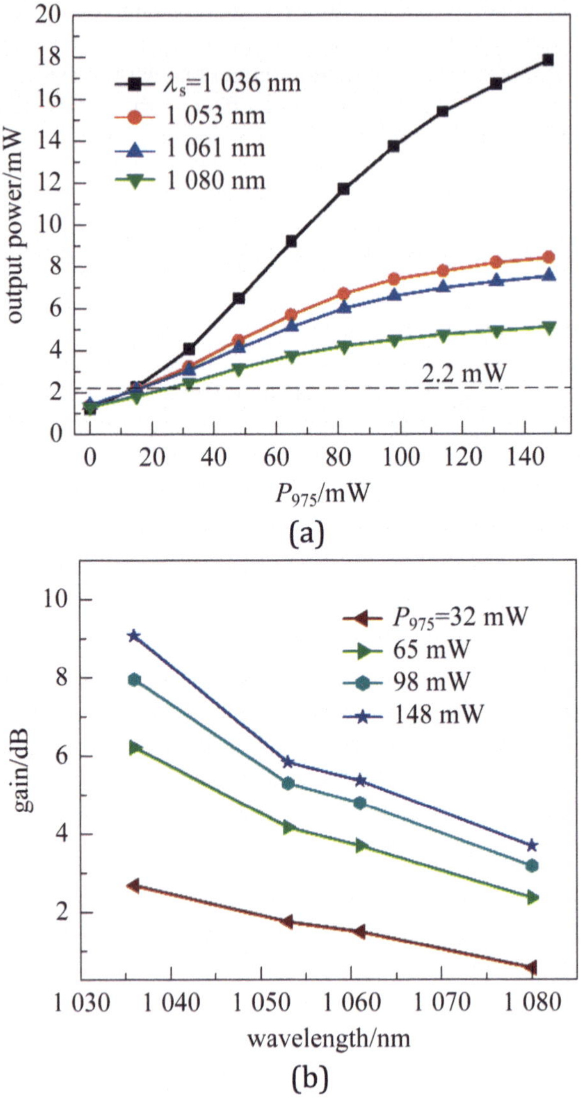

Fig. 8.30 (**a**) Relationship between amplified output power and P_{975} under 975 nm pumping; (**b**) Relationship between gain and wavelength

fiber and the 6/125 μm fiber to reduce insertion loss. The reflection end of the wavelength division multiplexer serves as the signal output.

First, the situation when 975 nm single-wavelength pumping is discussed. Figure 8.30a shows the relationship between the amplified output power of different wavelength seed lasers and P_{975}. When $P_{975} > 20$ mW, the amplifier can achieve laser amplification. Among them, the seed light with a wavelength of 1036 nm has the highest output power, and an output power of 18 mW is obtained under 148 mW pumping. For seed with wavelengths of 1053 nm, 1064 nm, and 1080 nm, the amplified output power is much lower, and the longer the wavelength, the lower the output power. Using the amplified output power to calculate the gain, as shown in Figure 8.30b, the relationship between the gain and the seed light wavelength is obtained.

Under different pumping powers, the gain curve is slightly different, but the gain of each seed laser generally decreases with the increase of wavelength. Obviously, this is consistent with the change of Yb^{3+} emission cross-section with wavelength.

Next, the case of 808 nm and 975 nm dual-wavelength pumping were discussed. Figure 8.31a shows the relationship between the amplified power of the 1061 nm

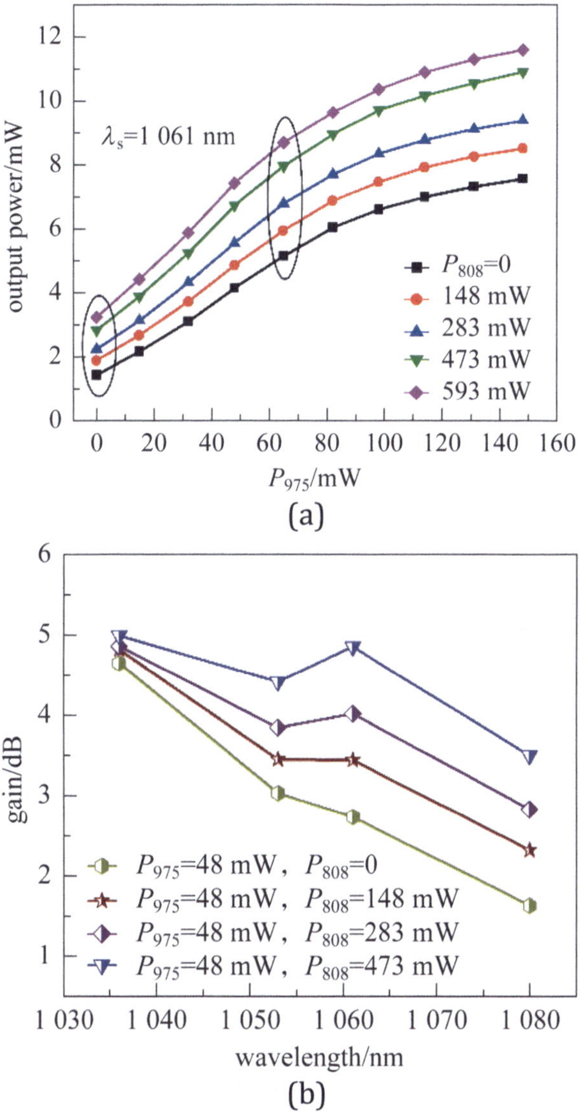

Fig. 8.31 (**a**) Amplified output power of 1061 nm under 808 nm and 975 nm dual-wavelength pumping and its relationship with P_{975}; (**b**) The impact of P_{808} on the gain curve ($P_{975} = 48$ mW)

seed laser under dual-wavelength pumping and P_{975}, with P_{808} fixed at 0, 148 mW, 283 mW, 473 mW, and 593 mW, respectively. Among them, $P_{808} = 0$ is the case of single-wavelength pumping at 975 nm, which is used here for comparative analysis. The comparison shows that the output curves of each $P_{808} > 0$ are similar to single-wavelength pumping at 975 nm, and the output power shows saturation as P_{975} increases. But the addition of P_{808} increases the output power, and this increase effect is significantly enhanced as P_{975} increases, as shown by the elliptical mark in Figure 8.31a. In fact, this is related to the spectral overlap between Nd^{3+} and Yb^{3+}, which is analyzed and explained in the following discussion on laser efficiency.

Using the same test process, the seed lasers at 1036 nm, 1053 nm, and 1080 nm are amplified and the gain is calculated. Figure 8.31b shows the gain spectrum under the condition of $P_{975} = 48$ mW, with P_{808} fixed at 0, 148 mW, 283 mW, and 473 mW respectively. Here, $P_{808} = 0$ is used for comparative analysis. As can be seen that P_{808} mainly produces gain for seed lasers at wavelengths such as 1053 nm, 1061 nm, and 1080 nm, which are within the emission range of Nd^{3+}, indicating that the energy transfer from Nd^{3+} to Yb^{3+} in NYF-1 fiber is not significant, which is consistent with the results of the simulation calculation analysis in Sect. 8.3.1. In addition, P_{808} has the effect of enhancing the gain bandwidth, making the gain spectrum profile flat.

8.3.2.2 Laser Efficiency

Next, the laser efficiency of Nd^{3+}/Yb^{3+} co-doped fiber in the spectral overlap region were further explored. Figure 8.32a shows the experimental setup of 1061 nm laser generated by 808 nm and 975 nm dual-wavelength pumping. The gain fiber is NYF-2 and the length is 22 cm. The experiment uses a fixed P_{975} and changes P_{808} to study the relationship between output power and P_{808}. Figure 8.32b shows the measurement results when P_{975} is 0, 45 mW, 80 mW, 146 mW, and 211 mW, where $P_{975} = 0$ corresponds to the situation of 808 nm single-wavelength pumping. Since 975 nm can also be used to generate 1061 nm laser, the laser power under the pump condition of $P_{808} = 20$ mW increases with the increase of P_{975}. Through linear fitting, the slope efficiency of 808 nm pumping under different P_{975} conditions can be obtained, as shown in Figure 8.32c. Obviously, the slope efficiency increases with the increase of P_{975}, but when $P_{975} > 70$ mW, the increase in slope efficiency shows a saturation trend. Ultimately, under the condition of $P_{975} = 211$ mW, the slope efficiency increased to 50%, which is twice as much as the initial 24.1%.

The increase in slope efficiency is caused by the spectral overlap of Nd^{3+} and Yb^{3+}. Since Nd^{3+} and Yb^{3+} can generate gain at 1061 nm, especially Nd^{3+} has a four-level structure and large emission cross-section, therefore under 808 nm pumping, Nd^{3+} generates gain at 1061 nm, leading to an increase in power density in the resonant cavity and enhanced stimulated radiation. As can be seen from formulas (8.15) and (8.16), enhanced stimulated radiation will lead to a decrease in spontaneous radiation, non-radiative transitions, and other non-laser processes; this is the same principle as suppressing ASE by increasing the reflectivity of the low-reflectivity fiber Bragg grating. The difference is that in the Nd^{3+}/Yb^{3+} co-doped fiber, the

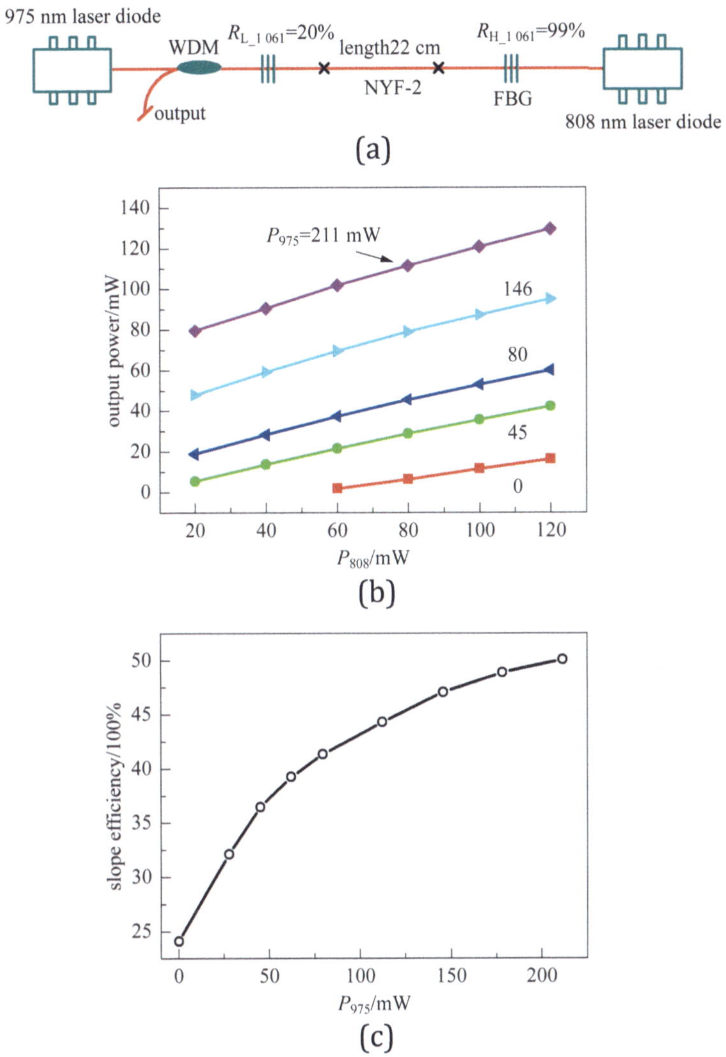

Fig. 8.32 (**a**) Experimental setup for 1061 nm lasing under 808 nm and 975 nm dual-wavelength pumping; (**b**) The relationship between output power and P_{808} under different P_{975} conditions; (**c**) The fitted 808 nm pump slope efficiency under different P_{975} conditions

reflectivity of the low-reflectivity fiber grating has not actually increased, so there will be a significant increase in slope efficiency. Obviously, this is the uniqueness of the Nd^{3+}/Yb^{3+} co-doped fiber.

It is impossible to distinguish that the laser is attributed by Nd^{3+} or Yb^{3+} due to their spectral overlap, so the characterization method of slope efficiency is not accurate. A better method is to convert the output power into optical-to-optical efficiency, which is also convenient for comparison with Yb^{3+}-doped fiber or Nd^{3+}-doped fiber. Figure 8.33 shows the optical-to-optical efficiency of two sets of experiments

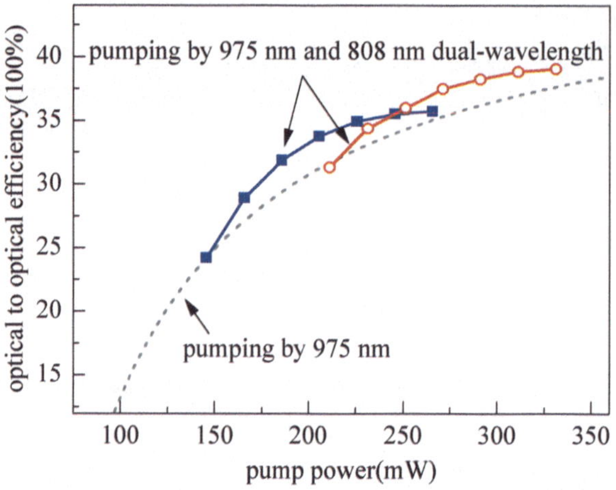

Fig. 8.33 Optical-to-optical efficiency under 975 nm and 808 nm dual-wavelength pumping (the square solid line and the circle solid line represent the conditions of P_{975} = 146 mW and 211 mW respectively, the horizontal axis is the total pump power of 975 nm and 808 nm dual-wavelength)

with P_{975} = 146 mW and P_{975} = 211 mW, where P_{808} ranges from 0 to 120 mW, and the horizontal axis represents the total pump power of 975 nm and 808 nm dual wavelengths. The dashed line in the figure shows the calculated result of the optical-to-optical efficiency under 975 nm single-wavelength pumping: based on the slope efficiency value of 48.2% for the 1061 nm laser generated by the NYF-2 fiber under 975 nm single-wavelength pumping. Using the slope efficiency fitting formula, for any P_{975}, the corresponding output power can be obtained, and finally, the optical-to-optical efficiency shown by the dashed line in the figure is obtained. It can be seen that the optical-to-optical efficiency of dual-wavelength pumping is higher than that of 975 nm single-wavelength pumping within a certain range. Moreover, when P_{975} = 211 mW, the range of dual-wavelength pumping with higher optical-to-optical efficiency is larger. Under the dual-wavelength pumping with P_{975} = 146 mW and P_{808} = 120 mW, the optical-to-optical efficiency is equivalent to the optical-to-optical efficiency of P_{975} = 266 mW single-wavelength pumping, while the slope efficiency of the 1061 nm laser generated by the NYF-2 fiber under 808 nm single-wavelength pumping is only 24%. It can be seen that the connection between Nd^{3+} and Yb^{3+} due to spectral overlap provides a beneficial role in improving laser performance.

Theoretically, the maximum efficiency of generating a 1061 nm laser through 975 nm pumping can reach up to 92%, while 808 nm pumping is only 76%. Therefore, in an ideal state, dual-wavelength pumping by 808 nm and 975 nm cannot achieve a higher laser efficiency than single-wavelength pumping at 975 nm. However, combined with the actual measurement results of Fig. 8.33, in some special cases, when the laser efficiency of Yb^{3+} is much lower than the theoretical limit or equivalent to that of Nd^{3+}, dual-wavelength pumping based on Nd^{3+}/Yb^{3+} co-doped fiber can achieve higher laser efficiency. The heat-resistant performance of the laser discussed below is an example.

8.3.2.3 Heat Resistance

Temperature is an important factor for Nd^{3+}/Yb^{3+} co-doped fiber, which is related to W_{ET} in eq. (8.15) and can affect the energy transfer process. The energy transfer from Yb^{3+} to Nd^{3+} is a process of absorbing phonons, which is very sensitive to temperature. Lin et al. [81] observed the energy transfer from Yb^{3+} to Nd^{3+} in highly doped Nd^{3+}/Yb^{3+} co-doped phosphate fiber using the thermal effect. Figure 8.34a, b are the fiber fluorescence spectra when the temperature is 25 °C, 90 °C, 150 °C, 210 °C, and 270 °C, respectively. The fiber used in the experiment is NYF-1, with a length of 5 cm, and $P_{975} = 48$ mW. The fiber is placed on a copper block, filled with thermal grease, fixed and then heated by a heater.

As shown, the 1336 nm emission band appears on the spectrum, which is a manifestation of the energy transfer from Yb^{3+} to Nd^{3+}. In the process of temperature increasing from 25 °C to 270 °C, the fluorescence intensity of Nd^{3+} at 1336 nm increased by 6 dB, while the fluorescence intensity of Yb^{3+} at 1030 nm decreased by 2 dB. When the temperature exceeds 150 °C, the emission of Nd^{3+} near 1060 nm gradually stands out on the spectrum profile. Although the $Yb^{3+} \rightarrow Nd^{3+}$ energy transfer is enhanced at high temperatures, the enhancement is not enough to have a substantial impact on the NYF-1 fiber like the phosphate fiber, which is largely related to the glass matrix.

Due to the self-absorption effect, Yb^{3+} is also sensitive to temperature. Figure 8.35a shows the experiment setup for characterizing the heat resistance of Nd^{3+}/Yb^{3+} co-doped silica fiber. The laser resonator consists of a 30 cm long NYF-1 fiber, a 1061 nm fiber Bragg grating, and a high-reflection dichroic mirror. During the experiment, the fiber is placed in the tube furnace and temperature-controlled.

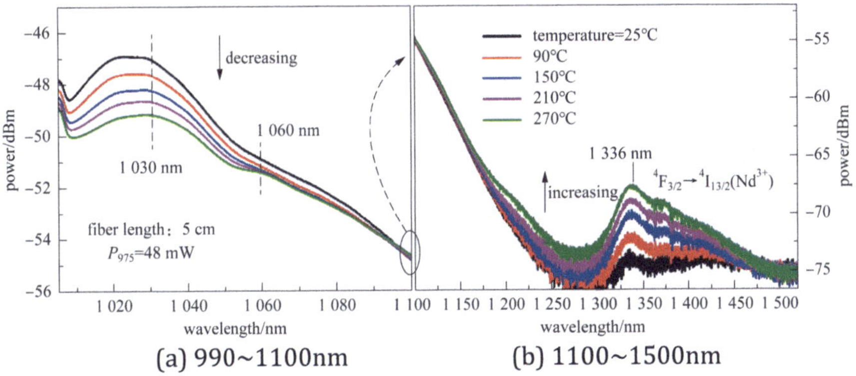

Fig. 8.34 Fluorescence spectrum under heating conditions. (**a**) 990 ~ 1100 nm. (**b**) 1100 ~ 1500 nm

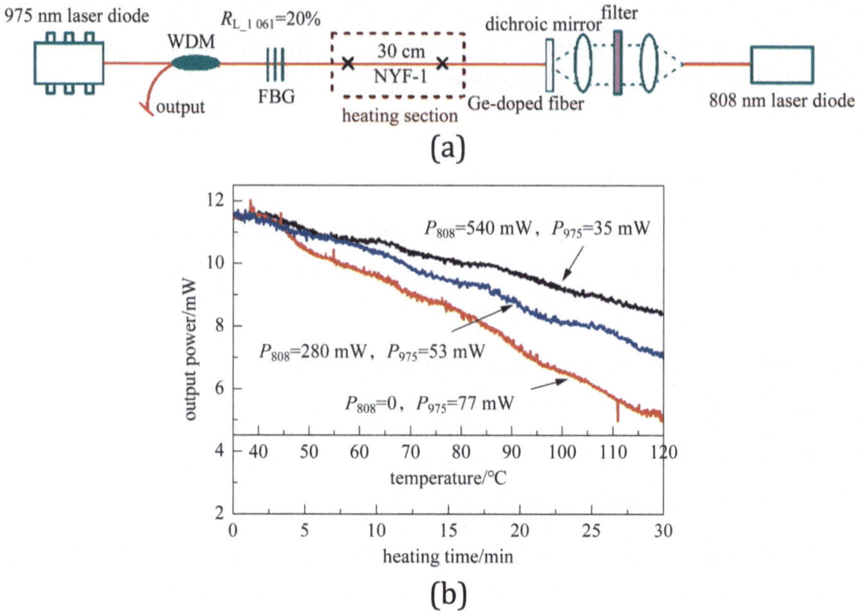

Fig. 8.35 (**a**) Schematic diagram of the variable temperature laser experiment; (**b**) The change of output power with heating time/fiber temperature

Figure 8.35b shows the power change of NYF-1 fiber in the temperature range of 35 ~ 120 °C, including three sets of experiments under different pumping conditions. The three sets of experiments are designed for the replacement of P_{808} with P_{975}, but all have the same initial power of 11.5 mW. Before the heating experiment, the fiber is first kept at 35 °C for 10 min to reach thermal equilibrium. The heating time is 30 min, that is, the temperature is heated to 120 °C within 30 min, and the heating rate is 2.83 °C/min. The result shown in the figure where P_{808} = 0 is the output power change of Yb^{3+} with temperature. When the temperature is heated to 120 °C, the output power is reduced to 5.3 mW, which is 54% less than the initial power. Obviously, after replacing with P_{808}, the heat resistance of the fiber is improved. Under the condition of 120 °C, the output power of the two sets of experiments with P_{808} = 280 mW and P_{808} = 540 mW are reduced by 37.4% and 26.0% respectively. According to the experimental results of 8.34a, for the highest heating temperature of 120 °C here, the role of $Yb^{3+} \rightarrow Nd^{3+}$ energy transfer can be ignored. Therefore, the improvement in the heat resistance of NYF-1 fiber caused by the replacement of P_{808} with P_{975} here is derived from Nd^{3+} itself. Nd^{3+} is a four-level structure, without self-absorption effect, so its heat resistance is better than Yb^{3+}.

Laser efficiency is an accurate reflection of the heat resistance of NYF fiber. Based on the experimental setup in Figure 8.32a, the impact of temperature on the laser performance was explored. The fiber temperature is controlled by a tube furnace, including temperatures of 20 °C, 50 °C, 75 °C, 100 °C, and 125 °C.

Figure 8.36a shows the change of slope efficiency with temperature under 975 nm and 808 nm single-wavelength pumping for NYF-2 fiber. As can be seen, as the temperature increases, the slope efficiency under both single-wavelength pumping conditions gradually decreases. Among them, the slope efficiency by 975 nm pumping decreases from the initial 49% to 43%, while the slope efficiency by 808 nm pumping decreases from the initial 23.9% to 21.2%, that is, the decrease in slope efficiency of Nd^{3+} is lower than that of Yb^{3+}, which is consistent with the results shown in Figure 8.34b. Figure 8.36b shows the optical-to-optical efficiency

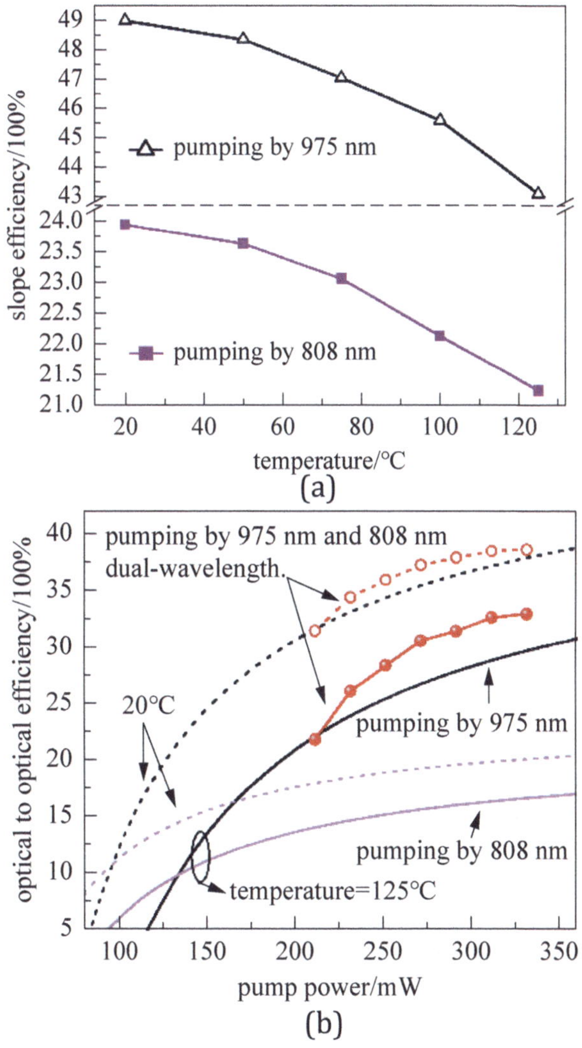

Fig. 8.36 (**a**) The relationship between slope efficiency and temperature; (**b**) The impact of temperature on optical-to-optical efficiency

under 808 nm and 975 nm dual-wavelength pumping at temperatures of 20 °C and 125 °C, where P_{975} = 211 mW. For comparison, the optical-to-optical efficiency under 975 nm and 808 nm single-wavelength pumping is also provided, which is calculated from the slope efficiency value of Figure 8.34a. Comparatively, the optical-to-optical efficiency under each pumping situation significantly decreases with the increase in temperature, but the decrease in the 975 nm single-wavelength pumping is the largest, which narrows the gap between the laser efficiency of Nd^{3+} and that of Yb^{3+}, and makes the effect of the dual-wavelength pumping on the improving optical-to-optical efficiency more significant.

8.3.2.4 Photodarkening

Figure 8.37a shows the experimental setup for characterizing the photodarkening of Nd^{3+}/Yb^{3+} co-doped silica fiber. 975 nm core pumping is used to excite Yb^{3+} to a saturation state, thereby accelerating the photodarkening process. The power change of 633 nm laser diode reflects the degree of photodarkening of the fiber. The incident power at 633 nm is set to 55 μW in the test. It should be emphasized that Nd^{3+} does not have an absorption peak at 633 nm, so the absorption of Nd^{3+}/Yb^{3+} co-doped silica fiber at 633 nm can be ignored. Before the test of photodarkening, the transmittance of the fiber under 975 nm pumping is first measured to ensure that the used P_{975} can saturate Yb^{3+}.

Figure 8.37b gives the relationship between the transmittance of the fiber under test (NYF-1, NYF-2, and Yb-doped silica fiber) and P_{975}. The fiber length for NYF-1, NYF-2, and Yb-doped silica fiber are 7.7 cm, 3 cm, and 5 cm respectively. Another Yb-doped fiber is also prepared by the sol-gel method and used here for comparison. The size of the fiber is 10/100 and the core molar composition is $98.35SiO_2$–$1.5Al_2O_3$–$0.15Yb_2O_3$. Among the three test fibers, the transmittance of NYF-2 is slightly lower, mainly because the splicing loss with passive fibers is higher due to the fiber eccentricity. As can be seen from the transmittance curve, when $P_{975} > 100$ mW, the three test fibers are close to saturation. During the experiment, the pump power of the NYF-1 fiber was 170 mW, and the pump power of the other two fibers was 244 mW. Figure 8.37c shows the online monitoring results of the photodarkening of the three test fibers, with continuous monitoring time exceeding 17 h. Among them, the monitoring power of NYF-1 fiber and NYF-2 fiber did not decrease significantly; while the monitoring power of Yb-doped fiber significantly decreased, from the initial 34 μW to 20 μW. It can be seen that the photodarkening effect of Nd^{3+}/Yb^{3+} co-doped fiber is not significant and can be ignored. Therefore, there is reason to believe that Nd^{3+} plays a positive role in suppressing the photodarkening of Nd^{3+} /Yb^{3+} co-doped silica fiber, but further experimental results are needed to verify this.

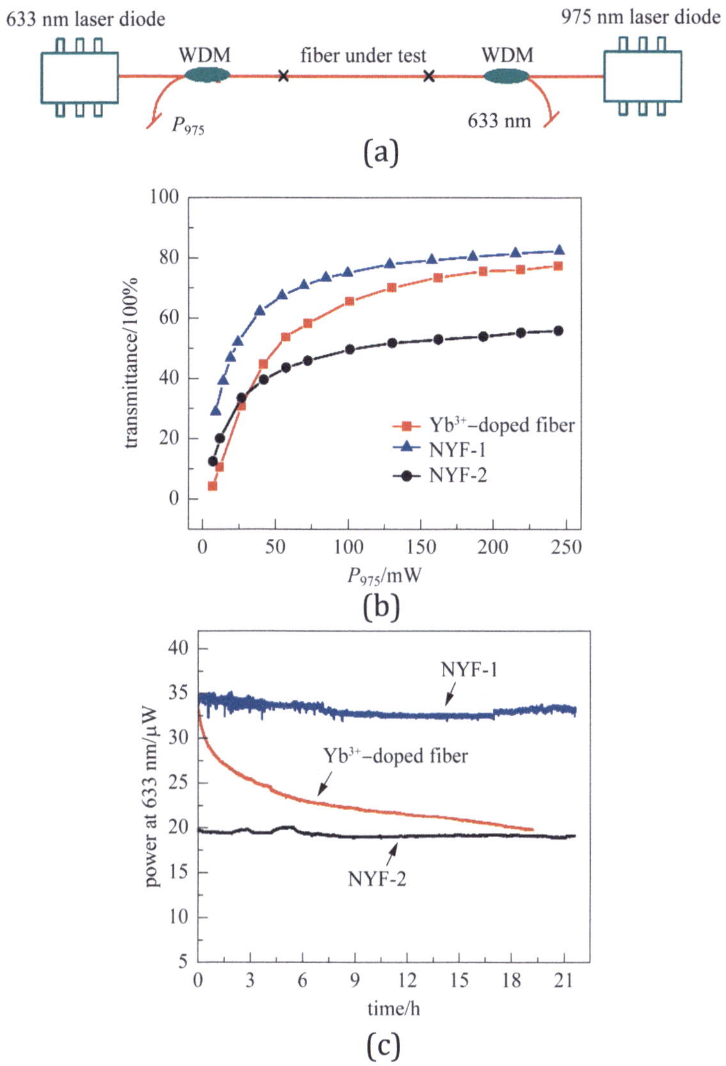

Fig. 8.37 (**a**) Experimental setup for the photodarkening test; (**b**) Transmittance of the fiber under 975 nm pumping; (**c**) Changes in power at 633 nm

8.3.3 *Dual-Wavelength Nd^{3+}/Yb^{3+} Co-Doped Silica Fiber Laser*

Dual-wavelength lasers have important applications in distributed fiber sensing, gas detection, terahertz radiation, etc [82] Dual-wavelength lasers are a common experimental phenomenon in rare-earth-doped fiber lasers, originating from the gain balance of the laser at two wavelengths. Methods for generating dual-wavelength lasers include fiber bending [83], polarization control [84], interference method [85], etc. In the discussion of Sect. 8.3.2, the characteristics of Nd^{3+}/Yb^{3+} co-doped fiber in gain bandwidth control have been introduced, which can be used to generate dual-wavelength lasers. The following part would introduce the generation methods and characteristics of dual-wavelength Nd^{3+}/Yb^{3+} co-doped fiber lasers.

8.3.3.1 Wavelength-Switchable Dual-Wavelength Laser

Figure 8.38a shows the experimental setup of 1036 nm and 1061 nm dual-wavelength lasers by the NYF-1 fiber. The resonator is composed of a 1036 nm, a 1061 nm fiber

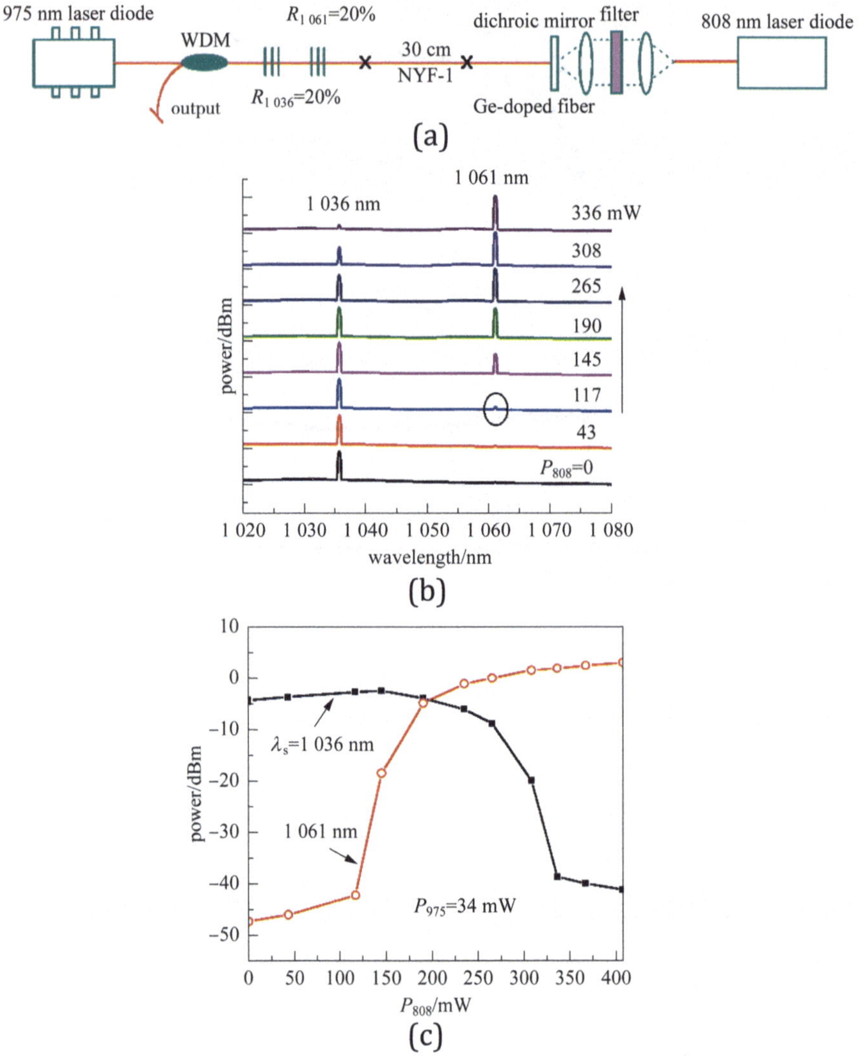

Fig. 8.38 (**a**) Experimental setup for the dual-wavelength laser; (**b**) The change of dual-wavelength laser with P_{808} ($P_{975} = 34$ mW); (**c**) The relationship between the intensity of 1036 nm and 1061 nm laser and P_{808}

Bragg grating with 20% reflectivity and a broadband high-reflective dichroic mirror. The length of the gain fiber is 30 cm, pumped by 808 nm and 975 nm pump lasers. Figure 8.38b shows the variation of 1036 nm and 1061 nm dual-wavelength lasers with P_{808}, where P_{975} is fixed at 43 mW. When $P_{808} = 0$, the 975 nm pump light at 43 mW can generate a 1036 nm laser, which is clearly produced by Yb^{3+}. When P_{808} increases to 145 mW, the laser spectrum shows a significant change, with a 1061 nm laser peak appearing. As P_{808} increases, the intensity of the 1061 nm laser peak gradually increases, while the intensity of the 1036 nm laser peak gradually decreases. Finally, when $P_{808} = 336$ mW, the 1036 nm laser peak disappears and becomes a single wavelength operation at 1061 nm. Figure 8.38c shows the details of the changes. It can be seen that by changing the P_{808}, the intensity ratio of the two laser peaks can be controlled to achieve equal intensity output or laser wavelength switching, etc. Here, the dual-wavelength laser output is achieved by controlling the pump power, which is an active gain control method, thanks to the gain bandwidth control characteristics of Nd^{3+}/Yb^{3+} co-doped fiber.

8.3.3.2 Wavelength Tunability of Dual-Wavelength Laser

Due to the large gain bandwidth of Nd^{3+}/Yb^{3+} co-doped fiber and its controllable characteristics, Nd^{3+}/Yb^{3+} co-doped fiber has good wavelength tunability. Based on Figure 8.38a, using fiber Bragg grating with working wavelengths at 1036 nm, 1053 nm, 1061 nm and 1080 nm to explore the wavelength tunability. Figure 8.39a shows a dual-wavelength laser with a fixed 1036 nm laser peak. Under the condition of $P_{975} = 31$ mW, by controlling P_{808}, dual-wavelength lasers of 1036 nm and 1053 nm, 1036 nm and 1061 nm, and 1036 nm and 1080 nm can be generated, with the corresponding P_{808} being 162 mW, 234 mW, and 436 mW respectively; the laser wavelength ranges from 1053 nm to 1080 nm, achieving a tuning range of 27 nm. Figure 8.39b shows a dual-wavelength laser with a fixed 1061 nm laser peak. Under the condition of $P_{808} = 412$ mW, a dual-wavelength laser of 1053 nm and 1061 nm can be obtained; the laser wavelength ranges from 1053 nm to 1036 nm, with a tuning range of 17 nm. By using fiber Bragg gratings with shorter wavelengths, the wavelength tunability range can be further expanded.

In summary, neodymium ions were the first rare earth ions that can achieve four-level laser output at room temperature. After decades of development, Nd^{3+}-doped laser glasses have played an indispensable role in large-scale laser devices. This chapter, with the background of the demand for high-energy, high-repetition-rate lasers for Nd^{3+}-doped silica glass, first introduces the spectral properties and manipulation methods of Nd^{3+}-doped silica glass in detail, and elaborates on the effects of Al, P, S, I and other elements on the structure and spectral properties, providing important references for the development of new Nd^{3+}-doped silica glass and fibers. Secondly, focusing on the unique needs of underwater communication, biological imaging, distributed fiber sensing and other fields for Nd^{3+}-doped silica fiber lasers at 0.9 μm and dual-wavelength Nd^{3+}/Yb^{3+} co-doped silica fiber lasers, it introduces the research progress of 0.9 μm fiber lasers in detail, sorts out the problems and

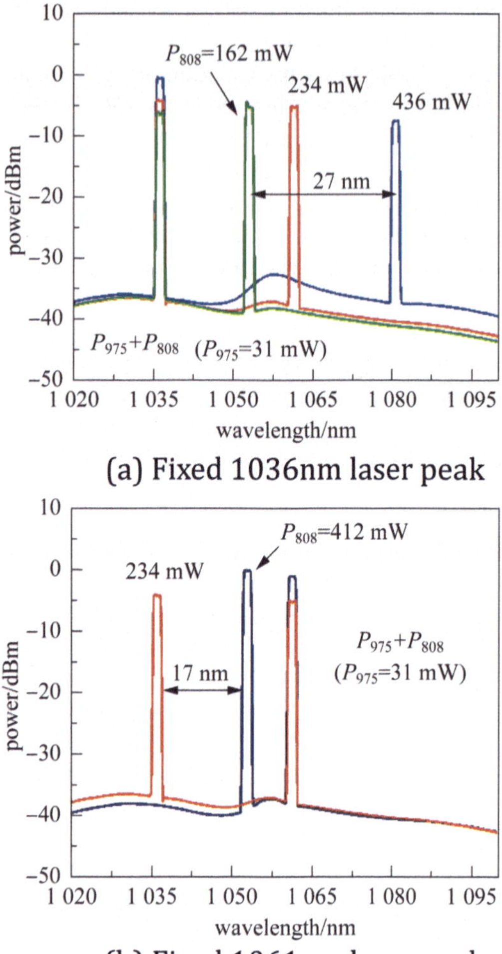

Fig. 8.39 Wavelength-tunable dual-wavelength laser. (**a**) Fixed 1036 nm laser peak. (**b**) Fixed 1061 nm laser peak

improvement measures, and deeply analyzes the rate equations and wavelength tuning characteristics of dual-wavelength Nd^{3+}/Yb^{3+} co-doped fiber lasers. It is believed that with further research on fluorescence branching ratio control methods, fiber structure parameters, and laser technologies, the gain, output power, efficiency and laser wavelength tunability of new Nd^{3+} and Nd^{3+}/Yb^{3+} co-doped silica fibers will be significantly improved, and they will play an important role in more application fields beyond high-energy lasers.

References

1. Snitzer E. Optical maser action of Nd^{3+} in barium crown glass [J]. Physical Review Letters, 1961,7(12):444-446.

2. Cellard C, Retraint D, Francois M, et al. Laser shock peening of Ti-17 titanium alloy: influence of process parameters [J]. Materials Science and Engineering: A, 2012(532):362-372.
3. Fujimoto Y, Murata K, Nakatsuka M. New material for high average power laser based on silica glass [J]. Fusion Engineering and Design, 1999,44(1-4):431-435.
4. Dragic P D, Little L M, Papen G C. Fiber amplification in the 940-nm water vapor absorption band using the $^4F_{3/2} \rightarrow {}^4I_{9/2}$ transition in Nd [J]. IEEE Photonics Technology Letters, 1997,9(11):1478-1480.
5. Anthony L. Cook H D H. Diode-laser-pumped tunable 896-939.5-nm neodymium-doped fiber laser with 43-mW output power [J]. Applied Optics, 1998,37(15):3276-3281.
6. Schirripa Spagnolo G, Cozzella L, Leccese F. Underwater optical wireless communications: overview [J]. Sensors (Basel), 2020,20(8):2261.
7. Li Fengqin, Zhao Biao, Wei Jiao, et al. Continuously tunable single-frequency 455nm blue laser for high-state excitation transition of cesium [J]. Optics Letters, 2019,44(15):3785-3788.
8. Chen Bingying, Rong Hao, Huang Xiaoshuai, et al. Robust hollow-fiber-pigtailed 930nm femtosecond Nd:fiber laser for volumetric two-photon imaging [J]. Optics Express, 2017,25(19):22704-22709.
9. Corre K L, Gilles H, Girard S, et al. Large core, low-NA neodymium-doped fiber for high power CW and pulsed laser operation near 900nm [C]. OSA Laser Congress 2019 (ASSL, LAC, LS&C), 2019.
10. Campbell J H, Suratwala T I. Nd-doped phosphate glasses for high-energy/high-peak-power lasers [J]. Journal of Non-Crystalline Solids, 2000(263-264):318-341.
11. Hu Lili, et al. Laser Glass and Its Applications [M]. Shanghai: Shanghai Scientific & Technical Publishers, 2019.
12. Judd B R. Optical absorption intensities of rare-earth ions [J]. Physical Review, 1962,127(3):750-761.
13. Ofelt G S. Intensities of crystal spectra of rare-earth ions [J]. Journal of Chemical Physics, 1962,37(3):511.
14. Walsh B M. Judd-Ofelt Theory: principles and practices; proceedings of the Advances in Spectroscopy for Lasers and Sensing, Dordrecht, F 2006[C]. Springer Netherlands.
15. Carnall W T, Fields P R, Rajnak K. Electronic energy levels in the trivalent lanthanide aquo ions. i. Pr^{3+}, Nd^{3+}, Pm^{3+}, Sm^{3+}, Dy^{3+}, Ho^{3+}, Er^{3+}, and Tm^{3+} [J]. The Journal of Chemical Physics, 1968,49(10):4424-4442.
16. Jørgensen C K, Judd B R. Hypersensitive pseudoquadrupole transitions in lanthanides [J]. Molecular Physics, 1964,8(3):281-290.
17. Takebe H, Nageno Y, Morinaga K. Compositional dependence of Judd-Ofelt parameters in silicate, borate and phosphate glasses [J]. Journal of the American Ceramic Society, 1995,78(5):1161-1168.
18. Tanabe S, Hanada T, Ohyagi T, et al. Correlation between ^{151}Eu Mossbauer isomer shift and Judd-Ofelt Ω_6 parameters of Nd^{3+} ions in phosphate and silicate laser glasses [J]. Physical Review B, 1993,48(14):10591-10594.
19. Jacobs R R, Weber M J. Dependence of the $^4F_{3/2} \rightarrow {}^4I_{11/2}$ induced emission cross section for Nd^{3+} on glass composition [J]. IEEE Journal of Quantum Electronics, 1976, QE-1(2):102-111.
20. Dantas N O, Serqueira E O, Bell M J V, et al. Influence of crystal field potential on the spectroscopic parameters of SiO_2 center dot B_2O_3 center dot PbO glass doped with Nd_2O_3 [J]. Journal of Luminescence, 2011,131(5):1029-1036.
21. Kaminskii A A, Li L. Analysis of spectral line intensities of TR^{3+} ions in crystal systems [J]. Physica Status Solidi (A), 1974(26):593-598.
22. Lomheim T S, Deshazer L G. New procedure of determining neodymium fluorescence branching ratios as applied to 25 crystal and glass hosts [J]. Optics Communications, 1978,24(1):89-94.
23. Arai K, Namikawa H, Kumata K, et al. Aluminum or phosphorus co-doping effects on the fluorescence and structural properties of neodymium-doped silica glass [J]. Journal of Chemical Physics, 1986,136(1):3430-3436.

24. Sen S. Atomic environment of high-field strength Nd and Al cations as dopants and major components in silicate glasses: a Nd L_{III} -edge and Al K-edge X-ray absorption spectroscopic study [J]. Journal of Non-Crystalline Solids, 2000,261(1-3):226-236.
25. Qiao Y, Da N, Chen D, et al. Spectroscopic properties of neodymium doped high silica glass and aluminum codoping effects on the enhancement of fluorescence emission [J]. Applied Physics B, 2007,87(4):717-722.
26. Funabiki F, Kajihara K, Kaneko K, et al. Characteristic coordination structure around Nd ions in sol-gel-derived Nd-Al-codoped silica glasses [J]. Journal of Physical Chemistry B, 2014,118(29):8792-8797.
27. Xu Wenbin, Wang Meng, Zhang Lei, et al. Effect of P^{5+}/Al^{3+} molar ratio on structure and spectroscopic properties of $Nd^{3+}/Al^{3+}/P^{5+}$ co-doped silica glass [J]. Journal of Non-Crystalline Solids, 2016(432):285-291.
28. Wang Shikai, Yu Chunlei, Chen Yinggang, et al. A method to effectively improve the fluorescence intensity of Nd^{3+} doped silica glass at 900nm: China, CN113024111A [P]. 2021.
29. Wang Siying, Li Yijun, Chen Yanchun, et al. Femtosecond all-polarization-maintaining Nd fiber laser at 920nm mode locked by a biased NALM [J]. Optics Express, 2021,29(23):38199-38205.
30. Peng Yu, Zhao Yang, Li Ye. Locking of 461nm external cavity frequency doubled laser using 3 methods [J]. Chinese Journal of Lasers, 2010,37(2):345-350.
31. Shen Quanhong, Xu Duanyi, Qi Guosheng, et al. High-density Blu-ray storage and its extension technology [J]. optical technique, 2005,31(6):921-927.
32. Gu Bo. High-power blue semiconductor lasers open a new door for metal processing [J]. Metalworking (Hot Working), 2021(3):1-6.
33. Gao Weinan, Xu Zuyan, Bi Yong, et al. Present development and tendency of laser display technology [J]. Chinese Journal of Engineering Science, 2020,22(3):85-91.
34. Corre K L, Robin T, Barnini A, et al. Linearly-polarized pulsed Nd-doped fiber MOPA at 905nm and frequency conversion to deep-UV at 226nm [J]. Optics Express, 2021,29(3):4240-4248.
35. Xin Guofeng, Feng Rongzhu, Chen Guoying, et al. 900nm high-power semiconductor laser array [J]. Laser Optoelectronics Progress, 2003,40(3):43-44.
36. Shayeganrad G, Cante S, Mosquera J P, et al. Highly efficient 110-W closed-cycle cryogenically cooled Nd:YAG laser operating at 946nm [J]. Optics Letters, 2020,45(19):5368-5371.
37. Sattayaporn S, Aka G, Loiseau P, et al. Optical spectroscopic properties, 0.946 and 1.074μm laser performances of Nd^{3+}-doped Y_2O_3 transparent ceramics [J]. Journal of Alloys and Compounds, 2017(711): 446-454.
38. Le Corre K, Robin T, Cadier B, et al. Mode-locked all-PM Nd-doped fiber laser near 910nm [J]. Optics Letters, 2021,46(15):3564-3567.
39. Wang Yafei, Li Xingyu, Wu Jiamin, et al. Three-level all-fiber laser at 915nm based on polarization-maintaining Nd^{3+}-doped silica fiber [J]. Chinese Optics Letters, 2020,18(1):011401.
40. Bufetov I A, Dudin V V, Shubin A V, et al. Efficient 0.9-μm neodymium-doped single-mode fibre laser [J]. Quantum Electronics, 2003,33(12):1035-1037.
41. J Kim P D, D B S Soh, J K Sahu, et al. Nd Al-doped depressed clad hollow fiber laser at 930nm [J]. OSA/ASSP, 2005, MC5.
42. Soh D B S, Seongwoo Y, Nilsson J, et al. Neodymium-doped cladding-pumped aluminosilicate fiber laser tunable in the 0.9μm wavelength range [J]. IEEE Journal of Quantum Electronics, 2004,40(9):1275-1282.
43. Jay W Dawson, Raymond Beach, Alex Drobshoff, et al. Scalable 11W 938nm Nd^{3+} doped fiber laser [J]. OSA/ASSP, 2004.
44. Laroche M, Cadier B, Gilles H, et al. 20W continuous-wave cladding-pumped Nd-doped fiber laser at 910nm [J]. Optics Letters, 2013,38(16):3065-3067.
45. Leconte B, Cadier B, Gilles H, et al. Extended tunability of Nd-doped fiber lasers operating at 872-936nm [J]. Optics Letters, 2015,40(17):4098-4101.
46. Fu L B, Ibsen M, Richardson D J, et al. Compact high-power tunable three-level operation of double cladding Nd-doped fiber laser [J]. IEEE Photonics Technology Letters, 2005,17(2):306-308.

47. Chen Bingying, Rong Hao, Huang Xiaoshuai, et al. Robust hollow-fiber-pigtailed 930nm femtosecond Nd: fiber laser for volumetric two-photon imaging [J]. Optics Express, 2017,25(19):22704-22709.
48. Wei Xiaoming, Kong Cihang, Sy Samuel, et al. Ultrafast time-stretch imaging at 932nm through a new highly-dispersive fiber [J]. Biomedical Optics Express, 2016,7(12):5208-5217.
49. Wang A, George A K, Knight J C. Three-level neodymium fiber laser incorporating photonic bandgap fiber [J]. Optics Letters, 2006,31(10):1388-1390.
50. Wang Yafei, Chen Weiwei, Cao Jiangkun, et al. Boosting the branching ratio at 900nm in Nd^{3+} doped germanophosphate glasses by crystal field strength and structural engineering for efficient blue fiber lasers [J]. Journal of Materials Chemistry C, 2019,7(38):11824-11833.
51. Chen Yinggang, Lin Zhiquan, Wang Yafei, et al. Nd^{3+} -doped silica glass and fiber prepared by modified sol-gel method [J]. Chinese Optics Letters, 2022,20(9):091601.
52. Chen Yinggang, Dong Hehe, Lin Zhiquan, et al. Research progress on key laser materials in the 900nm band [J]. Laser Optoelectronics Progress, 2022,59(15):1516004.
53. Alcock I P, Ferguson A I, Hanna D C. Continuous-wave neodymium-doped monomode-fiber laser operating at 0.900-0.945 and 1.070-1.135μm [J]. Optics Letters, 1986,11(11):709-711.
54. Dawson J, Drobshoff A, Liao Z, et al. High power 938nm cladding pumped fiber laser [J]. Proceeding of SPIE, 2003(4974):75-82.
55. Kilian Le C, Hervé G, Sylvain G, et al. 83W efficient Nd-doped LMA fiber laser at 910nm [J]. Proceedings of the SPIE, 2022, PC11981.
56. Geng Jihong, Wang Qing, Luo Tao, et al. Single-frequency narrow-linewidth Tm-doped fiber laser using silicate glass fiber [J]. Optics Letters, 2009,34(22):3493-3495.
57. Zhu Xiushan, Shi Wei, Zong Jie, et al. 976nm single-frequency distributed Bragg reflector fiber laser [J]. Optics Letters, 2012,37(20):4167-4169.
58. Xu Shanhui, Yang Zhongmin, Zhang Weinan, et al. 400mW ultrashort cavity low-noise single-frequency Yb^{3+} -doped phosphate fiber laser [J]. Optics Letters, 2011,36(18):3708-3710.
59. Fang Qiang, Xu Yang, Fu Shijie, et al. Single-frequency distributed Bragg reflector Nd doped silica fiber laser at 930nm [J]. Optics Letters, 2016,41(8):1829-1832.
60. Fermann M E, Hartl I. Ultrafast fibre lasers [J]. Nature Photonics, 2013,7(11):868-874.
61. Martinez A, Yamashita S. 10 GHz fundamental mode fiber laser using a graphene saturable absorber [J]. Applied Physics Letters, 2012,101(4):04118.
62. Thapa R, Nguyen D, Zong J, et al. All-fiber fundamentally mode-locked 12 GHz laser oscillator based on an Er/Yb-doped phosphate glass fiber [J]. Optics Letters, 2014,39(6):1418-1421.
63. Cheng Huihui, Wang Wenlong, Zhou Yi, et al. 5 GHz fundamental repetition rate, wavelength tunable, all-fiber passively mode-locked Yb-fiber laser [J]. Optics Express, 2017,25(22):27646-27651.
64. Drobizhev M, Makarov N S, Tillo S E, et al. Two-photon absorption properties of fluorescent proteins [J]. Nature Methods, 2011,8(5):393-399.
65. Clowes J, Godfrey I, Grudinin A B, et al. High power ultrafast fibre laser in the 920nm spectral range; proceedings of the Conference on Lasers and Electro-Optics/Quantum Electronics and Laser Science and Photonic Applications Systems Technologies, Baltimore, Maryland, F 2005/05/22[C]. Optica Publishing Group.
66. Gao Xiang, Zong Weijian, Chen Bingying, et al. Core-pumped femtosecond Nd: fiber laser at 910 and 935nm [J]. Optics Letters, 2014,39(15):4404-4407.
67. Chen Bingying, Jiang Tongxiao, Zong Weijian, et al. 910nm femtosecond Nd-doped fiber laser for in vivo two-photon microscopic imaging [J]. Optics Express, 2016,24(15):16544-16549.
68. Alcock I P, Ferguson A I, Hanna D C, et al. Tunable, continuous-wave neodymium-doped monomode-fiber laser operating at 0.900-0.945 and 1.070-1.135μm [J]. Optics Letters, 1986,11(11):709-711.
69. Hakimi F, Po H, Tumminelli R, et al. Glass fiber laser at 1.36μm from SiO_2: Nd [J]. Optics Letters, 1989,14(19):1060-1061.
70. Grubb S G, Barnes W L, Taylor E R, et al. Diode-pumped 1.36μm Nd-doped fibre laser [J]. Electronics Letters, 1990,26(2):121-122.

71. Ishikawa E, Aoki H, Yamashita T, et al. Laser emission and amplification at 1.3μm in neodymium-doped fluorophosphate fibres [J]. Electronics Letters, 1992,28(16):1497-1499.
72. Htein L, Fan W, Watekar P R, et al. Amplification at 1400-1450nm of the large-core Nd-doped fiber by white LED pumping [J]. IEEE Photonics Technology Letters, 2013,25(11):1081-1083.
73. Dawson J W, Pax P H, Allen G S, et al. 1.2W laser amplification at 1427nm on the $^4F_{3/2}$ to $^4I_{13/2}$ spectral line in an Nd^{3+} doped fused silica optical fiber [J]. Optics Express, 2016,24(25):29138-29152.
74. Dawson J W, Kiani L S, Pax P H, et al. E-band Nd^{3+} amplifier based on wavelength selection in an all-solid micro-structured fiber [J]. Optics Express, 2017,25(6):6524-6538.
75. Miniscalco W J, Andrews L J, Thompson B A, et al. 1.3μm fluoride fibre laser [J]. Electronics Letters, 1988,24(1):28-29.
76. Miyajima Y, Komukai T, Sugawa T, et al. Rare earth-doped fluoride fiber amplifiers and fiber lasers [J]. Optical Fiber Technology, 1994,1(1):35-47.
77. Lin Zhiquan, Yu Chunlei, Hu Lili. Laser properties of Nd^{3+}/Yb^{3+} co-doped glass fiber around 1μm [J]. Journal of the Optical Society of America B, 2021,38(8):2443-2450.
78. Lin Zhiquan, Wang Fan, Wang Meng, et al. Maintaining broadband gain in a Nd^{3+}/Yb^{3+} co-doped silica fiber amplifier via dual-laser pumping [J]. Optics Letters, 2018,43(14):3361-3364.
79. Lou Liren, Yin Min, Li Qingting. Fundamentals of Luminescent Physics: Solid State Optical Transition Process[M]. Hefei: Press of University of Science and Technology of China, 2014.
80. Rivera-López F, Babu P, Basavapoornima C, et al. Efficient Nd^{3+}/Yb^{3+} energy transfer processes in high phonon energy phosphate glasses for 1.0μm Yb^{3+} laser [J]. Journal of Applied Physics, 2011,109(12):123514.
81. Lin Zhiquan, Yu Chunlei, He Dongbing, et al. Dual-wavelength laser output in Nd^{3+}/Yb^{3+} co-doped phosphate glass fiber under 970nm pumping [J]. IEEE Photonics Technology Letters, 2016,28(23):2673-2676.
82. Wang Tianshu. Multi-wavelength Fiber Laser Technology [M]. Beijing: Science Press, 2017.
83. Ding Yaqian, Qi Yunfeng, Liu Yuan, et al. Study on the polarization-maintaining fiber power amplification characteristics of dual-wavelength fiber lasers [J]. Chinese Journal of Lasers, 2014,41(3):0302002.
84. Feng Xinhuan, Liu Yange, Fu Shenggui, et al. Switchable dual-wavelength ytterbium-doped fiber laser based on a few-mode fiber grating [J]. IEEE Photonics Technology Letters, 2004,16(3):762-764.
85. Ahmad H, Salim M M, Azzuhri S R, et al. Tunable dual-wavelength ytterbium-doped fiber laser using a strain technique on microfiber MachZehnder interferometer [J]. Applied Optics, 2016,55(4):778-782.

Chapter 9
Radiation-Resistant Rare Earth-Doped Silica Fiber and Its Applications

In recent years, major projects in the field of space science such as the starlink proposed by SpaceX, Micius satellite for quantum science experiments, gravitational wave detection, and space-earth integration network have created urgent demands for lasers for space applications. Fiber lasers with advantages including high brightness, good robustness, and high power-to-weight ratio, and have important applications in space laser communication, radar, remote sensing, space debris treatment, etc. [1–3]. Rare earth ions (Yb^{3+}, Er^{3+}, Er^{3+}-Yb^{3+}, Tm^{3+}, etc.) doped silica fiber, also known as "active fiber", is the core gain medium that constitutes the fiber laser. Its function is to generate lasers and achieve gain amplification. Space is a complex environment that includes strong radiation, high vacuum, and extreme temperature fluctuations. When the fiber laser is subjected to space ionizing radiation, the optical loss and noise factor of the active fiber will increase sharply, and the laser slope efficiency or gain performance will drop significantly. This phenomenon is called the radiation-induced darkening (RD) effect [4, 5], the fundamental reason is related to the color centers produced by irradiation. In addition, the absorption loss of color centers will also increase the thermal effect of active fibers, leading to a significant decrease in the threshold of transverse mode instability (TMI), severely degrading the output performance of fiber lasers. How to effectively solve the RD effect of active fibers for space applications is a difficult problem faced by researchers at home and abroad. It is necessary to conduct systematic research from the perspectives of the RD generation mechanism, influencing factors, and suppression methods.

This chapter first briefly introduces the space radiation environment, the application requirements of silica fibers in space, and the challenges faced; then it focuses on the RD mechanism of rare-earth-doped silica fibers, elaborates on the influencing factors of the radiation resistance of rare-earth-doped silica fibers, and finally summarizes the methods to improve the radiation resistance of rare-earth-doped silica fibers.

L. Hu, *Rare Earth Doped Silica Fiber and Its Applications*,
https://doi.org/10.1007/978-981-96-5306-5_9

9.1 Applications and Challenges of Rare Earth-Doped Silica Fiber in Space

9.1.1 Space Radiation Environment

Space is a complex environment with microgravity, high vacuum, extreme temperature fluctuations (±200 °C), and strong radiation. Among them, strong radiation has the most serious impact on optical and electrical devices.

Radiation refers to the propagation of energy in space or medium. As shown in Fig. 9.1, radiation is usually classified in two ways: ① according to the carrier of energy, it can be divided into electromagnetic radiation and particle radiation; ② according to the level of energy and the ability to ionize matter, it can be divided into ionizing radiation and non-ionizing radiation.

As we all know, light and all microscopic particles have wave-particle duality. Among them, radiation characterized by volatility is called electromagnetic radiation. Electromagnetic radiation energy from low to high mainly includes radio waves, microwaves, infrared, visible light, ultraviolet, X-rays, and γ-rays. Radiation characterized by particle nature is called particle radiation. Particle radiation mainly includes protons (H nucleus), α particles (He nucleus), β particles (electrons), neutrons, and heavy ions (proton number > 2).

Because the energy of fast particles, X-rays, and γ-rays is very high, they can cause ionization of water or biological tissues, so they are called ionizing radiation. Among them, proton, α particle, β particle, heavy ion and other charged particle

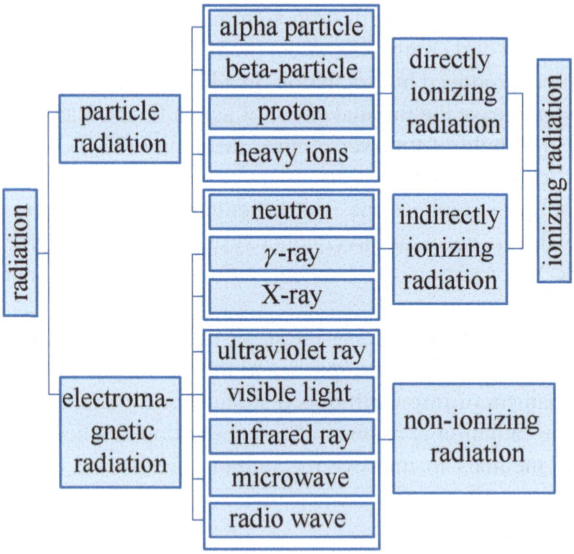

Fig. 9.1 Types of radiation

radiation can directly cause material ionization, which is called direct ionizing radiation; X-ray, γ-ray, neutron, and other uncharged radiation cause material ionization by producing secondary electrons, which is called indirect ionizing radiation. Because the energy of electromagnetic waves with a wavelength greater than X-rays (>100 nm) is low and cannot cause ionization of water or biological tissues, it is called non-ionizing radiation. In this chapter, unless specifically stated, all radiation refers specifically to ionizing radiation.

The penetration ability of different types of radiation varies greatly. Charged particles and electromagnetic waves will interact with the electrons or atomic nuclei in the material when penetrating the material, thereby consuming some energy. Charged particles will slow down as a result, and electromagnetic waves will be absorbed by the material they penetrate. Neutrons are not charged, and they only consume energy when they have inelastic collisions with atomic nuclei. The penetration ability of neutrons is closely related to their speed. As shown in Fig. 9.2, the penetration ability of α particles and other heavy ion beams is weak and can be blocked by thin sheet of paper. β particles can be blocked by aluminum plates with a thickness of millimeters or wooden boards with a thickness of centimeters. High-energy electromagnetic radiation has extremely strong penetration ability, especially X-rays and γ-rays. Studies have shown that a lead plate 15.6 mm thick can only shield about 50% of ^{60}Co γ-rays [6]. The penetration ability of high-energy neutron beams is the strongest, and it can penetrate lead plates several centimeters thick. It needs to be shielded with water or heavy concrete to be more thoroughly shielded.

The radiation belts around the Earth mainly contain galactic cosmic rays, geomagnetic capture radiation belts, and solar particle events. The components, energy, flux, and spatiotemporal distribution of each radiation belt are different, as shown in Table 9.1. Galactic cosmic rays are cosmic rays from within the Milky Way but

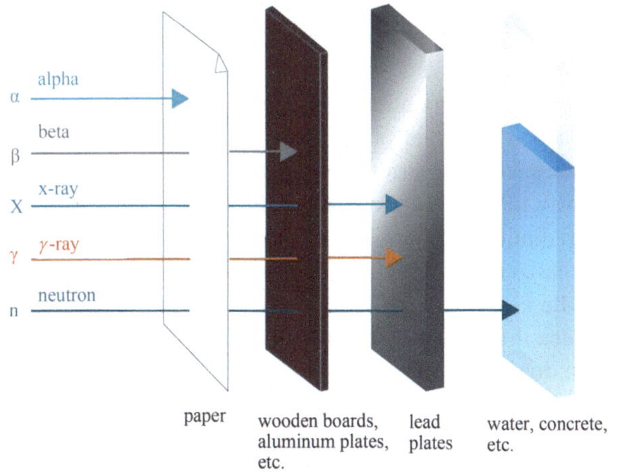

Fig. 9.2 Schematic diagram of the penetration ability of different types of radiation

Table 9.1 Components, time, and spatial distribution of radiation belts around the Earth

Radiation Belt	Galactic Cosmic Rays	Geomagnetic Capture Radiation	Solar Particle Events
Composition	83% protons 13% α particles 3% electrons and mesons 1% heavy ions	Protons Electrons α particles Heavy ions	99% protons <1% α particles <0.5% electrons
Energy	Hundreds of MeV ~ GeV	5 ~ 400 MeV	Hundreds of MeV
Characteristics	High energy, low flux; Flux related to solar activity	Low energy, high flux; Flux related to solar activity and magnetic field disturbances	Ultra-high flux; Flux and event occurrence time are random
Time/space distribution	Always present; Isotropic	Always present; Van Allen radiation belt; South Atlantic anomaly	Occasionally present; Anisotropic

outside the solar system. They may originate from supernova explosions and are accelerated by interstellar magnetic fields to reach the high-energy charged particle stream in Earth's space. The geomagnetic capture radiation belt refers to many high-energy particle streams captured by the Earth's magnetic field, including the inner radiation belt and the outer radiation belt. The inner radiation belt is located at a altitude of 10^5 ~ 10^7 m from the surface of the Earth, and the outer radiation belt is located at a distance from the surface of the Earth (1 ~ 6) × 10^7 m. Solar particle events refer to the large number of high-energy particle streams radiated by the sun during flare eruptions, with nearly 99% of its components being proton beams, hence also known as solar proton events.

It should be noted that although the components of the three radiation belts are mainly protons, α particles, and electrons, particle radiation will interact with spacecraft or the atmosphere to produce secondary radiation components, such asπ mesons, X-rays, γ-rays, etc. For the fiber lasers used inside the cabin, since most particle radiation can be shielded by the spacecraft cabin and laser housing, therefore γ-rays are the main factors affecting the radiation resistance characteristics of active fibers in space. The use of physical shielding methods will greatly increase the mass of the laser, severely restricting the application of lasers in space. Therefore, the fundamental way to improve the space radiation tolerance of active fiber lasers is to improve the radiation resistance characteristics of active fiber materials.

Table 9.2 summarizes the altitude, radiation environment, and uses of low earth orbit (LEO), middle earth orbit (MEO), and geosynchronous orbit (GSO) [7, 8]. Figure 9.3 illustrates the spatial distribution of LEO, MEO and GSO in space [7]. The total radiation dose in Table 9.2 refers to the space cabin skin screen.

Table 9.2 The altitude, radiation belt, and purpose corresponding to the three satellite orbits in space

Orbit Name	Altitude /km	Dose Rate/ (rad/min)	Total Dose /krad	Dominant Radiation Belt	Primary Purpose
LEO	<2000	<0.027	5 ~ 10	South Atlantic anomaly	Earth observation satellite
MEO	2000 ~ 36,000	<0.272	10 ~ 100	Van Allen Radiation Belt	Navigation system satellite
GSO	36,000	~0.135	~50	Galactic cosmic rays	Communication satellite

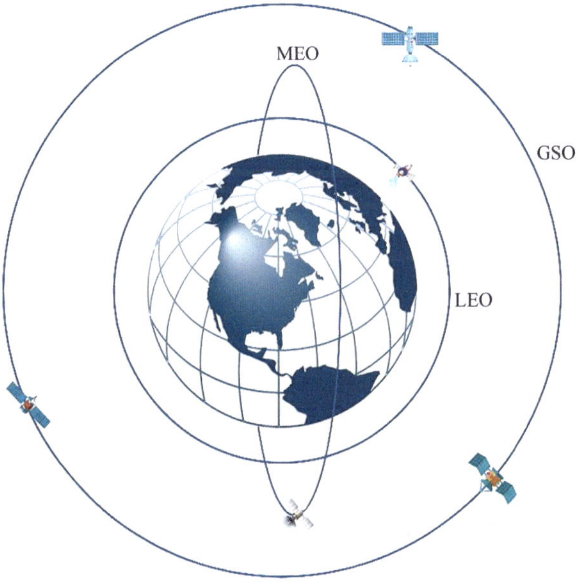

Fig. 9.3 Schematic diagram of Low Earth Orbit (LEO), Medium Earth Orbit (MEO), Geosynchronous Orbit (GSO)

The cumulative radiation dose under shielding conditions for 7 years of service, this data comes from the National Aeronautics and Space Administration (NASA) [7]. The dose rate is obtained by dividing the maximum value of the total radiation dose by time. The radiation dose refers to the average energy received by the unit mass of matter from radiation, and the unit is Gray (Gy) or rad, 1Gy = 1 J/kg = 100 rad. Because the magnetic field above the South Atlantic Anomaly is 30% ~ 50% weaker than other places, allowing more particle beams from outer space to penetrate, satellites operating in LEO will suffer more ionizing radiation when passing through this area.

9.1.2 *Applications and Challenges of Silica-Based Fiber in Space*

Silica fiber has the advantages of low loss, strong resistance to electromagnetic interference, wide transmission wavelength range (0.2 ~ 2.1 μm), easy coupling with laser diodes, etc., and has been widely used in information transmission, sensing, spectral analysis, and other fields. In addition, silica fiber also has high mechanical strength, good bending performance, excellent radiation resistance, high laser damage threshold, can be coated with hermetic high-temperature resistant (≥180 °C) coating (such as carbon coating, metal coating), etc., which is especially suitable for the complex environment of space with high vacuum, large temperature difference (±200 °C), and strong radiation. Table 9.3 summarizes the characteristic parameters of five different types of silica fibers and their applications and challenges in space.

The statistics show that from 1971 to 1986, there were 1589 accidents involving American satellites and their components, of which about 70% of the total number

Table 9.3 Characteristics, applications, and challenges in space of five different types of silica fibers

Fiber Type	Core/Cladding Size/μm	Core NA	Main Applications	Challenges	Refs.
Sensing/ communication single mode fiber	<10/125	<0.17	Temperature sensing Humidity sensing Pressure sensing Data transmission	Increase in loss Bragg wavelength drift Decrease in reflectivity	[12–15]
Sensing/ communication multimode fiber	50 ~ 62.5/125	0.18 ~ 0.23	Temperature sensing Humidity sensing Pressure sensing Data transmission	Increase in loss Bragg wavelength drift Decrease in reflectivity	[12–15]
Polarization maintaining fiber	<10/125	0.12 ~ 0.22	Fiber optic gyroscope	Increase in loss	[16]
Microstructure fiber	<20/125	<0.06	Large mode field/ single mode	Increase in loss	[17, 18]
Active fiber (Yb, Er, tm, etc.)	<10/125 <25/400	0.06 ~ 0.22	Fiber optic gyroscope Laser communication Laser radar Laser remote sensing Laser weapons Space debris treatment	Increase in loss Decrease in gain Decrease in efficiency Increase in noise	[19, 20]

of accidents were caused by space ionizing radiation [9]. Space ionizing radiation seriously restricts the normal operation of spacecraft and their components.

Figure 9.4 summarizes the effects of ionizing radiation on the performance of silica-based fibers, mainly including the following four aspects [10, 11]:

1. Radiation-induced attenuation (RIA). The formation of color centers can cause a sharp increase in the RIA of optical fibers, severely affecting the light-guiding and laser performance of the fibers.
2. Radiation-induced emission (RIE). RIE includes the emission of color centers and the Čerenkov emission. These spurious light signals cause the signal-to-noise ratio (SNR) of the light-guiding fiber or laser fiber to decrease.
3. Radiation-induced refractive index change (RIRIC). High-dose ionizing radiation induces an increase in the refractive index and density of silica glass, and the increase is related to the doping elements in the core and cladding. RIRIC causes changes in the numerical aperture (NA) of the optical fiber, thereby affecting the beam quality.
4. Radiation-induced coating degradation (RICD). High-dose ionizing radiation causes the polymer chains of the fiber coating to break (i.e., degrade). This is specifically manifested as the coating turning yellow and brittle, reducing mechanical strength of the fiber, and increasing interface loss between the coating and the outer cladding.

It should be noted that RIE, RIRIC, and RICD are usually only significantly manifested under high-dose radiation conditions (>1MGy). In low-dose radiation environments such as space (≤1 kGy), RIA is the main factor limiting the radiation resistance of optical fibers.

Figure 9.5 shows the RIA spectra of active fibers (Yb/Al/P/F co-doped silica fiber) and passive fibers (Ge single-doped silica fiber). The irradiation source is X-rays, the dose rate is 2.5 Gy/min, and the total dose is 500 Gy. Under the same irradiation conditions, the radiation-induced attenuation of ordinary active fibers (5228 dB/km@1200 nm) is three orders of magnitude larger than that of passive

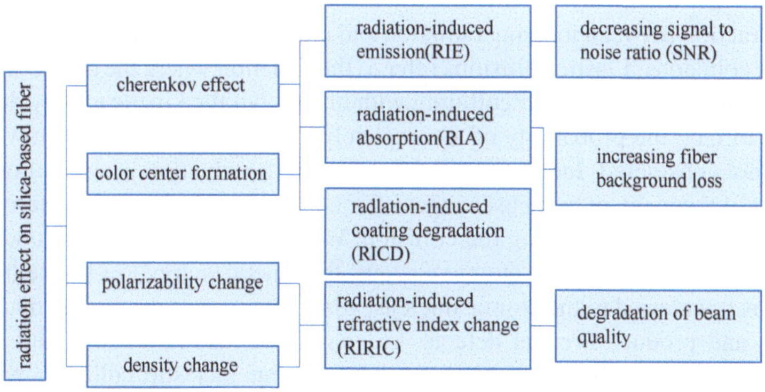

Fig. 9.4 The impact of ionizing radiation on the performance of silica-based optical fibers

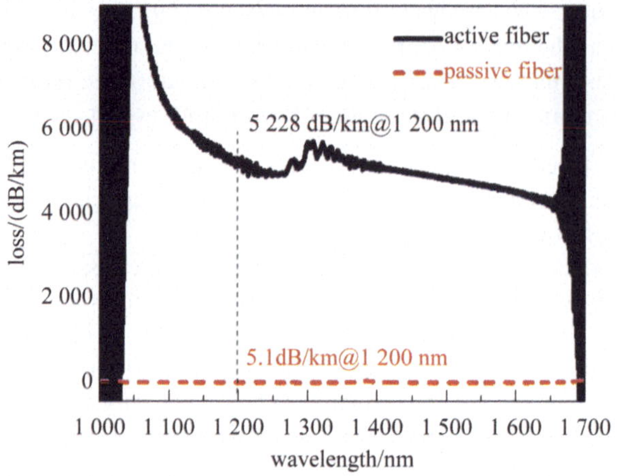

Fig. 9.5 RIA spectra of active and passive optical fibers

fibers (5.1 dB/km@1200 nm). The radiation sensitivity of active fibers is significantly higher than that of passive fibers.

The radiation resistance of passive fibers has basically met the current space application requirements [12]. However, how to effectively improve the radiation resistance of active fibers for space applications is a common problem faced by researchers at home and abroad. The following focuses on the RD mechanism, influencing factors, and methods to improve radiation resistance of active fibers.

9.2 Radiation-Induced Darkening Mechanism of Rare Earth-Doped Silica Fiber

9.2.1 Interaction of Ionizing Radiation with Silica Glass

The interaction between ionizing radiation and matter includes elastic collisions and inelastic collisions. Elastic collisions refer to the situation where the direction of the incident particle changes after colliding with matter, but the kinetic energy does not change. In fact, the probability of a completely elastic collision is very low and is usually not considered. Inelastic collisions refer to the situation where not only the direction of the incident particle changes after colliding with matter, but there is also a loss of kinetic energy during the collision. When the incident particle undergoes an inelastic collision with the atomic nucleus, the kinetic energy lost by the incident particle is transferred to the atomic nucleus, which may cause the atomic nucleus to displace and produce Frenkel defects, i.e., pairs of interstitial atoms and atomic vacancies. When the incident particle undergoes an inelastic collision with the extranuclear electron, the kinetic energy lost by the incident particle is transferred to the electron, causing the electron to be excited or ionized. If the energy gained by the electron only causes it to transition from a low energy level to a high energy level, this process is called excitation; if the energy gained by the electron is large enough to make it escape from the atomic nucleus and become a free electron, this process is called ionization.

Griscom and others from the U.S. Naval Research Laboratory [22–27] systematically studied the effects of different types of ionizing radiation on non-rare earth doped silica fibers (passive fibers) and explored their impact mechanisms at the atomic microscale.

Figure 9.6a is a model of the destruction of the Si-O-Si network in silica glass by ionizing radiation. From the microscopic atomic scale, the destruction of silica fiber by ionizing radiation mainly includes two aspects [10]: ① atomic displacement (knock on); ② ionization damage (radiolysis). Atomic displacement produces oxygen vacancies and interstitial oxygen atoms, while ionization damage produces electron-type color centers and hole-type color centers. Since the energy threshold required for ionization damage (≤8 eV) is much lower than the energy threshold required for atomic displacement (>10 eV), ionization damage is the main destruction mechanism of silica fiber by ionizing radiation.

Figure 9.6b is the RIA spectrum of aluminum single-doped silica glass. Studies have shown that the two absorption peaks in the ultraviolet band originate from the electron-type color centers (Si-E' and Al-E'), and the other two absorption peaks in the visible and near-infrared bands originate from the hole-type color centers (Al-OHC). Therefore, the electron-type color centers and hole-type color centers are the main reasons for the sharp increase in the RIA intensity of silica-based fibers.

Figure 9.7 shows a schematic diagram of the color center induced by different types of ionizing radiation in pure silica fiber [22, 23]. The generation of color centers mainly includes the following five stages:

1. Ionizing radiation stage.
2. Stage of atomic displacement and electron-hole pair generation. This process is usually transient (prompt occurrences), and whether it can occur mainly depends on the particle beam energy. For silica glass, the energy thresholds for oxygen atoms and silicon atoms to undergo atomic displacement are 10 eV and 18 eV, respectively [10]. Therefore, oxygen vacancy defects are usually much more than silicon vacancy defects. The bandgap width of high-purity SiO_2 glass is about 8 eV. When an electron transitions from the valence band to the conduction band, a hole and an electron are generated in the valence band and conduction band, respectively. Therefore, the energy threshold for pure silica glass to generate electron-hole pair color centers is about 8 eV. Since the energy of ultraviolet light is usually less than 8 eV, only photolytic defects can be generated.
3. Excited state relaxation and recombination stage. Atomic displacement leads to the generation of interstitial atoms and atomic vacancies, i.e., the generation of Frenkel defects. The recombination of interstitial atoms and atomic vacancies leads to the annihilation of Frenkel defects. Electron-hole pairs can recombine through radiative transition and non-radiative transition, the former produces a

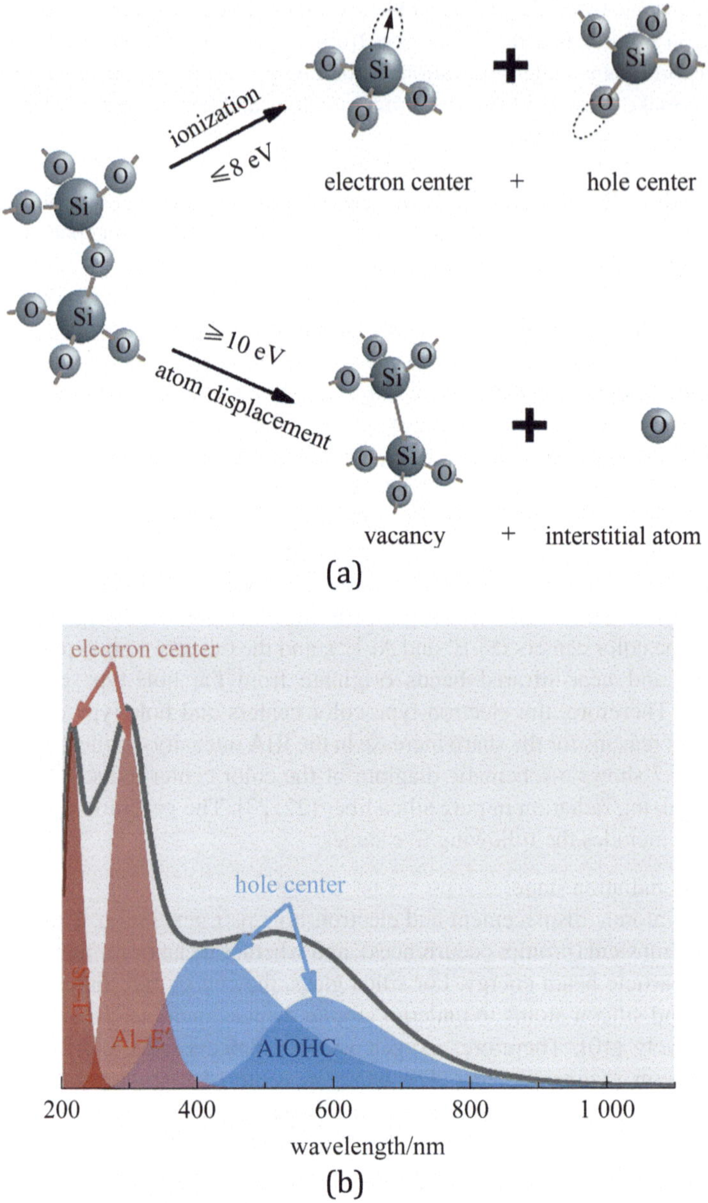

Fig. 9.6 (**a**) Model of ionizing radiation damage to the Si-O-Si network in silica glass; (**b**) RIA spectrum of aluminum single-doped silica glass

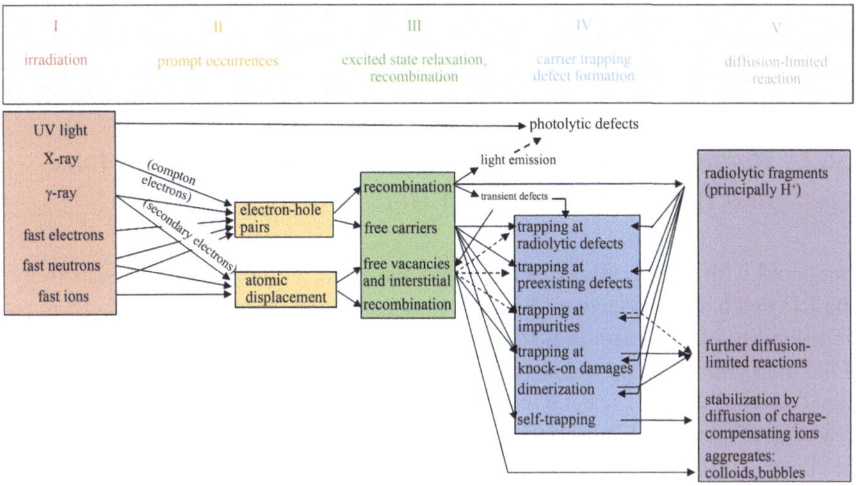

Fig. 9.7 Schematic of color center formation in pure silica fiber induced by ionizing radiation

luminescent signal, and the latter converts energy into heat transferred to the lattice.

4. Carrier capture and defect formation stage. Unrecombined holes and electrons can move freely during the ionizing radiation process, called carriers. Carriers may be captured by various lattice sites (such as radiation-induced defects, pre-existing defects, impurity ions, etc.) in SiO_2 glass, leading to the formation of electron-type and hole-type color centers. During the radiation process, the nucleus diffuses from one interstitial position to another, called interstitial diffusion; the nucleus diffuses from an equilibrium position to an atomic vacancy, leading to the generation of a new vacancy, called vacancy diffusion. The energy required for vacancy diffusion is much less than that required for interstitial diffusion. Therefore, vacancy diffusion is the main mechanism of nuclear diffusion. The mutual diffusion of the nucleus and vacancies often leads to the generation of new Frenkel defects.
5. Diffusion-limited reaction. Radiation-induced decomposition fragments (such as H^0 atoms) tend to undergo chemical reactions with various types of electron-type and hole-type color centers during the diffusion process, forming more stable defects (such as ≡Si-H, ≡Si-OH, etc.), or forming stable molecules (H_2, Cl_2, O_2 etc.), causing the carrier diffusion process to be pinned, i.e., diffusion is restricted. In addition, impurity-trapped defects, atomic displacement defects, and dimer reaction products (such as H_2) may further undergo chemical reactions during the radiation process, generating more stable defects, leading to further pinning of the diffusion process. Self-trapping hole (STH) defects are usually a metastable defect, which can rely on the charge compensation of free diffusion ions (such as H^+) to become stable defects. Interstitial atoms or vacancies tend to cluster together during high-dose irradiation, resulting in colloids or

bubbles (such as H_2, Cl_2, O_2 etc.), causing the diffusion process of vacancies and interstitial atoms to be pinned.

9.2.2 Radiation-Induced Valence Changes of Rare Earth Ions and Formation of Oxygen Hole Centers

Compared with passive fibers, active fibers have much higher radiation sensitivity. The RD mechanism of active fibers remains incompletely understood. To increase the solubility of rare earth ions in the silica matrix and control the refractive index of the fiber core, the core of active fibers usually needs to introduce a variety of co-doped ions, such as Al^{3+}, P^{5+}, Ge^{4+}, F^- etc. Research by Likhachev of the Russian Academy of Sciences [28] shows that the RIA of Er^{3+} doped silica fiber is related to the choice of co-doped ions and their co-doping ratio. Research by Lezius of Menlo Systems GmbH in Germany [29] shows that changes in Er^{3+} concentration do not significantly increase the RIA of fiber in the near-infrared band, and it is speculated that the main reason for the sharp increase in RIA of rare earth-doped silica fiber is the formation of color centers related to co-dopants (such as Ge, Al, P, etc.), but the nature and formation mechanism of the color centers have not been reported. Arai et al. of Fujikura Ltd. in Japan [30] used continuous wave electron paramagnetic resonance (CW-EPR) and optical absorption spectroscopy to show that the aluminum-related oxygen hole center (Al-OHC) is the main cause of radiation-induced darkening in Yb^{3+}/Al^{3+} co-doped silica fiber, but the formation mechanism of Al-OHC color center has not been reported. Deschamps et al. of École Polytechnique in France [31] used optical absorption, Raman, CW-EPR and other means to study that in $Yb^{3+}/Al^{3+}/P^{5+}$ In co-doped optical fiber preforms, the concentration of Al-OHC color centers decreases with the increase of P/Al co-doping ratio, and it is speculated that the $AlPO_4$ structure plays an important role in inhibiting the formation of Al-OHC color centers. Research by Fox et al. from the University of Arizona, USA [32] shows that Yb^{3+}/Er^{3+} co-doped optical fibers have better radiation resistance than Er^{3+} singly doped fibers. This is speculated to be related to the valence change ($Yb^{3+} \rightarrow Yb^{2+}$) of Yb^{3+} ions. Ollier et al. from the University of Paris-Saclay, France [33] used online fluorescence spectroscopy tests to show that during radiation, some Yb^{3+} ions in Yb^{3+}/Al^{3+} co-doped silica fibers were reduced to Yb^{2+} ions, and it is speculated that Yb^{2+} plays an important role in the radiation-induced darkening process.

Based on the above research, Shao et al. from the Shanghai Institute of Optics and Mechanics (SIOM), Chinese Academy of Sciences (CAS) [34, 36] systematically compared the radiation resistance behavior of Yb^{3+} singly doped, Yb^{3+}/Al^{3+} co-doped, Yb^{3+}/P^{5+} co-doped, and $Yb^{3+}/Al^{3+}/P^{5+}$ tri-doped silica glass and optical fibers with varying P/Al ratios. They systematically studied the RD mechanism of Yb^{3+} doped silica fibers from the atomic scale perspectives of Yb^{3+} local structure, glass network structure, and radiation-induced color centers. The following will illustrate with the example of Yb^{3+} singly doped silica glass.

Figure 9.8a shows the in-situ photoluminescence (PL) spectrum of Yb^{3+} singly doped silica glass, with a 193 nm ArF excimer laser as the excitation source. Figure 9.8b, c show the RIA spectra of pure silica glass (SiO_2) and Yb^{3+} single-doped silica glass (Yb:SiO_2), and their Gaussian peak fitting. For comparison with the RIA spectrum of SiO_2, the RIA spectrum of Yb:SiO_2 is added to Fig. 9.8b. Figure 9.8d is the CW-EPR spectra of SiO_2 and Yb:SiO_2 samples, and its theoretical simulation. The samples used in Figure 9.8b–d all refer to samples after 193 nm laser radiation for 100 minutes. From Fig. 9.8a, as the pump time increases, the luminescence intensity of Yb^{3+} (976 nm) gradually decreases, and the luminescence intensity of Yb^{2+} (530 nm) and Si-OHC (650 nm) color centers gradually increases. This indicates that some Yb^{3+} is reduced to Yb^{2+} during the radiation process. From Fig. 9.8b, the RIA intensity of Yb: SiO_2 is much higher than that of SiO_2 samples.

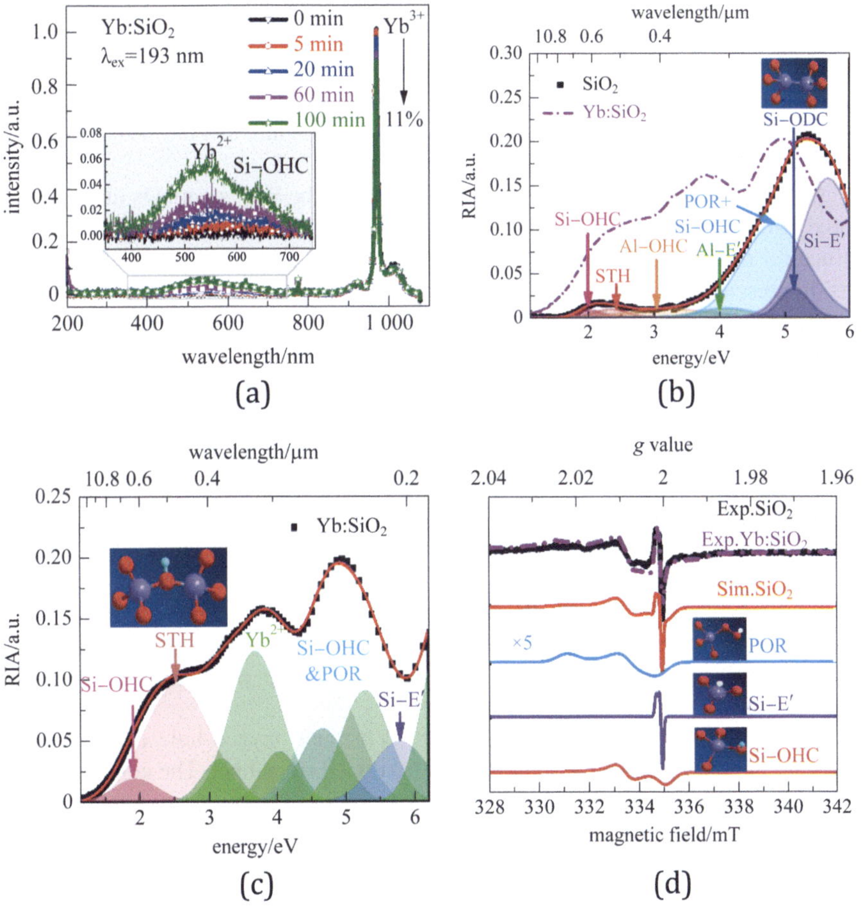

Fig. 9.8 (**a**) In-situ photoluminescence spectrum of Yb^{3+} singly doped silica glass (Yb: SiO_2), excited by a 193 nm laser; RIA spectra of SiO_2 glass (**b**) and Yb: SiO_2 glass (**c**); (**d**) CW-EPR spectra of SiO_2 and Yb:SiO_2 glasses of irradiation with a 193 nm laser after 100 min [35]

This indicates that co-doping with Yb^{3+} will reduce the radiation resistance of the glass, which is related to the formation of Yb^{2+} caused by irradiation, as shown in Fig. 9.8c. As shown in Fig. 9.8b, the RIA spectrum of SiO_2 samples is decomposed into seven Gaussian peaks, located at 2.0 eV, 2.4 eV, 3.2 eV, 4.1 eV, 4.8 eV, 5.1 eV, and 5.7 eV. The 3.2 eV and 4.1 eV absorption bands are attributed to Al-related defects. This is because a small amount of Al impurities is inevitably introduced from the corundum crucible wall during the high-temperature melting process of the glass. According to the research of Griscom et al. [24], the 2.0 eV, 2.4 eV, 5.1 eV, and 5.7 eV absorption bands are attributed to silicon oxygen hole centers (Si-OHC or NBOHC), self-trapped hole centers (STH), non-relaxed silicon oxygen vacancies (ODC(II)), and silicon dangling bond defects (Si-E'), respectively. The 4.8 eV absorption band is attributed to the absorption superposition of peroxide (POR) and Si-OHC. As shown in Fig. 9.8c, the RIA spectrum of Yb:SiO_2 samples is decomposed into ten Gaussian peaks, located at 1.96 eV, 2.5 eV, 3.1 eV, 3.7 eV, 4.0 eV, 4.7 eV, 4.8 eV, 5.3 eV, 5.8 eV and 6.5 eV. Among them, the absorption bands of 1.96 eV, 2.5 eV, and 5.8 eV are attributed to Si-OHC, STH, and Si-E' defects, respectively. The absorption bands of 3.1 eV, 3.7 eV, 4.0 eV, 4.7 eV, 5.3 eV, and 6.5 eV (corresponding to 400 nm, 335 nm, 310 nm, 265 nm, 234 nm, 190 nm) originates from Yb^{2+} ions.

As shown in Fig. 9.8d, the CW-EPR spectrum of the SiO_2 sample is divided into three parts, corresponding to the POR, Si-E', and Si-OHC defect centers. Their structural models can be represented as ≡Si—O—O·, ≡Si·, and ≡Si—O°, where "≡" represents three bridging oxygens, "·" represents a single electron, and "°" represents a hole. Due to the relatively low natural abundance of ^{29}Si (nuclear spin $I = 1/2$, natural abundance NA~4.7%), its hyperfine structure was not observed.

From Fig. 9.8b, c, it can be seen that compared to the SiO_2 sample, the RIA intensity of Yb^{2+}, Si-OHC, STH in the Yb:SiO_2 sample increases, while the RIA intensity of Si-E' decreases, and the oxygen vacancy center (ODC(II)) is completely suppressed. From Fig. 9.8d, it can also be seen that compared to the SiO_2 sample, the CW-EPR intensity of Si-OHC in the Yb: SiO_2 sample increases, while the CW-EPR intensity of Si-E' decreases. This indicates that the valence state change of $Yb^{3+/2+}$ plays an important role in the RD process of the Yb: SiO_2 sample. Compared to the SiO_2 sample, the Yb^{3+} ions in the Yb: SiO_2 sample captures many electrons to become Yb^{2+} ions under irradiation, suppressing the generation of electron-type color centers (Si-E'), while promoting the generation of hole-type color centers (Si-OHC, STH).

Further research shows [34, 35, 37, 38], Yb^{2+} and oxygen hole center (OHC) pairs are the fundamental cause of RD in Yb^{3+} doped silica fiber. The corresponding chemical reaction can be represented as:

$$Yb^{3+} - O - R \equiv \overset{h\nu}{\leftrightarrow} Yb^{2+} + \underset{(ROHC)}{{}^{\circ}OR \equiv} (R = Al, Si, P)$$

In the formula, "°" and "≡" represent a hole and three bridging oxygens connected to R, respectively.

Pulse EPR spectrum (HYSCORE projection) tests show [39]: In Yb^{3+} singly doped (SY) silica glass, there are many Yb-O-Si linkages. In Yb^{3+}/Al^{3+} doubly doped (SYA) silica glass, Yb^{3+} is in a mixed coordination environment of Yb-O-Al and Yb-O-Si. In Yb^{3+}/P^{5+} doubly doped (SYP) silica glass, it is mainly Yb-O-P linkages. In $Yb^{3+}/Al^{3+}/P^{5+}$ In tri-doped (SYAP) silica glass, as the P/Al co-doping ratio ($0.25 \rightarrow 2$) increases, Yb^{3+} ions gradually transition from a silicon-rich or aluminum-rich environment to a phosphorus-rich environment. It is worth noting that when P/Al ≈ 1 (SYAP1), Al^{3+} and P^{5+} preferentially form $[AlPO_4]^0$ units around Yb^{3+}; when P/Al < 1, Yb^{3+} is mainly in a silicon-rich or aluminum-rich environment; when P/Al > 1, Yb^{3+} is mainly in a phosphorus-rich environment.

In-situ ultraviolet excitation fluorescence spectra and CW-EPR tests show [34, 35], radiation-induced Yb^{2+} and OHC pairs are most abundant in Yb^{3+}/Al^{3+} co-doped silica glass, and least in Yb^{3+}/P^{5+} co-doped silica glass. This is related to the charge of coordination group of Yb^{3+} ion. The formation model of Yb^{2+} and OHC pairs is shown in Fig. 9.9 [34].

$$\left[AlO_{4/2}\right]^{-} + Yb^{3+} \overset{h\upsilon}{\leftrightarrow} \left[AlO_{4/2} + h^{+}\right]^{0} + Yb^{2+}$$

In Yb^{3+}/Al^{3+} co-doped silica glass, the surroundings of Yb^{3+} are mainly $[SiO_{4/2}]^0$ and $[AlO_{4/2}]^-$ groups. The electronegative $[AlO_{4/2}]^-$ easily captures a hole (i.e., releases an electron) during radiation to become an aluminum-oxygen hole center (Al-OHC), and the released electron is captured by Yb^{3+} ion, leading to a large amount of Yb^{3+} ions being reduced to Yb^{2+} ions. Therefore, the radiation resistance of Yb^{3+}/Al^{3+} co-doped silica glass is the worst.

In Yb^{3+} singly doped silica glass, Yb^{3+} is mainly coordinated to the $[SiO_{4/2}]^0$ group. The electrically neutral $[SiO_{4/2}]^0$ does not easily gain or lose electrons during radiation. Only a small amount of Yb^{3+} is reduced to Yb^{2+} during radiation. Therefore, compared with Yb^{3+}/Al^{3+} co-doped silica glass, the radiation resistance of Yb^{3+} singly doped silica glass is significantly improved.

In Yb^{3+}/P^{5+} co-doped silica glass, Yb^{3+} is mainly surrounded by the $[O{=}PO_{3/2}]^0$ group (i.e., $P^{(3)}$ group). The P=O double bond in the $P^{(3)}$ group is very easy to break

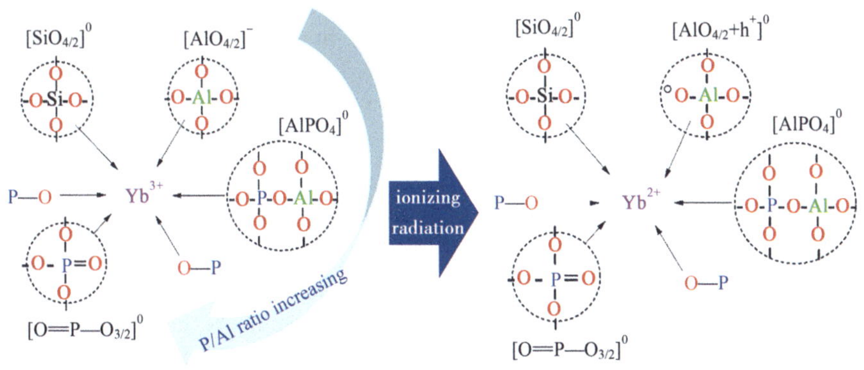

Fig. 9.9 Formation model of color centers caused by irradiation in Yb^{3+} doped silica glass/fiber

during radiation, which absorbs radiation energy. In addition, the P=O double bond itself plays the role of both electron capture center and hole capture center during radiation. The radiation-induced Yb^{2+} And the steady-state POHC color center is significantly reduced. Therefore, compared with Yb^{3+} singly doped or Yb^{3+}/Al^{3+} doubly doped silica glass, the radiation resistance of Yb^{3+}/P^{5+} doubly doped silica glass is the best. It is worth noting that the absorption peak of the P_1 color center generated by the ionization of the P=O double bond is located at 1.6 μ m, which severely deteriorates the luminescence properties of Er^{3+} ions.

In $Yb^{3+}/Al^{3+}/P^{5+}$ triply doped silica glass, as the P/Al co-doping ratio (0.25 → 2) increases, the radiation-induced Yb^{2+} and OHC color centers gradually decrease, and the radiation resistance of the glass gradually improves.

When P/Al < 1, aluminum and phosphorus preferentially form electrically neutral $[AlPO_4]^0$ units surrounding Yb^{3+}, and the excess aluminum mainly surrounds Yb^{3+} in the form of negatively charged $[AlO_{4/2}]^-$ units. Yb^{3+} and $[AlO_{4/2}]^-$ capture electrons and holes respectively during radiation, leading to the formation of a large amount of Yb^{2+} and Al-OHC.

When P/Al = 1, the surroundings of Yb^{3+} are mainly electrically neutral $[AlPO_4]^0$ groups. $[AlPO_4]^0$ is similar to the $[SiO_{4/2}]^0$ structure and is not easy to gain or lose electrons during radiation. Therefore, doping with equimolar amounts of aluminum and phosphorus can improve the radiation resistance of Yb^{3+} doped silica glass to a certain extent.

When P/Al > 1, most of the aluminum and phosphorus preferentially form electrically neutral $[AlPO_4]^0$ structural units, located far from Yb^{3+}. And the excess phosphorus mainly surrounds Yb^{3+} in the form of electrically neutral $[O{=}P\text{-}O_{3/2}]^0$ (i.e., $P^{(3)}$) structures. In this case, the local structure of Yb^{3+} is similar to that in Yb^{3+}/P^{5+} co-doped silica glass. Therefore, the radiation-induced Yb^{2+} and P-OHC color centers are significantly reduced relative to the P/Al ≤ 1 samples. The radiation resistance of the glass is significantly improved.

Under irradiation, the positively charged rare earth (RE) ions usually capture electrons, causing themselves to be reduced, and the negatively charged ligand groups usually lose electrons, that is, they capture holes to form oxygen hole centers (OHCs). At present, the valence change of RE ions (RE^{3+} → RE^{2+}) and the formation of OHCs are generally considered to be the fundamental reasons for the radiation darkening of active fibers [35, 38, 40, 42]. In addition to $Yb^{2+/3+}$, ionizing radiation induces $Ce^{4+/3+}$, $Sm^{3+/2+}$, $Eu^{2+/3+}$, $Er^{2+/3+}$, $Tm^{2+/3+}$ and other rare earth ions have been publicly reported to undergo valence changes [41, 43–47].

9.2.3 *Common Point Defects in Doped Silica Glass*

In addition to RE^{2+} and OHC color centers, ionizing radiation can also induce some color centers related to dopants (Al, P, Ge, etc.) in active fibers. It is worth noting that although fluorine is a common dopant in active silica optical fibers, there are currently no reports of fluorine-related color centers in silica glass. The characterization methods of radiation-induced defects mainly include absorption spectra, fluorescence spectra, thermoluminescence spectra, CW-EPR spectra, etc. Table 9.4

Table 9.4 Structural models, spectral characteristics, and CW-EPR parameters of point defects in pure and doped silica glasses

Defect	Structural Model	Spectral Parameters			CW-EPR Parameters		References
		Absorption Peak (eV)	FWHM of absorption peak (eV)	Emission Peak (eV)	(g_1, g_2, g_3) [a] (Dimensionles)	(A_1, A_2, A_3) [b] (G)	
ODC (I)	≡Si-Si≡	7.6	0.5	2.7 / 4.4	None	None	[48–50]
Si-E´	≡Si•	5.8	0.8	None	(2.0018, 2.0006, 2.0003)	Not observed	[23, 24, 50]
ODC (II)	≡Si..Si≡	4.95 ~ 5.05	0.3	2.7 /4.4	None	None	[48–50]
NBOHC	≡Si-O°	4.8/2.0	1.05/0.18	1.85 ~ 1.95	(1.9999, 2.0095, 2.078)	Not observed	[23, 24, 50]
POR	≡Si-O-O•	4.8/1.97	0.8/0.175	None	(2.0018, 2.0078, 2.067)	Not observed	[23, 24, 50]
POL	≡Si-O-O-Si≡	3.8		None	None	None	[23, 50]
Cl_2	Cl-Cl	3.8	0.7	None	None	None	[50]
STH	≡Si-O°-Si≡	2.6/2.16	1.5/1.2	None	(2.0054, 2.0078, 2.0125)	Not observed	[23–25, 50]
O_2	O=O	1.62 / 0.97	0.012/0.011	0.97	None	None	[50]
Al-ODC	≡Al..Si≡	4.96	0.47	2.6/3.4	None	None	[34]
Al-E´	≡Al•	4.1	1.02	None	2.0023	50	[34, 51, 52]
Al-OHC	≡Al-O°	3.2/2.3	1.0/0.9	None	(2.0402, 2.017, 2.0039)	(4.7, 10.3, 12.7)	[34, 52, 53]
P_4	-P•-	4.8	0.41	None	(2.0014, 1.9989, 1.9989)	300	[27]
P-ODC	≡P..Si≡	4.8/6.4	0.7/0.6	3.0	None	None	[35]
P_2	=P•=	4.5	1.27	None	(2.002, 1.999, 1.999)	800–1600	[27]
l-POHC	≡P-O°	3.1	0.73	None	(2.0039, 2.0027, 2.0026)	(50, 41, 48)	[27]

(continued)

Table 9.4 (continued)

Defect	Structural Model	Spectral Parameters			CW-EPR Parameters		References
		Absorption Peak (eV)	FWHM of absorption peak (eV)	Emission Peak (eV)	(g_1, g_2, g_3) [a] (Dimensionles)	(A_1, A_2, A_3) [b] (G)	
r-POHC	$=P\text{-}O_2^{\Upsilon}$	2.2, 2.5	0.35, 0.63	None	(2.0179, 2.0097, 2.0075)	(54, 52, 48)	[27]
P_1	≡P•	0.79	0.29	None	(2.002, 1.999, 1.999)	910	[27]
GLPC	=Ge:	5.15	0.42	4.3	None	None	[54, 55]
$Ge_{(1)}$	=Ge•=	4.4	1.2	None	(2.0006, 2.0000, 1.9930)	Not observed	[55, 56]
Ge-E´	≡Ge•	6.2	0.7	None	(2.0012, 1.9951, 1.9941)	Not observed	[54, 56]
$Ge_{(2)}$	≡Ge•-Ge≡	5.8	0.7	None	(2.0010, 1.9989, 1.9867)	Not observed	[54, 56]
Ge-OHC	≡Ge-O°	2.1	1	1.85	Not reported	Not reported	[55]
Ge-STH	Not clear	0.54	0.5	None	Not reported	Not reported	[55]

[a] *g* value refers to the Landé g-factor, which is related to the orbitals occupied by unpaired electrons and the local structural symmetry

[b] *A* value refers to the hyperfine coupling constant, which is related to the abundance of magnetic nuclei around unpaired electrons and the nuclear magnetic moment

summarizes the structural models, absorption, and emission spectra, and CW-EPR characteristic values of common point defects in pure silica glass, aluminum single-doped, phosphorus single-doped, and germanium single-doped silica glasses. As can be seen from Table 9.4, defects with luminescent signals (PL) usually do not have paramagnetic signals (CW-EPR), and vice versa.

Based on the data in Table 9.4, Shao [21] simulated the absorption spectra and CW-EPR spectra of common point defects in silica glass using MATLAB®. Among them, Figs. 9.10a, b are the absorption spectra and CW-EPR spectra of common point defects in pure silica glass, respectively. Figure 9.11a, b are the absorption spectra and CW-EPR spectra of common point defects in aluminum-doped silica glass, respectively. Figure 9.12a, b are the absorption spectra and CW-EPR spectra of common point defects in phosphorus-doped silica glass, respectively. Figure 9.13a, b are the absorption spectra and CW-EPR spectra of common point defects in germanium-doped silica glass, respectively.

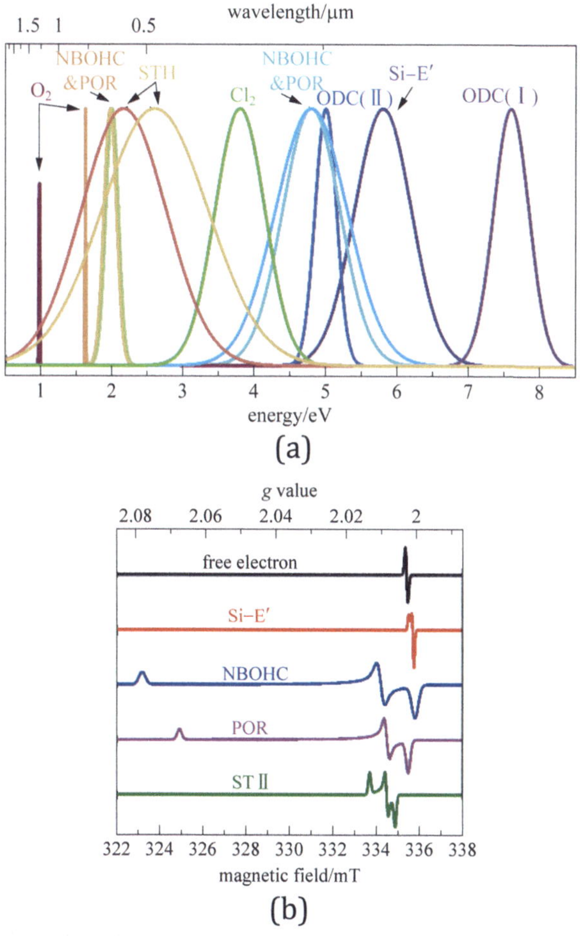

Fig. 9.10 Absorption (**a**) and CW-EPR spectra (**b**) of common point defects in pure silica glass

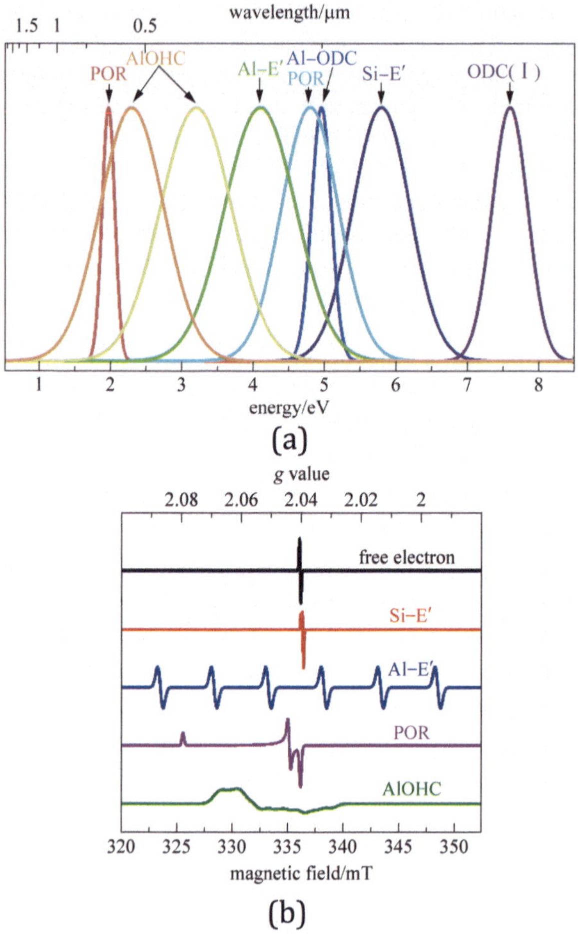

Fig. 9.11 Absorption (**a**) and CW-EPR spectra (**b**) of common point defects in aluminum-doped silica glass

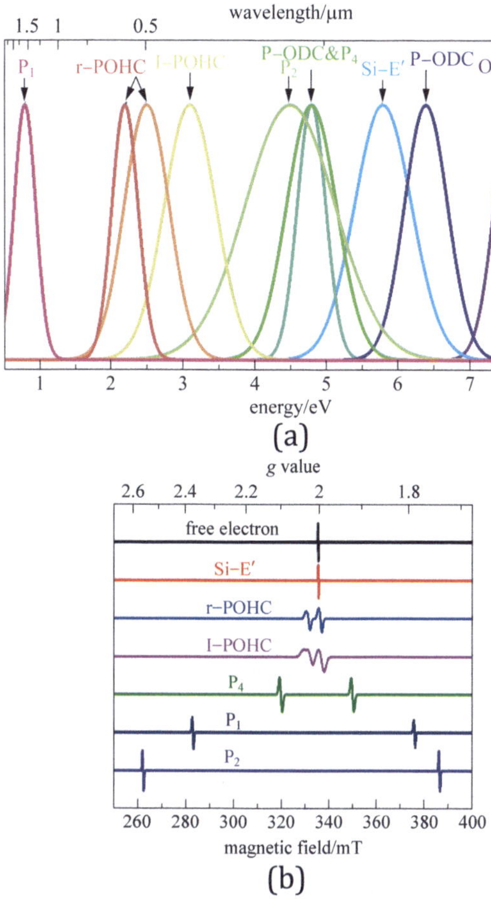

Fig. 9.12 Absorption (**a**) and CW-EPR spectra (**b**) of common point defects in phosphorus-doped silica glass

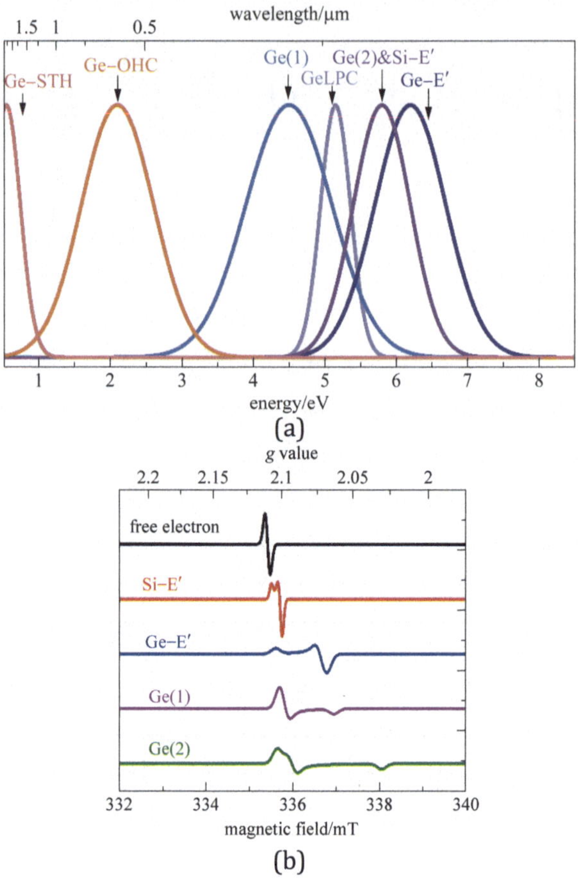

Fig. 9.13 Absorption (**a**) and CW-EPR spectra (**b**) of common point defects in germanium-doped silica glass

9.3 Factors Affecting the Radiation Resistance of Rare Earth-Doped Silica Fiber

Figure 9.14 summarizes the three main factors affecting the radiation resistance of active fibers: fiber parameters, irradiation environment parameters, and laser construction parameters [8]. Among them, fiber parameters are the main factors affecting the radiation resistance of active fibers, which include the composition of the fiber core, cladding and coating, the geometric parameters of the fiber, and the fiber preparation process; irradiation environment parameters include the type of radiation particles, total dose and dose rate, and environmental temperature; laser structure parameters include fiber pump structure, pump wavelength and power, fiber length, and fiber usage history. The following introduces these three major influencing factors respectively.

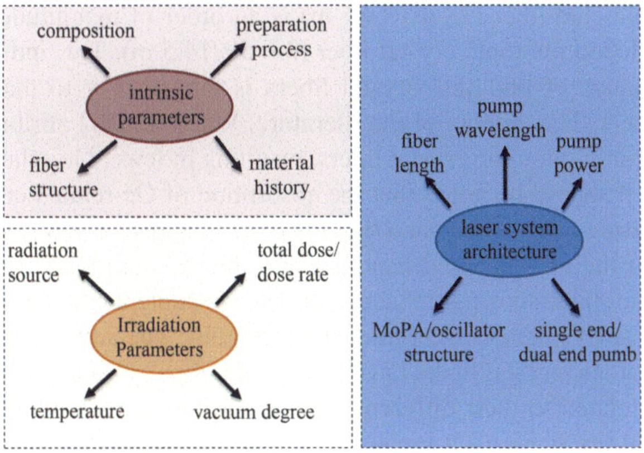

Fig. 9.14 The three main factors affecting the radiation resistance of active fibers

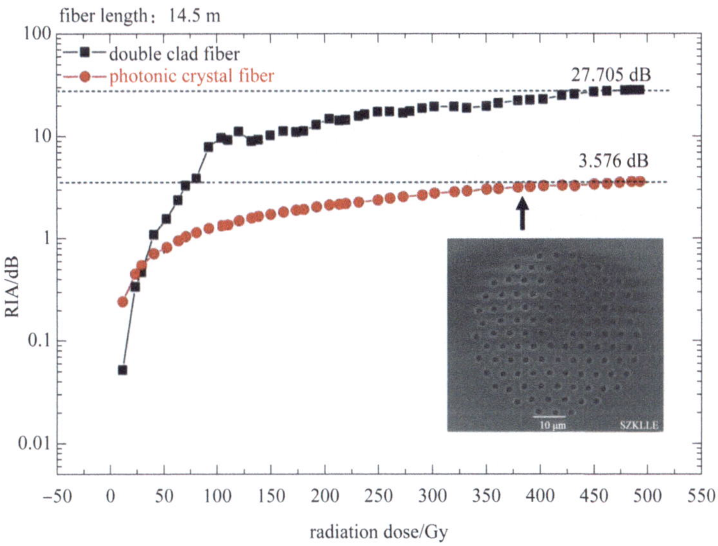

Fig. 9.15 The change in RIA intensity of Er^{3+} doped double-clad fiber and Er^{3+} doped photonic crystal fiber at wavelength of 1550 nm with radiation dose

9.3.1 *Fiber Parameters*

Studies have shown that the structure of the fiber has a great impact on the radiation resistance of the fiber. Figure 9.15 shows the change in RIA intensity at a wavelength of 1550 nm with radiation dose for Er-doped double-clad fiber and Er-doped photonic crystal fiber [57]. The illustration in Fig. 9.15 is a cross-section of a photonic crystal fiber. As the radiation dose increases, the RIA intensity of both fibers tends to saturate. When the radiation dose reaches 494Gy, the RIA intensity of the

Er-doped double-clad fiber (28 dB/14.5 m) is an order of magnitude greater than that of the Er-doped photonic crystal fiber (3.6 dB/14.5 m). This indicates that the radiation resistance of photonic crystal fibers is far superior to that of ordinary double-clad fibers. The authors of the literature, Wu et al. [57] attribute this to the lower Ge content in photonic crystal fibers, resulting in fewer Ge-related color centers. However, it should be noted that the absorption of Ge-related color centers is mainly in the ultraviolet-visible band (see Figs. 9.13), and the Ge-STH and Ge-OHC color centers in the near-infrared band are unstable at room temperature. In addition, early studies have shown that increasing the Ge content can improve the radiation resistance of Er fibers to a certain extent [58, 59]. The author believes that the reason why photonic crystal fibers have better radiation resistance than double-clad fibers may be related to their different fiber structures. The cladding thickness of photonic crystal fibers is much greater than that of double-clad fibers, which can shield the core from external radiation to a certain extent. On the other hand, the outer cladding of double-clad silica fibers usually uses low-refractive-index polymer materials, and its outer surface is coated with a layer of high-refractive-index polymer material as a protective layer. The outer cladding of photonic crystal fibers is usually pure silica glass, and the outer surface is no longer coated with a polymer material protective layer. Polymer materials are prone to degradation and aging during irradiation, which leads to a sharp increase in interface loss between the inner and outer cladding and between the outer cladding and the protective layer. Girard et al. [18, 60] have shown that the radiation resistance of pure silica core microstructure fibers (such as photonic crystal fibers and hollow fibers) is superior to that of conventional silica core double-clad pure fibers. This result rules out the difference in core compositions and confirms once again that the fiber structure does indeed significantly affect the radiation resistance of the fiber.

However, it must be pointed out that there are relatively few reports on the impact of fiber structure on the radiation resistance of active fibers. Whether the radiation resistance of microstructure active fibers is superior to that of double-clad structure active fibers is still a subject for further research.

The composition of the fiber core and cladding, as well as the coating material, can affect the radiation resistance of the optical fiber, with the core composition having the greatest impact. To regulate the refractive index of the core or increase the solubility of rare earth (RE) ions in silica glass, the SiO_2 core glass usually needs to be co-doped with Al, P, Ge, and F elements. These co-doped elements can form corresponding color centers during radiation, leading to a sharp increase in the loss of silica fiber. The Si, Al, P, Ge related color centers formed by radiation-induced doped silica glass are shown in Figs. 9.10, 9.11, 9.12 and 9.13. RE-related color centers are mainly related to the valence change of RE ions. There are few reports on F-related color centers.

Figure 9.16 shows the effect of core doping elements (Ge, P, Al) on the RIA of silica fiber at 1550 nm [55]. Under the same irradiation conditions, the radiation resistance of Al single-doped silica fiber is the worst, and that of pure silica fiber is the best. The radiation resistance of Ge single-doped silica fiber and P single-doped silica fiber is between the two.

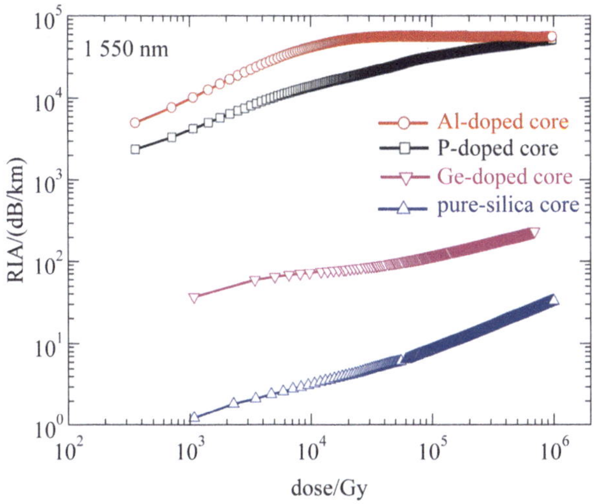

Fig. 9.16 The effect of core doping elements (Al, P, Ge) on the RIA of silica fiber at 1550 nm

The composition of the fiber cladding and the coating material can also affect the radiation resistance of the optical fiber to a certain extent. Studies have shown that using fluorine-doped silica glass as the outer cladding can improve the radiation resistance of silica fiber. Compared with pure silica cladding fiber, when the fiber cladding material contains elements such as P, Ge, the radiation tolerance of the fiber will drastically degrades [10]. Polyimide coating has better radiation resistance than acrylate coating [61]. In addition, the acrylate coating will release H^+ or H_2 during radiation and diffuse to the fiber cladding and core. H^+ or H_2 reacts with SiO_2 to form OH, causing an increase in fiber loss at 1.39 μm [62].

The preparation process of the optical fiber is also an important factor affecting the radiation resistance of the optical fiber. Hanafusa and others from the NTT Electrical Communication Laboratory in Japan first reported in 1986 the influence of drawing process parameters on the radiation resistance of optical fibers [63]. Their research shows that the radiation-induced loss of optical fiber decreases with the increase of drawing speed and the decrease of drawing temperature. Girard and others from the University of Saint-Etienne in France found that reducing the deposition temperature of the preform rod and the drawing tension can to some extent reduce the transient induced loss of germanosilicate fiber at 1310 nm and 1550 nm induced by pulsed X-rays [64]. The reasons may be multifaceted, for example: preform rods may produce peroxide (such as POL) or oxygen deficiency (such as ODC) defects due to different oxygen partial pressures during the preparation process. Optical fibers may also have some dangling bond defects (such as Si-E ‘) due to the influence of drawing tension during the drawing process. These pre-existing defects can affect the radiation resistance characteristics of the drawn optical fibers. In addition, the preform preparation and drawing process are different, the fictive temperature (T_f) of the drawn fiber may be different. Wang et al. from Toyota

Technological Institute in Japan [65] showed that the concentration of radiation-induced self-trapped hole centres (STHs) in pure silica glass is proportional to the T_f of glass. In addition, many studies have shown that the lower the T_f of silica glass, the better its radiation resistance [66, 67].

9.3.2 Radiation Environment Parameters

The radiation environment parameters include the type of radiation particles (protons, electrons, X-rays, γ-rays, etc.), radiation mode (pulse or continuous radiation), total dose and dose rate, and environmental temperature.

As shown in Fig. 9.17a, the effects of 50 MeV and 105 MeV protons and γ-rays on the RIA spectrum of Er^{3+}/Al^{3+} co-doped silica fiber are basically consistent. This indicates that the types and quantities of color centers induced by protons and γ-rays, two different types of radiation, are basically the same [68].

As shown in Fig. 9.17b, the effects of steady-state γ-rays and transient X-rays on the RIA of Ge-doped and N-doped core fibers are basically the same [69]. However, the effects of steady-state and transient radiation on the RIA of pure silica fiber are very different, the main reason is that the STH color center is unstable at room temperature, while NBOHC can stably exist at room temperature. Therefore, the STH color center contributes the most to the radiation-induced attenuation spectrum in transient radiation, while the NBOHC color center contributes the most in steady-state radiation [70].

As shown in Fig. 9.18a, when the total dose is kept constant, the RIA of fluorine-doped silica fibers at 1310 nm increases with the increase of dose rate [71].

As shown in Fig. 9.18b, when the radiation dose rate (5Gy/s) and total dose (1 kGy) are kept consistent, the RIA of phosphorus-doped silica fibers decreases with the rise of environmental temperature [72].

9.3.3 Laser Construction Parameters

Laser construction parameters include fiber pump structure, fiber length, pump light wavelength and power, fiber usage history, etc. Research by Ladaci and others from the University of Lyon, France [73] shows that optimizing the parameters of Er^{3+} doped fiber amplifiers (such as fiber length, pumping method, etc.) using particle swarm algorithm can further improve their radiation resistance.

Figure 9.19 shows the effects of 980 nm pump light power on the RIA spectra of Er^{3+}/Al^{3+} co-doped fibers [74]. The radiation source is γ-rays, and the total radiation dose is 330krad. As the pump light power increases, the RIA intensity gradually decreases. Under the action of pump light, the recombination of electrons and holes leads to the annihilation of color centers, known as color center bleaching. Usually, the shorter the wavelength and the higher the power of the pump light, the higher its

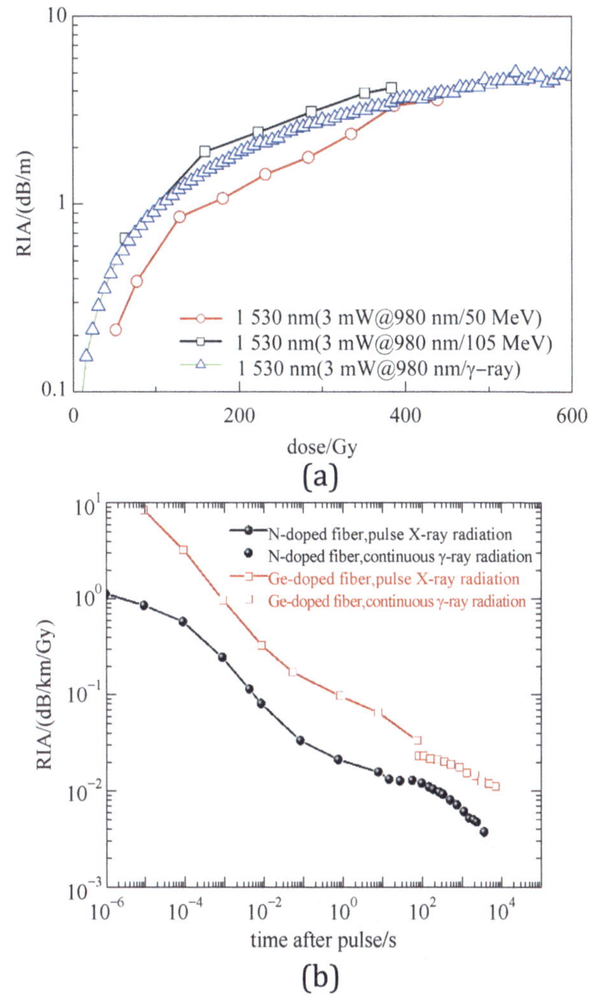

Fig. 9.17 (**a**) The effects of gamma and proton radiation on the RIA of Er^{3+}/Al^{3+} co-doped fiber [68]; (**b**) The impacts of steady-state γ-rays and transient X-rays on the RIA of passive optical fibers [69]

bleaching efficiency for radiation-induced color centers. Research by Zotov and others from the Russian Academy of Sciences [75] shows that 980 nm laser has a higher bleaching efficiency for Er^{3+} doped fiber after radiation darkening than 1480 nm laser, and it is believed that using 980 nm laser to pump Er^{3+} doped fiber can to some extent inhibit its radiation darkening effect.

In addition, the irradiation history and heat treatment history of the optical fiber will also affect its radiation resistance. As early as 1995, Griscom and others from the U.S. Naval Research Laboratory [76, 77] found that pre-irradiation treatment of pure silica core optical fibers with ultra-high dose rate (>1MGy) of γ-rays can

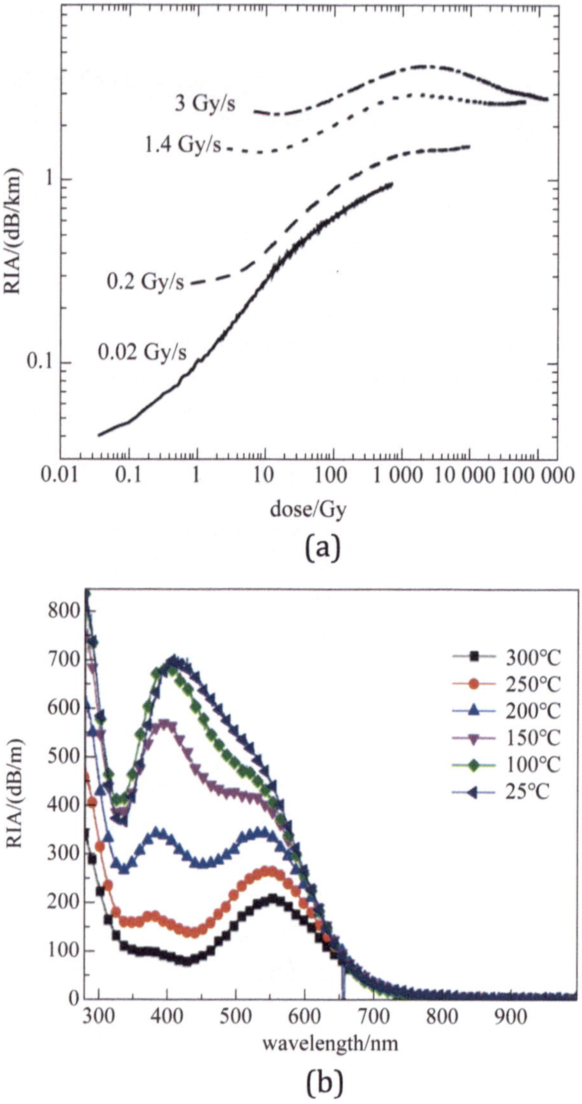

Fig. 9.18 (**a**) The effect of dose rate on the RIA of fluorine-doped silica fibers (Fujikura RRSMFB); (**b**) The effect of temperature on the RIA of phosphorus-doped silica fibers

improve the radiation resistance of the optical fibers, but the related mechanism has not been reported. Yeniay and others from the American Photon-X LLC company [78] have shown that the method of γ-ray pre-irradiation combined with thermal annealing (165 °C) can improve the radiation resistance of Yb^{3+}/Er^{3+} doped optical fibers.

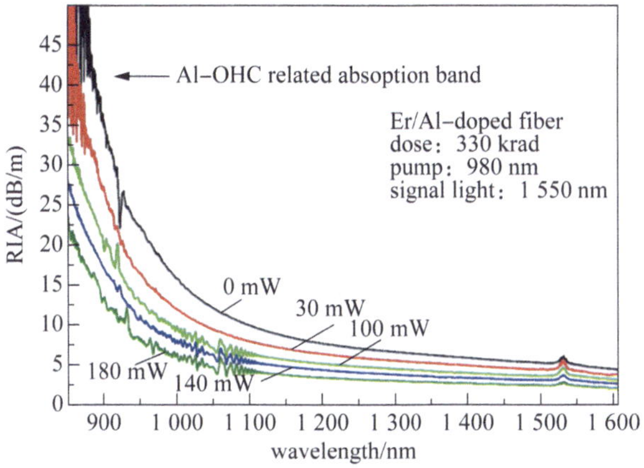

Fig. 9.19 The effect of pump power on the RIA of Er^{3+}/Al^{3+} co-doped fiber

9.4 Methods to Improve the Radiation Resistance of Rare Earth-Doped Silica Fiber

As shown in Fig. 9.20, the methods to improve the radiation resistance of optical fibers mainly include composition optimization, pre-treatment, post-treatment, and system optimization. The following will introduce them separately.

9.4.1 Compositions Optimization

Composition optimization mainly refers to the optimization of the core material compositions. Al, P, Ge and other co-doped elements are mainly used to regulate the refractive index of the core and increase the solubility of rare earth ions in silica glass. However, these co-doped elements will cause the radiation resistance of the optical fiber to drop sharply. Co-doping F can reduce the refractive index of the core and inhibit color center generation, but high doping F has difficulties in the process. Therefore, the optimization of the core compositions mainly starts from the following three aspects:

1. Reduce radiation-sensitive elements. The University of Montpellier II in France and Draka Comteq BV company have jointly developed a new type of rare earth ion doping technology—nanoparticle doping [79]. Compared with the traditional MCVD solution soaking method, this technology can reduce the doping concentration of Al or P and other co-dopants as much as possible under the premise that Er^{3+} does not undergo concentration quenching, and even Al or P and other radiation-sensitive elements can be not doped. Therefore, the optical fibers prepared by this technology often have better radiation resistance.

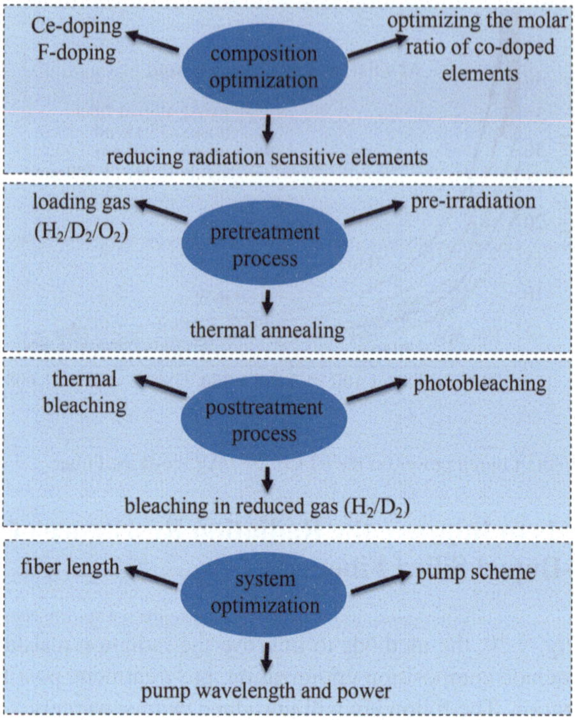

Fig. 9.20 Methods to improve the radiation resistance of active fibers

However, relying solely on nanoparticle doping technology cannot effectively inhibit the clustering of rare earth ions. To avoid the concentration quenching effect, this method requires that the co-doping concentration of rare earth ions be less than 1000 ppm.

2. Optimize the ratio of co-doped elements. As shown in Fig. 9.21, Likhachev and others from the Russian Academy of Sciences [28] systematically studied the effects of single and co doping of Er, Al, P, $AlPO_4$, Ge, etc. in different forms on the RIA of silica fibers. Among them, $AlPO_4$ refers to the structural unit formed by equimolar doping of Al and P. The research results show that the radiation resistance of Er/Al double-doped fiber is the worst, and the radiation resistance of Er-$AlPO_4$-Ge co-doped fiber is the best. The research by Jiao and others from the SIOM, CAS [80] shows that the radiation resistance of Er/Al/Ge co-doped fiber is far superior to that of Er/Al co-doped fiber, and the larger the Ge/Al co-doping ratio, the better the radiation resistance of Er/Al/Ge co-doped fiber. Shao et al., [34] shows that the radiation resistance of Yb/Al/P co-doped silica glass is far superior to that of Yb/Al co-doped silica glass, and when the P/Al ratio is slightly greater than 1, the radiation resistance of the glass is the best.
3. Co-doped with valence-changing ion $Ce^{3+/4+}$. $Ce^{3+/4+}$ was first used to enhance the darkening resistance of X-ray silica tubes, and later used to enhance the photo-

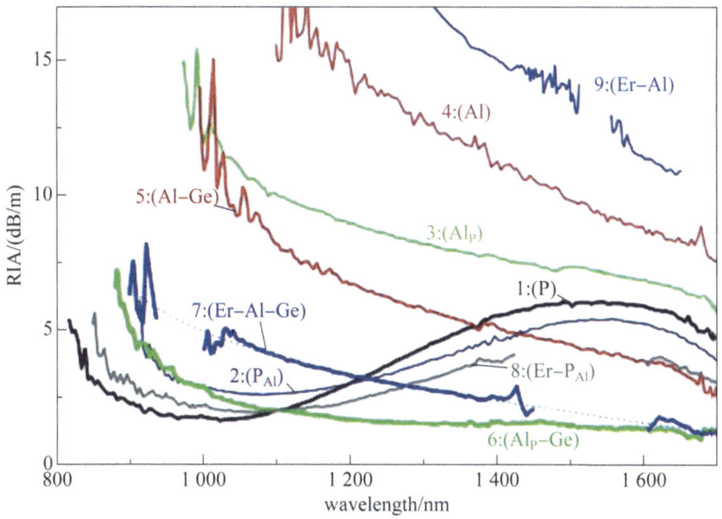

Fig. 9.21 The influence of core composition on the RIA spectra of optical fibers

darkening resistance and radiation resistance of rare earth-doped silica fibers [81]. The research group of Girard [82–85] shows that co-doping with a certain amount of $Ce^{3+/4+}$ does not significantly affect the laser performance of Er^{3+} doped silica fiber, and can significantly improve the radiation resistance of the fiber. The following will introduce further through specific examples.

Figure 9.22a, b show the effects of Ce content changes on the absorption coefficient and laser slope efficiency of P/Er/Yb/Ce doped silica fiber [82]. Compared with P/Er/Yb (without Ce) samples, the luminescence intensity and laser slope efficiency of P/Er/Yb/Ce (low Ce content) and P/Er/Yb/Ce + (high Ce content) samples only slightly decrease. This indicates that co-doping with a certain amount of Ce does not significantly deteriorate the luminescence and laser performance of Er-doped silica fiber.

Figure 9.22c, d show the effects of Ce content changes on the gain performance of P/Er/Yb/Ce doped silica fiber under online radiation and offline radiation conditions [82]. Online radiation refers to the in-situ laser amplification experiment conducted simultaneously on Er^{3+} doped fibers during the irradiation process. Offline radiation refers to the laser amplification experiment conducted after the irradiation of Er^{3+} doped fiber is completed. Whether it is online or offline radiation, compared to P/Er/Yb samples, the gain reduction of P/Er/Yb/Ce and P/Er/Yb/Ce + samples are significantly smaller. This indicates that co-doping with Ce can significantly improve the radiation resistance of Er fiber. Under the same radiation dose conditions, compared to offline radiation, the reduction in gain induced by online radiation in Er^{3+} doped fiber is smaller. This indicates that during the online irradiation process, there are photo-bleaching and thermal bleaching effects.

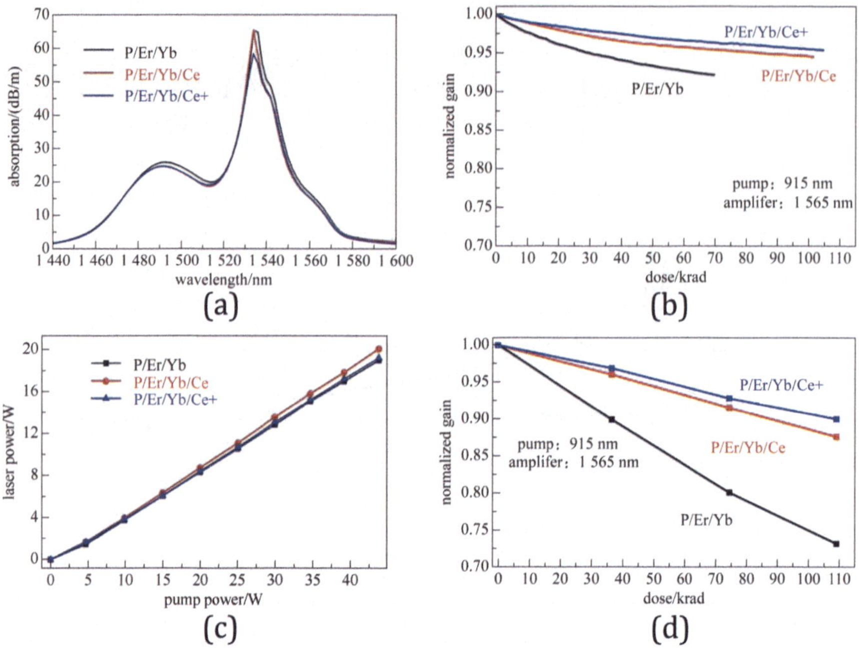

Fig. 9.22 The effect of cerium content on the absorption coefficient (**a**), laser slope efficiency (**b**), and gain performance under online radiation (**c**) and offline radiation (**d**) conditions for P/Er/Yb/Ce doped silica fibers

Figure 9.23a, b show the effects of four types of radiation sources: X-rays, γ-rays, electron beams, and proton beams on the fluorescence lifetime of the Er^{3+} ion's $^4I_{13/2}$ energy level in P/Er/Yb and P/Er/Yb/Ce co-doped silica fibers [84]. For P/Er/Yb co-doped fiber, the fluorescence lifetime of Er^{3+} ion's $^4I_{13/2}$ energy level shows a clear downward trend with the increase of radiation dose. For P/Er/Yb/Ce co-doped fiber, as the radiation dose increases, the fluorescence lifetime of Er^{3+} ion's$^4I_{13/2}$ energy level remains basically unchanged. This further indicates that co-doping with Ce can significantly improve the radiation resistance of Er fiber. In addition, the $^4I_{13/2}$ fluorescence lifetime of Er^{3+} ions in the unirradiated P/Er/Yb/Ce sample (~7.2 ms) is significantly shorter than that of the Er^{3+} ions in the P/Er/Yb sample (~9 ms). This indicates that there is energy transfer between the $^4I_{13/2}$ energy level of Er^{3+} and the 4f-5d energy level of Ce^{3+} or the 4f-4f energy level of Ce^{4+}. If the co-doped Ce^{3+} content is too high, it will lead to the deterioration of the luminescence and laser performance of the Er^{3+}-doped fiber.

Figure 9.24 shows the RIA spectra of P single-doped and P/Ce co-doped silica fibers after 300Gy gamma ray irradiation [85]. After irradiation, the P single-doped fiber has a strong broadband absorption peak at 1.6 μm, which corresponds to a phosphorus-related defect, commonly referred to as the P_1 color center, and its structural model can be represented as "≡P·". Compared with P single-doped fibers,

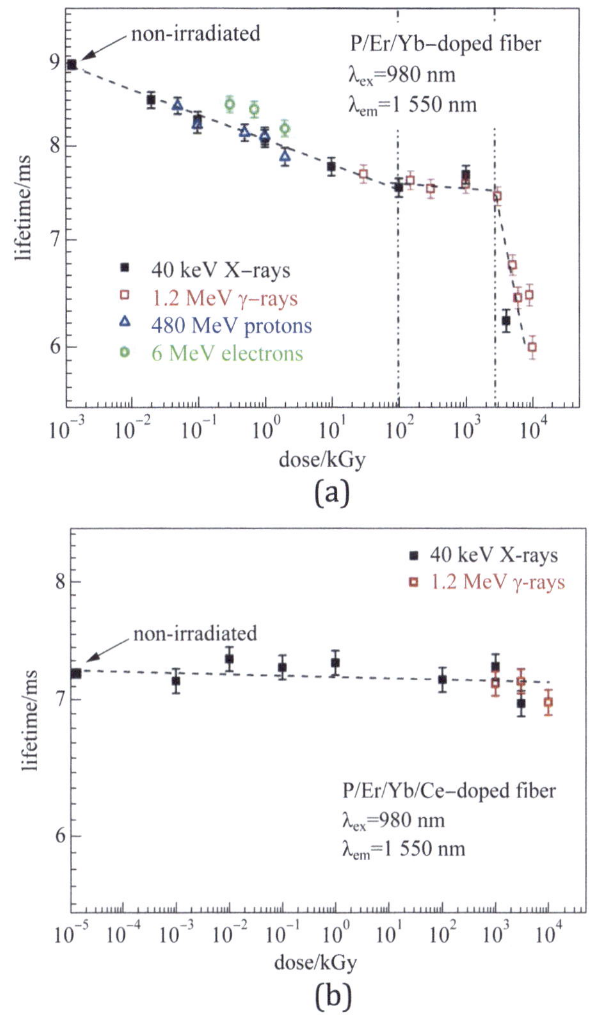

Fig. 9.23 The effect of four irradiation sources on the fluorescence lifetime of Er^{3+} ion's $^4I_{13/2}$ energy level in P/Er/Yb (**a**) and P/Er/Yb/Ce (**b**) co-doped silica fibers

no P_1 color center absorption peak was observed in irradiated P/Ce co-doped silica fibers. This indicates that co-doping with Ce can suppress the generation of the P_1 color center. The absorption peak position of P_1 center overlaps with the luminescence peak of Er^{3+}, the presence of P_1 center will cause the luminescence of Er^{3+} to be quenched.

Shao et al., [37] systematically studied the effect of Ce content changes on the radiation resistance of Yb/Al/Ce co-doped silica glass and its mechanism by combining absorption, fluorescence, X-ray photoelectron spectroscopy (XPS), and

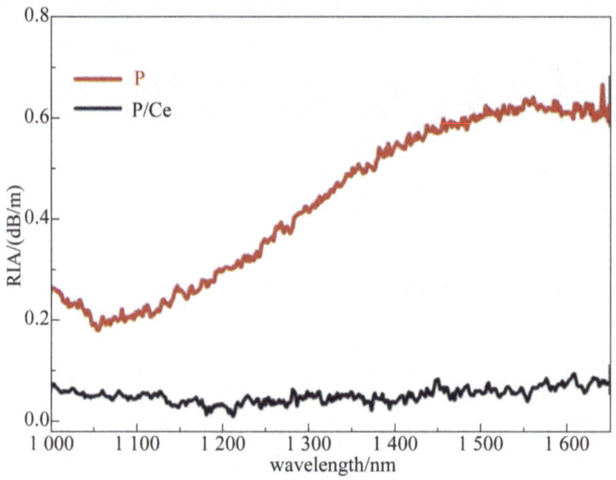

Fig. 9.24 RIA spectra of P single-doped and P/Ce co-doped silica fibers

CW-EPR. The study shows that the radiation resistance of the glass is positively correlated with the Ce content, and its radiation resistance mechanism is related to the valence state change of $Ce^{3+/4+}$. Ce^{3+} By capturing holes and becoming Ce^{4+}, it inhibits the production of hole-type color centers (Al-OHC). Ce^{4+} captures electrons and becomes Ce^{3+}, inhibiting the production of electron-type color centers (Yb^{2+}/Al-E'/Si-E'). Research by Liu and others at Huazhong University of Science and Technology (HUST), China [86] shows that co-doping with $Ce^{3+/4+}$ can promote the bleaching of Yb^{3+} fiber after darkening. However, it should be noted that co-doping with $Ce^{3+/4+}$ will increase the refractive index and numerical aperture of the fiber core, affecting the quality of the laser beam. Further research by Shao and others [87] shows that introducing F on the basis of Yb/Al/Ce co-doped silica glass can not only effectively reduce the refractive index of the fiber core, but also further improve the radiation resistance of the glass.

9.4.2 *Pretreatment*

Pretreatment refers to the radiation-resistant reinforcement treatment of the fiber or preform before irradiation, which includes preloading gas, pre-irradiation, and pre-annealing. The following will introduce each method.

1. Preloading gas. Loading gas refers to promoting a certain gas to diffuse into the fiber or preform under high-temperature or high-pressure conditions. The gas used is usually hydrogen or deuterium. As early as 1967, American scientists first discovered that loading hydrogen (H_2) can improve the radiation resistance of glass [88]. In 1985, Japanese scientists confirmed that loading H_2 can improve the radiation resistance of pure silica fiber [89]. In 2007, Zotov and others in

Russia [90, 91] first confirmed that loading H_2 can improve the radiation resistance of Er^{3+} doped silica fiber. In 2018, the research group of Li Jinyan at HUST, China [92] showed that loading deuterium (D_2) can effectively improve the radiation resistance of Tm^{3+} doped silica fiber. However, the disadvantage of improving the radiation resistance of active fibers by loading H_2 or D_2 is that: ①H_2 or D_2 entering the fiber in molecular form, it almost completely diffuses out of the fiber in just three months in the atmospheric environment, leading to a decrease in the radiation resistance of the fiber. Although the use of hermetic carbon coatings or metal coatings can effectively solve the problem of gas leakage [3], it increases the complexity of the process, and the sealed coating fiber needs higher temperature and pressure for gas loading. ② If the concentration of gas loaded into the fiber is too high, the strong absorption of H_2 (1.24 μm) or D_2 (1.71 μm) molecules in the near-infrared band will significantly reduce the laser slope efficiency of the active fiber. ③ High-temperature gas loading can severely damage the mechanical properties of the polymer-based coating layer. To solve the above three problems, Girard et al. [93, 94] designed a new fiber structure, called hole-assisted carbon-coated (HACC) fiber. The core of this fiber contains four elements: Er, Al, Ge, and Ce. The inner cladding has six air holes, and the coating is a sealed carbon coating, as shown in Fig. 9.25a. This HACC fiber can load H_2 or D_2 into the fiber at low temperatures through the air holes. Then, by controlling the gas content in the air holes, the gain and radiation resistance of the Er-doped fiber are optimized. Finally, sealed guiding fibers are fused at both ends of the fiber to prevent gas leakage. Studies have shown that the gas leakage rate of this HACC fiber is more than 100 times lower than that of ordinary double-clad hydrogen-loaded fibers. The radiation-induced loss of HACC fiber is more than 20 times lower than that of radiation tolerant acrylate coated (RTAC) fiber, as shown in Fig. 9.25b. After irradiation with 100krad γ-rays, the gain reduction of HACC fiber is less than 5%, while the gain reduction of other fibers is more than 15%, as shown in Fig. 9.25c. However, the disadvantages of this method are: ① the process is complex; ② it has special requirements for the fiber structure and is not applicable to ordinary double-clad fibers.

2. Pre-irradiation. In 1995, Griscom and others from the U.S. Naval Research Laboratory [76, 77] showed that pre-irradiation treatment of pure silica core fibers with a large dose ($\geq 10^7$ rad) can ensure that the RIA in 400 ~ 700 nm of pure silica fibers in the subsequent radiation process does not exceed 30 dB/km. In 2013, Yeniay and others from the American Photon-X LLC [78] showed that using γ-ray pre-irradiation combined with thermal annealing (maintained at 165 °C for 365 h) can improve the radiation resistance of Yb^{3+}/Er^{3+} co-doped silica fibers to some extent. However, it must be pointed out that due to the poor heat resistance of the polymer-based coating ($\leq$180 °C), both photobleaching and thermal bleaching cannot be too high in temperature, thus it is impossible to completely bleach the color centers and restore the loss of irradiated fibers to their original state. In addition, the mechanism of improving the radiation resistance of fibers by pre-irradiation may be related to the diffusion of H^+. Studies have shown that the acrylate coating releases H^+ or H_2 and diffuses to the fiber

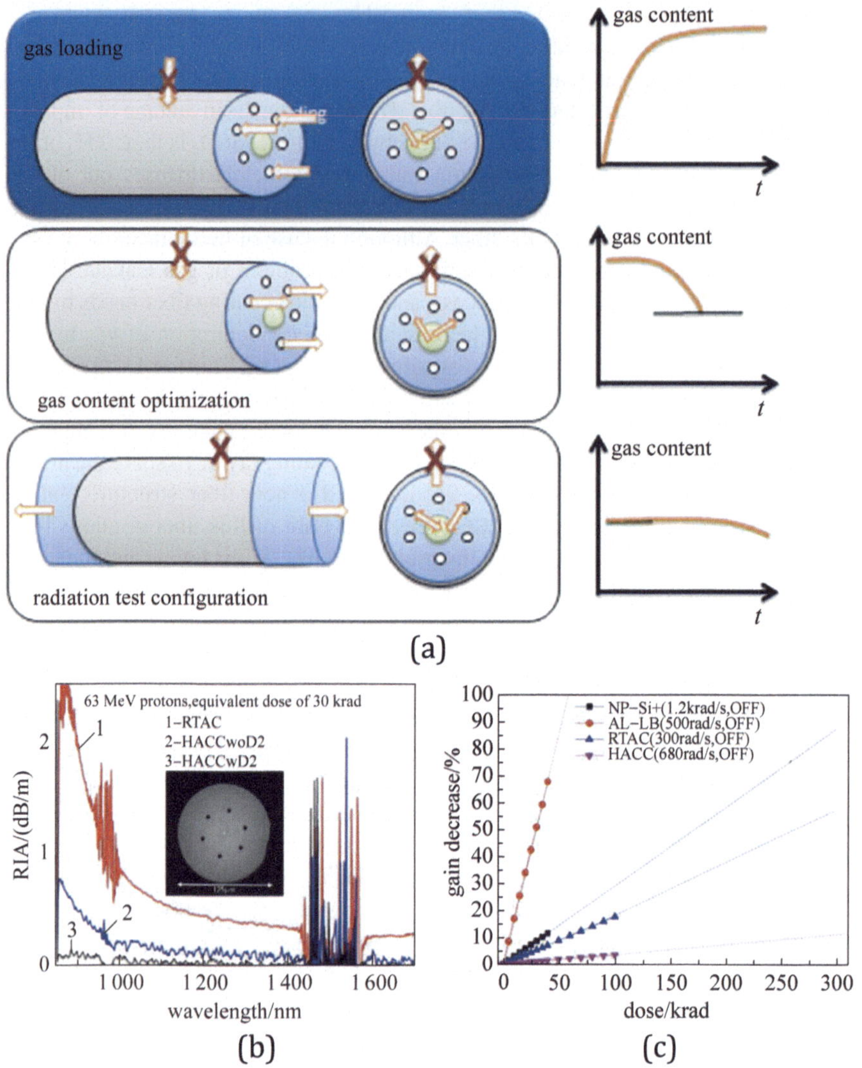

Fig. 9.25 Schematic diagram of the structure of a hole-assisted carbon coating (HACC) fiber [93] (**a**); RIA spectrum [93] (**b**); Gain reduction with irradiation dose [94] (**c**)

cladding and core during the radiation process, and H^+ or H_2 reacts with SiO_2 to form OH, causing an increase in fiber loss at 1.39 μm [62]. Liu and others from the SIOM, CAS [95] showed that increasing OH content can improve the radiation resistance of Yb^{3+} doped silica glass to some extent. In 2015, Griscom and others [96] published a correction stating that when he pre-irradiated fibers with a large dose in 1995, some of the fibers were soaked in water. During the radiation process, the H^+ in the water and the H^+ produced by the radiation-induced degradation of the acrylate coating are likely to diffuse into the fiber core and react with SiO_2. The reaction produces Si-OH, thereby improving the radiation resistance of the optical fiber. In 2016, Kim et, al. [97] found that pre-irradiation treatment does not improve the radiation resistance of fluorine-doped silica optical fibers in the near-infrared band during subsequent radiation processes.

3. Pre-annealing. In 2001, Hosono and others from Tokyo Institute of Technology in Japan [98] performed variable temperature (1400/1200/1100/900 °C) thermal annealing treatment on pure silica glass, aiming to change the T_f of glass. The results showed that reducing the T_f of pure silica glass can significantly improve the radiation resistance of the glass. In 2016, the University of Paris X in France [66] performed thermal annealing treatment on Er^{3+} single-doped silica optical fiber preforms at different temperatures (900/1000/1100/1200/1300 °C) to change its T_f, and found that reducing the T_f of the preform can effectively improve its radiation resistance.

The combination of the above pre-loading gas, pre-irradiation, and pre-annealing treatments can not only further improve the radiation resistance of the optical fiber, but also effectively prevent gas leakage. However, this combination method is not suitable for direct pre-treatment of optical fibers, mainly for two reasons: ① High-temperature loading gas or large-dose pre-irradiation will accelerate the degradation of the polymer-based coating of the optical fiber (turning yellow and brittle), affecting the mechanical strength and service life of the optical fiber; ② Large-dose pre-irradiation induces color centers in the optical fiber, leading to an increase in the background loss of optical fiber. Limited by the poor heat resistance of the polymer-based coating of optical fiber, both photobleaching and thermal bleaching can partially reduce its RIA.

In 2019, Shao et al. (SIOM,CAS) [99, 100] proposed a radiation-resistant reinforcement method for active fiber preforms by sequentially carrying out deuterium loading, pre-irradiation, and thermal annealing treatments. Taking the ytterbium-doped silica optical fiber preform as an example, its pre-treatment conditions and process are listed, as shown in Fig. 9.26. CW-EPR tests show that under the same radiation conditions, the concentration of radiation-induced color centers in the core rod of the preform treated by this method is more than an order of magnitude lower than that in the untreated core rod. The core rods obtained by this method can be used to prepare radiation-resistant rare-earth-doped silica optical fibers, which have the advantages of high laser slope efficiency, low background loss, and long-term stable use in a vacuum environment. The following will further introduce this pre-treatment method.

Figure 9.27a, d show the loss spectrum and laser slope efficiency of the original Yb^{3+}-doped silica fiber preform (without pre-treatment) before and after gamma irradiation. Figure 9.27b, e show the loss spectrum and laser slope efficiency of the fiber drawn from the hydrogen pre-treated preform before and after gamma irradiation. Figure 9.27c, f show the loss spectrum and laser slope efficiency of the fiber drawn from the deuterium pre-treated preform before and after gamma irradiation. The laser slope efficiency was tested using a spatial coupling method, with a 976 nm

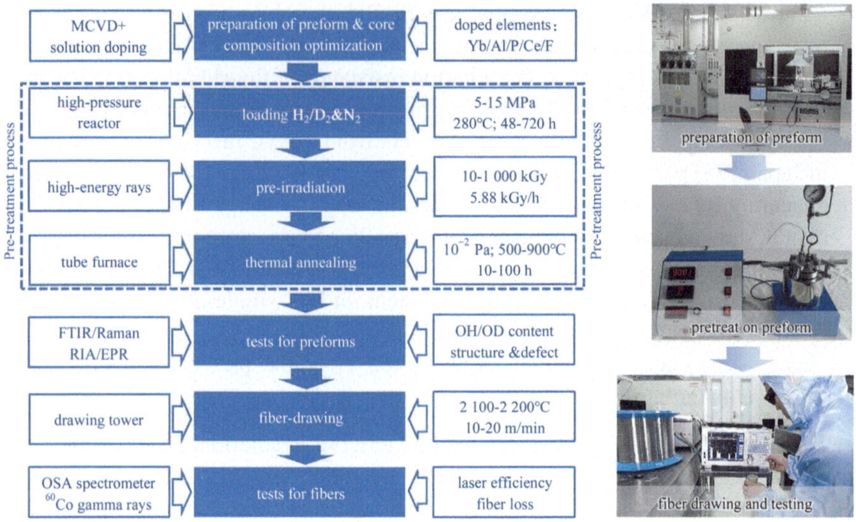

Fig. 9.26 Pre-treatment of active fiber preform and its fiber performance evaluation process [99, 100]

laser diode as the pump source, and the length of the Yb^{3+}-doped fiber used for testing was 25 m.

For comparison, Table 9.5 lists the specific values of background loss and laser slope efficiency at a wavelength of 1200 nm before and after gamma irradiation for three fibers. The total dose of gamma irradiation is about 700Gy, which corresponds to the cumulative irradiation dose absorbed by a satellite in geosynchronous orbit (GSO) over a period of 10 years.

Before gamma irradiation (0Gy), the background losses of the pristine, hydrogen-loaded, and deuterium-loaded fibers at a wavelength of 1200 nm are approximately 6 dB/km, 83 dB/km, and 20 dB/km respectively; the laser slope efficiencies of the three fibers are 79%, 45%, and 75% respectively.

After gamma irradiation (700Gy), the background losses of the pristine, hydrogen-loaded, and deuterium-loaded fibers at a wavelength of 1200 nm are approximately 533 dB/km, 130 dB/km, and 70 dB/km respectively; the laser slope efficiencies of the three fibers are 0, 32%, and 59% respectively.

As can be seen from Figs. 9.27 and Table 9.5, hydrogen or deuterium loading can significantly improve the radiation resistance of Yb^{3+} doped silica fibers. However, hydrogen loading severely deteriorates the laser performance of the unirradiated Yb^{3+} doped fiber, reducing the laser slope efficiency from 79% to 45%, and increasing the fiber loss from 6 dB/km to 83 dB/km; Deuterium loading does not have a significant negative impact on the laser performance of unirradiated Yb^{3+} doped fiber, with the laser slope efficiency decreasing from 79% to 75%, and the fiber loss increasing from 6 dB/km to 20 dB/km. The fundamental reason is correlated to the absorption peak positions and absorption intensities of the OH and OD groups.

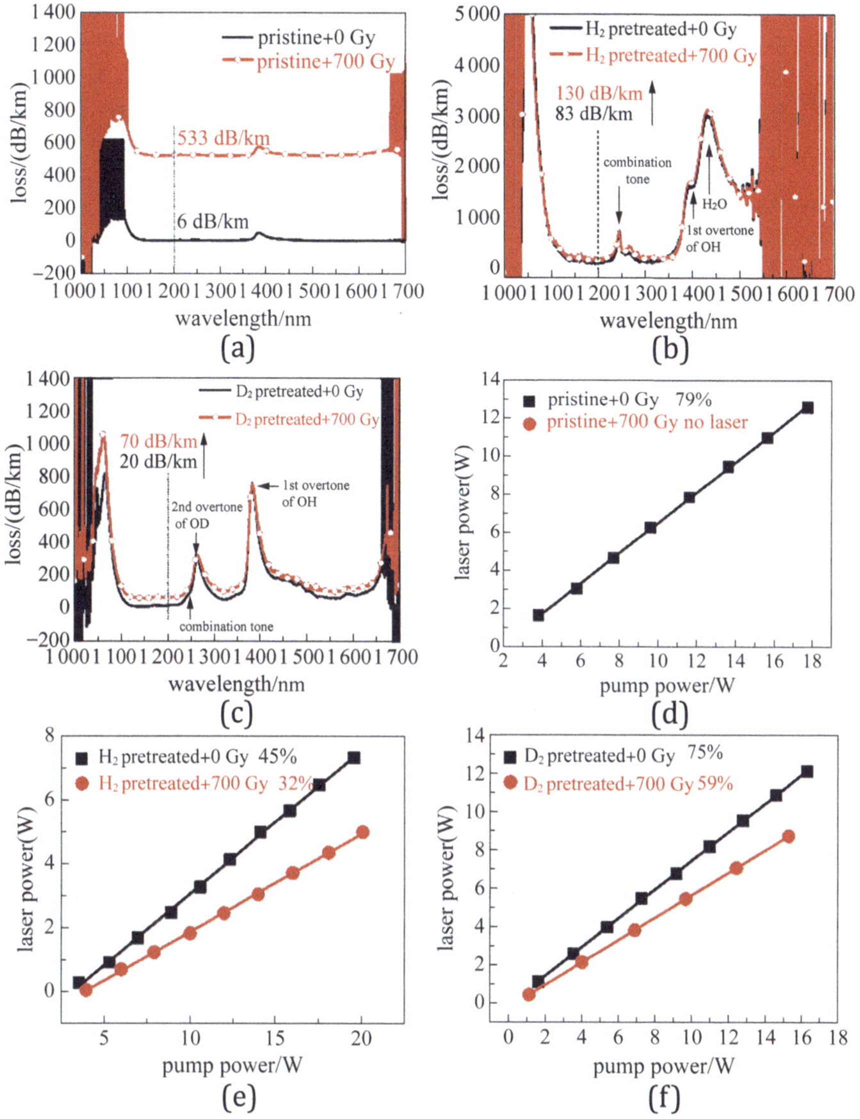

Fig. 9.27 Loss spectrum and laser efficiency of Yb^{3+}-doped silica fibers drawn from original (**a**, **d**), H_2 pre-treated (**b**, **e**), and D_2 pre-treated (**c**, **f**) preforms [101]

As is well known, the absorption of glass in the infrared band mainly comes from molecular vibrations. The frequency (ν) of molecular vibration depends on the reduced mass (μ) and bond force constant (K) in the vibrating group of anions and cations, and their relationship is as follows [102]:

Table 9.5 Laser slope efficiency and background loss at 1200 nm before and after irradiation of Yb^{3+} doped silica fibers prepared from original, H_2 pre-treated, and D_2 pre-treated preforms

Dose	Fiber Parameters	Pristine	H_2 pretreated	D_2 pretreated
0Gy	Loss@1200 nm (dB/km)	6	83	20
	Slope efficiency	79%	45%	75%
	Efficiency reduction caused by pretreatment	0	43%	5%
700Gy	Loss@1200 nm(dB/km)	533	130	70
	Slope efficiency	0	32%	59%
	Efficiency reduction caused by radiation	100%	29%	21%

$$\nu = \frac{1}{2\pi c}\sqrt{\frac{K}{\mu}} \tag{9.1}$$

$$\mu = \frac{m_1 m_2}{m_1 + m_2} \tag{9.2}$$

where, c is the speed of light; m_1 and m_2 are the molar masses of the anion and cation, respectively. According to the above two formulas, the vibration frequency of the OH group is about 1.37 times that of the OD vibration frequency. It is known that the fundamental vibration peak of OH is located at 2.7 μm, and the first and second overtone peaks are located at 1.38 μm and 0.95 μm, respectively [103]. It can be inferred that the fundamental vibration peak of OD is located at 3.8 μm, and the first and second overtone peaks are located at 1.87 μm and 1.26 μm, respectively. Studies have shown that the intensity of the OH fundamental vibration peak is about 50 ~ 60 times that of the first overtone peak, and more than 2500 times that of the second overtone peak [104].

Based on the above conclusions, Fig. 9.28 shows the main absorption peak positions and intensities of the OD and OH groups in the 800 ~ 4000 nm band [104]. As can be seen from the figure, compared to the OD group, the absorption peak wavelength of the OH group is closer to the pump and laser wavelength of Yb^{3+}. Therefore, compared with the same content of OD group, the OH group has a greater negative impact on the fluorescence and laser performance of Yb^{3+} doped fiber.

Figure 9.29 shows the normalized gain variation with irradiation dose for four Erbium-doped silica fibers (EDF) [105]. The original fiber and the deuterium pretreated fiber come from the same preform rod, with the core composition of the preform rod being co-doped with four elements: Er-Al-La-Ge. The same preform rod is divided into two sections, one of which is not treated at all, and the fiber drawn is called the SIOM pristine EDF; the other section undergoes deuterium pretreatment, and the fiber drawn is called the SIOM pre-treated EDF. The pre-treatment method refers to Fig. 9.27. The commercial ordinary EDF comes from OFS company, and the commercial radiation-resistant EDF comes from iXblue company, these two fibers are mainly used as reference samples. The radiation-induced gain variation (RIGV) is approximately linear with the irradiation dose, and by performing a linear fit, the RIGV value after larger dose radiation can be predicted.

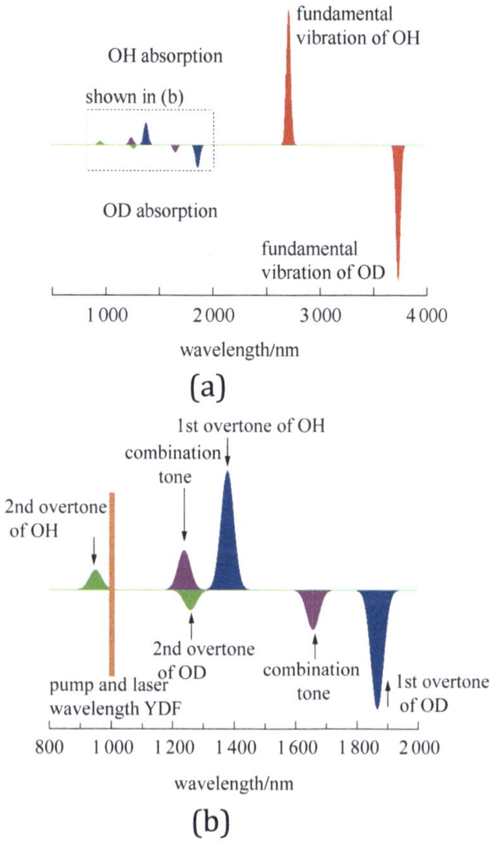

Fig. 9.28 Schematic diagram of the main vibrational absorption peaks of OH and OD groups in 500 ~ 4000 nm (**a**) and 800 ~ 2000 nm (**b**)

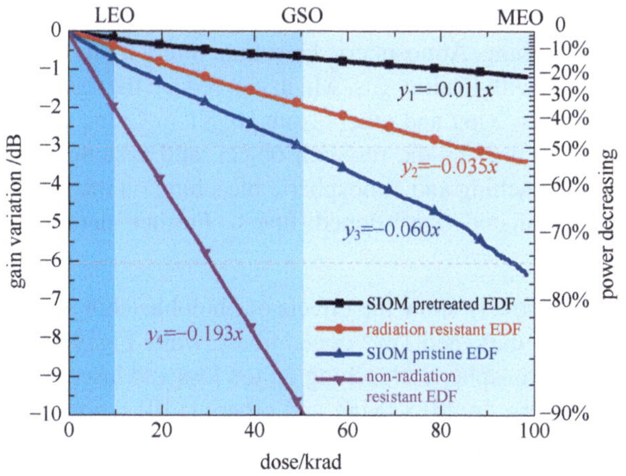

Fig. 9.29 Normalized gain variation with irradiation dose for four Erbium-doped silica fibers

According to the fitting results, after 100krad radiation, the RIGV values of commercial ordinary EDF, SIOM pristine EDF, commercial radiation-resistant EDF, and SIOM pre-treated EDF are 19.3 dB, 6 dB, 3.5 dB, and 1.1 dB, respectively. Obviously, the radiation resistance of SIOM pre-treated EDF is the best.

9.4.3 Post-Treatment

Post-treatment refers to the bleaching treatment of the fiber after irradiation, which includes thermal bleaching, photobleaching, and atmosphere bleaching. The following is a brief introduction:

1. Thermal bleaching. Within a certain temperature range, raising the temperature will cause the RIA intensity of fiber to decrease, or even fully recover to the level before radiation, this phenomenon is called thermal bleaching. In 1997, research by Ramsey et al. from Princeton University [106] showed that heating can bleach the irradiated silica fiber, and the higher the temperature, the better the bleaching effect. Research by Söderlund et al. from Aalto University in Finland [107] showed that to completely bleach the radiation-induced loss of Yb^{3+}-doped fiber, the temperature needs to be raised above 600 °C. However, high-temperature thermal bleaching poses a significant challenge to the coating of ordinary double-clad fibers.
2. Photobleaching. In 1981, Friebele and others from the U.S. Naval Research Laboratory [108] showed that by using a 0.85 μm laser diode to pump γ-ray irradiated optical fibers, the radiation-induced loss of the optical fibers could be observed to gradually decrease over time, a phenomenon known as photobleaching. Further research showed that the higher the pump power, the shorter the pump wavelength, and the longer the pump time, the better bleaching effect of the darkened optical fibers [74, 75].
3. Atmospheric bleaching. Atmospheric bleaching refers to the treatment of irradiated optical fibers with loading gas, which can reduce the radiation-induced loss of the optical fibers. Xing and others from HUST in China [92, 109–111] used Tm^{3+}-doped silica fiber as the research object, and systematically studied the effects of photobleaching and atmospheric bleaching on the loss and laser slope efficiency of irradiated Tm^{3+}-doped fibers. Further details will be introduced below.

Figure 9.30a, b respectively show the effects of photobleaching on the loss spectra and laser efficiency of darkened Tm^{3+} doped fibers, while Fig. 9.30c, d respectively show the effects of atmospheric bleaching on the loss and laser efficiency of darkened Tm^{3+} doped fibers. In 2015, Xing and others [110] reported that by using a 793 nm laser diode (LD) to pump Tm^{3+} doped fibers irradiated with different doses of γ-rays for a long time, the RIA intensity of the darkened fibers could be reduced and their laser efficiency could be increased, as shown in Fig. 9.30a, b. The research results show that the pump light can only partially bleach the darkened fiber, and the

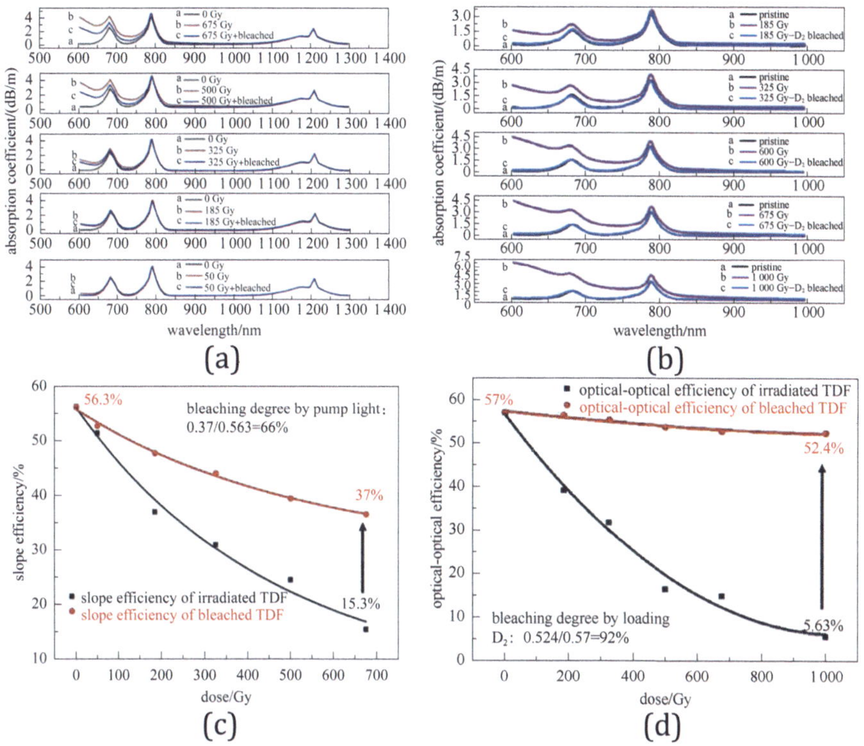

Fig. 9.30 Photobleaching [110] (**a**, **b**) and atmospheric bleaching [111] (**c**, **d**) on the loss spectra (**a**, **c**) and laser efficiency (**b**, **d**) of Tm^{3+} doped silica fiber after different doses of gamma ray irradiation

larger the radiation dose, the worse the photobleaching effect. When the radiation dose is 700 Gy, the degree of bleaching of the darkened fiber is only 66%, that is, the laser efficiency of the bleached fiber (37%) is only equivalent to 66% of the laser efficiency of the unirradiated fiber (56.3%). It must be pointed out that the efficiency of photobleaching is low, the time required is long (>70 h), the bleaching effect is poor, and the distribution along the length of the fiber is uneven, the pump end is thoroughly bleached, and the far end from the pump is not completely bleached.

To solve the above problems, Xing et al. [111] proposed in 2018 to bleach Tm^{3+} doped fiber in the atmosphere. The specific plan is to treat the irradiated fiber with high-pressure loading gas. The gas is a mixture of 95% N_2 and 5% D_2, or 95% N_2 and 5% H_2, or 100% N_2. The loading gas pressure and time are 0.3 MPa and 48 h respectively. The loading gas temperature is room temperature. The experimental results show that high-pressure nitrogen has almost no bleaching effect, and the bleaching effect of high-pressure deuterium is far better than that of hydrogen. Figure 9.30c, d show the effects of high-pressure deuterium on the loss spectrum and laser efficiency of darkened Tm^{3+} doped fiber. The loss spectrum and laser efficiency of the bleached fiber are nearly identical from those of the unirradiated fiber. And as the radiation dose increases, the bleaching effect does not significantly worsen. When the radiation dose is 1000 Gy, the degree of atmospheric bleaching of the darkened fiber is as high as 92%, that is, the laser efficiency of the bleached fiber (52.4%) is equivalent to 92% of the laser efficiency of the unirradiated fiber (57%). Compared with photobleaching, atmospheric bleaching has better effects and higher efficiency.

9.4.4 System Optimization

As described in Sect. 9.3, the factors affecting the radiation resistance of optical fibers include three aspects: fiber parameters, radiation environment, and application parameters. The above three methods to improve the radiation resistance of optical fibers (i.e., composition optimization, pretreatment, and post-treatment) all involve the optimization of fiber components. However, in practical applications, the effects of the radiation environment and application parameters on the stability of fiber laser or amplifier systems should also be considered. The Girard research group at the University of Lyon in France [74, 112–116] was the first to propose enhancing the radiation resistance of fiber lasers or amplifiers from a system optimization perspective. The overall approach is shown in Fig. 9.31 [74]: ① Improve the radiation resistance of fiber components by optimizing fiber structure, glass composition, and preparation process. Obtain the spectral parameters of the activated ions (such as absorption, emission cross-section, fluorescence lifetime, etc.) to create conditions for subsequent theoretical modeling. ② Define the service conditions of the laser or amplifier, such as total radiation dose, dose rate, temperature, etc. ③ Based on the above, optimize the system parameters of the laser or amplifier

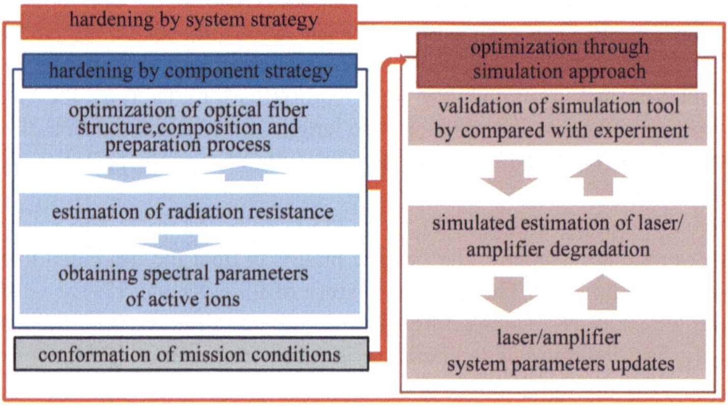

Fig. 9.31 Overall approach to improving the radiation resistance of fiber lasers or amplifiers through system optimization

such as fiber length, pumping method, pumping wavelength, pumping power, etc., through theoretical simulation, and verify the theoretical simulation results with specific experiments. Further introduction is given below.

Figure 9.32a–c show the effects of hydrogen loading, fiber length, and pumping method on the radiation resistance of Er^{3+} doped fiber amplifiers (EDFA), respectively [116]. Figure 9.32d compares the effects of fiber component optimization and system optimization on the radiation resistance of EDFA [116]. The experimental conditions in Fig. 9.32a are an active fiber length of 8 m and a forward pumping method (i.e., the pumping light and signal light are at the same end of the fiber). As can be seen from Fig. 9.32a, the gain reduction induced by radiation in hydrogen-loaded EDFA is much less than that in non-hydrogen-loaded EDFA. Whether it is hydrogen-loaded or non-hydrogen-loaded EDFA, the maximum error between the experimental results and the simulation results does not exceed 3%. This can confirm the correctness and reliability of the theoretical simulation. Figure 9.32b optimizes the length of the erbium-doped fiber in the hydrogen-loaded EDFA device

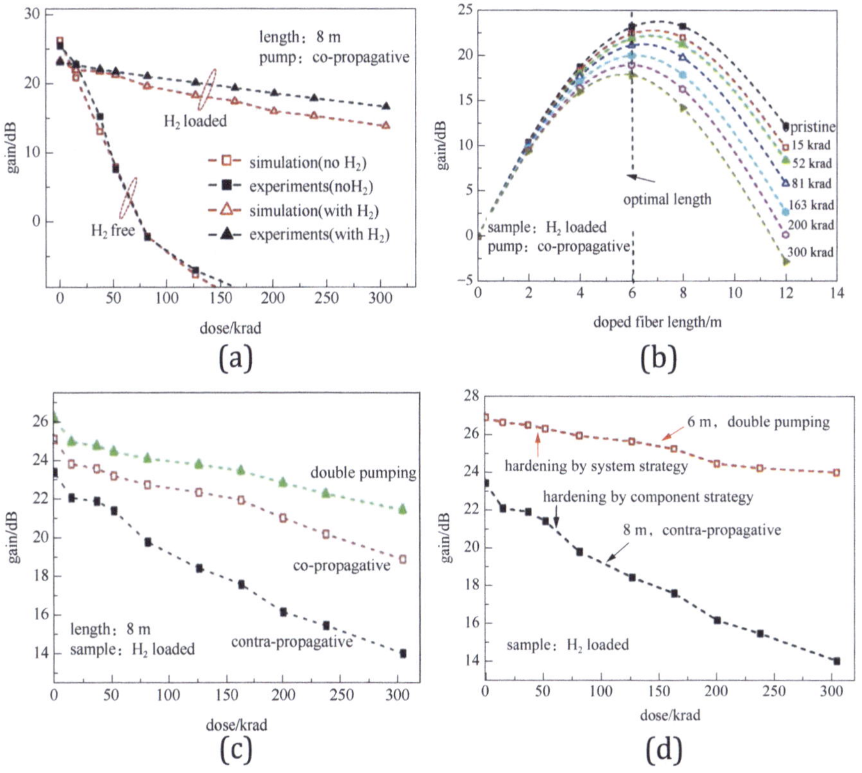

Fig. 9.32 Simulation of the impacts of hydrogen loading (**a**), fiber length (**b**), and pumping method (**c**) on the radiation resistance of EDFA; (**d**) Experimental comparison of the impacts of component optimization and system optimization on the radiation resistance of hydrogen-loaded EDFA [116]

through theoretical modeling. The pumping method is assumed to be co-directional pumping. The simulation results show that in the non-irradiated EDFA, the gain is maximum when the active fiber length is 7 m. As the radiation dose increases, the optimal length of the active fiber gradually shortens. When the radiation dose is 300krad, the optimal length of the active fiber is 6 m. Fig. 9.32c demonstrates the impact of three types of pumping methods (bidirectional pumping, forward pumping, and backward pumping) on the radiation resistance of hydrogen-loaded EDFA through theoretical simulation. The fiber length is assumed to be 8 m. In bidirectional pumping, it is assumed that two pump lights (~45 mW) are injected from both ends of the fiber, with the signal light located at one end of the fiber. In forward pumping, it is assumed that only one pump light (~90 mW) and the signal light are injected from the same end of the fiber. In backward pumping, it is assumed that only one pump light (~90 mW) and one signal light are injected from both ends of the fiber. The simulation results show that when bidirectional pumping is used, the gain of the EDFA is the highest, and the radiation-induced gain reduction is the smallest; when forward pumping is used, the gain of the EDFA is the smallest, and the radiation-induced gain reduction is the largest; the effect of forward pumping is between the two. Obviously, bidirectional pumping is most beneficial to improve the radiation resistance of hydrogen-loaded EDFA. Figure 9.32d compares the impact of fiber component optimization and laser/ amplifiers system optimization on the radiation resistance of hydrogen-loaded EDFA through specific experiments. If only the fiber components are optimized, when the fiber length is 8 m and the pumping method is forward pumping, the gain of the hydrogen-loaded EDFA gradually decreases from 23.5 dB before irradiation to 14 dB after 300krad irradiation, a decrease of 9.5 dB. If system optimization is considered, when the fiber length is 6 m and the pumping method is bidirectional pumping, the gain of the hydrogen-loaded EDFA gradually decreases from 27 dB before irradiation to 24 dB after 300krad irradiation, a decrease of only 3 dB. System optimization can effectively improve the radiation resistance of fiber lasers or amplifiers.

In summary, this chapter first introduces the applications and challenges of rare-earth-doped silica fibers in space, and then details the latest research progress in radiation-resistant rare-earth-doped silica fibers for space applications from three aspects: ① the radiation-induced darkening mechanism of active fibers; ② the influencing factors of the radiation resistance of active fibers; ③ the methods to improve the radiation resistance of active fibers. Future research on radiation-resistant active fibers can be conducted from the following five aspects: (1) Currently, the most research has been done on the radiation resistance of erbium-doped and ytterbium-doped silica fibers, and the radiation resistance reinforcement technology is relatively mature. Due to the different glass components, working wavelengths, and numerical apertures required for the single-mode cutoff frequency of silica fibers doped with different rare earth ions, the radiation resistance reinforcement technology of erbium-doped or ytterbium-doped fibers may not necessarily be applicable to other rare earth ions (such as Tm^{3+}) doped fibers. Therefore, it is necessary to carry out systematic research on the radiation resistance of silica fibers doped with different rare earth ions. (2) The factors affecting the radiation resistance of active fibers are extremely complex, including both intrinsic parameters of the fiber and environmental and application parameters. At present, there is no complete theory that can accurately simulate and predict the radiation resistance of active fibers, and accelerated radiation experiments are still the best method for evaluating the radiation resistance of active fibers. Future work should strengthen theoretical research and develop predictive models for the radiation resistance of active fibers. (3) Although many methods of radiation resistance reinforcement for active fibers have been proposed, the mechanisms and long-term effectiveness of these methods are not yet fully understood. For example, why can reducing the T_f of silica glass and fibers improve their radiation resistance? What is the mechanism? Can it be extended to other glass systems? For instance, increasing the hydroxyl content in active fibers can improve their radiation resistance, but what is the mechanism? Does it have long-term radiation stability? Therefore, future work should systematically study the physical origins behind different radiation resistance reinforcement methods, evaluate their long-term radiation stability, and integrate different radiation resistance reinforcement methods for comprehensive application. (4) The darkening and bleaching of fibers are a dynamic equilibrium process. In a non-radiation environment, pump light can induce photo-darkening effects in active fibers and can also bleach darkened fibers to a certain extent. Correspondingly, in a radiation environment, pump light will also affect the degree of darkening of active fibers. Therefore, conducting in-situ online testing of active fibers during the pumping process, while in a radiation environment, is closer to the actual working environment and has greater practical significance. (5) Fiber lasers for space applications are in a complex environment with strong vibrations, high vacuum, extreme temperature fluctuations (±200 °C), and harsh radiation. The spacecraft undergoes intense vibrations during ascent and landing, which requires the laser system to adopt an all-fiber structure. Under vacuum conditions, the laser cools down slowly, and some small molecule gases (such as H_2/Cl_2) can easily diffuse out of the fiber. The probability of 4f orbital electrons at different Stark levels of rare earth ion follows the Boltzmann distribution, which is closely related to temperature. The stability of radiation-induced color centers also heavily depends on the external temperature. Therefore, changes in external temperature will inevitably cause changes in the laser performance and radiation resistance of active fibers. Unlike the single, stable gamma acceleration radiation experiment conditions, there are multiple radiation sources in space with low dose rates, and dose rates that vary over time and space. The transient strong radiation may occur when encountering solar particle events. Therefore, future research needs to comprehensively assess the impacts of vibration intensity, vacuum degree, temperature change, and radiation condition on the laser performance and radiation resistance of active fibers.

References

1. Powell D. Lasers boost space communications [J]. Nature, 2013,499(7458):266-267.
2. Wright M W, Valley G C. Yb-doped fiber amplifier for deep-space optical communications [J]. Journal of Lightwave Technology, 2005,23(3):1369.
3. Huang Jianping, Zhang Ge, Wang Pupu, et al. Research of radiation resistant Er doped fiber for space detection[C]. SPIE, 2016.
4. Duchez J, Mady F, Mebrouk Y, et al. Interplay between photo- and radiation-induced darkening in ytterbium-doped fibers [J]. Optics Letters, 2014,39(20):5969-5972.
5. Williams G L M, Friebele E J. Space radiation effects on erbium-doped fiber devices: Sources, amplifiers, and passive measurements [J]. IEEE Transactions on Nuclear Science, 1998,45(3Pt3):399-404.
6. Smith D S, Stabin M G. Exposure rate constants and lead shielding values for over 1,100 radionuclides [J]. Health Physics, 2012,102(3):271-291.
7. Ott M N, Jin X L, Chuska R, et al. Space flight requirements for fiber optic components: qualification testing and lessons learned [C]. Society of Photo-Optical Instrumentation Engineers (SPIE) Conference Series, 2006.
8. Girard S, Morana A, Ladaci A, et al. Recent advances in radiation-hardened fiber-based technologies for space applications [J]. Journal of Optics, 2018,20(9):93001.
9. Velazco R, Fouillat P, Reis R. Radiation effects on embedded systems [M]. [S.l.]: Springer Science and Business Media, 2007.
10. Girard S, Kuhnhenn J, Gusarov A, et al. Radiation effects on silica-based optical fibers: recent advances and future challenges [J]. IEEE Transactions on Nuclear Science, 2013,60(3):2015-2036.
11. Deng Tao, Xie Junlin, Luo Jie, et al. Review and prospect of research on radiation resistance of optical fibers [J]. Optical Communication Technology, 2007(9):58-61.
12. Ott M N. Radiation effects data on commercially available optical fiber: database summary [C]. IEEE, 2002.
13. Gusarov A, Hoeffgen S K. Radiation effects on fiber gratings [J]. IEEE Transactions on Nuclear Science, 2013,60(3):2037-2053.
14. Berghmans F, Brichard B, Fernandez A F, et al. An introduction to radiation effects on optical components and fiber optic sensors [M]. [S.l.]: Springer Netherlands, 2008.
15. Perry M, Niewczas P, Johnston M. Effects of neutron-gamma radiation on fiber Bragg grating sensors: a review [J]. IEEE Sensors Journal, 2012,12(11):3248-3257.
16. Friebele E J, Gingerich M E, Brambani L A, et al. Radiation effects in polarization-maintaining fibers [C]//Proceedings of the SPIE — The International Society for Optical Engineering, 1990(1314):146-154.
17. Olanterä L, Sigaud C, Troska J, et al. Gamma irradiation of minimal latency hollow-core photonic bandgap fibres [J]. Journal of Instrumentation, 2013,8(12):C12010.
18. Girard S, Yahya A, Boukenter A, et al. γ-radiation-induced attenuation in photonic crystal fibre [J]. Electronics Letters, 2002,38(20),1169-1171.
19. Girard S, Ouerdane Y, Tortech B, et al. Radiation effects on ytterbium-and ytterbium/erbium-doped double-clad optical fibers [J]. IEEE Trans. Nucl. Sci., 2009,56(6):3293-3299.
20. Girard S, Ouerdaneb Y, Vivonab M, et al. Radiation effects on rare-earth doped optical fibers [C]. Proc. SPIE 7817, Nanophotonics and Macrophotonics for Space Environments IV, 2010:781701.https://doi.org/10.1117/12.862706.
21. Shao Chongyun. Study on the structure, spectrum, radiation resistance and radiation-induced darkening mechanism of Yb^{3+} doped silica glass [D]. Beijing: University of Chinese Academy of Sciences, 2019.
22. Griscom D L. A minireview of the natures of radiation-induced point defects in pure and doped silica glasses and their visible/near-IR absorption bands, with emphasis on self-trapped holes and how they can be controlled [J]. Physics Research International, 2013(2013):1-14.

23. Griscom D L. Nature of defects and defect generation in optical glasses: 1985 Albuquerque Conferences on Optics, 1985 [C]. International Society for Optics and Photonics.
24. Griscom D L. The natures of point defects in amorphous silicon dioxide [M]//Defects in SiO_2 and related dielectrics: science and technology. Springer, 2000:117-159.
25. Griscom D L. Optical properties and structure of defects in silica glass [J]. Nippon Seramikkusu Kyokai Gakujutsu Ronbunshi, 1991,99(10):923-942.
26. Griscom D L. Defect structure of glasses: some outstanding questions in regard to vitreous silica [J]. Journal of Non-Crystalline Solids, 1985,73(1-3):51-77.
27. Griscom D L, Friebele E J, Long K J, et al. Fundamental defect centers in glass: electron spin resonance and optical absorption studies of irradiated phosphorus-doped silica glass and optical fibers [J]. Journal of Applied Physics, 1983,54(7):3743-3762.
28. Likhachev M E, Bubnov M M, Zotov K V, et al. Radiation resistance of Er-doped silica fibers: effect of host glass composition [J]. Lightwave Technology, 2013,31(5):749-755.
29. Lezius M, Predehl K, Stöwer W, et al. Radiation induced absorption in rare earth doped optical fibers [J]. Nuclear Science, IEEE Transactions on, 2012,59(2):425-433.
30. Arai T, Ichii K, Tanigawa S, et al. Gamma-radiation-induced photodarkening in ytterbium-doped silica glasses [J]. Fiber Lasers VII: Technology, Systems, and Applications, 2011,7914(79140K).
31. Deschamps T, Vezin H, Gonnet C, et al. Evidence of AlOHC responsible for the radiation-induced darkening in Yb doped fiber [J]. Optics Express, 2013,21(7):8382.
32. Fox B P, Simmons-Potter K, Thomes W J, et al. Gamma-radiation-induced photodarkening in unpumped optical fibers doped with rare-earth constituents [J]. Ieee Transactions on Nuclear Science, 2010,57(33):1618-1625.
33. Ollier N, Corbel C, Duchez J, et al. In situ observation of the Yb^{2+} emission in the radiodarkening process of Yb-doped optical preform [J]. Optics Letters, 2016,41(9):2025-2028.
34. Shao Chongyun, Ren Jinjun, Wang Fan, et al. Origin of radiation-induced darkening in Yb^{3+}/Al^{3+}/P^{5+}-doped silica glasses: effect of the P/Al ratio [J]. The Journal of Physical Chemistry B, 2018,122(10):2809-2820.
35. Shao Chongyun, Guo Mengting, Zhang Yang, et al. 193nm excimer laser induced color centers in Yb^{3+}/Al^{3+}/P^{5+}-doped silica glasses [J]. Journal of Non-Crystalline Solids, 2020(544):120198.
36. Wang Fan, Shao Chongyun, Yu Chunlei, et al. Effect of $AlPO_4$ join concentration on optical properties and radiation hardening performance of Yb-doped Al_2O_3-P_2O_5-SiO_2 glass [J]. Journal of Applied Physics, 2019,125(17):173104.
37. Shao Chongyun, Xu Wenbin, Ollier N, et al. Suppression mechanism of radiation-induced darkening by Ce doping in Al/Yb/Ce-doped silica glasses: Evidence from optical spectroscopy, EPR and XPS analyses [J]. Journal of Applied Physics, 2016,120(15):153101.
38. Shao Chongyun, Xie Fenghou, Wang Fan, et al. UV absorption bands and its relevance to local structures of ytterbium ions in Yb^{3+}/Al^{3+}/P^{5+}-doped silica glasses [J]. Journal of Non-Crystalline Solids, 2019(512):53-59.
39. Hu Lili, et al. Laser Glass and Its Applications (in chinese) [M]. Shanghai Scientific and Technical Publishers, 2019.
40. Ollier N, Corbel C, Duchez J, et al. In situ observation of the Yb^{2+} emission in the radiodarkening process of Yb-doped optical preform [J]. Optics Letters, 2016,41(9):2025-2028.
41. Mebrouk Y, Mady F, Benabdesselam M, et al. Experimental evidence of Er^{3+} ion reduction in the radiation-induced degradation of erbium-doped silica fibers [J]. Optics Letters, 2014,39(21):6154-6157.
42. Hari Babu B, Ollier N, León Pichel M, et al. Radiation hardening in sol-gel derived Er^{3+}-doped silica glasses [J]. Journal of Applied Physics, 2015,118(12):123107.
43. Malchukova E, Boizot B. Reduction of Eu^{3+} to Eu^{2+} in aluminoborosilicate glasses under ionizing radiation [J]. Materials Research Bulletin, 2010,45(9):1299-1303.
44. Zhang Jun, Riesen H. Controlled generation of Tm^{2+} ions in nanocrystalline BaFCl: Tm^{3+} by X-ray irradiation [J]. The Journal of Physical Chemistry A, 2017,121(4):803-809.

45. Qiu J, Hirao K. γ-ray induced reduction of Sm^{3+} to Sm^{2+} in sodium aluminoborate glasses [J]. Journal of Materials Science Letters, 2001,20(8):691-693.
46. Vahedi S, Okada G, Morrell B, et al. X-ray induced Sm^{3+} to Sm^{2+} conversion in fluorophosphate and fluoroaluminate glasses for the monitoring of high-doses in microbeam radiation therapy [J]. Journal of Applied Physics, 2012,112(7):73108.
47. Singh G P, Kaur P, Kaur S, et al. Conversion of Ce^{3+} to Ce^{4+} ions after gamma ray irradiation on CeO_2-PbO-B_2O_3 glasses [J]. Physica B: Condensed Matter, 2013(408):115-118.
48. Imai H, Arai K, Imagawa H, et al. Two types of oxygen-deficient centers in synthetic silica glass [J]. Phys. Rev. B Condens. Matter., 1988,38(17):12772-12775.
49. Amossov A V, Rybaltovsky A O. Oxygen-deficient centers in silica glasses: a review of their properties and structure [J]. Journal of Non-Crystalline Solids, 1994(179):75-83.
50. Skuja L. Optically active oxygen-deficiency-related centers in amorphous silicon dioxide [J]. Journal of Non-Crystalline Solids, 1998,239(1-3):16-48.
51. Brower K L. Electron paramagnetic resonance of AlE'_1 centers in vitreous silica [J]. Physical Review B, 1979,20(5):1799.
52. Hideo H, Hiroshi K. Radiation-induced coloring and paramagnetic centers in synthetic SiO_2 :Al glasses [J]. Nuclear Instruments and Methods in Physics Research Section B: Beam Interactions with Materials and Atoms, 1994,91(1-4):395-399.
53. Chah K, Boizot B, Reynard B, et al. Micro-Raman and EPR studies of β-radiation damages in aluminosilicate glass [J]. Nuclear Instruments and Methods in Physics Research Section B: Beam Interactions with Materials and Atoms, 2002,191(1-4):337-341.
54. Fujimaki M, Watanabe T, Katoh T, et al. Structures and generation mechanisms of paramagnetic centers and absorption bands responsible for Ge-doped SiO_2 optical-fiber gratings [J]. Physical Review B, 1998,57(7):3920.
55. Girard S, Alessi A, Richard N, et al. Overview of radiation induced point defects in silica-based optical fibers [J]. Reviews in Physics, 2019,4(11):100032.
56. Alessi A, Agnello S, Gelardi F M, et al. Influence of Ge doping level on the EPR signal of Ge(1), Ge(2) and E'Ge defects in Ge-doped silica [J]. Journal of Non-Crystalline Solids, 2011,357(8):1900-1903.
57. Wu Xu, Liu Chengxiang, Wu Dong, et al. Radiation resistance of an Er/Ce codoped superfluorescent source of conventional fiber and photonic crystal fiber [J]. Optical Engineering, 2017,56(12):1.
58. Kobayashi Y, Sekiya E H, Saito K, et al. Effects of Ge co-doping on P-related radiation-induced absorption in Er/Yb-doped optical fibers for space applications [J]. Journal of Lightwave Technology, 2018,36(13):2723-2729.
59. Leon M, Lancry M, Ollier N, et al. Influence of Al/Ge ratio on radiation-induced attenuation in nanostructured erbium-doped fibers preforms[C]. OSA, 2015.
60. Girard S, Ouerdane Y, Bouazaoui M, et al. Transient radiation-induced effects on solid core microstructured optical fibers [J]. Opt. Express, 2011,19(22):21760-21767.
61. Barnes C E, Greenwell R A, Nelson G W. The effect of fiber coating on the radiation response of fluorosilicate clad, pure silica core step index fibers [C]. Technical Symposium Southeast, 1987.
62. Brichard B, Fernandez A F, Berghmans F, et al. Origin of the -doped optical fibers [J]. Journal of Lightwave Technology, 2018,36(4):972-978.
63. Hanafusa H, Hibino Y, Yamamoto F. Drawing condition dependence of radiation-induced loss in optical fibres[J]. Electronics Letters, 1986,22(2):106-108.
64. Girard S, Ouerdane Y, Boukenter A, et al. Transient radiation responses of silica-based optical fibers: Influence of modified chemical-vapor deposition process parameters[J]. Journal of Applied Physics, 2006,99(2):23104.
65. Wang R P, Tai N, Saito K, et al. Fluorine-doping concentration and fictive temperature dependence of self-trapped holes in SiO_2 glasses[J]. Journal of Applied Physics, 2005,98(2):23701.
66. Hari B B, Lancry M, Ollier N, et al. Radiation hardening of sol gel-derived silica fiber preforms through fictive temperature reduction[J]. Appl Opt, 2016,55(27):7455-7461.

67. Lancry M, Babu B H, Ollier N, et al. Radiation hardening of silica glass through fictive temperature reduction[J]. International Journal of Applied Glass Science, 2017,8(3):285-290.
68. Girard S, Tortech B, Regnier E, et al. Proton- and Gamma-Induced Effects on Erbium-Doped Optical Fibers[J]. IEEE Transactions on Nuclear Science, 2007,54(6):2426-2434.
69. GIRARD S, KEURINCK J, BOUKENTER A, et al. Gamma-rays and pulsed X-ray radiation responses of nitrogen-, germanium-doped and pure silica core optical fibers[J]. Nuclear Instruments and Methods in Physics Research Section B: Beam Interactions with Materials and Atoms, 2004,215(1-2):187-195.
70. S. G, B. B, J. B, et al. Comparative study of pulsed X-ray and γ-ray radiation-induced effects in pure-silica-core optical fibers: 2005 8th European Conference on Radiation and Its Effects on Components and Systems, 2005[C].2005, 19-23 Sept. 2005.
71. Wijnands T, Aikawa K, Kuhnhenn J, et al. Radiation Tolerant Optical Fibers: From Sample Testing to Large Series Production[J]. Journal of Lightwave Technology, 2011,29(22):3393-3400.
72. Girard S, Marcandella C, Morana A, et al. Combined High Dose and Temperature Radiation Effects on Multimode Silica-Based Optical Fibers[J]. IEEE Transactions on Nuclear Science, 2013,60(6):4305-4313.
73. Ladaci A, Girard S, Mescia L, et al. Optimized radiation-hardened erbium doped fiber amplifiers for long space missions[J]. 2017,121(16):163104.
74. Ladaci A. Rare earth doped optical fibers and amplifiers for space applications[D]. Université de Lyon; Politecnico di Bari. Dipartimento di Ingegneria ..., 2017.
75. Zotov K V, Likhachev M E, Tomashuk A L, et al. Radiation Resistant Er-Doped Fibers: Optimization of Pump Wavelength[J]. IEEE Photonics Technology Letters, 2008,20(17):1476-1478.
76. Griscom D L. Radiation hardening of pure silica core optical fibers and their method of making by ultra-high-dose gamma ray pre-irradiation[Z]. Google Patents, 1996.
77. Griscom D L. Radiation hardening of pure-silica-core optical fibers by ultra-high-dose γ-ray pre-irradiation[J]. Journal of applied physics, 1995, 77(10): 5008-5013.
78. Yeniay A, Gao R F. Radiation induced loss properties and hardness enhancement technique for ErYb doped fibers for avionic applications[J]. OPTICAL FIBER TECHNOLOGY, 2013,19(2):88-92.
79. Thomas J, Myara M, Troussellier L, et al. Radiation-resistant erbium-doped-nanoparticles optical fiber for space applications.[J]. Optics Express, 2012,20(3):2435-2444.
80. Jiao Y, Yang Q, Guo M, et al. Effect of the GeO_2 content on the radiation resistance of Er^{3+}-doped silica glasses and fibers[J]. Optical Materials Express, 2021,11(7):1885-1897.
81. Engholm M, Jelger P, Laurell F, et al. Improved photodarkening resistivity in ytterbium-doped fiber lasers by cerium codoping [J]. Optics Letters, 2009,34(8):1285-1287.
82. Ladaci A, Girard S, Mescia L, et al. Radiation hardened high-power Er^{3+}/Yb^{3+}-codoped fiber amplifiers for free-space optical communications [J]. Optics Letters, 2018,43(13):3049-3052.
83. Girard S, Vivona M, Laurent A, et al. Radiation hardening techniques for Er/Yb doped optical fibers and amplifiers for space application [J]. Optics Express, 2012,20(8):8457-8465.
84. Ladaci A, Girard S, Mescia L, et al. X-rays, γ-rays, electrons and protons radiation-induced changes on the lifetimes of Er^{3+} and Yb^{3+} ions in silica-based optical fibers [J]. Journal of Luminescence, 2018(195):402-407.
85. Vivona M, Girard S, Marcandella C, et al. Radiation hardening of rare-earth doped fiber amplifiers[C]. SPIE, 2017 .
86. Liu Xiaoxia, Liu Chaoping, Chen Gui, et al. Influence of cerium ions on thermal bleaching of photo-darkened ytterbium-doped fibers [J]. Frontiers of Optoelectronics, 2018,11(4),394-399.
87. Shao Chongyun, Wang Fan, Guo Mengting, et al. Structure and property of Yb^{3+}/Al^{3+}/Ce^{3+}/F^--doped silica glasses [J]. J. Chin. Ceram. Soc., 2019,47(1):120-131.
88. Faile S P, Schmidt J J, Roy D M. Irradiation effects in glasses: suppression by synthesis under high-pressure hydrogen [J]. Science, 1967,156(3782):1593-1595.

89. Nagasawa K, Hoshi Y, Ohki Y, et al. Improvement of radiation resistance of pure silica core fibers by hydrogen treatment [J]. Japanese Journal of Applied Physics.pt Regular Papers and Short Notes, 1985,24(9):1224-1228.
90. Zotov K V, Likhachev M E, Tomashuk A L, et al. Radiation-resistant erbium-doped fiber for spacecraft applications [J]. Radecs 2007: Proceedings of the 9th european Conference on Radiation and Its Effects on Components and Systems, 2007:450-453.
91. Zotov K V, Likhachev M E, Tomashuk A L, et al. Radiation-resistant erbium-doped silica fibre [J]. Quantum Electronics, 2007,37(10):946-949.
92. Xing Yingbin, Liu Yinzi, Zhao Man, et al. Radical passive bleaching of Tm-doped silica fiber with deuterium [J]. Optics Letters, 2018,43(5):1075-1078.
93. Girard S, Laurent A, Pinsard E, et al. Proton irradiation response of hole-assisted carbon coated erbium-doped fiber amplifiers [J]. IEEE Transactions on Nuclear Science, 2014,61(6),3309-3314.
94. Girard S, Laurent A, Pinsard E, et al. Radiation-hard erbium optical fiber and fiber amplifier for both low- and high-dose space missions [J]. Optics Letters, 2014,39(9):2541-2544.
95. Liu Shuang, Zheng Shupei, Yang Ke, et al. Radiation-induced change of OH content in Yb-doped silica glass [J]. Chinese Optics Letters, 2015,13(6):60602.
96. Griscom D L. Radiation hardening of pure-silica-core optical fibers by ultra-high-dose gamma-ray pre-irradiation [J]. Journal of Applied Physics, 1995,77(10):5008-5013.
97. Kim Y, Ju S, Jeong S, et al. Gamma-ray radiation response at 1550nm of fluorine-doped radiation hard single-mode optical fiber [J]. Optics Express, 2016,24(4):3910-3920.
98. Hosono H, Ikuta Y, Kinoshita T, et al. Physical disorder and optical properties in the vacuum ultraviolet region of amorphous SiO_2 [J]. Physical Review Letters, 2001,87(17):175501.
99. Shao Chongyun, Jiao Yan,Lou Fengguang, et al. Enhanced radiation resistance of ytterbium-doped silica fiber by pretreating on a fiber preform [J]. Optical Materials Express, 2020,10(2):408.
100. Shao Chongyun, Hu Lili, Yu Chunlei, et al., Radiation-resistant laser optical fiber preform core rod and preparation method therefor: U.S. Patent 12,001,052[P]. 2024-6-4.
101. Shao Chongyun, Yu Chunlei, Hu Lili. Radiation-Resistant Active Fibers for Space Applications (in chinese) [J]. Chinese Journal of Lasers, 2020,47(5):233-261.
102. Staurt B. Infrared spectroscopy: fundamentals and applications [J]. John Wiley and Sons, Ltd., West Sussex, England. DOI, 2004,10:470011149.
103. Humbach O, Fabian H, Grzesik U, et al. Analysis of OH absorption bands in synthetic silica [J]. Journal of Non-Crystalline Solids, 1996(203):19-26.
104. Stone J. Interactions of hydrogen and deuterium with silica optical fibers: a review [J]. Journal of Lightwave technology, 1987,5(5):712-733.
105. Jiao Y, Yang Q, Zhu Y, et al. Improved radiation resistance of an Er-doped silica fiber by a preform pretreatment method [J]. Optics Express, 2022,30(4):6236.
106. Ramsey A T, Tighe W, Bartolick J, et al. Radiation effects on heated optical fibers [J]. Review of scientific instruments, 1997,68(1):632-635.
107. Söderlund M J, I Ponsoda J J M, Koplow J P, et al. Thermal bleaching of photodarkening in ytterbium-doped fibers: SPIE LASE [C]. International Society for Optics and Photonics, 2010.
108. Friebele E J, Gingerich M E. Photobleaching effects in optical fiber waveguides [J]. Applied Optics, 1981,20(19):3448-3452.
109. Xing Yingbin, Huang Hongqi, Zhao Man, et al. Pump bleaching of Tm-doped fiber with 793nm pump source [J]. Optics Letters, 2015,40(5):681-684.
110. Xing Yingbin, Zhao Man, Liao Lei, et al. Active radiation hardening of Tm-doped silica fiber based on pump bleaching [J]. Optics Express, 2015,23(19):24236.
111. Xing Yingbin, Liu Yinzi, Cao Ruiting, et al. Elimination of radiation damage in Tm-doped silica fibers based on the radical bleaching of deuterium loading [J]. OSA Continuum, 2018,1(3):987.

112. Girard S, Ouerdane Y, Origlio G, et al. Radiation Effects on Silica-Based Preforms and Optical Fibers — I: Experimental Study With Canonical Samples [J]. IEEE Transactions on Nuclear Science, 2008,55(6):3473-3482.
113. Girard S, Richard N, Ouerdane Y, et al. Radiation Effects on silica-based preforms and optical fibers-II: coupling Ab initio simulations and experiments [J]. IEEE Transactions on Nuclear Science, 2008,55(6):3508-3514.
114. Girard S, Mescia L, Vivona M, et al. Design of radiation-hardened rare-earth doped amplifiers through a coupled experiment/simulation approach [J]. Journal of Lightwave Technology, 2013,31(8):1247-1254.
115. Mescia L, Girard S, Bia P, et al. Optimization of the design of high power Er^{3+} /Yb^{3+} -codoped fiber amplifiers for space missions by means of particle swarm approach [J]. IEEE Journal of Selected Topics in Quantum Electronics, 2014,20(5),484-491.
116. Ladaci A, Girard S, Mescia L, et al. Optimized radiation-hardened erbium doped fiber amplifiers for long space missions [J]. Journal of Applied Physics, 2017,121(16):163104.

Index

L. Hu, *Rare Earth Doped Silica Fiber and Its Applications*,
https://doi.org/10.1007/978-981-96-5306-5

The manufacturer's authorised representative in the EU is Springer Nature Customer Service Centre GmbH, Europaplatz 3, 69115 Heidelberg, Germany. If you have any concerns regarding our products, please contact ProductSafety@springernature.com

Printed and bound by CPI Group (UK) Ltd, Croydon, CR0 4YY
01/10/2026
02225923-0020